INSTRUCTOR'S MANUAL WITH SOLUTIONS TO ACCOMPANY

COLLEGE PHYSICS
THIRD EDITION

JERRY S. FAUGHN & CHARLES D. TEAGUE

Eastern Kentucky University

SAUNDERS GOLDEN SUNBURST SERIES

Saunders College Publishing
A Harcourt Brace Jovanovich College Publisher
Fort Worth Philadelphia San Diego New York Orlando Austin San Antonio
Toronto Montreal London Sydney Tokyo

Copyright ©1992 by Saunders College Publishing

All rights reserved. No part of this publication may be reproduced or transmitted in any form or by any means, electronic or mechanical, including photocopy, recording or any information storage and retrieval system, without permission in writing from the publisher.

Requests for permission to make copies of any part of the work should be mailed to: Permissions, Holt, Rinehart and Winston, Inc., Orlando, Florida 32887.

Printed in the United States of America.

Serway: Instructor's Manual with Solutions to accompany COLLEGE PHYSICS, 3/E.

ISBN 0-03-075012-1

34 082 9876543

PREFACE

This guide has been written to accompany the textbook "College Physics" (3rd edition) by Serway and Faughn. It is divided into two parts:

I. solutions to all the problems in the text, and
II. answers to all even-numbered problems.

We welcome your comments on the accuracy of the solutions as presented here, as well as suggestions for alternative approaches.

This manual was written on a Macintosh SE/30 using Microsoft Word 4.00 for the text and MacDraw II for the figures. If you would like to modify the solutions to satisfy your own tastes, the manual is available on discs that you can obtain by writing either author and enclosing three double-sided, formatted discs.

Jerry Faughn
Charles Teague

Physics Department
Eastern Kentucky Univ.
Richmond, Ky 40475

Table of Contents

Chapter One Solutions	1
Chapter Two Solutions	7
Chapter Three Solutions	19
Chapter Four Solutions	34
Chapter Five Solutions	57
Chapter Six Solutions	74
Chapter Seven Solutions	91
Chapter Eight Solutions	105
Chapter Nine Solutions	128
Chapter Ten Solutions	143
Chapter Eleven Solutions	155
Chapter Twelve Solutions	171
Chapter Thirteen Solutions	182
Chapter Fourteen Solutions	195
Chapter Fifteen Solutions	209
Chapter Sixteen Solutions	224
Chapter Seventeen Solutions	242
Chapter Eighteen Solutions	257
Chapter Nineteen Solutions	268
Chapter Twenty Solutions	287
Chapter Twenty-One Solutions	300
Chapter Twenty-Two Solutions	311
Chapter Twenty-Three Solutions	325
Chapter Twenty-Four Solutions	331
Chapter Twenty-Five Solutions	346
Chapter Twenty-Six Solutions	363

Chapter Twenty-Seven Solutions	377
Chapter Twenty-Eight Solutions	392
Chapter Twenty-Nine Solutions	404
Chapter Thirty Solutions	417
Chapter Thirty-One Solutions	431
Chapter Thirty-Two Solutions	440
Chapter Thirty-Three Solutions	454
Chapter Thirty-Four Solutions	463
Even-Numbered Answers	473

CHAPTER ONE SOLUTIONS

1.1 $v^2 = v_0^2 + 2ax$ has units of $(L/T)^2 = (L/T)^2 + (L/T^2)(L)$. All terms have units of
$L^2 T^{-2}$.

1.2 (a) From $x = ct^2$, we have $c = x/t^2$. Thus, c has units of L/T^2.
(b) $x = A\sin(2\pi t)$ has units of $L = A$(pure number). Thus, A has units of L.

1.3 Given that $a \propto F/m$, we have $F \propto ma$. Therefore, the units of force are those of ma.
$F = M(L/T^2)$.

1.4. (a) Both v and v_0 have units of L/T, but the term ax has units of $(L/T^2)(L) = \dfrac{L^2}{T^2}$. Thus, this equation is not dimensionally correct.

(b) The argument of the cosine term has no dimensions, and the units of the number 2 are those of m. Thus, the units are L on each side, and the equation is dimensionally correct.

1.5 We know volume of cylindrical column = volume of spherical drop
Thus, $\pi r^2 h = \dfrac{4}{3}\pi R^3$ where r = radius of cylinder, R = radius of drop, and h = height of cylinder.
So, $h = \dfrac{4R^3}{3r^2}$

As a dimension check, we have $(L) = \dfrac{(L^3)}{(L^2)} = (L)$

1.6 Substituting in dimensions, we have
$(T) = \sqrt{\dfrac{(L)}{(L/T^2)}} = \sqrt{T^2} = (T)$
Thus, the dimensions are consistent.

1.7 $c = 2.997924574 \times 10^8$ m/s
(a) Rounded to 3 significant figures: $c = 3.00 \times 10^8$ m/s
(b) Rounded to 5 significant figures: $c = 2.9979 \times 10^8$ m/s
(c) Rounded to 7 significant figures: $c = 2.997925 \times 10^8$ m/s

---

1.8 (a) 78.9 ± 0.2 has 3 significant figures.
(b) 3.788×10^9 has 4 significant figures.
(c) 2.46×10^{-6} has 3 significant figures.
(d) $0.0032 = 3.2 \times 10^{-3}$ has 2 significant figures.

1.9 (a) The sum is rounded to 797 because 756 in the terms to be added has no positions beyond the decimal.

CHAPTER ONE SOLUTIONS

(b) 3.2 X 3.563 must be rounded to 11 because 3.2 has only two significant figures.
(c) 5.67 X π must be rounded to 17.8 because 5.67 has only three significant figures.

1.10 (a) The answer is 719 because the number 756 carries no information beyond the decimal.
(b) 3.2/1.4577 must be rounded to 2.2 because 3.2 has only two significant figures.

1.11 (a) $c = 2\pi r = 2\pi(3.5 \text{ cm})$ must be rounded to 22 cm. The number 2 and π are considered as known to many significant figures, while 3.5 is known only to two.
(b) $A = \pi r^2 = \pi(4.65 \text{ cm})^2$ must be rounded to the number of significant figures in 4.65, which gives an answer of 67.9 cm^2.

1.12 The distance around is 38.44 m + 19.5 m + 38.44 m + 19.5 m = 115.88 m, but this answer must be rounded to 115.9 m because the distance 19.5 m carries information to only one place past the decimal.

1.13 100 yds = (100 yds)$\left(\dfrac{3 \text{ ft}}{1 \text{ yd}}\right)\left(\dfrac{1 \text{ m}}{3.28 \text{ ft}}\right)$ = 91.4 m

---

1.14 Volume of cube = L^3 = 1 quart (Where L is the length of one side of the cube.)

Thus, L^3 = (1 quart)$\left(\dfrac{1 \text{ gallon}}{4 \text{ quarts}}\right)\left(\dfrac{3.786 \text{ liters}}{1 \text{ gallon}}\right)\left(\dfrac{1000 \text{ cm}^3}{1 \text{ liter}}\right)$ = 946.5 cm^3
and   L = 9.82 cm

1.15 Age of earth = (1 X 10^{17}s)(1 y/3.156 X 10^7 s) = 3.16 X 10^9 y.

1.16 Distance = (4 X 10^{16} m)(3.281 ft/1 m) = 1 X 10^{17} ft

---

1.17 Volume of house = (50 ft)(26 ft)(8 ft) = 1.04 X 10^4 ft^3
= (1.04 X 10^4 ft^3)(2.832 X 10^{-2} m^3/ft^3) = 295 m^3
= (2.95 X 10^2 m^3)(10^2 cm/m)^3 = 2.95 X 10^8 cm^3

1.18 (a) 1 yr = (1 yr)(365.242 days/1yr)(86400 s/1day) = 3.16 X 10^7 s.
(b)   Let us consider a segment of the surface of the moon which has an area of 1 m^2 and a depth of 1 m. This imaginary box on the moon has a volume of 1.00 m^3. The volume of a meteorite is found as

Vol/meteorite = $\dfrac{4}{3}\pi r^3 = \dfrac{4}{3}\pi(0.5 \times 10^{-6} \text{ m})^3$ = 5.24 X 10^{-19} m^3. Since this volume of meteorites strikes the 1 m^2 area of the moon each second, the time to fill the volume is

time = (volume of box)/(volume of meteorites striking box per second)
time = (1 m^3)/(5.24 X 10^{-19} m^3/s) = 1.91 X 10^{18} s = 6.05 X 10^{10} yr

CHAPTER ONE SOLUTIONS

1.19 area = 4(area of one wall) = (4)(8 ft)(12 ft) = 384 ft^2. This is converted to square meters as
384 ft^2(1m/3.281ft)^2 = 35.7 m^2

1.20 Volume of oil = 1.0 m^3
(area)(thickness) = 1.0 m^3
(area)(1 μm) = 1.0 m^3
area = $\dfrac{1.0 \text{ m}^3}{1 \text{ μm}}$ = 10^6 m^2 $\left(\dfrac{1 \text{ mile}}{1609 \text{ m}}\right)^2$ = 0.39 sq. miles

---

1.21 We require (mass)$_{al}$ = (mass)$_{iron}$

Thus, (density)$_{al}\left(\dfrac{4}{3}\pi r^3\right)$ = (density)$_{iron}\left(\dfrac{4}{3}\pi(2.0 \text{ cm})^3\right)$

or, r^3 = $\left(\dfrac{(\text{density})_{iron}}{(\text{density})_{al}}\right)$(2.0 cm)^3 = $\left(\dfrac{7.86 \text{ kg/m}^3}{2.70 \text{ kg/m}^3}\right)$(2.0 cm)^3 = 23.3 cm^3

and r = 2.86 cm, or 2.9 cm to two significant figures.

1.22 (a) mass = (density)(volume) = $\left(\dfrac{1.0 \times 10^{-3} \text{ kg}}{1.0 \text{ cm}^3}\right)$(1 m^3)

= (1.0 × 10^{-3} kg/cm^3)(1 m^3)$\left(\dfrac{10^2 \text{ cm}}{1 \text{ m}}\right)^3$ = 1000 kg

(b) As rough calculation, treat as if 100% water.

cell: mass = density × volume = $\left(\dfrac{10^3 \text{ kg}}{1 \text{ m}^3}\right)\dfrac{4}{3}\pi$ (0.5 × 10^{-6} m)^3 = 5.2 × 10^{-16} kg

kidney: mass = density × volume = $\left(\dfrac{10^3 \text{ kg}}{1 \text{ m}^3}\right)\dfrac{4}{3}\pi$ (4 × 10^{-2} m)^3 = 0.27 kg

fly: mass = density × vol = (density)($\pi r^2 h$)
= $\left(\dfrac{10^3 \text{ kg}}{1 \text{ m}^3}\right)\pi$ (10^{-3} m)^2(4 × 10^{-3} m) = 1.3 × 10^{-5} kg

---

1.23 (a) 1 mi/h = (1 mi/h)(1.609 km/mi) = 1.609 km/h
Thus, 1 mi/h = 1.609 km/h
(b) 55 mi/h = (55 mi/h)(1.609 km per hr/1 mi per hr) = 88.5 km/h
(c) increase = 10 mi/h = (10 mi/h)(1.609 km per hr/1 mi per hr) = 16.1 km/h

---

1.24 c = (3 × 10^8 m/s)(3600 s/h)(1 km/10^3 m)(1mi/1.609 km) = 6.71 × 10^8 mi/h

---

1.25 distance = (speed)(time)
= (3 × 10^8 m/s)(3600 s/1h)(1 km/10^3 m)(1mi/1.609 km)
= 6.71 × 10^8 mi.

1.26 Volume of pyramid = $\dfrac{1}{3}$ (base)(height)

= $\dfrac{1}{3}$ [(13 acres)(43,560 ft^2/acre)](481 ft) = 9.08 × 10^7 ft^3,

= (9.08 × 10^7 ft^3)(2.832 × 10^{-2} m^3)/(1 ft^3) = 2.57 × 10^6 m^3.

CHAPTER ONE SOLUTIONS

1.27 weight = (number of blocks)(weight per block)
= $(2 \times 10^6)(2.5 \text{ tons}) = 5 \times 10^6$ tons
or, in lbs
$(5 \times 10^6 \text{ tons})(2 \times 10^3 \text{ lb/ton}) = 1 \times 10^{10}$ lb
in newtons, we have
$(1 \times 10^{10} \text{ lb})(1 \text{ N}/0.2248 \text{ lb}) = 4.45 \times 10^{10}$ N

1.28 number of balls needed = (number lost per hitter)(number hitters)(games) = (1 ball per hitter)(10 hitters per inning)(9 innings per game)(81 games) = 7300 balls
Assumptions are 1 ball lost per hitter, 10 hitters per inning, 9 innings per game, and 81 games per season.

1.29 number of pounds = (number of burgers)(weight/burger)
= $(5 \times 10^{10} \text{ burgers})(0.25 \text{ lb/burger}) = 1.25 \times 10^{10}$ lb
number of head of cattle = (weight needed)/(weight per head)
= $(1.25 \times 10^{10} \text{ lb})/(300 \text{ lb/head}) = 4.17 \times 10^7$ head
Assumptions are 0.25 lb of meat per burger and 300 lb of meat per head of cattle

---

1.30 The number of tuners is found by dividing the number of residents of the city by the number of residents serviced by one tuner. We shall assume 1 tuner per 10000 residents and a population of 7.5 million. Thus,
number of tuners = $(7.5 \times 10^6)/(10,000) = 750$

1.31 number of pitches = (number pitches per inning)(innings per game)(games)
= (30 pitches per inning)(9 innings per game)(162 games)
= 4.4×10^4 pitches per season
Assumptions 30 pitches per inning and 9 innings per game

1.32 The x coordinate is found as
$x = r \cos\theta = (2.5 \text{ m})(\cos 35°) = 2.05$ m
and the y coordinate is
$y = r \sin\theta = (2.5 \text{ m})(\sin 35°) = 1.43$ m

1.33 The x distance out to the fly is 2 m and the y distance up to the fly is 1 m. Thus, we can use the Pythagorean theorem to find the distance from the origin to the fly as,
distance = $\sqrt{x^2 + y^2} = \sqrt{(2.0 \text{ m})^2 + (1.0 \text{ m})^2} = \sqrt{5.0 \text{ m}^2} = 2.24$ m

---

1.34 The distance from the origin to the fly is r in polar coordinates, and this was found to be 2.24 m in problem 33. The angle θ is the angle between r and the horizontal reference line (the x axis in this case). Thus, the angle can be found as
$\tan\theta = y/x = (1.0 \text{ m})/(2.0 \text{ m}) = 0.5$
Thus, $\theta = 26.6°$. The polar coordinates are r = 2.24 m and $\theta = 26.6°$.

CHAPTER ONE SOLUTIONS

1.35 The x distance between the two points is 8 cm and the y distance between them is 1 cm. The distance beween them is found from the Pythagorean theorem.

distance = $\sqrt{x^2 + y^2} = \sqrt{(8\text{ cm})^2 + (1\text{ cm})^2} = \sqrt{65\text{ cm}^2} = 8.06$ cm.

1.36 (a) The side opposite $\theta = 3$.
(b) The side adjacent to $\phi = 3$
(c) $\cos\theta = \dfrac{4}{5}$
(d) $\sin\phi = \dfrac{4}{5}$
(e) $\tan\phi = \dfrac{4}{3}$

1.37 From the Pythagorean theorem,
$c = \sqrt{(5\text{ m})^2 + (7\text{ m})^2} = 8.60$ m

1.38 $\tan\theta = \dfrac{5}{7}$, so $\theta = 35.5°$

1.39 (a) $\sin\theta = \dfrac{\text{side opposite}}{\text{hypotenuse}}$
so side opposite = $(\sin 30°)(3\text{ m}) = 1.5$ m
(b) $\cos\theta = \dfrac{\text{adjacent side}}{\text{hypotenuse}}$
so, adjacent side = $(\cos 30°)(3\text{ m}) = 2.60$ m

1.40 (a) The length of the unknown side is $b = \sqrt{c^2 - a^2} = \sqrt{(9)^2 - (6)^2} = 6.71$ m
(b) $\tan\theta = \dfrac{6}{6.71} = 0.894$
(c) $\sin\phi = \dfrac{6.71}{9} = 0.746$

1.41 (a) 10^{12} microphones = $10^{12}(10^{-6})$ phones = 10^6 phones = 1 mega-phone
(b) 10^{21} picolos = $(10^{21})(10^{-12})$ los = 10^9 los = 1 giga-lo
(c) 10 rations = (10^1) rations = 1 deca-ration
(d) 10^6 bicycles = $(10^6)(2\text{ cycles})$ = 2 (mega) cycles = 2 mega-cycles
(e) 10^{12} pins = (tera) pins = 1 tera-pin
(f) $3\tfrac{1}{3}$ tridents = $(3\tfrac{1}{3})(3\text{ dents})$ = 10 dents = 1 deca-dent
(g) 2000 mockingbirds = 2(1000)mockingbirds = 2 kilo-mockingbirds
(h) 10^{-12} boo = 1 pico-boo
(i) 10^{-9} goat = 1 nano-goat

1.42 (a) The volume of Saturn is
$V = \dfrac{4}{3}\pi r^3 = \dfrac{4}{3}\pi(5.85 \times 10^7\text{ m})^3 = 8.39 \times 10^{23}\text{ m}^3$
and the density is

CHAPTER ONE SOLUTIONS

$$\rho = \frac{m}{V} = \frac{(5.68 \times 10^{26} \text{ kg})}{(8.39 \times 10^{23} \text{ m}^3)} = 677 \text{ kg/m}^3 = 0.677 \text{ g/cm}^3. \text{ (less dense, on average, than water)}$$

(b) The surface area of Saturn is
$A = 4\pi r^2 = 4\pi (5.85 \times 10^7 \text{ m})^2 = 4.30 \times 10^{16} \text{ m}^2$

1.43 The constants must have units of $1.5 \frac{\text{million ft}^3}{\text{month}}$ and $0.008 \frac{\text{million ft}^3}{(\text{month})^2}$.

Thus $V = \left(1.5 \times 10^6 \frac{\text{ft}^3}{\text{month}}\right) t + \left(0.008 \times 10^6 \frac{\text{ft}^3}{(\text{month}^2)}\right) t^2$

To convert, use 1 month = 2.59×10^6 sec, to obtain

$V = \left(0.579 \frac{\text{ft}^3}{\text{sec}}\right) t + \left(1.19 \times 10^{-9} \frac{\text{ft}^3}{(\text{sec}^2)}\right) t^2$

1.44 Assume an average of 1 can per person each week and a population of 250 million.
number cans per yr
  = (number cans per person each week)(population)(weeks per yr)
  =(1 can per person each week)(2.5×10^8 people)(52 weeks per yr)
  = 1.3×10^{10} can/yr
 number of tons = (weight per can)(number of cans per yr)
  = (0.5 oz per can)(1.3×10^{10} cans per yr) = 2×10^5 tons/yr
Assumes an average weight of 0.5 oz of aluminum per can.

1.45 Assume 200 bills per inch, and the height is found as
height = (number of bills)/(number of bills per inch) = (1×10^6)/(200/in) = 5,000 in, or 417 ft. (Note: If the bills are placed end-to-end, they will stretch for about 100 miles.)

1.46 The term s has dimensions of L, a has dimensions of LT^{-2}, and t has dimensions of T. Therefore, the equation, $s = ka^m t^n$ has dimensions of
  $L = (LT^{-2})^m (T)^n$
or
  $L^1 T^0 = L^m T^{n-2m}$,
The powers of L and T must be the same on each side of the equation. Therefore, $L^1 = L^m$ and m = 1.
Likewise, equating terms in T, we see that n - 2m must equal 0. Thus, n = 2m = 2. The value of k, a dimensionless constant, cannot be obtained by dimensional analysis.

1.47 The x distance between these points is 5.0 m, and the y distance between these points is 7.0 m. The distance between the points is found from the Pythagorean theorem as
  $\sqrt{(5.0 \text{ m})^2 + (7.0 \text{ m})^2} = 8.60$ m

CHAPTER TWO SOLUTIONS

2.1 $\bar{v} = \Delta x/\Delta t = (100 \text{ yds}/15 \text{ s})(3 \text{ ft/ yd})(1 \text{ m}/3.28 \text{ ft}) = 6.10 \text{ m/s}$

2.2 The radius of the earth is 6.38×10^6 m = 3965 mi. From $\bar{v}$ = dist/time, the time to circle the earth is its circumference divided by the velocity.
$$t = 2\pi R_e/\bar{v} = 2\pi(3965 \text{ mi})/(19,800 \text{ mi/h}) = 1.26 \text{ h} = 75.5 \text{ min}$$

2.3 $t = \dfrac{x}{v} = \dfrac{8.4 \times 10^{-2} \text{ m}}{3.5 \times 10^{-6} \text{ m/s}} = 2.4 \times 10^4 \text{ s} = 6.67 \text{ h}$

2.4 Distances traveled are
$x_1 = v_1 t_1 = (80 \text{ km/h})(0.5 \text{ h}) = 40 \text{ km}$
$x_2 = v_2 t_2 = (100 \text{ km/h})(0.2 \text{ h}) = 20 \text{ km}$
$x_3 = v_3 t_3 = (40 \text{ km/h})(0.75 \text{ h}) = 30 \text{ km}$
Thus, the total elapsed time is 1.7 h, and the total distance traveled is 90 km.
(a) $\bar{v} = \dfrac{x}{t} = \dfrac{90 \text{ km}}{1.7 \text{ h}} = 52.9 \text{ km/h}$
(b) x = 90 km (see above)

2.5 (a) In the first half of the trip, the average velocity is
$\bar{v} = (x_2 - x_1)/20 \text{ s} = +50 \text{ m}/20 \text{ s} = +2.50 \text{ m/s}$
(b) On the return leg, we have
$\bar{v} = (x_3 - x_2)/22 \text{ s} = (0 - 50 \text{ m})/22 \text{ s} = -2.27 \text{ m/s}$
(c) For the entire trip,
$\bar{v} = (x_3 - x_1)/42 \text{ s} = 0/42 \text{ s} = 0$

2.6 The total time for the trip is $t = t_1 + 22$ min $= t_1 + 0.367$ h, where t_1 is the time spent traveling at 89.5 km/h. Thus, the distance traveled is
$x = \bar{v} t = (89.5 \text{ km/h})t_1 = (77.8 \text{ km/h})(t_1 + 0.367 \text{ h})$
or $(89.5 \text{ km/h})t_1 = (77.8 \text{ km/h})t_1 + 28.5 \text{ km}$
From which, $t_1 = 2.44$ h
for a total time of $t = t_1 + 0.367$ h $= 2.81$ h
Therefore, $x = \bar{v} t = (77.8 \text{ km/h})(2.81 \text{ h}) = 218 \text{ km}$

2.7 (a) The displacement in the first 5 minutes is
$\Delta x_1 = \bar{v}_1 (\Delta t)_1 = (2 \text{ m/s})(300 \text{ s}) = 600 \text{ m}$
and the displacement in the next 2 min is
$\Delta x_2 = \bar{v}_2 (\Delta t)_2 = (1.5 \text{ m/s})(120 \text{ s}) = 180 \text{ m}$
The total displacement $= \Delta x_1 + \Delta x_2 = 780$ m
(b) The average velocity
$\bar{v} = (\Delta x)_{total}/(\Delta t)_{total} = 780 \text{ m}/420 \text{ s} = 1.86 \text{ m/s}$

CHAPTER TWO SOLUTIONS

2.8 The distance traveled by A (the runner traveling at 6 mi/h) when they meet is $x_1 = v_A t$, and the distance traveled by B (the runner moving at 5 mi/h) when they meet is $x_2 = v_B t$. We also know that $x_1 + x_2 = 7$ mi. From this last equation, we find, (6 mi/h) t + (5 mi/h)t = 7 mi, or (11 mi/h)t = 7 mi, and

$$t = \frac{7}{11} \text{ h}$$

Thus, the distance traveled by A when they meet is

$$x_1 = (6 \text{ mi/h})(\frac{7}{11} \text{ h}) = 3.818 \text{ mi} \quad \text{and} \quad x_2 = (5 \text{ mi/h})(\frac{7}{11}\text{h}) = 3.182 \text{ mi}$$

They meet 2/11 mi or 0.182 miles west of the flag pole.

2.9 Consider the slope of the line tangent to the graph at the point of interest.
(a) v < 0 (slope is negative)
(b) v > 0 (slope positive)
(c) v = 0 (zero slope)
(d) v = 0 (zero slope)

2.10

time interval (s)	Δt (s)	Δx (m)	v (m/s)
2 to 2.01	0.01	0.014	1.40
2 to 2.2	0.20	0.27	1.35
2 to 2.5	0.50	0.66	1.32
2 to 3.0	1.00	1.26	1.26
2 to 4.0	2.00	2.34	1.17

(a) The average velocity for the complete interval = $\Delta x_t/\Delta t$ = 2.34 m/2.00 s = 1.17 m/s.
(b) The velocity at t = 2s is approximately equal to 1.40 m/s.

2.11 (a) A few typical values are

t(s)	x(m)
1.0	5.75
2.0	16.0
3.0	35.3
4.0	68.0
5.0	118.8

(b) We will use a 0.4 s interval centered at t = 4 s. We find
at t = 3.8 s, x = 60.154 m
at t = 4.2 s, x = 76.566 m

So, $v = \frac{\Delta x}{\Delta t} = \frac{16.412 \text{ m}}{0.4 \text{ s}} = 41.03$ m/s

Using a time interval of 0.2 s, we find the corresponding values to be
at t = 3.9 s, x = 63.989 m
at t = 4.1 s, x = 72.191 m

and $v = \frac{\Delta x}{\Delta t} = \frac{8.201 \text{ m}}{0.2 \text{ s}} = 41.008$ m/s

For a time interval of 0.1 s, the values are
at t = 3.95 s, x = 65.972 m
at t = 4.05 s, x = 70.073 m

and $v = \frac{\Delta x}{\Delta t} = \frac{4.1002 \text{ m}}{0.1 \text{ s}} = 41.002$ m/s

CHAPTER TWO SOLUTIONS

(c) at t = 4 s, x = 68 m. Thus, for the first 4 s,
$$\bar{v} = \frac{\Delta x}{\Delta t} = \frac{68 \text{ m}}{4 \text{ s}} = 17 \text{ m/s}$$
This value is considerably less than the instantaneous velocity at t = 4 s.

2.12 The plot of $x = 2t^2$ looks somewhat like that pictured at the right.

(a) The position of the plane at t = 0 is x = 0.
At t = 3 s, the position is $x = 2(3)^2 = 18$ m.
The average velocity is
$\Delta x/\Delta t = 18$ m/3 s = + 6 m/s

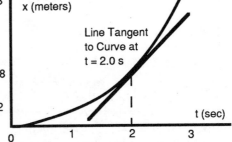

(b) The slope of the line tangent to the curve at t = 2 s gives the instantaneous velocity at that time. This slope is +8 m/s.

2.13 (a) $v_{0,1} = (x_1 - x_0)/(\Delta t) = (4 \text{ m} - 0)/1 \text{ s} = +4$ m/s
(b) $v_{0,4} = (x_4 - x_0)/\Delta t = (-2 \text{ m} - 0)/ 4 \text{ s} = - 0.5$ m/s
(c) $v_{1,5} = (x_5 - x_1)/\Delta t = (0 - 4 \text{ m})/4 \text{ s} = - 1$ m/s
(d) $v_{0,5} = (x_5 - x_0)/\Delta t = (0 - 0)/5 \text{ s} = 0$

2.14 (a) v(t = 0.5 s) = [x(t = 1 s) - x (t = 0)] /(1 s)= +4 m/1 s = +4 m/s
(b) v(t = 2s) = [x(t = 2.5 s) - x(t = 1.5 s)] /(2.5 s - 1 s)= (-2 m - 4 m)/1.5 s
= -6 m/1.5 s = -4 m/s
(c) v(t=3 s) = [x(t = 4 s) - x(t = 2.5 s)]/(4s - 2.5 s) = (-2m -(-2m))/1.5 s = 0
(d) v(t = 4.5 s) = [x(t = 5 s) - x(t = 4s)]/(5 s - 4 s) = (0 - (-2 m))/1 s = + 2 m/s

---

2.15 The position-time graph should look somewhat like that sketched below.

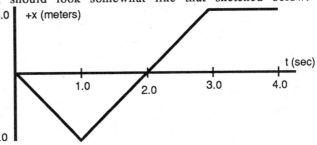

The average velocity during the total time interval is the total displacement divided by the total elapsed time.
$\bar{v}$ = + 3 m/ 4 s = 0.75 m/s
The instantaneous velocity at t = 0.5 s is the slope of the line between t = 0 and t = 1 s.
v = (-3 m - 0)/(1 s - 0) = - 3 m/s
At t = 2 s, the instantaneous velocity is the slope of the line between t = 1 s and t = 3 s.
v = (+3m - (-3 m))/2 s = +3 m/s
At t = 3.5 s, the instantaneous velocity is the slope of the line between t = 3 s and t = 4 s. This line has zero slope; thus, v = 0.

2.16 The average acceleration is found as

9

CHAPTER TWO SOLUTIONS

$\bar{a} = \Delta v/\Delta t = (+8 \text{ m/s} - 5 \text{ m/s})/(4 \text{ s}) = 0.75 \text{ m/s}^2.$

2.17 $\bar{a} = \Delta v/\Delta t = (v_f - v_i)/\Delta t = (6 \text{ m/s} - 10 \text{ m/s})/3 \text{ s} = (-4 \text{ m/s})/3 \text{ s} = -1.33 \text{ m/s}^2.$

2.18 $\bar{a} = \Delta v/\Delta t = \dfrac{(-8 \text{ m/s}) - (10 \text{ m/s})}{12 \times 10^{-3} \text{ s}} = -1500 \text{ m/s}^2$

2.19 From the definition of acceleration, $\Delta v = a(\Delta t) = (0.8 \text{ m/s}^2)(2 \text{ s}) = 1.6 \text{ m/s}.$
From this, the final velocity is
  $V_f = 7 \text{ m/s} + 1.6 \text{ m/s} = 8.6 \text{ m/s}$

2.20 (a) $\bar{a}(0 \text{ to } 5 \text{ s}) = \Delta v/\Delta t = (0 - 0)/5 \text{ s} = 0$
$\bar{a}(5 \text{ s to } 15 \text{ s}) = (+8 \text{ m/s} - (-8 \text{ m/s}))/(10 \text{ s}) = +1.6 \text{ m/s}^2.$
$\bar{a}(0 \text{ to } 20 \text{ s}) = (+8 \text{ m/s} - (-8 \text{ m/s}))/20 \text{ s} = +0.80 \text{ m/s}^2.$
(b) At $t = 2$ s, the slope of the tangent line to the curve is 0.
At $t = 10$ s, the slope of the tangent line is $+1.6 \text{ m/s}^2$.
At $t = 18$ s, the slope of the tangent line is 0

2.21 The change in velocity during the first 15 s is $\Delta v = a \Delta t = (2.77 \text{ m/s}^2)(15 \text{ s}) = 41.55 \text{ m/s}.$ Thus, the velocity at $t = 15$ s is 41.55 m/s.
For the last 4.39 s, we have
  $\Delta v = a \Delta t = v_f - 41.55 \text{ m/s} = (-9.47 \text{ m/s}^2)(4.39 \text{ s}) = -41.57 \text{ m/s}$
Thus, the final velocity is -0.002 m/s
(a) $x = x_1 + x_2 + x_3$
$= (0 + \dfrac{1}{2}(2.77 \text{ m/s}^2)(15 \text{ s})^2) + (41.55 \text{ m/s }(123 \text{ s}) + 0)$

  $+ (41.55 \text{ m/s }(4.39 \text{ s}) + \dfrac{1}{2}(-9.47 \text{ m/s}^2)(4.39 \text{ s})^2)$

$= 311.6 \text{ m} + 5110.6 + 91.2 \text{ m} = 5513.4 \text{ m}$
(b) $\bar{v}_1 = \dfrac{311.6 \text{ m}}{15 \text{ s}} = 20.8 \text{ m/s}, \quad \bar{v}_2 = \dfrac{5110.6 \text{ m}}{123 \text{ s}} = 41.6 \text{ m/s}, \quad \bar{v}_3 = \dfrac{91.2 \text{ m}}{4.39 \text{ s}} = 20.8 \text{ m/s}$
and $\bar{v}_{total} = \dfrac{(\Delta x)_{total}}{t_{total}} = \dfrac{5513.4 \text{ m}}{142.4 \text{ s}} = 38.7 \text{ m/s}$

2.22 $v_i = 55 \dfrac{\text{mi}}{\text{h}} \dfrac{(0.447 \text{ m/s})}{(1 \text{ mi/h})} = 24.58 \text{ m/s}$;
and by the same method, $v_f = 26.82$ m/s. Thus, $\Delta v = 2.24$ m/s, and from the definition of acceleration
  $\Delta t = \Delta v/a = (2.24 \text{ m/s})/(0.6 \text{ m/s}^2) = 3.73 \text{ s}$

2.23 (a) The average acceleration can be found from the curve, and its value will be
  $\bar{a} = \Delta v/\Delta t = (16 \text{ m/s})/2 \text{ s} = +8.00 \text{ m/s}^2.$
(b) The instantaneous acceleration at $t = 1.5$ s equals the slope of the tangent line to the curve at that time. The line will have a slope of about $+11.0 \text{ m/s}^2$.

2.24 (a) From $v^2 = v_0^2 + 2ax$ we have
  $a = \dfrac{v^2 - v_0^2}{2x} = \dfrac{(30 \text{ m/s})^2 - (20 \text{ m/s})^2}{2(200 \text{ m})} = +1.25 \text{ m/s}^2.$

CHAPTER TWO SOLUTIONS

(b) From $\Delta x = \bar{v} \Delta t$ and $\bar{v} = (v_0 + v_f)/2$, we have

$$\Delta t = \frac{(2 \Delta x)}{(v_0 + v_f)} = \frac{(2(200 \text{ m}))}{(20 \text{ m/s} + 30 \text{ m/s})} = 8 \text{ s}$$

2.25 Use $v^2 = v_0^2 + 2ax$. We have
$$v^2 = 0 + 2(0.3 \text{ m/s}^2)(25 \text{ m})$$
and $v = 3.87$ m/s

2.26 $v_f = 40$ mi/h = 17.9 m/s and $v_0 = 0$
(a) To find the distance traveled, we first find the acceleration as
$$a = \frac{v - v_0}{t} = \frac{17.9 \text{ m/s} - 0}{12 \text{ s}} = 1.49 \text{ m/s}^2.$$
and $x = v_0 t + \frac{1}{2} at^2 = \frac{1}{2}(1.49 \text{ m/s}^2)(12 \text{ s})^2 = 107$ m

(b) The acceleration was found in part (a) as 1.49 m/s^2

2.27 The final velocity of the car is found as
$$v^2 = v_0^2 + 2ax = (20 \text{ m/s})^2 + 2(4 \text{ m/s}^2)(50 \text{ m})$$
From which, $v = 28.3$ m/s
Then from the definition of acceleration, we find t as
$$t = \frac{v - v_0}{a} = \frac{28.3 \text{ m/s} - 20 \text{ m/s}}{4 \text{ m/s}^2} = 2.07 \text{ s}$$

2.28 (a) From the definition of acceleration, we have
$$a = \frac{v - v_0}{t} = \frac{0 - 40 \text{ m/s}}{5 \text{ s}} = -8 \text{ m/s}^2.$$
(b) From $x = v_0 t + \frac{1}{2} at^2$, we have
$$x = (40 \text{ m/s})(5 \text{ s}) + \frac{1}{2}(-8 \text{ m/s}^2)(5 \text{ s})^2 = 100 \text{ m}$$

2.29 Use $v^2 = v_0^2 + 2ax$.
$$a = \frac{v^2 - v_0^2}{2x} = \frac{(8.0 \text{ m/s})^2 - (2.0 \text{ m/s})^2}{2(12.0 \text{ m})} = +2.5 \text{ m/s}^2.$$

2.30 (a) From $v = v_0 + at$, we have
$$t = \frac{v - v_0}{a} = \frac{5.4 \times 10^5 \text{ m/s} - 3.0 \times 10^5 \text{ m/s}}{8.0 \times 10^{14} \text{ m/s}^2} = 3 \times 10^{-10} \text{ s}$$
(b) From $x = v_0 t + \frac{1}{2} at^2$
$$x = (3.0 \times 10^5 \text{ m/s})(3 \times 10^{-10} \text{ s}) + \frac{1}{2}(8.0 \times 10^{14} \text{ m/s}^2)(3 \times 10^{-10} \text{ s})^2 = 1.26 \times 10^{-4} \text{ m}$$

2.31 (a) $t = \dfrac{v - v_0}{a} = \dfrac{0 - 100 \text{ m/s}}{-5 \text{ m/s}^2} = 20$ s

(b) $x = \bar{v} t = \left(\dfrac{v + v_0}{2}\right) t = \dfrac{100 \text{ m/s} + 0}{2} 20 \text{ s} = 1000 \text{ m} = 1.0$ km

CHAPTER TWO SOLUTIONS

Therefore, the minimum distance to stop exceeds the length of the runway, so it cannot land safely.

2.32 The initial velocity of the train is v_i = 82.4 km/h = 22.9 m/s and the final velocity is v_f = 16.4 km/h = 4.56 m/s. We also know that $\Delta x = \bar{v}\, t$, so

$$t = \frac{\Delta x}{\bar{v}} \quad \text{and} \quad \bar{v} = \frac{v_i + v_f}{2} = 13.73 \text{ m/s}$$

so $\quad t = \dfrac{400 \text{ m}}{13.73 \text{ m/s}} = 29.1 \text{ s}$

2.33 The velocity when the brakes are applied is
$\quad v_i = v_0 + at = 0 + (1.5 \text{ m/s}^2)(5 \text{ s}) = 7.5 \text{ m/s}$
(a) After braking
$\quad v_f = v_i + at = 7.5 \text{ m/s} + (-2 \text{ m/s}^2)(3 \text{ s}) = 1.5 \text{ m/s}$
(b) $x = x_1 + x_2 = \bar{v}_1 t_1 + \bar{v}_2 t_2$
$\quad \bar{v}_1 = \dfrac{0 + 7.5 \text{ m/s}}{2} = 3.75 \text{ m/s}, \quad \bar{v}_2 = \dfrac{7.5 \text{ m/s} + 1.5 \text{ m/s}}{2} = 4.5 \text{ m/s}$
Thus, $x = (3.75 \text{ m/s})(5 \text{ s}) + (4.5 \text{ m/s})(3 \text{ s}) = 32.3 \text{ m}$

2.34 Let us first find the speed of the student at the beginning of the final 500 yd dash. For the first 10 min, she had run a distance of one mile (1760 yd) less 500 yd for a total distance of 1260 yd. Thus her average velocity was
$\quad \bar{v}$ = 1260 yd/600 s, which when converted to m/s is 1.92 m/s.
We shall assume that this was her speed at the beginning of the final 500 yd dash. Thus, she must cover 500 yd (457 m) in 2 min (120 s) starting with 1.92 m/s initial velocity.
The average velocity that she must have to make the last 500 yd in two minutes is
$\quad \bar{v} = \Delta x/\Delta t = (457 \text{ m})/(120 \text{ s}) = 3.81 \text{ m/s}$
From $\bar{v} = \dfrac{v_0 + v}{2}$, we find the required final velocity to be
$\quad v = 2\bar{v} - v_0 = 2(3.81 \text{ m/s}) - 1.92 \text{ m/s} = 5.70 \text{ m/s}$
and the needed acceleration is
$\quad a = \dfrac{v - v_0}{t} = \dfrac{5.70 \text{ m/s} - 1.92 \text{ m/s}}{120 \text{ s}} = 0.032 \text{ m/s}^2.$
Thus, her maximum acceleration of 0.15 m/s² is more than sufficient.

2.35 (a) Take $t = 0$ at the time when the first player starts to chase the second player. At this time, the second player is 36 m in front of the first player. Let us write down $x = v_0 t + \frac{1}{2} at^2$ for both players. For the first player, we have

$\quad x_1 = v_0 t + \frac{1}{2} at^2 = 0 + \frac{1}{2}(4 \text{ m/s}^2)t^2, \qquad (1)$

and for the second player,

$\quad x_2 = v_0 t + \frac{1}{2} at^2 = (12 \text{ m/s})t + 0 \qquad (2)$

When the players are side-by-side,
$\quad x_1 = x_2 + 36 \text{ m} \qquad (3)$
From Equations (1), (2), and (3), we find

CHAPTER TWO SOLUTIONS

$t^2 - 6t - 18 = 0$

The roots of this equation are $t = -2.2$ s and $t = +8.2$ s. We must choose the 8.2 s answer since the time must be greater than zero.

(b) $\Delta x_1 = v_{01}t + \frac{1}{2}a_1 t^2 = 0 + \frac{1}{2}(4 \text{ m/s}^2)(8.2 \text{ s})^2 = 134$ m

---

2.36 (a) Use $v^2 = v_0^2 + 2ay$ with $v = 0$. We have

$0 = (25 \text{ m/s})^2 + 2(-9.8 \text{ m/s}^2)y_m$.

This gives the maximum height, y_m, as

$y_m = 31.9$ m.

(b) The time to reach the highest point is found from the definition of acceleration as

$t = \dfrac{0 - 25 \text{ m/s}}{(-9.8 \text{ m/s}^2)} = 2.55$ s

(c) From the symmetry of the motion, the ball takes the same amount of time to reach the ground from its highest point as it does to move from the ground to its highest point. Thus, $t = 2.55$ s.

(d) We can use $v = v_0 + at$, with the position of the ball at its highest point as the origin of our coordinate system. Thus, $v_0 = 0$, and t is the time for the ball to move from its maximum height to ground level. This was found in part (c) to be 2.55 s. Thus,

$v = 0 + (-9.8 \text{ m/s}^2)(2.55 \text{ s}) = -25$ m/s

2.37 $y = v_{0y}t + \frac{1}{2}at^2$ at $t = 3$ s becomes

$y = 0 + \frac{1}{2}(-9.8 \text{ m/s}^2)(3 \text{ s})^2 = -44.1$ m

---

2.38 $v^2 = v_0^2 + 2gy$ becomes

$0 = v_0^2 + 2(-9.8 \text{ m/s}^2)(3.43 \times 10^{-2} \text{ m})$

From which, $v_0 = 0.82$ m/s

2.39 We shall use $v^2 = v_0^2 + 2ay$ in the vertical direction, and we shall select the origin of our coordinate system at the position of the maximum height reached by the ball. At this point, the initial velocity in the vertical direction is zero. The displacement from the maximum height, 40 m, to the point where it is caught, 30 m, is -10 m. Thus,

$v^2 = 0 + 2(-9.8 \text{ m/s}^2)(-10 \text{ m})$

and

$v = -14$ m/s

2.40 (a) To find the maximum upward speed, we shall use $v^2 = v_0^2 + 2ay$, and require that $v = 0$ just as the ball reaches ceiling height. Thus the initial speed is found as

$0 = v_0^2 + 2(-9.8 \text{ m/s}^2)(2 \text{ m})$

From which,

$v_0 = 6.26$ m/s

(b) The time to reach the maximum height is found from the definition of acceleration as

CHAPTER TWO SOLUTIONS

$$t = \frac{0 - 6.26 \text{ m/s}}{(-9.8 \text{ m/s}^2)} = 0.639 \text{ s}.$$

Thus, the total time of flight is two times this value, or 1.28 s.

2.41 Use $y = v_{0y}t + \frac{1}{2}at^2$

$$-76 \text{ m} = 0 + \frac{1}{2}(-9.8 \text{ m/s}^2)t^2$$

$$t = 3.94 \text{ s}$$

2.42 (a) Choose the origin of the coordinate system at the location of the parachutist. We use
$$v^2 = v_0^2 + 2ay$$
$$v^2 = (-10 \text{ m/s})^2 + 2(-9.8 \text{ m/s}^2)(-50 \text{ m})$$
From which $v = -32.9$ m/s. The time to reach the ground is found from $v = v_0 + at$

$$t = \frac{v - v_0}{a} = \frac{-32.9 \text{ m/s} - (-10 \text{ m/s})}{(-9.8 \text{ m/s}^2)} = 2.33 \text{ s}$$

(b) The velocity was found to be -32.9 m/s in part (a)

2.43 We shall first find the height of the rocket and its velocity at the instant it runs out of fuel. The height of the rocket is found from $y = v_0t + \frac{1}{2}at^2$.

$$y_1 = 0 + \frac{1}{2}(29.4 \text{ m/s}^2)(4 \text{ s})^2 = 235 \text{ m}$$

The velocity is found from $v = v_0 + at$. We have

$$v_1 = 0 + (29.4 \text{ m/s}^2)(4 \text{ s}) = 117.6 \text{ m/s}$$

At this point, the rocket begins to behave as a freely falling body. We shall now find how much higher it rises once its fuel is exhausted by the use of $v^2 = v_0^2 + 2ay$.

$$0 = (117.6 \text{ m/s})^2 + 2(-9.8 \text{ m/s}^2)h.$$

This gives
$$h = 706 \text{ m}$$

Thus, the total height reached is
$$y_{max} = y_1 + h = 235 \text{ m} + 706 \text{ m} = 941 \text{ m}$$

2.44 (a) The time to fall a distance of 24 km is found from $y = v_0t + \frac{1}{2}at^2$.

$$-2.4 \times 10^4 \text{ m} = 0 + \frac{1}{2}(0.38)(-9.8 \text{ m/s}^2)t^2$$

From which
$$t = 113.5 \text{ s}.$$

(b) The velocity at this time is found from $v = v_0 + at$.

$$v = 0 + (0.38)(-9.8 \text{ m/s}^2)(113.5 \text{ s}) = -423 \text{ m/s}$$

(Air resistance increases as the speed of an object increases. As a result, air resistance should become significant at these speeds.)

2.45 (a) The time to reach the ground from the initial position of the ball is found as

$$y = v_0t + \frac{1}{2}at^2$$

CHAPTER TWO SOLUTIONS

$$-2\text{ m} = 0 + \frac{1}{2}(-9.8\text{ m/s}^2)t^2$$

This gives
$$t = 0.639\text{ s}$$

The velocity just before striking the ground is now found from $v = v_0 + at$.
$$v = 0 + (-9.8\text{ m/s}^2)(0.639\text{ s}) = -6.26\text{ m/s}$$

(b) The initial velocity of the ball on its upward flight after collision with the ground is found from
$$v^2 = v_0^2 + 2ay$$
$$0 = v_0^2 + 2(-9.8\text{ m/s}^2)(1.85\text{ m})$$
$$v_0 = 6.02\text{ m/s}$$

(c) The time to rise to its maximum height is found from the definition of acceleration.
$$t = \frac{0 - 6.02\text{ m/s}}{-9.8\text{ m/s}^2} = 0.614\text{ s}$$

The time to fall from its initial position was found in part (a) to be 0.639 s. Thus the total time of movement is 0.639 s + 0.614 s = 1.25 s

---

2.46 (a) The acceleration of the bullet is found from $v^2 = v_0^2 + 2ax$.
$$(300\text{ m/s})^2 = (400\text{ m/s})^2 + 2a(0.1\text{ m})$$
From which, we find $a = -3.5 \times 10^5\text{ m/s}^2$.

(b) The time of contact with the board is found from $v = v_0 + at$.
$$t = \frac{v - v_0}{a} = \frac{300\text{ m/s} - 400\text{ m/s}}{-3.5 \times 10^5\text{ m/s}^2} = 2.86 \times 10^{-4}\text{ s}$$

2.47 The acceleration of the ball is -9.8 m/s^2 even though it was thrown downward. The velocity of the ball is given by
$$v = v_0 + at = (-10\text{ m/s}) - (9.8\text{ m/s}^2)(2\text{ s}) = -29.6\text{ m/s}$$

2.48 We use $\bar{a} = \Delta v / \Delta t$

(a) 0 to 1 s: $\bar{a} = \frac{((-3\text{ m/s}) - 0)}{(1\text{ s})} = -3\text{ m/s}^2$.

(b) 1 s to 3 s: $\bar{a} = \frac{((3\text{ m/s}) - (-3\text{ m/s}))}{(2\text{ s})} = 3\text{ m/s}^2$

(c) 3 s to 4 s: $\bar{a} = \frac{((3\text{ m/s}) - 3\text{ m/s})}{(1\text{ s})} = 0$

(d) The instantaneous acceleration is the slope of the tangent line drawn to the v vs t curve at the time of interest. At t = 0.5s, the slope of this line is -3 m/s^2; at t = 2 s, the slope is 3 m/s^2, and at t = 3.5 s, the slope is 0.

2.49 The initial distance the leader is ahead of the second swimmer equals the distance the leader swims in 0.5 s, which is (4 m/s)(0.5 s) = 2 m. Thus, the leader has 50 m to go and the second swimmer has 52 m to the end of the pool. The time for the leader to reach the end of the pool is
$$t = \frac{x}{v} = \frac{50\text{ m}}{4\text{ m/s}} = 12.5\text{ s}$$

In this same time, the second swimmer must swim 52 m. The minimum velocity is

CHAPTER TWO SOLUTIONS

$$v = \frac{x}{t} = \frac{52 \text{ m}}{12.5 \text{ s}} = 4.16 \text{ m/s}$$

2.50 (a) The velocity of the woman just before striking the box is found from
$$v^2 = v_0^2 + 2ay$$
$$v^2 = 0 + 2(-9.8 \text{ m/s}^2)(-44 \text{ m})$$
$$v = -29.4 \text{ m/s}$$
(b) While the box crumpled, she went from a velocity of -29.4 m/s to 0 in 0.46 m. Her deceleration is found from $v^2 = v_0^2 + 2ay$.
$$a = \frac{v^2 - v_0^2}{2x} = \frac{0 - (-29.4 \text{ m/s})^2}{2(-0.46 \text{ m})} = 939 \text{ m/s}^2.$$
(c) and the time is found from the definition of acceleration as
$$t = \frac{v - v_0}{a} = \frac{0 - (-29.4 \text{ m/s})}{939 \text{ m/s}^2} = 3.13 \times 10^{-2} \text{ s}$$

2.51 The falling ball moves a distance of (15 m - h) before they meet, where h is the height above the ground at which they pass. Apply $y = v_0 t + \frac{1}{2} at^2$
$$-(15 \text{ m} - h) = 0 - \frac{1}{2} gt^2.$$
from which
$$h = 15 \text{ m} - \frac{g}{2} t^2. \quad (1)$$
Applying $y = v_0 t + \frac{1}{2} at^2$ to the rising ball gives
$$h = (25 \text{ m/s}) t - \frac{1}{2} gt^2 \quad (2)$$
Equating the expressions for h in (1) and (2) and solving for t gives t = 0.6 s.

2.52 35 mi/h = 51.3 ft/s
At this speed, during the reaction time t_r, the car will travel a distance given by
$$x_1 = (51.3 \text{ ft/s}) t_r \quad (1)$$
The distance required for the car to come to rest from a speed of 51.3 ft/s is
$$v^2 = v_0^2 + 2ax$$
$$0 = (51.3 \text{ ft/s})^2 + 2(-9.0 \text{ ft/s}^2) x$$
$$x = 146 \text{ ft}$$
Thus, the distance the car can travel during the reaction time is 200 ft - 146 ft = 54 ft.
$$t_r = \frac{54 \text{ ft}}{51.3 \text{ ft/s}} = 1.05 \text{ s}$$

2.53 (a) The velocity with which the first stone hits the water is
$$v_1^2 = v_0^2 + 2ay$$
$$v_1^2 = (-2 \text{ m/s})^2 + 2(-9.8 \text{ m/s}^2)(-50 \text{ m})$$
$$v_1 = -31.3 \text{ m/s}$$
And the time for the first stone to reach the water is
$$v = v_0 + at$$
$$-31.37 \text{ m/s} = -2 \text{ m/s} - (9.8 \text{ m/s}^2) t_1$$
$$t_1 = 3.0 \text{ s}$$

CHAPTER TWO SOLUTIONS

(b) Since they hit simultaneously, the second stone which is released 1 s later, will hit the water after an elapsed time of 2.0 s. Thus,
$$y = v_0 t + \frac{1}{2} a t^2$$
$$(-50 \text{ m}) = v_0(2.0 \text{ s}) + \frac{1}{2}(-9.8 \text{ m/s}^2)(2.0 \text{ s})^2$$
Which yields, $v_0 = -15.2$ m/s

(c) We found the velocity of the first stone in part (a) above. The velocity of the second is found as
$$v_2^2 = v_0^2 + 2ay$$
$$v_2^2 = (-15.2 \text{ m/s})^2 + 2(-9.8 \text{ m/s}^2)(-50 \text{ m})$$
$$v_2 = -34.8 \text{ m/s}$$

2.54 (a) $\Delta x = v \Delta t = (110 \text{ ft/s})(5 \text{ s}) = 5500$ ft
(b) For the airplane, $v = \Delta x / \Delta t = (5500 \text{ ft})(5 + 10)\text{s} = 367$ ft/s
(c) Light would require a time $\Delta t = 5500 \text{ ft}/9.9 \times 10^8$ ft/s $= 5.56$ μs to travel from plane to observer. During this time the plane would travel only a distance of 0.002 ft.

2.55 (a) The distance the stock car travels before the sports car starts is
$$x = v_0 t + \frac{1}{2} a t^2 = 0 + \frac{1}{2}(12 \text{ ft/s}^2)(1 \text{ s})^2 = 6 \text{ ft.}$$
The speed of the stock car when the sports car starts is
$$v = v_0 + at = 0 + (12 \text{ ft/s}^2)(1 \text{ s}) = 12 \text{ ft/s}$$
After the sports car starts, the stock car travels a distance of
$$x_1 = v_0 t + \frac{1}{2} a t^2 = (12 \text{ ft/s})t + \frac{1}{2}(12 \text{ ft/s}^2)t^2$$
In this same time, t, the sports car must travel a distance $x_2 = x_1 + 6$ ft. We find x_2 as
$$x_2 = v_0 t + \frac{1}{2} a t^2 = 0 + \frac{1}{2}(16 \text{ ft/s}^2)t^2$$
Now use $x_2 = x_1 + 6$ to get a quadratic equation for t. This can be solved to find
$t = 6.46$ s.

(b) At the time $t = 6.46$ s, the sports car's position is
$$D = v_0 t + \frac{1}{2} a t^2 = 0 + \frac{1}{2}(16 \text{ ft/s}^2)(6.46 \text{ s})^2 = 334 \text{ ft.}$$
(c) The velocity of the sports car is
$$v = v_0 + at = 0 + (16 \text{ ft/s}^2)(6.46 \text{ s}) = 103 \text{ ft/s}$$
and the velocity of the stock car is
$$v = v_0 + at = 12 \text{ ft/s} + (12 \text{ ft/s}^2)(6.46 \text{ s}) = 89.6 \text{ ft/s}$$

2.56 The total distance traveled is: $(d + 4)$ and
$$(d + 4) = v_0(10) - \frac{1}{2} a(10)^2$$
The distance to the dog is d, and $d = v_0(8) - \frac{1}{2} a(8)^2$

When the trolley is finally stopped, $v = 0$ and $0 - v_0^2 = -2a(d + 4)$.
Solving these three equations, we eliminate v_0 and a and find $d = 96$ m.

2.57 Consider the motion of the sled in three time intervals:
1) Constant acceleration for t_1.

CHAPTER TWO SOLUTIONS

2) Constant velocity for t_2, and
3) Constant (negative) acceleration for t_3.

(a) $x_1 = v_0 t_1 + \frac{1}{2} a_1 t_1^2$ and $x_2 = v_{02} t_2 + \frac{1}{2} a_2 t_2^2$

But $t_2 = (90 - t_1)$, $a_2 = 0$, and $v_{02} = v_1 = a_1 t_1$
so that $x_2 = (a_1 t_1)(90 - t_1)$

$x_{(1+2)} = \frac{1}{2} a_1 t_1^2 + (a_1 t_1)(90 - t_1)$ or

$17500 \text{ ft} = \frac{1}{2} (40 \text{ ft/s}^2)(t_1^2) + (40 \text{ ft/s}^2)(90 \text{ s}) t_1 - (40 \text{ ft/s}^2)(t_1^2)$

This becomes $t_1^2 + 180 t_1 - 875 = 0$ and $t_1 = 4.74$ s and $t_2 = 90 - t_1 = 85.26$ s

(b) $v = d_2/t_2$ when $d_2 = 17{,}500 \text{ ft} - d_1$ and

$d_1 = \frac{1}{2} a_1 t_1^2 = \frac{1}{2} (40 \text{ ft/s}^2)(4.64 \text{ s})^2 = 449$ ft

Therefore, $v = (17{,}500 - 449) \text{ft}/(85.26 \text{ s}) = 200$ ft/s

(c) For the third time interval
$v^2 = v_1^2 + 2ax$
$x_3 = v_3^2 - v_{03}^2 / 2a_3 = 0 - (200 \text{ ft/s})^2/(2)(-20 \text{ ft/s}^2) = 1000$ ft

Final position of the sled is 18,500 ft from the starting point.

(d) $T = (v - v_0)/a = 0 - (200 \text{ ft/s})/(20 \text{ ft/s}^2) = 10$ s

Sled stops 10 s after the end of the 90 s interval or a total time of travel of 100 s.

CHAPTER THREE SOLUTIONS

3.1 The displacement from the 260 mi marker to the 150 mi marker is 110 mi. The displacement vector from the 150 mi marker back to the 175 mi marker is 25 mi in the opposite direction to the first displacement. The vectors must be subtracted to give a resultant of 85 mi.

3.2 When displacements are in the same direction, the vectors are added as algebraic quantities. Thus, an 80 m walk east followed by a 125 m walk also eastward gives a resultant displacement of 205 m eastward.
(b) The displacements are in opposite directions in this case, and thus, the vectors are subtracted to give a resultant of 45 m westward.

3.3 (a) Carefully draw, to scale, a vector 3 units long along the x direction, and from the tip of this vector, draw another of length 4 units in the negative y direction. The resultant is the length, to scale, of the vector drawn from the tail of the first to the tip of the second. This will be a vector 5 units long and at an angle of 53° below the x axis.
(b) In this case, the second vector, B, will be in the + y direction. The resultant will still be 5 units long, but at an angle of 53° above the + x direction.

3.4 (a) Drawing these to scale and maintaining their respective directions yields a resultant of 5.2 m at an angle of 60° above the x axis.
(b) Maintain the direction of **A**, but reverse the direction of **B** by 180°. The resultant is 3.0 m at an angle of 30° below the x axis.
(c) Maintain the direction of **B**, but reverse the direction of **A**. The resultant is 3.0 m at an angle of 150° with respect to the + x axis.
(d) Maintain the direction of A, reverse the direction of B, and multiply its magnitude by two. The resultant is 5.2 m at an angle of 60° below the + x axis.

3.5 Your sketch when drawn to scale should look somewhat like the one at the right. The distance R and the angle θ can be measured to give, upon use of your scale factor, the values of R = 421 ft at about 3° below the horizontal.

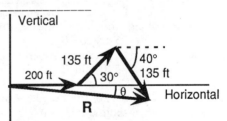

3.6 Your sketch should be drawn to scale, and should look somewhat like that pictured at the right. The angle from the westward direction, θ, can be measured to be 4.34° north of west, and the distance from your sketch can be converted according to the scale used to be 7.92 m.

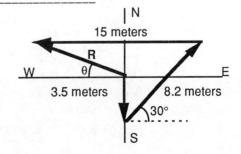

19

CHAPTER THREE SOLUTIONS

3.7 The vector diagram sketched for this problem should look like the one shown at the right. The initial displacement **A** = 100 m and the resultant **R** = 175 m are both known. In order to reach the end point of the run following the initial displacement, the jogger must follow the path shown as **B**. The distance can be found from the scale used for your sketch and the angle θ measured. The results should be about 83 m at 33° north of west.

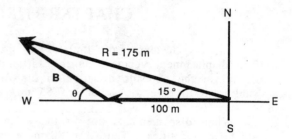

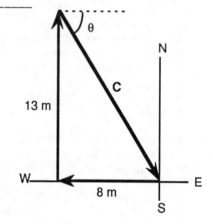

3.8 The displacement vectors **A** = 8 m westward and **B** = 13 m north can be drawn to scale as at the right. The vector **C** represents the displacement that the man in the maze must undergo to return to his starting point. The scale used to draw the sketch can be used to find **C** to be 15.3 m and the angle θ can be measured to be about 58°.

3.9 Moving for 135 ft an an angle of 40° above the horizontal moves the roller coaster along the horizontal a distance x given by
　　x = (135 ft)(cos40°) = 103 ft
and the coaster rises vertically a distance y given by
　　y = (135 ft)(sin40°) = 86.8 ft

3.10 Diving at 30° below the horizontal for a distance of 50 m brings the sub to a distance h below the surface given by
　　h = (50 m)(sin30°) = 25 m

3.11 Let **A** be the vector corresponding to the 10 yd run, **B** to the 15 yd run, and **C** to the 50 yd pass. Also, we choose a coordinate system with the +y direction downfield, and the +x direction toward the sideline to which the player runs. The components of the vectors are then
　　$A_x = 0$　　　$A_y = -10$ yd
　　$B_x = 15$ yds　$B_y = 0$
　　$C_x = 0$　　　$C_y = +50$ yds
From these, $R_x = 15$ yds, and $R_y = 40$ yds, and the Pythagorean theorem gives
　　$R = \sqrt{(R_x)^2 + (R_y)^2} = \sqrt{(15 \text{ yds})^2 + (40 \text{ yds})^2} = 42.7$ yds

CHAPTER THREE SOLUTIONS

3.12 Let **A** be the vector corresponding to the 100 m displacement, **B** to the 300 m displacement, **C** to 150 m, and **D** to 200 m. The components are
$A_x = 100$ m $A_y = 0$
$B_x = 0$ $B_y = -300$ m
$C_x = -129.9$ m $C_y = -75.0$ m
$D_x = -100$ m $D_y = 173.2$ m

The resultant x component is -129.9 m, and the resultant y component is -201.8 m. From the Pythagorean theorem, we find
$$R = \sqrt{(R_x)^2 + (R_y)^2} = \sqrt{(-129.9 \text{ m})^2 + (-201.8 \text{ m})^2} = 240 \text{ m}$$
and $\theta = \tan^{-1}\dfrac{R_y}{R_x} = \tan^{-1}\dfrac{-201.8}{-129.9} = \tan^{-1}(1.553)$
$\theta = 237°$

3.13 The two basic situations are shown in the figure below. In case (a) the net displacement in the x direction is $R_x = 40$ m + 20 m = 80 m, and the net displacement in the y direction is 15 m. The resultant is found by the Pythagorean theorem as
$R = \sqrt{(R_x)^2 + (R_y)^2} = \sqrt{(60 \text{ m})^2 + (15 \text{ m})^2} = 61.8$ m
and the angle θ is found as
$\tan\theta = \dfrac{15 \text{ m}}{60 \text{ m}} = 0.25$, from which $\theta = 14.0°$.

(case b) For the situation depicted in case (b) the resultant x displacement is 20 m and the resultant y displacement is 15 m. Using the Pythagorean theorem and trig functions as above, R = 25 m and $\theta = 36.9°$.

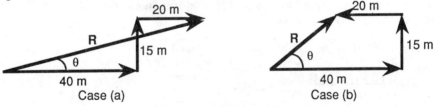

Case (a)   Case (b)

3.14 The x component of the initial movement from city A to city B is 800 mi and the y component is zero. (The x axis is selected as positive toward the east and the y axis is positive northward.) The x component of the displacement from city B to city C is
x(B to C) = (600 mi)(cos40°) = 460 mi.
The y component of the displacement from city B to C is
y(B to C) = (600 mi)(sin40°) = 386 mi.
The resultant displacement in the x direction is
$R_x = 1260$ mi,
and the resultant y displacement is
$R_y = 386$ mi.
The Pythagorean theorem then gives the resultant displacement as
$$R = \sqrt{(R_x)^2 + (R_y)^2} = \sqrt{(1260 \text{ mi})^2 + (386 \text{ mi})^2} = 1320 \text{ mi}$$
The angle between R and the eastward direction can be found from
$\tan\theta = \dfrac{R_y}{R_x} = \dfrac{386}{1260} = 0.306$
from which θ is found to be 17.0° north of east.

CHAPTER THREE SOLUTIONS

3.15 (a) Let us call $\mathbf{F}_1$ the 120 N force and $\mathbf{F}_2$ the 80 N force. The components of these forces are

$F_{1x} = 60$ N $\quad F_{1y} = 103.9$ N
$F_{2x} = -20.7$ N $\quad F_{2y} = 77.3$ N

The resultant x component is 39.3 N, and the resultant y component is 181.2 N. The resultant is

$$R = \sqrt{(R_x)^2 + (R_y)^2} = 185.4 \text{ N}$$

also $\tan\theta = \dfrac{181.2}{39.3} = 4.61$, from which $\theta = 77.8°$.

(b) To have a resultant of zero on the mule, the net force above must be cancelled by a force equal in magnitude to it and oppositely directed. Thus, the required force is 185.4 N at an angle of 257.8°

3.16 We shall use the vector equation

$\mathbf{A} + \mathbf{B} = \mathbf{R}$

where $\mathbf{A}$ is the 150 cm displacement at 120°, $\mathbf{R}$ is the resultant displacement and $\mathbf{B}$ is the second displacement that we are looking for. The equation above can be solved for $\mathbf{B}$ to give

$\mathbf{B} = \mathbf{R} - \mathbf{A} = \mathbf{R} + (-\mathbf{A})$

The vector $\mathbf{R}$ has the following components

$R_x = (140 \text{ cm})(\cos 35°) = 114.7$ cm $\quad R_y = (140 \text{ cm})(\sin 35°) = 80.3$ cm

The vector $-\mathbf{A}$ has components

$A_x = 75$ cm $\quad A_y = -130$ cm

Thus, the vector B has the following components

$B_x = 190$ cm $\quad B_y = -49.7$ cm

From the Pythagorean theorem, $B = 196$ cm, and $\tan\theta$ gives $\theta = -14.7°$ with respect to the positive x axis.

3.17 The time of flight of the football is found from $v = v_0 + at$, applied in the vertical direction

$$t = \dfrac{-10 \text{ m/s} - 10 \text{ m/s}}{-9.8 \text{ m/s}^2} = 2.04 \text{ s}$$

The initial velocity in the x direction is found from

$v_{0x} = v_0 \cos\theta = (20 \text{ m/s})\cos 30° = 17.3$ m/s

The horizontal distance moved during this time can now be found as

$x = v_{0x}t = (17.3 \text{ m/s})(2.04 \text{ s}) = 35.3$ m

3.18 We have $v_{0x} = 5$ m/s and $v_{0y} = 0$. The time of flight for Tom is found from

$$y = v_{0y}t + \dfrac{1}{2}at^2$$

$-1.5 \text{ m} = 0 + \dfrac{1}{2}(-9.8 \text{ m/s}^2)t^2$

$t = 0.553$ s

The distance moved in the x direction during this time is

$x = v_{0x}t = (5 \text{ m/s})(0.553 \text{ s}) = 2.77$ m

The horizontal component of velocity does not change during the flight, so Tom strikes the floor with a horizontal component of velocity of $v_{0x} = 5$ m/s. The vertical component of velocity is found as

$v_y = v_{0y} + at = 0 - (9.8 \text{ m/s}^2)(0.553 \text{s}) = -5.42$ m/s

CHAPTER THREE SOLUTIONS

3.19 We use $y = v_{0y}t + \frac{1}{2}at^2$ to find the initial velocity in the vertical direction. At the end of the flight, $y = 0$ and $t = 5$ s. Thus,

$$0 = v_{0y}(5 \text{ s}) + \frac{1}{2}(-9.8 \text{ m/s}^2)(5 \text{ s})^2$$

$$v_{0y} = 24.5 \text{ m/s}$$

but, $v_{0y} = v_0 \sin 45°$. From which $v_0 = 34.6$ m/s

3.20 We choose our coordinate system at the initial position of the projectile. After 3 s, it is at ground level, -H. To find H, we use $y = v_{0y}t + \frac{1}{2}at^2$.

$$-H = (15 \text{ m/s})(\sin 25°)(3 \text{ s}) + \frac{1}{2}(-9.8 \text{ m/s}^2)(3 \text{ s})^2 = -25.1 \text{ m, or } H = 25.1 \text{ m}$$

3.21 We shall first find the initial velocity of the ball thrown vertically upward. At its maximum height, $v = 0$ and $t = 1.5$ s.

$$v = v_0 + at$$

$$0 = v_{0y} + (-9.8 \text{ m/s}^2)(1.5 \text{ s})$$

$$v_{0y} = 14.7 \text{ m/s}$$

In order for the second ball to reach the same vertical height as the first, the second must have the same initial vertical velocity. Thus, we can find v_0 as

$$v_0 = \frac{v_{0y}}{\sin 30°} = 29.4 \text{ m/s}$$

3.22 Let us first find the time for an object to free fall 20 m starting with an initial velocity of 0.

$$y = v_{0y}t + \frac{1}{2}a_y t^2$$

becomes $\quad -20 \text{ m} = 0 + \frac{1}{2}(-9.8 \text{ m/s}^2)t^2$

which gives $t = 2.02$ s

Now, let us look at the horizontal motion of the ball moving as a projectile.

$$x = vt = (2.5 \text{ m/s})(2.02 \text{ s}) = 5.05 \text{ m}$$

3.23 Consider first the horizontal motion of the projectile. We have

$$x = v_{0x}t$$

or, $\quad 500 \text{ m} = (v_0 \cos 35)t$

which gives, $\quad t = \dfrac{500 \text{ m}}{v_0 \cos 35}$ (1)

Now, consider the vertical motion.

$$y = v_{0y}t + \frac{1}{2}a_y t^2$$

becomes, $\quad 300 \text{ m} = (v_0 \sin 35)\left(\dfrac{500 \text{ m}}{v_0 \cos 35}\right) + \frac{1}{2}(-9.8 \text{ m/s}^2)\left(\dfrac{500 \text{ m}}{v_0 \cos 35}\right)^2$

which reduces to $\quad 300 \text{ m} = 350.1 \text{ m} - (4.9 \text{ m/s}^2)\left(\dfrac{610.4 \text{ m}}{v_0}\right)^2$

This can be solved for v_0 to give $v_0 = 191$ m/s

3.24 (a) First, find the speed of the car when it reaches the edge of the cliff from

CHAPTER THREE SOLUTIONS

$v^2 = v_0^2 + 2ax = 0 + 2(4 \text{ m/s}^2)(50 \text{ m})$
$v = 20 \text{ m/s}$

Now, consider the projectile phase of the car's motion. We shall first find the vertical velocity with which the car strikes the water as

$v_y^2 = v_{0y}^2 + 2ay$
$v_y^2 = (-20\sin37° \text{ m/s})^2 + 2(-9.8 \text{ m/s}^2)(-30 \text{ m})$
$v_y = -27.1 \text{ m/s}$

and the time of flight is found from $v_y = v_{0y} + at$

$-27.1 \text{ m/s} = (-20 \text{ m/s})(\sin37°) + (-9.8 \text{ m/s}^2)t$

This gives $t = 1.53$ s.
The horizontal motion of the car during this time is

$x = v_{0x}t = (20 \text{ m/s})(\cos37°)(1.53 \text{ s}) = 24.5 \text{ m}$

(b) The time of flight of the car has already been found in part (a).

3.25 We have $v_{1s} = 0.511$ km/h (velocity of boat 1 relative to the shore), and $v_{2s} = -0.433$ km/h (velocity of boat 2 relative to the shore).
Let v_{21} be the velocity of boat 2 relative to boat 1. Then

$v_{2s} = v_{21} + v_{1s}$

or $v_{21} = v_{2s} - v_{1s} = -0.433$ km/h $-$ (0.511 km/h) $= -0.944$ km/h.

3.26 (a) We have $v_{pe} = 0.511$ km/h (velocity of police relative to the earth), and $v_{me} = 80$ km/h (velocity of motorist relative to the earth).
We have $v_{me} = v_{mp} + v_{pe}$
or $v_{mp} = v_{me} - v_{pe} = 80$ km/h $- 95$ km/h $= -15$ km/h
(b) $v_{pe} = v_{pm} + v_{me}$ gives $v_{pm} = v_{pe} - v_{me} = 95$ km/h $- 80$ km/h $= 15$ km/h

---

3.27 We have $v_{eg} = \dfrac{20.0 \text{ m}}{50 \text{ s}} = 0.4$ m/s (velocity of escalator relative to the ground), and $v_{pe} = \pm 0.5$ m/s (velocity of person relative to the escalator). (The plus sign is used when the person is going up; the negative sign when the person is going down.) Also, v_{pg} is the velocity of the person relative to ground.
We have $v_{pg} = v_{pe} + v_{eg}$
(a) When he walks up the escalator:

$v_{pg} = 0.5 \text{ m/s} + 0.4 \text{ m/s} = 0.9 \text{ m/s}$

and the time is $t = \dfrac{20 \text{ m}}{0.9 \text{ m/s}} = 22.25$ s

(b) When he walks down the escalator

$v_{pg} = -0.5 \text{ m/s} + 0.4 \text{ m/s} = -0.1 \text{ m/s}$

and the time is $t = \dfrac{-20 \text{ m}}{-0.1 \text{ m/s}} = 200$ s

---

3.28 We have v_{bs} = velocity of boat relative to the shore
v_{bw} = velocity of boat relative to the water
and v_{ws} = velocity of water relative to the shore
$v_{bs} = v_{bw} + v_{ws}$

Take downstream as the positive direction. Then $v_{ws} = 1.5$ m/s for both parts of the trip.
Going downstream: $v_{bw} = 10$ m/s
Therefore, $v_{bs} = 10$ m/s $+ 1.5$ m/s $= 11.5$ m/s

CHAPTER THREE SOLUTIONS

and the time required is $t_1 = \dfrac{300 \text{ m}}{11.5 \text{ m/s}} = 26.1$ s

Going upstream, $v_{bw} = -10$ m/s

Thus, $v_{bs} = -10$ m/s $+ 1.5$ m/s $= -8.5$ m/s

and the time is $t_2 = \dfrac{-300 \text{ m}}{-8.5 \text{ m/s}} = 35.3$ s

The time for the round trip is $t = t_1 + t_2 = 61.4$ s

3.29 We have $v_{wg} = -0.50$ m/s = velocity of water relative to the ground

$v_{sg} = \dfrac{0.56 \text{ m}}{0.8 \text{ s}} = 0.7$ m/s = velocity of skater relative to the ground

$v_{sg} = v_{sw} + v_{wg}$

(a) (i) $v_{sg} = v_{sw} + v_{wg}$

is   0.7 m/s $= v_{sw} + (-0.5$ m/s$)$

so $v_{sw} = 1.2$ m/s

(ii) $v_{sg} = -0.5$ m/s (same as the water)

$v_{sg} = v_{sw} + v_{wg}$

is -0.5 m/s $= v_{sw} + (-0.5$ m/s$)$

$v_{sw} = 0$

(b) $d_{sw} = v_{sw} \, t = (1.2$ m/s$)(0.8$ s$) = 0.96$ m

(c) time to go upstream = 0.8 s

time to drift back downstream = $\dfrac{0.56 \text{ m}}{0.5 \text{ m/s}} = 1.12$ s

for a total time of 1.92 s

Therefore, $\bar{v} = \dfrac{d_{sw}}{\text{time}} = \dfrac{0.96 \text{ m}}{1.92 \text{ s}} = 0.5$ m/s

---

3.30 $v_{1w} = -v_{2w}$ (The canoes have the same speed relative to the water, but go in opposite directions.)

(1e = number one relative to the earth)
(1w = number one relative to the water)
(we = water relative to the earth)

$v_{1e} = v_{1w} + v_{we}$

and $v_{2e} = v_{2w} + v_{we}$

We are given that $v_{1e} = 2.9$ m/s and $v_{2e} = -1.2$ m/s

We have   2.9 m/s $= -v_{2w} + v_{we}$ (1)

            -1.2 m/s $= v_{2w} + v_{we}$ (2)

Adding, gives $2v_{we} = 1.7$ m/s

So, $v_{we} = 0.85$ m/s

Then (1) gives $v_{1w} = 2.05$ m/s

and (2) gives $v_{2w} = -2.05$ m/s

CHAPTER THREE SOLUTIONS

3.31 We have $v_{bw} = 10$ m/s is the velocity of the boat relative to the water, and is a vector directed northward.
$v_{ws} = 1.5$ m/s, the velocity of the water relative to the shore, and is directed east.
v_{bs} = the velocity of the boat relative to the shore, and is a vector directed at an angle of θ, relative to the northward direction.

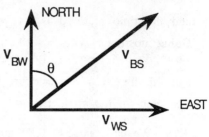

$$v_{bs} = v_{bw} + v_{ws}$$
Northward component of v_{bs} is $v_{bs} \cos\theta = 10$ m/s (1)
Eastward component is $v_{bs} \sin\theta = 1.5$ m/s (2)
Dividing (2) by (1) gives $\tan\theta = 0.15$ and $\theta = 8.53°$
Then, from (1) $v_{bs} = \dfrac{10 \text{ m/s}}{\cos 8.53} = 10.1$ m/s
The time to cross the river is $t = \dfrac{300 \text{ m}}{v_{bs} \cos\theta} = \dfrac{300 \text{ m}}{10 \text{ m/s}} = 30$ s
The eastward drift = $(v_{bs} \sin\theta)t = (1.5 \text{ m/s})(30 \text{ s}) = 45$ m

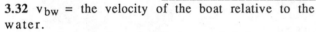

3.32 v_{bw} = the velocity of the boat relative to the water.
v_{ws} = the velocity of the water relative to the shore, and is directed east.
v_{bs} = the velocity of the boat relative to the shore.

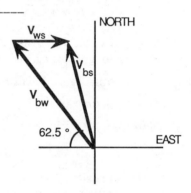

$$v_{bs} = v_{bw} + v_{ws}$$
The northward components of this equation (where north is across the stream) are
$v_{bs})_N = v_{bw})_N + v_{ws})_N$
$v_{bs})_N = (3.3 \text{ mi/h}) \sin 62.5 + 0 = 2.93$ mi/h
And, the time to cross the stream is
$t = \dfrac{0.505 \text{ mi}}{2.93 \text{ mi/h}} = 0.172$ h
The eastward components of the relative velocity equation (where east is parallel to the current) are
$v_{bs})_E = v_{bw})_E + v_{ws})_E = -(3.3 \text{ mi/h}) \cos 62.5 + 1.25 \text{ mi/h} = -0.274$ mi/h
and the distance traveled parallel to the shore = $(v_{bs})_E \, t = (-0.274 \text{ mi/h})(0.172 \text{ h}) = -0.0472$ mi = -249 ft = 249 ft upstream.

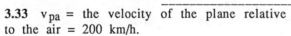

3.33 v_{pa} = the velocity of the plane relative to the air = 200 km/h.
v_{ag} = the velocity of the air relative to the ground = 50 km/h (south).
v_{pg} = the velocity of the plane relative to the ground (to be due east).

$$v_{pg} = v_{pa} + v_{ag}$$
For v_{pg} to have zero northward component, we must have
$v_{pa} \sin\theta = v_{ag}$
or $\sin\theta = \dfrac{50 \text{ km/h}}{200 \text{ km/h}} = 0.25$ and $\theta = 14.5°$ Thus, the plane should head at 14.5° north of west.
Since v_{pg} has zero northward component,

CHAPTER THREE SOLUTIONS

$v_{pg} = v_{pa}\cos\theta = (200 \text{ km/h})\cos 14.5°$
Which gives, $v_{pg} = 194$ m/s, the plane's ground speed.

3.34 v_{wc} = the velocity of the water relative to the car.
v_{we} = the velocity of the water relative to the earth.
v_{ce} = the velocity of the car relative to the earth.

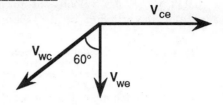

$v_{we} = v_{wc} + v_{ce}$
(a) v_{we} has zero horizontal component. Thus,
$v_{wc}\sin 60° = v_{ce} = 50$ km/h
or, $v_{wc} = 57.7$ km/h at 60° west of vertical.
(b) Since v_{ce} has zero vertical component,
$v_{we} = v_{wc}\cos 60° = (57.7 \text{ km/h})(0.5) = 28.9$ km/h downward

---

3.35 v_{bc} = the velocity of the ball relative to the car.
v_{be} = the velocity of the ball relative to the earth.
v_{ce} = velocity of the car relative to the earth = 10 m/s.

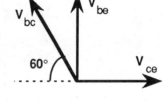

$v_{be} = v_{bc} + v_{ce}$
Since v_{be} has zero horizontal component, we have
$0 = -v_{bc}\cos 60 + v_{ce}$
So, $v_{bc} = \dfrac{10 \text{ m/s}}{0.5} = 20$ m/s
and the vertical components give
$v_{be} = v_{bc}\sin 60 + 0 = (20 \text{ m/s})(0.866) = 17.3$ m/s
This is the initial velocity of the ball relative to the earth.
Now from $v_y^2 = v_{0y}^2 + 2ay$, we have
$0 = (17.3 \text{ m/s})^2 + 2(-9.8 \text{ m/s}^2)h$
to give h = 15.3 m, the maximum height the ball rises.

3.36 (a) The horizontal component is $v_{0x} = (20 \text{ m/s})(\cos 35°) = 16.4$ m/s, and
(b) the vertical component is $v_{0y} = (20 \text{ m/s})(\sin 35°) = 11.5$ m/s.

3.37 v_{se} = the velocity of the swimmer relative to the earth.
v_{sw} = the velocity of the swimmer relative to the water.
v_{we} = the velocity of the water relative to the earth.
$v_{sw} = 1.2$ m/s , when going downstream
$v_{se} = 1.2$ m/s + 0.5 m/s = 1.7 m/s
and the time $t_1 = \dfrac{1000 \text{m}}{1.7 \text{ m/s}} = 588.2$ s
$v_{sw} = -1.2$ m/s , when going upstream
$v_{se} = -1.2$ m/s + 0.5 m/s = -0.7 m/s
and the time $t_2 = \dfrac{-1000 \text{ m}}{-0.7 \text{ m/s}} = 1428.6$ s
Thus the time for the round trip = $t_1 + t_2 = 2017$ s (with flowing water).
If the water were still, the time for the trip would be
$t' = \dfrac{2(1000 \text{ m})}{1.2 \text{ m/s}} = 1667$ s
Thus, the percent increase in time due to the moving water is

CHAPTER THREE SOLUTIONS

$$\frac{2017\text{ s} - 1667\text{ s}}{1667\text{ s}} \; 100\% = 21\% \text{ increase}$$

3.38 v_{sg} = the velocity of the shopper relative to the ground.
v_{se} = the velocity of the shopper relative to the escalator.
v_{eg} = the velocity of the escalator relative to the ground.
If $v_{eg} = 0$ (the escalator is stalled), $v_{sg} = v_{se}$ and

$$t_1 = \frac{L}{v_{sg}} = \frac{L}{v_{se}} = 30\text{ s, or} \quad v_{se} = \frac{L}{30\text{ s}}$$

If $v_{se} = 0$ (shopper rides up), $v_{sg} = v_{eg}$ and

$$t_2 = \frac{L}{v_{sg}} = \frac{L}{v_{eg}} = 20\text{ s, or} \quad v_{eg} = \frac{L}{20\text{ s}}$$

If the escalator is running and the shopper walks up:
$$v_{sg} = v_{se} + v_{eg}$$

and the time required is $t = \frac{L}{v_{sg}}$, or $v_{sg} = \frac{L}{t}$

So, $\frac{L}{t} = \frac{L}{30\text{ s}} + \frac{L}{20\text{ s}}$, from which $t = 12$ s

---

3.39 The components are

	east (x component)	north (y component)
10 km:	0	-10 km
6 km:	5.20 km	+3 km
	R_x = +5.20 km	R_y = -7 km

and from $\sqrt{(R_x)^2 + (R_y)^2}$, we find $R = 8.72$ km.
From $\tan\theta = \frac{R_x}{R_y} = -1.35$, we find $\theta = -53.4°$ or $\theta = 53.4°$ south of east.

3.40 (a) The projected distance on the seventh line is very nearly equal to the distance fallen. Thus,

$$7(0.15\text{ m}) = 1.05\text{ m} = \frac{1}{2}gt^2 \text{ and and } t \text{ is approximately equal to } 0.45\text{ s.}$$

Thus, the projected velocity is $v = \frac{1.05\text{ m}}{45\text{ s}} = 2.3$ m/s.

(b) All objects fall vertically at the same acceleration, $g = 9.8$ m/s^2.

3.41 An expression for the horizontal range can be found from $x = v_{0x}t$. The time of flight is found from $y = v_{0y}t + \frac{1}{2}at^2$ with $y = 0$, as $t = \frac{2v_{0y}}{g}$. This gives the range as

$x = v_{0x}\left(\frac{2v_{0y}}{g}\right)$. On earth this becomes $x_e = v_{0x}\left(\frac{2v_{0y}}{g_e}\right)$ and on the moon,

$x_m = v_{0x}\left(\frac{2v_{0y}}{g_m}\right)$. Dividing x_m by x_e, we have $x_m = \left(\frac{g_e}{g_m}\right)(x_e)$. With $g_m = \frac{1}{6}g_e$, we find $x_m = 18$ m. For Mars, $g_{mars} = 0.38g_e$, and we find $x_{mars} = 7.89$ m.

3.42 The time to reach the fence is found from $x = v_{0x}t$

$$t = \frac{130\text{ m}}{v_0\cos 35°}$$

CHAPTER THREE SOLUTIONS

At this time, the ball must be 2 m above its launch position. Use $y = v_{0y}t + \frac{1}{2}at^2$

$$2 \text{ m} = (v_0 \sin 35°)\frac{130\text{m}}{v_0 \cos 35°} - \frac{9.8 \text{ m/s}^2}{2}\frac{(130 \text{ m})^2}{(v_0 \cos 35°)^2}$$

From which, $v_0 = 37.2$ m/s.

3.43 $AC = v_1 t = (90 \text{ km/h})(2.5 \text{ h}) = 225$ km
$AD = AC \cos 40 = 172.4$ km, and
$CD = AD \sin 40 = 144.6$ km
$BD = AD - AB = 172.4$ km $- 80$ km $= 92.4$ km
From the triangle BCD, $BC = \sqrt{(BD)^2 + (CD)^2} =$
$\sqrt{(92.4 \text{ km})^2 + (144.6 \text{ km})^2} = 171.6$ km
Since car 2 travels this distance in 2.5 h, its constant speed is
$$v_2 = \frac{171.6 \text{ km}}{2.5 \text{ h}} = 68.6 \text{ km/h}$$

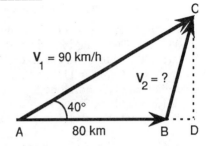

3.44 We know, $v_{0x} = v_0 \cos 45 = 17.68$ m/s, and $v_{0y} = v_0 \sin 45 = 17.68$ m/s
The time to go 50 m horizontally is $t = \frac{\Delta x}{v_{0x}} = \frac{50 \text{ m}}{17.68 \text{ m/s}} = 2.83$ s
and the height at this time is
$$y = v_0 t + \frac{1}{2}at^2 = (17.68 \text{ m/s})(2.83 \text{ s}) + \frac{1}{2}(-9.8 \text{ m/s}^2)(2.83 \text{ s})^2 = 10.8 \text{ m}.$$ Thus, the net should be placed 10.8 m above the ground.

3.45 The time to reach the opposite bank is found from $x = v_{0x}t$ as
$$t = \frac{10 \text{ m}}{v_0 \cos 15°} \quad (1)$$
Now use $y = v_{0y}t + \frac{1}{2}at^2$ with $y = 0$ at the opposite bank. Thus, this equation reduces to
$$0 = v_{0y}t - \frac{1}{2}gt^2 \text{ or } v_{0y} = \frac{gt}{2} = v_0 \sin 15° \quad (2)$$
Now eliminate t from (2) by use of (1) and solve for v_0.
The result is $v_0 = 14.0$ m/s.

3.46 (a) The distance, s, moved in the first three seconds is given by
$$s = v_0 t + \frac{1}{2}at^2 = (100 \text{ m/s})(3 \text{ s}) + \frac{1}{2}(30 \text{ m/s}^2)(3 \text{ s})^2 = 435 \text{ m}$$
At this time its x location is $x_1 = s \cos 53° = 261.8$ m, its vertical height is
$h = s \sin 53° = 347.4$ m and its velocity is
$$v = v_0 + at = 100 \text{ m/s} + (30 \text{ m/s}^2)(3 \text{ s}) = 190 \text{ m/s}.$$
This is the initial velocity for the next phase of its motion. Thus, the rocket begins its projectile motion with an x component of velocity of 114.3 m/s and a y component of 151.7 m/s. We find the maximum height during the projectile phase from
$$v_y = v_{0y} + at$$
$$0 = 151.7 \text{ m/s} - (9.8 \text{ m/s}^2)t_2$$
which gives $t_2 = 15.48$ s as the time to reach the top of its arc after free-fall starts. During this time, it rises a distance

CHAPTER THREE SOLUTIONS

$\Delta y = v_{0y}t + \frac{1}{2} at_2^2 = (151.7 \text{ m/s})(15.48 \text{ s}) + \frac{1}{2}(-9.8 \text{ m/s}^2)(15.48 \text{ s})^2 = 1174.1 \text{ m}$

Thus, the maximum height reached is $H = h + \Delta y = 347.4 \text{ m} + 1174.1 \text{ m} = 1521.5 \text{ m}$.

(b) The time to fall a distance H vertically, starting with $v_{0y} = 0$ is found from

$y = v_{0y}t + \frac{1}{2} at^2$;

$-1521.5 \text{ m} = 0 + \frac{1}{2}(-9.8 \text{ m/s}^2)t_3^2$

which gives $t_3 = 17.62$ s, and a total time of free-fall flight of $15.48 \text{ s} + 17.62 \text{ s} = 33.1$ s. Thus, the total time of flight is $T = 3 \text{ s} + 33.1 \text{ s} = 36.1$ s.

(c) The horizontal range during free-fall is

$x = v_{0x}t = (114.3 \text{ m/s})(33.1 \text{ s}) = 3783.3 \text{ m}$

and the total horizontal range is the sum of the range during powered flight, 261.8 m, plus 3783.3 m.

Range = 261.8 m + 3783.3 m = 4045 m

3.47 Find initial velocity of dart when shot at rest, horizontally, one meter above the ground.

From $y = v_{y0}t + \frac{1}{2} at^2$ we have $t = \left(\frac{-2y}{g}\right)^{1/2}$

and $x = v_{x0}t$ thus $v_{x0} = x/t = x/(-2y/g)^{1/2} = x(g/(-2y))^{1/2} = 5(9.8/2)^{1/2}$

$v_{x0} = 11.1$ m/s

Find how far the dart will go if it is shot horizontally, one meter above the ground while sliding down board at 2 m/s.

$y = v_{y0}t + \frac{1}{2} at^2$

$0 = gt^2/2 + v_{y0}t - y = 4.9t^2 + 2(0.707)t - 1$ and thus

$t = -1.414 + [(1.414)^2 + 4(4.90)]^{1/2}/2(4.90) = 0.32995$ s

$x = v_{x0}t = [11.07 + 2(0.707)](0.32995) = 4.12$ m

3.48 (a) First, find the time for the coyote to travel the 70 m to the edge of the cliff.

$x = v_0t + \frac{1}{2} at^2$

$70 \text{ m} = 0 + \frac{1}{2}(15 \text{ m/s}^2)t^2$

which gives $t = 3.06$ s.

The minimum speed of the roadrunner is

$v = \frac{x}{t} = \frac{70 \text{ m}}{3.06 \text{ s}} = 22.9$ m/s

(b) Find the horizontal velocity of the coyote when he reaches the edge of the cliff.

$v_x = v_0 + at = 0 + (15 \text{ m/s}^2)(3.06 \text{ s}) = 45.8$ m/s

Now, find the time to drop 100 m vertically starting with $v_{0y} = 0$.

$y = v_{0y}t + \frac{1}{2} at^2$ $-100 \text{ m} = 0 + \frac{1}{2}(-9.8 \text{ m/s}^2)t^2$

From which, $t = 4.52$ s.

At this time, the horizontal position can be found as

$x = v_0t + \frac{1}{2} at^2$ $x = (45.8 \text{ m/s})(4.52 \text{ s}) + \frac{1}{2}(15 \text{ m/s}^2)(4.52 \text{ s})^2 = 360$ m

CHAPTER THREE SOLUTIONS

3.49 (a) At the top of the arc $v_y = 0$, and from $v_y = v_{0y} - gt$, we find the time to reach the top of the arc to be

$$t = \frac{v_0 \sin\theta}{g} \quad (1)$$

the vertical height, h, reached in this time is found from $y = v_{0y}t - \frac{1}{2}gt^2$ as

$$h = (v_0\sin\theta)t - \frac{1}{2}gt^2 \quad (2)$$

Substitute into (2) for t from (1), and we find

$$h = \frac{v_0^2 \sin^2\theta}{2g}$$

(b) The total time of flight, T, is twice the time given in (1) above. The horizontal range, R, is found from $x = v_{0x}t$ as

$$R = (v_0\cos\theta)\frac{2v_0\sin\theta}{g}$$

or

$$R = \frac{v_0^2 \sin(2\theta)}{g}$$

3.50 Note that $\tan\theta_0 = h_0/x$ where h_0 is the intial target height and x is the horizontal displacement.

For the projectile $y_p = v_0\sin\theta_0 t - \frac{1}{2}gt^2$ and $x = v_0\cos\theta_0 t$

Combining and eliminating t in first term on right side gives

$$y_p = \tan\theta_0 x - \frac{1}{2}gt^2$$

Substituting h_0/x for $\tan\theta_0$ gives $\quad y_p = h_0 - \frac{1}{2}gt^2 \quad (1)$

For the falling target $\quad y_t = h_0 - \frac{1}{2}gt^2 \quad (2)$

From (1) and (2) it can be inferred that $\quad y_t = y_p$

3.51 Find highest elevation, θ_H, that will clear the mountain peak; this will yield the range of the closest point of bombardment. Next find lowest elevation, θ_L, that will clear the mountain peak; this will yield the maximum range under these condition if both θ_H and θ_L, are $> 45°$:
$x = 2500$ m, $y = 1800$ m, $v_0 = 250$ m/s.

$$y = v_{y0}t - \frac{1}{2}gt^2 = v_0(\sin\theta)t - \frac{1}{2}gt^2; \quad x = v_{x0}t = v_0(\cos\theta)t$$

thus $t = \frac{x}{v_0(\cos\theta)}$

Substitute into expression for y.

$y = v_0(\sin\theta)x/(v_0(\cos\theta)) - \frac{1}{2}g(x/v_0(\cos\theta))^2 = x\tan\theta - [gx^2/2v_0^2][\tan^2\theta + 1]$ and

$0 = gx^2/2v_0^2 \tan^2\theta - x\tan\theta + gx^2/2v_0^2 + y$

Substitute values, use the quadratic formula, and find
$\tan\theta = 3.905$ or 1.197 which gives $\theta_H = 75.6°$ and $\theta_L = 50.1°$
Range(at θ_H) = $v_0^2 \sin 2\theta_H/g$ = 3065 m from enemy ship
3065 - 2500 - 300 = 265 m from shore.
Range(at θ_L) = $v_0^2 \sin 2\theta_L/g$ = 6276 m from enemy ship

CHAPTER THREE SOLUTIONS

6276 - 2500 -200 = 3476 m from shore
Therefore, safe distance is < 265 m or > 3476 m from shore.

3.52 The components of **R** are found from
$R_x = R \cos\theta$ and $R_y = R \sin\theta$
These components are given for the various angles below.

	θ	R_x	R_y
(a)	60.°	+50.0	+86.6
(b)	130°	-64.3	+76.6
(c)	200°	-94.0	-34.2
(d)	290°	+34.2	-94.0

3.53 The components of the displacements are given below

	x comp	y comp
75 m	0.0	+75.0 m
250 m	+250.0	0.0
125 m	+108.0	+62.5
150 m	0.0	-150.0
	$R_x = +358$ m	$R_y = -12.5$ m

and from the Pythagorean theorem

$R = \sqrt{(358 \text{ m})^2 + (-12.5 \text{ m})^2} = 358$ m

and $\tan\theta = R_y/R_x = -0.004$ from which, $\theta = -2.00°$.
Thus, **R** = 358 m at 2° south of east.

3.54 (a) For the male anatomy, we have

x-component   y component
$d_{1x} = 0$ $d_{1y} = 104$ cm
$d_{2x} = 46$ cm $d_{2y} = 19.5$ cm

For an resultant component of $d_x = 46$ cm, and a resultant y component $d_y = 123.5$ cm.

So, $d = \sqrt{(46 \text{ cm})^2 + (123.5 \text{ cm})^2} = 131.8$ cm ; $\tan\theta = 2.68$, and $\theta = 69.6$.

For the female anatomy, we have

x-component   y component
$d_{1x} = 0$ $d_{1y} = 84$ cm
$d_{2x} = 38$ cm $d_{2y} = 20.2$ cm

For an resultant component of $d_x = 38$ cm, and a resultant y component $d_y = 104.2$ cm.

So, $d = \sqrt{(38 \text{ cm})^2 + (104.2 \text{ cm})^2} = 110.9$ cm ; $\tan\theta = 2.74$, and $\theta = 70$.

(b) To normalize, multiply all distances by the appropriate scale factors which are:

$s_m = \dfrac{200 \text{ cm}}{180 \text{ cm}} = 1.111$ and $s_f = \dfrac{200 \text{ cm}}{168 \text{ cm}} = 1.190$

Multiplying all distances by these scale factors and recomputing the sums, yields

$\mathbf{d_m}' = 146.4$ cm and the angle is 69.6°
$\mathbf{d_f}' = 132.0$ cm and the angle is 70°

To compute the vector difference
$\Delta\mathbf{d} = \mathbf{d_m}' - \mathbf{d_f}' = \mathbf{d_m}' + (-\mathbf{d_f}')$, we have

x-component   y component
$d_{mx}' = 51.0$ cm $d_{my}' = 137.2$ cm

CHAPTER THREE SOLUTIONS

$-d_{fx}' = -45.1$ cm $-d_{fx}' = -124.0$ cm
For a resultant x component of 5.9 cm, and a resultant y component of 13.2 cm
The Pythagorean theorem yields, $\Delta d = 14.5$ cm
and the $\tan\theta = \dfrac{13.2 \text{ cm}}{5.9 \text{ cm}} = 2.24$, from which $\theta = 65.9°$

3.55

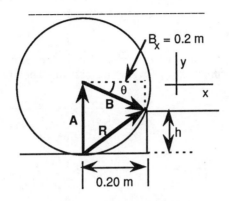

First, find the angle θ: $B_x = B\cos\theta$, so $\cos\theta = \dfrac{0.20 \text{ m}}{1.50 \text{ m}} = 0.133$, and $\theta = 82.33°$.
Then, realize that $h = R_y = A_y - B_y = 1.50$ m $- (1.50$ m$)(\sin\theta)$
           $= 1.50$ m$(1 - \sin 82.33°) = 0.0134$ m $= 1.34$ cm.

3.56

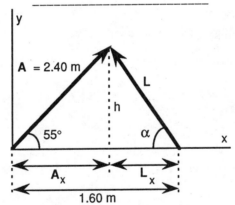

Recognize that $h = A_y = (2.40$ m$)\sin 55 = 1.97$ m $= h = L_y$.
  $L_x = 1.60$ m $- A_x = 1.60$ m $- (2.40$ m$)\cos 55 = 0.22$ m.
Then from the Pythagorean theorem,
  $L = \sqrt{(L_x)^2 + (L_y)^2} = \sqrt{(L_x)^2 + (h)^2} = 1.98$ m
and $\tan\alpha = \dfrac{h}{L_x} = \dfrac{1.97}{0.22} = 8.95$, and $\alpha = 83.6°$

CHAPTER FOUR SOLUTIONS

4.1 $w = mg = (2 \text{ kg})(9.8 \text{ m/s}^2) = 19.6 \text{ N}$
$w = (19.6 \text{ N})(0.225 \text{ lb}/ 1\text{N}) = 4.41 \text{ lb}$

4.2 Since the car is moving with a constant speed and in a straight line, the resultant force on it must be zero regardless of whether it is moving toward the right or the left.

4.3 Using $m = \dfrac{w}{g}$, we have $\dfrac{w_{Jupiter}}{g_{Jupiter}} = \dfrac{w_{earth}}{g_{earth}}$ (The mass is the same on both planets.)

So, $w_{Jupiter} = w_{earth}\left(\dfrac{g_{Jupiter}}{g_{earth}}\right) = 900 \text{ N}\left(\dfrac{25.9 \text{ m/s}^2}{9.8 \text{ m/s}^2}\right) = 2379 \text{ N}$

4.4 On earth $w = mg$. If g changes to g', the new weight would be $w' = mg'$. Therefore,

$$\dfrac{w'}{w} = \dfrac{mg'}{mg} = \dfrac{g'}{g} \quad \text{or} \quad w' = \dfrac{g'}{g} w$$

On the moon, $g' = \dfrac{1}{6} g$. Thus, $w' = \dfrac{1}{6} w = \dfrac{5}{6}$ lb. This is equivalent to 3.71 N.

On Jupiter, $g' = 2.64$ g. The techniques above give the weight on Jupiter to be 58.7 N.

The mass is the same at all three locations. The weight on the earth is 22.2 N, and

$$m = \dfrac{w}{g} = \dfrac{22.2 \text{ N}}{9.8 \text{ m/s}^2} = 2.27 \text{ kg}.$$

4.5 (a) $w = (120 \text{ lb})(4.45 \text{ N}/1 \text{ lb}) = 534 \text{ N}$
(b) $m = w/g = 534 \text{ N}/(9.8 \text{ m/s}^2) = 54.5 \text{ kg}$

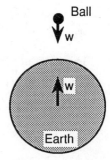

4.6 (a) The only force acting on the freely-falling baseball is its weight, w, as shown at the right. The reaction to this force is a force of magnitude w directed upward and acting on the earth at its center.

(b) The only force acting on the projectile is its weight. Thus, the answer to (b) is the same as the answer to part (a).

4.7 (a) $F_R = ma = (6 \text{ kg})(2 \text{ m/s}^2) = 12 \text{ N}$
(b) $a = \dfrac{F_R}{m} = \dfrac{12 \text{ N}}{4 \text{ kg}} = 3 \text{ m/s}^2$.

CHAPTER FOUR SOLUTIONS

4.8 The mass of the object is $m = w/g = (20 \text{ N})/(9.8 \text{ m/s}^2) = 2.04$ kg. Thus, the acceleration is given by $a = F_R/m = (5 \text{ N})/(2.04 \text{ kg}) = 2.45 \text{ m/s}^2$.

4.9 The acceleration given to the football is found from $v = v_0 + at$ as
$$a = \frac{v - v_0}{t} = \frac{10 \text{ m/s} - 0}{0.2 \text{ s}} = 50 \text{ m/s}^2.$$
Then, from Newton's Second Law, we find
$$F = ma = (0.5 \text{ kg})(50 \text{ m/s}^2) = 25 \text{ N}$$

4.10 Summing the forces on the plane shown gives
$\Sigma F_x = F - f = 10 \text{ N} - f = (0.2 \text{ kg})(2 \text{ m/s}^2)$
From which
$f = 9.6 \text{ N}$

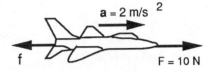

4.11 First, find the acceleration of the bullet from the equations of motion with constant acceleration, as
$$a = \frac{v^2 - v_0^2}{2x} = \frac{(320 \text{ m/s})^2 - 0}{2(0.82 \text{ m})} = 6.24 \times 10^4 \text{ m/s}^2$$
Then, $F = ma = (5 \times 10^{-3} \text{ kg})(6.24 \times 10^4 \text{ m/s}^2) = 312 \text{ N}$

4.12 The resultant of the two forces is found from the Pythagorean theorem as
$$F_R = \sqrt{(390 \text{ N})^2 + (180 \text{ N})^2} = 430 \text{ N}$$
and $\tan \theta = \dfrac{390 \text{ N}}{180 \text{ N}} = 2.166 \qquad \theta = 65.2°$

The resultant acceleration is $a = \dfrac{F_R}{m} = \dfrac{430 \text{ N}}{270 \text{ kg}} = 1.59 \text{ m/s}^2$ at 65.2° north of east

---

4.13 (a) We resolve the forces shown into their components as

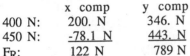

	x comp	y comp
400 N:	200. N	346. N
450 N:	-78.1 N	443. N
F_R:	122 N	789 N

The magnitude of the resultant force is found from the Pythagorean theorem as
$$F_R = \sqrt{(\Sigma F_x)^2 + (\Sigma F_y)^2} = \sqrt{(122 \text{ N})^2 + (789 \text{ N})^2}$$
$= 798 \text{ N}$

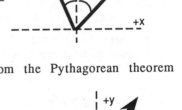

and $\tan \theta = \dfrac{\Sigma F_y}{\Sigma F_x} = \dfrac{789}{122} = 6.47 \quad$ from which $\theta = 81.2°$

Thus, the resultant force is at an angle of 8.8° to the right of the forward direction.
(b) The acceleration is in the same direction as F_R and is given by
$$a = F_R/m = 798 \text{ N}/ 3000 \text{ kg} = 0.266 \text{ m/s}^2.$$

4.14 (a) From the second law, the acceleration of the boat is

35

CHAPTER FOUR SOLUTIONS

$a = \dfrac{F_R}{m} = \dfrac{F - f}{m} = \dfrac{2000 \text{ N} - 1800 \text{ N}}{1000 \text{ kg}} = 0.20$ m/s².

(b) The distance moved is found from
$$x = v_0 t + \tfrac{1}{2} a t^2 = 0 + \tfrac{1}{2}(0.2 \text{ m/s}^2)(10 \text{ s})^2 = 10 \text{ m}$$
(c) Its velocity is found as
$$v = v_0 + at = 0 + (0.2 \text{ m/s}^2)(10 \text{ s}) = 2 \text{ m/s}$$

4.15 $\Sigma F_x = 0$ becomes
$$T_1 \cos 8° - T_2 \cos 8° = 0$$
or $\quad T_1 = T_2$
$\Sigma F_y = 0$ becomes
$$T_1 \sin 8° + T_2 \sin 8° - 600 \text{ N} = 0$$
or $\quad 2 T_2 \sin 8° = 600 \text{ N}$
$\quad T_2 = T_1 = 2156 \text{ N}$

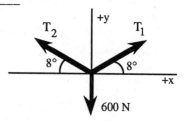

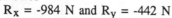

4.16 The resultant force in the x direction on the sled is given by
$$(800 \text{ N})\cos 50° + (500 \text{ N})\cos 20° = 984 \text{ N}$$
and the resultant force in the y direction is
$$(800 \text{ N})\sin 50° - (500 \text{ N})\sin 20° = 442 \text{ N}$$
In order for the sled to move at constant velocity, the farmer will have to exert a force that will cancel these resultant component forces. Thus, his force will have components
$$R_x = -984 \text{ N} \text{ and } R_y = -442 \text{ N}$$
The magnitude of the farmer's force is found from the Pythagorean theorem as
$$\sqrt{(984 \text{ N})^2 + (442 \text{ N})^2} = 1079 \text{ N}$$
The angle at which this force must be exerted is
$$\tan \theta = \dfrac{442}{984} = .449 \text{ and } \theta = 24.2°$$
The angle is counterclockwise from the -x direction.
The force he must exert is 1079 N at 24.2° + 180° = 204.2°.

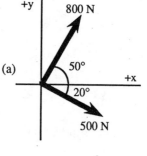

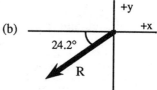

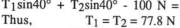

4.17 Use $\Sigma F_x = 0$ as
$$T_1 \cos 40° - T_2 \cos 40° = 0$$
or $\quad T_1 = T_2$
From $\Sigma F_y = 0$, we have
$$T_1 \sin 40° + T_2 \sin 40° - 100 \text{ N} = 0$$
Thus, $\quad T_1 = T_2 = 77.8 \text{ N}$

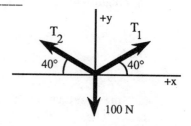

CHAPTER FOUR SOLUTIONS

4.18 From $\Sigma F_x = 0$, we have
$$T_1 \cos 30° - T_2 \cos 60° = 0$$
Thus, $T_1 = 0.577\, T_2$ (1)
From $\Sigma F_y = 0$,
$$T_1 \sin 30° + T_2 \sin 60° - 150 \text{ N} = 0 \quad (2)$$
Solving (1) and (2) simultaneously,
$$T_2 = 130 \text{ N and } T_1 = 75 \text{ N}$$

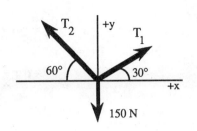

4.19 From $\Sigma F_x = 0$, we have
$$T_1 \cos 32° + F_x - T_2 \cos 24° = 0$$
or
$$F_x = -200 \text{ N} \cos 32° + 100 \text{ N} \cos 24°$$
$$= -78.3 \text{ N}$$
From $\Sigma F_y = 0$,
$$T_1 \sin 32° + T_2 \sin 24° - F_y = 0$$
or
$$F_y = 200 \text{ N} \sin 32° + 100 \text{ N} \sin 24°$$
$$= 146.7 \text{ N}$$
The force of the hand on the cord is found from the Pythagorean theorem as
$$\sqrt{(-78.3 \text{ N})^2 + (146.7 \text{ N})^2} = 166 \text{ N}$$
and $\tan \theta = \dfrac{F_x}{F_y} = 0.534 \qquad \theta = 28.1°$

Thus the resultant force is 166 N, downward and to the left of vertical at an angle of 28.1°.

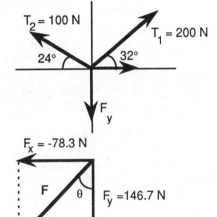

4.20 From $\Sigma F_x = 0$, we have
$$W_2 \cos \alpha - (110 \text{ N}) \cos 40° = 0 \quad (1)$$
From $\Sigma F_y = 0$, we have
$$W_2 \sin \alpha + (110 \text{ N}) \sin 40° - 220 \text{ N} = 0 \quad (2)$$
Dividing (2) by (1) yields
$$\tan \alpha = \frac{149.3}{84.26} = 1.772 \qquad \alpha = 60.55°$$
Then, from either (1) or (2), $W_2 = 171.4$ N.

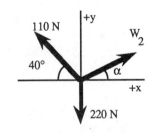

4.21 From the second law, we find the acceleration as
$$a = \frac{F}{m} = \frac{10 \text{ N}}{30 \text{ kg}} = 0.333 \text{ m/s}^2 \,.$$
To find the distance moved, we use
$$x = v_0 t + \frac{1}{2} a t^2 = 0 + \frac{1}{2}(0.333 \text{ m/s}^2)(3 \text{ s})^2 = 1.50 \text{ m}$$

4.22 In this case, the total mass will be $m = m_{cart} + m_{child}$. The mass of the child is
$$m_{child} = \frac{w_{child}}{g} = \frac{30 \text{ N}}{9.8 \text{ m/s}^2} = 3.1 \text{ kg}$$
and, thus, the total mass, m, is 30 kg + 3.1 kg = 33.1 kg. The acceleration is

CHAPTER FOUR SOLUTIONS

$$a = \frac{F}{m} = \frac{10 \text{ N}}{33.1 \text{ kg}} = 0.302 \text{ m/s}^2 .$$

and the distance moved is found as

$$x = v_0 t + \frac{1}{2} a t^2 = 0 + \frac{1}{2} (0.302 \text{ m/s}^2)(3 \text{ s})^2 = 1.36 \text{ m}$$

4.23 (a) We first find the acceleration of the ball. We know the final velocity = 0, so

$$v^2 = v_0^2 + 2ax$$
$$0 = (20 \text{ m/s})^2 + 2a(0.08 \text{ m})$$

From which, $a = -2500 \text{ m/s}^2$.
Thus, the force on the ball is

$$F = ma = (0.15 \text{ kg})(-2500 \text{ m/s}^2) = -375 \text{ N}$$

(b) The force on the ball exerted by the glove and the force of the glove by the ball are action-reaction pairs. Thus, from the third law,

$$F_{glove} = +375 \text{ N}$$

4.24 $v_0 = 90 \text{ km/h} = 25 \text{ m/s}$

and $a = \frac{F}{m} = \frac{-1.87 \times 10^6 \text{ N}}{5.22 \times 10^6 \text{ kg}} = -0.358 \text{ m/s}^2 .$

(a) $v = v_0 + at$ gives

$$v = 25 \text{ m/s} + (-0.358 \text{ m/s}^2)(30 \text{ s}) = 14.3 \text{ m/s}$$

(b) $x = \bar{v} t = \left(\frac{25.0 \text{ m/s} + 14.3 \text{ m/s}}{2} \right) (30 \text{ s}) = 589 \text{ m}$

4.25 Let us choose the positive direction of the x axis down the incline. The acceleration along the 135 ft (41.1 m) incline is then found from the second law.

$$a_x = \frac{F_x}{m} = \frac{mg\sin\theta}{m} = g \sin\theta = (9.8 \text{ m/s}^2)(\sin 40°) = 6.30 \text{ m/s}^2 .$$

and

$$v_x^2 = v_{0x}^2 + 2a_x x = (4 \text{ m/s})^2 + 2(6.30 \text{ m/s}^2)(41.1 \text{ m})$$

giving

$$v_x = 23.1 \text{ m/s}$$

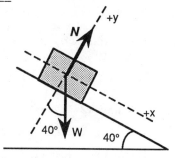

4.26 The acceleration of the mass down the incline is given by

$$x = v_0 t + \frac{1}{2} a t^2$$

$$0.8 \text{ m} = 0 + \frac{1}{2} a (0.5 \text{ s})^2, \text{ giving}$$

$$a = 6.4 \text{ m/s}^2 .$$

Thus, the force down the incline is

$$F = ma = (2 \text{ kg})(6.4 \text{ m/s}^2) = 12.8 \text{ N}$$

4.27 Choosing our x axis along the incline with the positive direction up the incline, we have

$$\Sigma F_x = ma_x =$$
$$T - w \sin 18.5° = (40 \text{ kg})a_x$$

38

CHAPTER FOUR SOLUTIONS

which gives $a_x = \dfrac{140 \text{ N} - 124.4 \text{ N}}{40 \text{ kg}} = 0.390 \text{ m/s}^2$,

Since we have constant acceleration, we find
$$v^2 = v_0^2 + 2ax$$
becomes $\quad v^2 = 0 + 2(0.39 \text{ m/s}^2)(80 \text{ m})$
and $\quad v = 7.90 \text{ m/s}$

4.28 We select the positive direction of the x axis up the incline. The component of the weight acting down the incline is
$$w_t = w\sin 20° = mg(0.342)$$
From the second law, we have
$$\Sigma F_x = F - w_t = ma_x,$$
or
$$F = w_t + ma_x = mg(0.342) + m(0.5 \text{ m/s}^2)$$
$$= m(0.342g + 0.5 \text{ m/s}^2) = 40 \text{ kg}(3.85 \text{ m/s}^2)$$
$$F = 154 \text{ N}$$

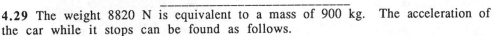

4.29 The weight 8820 N is equivalent to a mass of 900 kg. The acceleration of the car while it stops can be found as follows.
$$v^2 = v_0^2 + 2ax$$
$$0 = (35 \text{ m/s})^2 + 2a(1000 \text{ m})$$
so $\quad a = -0.6125 \text{ m/s}^2$.
and the retarding force on the car is
$$F = ma = (900 \text{ kg})(-0.6125 \text{ m/s}^2) = -551 \text{ N}$$

4.30 The forces on the bucket are the tension in the rope and the weight, 49 N, of the bucket. We call the positive direction upward, and use the second law.

$$\Sigma F_y = ma_y$$
$$T - 49 \text{ N} = (5 \text{ kg})(3 \text{ m/s}^2)$$
$$T = 64 \text{ N}$$

4.31 Let us call the forces exerted by each of the men F_1 and F_2. Thus, when pulling in the same direction, Newton's second law becomes
$$F_1 + F_2 = (200 \text{ kg})(1.52 \text{ m/s}^2)$$
or $\quad F_1 + F_2 = 304 \text{ N} \quad (1)$
When pulling in opposite directions,
$$F_1 - F_2 = (200 \text{ kg})(-0.518 \text{ m/s}^2)$$
or $\quad F_1 - F_2 = -103.6 \text{ N} \quad (2)$
Solving simultaneously, we find
$\quad F_1 = 100.2 \text{ N}$, and $\quad F_2 = 203.8 \text{ N}$

4.32 In the vertical direction, we have
$$(8000 \text{ N}) \sin 65° - w = 0$$
so $\quad w = 7250 \text{ N}$

39

CHAPTER FOUR SOLUTIONS

$$m = \frac{w}{g} = \frac{7250 \text{ N}}{9.8 \text{ m/s}^2} = 739.8 \text{ kg}$$

and, along the horizontal, the second law becomes,
$(8000 \text{ N}) \cos 65° = (739.8 \text{ kg})a_x$

so $a_x = 4.57 \text{ m/s}^2$

4.33 First consider the system as a whole and apply the second law
$\Sigma F = ma$

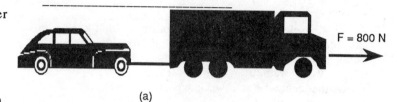

$800 \text{ N} = (11{,}000 \text{kg})a$
$a = 7.27 \times 10^{-2} \text{m/s}^2$.

Now consider the tow truck alone. The horizontal forces are the 800 N push of the motor and the tension in the cable. We have

$\Sigma F = ma$

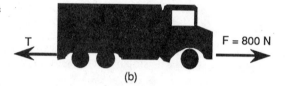

$800 \text{ N} - T = (8000 \text{ kg})(7.27 \times 10^{-2} \text{ m/s}^2)$
$T = 218 \text{ N}$

4.34

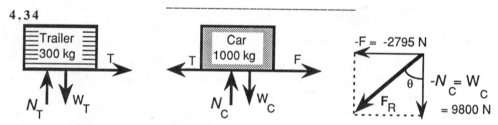

For the trailer, the second law becomes
$T = (300 \text{ kg})(2.15 \text{ m/s}^2)$ or $T = 645 \text{ N}$

For the car $F - T = 1000 \text{ kg}(2.15 \text{ m/s}^2)$, which gives F, since T is known.
$F = 2795 \text{ N}$

(a) The net force on the car $= F - T = m_{car}a = 2150 \text{ N}$ in the forward direction.
(b) From the above, we know that $T = 645 \text{ N}$ in forward direction.
(c) Using the third law, the force the trailer exerts on the car is $= -T = 645 \text{ N}$ in rearward direction.
(d) The force exerted on the road by the car is the resultant of the reaction force -F and the weight of the car as show above. The magnitude and direction of this force is found to be $F_R = 10{,}190 \text{ N}$ at $\theta = 15.9°$ to the left of the vertical.

CHAPTER FOUR SOLUTIONS

4.35 First, consider the forces on the upper block. They are T_1, the tension in the cable connecting the block to the ceiling of the elevator, T_2, the tension connecting the blocks, and the weight of the block, 98 N.

$\Sigma F_y = ma_y$

$T_1 - T_2 - 98 \text{ N} = (10 \text{ kg})(2 \text{ m/s}^2)$ (1)

The forces on the lower block are T_2 and its weight, 98 N.

$\Sigma F_y = ma_y$

$T_2 - 98 \text{ N} = (10 \text{ kg})(2 \text{ m/s}^2)$ (2)

From (2), we find
$T_2 = 118 \text{ N}$
and using this in (1), we find
$T_1 = 236 \text{ N}$

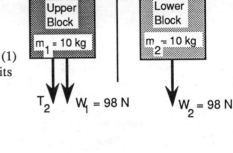

4.36 First consider the block moving along the horizontal. The only force in the direction of movement is T. Thus

$\Sigma F_x = ma$

$T = (5 \text{ kg})a$ (1)

Next consider the block which moves vertically. The forces on it are the tension T and its weight, 98 N. We have

$\Sigma F_y = ma$

$98 \text{ N} - T = (10 \text{ kg})a$ (2)

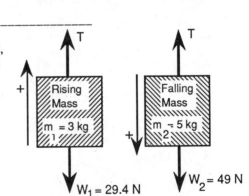

Note that both blocks must have the same magnitude of acceleration. Equations (1) and (2) can be solved simultaneously to give.

$a = 6.53 \text{ m/s}^2$ and $T = 32.7 \text{ N}$

4.37 First, consider the 3 kg rising mass. The forces on it are the tension, T, and its weight, 29.4 N. With the upward direction as positive, the second law becomes

$\Sigma F_y = ma_y$

$T - 29.4 \text{ N} = (3 \text{ kg})a$ (1)

The forces on the falling 5 kg mass are its weight and T, and its acceleration is the same as that of the rising mass. Calling the positive direction down for this mass, we have

$F = ma$

$49 \text{ N} - T = (5 \text{ kg})a$ (2)

Equations (1) and (2) can be solved simultaneously to give
(a) the tension as $T = 36.8 \text{ N}$,
(b) and the acceleration as $a = 2.45 \text{ m/s}^2$.
(c) Consider the 3 kg mass. We have

41

CHAPTER FOUR SOLUTIONS

$$y = v_0 t + \frac{1}{2} a t^2 = 0 + \frac{1}{2} (2.45 \text{ m/s}^2)(1 \text{ s})^2 = 1.23 \text{ m}$$

4.38 (a) When motion is impending, we know that
$$f_{max} = \mu_s N = 0.4(98 \text{ N}) = 39.2 \text{ N}$$
Thus, the maximum value of the force that can be exerted without causing movement is
$$F = 39.2 \text{ N}$$
(b) When slipping occurs,
$$f = \mu_k N$$
but $\mu_k = 0.8 \mu_s = 0.8(0.4) = 0.32$
Thus,
$$f = 0.32(98 \text{ N}) = 31.4 \text{ N}$$
Thus, if the applied force is maintained at 39.2 N as found in (a), we have
$$\Sigma F_x = m a_x$$
$$39.2 \text{ N} - 31.4 \text{ N} = (10 \text{ kg})a$$
and $a = 0.78 \text{ m/s}^2$.

4.39 The mass of the bobsled is 102 kg, and the normal force N is equal to 1000 N, its weight. Let us first find the acceleration of the bobsled. (Choose a coordinate system with +x in the direction of motion of the sled.)
$$v^2 = v_0^2 + 2ax$$
$$0 = (30 \text{ m/s})^2 + 2a(150 \text{ m})$$
From which, $a = -3 \text{ m/s}^2$.
The negative sign indicates that the acceleration is opposite the direction of the velocity. We now write down the second law, with the positive direction selected to be in the direction of the force of friction, f. We have
$$\Sigma F_x = m a_x$$
$$f = (102 \text{ kg})(3 \text{ m/s}^2) = 306 \text{ N}$$
and we find μ_k as
$$\mu_k = \frac{f}{N} = \frac{306 \text{ N}}{1000 \text{ N}} = 0.306$$

4.40 (a) To bring the block to the verge of motion, we exert a force of 75 N, but from the first law, we know
$$\Sigma F_x = 0$$
$$75 \text{ N} - f = 0 \quad ; \quad f = 75 \text{ N}$$
where f is the force of friction. We know, since motion is imminent, that
$$f = \mu_s N$$
where μ_s is the coefficient of static friction and N is the normal force. For our case, $N = mg$
Thus, $f = \mu m g$
and $$\mu_s = \frac{f}{m g} = \frac{75 \text{ N}}{196 \text{ N}} = 0.383$$
(b) When there is motion, the analysis is virtually the same as that above. We find
$$\mu_k = \frac{60 \text{ N}}{196 \text{ N}} = 0.306$$

4.41 In the vertical direction, we have
$$N - 300 \text{ N} - (400 \text{ N})\sin 35.2° = 0$$

CHAPTER FOUR SOLUTIONS

from which, $N = 530.6$ N
Therefore, $f = \mu_k N = (0.57)(530.6 \text{ N}) = 302.4$ N
Then, from the second law, applied along the horizontal direction, we have
 $(400 \text{ N}) \cos 35.2° - 302.4 \text{ N} = (30.6 \text{ kg})a_x$ from which $a_x = 0.80$ m/s^2
Then, from $x = v_{0x}t + \frac{1}{2} a_x t^2$, we have
$$4 \text{ m} = 0 + \frac{1}{2} (0.8 \text{ m/s}^2)t^2$$
which gives $t = 3.16$ s

4.42 Use $\Sigma F_x = 0$
   $(35 \text{ N})\cos 30° - f = 0$
which gives, $f = 30.3$ N
Now use $\Sigma F_y = 0$ to find the normal force N.
   $(35 \text{ N})\sin 30° + N - 150 \text{ N} = 0$
   $N = 132.5$ N
Then $f = \mu_k N$ gives $\mu_k = f/N = 30.3/132.5 = 0.229$

4.43 The friction force is
   $f = \mu_k N = 0.2 N$ (1)
and from $\Sigma F_y = 0$, we have
   $F\sin 35° - (100 \text{ N})\cos 20° + N = 0$
Thus, $N = 94 \text{ N} - 0.574F$ (2)
and from $\Sigma F_x = 0$, we have
   $F\cos 35° - f - (100 \text{ N})\sin 20° = 0$
or $0.819F - f - 34.2 \text{ N} = 0$ (3)
From (1) and (2), $f = 0.2(94 \text{ N} - 0.574F)$
This can be substituted into (3) for f to give F as
   $F = 56.7$ N

4.44 $\Sigma F_x = 0$ gives
   $T - mg\sin 53° = 0$
or, $7000 \text{ N} - 0.799mg = 0$
From which we find $m = 894$ kg

4.45 The forces on the center board are shown at the right. In the horizontal direction, we see that
   $N_1 = N_2 = N$
Therefore, $f_1 = \mu_s N_1$ and $f_2 = \mu_s N_2$
So, $f_1 = f_2 = f$
In the vertical direction, we have $\Sigma F_y = 0$, which gives
   $2f - 95.5 \text{ N} = 0$
Thus, $f = 47.75$ N
$$N = \frac{f}{\mu_s} = \frac{47.75 \text{ N}}{0.663} = 72.0 \text{ N}$$

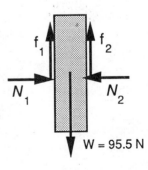

CHAPTER FOUR SOLUTIONS

4.46 From the free-body diagram of the person, find the compression force acting along the line of each crutch. $\Sigma F_x = 0$ gives $F_1 \sin 22° - F_2 \sin 22° = 0$
or $F_1 = F_2 = F$
Then $\Sigma F_y = 0$ gives
  $2F\cos 22° - 85$ lb $= 0$
From which, $F = 45.8$ lb

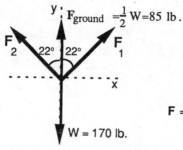

Free-Body Diagram of Person

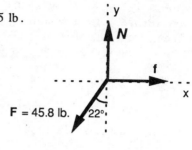

Free-Body Diagram of Crutch Tip

Now consider a crutch tip as shown. $\Sigma F_x = 0$ becomes $f - (45.8 \text{ lb}) \sin 22° = 0$
or $f = 17.2$ lb

$\Sigma F_y = 0$ becomes $N - (45.8 \text{ lb})\cos 22° = 0$ or $N = 42.5$ lb
Assuming it is on the verge of slipping, the minimum possible coefficient of friction is
$$\mu_s = \frac{f}{N} = \frac{17.2 \text{ lb}}{42.5 \text{ lb}} = 0.404$$

4.47 (a) First, let us use $\Sigma F_y = 0$ to find the normal force N on the 25 kg (245 N) box.
  $\Sigma F_y = 0 = N + (80 \text{ N}) \sin 25° - 245$ N
which gives $N = 211$ N
Now find the friction force, f, as
  $f = \mu_k N = 0.3(211 \text{ N}) = 63.3$ N
From the second law, we have
  $\Sigma F_x = ma$
  $(80 \text{ N})\cos 25° - 63.4 \text{ N} = (25 \text{ kg})a$
or $a = 0.366$ m/s^2.

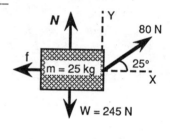

(b) When the box is on the ramp, the normal force is found as
  $\Sigma F_y = 0 = N + (80 \text{ N})\sin 25° - (245 \text{ N})\cos 10°$
which gives $N = 208$ N
and the friction force is
  $f = \mu_k N = 0.3(208 \text{ N}) = 62.2$ N
The net force up the incline is
  $(80 \text{ N})\cos 25° = 72.5$ N
the net force down the ramp is the sum of f and the component of weight down the ramp.
  $62.2 \text{ N} + (245 \text{ N})\sin 10° = 105$ N
Thus, the retarding force exceeds the force up the incline. The box will not make it up the ramp.

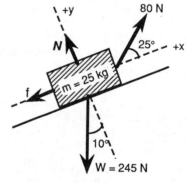

4.48 First, let us find the normal force acting on the box by use of $\Sigma F_y = 0$. We have

44

CHAPTER FOUR SOLUTIONS

$N - w\cos 30° = 0$
or $N = mg\cos 30°$
Now use the second law, applied along the ramp.
   $\Sigma F_x = mg\sin 30° - f = ma_x$.
But, $f = \mu_k N = \mu_k mg\cos 30°$
Thus, the second law becomes
   $mg\sin 30° - \mu_k mg\cos 30° = ma_x$.
or

$$\mu_k = \frac{\sin 30°}{\cos 30°} - \frac{a_x}{g\cos 30°} = \tan 30° - \frac{a_x}{g\cos 30°}$$

$$= 0.577 - \frac{1.2 \text{ m/s}^2}{(9.8 \text{ m/s}^2)(0.866)} = 0.436$$

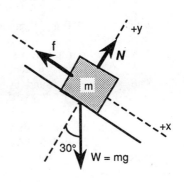

4.49 (a) The force of friction is found as
   $f = \mu_k N = \mu_k mg$
Now, choose the positive direction of the x axis in the direction of f and apply the second law. We have
   $f = ma_x$
or $a_x = f/m = \mu_k g$
Now choose the positive direction of the x axis to be in the direction of motion of the car and use $v^2 = v_0^2 + 2ax$, with $v = 0$, $v_0 = 50$ km/h $= 13.9$ m/s. We have
   $0 = (13.9 \text{ m/s})^2 + 2(-\mu_k g)x$ (1)
With $\mu_k = 0.1$, this gives a value for x of
   $x = 98.6$ m
(b) With $\mu_k = 0.6$, (1) above gives
   $x = 16.4$ m

4.50 The final velocity of the car $= 80$ mph $= 35.8$ m/s
   $a = \frac{v_f - v_0}{t} = \frac{35.8 \text{ m/s} - 0}{8 \text{ s}} = 4.47 \text{ m/s}^2$
We have equilibrium in the vertical direction, so $N = mg$.
Newton's second law applied along the horizontal gives
   $f = m(4.47 \text{ m/s}^2)$
For the minimum coefficient of friction, we are on the verge of slipping, so
   $f = \mu_s N$
   $m(4.47 \text{ m/s}^2) = \mu_s(m(9.8 \text{ m/s}^2))$
so $\mu_s = 0.456$

4.51 While skidding $f = \mu_k N = 0.153 N$
We have equilibrium in a direction perpendicular to the incline, so
   $N - w \cos 10.5° = 0$
or $N = m(9.64 \text{ m/s}^2)$ (1)
and $f = 0.153(m(9.64 \text{ m/s}^2)) = (1.474 \text{ m/s}^2)m$ (2)
Along the incline, the second law gives
   $-f - w \sin 10.5° = ma_x$
Using (1) and (2), this gives
   $a_x = -3.26 \text{ m/s}^2$
Also, $v_{0x} = 40$ km/h $= 11.1$ m/s and $v_f = 0$
so $v_x^2 = v_{0x}^2 + 2a_x x$
becomes $0 = (11.1 \text{ m/s})^2 + 2(-3.26 \text{ m/s}^2)x$

CHAPTER FOUR SOLUTIONS

and   x = 19.0 m   measured along the inclined surface

4.52 The normal force is
   $N = mg\cos\theta$ and $f = \mu_k N$.
The angle θ is found as
   $\sin\theta = 1.3/7.5 = 0.173$
From which, $\theta = 10°$
Let +x be up the incline, and use $\Sigma F_x = 0$
   $F - f - w\sin\theta = 0$
Where F is the force required to pull the box up the incline.
With $\mu_k = 0.6$, $\theta = 10°$, and w = 5400 N, we find
   F = 4130 N

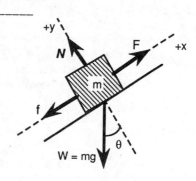

4.53 See the solution to problem 4.52. The approach is identical, and we find
   $F = f + w\sin\theta$
with m = 60 kg, $\mu_k = 0.2$, and $\theta = 25°$, we have
   F = 355 N

4.54 We use $\Sigma F_x = 0$. The forces along the incline are the force of friction f and the component of weight down the plane. We have
   $f = w\sin\theta$ (1)
and the normal force is given by
   $N = w\cos\theta$ (2)
Divide (1) by (2) to give
   $\dfrac{f}{N} = \tan\theta$
When the block is on the verge of motion $f = \mu_s N$ and therefore
   $\mu_s = \tan 36° = 0.727$
When the block is in motion $f = \mu_k N$, and $\mu_k = \tan 30° = 0.577$.

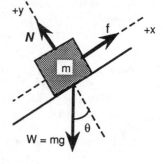

4.55 Consider the 5 kg mass. The normal force is 49 N, and the force of friction is
   $f = \mu_k N = 0.25(49\text{ N}) = 12.3\text{ N}$
Applying the second law, to the 5 kg mass gives
   $T - f = (5\text{ kg})a$
or
   $T = 12.3\text{ N} + (5\text{ kg})a$ (1)
For the 10 kg mass, the second law gives
   $98\text{ N} - T = (10\text{ kg})a$ (2)
Solving (1) and (2) simultaneously gives
   $a = 5.72\text{ m/s}^2$.

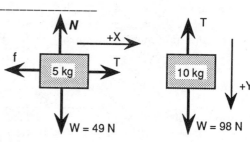

CHAPTER FOUR SOLUTIONS

4.56 First, apply the second law to the 10 kg (98 N) hanging block.
$$98\text{ N} - T = (10\text{ kg})a \quad (1)$$
For the 5 kg block, we have
$$N = w\cos 37° = 39.1\text{ N}$$
and
$$f = \mu_k N = 0.25(39.1\text{ N}) = 9.78\text{ N}$$
Apply the second law to the 5 kg block. We have
$$T - f - w\sin 37° = (5\text{ kg})a \quad (2)$$
Solving (1) and (2) simultaneously gives
$$a = 3.92\text{ m/s}^2.$$

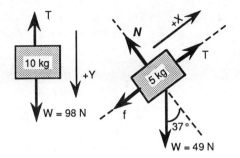

4.57 (a) $a = \dfrac{v_f - v_0}{t} = \dfrac{6\text{ m/s} - 12\text{ m/s}}{5\text{ s}} = -1.20\text{ m/s}^2$

We also know that $N = mg$. Now apply the second law:
$$-f = m(-1.2\text{ m/s}^2) \quad \text{or} \quad f = m(1.2\text{ m/s}^2) \quad (1)$$
But, also, $f = \mu_k N = \mu_k mg$. So, from (1), we have
$$\mu_k mg = m(1.2\text{ m/s}^2)$$
so, $\mu_k = \dfrac{1.2\text{ m/s}^2}{9.8\text{ m/s}^2} = 0.122$

(c) and $x = \bar{v}t = \left(\dfrac{v_0 + v_f}{2}\right)t = \left(\dfrac{12\text{ m/s} + 6\text{ m/s}}{2}\right)(5\text{ s}) = 45\text{ m}$

4.58 Consider the free body diagram of log 1. It moves with a constant velocity, so

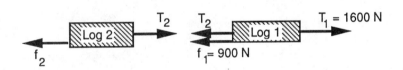

$T_1 - T_2 - f_1 = 0$, or
$T_2 = 1600\text{ N} - 900\text{ N} = 700\text{ N}$ (This is the tension in the chain between logs.)
Now, consider the free body diagram of log 2:
$T_2 - f_2 = 0$, so, $f_2 = 700\text{ N}$ (This is the force of friction on the back log.)

4.59 From Fig. (a), the normal force on the penguin is $N = 70\text{ N}$, and the second law applied to the penguin gives
$$f = (7.14\text{ kg})a$$
or $a = f/(7.14\text{ kg})$
When slipping is about to occur
$$f = f_{max} = \mu_s N = 0.7(70\text{ N}) = 49\text{ N}$$
Thus

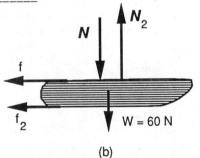

$a_{max} = f_{max}/(7.14\text{ kg}) = 49\text{ N}/7.14\text{ kg} = 6.86\text{ m/s}^2$.
From the free-body diagram of the sled [Fig. (b) above], $N_2 = 70\text{ N} + 60\text{ N} = 130\text{ N}$
Thus, the force of friction between sled and snow is
$$f_2 = \mu_k N_2 = 0.1(130\text{ N}) = 13\text{ N}$$

CHAPTER FOUR SOLUTIONS

Applying the second law to the sled gives
$$F - f - f_2 = (6.12 \text{ kg})a$$
or $\quad F = f + 13 \text{ N} + (6.12 \text{ kg})a$
F has its maximum value when f and a have their maximum values. Thus,
$$F_{max} = 49 \text{ N} + 13 \text{ N} + (6.12 \text{ kg})(6.86 \text{ m/s}^2) = 104 \text{ N}$$

4.60 $F_x = (70 \text{ N})\cos 30° = 60.6 \text{ N}$
$F_y = (70 \text{ N})\sin 30° = 35.0 \text{ N}$

4.61 The x and y components of the forces on the boat are

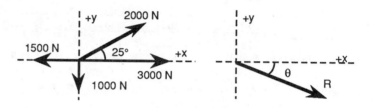

	x comp	y comp
3000 N	3000 N	0
2000 N	1813 N	845 N
1500 N	-1500 N	0
1000 N	0	-1000 N
	3313 N	-155 N

and from the Pythagorean theorem, $\quad R = \sqrt{(3313 \text{ N})^2 + (-155 \text{ N})^2} = 3317 \text{ N}$
and $\quad \tan\theta = 155/3313 = 4.68 \times 10^{-2}$
$\quad\quad \theta = 2.68°$ clockwise from the 3000 N force.

4.62 The resultant x component of the two forces exerted on the traffic light by the cables is 0, so
$$(60 \text{ N})\cos 45° - (60 \text{ N})\cos 45° = 0$$
The resultant y component is
$$(60 \text{ N})\sin 45° + (60 \text{ N})\sin 45° = 84.9 \text{ N}$$
The resultant force is, thus, 84.9 N vertically upward.
(T = 42.45 N in each cable.)
The forces on the traffic light are the 84.9 vertical force exerted by the cables and the weight. The resultant of these two must equal zero, from the first law. Thus,
$$w = 84.9 \text{ N}$$

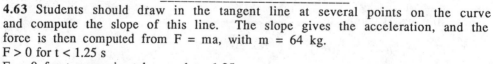

4.63 Students should draw in the tangent line at several points on the curve and compute the slope of this line. The slope gives the acceleration, and the force is then computed from F = ma, with m = 64 kg.
F > 0 for t < 1.25 s
F = 0 for t approximately equal to 1.25 s
F < 0 for t > 1.25 s

4.64 To find the acceleration, we use $v^2 = v_0^2 + 2ax$, which will be applied during the period of time the ball is in the thrower's hand. We have v = 100 mi/h = 44.7 m/s, and the initial velocity is zero.

48

CHAPTER FOUR SOLUTIONS

$v^2 = v_0^2 + 2ax$

$(44.7 \text{ m/s})^2 = 0 + 2a(1.5 \text{ m})$

which gives $a = 666 \text{ m/s}^2$.
and from the second law

$F = ma = (0.15 \text{ kg})(666 \text{ m/s}^2) = 99.9 \text{ N}$

4.65 The acceleration of the ball is found as

$v^2 = v_0^2 + 2ax$

$(20 \text{ m/s})^2 = 0 + 2a(1.5 \text{ m})$

From which, $a = 133.3 \text{ m/s}^2$.
The resultant force on the ball is the upward force, F, exerted by the thrower, less the weight of the ball, 1.47 N, downward. The second law becomes

$F - 1.47 \text{ N} = (.15 \text{ kg})(133.3 \text{ m/s}^2)$

$F = 21.5 \text{ N}$

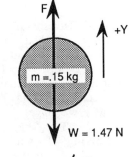

4.66 (a) The second law becomes

$w\sin 20° = ma$

or $a = g\sin 20° = 3.35 \text{ m/s}^2$.

(b) We find the time as

$x = v_0 t + \frac{1}{2} at^2$

$10 \text{ m} = 0 + \frac{1}{2}(3.35 \text{ m/s}^2)t^2$

From which $t = 2.44 \text{ s}$.

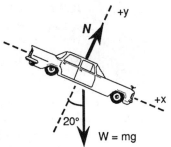

4.67 (a) The average acceleration is given by

$\bar{a} = \Delta v/\Delta t = (v_f - v_i)/\Delta t = (5 \text{ m/s} - 20 \text{ m/s})/4 \text{ s} = -3.75 \text{ m/s}^2$.

The average force is found from the second law as

$\bar{F} = m\bar{a} = (2 \times 10^3 \text{ kg})(3.75 \text{ m/s}^2) = 7500 \text{ N}$

(b) The distance traveled is

$x = \bar{v} t = \frac{(5 \text{ m/s} + 20 \text{ m/s})}{2}(4 \text{ s}) = 50 \text{ m}$

4.68 On the level surface the normal force, N, equals 600 N, and the force of friction is

$f = \mu_k N = (0.05)(600 \text{ N}) = 30 \text{ N}$

The 600 N weight is equivalent to a mass of 61.2 kg, and from the second law, the acceleration is

$a = f/m = (-30 \text{ N})/(61.2 \text{ kg}) = -0.49 \text{ m/s}^2$.

The distance traveled before coming to rest is found as

$v^2 = v_0^2 + 2ax$

$0 = (7 \text{ m/s})^2 + 2(-0.49 \text{ m/s}^2)x$

From which, $x = 50 \text{ m}$.

CHAPTER FOUR SOLUTIONS

4.69 The forces on the elevator are the tension in the cable and its weight, 1960 N. The tension is found from the second law.
  T - 1960 N = (200 kg)(1.5 m/s^2)
  T = 2260 N

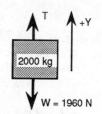

4.70 Horizontal forces on the 1st barrel are the 30 N pull and tension in the cable attached to the second barrel. Newton's second law becomes
  30 N - T = (7 kg)a   (1)
The only horizontal force on the second barrel is T. From the second law, we have
  T = (4 kg)a   (2)
Solving (1) and (2) simultaneously, we have
  T = 10.9 N   and   a = 2.73 m/s^2.

4.71 (a) From $\Sigma F_x = 0$, we have
  -f + (750 N)cos30° = 0
gives f = 650 N
From $\Sigma F_y = 0$, we have
  N - 2000 N + (750 N)sin30° = 0
From which,   N = 1625 N
Thus, $\mu_k = \frac{f}{N} = \frac{650}{1625} = 0.4$

4.72 Use $\Sigma F_x = 0$.
  Tcos10° - (1200N)sin25° = 0
  T = 515 N

4.73 (a) From $\Sigma F_x = 0$, we have
  T_1cos10° - T_2cos10° = 0
Thus, $T_1 = T_2 = T$ (the tension in the rope)
Now, use $\Sigma F_y = 0$.
  2Tsin10° - 600 N = 0   (1)
Giving,   T = 1728 N
(b) Following the same procedure outlined in part (a), we would find that Eq. (1) would become
  2Tsinθ - 600 N = 0
When T = 2000 N,   θ = 8.63°

CHAPTER FOUR SOLUTIONS

4.74 Since the surfaces are frictionless, the forces they exert must be perpendicular to those surfaces. The free body diagram is shown at the right.

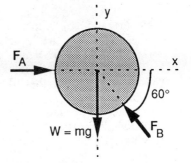

$\Sigma F_y = 0$ gives $\qquad F_B \sin 60° - w = 0$, or

$$F_B = \frac{w}{\sin 60°} = 1.155 \, Mg$$

and $\Sigma F_x = 0$ gives $\quad F_A - F_B \cos 60° = 0$
$\qquad$ So, $\quad F_A = 0.5775 \, Mg$

4.75 First, use $\Sigma F_y = 0$.
$\qquad T_2 \sin 37° - 600 \, N = 0$
$\qquad$ From which, $\quad T_2 = 997 \, N$
$\qquad$ Now, use $\Sigma F_x = 0$
$\qquad T_2 \cos 37° - T_1 = 0$
$\qquad$ Since $T_2 = 997 \, N$, we find $T_1 = 796 \, N$.

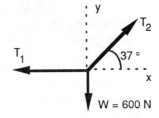

4.76 (a) Consider the weight on the table. Apply $\Sigma F_y = 0$, as
$\qquad N - w_1 = 0 \quad (1)$
where N is the normal force.
From $\Sigma F_x = 0$, we have
$\qquad T - f = 0 \quad (2)$
For the hanging weight, $\Sigma F_y = 0$ becomes
$\qquad w_2 - T = 0 \quad (3)$
With $w_1 = 100 \, N$ and $w_2 = 50 \, N$, Equation (3) gives
$\qquad T = w_2 = 50 \, N$. Then, from eq. (2), we find
$\qquad f = 50 \, N$.
(b) If the system is on the verge of slipping,
$\qquad f = \mu_s N$.
With $f = 50 \, N$ and $N = 100 \, N$ from Equation (1), we find
$\qquad \mu_s = 0.5$
(c) If $\mu_k = 0.25$, then $f = (0.25)(100 \, N) = 25 \, N$, and from Equation (2) $T = 25 \, N$. Finally, from Equation (3), $w_2 = T = 25 \, N$.

4.77 (a) The friction force between the box and the truck bed causes the box to move with the truck.
(b) The maximum value of the acceleration the truck can have before the box slides can be found by finding the maximum value of the friction force on the box. This is
$\qquad f_{max} = \mu_s N = \mu_s mg$
Thus, from Newton's Second Law
$\qquad a_{max} = f_{max}/m = \mu_s g = 0.3(9.8 \, m/s^2) = 2.94 \, m/s^2$.

4.78 Newton's second law applied along the incline becomes
$\qquad T - w \sin 30° = ma$
$\qquad 10{,}000 \, N - (1500 \, kg)(9.8 \, m/s^2) \sin 30° = (1500 \, kg)a$
and $\quad a = 1.77 \, m/s^2$.

CHAPTER FOUR SOLUTIONS

4.79 (a) and (b) Applying the second law to the car along the x direction, we have
$$T - mg\sin 30° = ma$$
$$T - (14{,}700 \text{ N})\sin 30° = (1500 \text{ kg})a \quad (1)$$
The second law for the 10,000 N (1020 kg) weight is
$$Mg - T = Ma$$
$$10{,}000 \text{ N} - T = (1020 \text{ kg})a \quad (2)$$
Solving (1) and (2) simultaneously, we have
$$a = 1.05 \text{ m/s}^2 \text{ and } T = 8.93 \times 10^3 \text{ N}$$
(c) Let us call the positive direction down the incline for the car, and upward for the rising weight. Newton's second law for the car is
$$(14{,}700 \text{ N})\sin 30° - T = (1500 \text{ kg})(2 \text{ m/s}^2) \quad (3)$$
and for the rising weight, we have
$$T - M(9.8 \text{ m/s}^2) = M(2 \text{ m/s}^2) \quad (4)$$
Solve (3) for T and substitute into (4) to give
$$M = 369 \text{ kg}$$

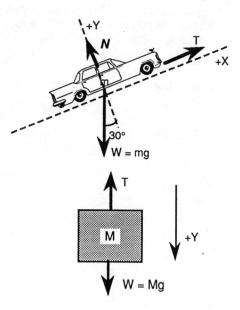

4.80 (a) $x = v_0 t + \frac{1}{2} a t^2$
$$2 \text{ m} = 0 + \frac{1}{2} a_x (1.5 \text{ s})^2$$
gives $a_x = 1.78 \text{ m/s}^2$
(b) Newtons second law along the incline gives
$$(29.4 \text{ N})\sin 30° - f = (3 \text{ kg})(1.78 \text{ m/s}^2)$$
We find $f = 9.37 \text{ N}$
Perpendicular to the plane, we have equilibrium, so
$$N = (29.4 \text{ N}) \cos 30° = 25.5 \text{ N}$$
Then $f = \mu_k N$ or $\mu_k = \frac{f}{N} = \frac{9.37 \text{ N}}{25.5 \text{ N}} = 0.368$
and finally $v^2 = v_0^2 + 2ax$
becomes $v^2 = 0 + 2(1.78 \text{ m/s}^2)(2 \text{ m})$, from which $v = 2.67 \text{ m/s}$

4.81 First, we will compute the needed accelerations:
(1) Before it starts to move, $a_y = 0$
(2) During the first 0.8 s $a_y = \frac{v_y - v_{0y}}{t} = \frac{1.2 \text{ m/s} - 0}{0.8 \text{ s}} = 1.5 \text{ m/s}^2$
(3) While moving at constant velocity $a_y = 0$
(4) During the last 1.5 s $a_y = \frac{v_y - v_{0y}}{t} = \frac{0 - 1.2 \text{ m/s}}{1.5 \text{ s}} = -0.8 \text{ m/s}^2$
Newton's second law is
$$T - 706 \text{ N} = (72 \text{ kg})a_y$$
or $T = 706 \text{ N} + (72 \text{ kg})a_y$
(a) When $a_y = 0$ $T = 706 \text{ N}$
(b) $a_y = 1.5 \text{ m/s}^2$ $T = 814 \text{ N}$
(c) $a_y = 0$ $T = 706 \text{ N}$

CHAPTER FOUR SOLUTIONS

(d) $a_y = -0.8$ m/s^2 $T = 648$ N

4.82 In the vertical direction, we have
$T\cos 4° - mg = 0$
or $T\cos 4° = 29.4$ N
Thus, $T = 29.5$ N
In the horizontal direction, the second law becomes
$T\sin 4° = (3 \text{ kg})a$
Since $T = 29.5$ N, we have
$a = 0.685$ m/s^2.

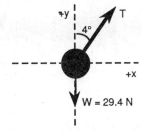

4.83 (a) The 70 N object has a mass of 7.14 kg. When the car accelerates upward at 3 m/s^2, the second law becomes
$T - mg = ma$
$T - 70 \text{ N} = (7.14 \text{ kg})(3 \text{ m/s}^2)$
and $T = 91.4$ N
(b) When accelerating downward at 3 m/s^2, we have
$mg - T = ma$
$70 \text{ N} - T = (7.14 \text{ kg})(3 \text{ m/s}^2)$
$T = 48.6$ N

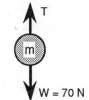

4.84 (a) Apply the second law to the 10 kg block
$T = (10 \text{ kg})a$ (1)
and then to the 20 kg block
$50 \text{ N} - T = (20 \text{ kg})a$ (2)
Solving (1) and (2) simultaneously,
$T = 16.7$ N and $a = 1.67$ m/s^2.
(b) The friction force on the 10 kg block is
$f_1 = \mu_k N = 0.1(98 \text{ N}) = 9.8$ N
and on the 20 kg block, we have
$f_2 = 0.1(196 \text{ N}) = 19.6$ N
Thus, the second law for the 10 kg block is
$T - 9.8 \text{ N} = (10 \text{ kg})a$ (3)
and for the 20 kg block
$50 \text{ N} - T - 19.6 \text{ N} = (20 \text{ kg})a$ (4)
Solving (3) and (4) simultaneously, we have
$T = 16.7$ N and $a = 0.687$ m/s^2.

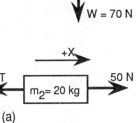

4.85 (a) Force diagrams for penguin and sled are shown. The primed forces are reaction forces for the corresponding unprimed forces.

CHAPTER FOUR SOLUTIONS

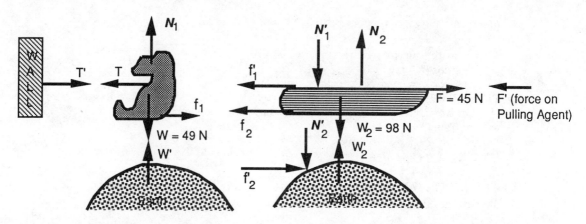

(b) The weight of the penguin is 49 N, and hence the normal force exerted on him by the sled, N_1, is also 49 N. Thus, the friction force acting on the penguin is
$$f_1 = \mu_k(49 \text{ N}) = 0.2(49 \text{ N}) = 9.8 \text{ N}$$
The penguin does not accelerate. Thus, the tension in the cord attached to the wall and the friction force must be equal.
$$T = 9.8 \text{ N}$$
(c) The normal force exerted on the sled plus penguin is the weight of the penguin (49 N) plus the weight of the sled (98 N). Thus, the net normal force, N_2 equals 147 N, and the friction force between sled and ground is
$$f_2 = \mu_k(147 \text{ N}) = 0.2(147 \text{ N}) = 29.4 \text{ N}$$
Applying the second law to the sled along the horizontal direction gives
$$45 \text{ N} - 9.8 \text{ N} - 29.4 \text{ N} = (10 \text{ kg})a$$
from which
$$a = 0.580 \text{ m/s}^2.$$

4.86 (a) The book has a weight of 8.82 N. Hence, the normal force on it while it is on the ramp is
$$N = (8.82 \text{ N})\cos 30° = 7.64 \text{ N}$$
and the friction force acting on it is
$$f = \mu_k(7.64 \text{ N}) = 0.2(7.64 \text{ N}) = 1.53 \text{ N}$$
With the positive direction up the incline, the second law is $\quad -f - w\sin 30° = ma$
$$-1.53 \text{ N} - (8.82 \text{ N})\sin 30° = (0.9 \text{ kg})a$$
which gives $a = -6.6 \text{ m/s}^2$.
The distance up the incline is found from
$$v^2 = v_0^2 + 2ax$$
$$0 = (3 \text{ m/s})^2 + 2(-6.6 \text{ m/s}^2)x. \quad \text{Thus,} \quad x = 0.682 \text{ m}$$

(a)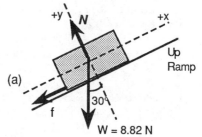

CHAPTER FOUR SOLUTIONS

(b) As in the case above, while the book moves down the ramp, the normal force is 7.64 N and the friction force is 1.53 N. However, the friction force is now up the incline. With the positive direction down the ramp, we have

$$w\sin\theta - f = ma$$
$$(8.82 \text{ N})\sin 30° - 1.53 \text{ N} = (0.9 \text{ kg})a$$

Hence, $a = 3.2 \text{ m/s}^2$.

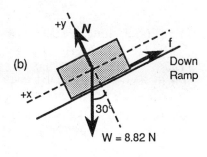

(c) The speed at the bottom is found as
$$v^2 = v_0^2 + 2ax$$
$$v^2 = 0 + 2(3.2 \text{ m/s}^2)(0.682 \text{ m}). \qquad v = 2.09 \text{ m/s}$$

4.87 (a)

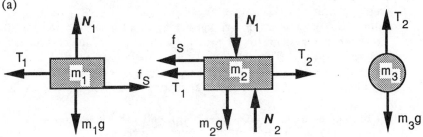

(b) and (c) $\Sigma F = ma$, Therefore

#1: $T_1 - \mu m_1 g = m_1 a$
$T_1 - (0.3)(2)(9.8) = 2a$

#2: $T_2 - T_1 - \mu m_1 g = m_2 a$
$T_2 - T_1 - 2(9.8)(0.3) = 3a$

#3: $T_2 - m_3 g = -m_3 a$
$T_2 - 10(9.80) = -10a$

Solving the above gives: $T_1 = 17.4 \text{ N}$, $T_2 = 40.5 \text{ N}$, and $a = 5.75 \text{ m/s}^2$.

4.88 (a) For every 2 meters of string passing over m_2, m_1 travels 2 meters and m_2 travels only 1 meter. Hence $2a_2 = a_1$.

(b) and (c) $\Sigma F = ma$, Therefore for the block we have
$$T = m_1 a_1 = 2 m_1 a_2.$$
and for the pulley
$$F - 2T = m_2 a_2.$$
Solving the above gives

$$T = \frac{2 m_1 F}{(4m_1 + m_2)}, \quad a_1 = \frac{2F}{(4m_1 + m_2)}, \text{ and } a_2 = \frac{F}{(4m_1 + m_2)}$$

4.89 (a) $a = \dfrac{F}{m_r} = \dfrac{18 \text{ N}}{2 \text{ kg} + 3 \text{ kg} + 4 \text{ kg}} = 2 \text{ m/s}^2$.

(b) $F_1 = m_1 a = (2 \text{ kg})(2 \text{ m/s}^2) = 4 \text{ N}$
$F_2 = m_2 a = (3 \text{ kg})(2 \text{ m/s}^2) = 6 \text{ N}$
$F_3 = m_3 a = (4 \text{ kg})(2 \text{ m/s}^2) = 8 \text{ N}$

(c) The force exerted by m_2 on $m_3 = 8$ N, and the force exerted by m_3 on $m_2 = -8$ N
The force exerted by m_1 on $m_2 = 6$ N $-$ (-8 N) $= 14$ N, and the force exerted by m_2 on $m_1 = -14$ N

CHAPTER FOUR SOLUTIONS

4.90 (a) $F - \mu mg = ma$
   $18 \text{ N} - (2 \text{ kg} + 3 \text{ kg} + 4 \text{ kg})(9.8 \text{ m/s}^2)(0.1) = (2 \text{ kg} + 3 \text{ kg} + 4 \text{ kg})a$
Thus, $a = 1.02 \text{ m/s}^2$
(b) $F_1 = m_1 a = 2.04 \text{ N}$
$F_2 = m_2 a = 3.06 \text{ N}$ and $F_3 = m_3 a = 4.08 \text{ N}$
(c) $18 - F_{12} - (2 \text{ kg})(9.80 \text{ m/s}^2)(0.1) = 2.04 \text{ N}$ $F_{12} = 14 \text{ N}$
$F_{12} - F_{23} - (3 \text{ kg})(9.80 \text{ m/s}^2)(0.1) = 3.06 \text{ N}$ $F_{23} = 8 \text{ N}$

4.91 We use $\Sigma F = ma$ with $a = 4 \text{ m/s}^2$ at all points.
Applying the second law to the entire system, we find the tension at the ceiling, point A, as
   $T - mg = ma$, or $T = m(g + a)$ (1)
Thus, T(ceiling) $= (10 \text{ kg} + 10 \text{ kg} + 1 \text{ kg} + 1 \text{ kg})(9.80 \text{ m/s}^2 + 4 \text{ m/s}^2) = 304 \text{ N}$
(b) Now, consider the system to be the upper block, the lower block, and the rope that connects them. The tension, from equation 1, is
   T(top of top block) $= (10 \text{ kg} + 10 \text{ kg} + 1 \text{ kg})(9.80 \text{ m/s}^2 + 4 \text{ m/s}^2) = 290 \text{ N}$
(c) Consider the system to be the bottom rope and the bottom block and apply (1).
   T(top of bottom rope) $= (10 \text{ kg} + 1 \text{ kg})(9.80 \text{ m/s}^2 + 4 \text{ m/s}^2) = 152 \text{ N}$
(d) Finally, apply (1) at the top of the bottom block.
   T(top of bottom block) $= (10 \text{ kg})(9.80 \text{ m/s}^2 + 4 \text{ m/s}^2) = 138 \text{ N}$

4.92 We use $\Sigma F = ma$ with a having the same magnitude for all blocks.
For the 10 kg block, we have $T_1 - (10 \text{ kg})(9.80 \text{ m/s}^2) = -(10 \text{ kg})(2 \text{ m/s}^2)$
For the 5 kg block, $T_2 - T_1 + f_5 = -(5 \text{ kg})(2 \text{ m/s}^2)$
and for the 3 kg block, $T_2 - (3 \text{ kg})(9.80 \text{ m/s}^2)\sin 25° - f_3 = (3 \text{ kg})(2 \text{ m/s}^2)$
$f_5 = \mu(5 \text{ kg})(9.80 \text{ m/s}^2)$ and $f_3 = \mu(3 \text{ kg})(9.80 \text{ m/s}^2)\cos 25°$
Solving these gives $T_1 = 78 \text{ N}$, $T_2 = 35.9 \text{ N}$, and $\mu = 0.655$

4.93 We use $\Sigma F = ma$ with a having the same magnitude for all blocks.
For block m_1, we have $T - m_1 g \sin\theta + f_1 = -m_1 a$
For block m_2, $T - m_2 g \sin\theta - f_2 = m_2 a$
where $f_1 = \mu m_1 g \cos\theta$ and $f_2 = \mu m_2 g \cos\theta$
Solving the above gives
$T = \dfrac{2m_1 m_1 g \sin\theta}{(m_1 + m_2)}$ and $a = g\sin\theta \left(\dfrac{m_1 - m_2}{m_1 + m_2}\right) - \mu g \cos\theta$

CHAPTER FIVE SOLUTIONS

5.1 $W = (F\cos\theta)s = (5000 \text{ N})(\cos 0°)(3000 \text{ m}) = 1.5 \times 10^7 \text{ J}$

5.2 If the weights are to move at constant velocity, the net force on them must be zero. Thus, the lifting force is equal to 350 N, and the work done against gravity is
$W = (F\cos\theta)s = (350 \text{ N})(\cos 0°)(2 \text{ m}) = 700 \text{ J}$

5.3 $w = mg = (20 \text{ kg})(9.8 \text{ m/s}^2)$ and 196 N = the applied force to lift the bucket at constant speed.
$W = (F\cos\theta)s = 6 \times 10^3 \text{ J} = Fs = (196 \text{ N})s$
So, $s = 30.6$ m

5.4 $f = \mu_k N = (0.9)(5.0 \text{ N}) = 4.5 \text{ N}$
Total distance moved $= (20)(0.75 \times 10^{-2} \text{ m}) = 0.15$ m
$W = (F\cos\theta)s = (4.5 \text{ N})(\cos 0°)(0.15 \text{ m}) = 0.675$ J

5.5 The component of force along the direction of motion is
$F\cos\theta = (35 \text{ N})\cos 25° = 31.7$ N
The work done by this force is
$W = (F\cos\theta)s = (31.7 \text{ N})(50 \text{ m}) = 1.59 \times 10^3$ J

5.6 $W = (F\cos\theta)s = (80 \text{ N})\cos 50°(s) = 450$ J
$s = 8.75$ m

5.7 Batman's weight = $mg = (80 \text{ kg})(9.8 \text{ m/s}^2) = 784$ N
The work done is equal to the weight of Batman times the vertical height his weight is raised.
This height, h, is
$h = 12 \text{ m} - (12 \text{ m})\cos 60° = 6$ m
Thus, $W = (784 \text{ N})(6 \text{ m}) = 4704$ J

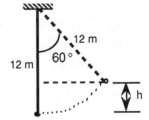

5.8 (a) We use the work-energy theorem to find the work.
$W = \frac{1}{2}mv^2 - \frac{1}{2}mv_0^2$
$W = 0 - \frac{1}{2}(70 \text{ kg})(4 \text{ m/s})^2 = -560$ J

(b) $W = (F\cos\theta)s = (f\cos 180°)s = -fs$
But, the frictional force f is found as, $f = \mu_k N = (0.7)(70 \text{ kg})(9.8 \text{ m/s}^2) = 480$ N.
Thus, $W = -fs = -(480 \text{ N})s = -560$ J
$s = 1.17$ m

5.9 (a) The work done by the 150 N force is
$W = (F\cos\theta)s = (150 \text{ N})(6 \text{ m}) = 900$ J
(b) If the object is to move at constant velocity, the friction force and the 150 N applied force must be equal. Therefore,

CHAPTER FIVE SOLUTIONS

W_f = (150 N)(cos180°)(6 m) = -900 J
(c) μ_k = f/N = (150 N)/(392 N) = 0.383.

5.10 (a) The work done by the stewardess is
$$W = (F\cos\theta)s = (40 \text{ N})(\cos 50°)(200 \text{ m}) = 5.14 \times 10^3 \text{ J}$$
(b) In order for the flight bag to move at constant velocity, the force of friction, f, must be equal in magnitude and opposite in direction to the horizontal component of the 40 N pull. Thus,
$$W_f = (F\cos\theta)s = -(40 \text{ N})\cos 50°(200 \text{ m}) = -5.14 \times 10^3 \text{ J}$$
(c) The magnitude of the force of friction is f = (40 N)(cos50°) = 25.7 N.
The magnitude of the normal force, N, is found from $\Sigma F_y = 0$.
$$N + (40 \text{ N})\sin 50° - 70 \text{ N} = 0$$
$$N = 39.4 \text{ N}$$
Therefore, μ_k = f/N = 25.7 N/39.4 N = 0.653

5.11 The initial speed of the car is 55 mi/h = 24.6 m/s, and the initial kinetic energy is
$$KE_i = \frac{1}{2} mv^2 = \frac{1}{2} (3000 \text{ kg})(24.6 \text{ m/s})^2 = 9.07 \times 10^5 \text{ J}$$
The initial kinetic energy of the car must be dissipated in the brake linings to bring the car to a stop.
energy lost = 9.07×10^5 J

5.12 The speed of the ball thrown by the very good pitcher is 100 mi/h = 44.7 m/s. Similarly, the average pitcher's ball has a speed of 80 mi/h = 35.8 m/s. The kinetic energy of the fast pitch is
$$KE = \frac{1}{2} mv^2 = \frac{1}{2} (0.15 \text{ kg})(44.7 \text{ m/s})^2 = 150 \text{ J}$$
Using the same approach as above, the kinetic energy of the slower pitch is
$$KE = \frac{1}{2} mv^2 = \frac{1}{2} (0.15 \text{ kg})(35.8 \text{ m/s})^2 = 96.0 \text{ J}$$

5.13 (a) $KE_A = \frac{1}{2} mv_A^2 = \frac{1}{2} (0.6 \text{ kg})(2 \text{ m/s})^2 = 1.2 \text{ J}$
(b) $KE_B = \frac{1}{2} mv_B^2 = 7.5 \text{ J} = \frac{1}{2} (0.6 \text{ kg})v_B^2$
Thus, v_B = 5.0 m/s
(c) Work = $\Delta KE = KE_B - KE_A$ = 7.5 J - 1.2 J = 6.3 J

5.14 net work = $\Delta KE = \frac{1}{2} mv_f^2 - \frac{1}{2} mv_i^2$
The net upward force = 2(355 N) - 700 N
Thus, $(2(355 \text{ N}) - 700 \text{ N})0.25 \text{ m} = \frac{1}{2} (71.4 \text{ kg})v_f^2 - 0$
and v_f = 0.265 m/s

5.15 The kinetic energy of the gas molecule is found as
$$KE = \frac{1}{2} mv^2 = \frac{1}{2} (5 \times 10^{-26} \text{ kg})(500 \text{ m/s})^2 = 6.25 \times 10^{-21} \text{ J}$$

5.16 The final kinetic energy of the bullet is

CHAPTER FIVE SOLUTIONS

$$KE_f = \frac{1}{2}mv^2 = \frac{1}{2}(2 \times 10^{-3} \text{ kg})(3 \times 10^2 \text{ m/s})^2 = 90 \text{ J}$$

(b) We know that
$$W = \Delta KE = 90 \text{ J} - 0 = 90 \text{ J}$$
and that $W = (F\cos\theta)s = (F)(\cos 0°)(0.5 \text{ m}) = 90$ J.
Thus, $F = 180$ N.

5.17 The net work is equal to the change in kinetic energy, $W_{net} = \Delta KE$. Thus,
$$W_{net} = KE_f - KE_i = 0 - KE_i = -\frac{1}{2}mv_i^2 = -\frac{1}{2}(1200 \text{ kg})(20 \text{ m/s})^2 = -2.40 \times 10^5 \text{ J}.$$

5.18 (a) $W = \Delta KE$
$$5000 \text{ J} = \frac{1}{2}(2500 \text{ kg})v_f^2 - 0$$
$$v_f = 2 \text{ m/s}$$
(b) $W = (F\cos\theta)s = F(25 \text{ m}) = 5000$ J
$F = 200$ N

5.19 We have equilibrium in a direction perpendicular to the incline. Thus, from $\Sigma F_y = 0$, we find the normal force to be
$$N = w \cos 20° = (98 \text{ N})(0.94) = 92.1 \text{ N}$$
and $f = \mu_k N = (0.4)(92.1 \text{ N}) = 36.8$ N
(a) The work done by the gravitational force is equal to the product of the weight of the object times the vertical height through which it has been raised. The work is also negative because the force and displacement are in opposite directions ($\cos\theta = \cos 180°$).
$$W = -ws\sin 20° = -(98 \text{ N})(5 \text{ m})\sin 20° = -168 \text{ J}$$
(b) W(by friction) = $fs\cos 180° = (36.8 \text{ N})(5 \text{ m})(-1) = -184$ J
(c) W(applied force) = $Fs = (100 \text{ N})(5 \text{ m}) = 500$ J
(d) W(net) = ΔKE = W(grav) + W(friction) + W(applied force)
   = -168 J $- 184$ J $+ 500$ J $= 148$ J
(e) $\Delta KE = \frac{1}{2}mv_f^2 - \frac{1}{2}mv_i^2 = \frac{1}{2}(10 \text{ kg})v_f^2 - \frac{1}{2}(10 \text{ kg})(1.5 \text{ m/s})^2 = 148$ J
$v_f = 5.64$ m/s

5.20 The work energy theorem is $W_{net} = KE_f - KE_i$. The net work done on the car is the work done by the motor less the work done against friction. We have
$$W_{net} = (10^3 \text{ N})(20 \text{ m}) - (950 \text{ N})(20 \text{ m}) = 1000 \text{ J}$$
We are given that $KE_i = 0$. Thus,
$$KE_f = 1000 \text{ J} = \frac{1}{2}(2000 \text{ kg})v^2.$$
From which, $v = 1$ m/s

5.21 The initial kinetic energy of the sled is $KE_i = \frac{1}{2}(10 \text{ kg})(2 \text{ m/s})^2 = 20$ J
and the force of friction = $f = \mu_k N = (0.1)(98 \text{ N}) = 9.8$ N
$W = fs\cos 180° = KE_f - KE_i$; $(9.8 \text{ N})s(-1) = 0 - 20$ J
$s = 2.04$ m

5.22 PE = $mgy = (0.15 \text{ kg})(9.8 \text{ m/s}^2)(100 \text{ m}) = 147$ J

CHAPTER FIVE SOLUTIONS

5.23 (a) Relative to the ceiling, h = -1 m
Thus, PE = mgy = (2 kg)(9.8 m/s^2)(-1 m) = -19.6 J
(b) Relative to the floor, h = 2 m
    PE = mgy = (2 kg)(9.8 m/s^2)(2 m) = 39.2 J
(c) Relative to the height of the ball, h = 0, so PE = 0

5.24 (a) With our choice for the zero level for potential energy at point B, PE$_B$ = 0. At point A, the potential energy is given by
PE$_A$ = mgy where y is the vertical height above the zero level. With 135 ft = 41.1 m, this height is found as:
    y = (41.1 m)sin40° = 26.4 m
Thus, PE$_A$ = (1000 kg)(9.8 m/s^2)(26.4 m) = 2.59 X 10^5 J

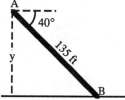

The difference in potential energy between points A and B is
    PE$_A$ - PE$_B$ = 2.59 X 10^5 J - 0 = 2.59 X 10^5 J
(b) With our choice of the zero level at point A, we have PE$_A$ = 0. The potential energy at B is given by
    PE$_B$ = mgy where, y is the vertical distance of point B below point A. In part (a) we found the magnitude of this distance to be 26.5 m. Because this distance is now below the zero reference level, it is a negative number. Thus,
    PE$_B$ = (1000 kg)(9.8 m/s^2)(-26.5 m) = - 2.59 X 10^5 J
The difference in potential energy is
    PE$_A$ - PE$_B$ = 0 - (-2.59 X 10^5 J) = 2.59 X 10^5 J

5.25 The net force in the direction of motion is the component of weight down the incline, wsin40° = (9800 N)sin40° = 6.3 X 10^3 N
The work done by this force during the 135 ft (41.1 m) displacement is
    W = Fs = (6.3 X 10^3 N)(41.1 m) = 2.59 X 10^5 J
Thus, we see that the work done is equal to the change in potential energy between these points.

5.26 (a) We take the zero level of potential energy at the lowest point of the arc. When the string is held horizontal initially, the initial position is 2 m above the zero level.
Thus, PE = mgy = wy
       = (40 N)(2 m) = 80 J
(b) From the sketch, we see that at an angle of 30° the ball is at a vertical height of (2 m)(1 - cos30°) above the lowest point of the arc.
Thus,
   PE = mgy = wy
      = (40 N)(2 m)(1 - cos30°)
      = 10.7 J

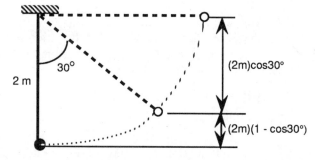

(c) The zero level has been selected at the lowest point the arc. Therefore, PE = 0 at this location.

5.27 w = mg = (3 kg)(9.8 m/s^2) = 29.4 N
(a) Work along OAC = work along OA + work along AC
      = w(OA)cos 90° + w(AC)cos 180°

CHAPTER FIVE SOLUTIONS

$\qquad = (29.4 \text{ N})(5 \text{ m})(0) + (29.4 \text{ N})(5 \text{ m})(-1) = -147 \text{ J}$
(b) W along OBC = W along OB + W along BC
$\qquad = (29.4 \text{ N})(5 \text{ m})\cos 180° + (29.4 \text{ N})(5 \text{ m})\cos 90° = -147 \text{ J}$
(c) Work along OC = w(OC)$\cos 135° = (29.4 \text{ N})(5 \times \sqrt{2} \text{ m})(-\frac{1}{\sqrt{2}}) = -147 \text{ J}$

The results should all be the same since gravitational forces are conservative.

5.28 (a) $W = W_{OA} + W_{AO} = (3 \text{ N})(5 \text{ m})\cos 180° + (3 \text{ N})(5 \text{ m})\cos 180° = -30 \text{ J}$
(b) $W = W_{OA} + W_{AC} + W_{CO}$
$\qquad = (3 \text{ N})(5 \text{ m})\cos 180° + (3 \text{ N})(5 \text{ m})\cos 180° + (3 \text{ N})(5 \times \sqrt{2} \text{ m})\cos 180°$
$\qquad = -15 \text{ J} - 15 \text{ J} - 21.2 \text{ J} = -51.2 \text{ J}$
(c) $W = W_{OC} + W_{CO} = (3 \text{ N})(5 \times \sqrt{2} \text{ m})\cos 180° + (3 \text{ N})(5 \times \sqrt{2} \text{ m})\cos 180° = -42.4 \text{ J}$
(d) The significance is that friction forces are shown to be non-conservative.

5.29 We use conservation of mechanical energy with the zero level selected at the base of the building. We also know that the initial speed of the ball is zero.

$$\frac{1}{2} mv_i^2 + mgy_i = \frac{1}{2} mv_f^2 + mgy_f$$

$$0 + mg(100 \text{ m}) = \frac{1}{2} mv_f^2 + 0$$

The mass cancels, and we solve for v_f to find
$\qquad v_f = 44.3 \text{ m/s}$

5.30 There are no non-conservative forces that do any work. As a result, we use conservation of mechanical energy, with the zero level for potential energy selected at the base of the hill. Also, note that $v_i = 0$, and we shall call the initial vertical height of the sled h.

$$\frac{1}{2} mv_i^2 + mgy_i = \frac{1}{2} mv_f^2 + mgy_f$$

$$0 + mgh = \frac{1}{2} m(3 \text{ m/s})^2 + 0$$

The mass cancels, and we solve for h to find
$\qquad h = 0.459 \text{ m}$

5.31 We use conservation of mechanical energy with the zero level for gravitational potential energy at ground level. The velocity at the maximum height, h, goes to zero, and we have

$$\frac{1}{2} mv_i^2 + mgy_i = \frac{1}{2} mv_f^2 + mgy_f$$

$$\frac{1}{2} m(6 \text{ m/s})^2 + 0 = 0 + mgh$$

Solve for h to find $\quad h = 1.84 \text{ m}$

5.32 (a) We choose the zero level for potential energy at ground level and note that at the maximum height, h, the velocity of the ball goes to zero. Thus, conservation of mechanical energy is written as

$$\frac{1}{2} mv_i^2 + mgy_i = \frac{1}{2} mv_f^2 + mgy_f$$

$$\frac{1}{2} m(20 \text{ m/s})^2 + 0 = 0 + mgh$$

Which gives $\qquad h = 20.4 \text{ m}$
(b) At half the maximum height, 10.2 m, we find the speed as

CHAPTER FIVE SOLUTIONS

$$\frac{1}{2} mv_i^2 + mgy_i = \frac{1}{2} mv_f^2 + mgy_f$$

$$\frac{1}{2} m(20 \text{ m/s})^2 + 0 = \frac{1}{2} mv^2 + mg(10.2 \text{ m})$$

giving, $v = 14.1$ m/s

5.33 (a) We choose the zero level for potential energy at the bottom of the arc. The initial height of Tarzan above this level is shown in the sketch to be
$(30\text{m})(1 - \cos 37°) = 6.04$ m
We use conservation of mechanical energy.

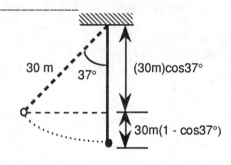

$$\frac{1}{2} mv_i^2 + mgy_i = \frac{1}{2} mv_f^2 + mgy_f$$

$$0 + mg(6.04 \text{ m}) = \frac{1}{2} mv^2 + 0$$

From which, $v = 10.9$ m/s
(b) In this case, conservation of mechanical energy becomes

$$\frac{1}{2} mv_i^2 + mgy_i = \frac{1}{2} mv_f^2 + mgy_f$$

$$\frac{1}{2} m(4 \text{ m/s})^2 + mg(6.04 \text{ m}) = \frac{1}{2} mv^2 + 0$$

and $v = 11.6$ m/s

5.34 (a) Neglecting air resistance $R = \dfrac{v_0^2 \sin 2\theta}{g}$ (See problem set chapter 2)
The maximum range occurs when $\theta = 45°$, so

$$R_{max} = \frac{v_0^2}{g} \quad \text{from which } v_0^2 = gR_{max}$$

Thus, $KE_i = \frac{1}{2} mv_0^2 = \frac{1}{2} mgR_{max}$. Now consider each of the objects in turn.

javelin $\quad \frac{1}{2} mgR_{max} = \frac{1}{2} (0.8 \text{ kg})g(89 \text{ m}) = 349$ J

discus $\quad = \frac{1}{2} (2.0 \text{ kg})g(69 \text{ m}) = 676$ J

shot $\quad = \frac{1}{2} (7.2 \text{ kg})g(21 \text{ m}) = 741$ J

(b) We use $W = \Delta KE$, with the initial velocity of the object equal to zero. The force, F, is applied over an assumed distance of 2 m while it is in the thrower's hand. We see,
$W = Fs = F(2 \text{ m}) = \Delta KE$
or $\quad F = \Delta KE/(2 \text{ m})$
javelin $\quad F = (349 \text{ J})/(2 \text{ m}) = 175$ N
discus $\quad F = (676 \text{ J})/(2 \text{ m}) = 338$ N
shot $\quad F = (741 \text{ J})/(2 \text{ m}) = 371$ N
(c) Yes, if the muscles of the body are capable of exerting a force of 371 N on an object (and giving it a kinetic energy of 741 J), one would expect the javelin to leave the hand with a speed of 43 m/s and to reach a range of 189 m. Since this predicted range is much larger than the range found in practice, air resistance must exert a considerable influence.

CHAPTER FIVE SOLUTIONS

5.35 (a) We choose the zero level for gravitational potential energy to be at the bottom of the slide. If the slide is frictionless, we can use conservation of mechanical energy, as

$$\frac{1}{2} mv_i^2 + mgy_i = \frac{1}{2} mv_f^2 + mgy_f$$

$$0 + mg(4 \text{ m}) = \frac{1}{2} mv^2 + 0$$

From which, $v = 8.85$ m/s

(b) When non-conservative forces are present, we must use

$$W_{nc} = \frac{1}{2} mv_f^2 - \frac{1}{2} mv_i^2 + mgy_f - mgy_i$$

$$W_{nc} = \frac{1}{2} m(6 \text{ m/s})^2 - 0 + 0 - mg(4 \text{ m})$$

The percentage of the energy lost, W_{nc}, to the total energy at the top of the slide, mgh, is

$$\frac{W_{nc}}{mg(4 \text{ m})} (100 \%) = 54.1 \%$$

5.36 The initial vertical height of the car above the zero reference level at the base of the hill is $h = (5 \text{ m})\sin 20° = 1.71$ m
The energy lost through friction is

$$W_{nc} = -fs = -(4000 \text{ N})(5 \text{ m}) = -2 \times 10^4 \text{ J}$$

We now use,

$$W_{nc} = \frac{1}{2} mv_f^2 - \frac{1}{2} mv_i^2 + mgy_f - mgy_i$$

$$-2 \times 10^4 \text{ J} = \frac{1}{2} (2000 \text{ kg})v^2 - 0 + 0 - (2000 \text{ kg})g(1.71 \text{ m})$$

and $v = 3.68$ m/s

5.37 The distance traveled by the ball from the top of the arc to the bottom, is
$s = \pi r = \pi(0.6 \text{ m}) = 1.885$ m
The work done by the non-conservative force, the force exerted by the pitcher, is

$$W_{nc} = (30 \text{ N})(1.885 \text{ m})\cos 0° = 56.55 \text{ J}$$

We shall choose the potential energy reference level to be at the bottom of the arc. We have

$$W_{nc} = \frac{1}{2} mv_f^2 - \frac{1}{2} mv_i^2 + mgy_f - mgy_i$$

$$56.55 \text{ J} = \frac{1}{2} (0.25 \text{ kg})v_f^2 - \frac{1}{2} (0.25 \text{ kg})(15 \text{ m/s})^2 + 0 - (0.25 \text{ kg})g(1.2 \text{ m})$$

giving $v_f = 26.5$ m/s

5.38 We shall take the zero level of potential energy to be at the lowest level reached by the diver under the water.

$$W_{nc} = \frac{1}{2} mv_f^2 - \frac{1}{2} mv_i^2 + mgy_f - mgy_i$$

$$\bar{F} (5 \text{ m})\cos 180° = 0 - 0 + 0 - (70 \text{ kg})g(15 \text{ m})$$

$$\bar{F} = 2060 \text{ N}$$

5.39 (a) The initial vertical height of the child above the zero level for gravitational energy at the bottom of the arc is

63

CHAPTER FIVE SOLUTIONS

(2 m)(1 - cos30°) = 0.268 m   (See problem 33.)
In the absence of friction, we use conservation of mechanical energy as

$$\frac{1}{2} mv_i^2 + mgy_i = \frac{1}{2} mv_f^2 + mgy_f$$

$$0 + mg(0.268 \text{ m}) = \frac{1}{2} mv^2 + 0$$

v = 2.29 m/s

(b) In the presence of friction, we use

$$W_{nc} = \frac{1}{2} mv_f^2 - \frac{1}{2} mv_i^2 + mgy_f - mgy_i$$

$$W_{nc} = \frac{1}{2} (25 \text{ kg})(2 \text{ m/s})^2 - 0 + 0 - (25 \text{ kg})g(0.268 \text{ m}) = -15.6 \text{ J}$$

5.40 The vertical distance dropped by the plane is
(500 m)sin30° = 250 m
Let us calculate the work done by non-conservative forces on the plane. The work done by the engine is
   $W_e = Fs = (7.5 \times 10^4 \text{ N})(500 \text{ m}) = 3.75 \times 10^7 \text{ J}$
and the work done by the resistive force, f, is
   $W_f = -fs = -(4 \times 10^4 \text{ N})(500 \text{ m}) = -2 \times 10^7 \text{ J}$
Thus, $W_{nc} = W_e + W_f = 1.75 \times 10^7$ J
Now, we use

$$W_{nc} = \frac{1}{2} mv_f^2 - \frac{1}{2} mv_i^2 + mgy_f - mgy_i$$

$1.75 \times 10^7 \text{ J} = \frac{1}{2} (1.5 \times 10^4 \text{ kg})v^2 - \frac{1}{2} (1.5 \times 10^4 \text{ kg})(60 \text{ m/s})^2$
                                        $- (1.5 \times 10^4 \text{ kg})g(250 \text{ m})$
Which gives   v = 104 m/s

5.41 The normal force, N, acting on the sled is found from $\Sigma F_y = 0$.
   $N - (20 \text{ kg})g\cos20° = 0$
   $N = 184$ N
The friction force is
   $f = \mu_k N = 0.2(184 \text{ N}) = 36.8$ N
and the work done by the force of friction as the sled moves a distance s up the hill is
   $W_{nc} = -fs = -(36.8 \text{ N})s$
Now use $W_{nc} = \frac{1}{2} mv_f^2 - \frac{1}{2} mv_i^2 + mgy_f - mgy_i$

$-(36.8 \text{ N})s = 0 - \frac{1}{2} (20 \text{ kg})(4 \text{ m/s})^2 + (20 \text{ kg})g(s)\sin20° - 0$

In the last term, s (sin20°) is the final vertical height of the sled above the zero reference level at the bottom of the hill.
Solving for s we find   s = 1.54 m

5.42 (a) The normal force, N, is found from $\Sigma F_y = 0$ (y direction is perpendicular to the hill) as
   $N - w\cos30° = 0$
so, $N = w\cos30°$, and the force of friction is
   $f = \mu_k N = 0.7(2 \text{ kg})g\cos30° = 11.9$ N

CHAPTER FIVE SOLUTIONS

We shall call the distance the block moves up the incline s. As a result, the final vertical height is s(sin30°).
The work done by friction over the distance s is
$$W_{nc} = -fs = -(11.9 \text{ N})s$$
Now, use $\quad W_{nc} = \frac{1}{2} mv_f^2 - \frac{1}{2} mv_i^2 + mgy_f - mgy_i$

$$-(11.9 \text{ N})s = 0 - \frac{1}{2}(2 \text{ kg})(3 \text{ m/s})^2 + (2 \text{ kg})g(s)\sin 30° - 0$$

Which yields $\quad s = 0.415$ m

(b) The work done by the force of friction is
$$W_{nc} = -fs = -(11.9 \text{ N})s = -(11.9 \text{ N})(0.415 \text{ m}) = -4.94 \text{ J}$$

(c) When s = 0.3 m, we find the final vertical height to be (0.3 m)sin30° = 0.15 m, and the potential energy at this height (with the zero level at the base of the incline) is
$$PE = mgy = (2 \text{ kg})g(0.15 \text{ m}) = 2.94 \text{ J}$$
Thus, the change in potential energy during this part of the motion is
$$\Delta PE = 2.94 \text{ J}$$
To find the change in kinetic energy, we use
$$W_{nc} = \frac{1}{2} mv_f^2 - \frac{1}{2} mv_i^2 + mgy_f - mgy_i, \text{ or } \quad W_{nc} = \Delta KE + \Delta PE$$
Thus, $\Delta KE = W_{nc} - \Delta PE = -3.57 \text{ J} - 2.94 \text{ J} = -6.51 \text{ J}$

5.43 $W_{nc} = \Delta KE + \Delta PE$

$$fs\cos 180° = 0 - \frac{1}{2} mv_i^2 + 0 - 0$$

$$(\mu_k mg)s(-1) = -\frac{1}{2} mv_i^2$$

$$(0.115)(9.8 \text{ m/s}^2)s = \frac{1}{2}(4.0 \text{ m/s})^2$$

$$s = 7.10 \text{ m}$$

5.44 We shall take the zero level for PE at the base of the hill. The skier starts at a vertical height above this level of $h_0 = (200 \text{ m})\sin 10.5° = 36.45$ m.
While on the hill, the normal force is $N_1 = w \cos 10.5° = mg \cos 10.5°$, and while on the level ground, $N_2 = w = mg$.

$$W_{nc} = \frac{1}{2} mv_f^2 - \frac{1}{2} mv_i^2 + mgy_f - mgy_i$$
$$-f_1(200 \text{ m}) - f_2(x) = 0 - 0 + 0 - mgh_0$$
$$-\mu_k N_1(200 \text{ m}) - \mu_k N_2(x) = -mgh_0$$
$$-(0.075)mg \cos 10.5°(200 \text{ m}) - (0.075)mg(x) = -mgh_0$$
$$x = 289 \text{ m}$$

5.45 (a) We use the work-kinetic energy theorem as
$$W = \frac{1}{2} mv^2 - \frac{1}{2} mv_0^2$$
so, $\quad W = \frac{1}{2} mv^2 - 0 = \frac{1}{2}(1500 \text{ kg})(10 \text{ m/s})^2 = 7.5 \times 10^4 \text{ J}$

(b) The average power is given by
$$\bar{P} = W/\Delta t = 7.5 \times 10^4 \text{ J}/3 \text{ s} = 2.5 \times 10^4 \text{ W} = 33.5 \text{ hp}$$

(c) $a = \dfrac{\Delta v}{\Delta t} = \dfrac{10 \text{ m/s} - 0}{3 \text{ s}} = 3.33 \text{ m/s}^2$

CHAPTER FIVE SOLUTIONS

$F = ma$ becomes $\quad F = (1500 \text{ kg})(3.33 \text{ m/s}^2) = 5000$ N
at $t = 2s$, $v = v_0 + at$ gives $\quad v = 0 + (3.33 \text{ m/s}^2)(2 \text{ s}) = 6.66$ m/s
so, $\quad P_{instantaneous} = Fv = (5000 \text{ N})(6.66 \text{ m/s}) = 3.33 \times 10^4$ W = 44.7 hp

---

5.46 (a) Let us first find the net work done on the car by use of
$$W_{net} = \frac{1}{2} mv^2 - \frac{1}{2} mv_0^2$$
$$W_{net} = \frac{1}{2} (1500 \text{ kg})(18 \text{ m/s})^2 - 0 = 2.43 \times 10^5 \text{ J}$$
The net work, W_{net}, is the work done by the engine, W_E, minus the work done by friction, W_f. Thus, the work done by the engine is
$$W_E = W_{net} + W_f = 2.43 \times 10^5 \text{ J} + (400 \text{ N})d$$
where d is the distance traveled by the car in 12 s. The power input by the engine is
$$\bar{P} = W_E/12 \text{ s} = (2.43 \times 10^5 \text{ J}/12 \text{ s}) + (400 \text{ N})d/12 \text{ s}$$, but d/12 s is the average velocity during this interval, which is 9 m/s.
Thus,
$$\bar{P} = 2.39 \times 10^4 \text{ W} = 32 \text{ hp}$$
(b) The constant acceleration of the car during the 12 seconds is
$$a = \Delta v/\Delta t = (18 \text{ m/s})/12 \text{ s} = 1.5 \text{ m/s}^2.$$
and the net force is given by $F = ma$ as
$\quad F_E - f = ma$ where F_E is the forward thrust by the engine.
so, $\quad F_E = f + ma = 400 \text{ N} + (1500 \text{ kg})(1.5 \text{ m/s}^2) = 2650$ N
and the instantaneous power at 12 s (when $v = 18$ m/s) is
$P_i = Fv = (2650 \text{ N})(18 \text{ m/s}) = 4.77 \times 10^4$ W = 63.9 hp

---

5.47 $W_{nc} = \Delta PE = mgh = (800 \text{ N})(6 \text{ m}) = 4.80 \times 10^3$ J
$\quad \bar{P} = W/\Delta t = 4.80 \times 10^3 \text{ J}/8 \text{ s} = 600 \text{ W} = 0.8$ hp

---

5.48 If the skier is to be pulled at constant speed, the force applied by the motor is equal to the force of friction. Thus,
$\quad \bar{P} = fv = \mu_k mgv = 0.15(80 \text{ kg})(g)(2 \text{ m/s}) = 235 \text{ W} = 0.315$ hp

---

5.49 $P = \Delta W/\Delta t = mgh/\Delta t = (300 \text{ kg})g(5 \text{ m})/8 \text{ s} = 1.84 \times 10^3 \text{ W} = 2.46$ hp

---

5.50 (a) $W_{nc} = \Delta KE + \Delta PE$, but $\Delta KE = 0$ because he moves at constant speed. The skier rises a vertical distance of $(60 \text{ m})\sin 30° = 30$ m. Thus,
$\quad W_{nc} = (70 \text{ kg})g(30 \text{ m}) = 2.06 \times 10^4$ J
(b) The time to travel 60 m at a constant speed of 2 m/s is 30 s. Thus,
$$P_{input} = \frac{W}{\Delta t} = \frac{2.06 \times 10^4 \text{ J}}{30 \text{ s}} = 686 \text{ W} = 0.919 \text{ hp}$$

5.51 (a) The work done by the student is
$\quad W = \Delta PE = mgy = (50 \text{ kg})g(5 \text{ m}) = 2.45 \times 10^3$ J
The time to do this work if she is to match the power output of a 200 W lightbulb is.
$\quad \Delta t = W/\bar{P} = 2.45 \times 10^3 \text{ J}/200 \text{ W} = 12.25$ s
Thus, the average velocity is

CHAPTER FIVE SOLUTIONS

$\bar{v} = 5\text{m}/12.25 \text{ s} = 0.408 \text{ m/s}$
(b) The work done has already been found in part (a).

5.52 To lift 60 lbs of water through a height of 80 ft (at constant speed) requires a minimum amount of work given by,
$$W = Fs = (60 \text{ lbs})(80 \text{ ft}) = 4.80 \times 10^3 \text{ ft lbs}$$
To do this much work each second requires a power input of
$$P = \Delta W/\Delta t = 4.80 \times 10^3 \text{ ft lbs/s} = 8.73 \text{ hp}$$

5.53 (a) The distance moved upward in the first 3 s is
$$\Delta y = \bar{v} t = \left(\frac{0 + 1.75 \text{ m/s}}{2}\right)(3 \text{ s}) = 2.625 \text{ m}$$
$$W_{nc} = \frac{1}{2} mv_f^2 - \frac{1}{2} mv_i^2 + mgy_f - mgy_i$$
$$W_{nc} = \frac{1}{2}(650 \text{ kg})(1.75 \text{ m/s})^2 - 0 + (650 \text{ kg})g(2.625 \text{ m}) = 1.772 \times 10^4 \text{ J}$$
Also, $W_{nc} = \bar{P} t$, so $\bar{P} = \dfrac{W_{nc}}{t} = \dfrac{1.772 \times 10^4 \text{ J}}{3 \text{ s}} = 5910 \text{ W} = 7.92 \text{ hp}$

(b) When moving upward at constant speed ($v = 1.75$ m/s), the applied force equals the weight = $(650 \text{ kg})(9.8 \text{ m/s}^2) = 6370 \text{ N}$.
Therefore, $P = Fv = (6370 \text{ N})(1.75 \text{ m/s}) = 11{,}200 \text{ W} = 14.1 \text{ hp}$

5.54 130 hp = 9.70×10^4 W
$P = Fv$, where F is the propulsive force required to overcome water resistance
$$F = \frac{P}{v} = \frac{9.70 \times 10^4 \text{ W}}{15 \text{ m/s}} = 6.47 \times 10^3 \text{ N} \quad \text{(approximately equal to 1500 lb)}$$

5.55 (a) The work done by a variable force equals the area under the force versus displacement graph. This area is in the shape of a triangle. Therefore,
$$W = \frac{1}{2}(\text{base})(\text{height}) = \frac{1}{2}(15 \text{ N})(3 \text{ m}) = 22.5 \text{ J}$$
(b) But we know from the work-kinetic energy theorem that
$$W = \frac{1}{2} mv^2 - \frac{1}{2} mv_0^2$$
$$22.5 \text{ J} = \frac{1}{2}(1 \text{ kg})v^2 - 0$$
Thus, $v = 6.71$ m/s

5.56 The work done by a variable force equals the area under the force versus displacement graph.
(a) Between $x = 0$ and $x = 5$ m, the area is that of a triangle, so
$$W = \frac{1}{2}(\text{base})(\text{height}) = \frac{1}{2}(5 \text{ m})(3 \text{ N}) = 7.5 \text{ J}$$
(b) Between $x = 5$ m and $x = 10$ m, the area is a rectangle, so
$$W = (\text{base})(\text{height}) = (5 \text{ m})(3 \text{ N}) = 15 \text{ J}$$
(c) Between $x = 10$ m and $x = 15$ m, the area is a triangle, so
$$W = \frac{1}{2}(\text{base})(\text{height}) = \frac{1}{2}(5 \text{ m})(3 \text{ N}) = 7.5 \text{ J}$$
(d) Between $x = 0$ and $x = 15$ m, the work is the total area under the curve, which is the sum of (a), (b), and (c).

CHAPTER FIVE SOLUTIONS

$$W = 7.5 \text{ J} + 15 \text{ J} + 7.5 \text{ J} = 30 \text{ J}$$

5.57 We know $W = \frac{1}{2}mv^2 - \frac{1}{2}mv_0^2 = KE_f - KE_i$, or $KE_f = W + KE_i$

The initial kinetic energy is $KE_i = \frac{1}{2}(3 \text{ kg})(0.5 \text{ m/s})^2 = 0.375 \text{ J}$

Thus, $KE_f = W + 0.375 \text{ J}$

(a) At $x = 5$ m, the work that has been done is $W = 7.5$ J, and the final value of the kinetic energy is

$$KE_f = 7.5 \text{ J} + 0.375 \text{ J} = 7.88 \text{ J} = \frac{1}{2}(3 \text{ kg})(v_f)^2$$

giving, $v_f = 2.29$ m/s

(b) The same approach is used for parts (b) and (c) as in part (a). For part (b), we find $KE_f = 22.9$ J and the final velocity is 3.91 m/s

(c) The final kinetic energy is 30.4 J, and the final velocity is 4.50 m/s.

5.58 Use conservation of mechanical energy, with the zero point of potential energy chosen to be at the top of the professor's head.

$$\frac{1}{2}mv_i^2 + mgy_i = \frac{1}{2}mv_f^2 + mgy_f$$

$$0 + mgh = \frac{1}{2}m(10 \text{ m/s})^2.$$

The mass cancels, and we find
$h = 5.10$ m

5.59 The work done against friction is $W = fs$

(a) For the circular path, $s = \frac{1}{2}$(circumference of the circle) $= \pi r$.

Since $r = 1$ m, $s = \pi$ m, and
$W = fs = (4 \text{ N})(\pi \text{ m}) = 4\pi$ J

(b) Along the triangular path, $s = 2\sqrt{2}$ m. Thus,
$W = fs = (4 \text{ N})(2\sqrt{2} \text{ m}) = 8\sqrt{2}$ J $= 11.3$ J

(c) Along the straight line path, $s = 2$ m, and
$W = fs = (4 \text{ N})(2 \text{ m}) = 8$ J

5.60 For uniformly accelerated motion, we have
$v^2 = v_0^2 + 2ay$
$v^2 = 0 + 2(1 \text{ m/s}^2)(30 \text{ m})$

So, $v^2 = 60$ m^2/s^2

Now, $W_{nc} = \frac{1}{2}mv_f^2 - \frac{1}{2}mv_i^2 + mgy_f - mgy_i$

$= \frac{1}{2}(8000 \text{ kg})(60 \text{ m}^2/\text{s}^2) - 0 + (8000 \text{ kg})g(30 \text{ m}) - 0 = 2.59 \times 10^6$ J

5.61 (a) We take the zero level for potential energy at the level of point B, and use

$$\frac{1}{2}mv_i^2 + mgy_i = \frac{1}{2}mv_f^2 + mgy_f$$

$$0 + (0.4 \text{ kg})g(5 \text{ m}) = \frac{1}{2}(0.4 \text{ kg})v_B^2 = 0$$

CHAPTER FIVE SOLUTIONS

We find, $v_B = 9.90$ m/s

(b) At point C, with the starting point at A, we have

$$\frac{1}{2} mv_i^2 + mgy_i = \frac{1}{2} mv_f^2 + mgy_f$$

$$0 + (0.4 \text{ kg})g(5 \text{ m}) = \frac{1}{2}(0.4 \text{ kg})v_C^2 + (0.4 \text{ kg})g(2 \text{ m})$$

and, $v_f = 7.67$ m/s

5.62 The velocity of the plane is 500 mi/h = 7.34×10^2 ft/s.
$P = Fv = (4000 \text{ lb})(7.34 \times 10^2 \text{ ft/s}) = 2.93 \times 10^6$ ft lb/s = 5.33×10^3 hp.

5.63 (a) The force of friction between block and surface is
$f = \mu_k N = 0.4(1960 \text{ N}) = 784$ N

If the crate is to move at constant speed, the force, F, exerted on it by the engine must equal f. Thus, the power is

$P = Fv = (784 \text{ N})(5 \text{ m/s}) = 3.92 \times 10^3$ W = 5.25 hp.

(b) The work done is $W = P(\Delta t) = (3.92 \times 10^3 \text{ W})(180 \text{ s}) = 7.06 \times 10^5$ J.

5.64 Convert 90 hp to ft lbs/s, and convert 60 mi/h to ft/s. Then substitute into $P = Fv$ to find
$F = 563$ lbs.

5.65 (a) Choose the zero level for potential energy at the level of B. Between A and B, we can use

$$\frac{1}{2} mv_i^2 + mgy_i = \frac{1}{2} mv_f^2 + mgy_f$$

$$0 + (0.4 \text{ kg})g(5 \text{ m}) = \frac{1}{2}(0.4 \text{ kg})v_B^2 + 0$$

$v_B = 9.90$ m/s

(b) We choose the starting point at B, the zero level at B, and the end point at C.

$$W_{nc} = \frac{1}{2} mv_f^2 - \frac{1}{2} mv_i^2 + mgy_f - mgy_i$$

$$W_{nc} = 0 - \frac{1}{2}(0.4 \text{ kg})(9.90 \text{ m/s})^2 + (0.4 \text{ kg})g(2 \text{ m}) - 0$$

$$W_{nc} = -11.8 \text{ J}$$

5.66 (a) Use $W = \frac{1}{2} mv_f^2 - \frac{1}{2} mv_0^2$ with W the work done on the baseball by the hand. (This is the negative of the work done on the hand by the baseball.)

$$W = -Fs = -F(.02 \text{ m}) = 0 - \frac{1}{2}(0.15 \text{ kg})(25 \text{ m/s})^2.$$

$F = 2.34 \times 10^3$ N.

From the third law this is also the magnitude of the force on the hand.

(b) The steps are identical to those above, except $s = 0.10$ m. The result is
$F = 469$ N

5.67 The mass of the cart is 10 kg, and we use

$$W = \frac{1}{2} mv_f^2 - \frac{1}{2} mv_0^2$$

$$(40 \text{ N})(12 \text{ m}) = \frac{1}{2}(10 \text{ kg})v_f^2 - 0$$

CHAPTER FIVE SOLUTIONS

$v_f = 9.80$ m/s

5.68 $W = \frac{1}{2}mv^2 - \frac{1}{2}mv_0^2$

$W = (40 \text{ N} - f)(12 \text{ m}) = \frac{1}{2}(10 \text{ kg})(4.90 \text{ m/s})^2 - 0$

Which gives, $f = 30$ N
and $\mu_k = f/N = 30$ N/98 N $= 0.306$

5.69 (a) $\frac{1}{2}mv_i^2 + mgy_i = \frac{1}{2}mv_f^2 + mgy_f$

$0 + (75 \text{ kg})g(1 \text{ m}) = \frac{1}{2}(75 \text{ kg})v_f^2 + 0$

Thus, the velocity with which he hits is, $v_f = 4.43$ m/s

(b) $W_{nc} = -Fs = -F(5 \times 10^{-3} \text{ m})$ where F is the force exerted on the man by the floor.

$W_{nc} = \frac{1}{2}mv_f^2 - \frac{1}{2}mv_i^2 + mgy_f - mgy_i$

$-F(5 \times 10^{-3} \text{ m}) = 0 - \frac{1}{2}(75 \text{ kg})(4.43 \text{ m/s})^2 + 0 - (75 \text{ kg})g(5 \times 10^{-3} \text{ m})$

Thus, $F = 1.47 \times 10^5$ N

5.70 The initial height of Tarzan above the zero level for potential energy is
$(5 \text{ m})(1 - \sin 30°) = 2.5$ m   (See problem 33.)
The velocity at the bottom of the arc is found from

$\frac{1}{2}mv_i^2 + mgy_i = \frac{1}{2}mv_f^2 + mgy_f$

$0 + mg(2.5 \text{ m}) = \frac{1}{2}mv^2 + 0$

The mass cancels, and we find $v^2 = 2g(2.5 \text{ m})$
On the upswing, Tarzan continues alone, and we use

$\frac{1}{2}mv_i^2 + mgy_i = \frac{1}{2}mv_f^2 + mgy_f$

$\frac{1}{2}m(2g)(2.5 \text{ m}) + 0 = 0 + mgh$

The mass again cancels, and we solve for h, the final height of Tarzan, to find
$h = 2.5$ m
He returns to the same height from which the two started.

5.71 (a) The 6 kg object has a weight of 58.8 N.   The work done by the tension force is
$W = (80 \text{ N})(5 \text{ m}) = 400$ J
(b) The work done by gravity is
$W_g = (58.8 \text{ N})\cos 180°(5 \text{ m}) = -294$ J
(c) $W_{nc} = \frac{1}{2}mv_f^2 - \frac{1}{2}mv_i^2 + mgy_f - mgy_i$

$400 \text{ J} = \frac{1}{2}(6 \text{ kg})v^2 - 0 + (6 \text{ kg})g(5 \text{ m})$

$v = 5.94$ m/s

5.72 $\frac{1}{2}mv_i^2 + mgy_i = \frac{1}{2}mv_f^2 + mgy_f$

CHAPTER FIVE SOLUTIONS

$\frac{1}{2} mv_i^2 + 0 = 0 + mg(2.5 \times 10^{-2} \text{ m})$

$v = 0.70$ m/s

5.73 (a) $W_{nc} = \frac{1}{2} mv_f^2 - \frac{1}{2} mv_i^2 + mgy_f - mgy_i$

$W_{nc} = mgy_f - mgy_i$ because $\Delta KE = 0$

$W_{nc} = (65 \text{ kg})g(205 \text{ m}) - 0$; 205 m is the vertical distance risen above the zero level of potential energy selected to be at the base of the hill.

$W_{nc} = 1.31 \times 10^5$ J

(b) $P = W/t = 1.31 \times 10^5$ J/150 s = 871 W = 1.17 hp

5.74 (a) $KE = \frac{1}{2} mv^2 = \frac{1}{2} (3 \text{ kg})[(5 \text{ m/s})^2 + (-3 \text{ m/s})^2] = 51$ J

(b) $KE_2 = \frac{1}{2} (3 \text{ kg})[(8 \text{ m/s})^2 + (4 \text{ m/s})^2] = 120$ J. Thus, $\Delta KE = 69$ J

5.75 (a) The work done on the projectile by gravity is the product of the weight and the vertical distance through which the projectile falls. Thus,

$W = mgy$

(b) $W = \Delta KE = mgy$

(c) $\Delta K = KE_f - KE_i = W$, so $KE_f = W + KE_i = mgy + \frac{1}{2} mv_0^2$

5.76 On the incline $N = w\cos 37° = 391$ N, and $f = \mu_k N = (0.25)(391 \text{ N}) = 97.8$ N. Now, we apply Newton's second law to the block on the incline in a direction along the incline

$T - 97.8 \text{ N} - (490 \text{ N}) \sin 37° = (50 \text{ kg})a$

or $T = 392.7 \text{ N} + (50 \text{ kg})a$ (1)

Now consider the hanging block, with the positive direction selected as downward.

$980 \text{ N} - T = (100 \text{ kg})a$ (2)

Solving (1) and (2) simultaneously,

$T = 588.5$ N

From the work energy theorem $(\Delta KE)_A$ = net work done on A by all forces =

$T \, s \, \cos 0° + N \, s \, \cos 90° + f \, s \, \cos 180° + w \, s \, \cos 127°$

$= (588.5 \text{ N})(20 \text{ m})(1) + 0 + (97.8 \text{ N})(20 \text{ m})(-1) + (490 \text{ N})(20 \text{ m})(-0.602) = 3914$ J

5.77 (a) $a = \frac{\Delta v}{t} = \left(\frac{97 \text{ km/h}}{10 \text{ s}}\right)\left(\frac{1000 \text{ m}}{\text{km}}\right)\left(\frac{1 \text{ h}}{3600 \text{ s}}\right) = 2.69$ m/s^2

(b) $F_{min} = f = \mu(m/2)g = ma$. We use m/2 for the 2 rear tires which provide the accelerating force. Therefore, $\mu = \frac{2a}{g} = \frac{2(2.69 \text{ m/s}^2)}{9.8 \text{ m/s}^2} = 0.55$

(c) $v_f = 97$ km/h = 26.94 m/s

$W = \Delta KE + \Delta PE = \frac{1}{2} (1500 \text{ kg})(26.94 \text{ m/s})^2 - 0 = 5.45 \times 10^5$ J

and $P = \frac{W}{t} = \frac{5.45 \times 10^5 \text{ J}}{10 \text{ s}} = 5.45 \times 10^4$ W = 73.1 hp

---

5.78 (a) The work done by the 40 N force is

$W = (F \cos 37°)s = (40 \text{ N} \cos 37°)2 \text{ m} = 63.9$ J

CHAPTER FIVE SOLUTIONS

(b) Only the component of the weight (29.4 N) which is tangent to the incline does any work. This is $(29.4 \text{ N})\sin 37° = 17.7$ N. Thus,
$$W_g = -(17.7 \text{ N})(2 \text{ m}) = -35.4 \text{ J}$$
(c) The work done by friction is
$$W_f = -fs$$
$f = \mu_k N$; we first calculate the normal force by considering $\Sigma F_y = 0$.
$$N - (40 \text{ N})\sin 37° - (29.4 \text{ N})(\cos 37°) = 0$$
$$N = 47.6 \text{ N}$$
Thus, $\quad f = 0.1(47.6 \text{ N}) = 4.76 \text{ N}'$
and the work done by friction is $\quad W_f = -(4.76 \text{ N})(2 \text{ m}) = -9.51$ J
(d) We use
$$W_{nc} = \frac{1}{2} mv_f^2 - \frac{1}{2} mv_i^2 + mgy_f - mgy_i$$
Where W_{nc} is the sum of the work done by the 40 N force and the work done by friction.
$$W_{nc} = 63.9 \text{ J} - 9.51 \text{ J} = 54.4 \text{ J}$$
We also note that the work done by gravity is equal to the negative of the change in potential energy.
$$54.4 \text{ J} = \Delta KE + 35.4 \text{ J}$$
$$\Delta KE = 19.0 \text{ J}$$

5.79 (a) We recognize that the normal force, N, is $(0.4 \text{ kg})g = 3.92$ N, and we use
$$W = \frac{1}{2} mv^2 - \frac{1}{2} mv_0^2$$

$$W = \frac{1}{2} (0.4 \text{ kg})(6 \text{ m/s})^2 - \frac{1}{2} (0.4 \text{ kg})(8 \text{ m/s})^2 = -5.6 \text{ J}$$
(b) Also, $\quad W = -fs = -5.6$ J
In one revolution, $s = 2\pi r$. So
$$f = 5.6 \text{ J}/(2\pi(1.5 \text{ m})) = 0.594 \text{ N}$$
and $\quad \mu_k = f/N = 0.594 \text{ N}/3.92 \text{ N} = 0.152$
(c) We have
$$\text{number of revolutions} = \frac{\text{initial energy}}{\text{energy lost per revolution}}$$
The initial energy $= \frac{1}{2} (0.4 \text{ kg})(8 \text{ m/s})^2 = 12.8$ J, and the energy lost per revolution is 5.6 J. Thus,
$$\text{number of revolutions} = 12.8 \text{ J}/5.6 \text{ J} = 2.29 \text{ revolutions}$$

5.80 (a) Choose the zero level for potential energy at point B, and we have
$$PE_A = mgh = (0.2 \text{ kg})(9.8 \text{ m/s}^2)(0.30 \text{ m}) = 0.588 \text{ J}$$
(b) From conservation of mechanical energy,
$$\Delta KE = \Delta U = 0.588 \text{ J} \quad \text{and since } KE_i = 0, \quad KE_f = 0.588 \text{ J}$$
(c) From $KE_f = \frac{1}{2} mv^2$, we have $\quad 0.588 \text{ J} = \frac{1}{2} (0.2 \text{ kg})v^2 \quad$ and $\quad v = 2.42$ m/s
(d) At point C, $\quad U = mgh = (0.2 \text{ kg})(9.8 \text{ m/s}^2)(0.20 \text{ m}) = 0.392$ J
and $\quad KE = 0.588 \text{ J} - 0.392 \text{ J} = 0.196$ J

5.81 (a) $KE_B = \frac{1}{2} mv^2 = \frac{1}{2} (0.2 \text{ kg})(1.5 \text{ m/s})^2 = 0.225$ J

CHAPTER FIVE SOLUTIONS

(b) In the preceding problem, we found the initial energy at point A to be 0.588 J. In the absence of friction, this should also be the total energy at point B. Thus, the energy lost is
$$0.588 \text{ J} - 0.225 \text{ J} = 0.363 \text{ J}$$
(c) Even though the energy lost through work against friction is known to be 0.363 J, we still cannot find μ. The reason is that the normal force and therefore f both change with position as the block slides on the inside of the bowl.

5.82 (a) $PE_s = \frac{1}{2} kx^2 = \frac{1}{2} (500 \text{ N/m})(4 \times 10^{-2} \text{ m})^2 = 0.400 \text{ J}$

(b) $PE_s = \frac{1}{2} kx^2 = \frac{1}{2} (500 \text{ N/m})(3 \times 10^{-2} \text{ m})^2 = 0.225 \text{ J}$

(c) $PE_s = \frac{1}{2} kx^2 = \frac{1}{2} (500 \text{ N/m})(0)^2 = 0 \text{ J}$

5.83 $\frac{1}{2} (5000 \text{ N/m})(0.1 \text{ m})^2 = (0.25 \text{ kg})(9.80 \text{ m/s}^2)h$

gives the maximum height $h = 10.2$ m

5.84 The velocity, v_0, at which the mass leaves the spring is found as
$$\frac{1}{2} (100 \text{ N/m})(10 \times 10^{-2} \text{ m})^2 = \frac{1}{2} (2 \text{ kg}) v_0^2$$
Thus, $v_0 = 0.707$ m/s
From $v^2 = v_0^2 + 2ax$, we find the acceleration,
$$0 = (0.707 \text{ m/s})^2 + 2a(0.25 \text{ m}) \quad \text{gives } a = -1.00 \text{ m/s}^2$$
and from Newtons second law $\mu mg = ma$ becomes
$$(0.25 \text{ kg})((9.80 \text{ m/s}^2) = (0.25 \text{ kg})(-1.00 \text{ m/s}^2)$$
so, $\mu = 0.102$

CHAPTER SIX SOLUTIONS

6.1 Use $p = mv$
(a) $p = (1.67 \times 10^{-27} \text{ kg})(5 \times 10^6 \text{ m/s}) = 8.35 \times 10^{-21}$ kg m/s
(b) $p = (1.5 \times 10^{-2} \text{ kg})(3 \times 10^2 \text{ m/s}) = 4.50$ kg m/s
(c) $p = (75 \text{ kg})(10 \text{ m/s}) = 750$ kg m/s
(d) $p = (5.98 \times 10^{24} \text{ kg})(2.98 \times 10^4 \text{ m/s}) = 1.78 \times 10^{29}$ kg m/s

---

6.2 The original momentum and kinetic energy are $p_0 = mv_0$, and $KE_0 = \frac{1}{2}mv_0^2$. If the speed becomes $v = 2v_0$, the new momentum and kinetic energy are
    (a) $p = mv = m(2v_0) = 2(mv_0) = 2p_0$ (momentum is doubled)
    (b) $KE = \frac{1}{2}mv^2 = \frac{1}{2}m(2v_0)^2 = 4KE_0$ (kinetic energy is quadrupled)

6.3 We are given $p_c = p_t$
Therefore, $m_c v_c = m_t v_t$
and $v_c = (5000 \text{ kg})(15 \text{ m/s})/1500 \text{ kg} = 50.0$ m/s

---

6.4 (a) At the maximum height, h, $v = 0$. Therefore, $p = 0$
(b) The maximum height is found from
     $v^2 = v_0^2 + 2ay$
     $0 = (15 \text{ m/s})^2 - 2(9.8 \text{ m/s}^2)h$
Thus, $h = 11.5$ m
We need the velocity at $h/2 = 5.75$ m
     $v^2 = v_0^2 + 2ay$
     $v^2 = (15 \text{ m/s})^2 - 2(9.8 \text{ m/s}^2)(5.75 \text{ m})$
     $v = 10.6$ m/s
So, $p = (0.10 \text{ kg})(10.6 \text{ m/s}) = 1.06$ kgm/s

6.5 We have $v_{0y} = v_0 \sin 30° = 10$ m/s
and $v_{0x} = v_0 \cos 30° = 17.3$ m/s
(a) At the top of the arc, the velocity is solely in the x direction at 17.3 m/s. Thus,
     $p = mv = (0.1 \text{ kg})(17.3 \text{ m/s}) = 1.73$ kgm/s   (in the horizontal direction)
(b) At ground level the velocity of the ball has the same magnitude as that with which it was projected (20 m/s). Thus,
     $p = mv = (0.1 \text{ kg})(20 \text{ m/s}) = 2.00$ kgm/s (at an angle of 30° below the horizontal.)

6.6 Use, $F\Delta t = \Delta p = mv_f - mv_i$.
         $F(0.3 \text{ s}) = 0 - (1500 \text{ kg})(15 \text{ m/s})$
         $F = -7.5 \times 10^4$ N (the negative sign indicates the direction of the force is opposite to that of the car's original motion.

6.7 Use, $F\Delta t = \Delta p = mv_f - mv_i$.

CHAPTER SIX SOLUTIONS

$F(0.5 \text{ s}) = 0 - (1.8 \times 10^4 \text{kg})(15 \text{ m/s})$
$F = -5.4 \times 10^5 \text{ N}$

6.8 (a) First, find the impulse delivered to the ball.
$F\Delta t = \Delta p = mv_f - mv_i = 0 - (0.5 \text{ kg})(15 \text{ m/s}) = -7.5 \text{ kg m/s}$
(b) $F\Delta t = -7.5 \text{ kg m/s}$, and $\Delta t = 0.02 \text{ s}$
so, $F = \dfrac{-7.5 \text{ kg m/s}}{0.02 \text{ s}} = -375 \text{ N}$
Thus, the force on the ball is 375 N in the direction opposite to the original direction of motion of the ball. The force on the receiver is also 375 N but in the direction the ball was moving.

6.9 (a) Impulse = area under curve = 2 triangular areas of altitude 18,000 N and base of 0.75 s.
$F\Delta t = 2(\tfrac{1}{2}(0.75 \text{ s})(18,000 \text{ N})) = 1.35 \times 10^4 \text{ N s}$
(b) $\bar{F} = \dfrac{\text{impulse}}{\text{time}} = \dfrac{1.35 \times 10^4 \text{ Ns}}{1.5 \text{ s}} = 9000 \text{ N}$ (approximately 2000 lb)

6.10 Impulse = area under curve = (two triangular areas each of altitude 4 N and base 2 s) + (one rectangular area of width 1 s and height of 4 N.)
(a) Thus, Impulse = $2(\tfrac{1}{2}(2 \text{ s})(4 \text{ N})) + (1 \text{ s})(4 \text{ N}) = 12 \text{ N s}$
(b) Δp = impulse = $mv_f - mv_i$ = 12 N s
$(2 \text{ kg})v_f - (2 \text{ kg})(0) = 12 \text{ N s}$
so $v_f = 6 \text{ m/s}$
(c) if $v_0 = -2 \text{ m/s}$
$(2 \text{ kg})v_f - (2 \text{ kg})(-2 \text{ m/s}) = 12 \text{ N s}$
and $v_f = 4 \text{ m/s}$

6.11 $F\Delta t$ = impulse = area under curve = 12 N s (See solution of problem 6.10.)
$\bar{F} = \dfrac{\text{impulse}}{\text{time}} = \dfrac{12 \text{ Ns}}{5 \text{ s}} = 2.4 \text{ N}$

6.12 (a) Use $F\Delta t = \Delta p = mv_f - mv_i$.
$F\Delta t = (0.15 \text{ kg})[(-22 \text{ m/s}) - (20 \text{ m/s})] = -6.30 \text{ kgm/s}$
The negative sign indicates that the impulse is in the opposite direction to that of the initial velocity.
(b) $F = (\text{impulse})/\Delta t = -(6.30 \text{ kgm/s})/2 \times 10^{-3} \text{ s} = -3.15 \times 10^3 \text{ N}$

6.13 From $F\Delta t = \Delta p = mv_f - mv_i$
we have $(3 \text{ N})(1.5 \text{ s}) = (0.5 \text{ kg})v - 0$
$v = 9 \text{ m/s}$
(b) Again use $F\Delta t = \Delta p = mv_f - mv_i$.
$(-4 \text{ N})(3 \text{ s}) = (0.5 \text{ kg})v - (0.5 \text{ kg})(9 \text{ m/s})$
$v = -15.0 \text{ m/s}$

CHAPTER SIX SOLUTIONS

6.14 First, we find the speed after the bag has dropped 25 m.
$$v^2 = v_0^2 + 2ay$$
$$v^2 = 0 + 2(-9.8 \text{ m/s}^2)(-25 \text{ m})$$
$$v = -22.1 \text{ m/s}$$
Now, find the time to travel a distance equal to its diameter (0.15 m).
$$t = d/v_y = 1.5 \times 10^{-1} \text{ m}/22.1 \text{ m/s} = 6.79 \times 10^{-3} \text{ s}$$
Finally, use $F\Delta t = \Delta p = mv_f - mv_i$.
$$F_{net}(6.79 \times 10^{-3} \text{ s}) = 0 - (2 \text{ kg})(-22.1 \text{ m/s})$$
$$F_{net} = 6.51 \times 10^3 \text{ N} \quad \text{(Net force experienced by bag as it slows down)}$$
But, $F_{net} = F_{head} - w$
So, $F_{head} = 6510 \text{ N} + 19.6 \text{ N} = 6530 \text{ N}$, and by Newton's third law, the bag exerts a force equal to 6530 N (downward) on the head.

6.15 The initial momentum = 0. Therefore, the final momentum, p_f, must also be zero. We have, (taking eastward as the positive direction),
$$p_f = (40 \text{ kg})(-v_c) + (0.5 \text{ kg})(5 \text{ m/s}) = 0$$
$$v_c = -6.25 \times 10^{-2} \text{ m/s} \quad \text{(The child recoils westward.)}$$

6.16 (a) From conservation of momentum, we have, choosing the direction of the bullet's motion as positive,
$$m_R v_R + m_b v_b = 0$$
$$v_R = -\frac{m_b}{m_R} v_b = -\frac{5 \times 10^{-3}}{3.06} \; 300 \text{ m/s} = -0.49 \text{ m/s}$$
(b) The mass of the man plus rifle is 74.5 kg. We use the same approach as in (a), to find
$$v = -\frac{5 \times 10^{-3}}{74.5} (300 \text{ m/s}) = -2.01 \times 10^{-2} \text{ m/s}$$

6.17 The momentum before the event is zero, and after, it is
$$2m(-v_2) + m(v_1) = 0 \quad \text{(where } v_2 \text{ is the mass of the larger piece)}$$
This reduces to $v_2 = \dfrac{v_1}{2}$
Or, the particle of twice the mass has a velocity which is 1/2 that of the less massive part. Also, the velocities have opposite directions.

6.18 First, determine the recoil speed, v_0, of the base of the swing immediately after the bird takes off. From conservation of momentum, we have
$$(52 \text{ g})(2 \text{ m/s}) + (153 \text{ g})(-v_0) = 0$$
and $v_0 = 0.68$ m/s
Now, apply conservation of energy to the rising swing.
$$KE_f + PE_f = KE_i + PE_i$$
$$0 + mgh = \frac{1}{2} mv_0^2 + 0$$
gives $h = \dfrac{v_0^2}{2g} = \dfrac{(0.68 \text{ m/s})^2}{2(9.8 \text{ m/s}^2)} = 2.36 \times 10^{-2}$ m = 2.36 cm

---

6.19 $p_{after} = p_{before}$
Thus, $(65 \text{ kg})v_{boy} + (40 \text{ kg})(4 \text{ m/s}) = 0$

CHAPTER SIX SOLUTIONS

v_{boy} = -2.46 m/s   (so the boy moves westward)

6.20 From conservation of momentum (after = before), we find the velocity of the man.
$$(74.5 \text{ kg})v_{man} + (1.2 \text{ kg})(5 \text{ m/s}) = 0$$
$$v_{man} = -8.05 \times 10^{-2} \text{ m/s} \quad \text{(direction is southward)}$$
The time to travel 5 m to shore is
$$t = 5 \text{ m}/8.05 \times 10^{-2} \text{ m/s} = 62.1 \text{ s}$$

6.21 We call the initial speed of the bowling ball v_0 and from momentum conservation,
$$(7 \text{ kg})(v_0) + (2 \text{ kg})(0) = (7 \text{ kg})(1.8 \text{ m/s}) + (2 \text{ kg})(3 \text{ m/s})$$
gives v_0 = 2.66 m/s

6.22 Consider the thrower first, with velocity after the throw of v_t
momentum after = momentum before
$$(65 \text{ kg})v_t + (0.045 \text{ kg})(30 \text{ m/s}) = (65.045 \text{ kg})(2.5 \text{ m/s})$$
$$v_t = 2.48 \text{ m/s}$$
Now, consider the catcher, with velocity after of v_c
$$(60.045 \text{ kg})v_c = (0.045 \text{ kg})(30 \text{ m/s})$$
$$v_c = 2.25 \times 10^{-2} \text{ m/s} = 2.25 \text{ cm/s}$$

6.23 p_{before} = M(5 m/s) + 0; p_{after} = 2Mv
From $p_{after} = p_{before}$
$$2Mv = M(5 \text{ m/s})$$
v = 2.50 m/s   (in same direction as the car was originally going.

6.24 p_{before} = M(.8 m/s) + 0; p_{after} = Mv + M(0.5 m/s)
Use $p_{after} = p_{before}$ and solve for v, the velocity of the first car, gives
$$v = 0.3 \text{ m/s}$$

6.25 $p_{xafter} = p_{xbefore}$
$$(1.2 \text{ kg} + 0.8 \text{ kg})v = (1.2 \text{ kg})(5 \text{ m/s})$$
$$v = 3 \text{ m/s}$$

6.26 (a) $p_{after} = p_{before}$
$$(3M)v = M(3 \text{ m/s}) + (2M)(1.2 \text{ m/s})$$ (M is the common mass of the cars, and v is the velocity of the combination after the collision.)
$$v = 1.8 \text{ m/s}$$
(b) KE(before) = $\frac{1}{2}(2 \times 10^4 \text{ kg})(3 \text{ m/s})^2 + \frac{1}{2}(4 \times 10^4 \text{ kg})(1.2 \text{ m/s})^2 = 1.19 \times 10^5$ J

KE(after) = $\frac{1}{2}(60000 \text{ kg})(1.8 \text{ m/s})^2 = 9.72 \times 10^4$ J

Thus, the kinetic energy lost = 2.16×10^4 J

6.27 $p_{after} = p_{before}$
$$(3 \text{ kg} + M)\frac{v}{3} = (3 \text{ kg})v$$
Cancel v and solve for M;   M = 6 kg

6.28 (a) $m_1v_{1i} + m_2v_{2i} = m_1v_{1f} + m_2v_{2f}$.

CHAPTER SIX SOLUTIONS

(Subscripts 1 refer to the 5 g ball, and subscripts 2 to the 10 g ball.) We have
$$(5\ g)(20\ cm/s) = (5\ g)v_{1f} + (10g)v_{2f}. \quad (1)$$
Also for a head-on elastic collision, we have
$$v_{1i} - v_{2i} = -(v_{1f} - v_{2f})$$
$$20\ cm/s - 0 = -(v_{1f} - v_{2f}) \quad (2)$$
Solve (1) and (2) simultaneously,
$$v_{1f} = -6.70\ cm/s \quad and \quad v_{2f} = 13.3\ cm/s$$

(b) $KE(before) = \frac{1}{2}(5 \times 10^{-3}\ kg)(20 \times 10^{-2}\ m/s)^2 = 10^{-4}\ J$

$KE(second\ object) = \frac{1}{2}(10 \times 10^{-3}\ kg)(13.3 \times 10^{-2})^2 = 8.89 \times 10^{-5}\ J$

$$\frac{KE(2)}{KE(before)} = 0.889$$

6.29 $m_1v_{1i} + m_2v_{2i} = m_1v_{1f} + m_2v_{2f}.$
$$(10g)(20\ cm/s) + (15\ g)(-30\ cm/s) = (10\ g)v_{1f} + (15\ g)v_{2f} \quad (1)$$
and
$$v_{1i} - v_{2i} = -(v_{1f} - v_{2f})$$
$$20\ cm/s - (-30\ cm/s) = -(v_{1f} - v_{2f}) \quad (2)$$
Solve (1) and (2) simultaneously
$$v_{1f} = -40\ cm/s; \qquad v_{2f} = 10\ cm/s$$

6.30 $m_1v_{1i} + m_2v_{2i} = m_1v_{1f} + m_2v_{2f}.$
$$(25\ g)(20\ cm/s) + (10\ g)(15\ cm/s) = (25\ g)v_{1f} + (10\ g)v_{2f} \quad (1)$$
and $v_{1i} - v_{2i} = -(v_{1f} - v_{2f})$
$$(20\ cm/s - 15\ cm/s) = -(v_{1f} - v_{2f}) \quad (2)$$
Solving (1) and (2) yields $v_{1f} = 17.1\ cm/s$ and $v_{2f} = 22.1\ cm/s$

6.31 (a) $m_1v_{1i} + m_2v_{2i} = m_1v_{1f} + m_2v_{2f}.$
$$(4\ \mu)(10^6\ m/s) + 0 = (4\mu)v_{1f} + (1\mu)v_{2f} \quad (1)$$
and $v_{1i} - v_{2i} = -(v_{1f} - v_{2f})$
$$10^6\ m/s - 0 = -(v_{1f} - v_{2f}) \quad (2)$$
Solving (1) and (2) simultaneously, we have
$$v_{1f} = 6 \times 10^5\ m/s \quad and \quad v_{2f} = 1.6 \times 10^6\ m/s$$
(b) The kinetic energy of the proton before is zero. The kinetic energy of the alpha is,

$KE(alpha\ after) = \frac{1}{2}(4 \times 1.67 \times 10^{-27}\ kg)(6 \times 10^5\ m/s)^2 = 1.20 \times 10^{-15}\ J$

$KE(proton\ after) = \frac{1}{2}(1.67 \times 10^{-27}\ kg)(1.6 \times 10^6\ m/s)^2 = 2.14 \times 10^{-15}\ J$

$KE(alpha\ before) = \frac{1}{2}(4 \times 1.67 \times 10^{-27}\ kg)(10^6\ m/s)^2 = 3.34 \times 10^{-15}\ J$

6.32 Applying conservation of mechanical energy from just after the collision until the end of the swing is reached, we have
$$\frac{1}{2}(M + m)V^2 = (M + m)gh$$
where M is the mass of the pendulum, m the mass of the bullet, h the vertical height through which the pendulum swings, and V is the velocity of the pendulum plus bullet immediately after the collision. The equation above reduces to

CHAPTER SIX SOLUTIONS

$$V = \sqrt{2gh} \quad (1)$$

Now apply conservation of momentum from just before to just after the collision. We have

$$mv_0 = (M + m)V \quad (2)$$

where v_0 is the velocity of the bullet just prior to collision.
We now solve (1) and (2) simultaneously to find the following equation for v_0.

$$v_0 = \frac{(M + m)}{m}\sqrt{2gh}$$

$$= \frac{(2.5 \text{ kg} + 8 \times 10^{-3} \text{ kg})}{8 \times 10^{-3} \text{ kg}}\sqrt{2g(6 \times 10^{-2} \text{ m})} = 3.40 \times 10^2 \text{ m/s}$$

6.33 Momentum is conserved only along the horizontal direction. We have

$$P_{after} = P_{before}$$
$$-(50 \text{ kg})V + (0.15 \text{ kg})(20 \text{ m/s})(\cos 30°) = 0$$
$$V = 5.20 \times 10^{-2} \text{ m/s} \text{ (V is the velocity of the machine horizontally across the ice.)}$$

6.34 (a) First, we conserve momentum in the x direction (the direction of travel of the fullback).

$$(90 \text{ kg})(5 \text{ m/s}) + 0 = (185 \text{ kg})V\cos\theta$$

Where θ is the angle between the direction of the final velocity V and the x axis. We find $\quad V\cos\theta = 2.43 \text{ m/s} \quad (1)$
Now consider conservation of momentum in the y direction (the direction of travel of the opponent.

$$(95 \text{ kg})(3 \text{ m/s}) + 0 = (185 \text{ kg})(V\sin\theta)$$

which gives, $\quad V\sin\theta = 1.54 \text{ m/s} \quad (2)$
Divide equation (2) by (1)
$\quad \tan\theta = 1.54/2.43 = 0.633 \quad$ From which $\quad \theta = 32.3°$
Then, either (1) or (2) gives $\quad V = 2.88 \text{ m/s}$

(b) $KE(before) = \frac{1}{2}(90 \text{ kg})(5 \text{ m/s})^2 + \frac{1}{2}(95 \text{ kg})(3 \text{ m/s})^2 = 1.55 \times 10^3 \text{ J}$

$KE(after) = \frac{1}{2}(185 \text{ kg})(2.88 \text{ m/s})^2 = 7.67 \times 10^2 \text{ J}$

Thus, the kinetic energy lost is 783 J

6.35 We choose the x direction along the direction of flight of the bird, and the y direction along the flight path of the bee. We have, from conservation of momentum in the x direction, with θ = the angle between the x axis and the final direction of flight.

$$(0.130 \text{ kg}) v\cos\theta = (0.125 \text{ kg})(0.6 \text{ m/s})$$
or $\quad v\cos\theta = 0.577 \text{ m/s} \quad (1)$
Conservation of momentum in the y direction yields,
$$(0.130 \text{ kg}) v\sin\theta = (5 \times 10^{-3} \text{ kg})(15 \text{ m/s})$$
or $\quad v\sin\theta = 0.577 \text{ m/s} \quad (2)$
Dividing (2) by (1) yields $\tan\theta = 1$ and $\theta = 45°$
Squaring (1) and (2) and adding them yields, $v = 0.816 \text{ m/s}$

6.36 The initial momentum of the system is 0. Thus,
$$(1.2m) v_{B0} = m(10 \text{ m/s})$$
and $\quad v_{B0} = 8.33 \text{ m/s}$

$$KE_{before} = \frac{1}{2}m(10 \text{ m/s})^2 + \frac{1}{2}(1.2 \text{ m})(8.33 \text{ m/s})^2 = \frac{1}{2}m(183.3 \text{ m}^2/\text{s}^2)$$

CHAPTER SIX SOLUTIONS

$$KE_{after} = \frac{1}{2}m(v_w)^2 + \frac{1}{2}(1.2\ m)(v_B)^2 = \frac{1}{2}(\frac{1}{2}\ m(183.3\ m^2/s^2))$$

or $\quad v_w^2 + 1.2\ v_B^2 = 91.67\ m^2/s^2 \quad (1)$

From conservation of momentum,
$$m v_w = (1.2\ m) v_B$$
or $\quad v_w = 1.2\ v_B \quad (2)$

Solving (1) and (2) simultaneously, we find
$\quad v_w = 7.07\ m/s$ (speed of white puck after collision)
and $\quad v_B = 5.89\ m/s$ (speed of black puck after collision)

6.37 We call east the positive x direction, and from conservation of momentum in the x direction, we have
$$(1500\ kg)(20\ m/s) = (4000\ kg) v \cos\theta\ .$$
From which, $\qquad v\cos\theta = 7.50\ m/s \quad (1)$

Now, we use conservation of momentum in the y direction with north positive.
$$-(2500\ kg)(15\ m/s) = -(4000\ kg) v \sin\theta$$
which gives, $\qquad v\sin\theta = 9.39\ m/s \quad (2)$

Divide (2) by (1) to give $\tan\theta = 1.25$ or $\theta = 51.3°$ (south of east)
Then use either (1) or (2) to find $\quad v = 12.0\ m/s$
Assuming uniformly accelerated motion, we have
$$v^2 = v_0^2 + 2ad$$
$$0 = (12\ m/s)^2 + 2a(6\ m)$$
$$a = -12.0\ m/s^2$$
and $\quad F = ma = (4000\ kg)(-12\ m/s^2) = -48{,}000\ N$

This force is in the direction opposite to v. Hence, it is at an angle of 51.3° west of north.

6.38 (a) Use $p_{after} = p_{before}$

$$m_1 v_{1f} + m_2 v_{2f} = m_1 v_0 \quad (1)$$

Where m_1 is the mass of the neutron and v_{1f} is its final velocity, and m_2 and v_{2f} refer to the carbon atom. For a head-on elastic collision, we use
$$v_{1i} - v_{2i} = -(v_{1f} - v_{2f})$$
$$v_0 = -(v_{1f} - v_{2f}) \quad (2)$$

Use $m_2 = 12 m_1$ in equations (1) and (2) and solve them simultaneously to give
$$v_{1f} = -\frac{11}{13} v_0 \quad \text{and} \quad v_{2f} = \frac{2}{13} v_0$$

Now, $KE(\text{neutron initial}) = \frac{1}{2} m_1 v_0^2$ and the kinetic energy of the carbon nucleus after the collision is
$$KE(\text{carbon after}) = \frac{1}{2}(12\ m_1)\left(\frac{2}{13}\right)^2 v_0^2$$

The ratio of KE(carbon after) to KE(neutron initial) = $\frac{48}{169} = 0.284$.

(b) If KE(neutron before) = 1.6×10^{-13} J, and with KE(carbon)/KE(neutron) = 0.284, we have
$$KE(\text{carbon}) = 4.54 \times 10^{-14}\ J$$
and the remaining energy $1.6 \times 10^{-13}\ J - 4.54 \times 10^{-14}\ J = 1.15 \times 10^{-13}\ J$ remains with the neutron.

CHAPTER SIX SOLUTIONS

6.39 We have found in the example problem associated with the ballistic pendulum that

$$V = \frac{m}{m+M} v_0 \quad (1) \quad \text{(See solution to problem 32)}$$

where V is the velocity of the pendulum-bullet combination immediately after the collision, and v_0 is the velocity of the bullet. M is the pendulum mass, and m is the bullet's mass. We have

$$\frac{KE_f}{KE_i} = \frac{(m+M)V^2}{mv_0^2}$$

Eliminate V by use of (1) in the preceding equation to find,

$$\frac{KE_f}{KE_i} = \frac{m}{m+M} = \frac{8 \times 10^{-3} \text{ kg}}{8 \times 10^{-3} \text{ kg} + 2 \text{ kg}} = 3.98 \times 10^{-3}, \text{ or } 0.398\%$$

6.40 Consider first conservation of momentum in the x direction.

$$mv_0 = mv_2\cos\theta + mv_1\cos30° \quad (1) \quad \text{(note that masses cancel)}$$

where v_1 is the velocity of the cue ball after the collision, v_2 that of the target, θ is the angle between the final velocity of the target and the x axis, and v_0 the initial velocity of the cue ball. Now, consider conservation of momentum in the y direction.

$$0 = mv_1\sin30° - mv_2\sin\theta \quad \text{(note the masses are equal and cancel)} \quad (2)$$

Since this is an elastic collision, kinetic energy is conserved. We have

$$\frac{1}{2}mv_0^2 = \frac{1}{2}mv_1^2 + \frac{1}{2}mv_2^2 \quad (3)$$

To solve (1), (2), and (3) simultaneously, rearrange (1) and (2) so the terms involving θ are isolated on the left sides, then square (1) and (2) and add them. After some reduction, the resulting equation is

$$v_2^2 = v_0^2 - (2v_0\cos30°)v_1 + v_1^2 \quad (4)$$

Now substitute (4) into (3) to eliminate v_2. The result is

$$v_1 = v_0\cos30° \quad (5)$$

With $v_0 = 4$ m/s, we have $v_1 = 3.464$ m/s. Then from (4), we find

$$v_2 = 2.00 \text{ m/s}$$

Finally, we find from (2) that $\sin\theta = 0.866$, or $\theta = 60.0°$.

6.41 The x location of the center of gravity is

$$x_{cg} = \frac{\sum m_i x_i}{\sum m_i} = \frac{(0+0+0+0)}{(2 \text{ kg} + 3 \text{ kg} + 2.5 \text{ kg} + 4 \text{ kg})} = 0$$

and the y location of the center of gravity is

$$y_{cg} = \frac{\sum m_i y_i}{\sum m_i}$$

$$= \frac{(2 \text{ kg})(3 \text{ m}) + (3 \text{ kg})(2.5 \text{ m}) + (2.5 \text{ kg})(0) + (4 \text{ kg})(-.5 \text{ m})}{2 \text{ kg} + 3 \text{ kg} + 2.5 \text{ kg} + 4 \text{ kg}}$$

$$y_{cg} = 1 \text{ m}$$

6.42 In order for $y_{cg} = 0$, we must have $\sum m_i y = 0$, so

$(2 \text{ kg})(3 \text{ m}) + (3 \text{ kg})(2.5 \text{ m}) + (2.0 \text{ kg})(0) + (4 \text{ kg})(-.5 \text{ m}) + (5 \text{ kg})(y_5) = 0$

where y_5 is the y location of the 5 kg mass.
We have $\quad y_5 = -2.3$ m

CHAPTER SIX SOLUTIONS

6.43 We have $\Sigma m_i = 3\text{ kg} + 1\text{ kg} + 2\text{ kg} = 6\text{ kg}$

Thus, $x_{cg} = \dfrac{\Sigma m_i x_i}{\Sigma m_i} = \dfrac{(3\text{ kg})(-2\text{ m}) + (1\text{ kg})(2\text{ m}) + (2\text{ kg})(3\text{ m})}{6\text{ kg}}$

From which, $x_{cg} = \dfrac{1}{3}\text{ m}$

and $y_{cg} = \dfrac{\Sigma m_i y_i}{\Sigma m_i} = \dfrac{(3\text{ kg})(4\text{ m}) + (1\text{ kg})(2\text{ m}) + (2\text{ kg})(-2\text{ m})}{6\text{ kg}}$

giving, $y_{cg} = \dfrac{5}{3}\text{ m}$

6.44 The center of gravity of a uniform object is at its geometrical center. With point O as origin, the center of gravity of piece 1 is at $x_1 = 0$ and $y_1 = -1.5$ in. The center of gravity of piece 2 is at $x_2 = 0$ and $y_2 = -17$ in. The mass of piece 1 is $m_1 = \sigma A$ and the mass of piece 2 is given by $m_2 = \sigma A_2$, where σ is the mass per unit area of the pieces.

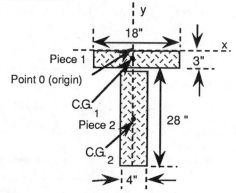

Since the center of gravity of both pieces lie along the y axis, we have
$x_{cg} = 0$ and y_{cg} is found as

$y_{cg} = \dfrac{\Sigma m_i y_i}{\Sigma m_i}$

$= \dfrac{\sigma(54\text{ in}^2)(-1.5\text{ in}) + \sigma(112\text{ in}^2)(-17\text{ in})}{\sigma(54\text{ in}^2) + \sigma(112\text{ in}^2)}$. σ cancels, and we have

$y_{cg} = -12.0$ in

6.45 The coordinates of the center of gravity of piece 1 are $x_1 = 2$ cm and $y_1 = 9$ cm. The coordinates for piece 2 are $x_2 = 8$ cm and $y_2 = 2$ cm. The area of each piece is $A_1 = 72$ cm^2 and $A_2 = 32$ cm^2. As shown in problem 44, the mass of each piece is proportional to the area. Thus,

CHAPTER SIX SOLUTIONS

$$x_{cg} = \frac{\sum m_i x_i}{\sum m_i}$$

$$= \frac{(72 \text{ cm}^2)(2 \text{ cm}) + (32 \text{ cm}^2)(8 \text{ cm})}{72 \text{ cm}^2 + 32 \text{ cm}^2} = 3.85 \text{ cm}$$

and $y_{cg} = \frac{\sum m_i y_i}{\sum m_i}$

$$= \frac{(72 \text{ cm}^2)(9 \text{ cm}) + (32 \text{ cm}^2)(2 \text{ cm})}{104 \text{ cm}^2} = 6.85 \text{ cm}$$

6.46 We use, with $x = 0$ at the center of the earth,

$$x_{cm} = \frac{\sum m_i x_i}{\sum m_i} = \frac{0 + (7.36 \times 10^{22} \text{ kg})(3.84 \times 10^8 \text{ m})}{5.98 \times 10^{24} \text{ kg} + 7.36 \times 10^{22} \text{ kg}} = 4.67 \times 10^6 \text{ m}$$

This is 0.732(radius of the earth).

6.47 Choose the origin at the proton, and use

$$x_{cm} = \frac{\sum m_i x_i}{\sum m_i} = \frac{0 + (9.1 \times 10^{-31} \text{ kg})(5.1 \times 10^{-11} \text{ m})}{1.67 \times 10^{-27} \text{ kg} + 9.1 \times 10^{-31} \text{ kg}} = 2.78 \times 10^{-14} \text{ m}$$

6.48 $x_{cm} = \frac{\sum m_i x_i}{\sum m_i} = \frac{(3 \text{ kg})(-2 \text{ m}) + (1 \text{ kg})(2 \text{ m}) + (2 \text{ kg})(3 \text{ m})}{3 \text{ kg} + 1 \text{ kg} + 2 \text{ kg}} = 0.333 \text{ m}$

$y_{cg} = \frac{\sum m_i y_i}{\sum m_i} = \frac{(3 \text{ kg})(4 \text{ m}) + (1 \text{ kg})(2 \text{ m}) + (2 \text{ kg})(-2 \text{ m})}{6 \text{ kg}} = 1.67 \text{ m}$

6.49 (a) No external force acts on the system, so the center of mass remains stationary at 100 m from the ship. Then, when the engine pack reaches the end of the rope, the man is at a distance of x_A from the center of mass, and the pack is a distance of x_e from the center of mass. We have

$x_A + x_e = 100 \text{ m}$ (1)

and, from the definition of the center of mass,

$(80 \text{ kg})x_A = (20 \text{ kg})x_e$

or $x_e = 4x_A$ (2)

Solving (1) and (2) simultaneously, we find

$x_A = 20 \text{ m}$ and $x_e = 80 \text{ m}$

Thus, the rocket stops 20 m short of the ship.

(b) He should unhook the engine from the rope and shove the engine straight out away from the ship. He will drift toward the ship with a speed of $\frac{1}{4}$ th that imparted to the engine.

CHAPTER SIX SOLUTIONS

6.50 Impulse = $F\Delta t = \Delta p = mv_f - mv_i$.
$\qquad$ = 0.4 kg(-22 m/s - 15 m/s) = -14.8 kg m/s

Thus, an impulse of 14.8 kg m/s in the direction opposite the initial velocity is delivered to the ball.

6.51 $p_{after} = p_{before}$
$\qquad$ (4 g)(0.5 m/s) + (2 g)v_{2f} = (4 g)(0.8 m/s) + 0
$\qquad$ V_{2f} = 0.60 m/s

6.52 $p_{after} = p_{before} = 0$ Thus, choosing eastward as positive, we have
$\qquad$ (80 kg)(-0.8 m/s) + (240 kg)v_{boat} = 0
$\qquad$ v_{boat} = 0.267 m/s (eastward)

6.53 (a) We use $m_1v_{1i} + m_2v_{2i} = m_1v_{1f} + m_2v_{2f}$.   (1)
and
$\qquad$ $v_{1i} - v_{2i} = -(v_{1f} - v_{2f})$ (2)
with $v_{2i} = 0$
(1) and (2) can be solved simultaneously to show that
$$v_{2f} = \frac{2m_1}{m_1 + m_2} v_{1i} \quad \text{and} \quad v_{1f} = \frac{m_1 - m_2}{m_1 + m_2} v_{1i}$$
With m_1 = 2 g, m_2 = 1 g, and v_{1i} = 8 m/s, we find
$\qquad$ $v_{2f} = \frac{32}{3}$ m/s and $v_{1f} = \frac{8}{3}$ m/s

(b) with m_1 = 2 g, m_2 = 10 g, and v_{1i} = 8 m/s, we have
$\qquad$ $v_{2f} = \frac{8}{3}$ m/s and $v_{1f} = \frac{-16}{3}$ m/s

(c) For case (a) KE(final) = $\frac{1}{2} m_1 v_{1f}^2 = \frac{1}{2}$ (2 X 10^{-3} kg)($\frac{8}{3}$ m/s)^2 = 7.11 X 10^{-3} J

and for case (b) KE(final) = $\frac{1}{2} m_1 v_{1f}^2 = \frac{1}{2}$ (2 X 10^{-3} kg)($\frac{16}{3}$ m/s)^2 = 2.84 X10^{-2} J

Since the incident kinetic energy is the same in case (a) and case (b), it is clear that the incident particle loses more KE in case (a).

6.54 (a) To find the velocity at pool level, we use conservation of mechanical energy with the zero level for potential energy at the level of the diving board. We have,
$$\frac{1}{2} mv_i^2 + mgy_i = \frac{1}{2} mv_f^2 + mgy_f$$
$$0 + 0 = \frac{1}{2} (70 \text{ kg})v^2 - (70 \text{ kg})g(3 \text{ m})$$
where v is the velocity before striking the water. We find,
$\qquad$ v = 7.67 m/s and from this, the momentum is
$\qquad$ p = (70 kg)(7.67 m/s) = 537 kg m/s

(b) Using, the same approach as above, but with h = 1.5 m, we find v = 5.42 m/s, and
$\qquad$ p = (70 kg)(5.42 m) = 380 kg m/s

6.55 We will first find the speed as he enters the water. We select the origin of the coordinate system at the initial position of the diver.

84

CHAPTER SIX SOLUTIONS

$$v^2 = v_0^2 + 2ax$$
$$v^2 = 0 + 2(-9.8 \text{ m/s}^2)(-3 \text{ m})$$
$$v = -7.67 \text{ m/s}$$

Now consider the deceleration period. From the impulse-momentum theorem, we have

$$F\Delta t = \Delta p = mv_f - mv_i.$$
$$F(2.0 \text{ s}) = 0 - (80 \text{ kg})(-7.67 \text{ m/s})$$
$$F = 3.07 \times 10^2 \text{ N}$$

This is the net force acting on the man. But

$$F = F_{water} - mg = F_{water} - 784 \text{ N} = 307 \text{ N}$$

Thus, $F_{water} = 1.09 \times 10^3$ N

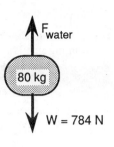

6.56 (a) Conserving momentum, we have
$$(2 \text{ g})(4 \text{ m/s}) + (3 \text{ g})(-3 \text{ m/s}) = (5 \text{ g})V$$
where V is the velocity of the combined mass,
$$V = -\frac{1}{5} \text{ m/s}$$

(b) If V is to equal zero, we see that
$$(2 \text{ g})(4 \text{ m/s}) + (3g)v = 0$$
where v is the required speed of the 3 g mass. This gives,
$$v = -\frac{8}{3} \text{ m/s}$$

6.57 First, consider conservation of momentum,
$$m_1v_{1i} + m_2v_{2i} = m_1v_{1f} + m_2v_{2f}.$$
Both balls have the same mass, so this equation becomes
$$v_{1i} + v_{2i} = v_{1f} + v_{2f}. \quad (1)$$
For an elastic head-on collision, we also have
$$v_{1i} - v_{2i} = -(v_{1f} - v_{2f})$$
Let us solve this equation for v_{1f}.
$$v_{1f} = v_{2f} + v_{2i} - v_{1i} \quad (2)$$
Use (2) to eliminate v_{1f} from (1). The result is
$$v_{2f} = v_{1i} \quad (3)$$
Now, eliminate v_{2f} from (1) by use of (3). We find
$$v_{1f} = v_{2i} \quad (4)$$
Thus, equations (3) and (4) show us that under the conditions of equal mass objects striking one another in a head-on elastic collision, the two objects exchange velocities. Thus, we may write the results of the various collisions as
(a) $v_{1f} = 0$, $v_{2f} = 1.5$ m/s
(b) $v_{1f} = -1.0$ m/s, $v_{2f} = 1.5$ m/s
(c) $v_{1f} = 1.0$ m/s, $v_{2f} = 1.5$ m/s

6.58 We equate momentum before to momentum just after, to get
$$(3 \times 10^{-2} \text{ kg})(200 \text{ m/s}) = (1.8 \times 10^{-1} \text{ kg})V$$
V is the velocity of the bullet-baseball combination immediately after the collision, and its value is found to be, V = 33.33 m/s.
The problem now becomes that of finding the height, H, reached by a projectile launched straight upward at 33.33 m/s.
$$v^2 = v_0^2 + 2ay$$
$$0 = (33.33 \text{ m/s})^2 + 2(-9.8 \text{ m/s}^2)H$$
$$H = 56.7 \text{ m}$$

CHAPTER SIX SOLUTIONS

6.59 No external force acts on the system (astronaut plus wrench), so the total momentum is unchanged. We have final momentum = initial momentum = 0.

$$m_{wrench}v_{wrench} = m_{astronaut}v_{astronaut} = 0$$

and $v_{astronaut} = -\dfrac{(0.5 \text{ kg})(20 \text{ m/s})}{80 \text{ kg}} = -0.125 \text{ m/s}$

or, the astronaut drifts toward the ship at 0.125 m/s. At this speed, the time to travel to the ship is

$$t = \dfrac{30 \text{ m}}{0.125 \text{ m/s}} = 240 \text{ s} = 4 \text{ minutes}$$

6.60 Let us call v_0 the velocity of the bullet before the collision and V the velocity of the bullet-block combination just after the collision. We apply the work energy theorem from just after the impact until the block slides to a stop.

ΔKE = work done = $KE_f - KE_i$ = $-f(7.5 \text{ m})$ with $f = \mu mg$, we have

$$KE_i = \tfrac{1}{2}mV^2 = (0.65)mg(7.5 \text{ m})$$

which gives,   V = 9.77 m/s

Now, apply conservation of momentum from just before the impact to just after.

$$(0.012 \text{ kg})v_0 = (0.112 \text{ kg})(9.77 \text{ m/s})$$

yielding, v_0 = 91.2 m/s

6.61 We use a coordinate system with the positive direction of the x axis toward the east and positive y toward the north. We set the initial velocity of the 3000 kg car as v_0, and use conservation of momentum. First, we consider the y direction.

$$(3000 \text{ kg})v_0 = (5000 \text{ kg})(5.22 \text{ m/s})\sin 40°$$

$$v_0 = 5.59 \text{ m/s}$$

A quick check will reveal that with this value of v_0, momentum is also conserved in the x direction.

6.62 We shall first use conservation of energy to find the velocity of the bead just before it strikes the ball. The zero level of potential energy is at the level of point B. We have,

$$\tfrac{1}{2}mv_i^2 + mgy_i = \tfrac{1}{2}mv_f^2 + mgy_f$$

$$0 + (0.4 \text{ kg})(g)(1.5 \text{ m}) = \tfrac{1}{2}(0.4 \text{ kg})v_{1i}^2 + 0$$

$$v_{1i} = 5.42 \text{ m/s}$$

We now treat the collision as an elastic head-on collision to find the velocity of the ball immediately after the collision. We find

$$v_{2f} = \dfrac{2m_1}{m_1 + m_2}v_{1i} = \dfrac{2(0.4 \text{ kg})}{1.0 \text{ kg}}(5.42 \text{ m/s}) = 4.34 \text{ m/s}$$

Now we use conservation of energy once again to find the height, H, that the 0.6 kg ball rises after the collision.

$$\tfrac{1}{2}mv_i^2 + mgy_i = \tfrac{1}{2}mv_f^2 + mgy_f$$

$$\tfrac{1}{2}(0.6 \text{ kg})(4.34 \text{ m/s})^2 + 0 = 0 + (0.6 \text{ kg})(g)H$$

$$H = 0.96 \text{ m}$$

CHAPTER SIX SOLUTIONS

6.63 We shall use conservation of energy, with the zero level for gravitational potential energy at the bottom of the arc, to find the velocity of Tarzan, v_0, just as he reaches Jane.

$$\frac{1}{2} mv_i^2 + mgy_i = \frac{1}{2} mv_f^2 + mgy_f$$

$$0 + (80 \text{ kg})g(3 \text{ m}) = \frac{1}{2} (80 \text{ kg})v_0^2 + 0$$

$$v_0 = 7.67 \text{ m/s}$$

Now, use conservation of momentum to find the velocity, V, of Tarzan + Jane just after he picks her up.

$$(80 \text{ kg})(7.67 \text{ m/s}) = (140 \text{ kg})V$$
$$V = 4.38 \text{ m/s}$$

Finally, we use conservation of mechanical energy from just after he picks her up to the end of their swing to determine the maximum height, H, reached.

$$\frac{1}{2} mv_i^2 + mgy_i = \frac{1}{2} mv_f^2 + mgy_f$$

$$\frac{1}{2} (140 \text{ kg})(4.38 \text{ m/s})^2 + 0 = 0 + (140 \text{ kg})gH$$

$$H = 0.98 \text{ m}$$

6.64 The mass of the truck is 2550 kg, and we shall call the mass of the pallet m. We use conservation of momentum in the horizontal direction, realizing that $v_f = 0.9v_0 = 0.9(7 \text{ m/s})$.

$$(2550 \text{ kg})(7 \text{ m/s}) = (2550 \text{ kg} + m)(0.9)(7 \text{ m/s})$$
$$m = 283.3 \text{ kg}.$$

Hence, the weight of the pallet is $w = mg = 2.78 \times 10^3$ N.

6.65 (a) The impulse equals the area under the F versus t graph. This area is the sum of the area of the rectangle plus the area of the triangle. Thus,

$$\text{Impulse} = (3 \text{ s})(2 \text{ N}) + \frac{1}{2}(2 \text{ s})(2 \text{ N}) = 8 \text{ N s}$$

(b) $F\Delta t = \Delta p = mv_f - mv_i$.
$$8 \text{ Ns} = (1.5 \text{ kg})v_f - 0$$
$$v_f = 5.33 \text{ m/s}$$

(c) $F\Delta t = \Delta p = mv_f - mv_i$.
$$8 \text{ Ns} = (1.5 \text{ kg})v_f - (1.5 \text{ kg})(-2 \text{ m/s})$$
$$v_f = 3.33 \text{ m/s}$$

6.66 (a) For each collision, momentum is conserved. For the first collision,
$$(1 \text{ kg})(5 \text{ m/s}) = (1 \text{ kg} + 6 \text{ kg}) v_1,$$
For the second collision, $(1 \text{ kg} + 6 \text{ kg})v_1 = (1 \text{ kg} + 6 \text{ kg} + 2 \text{ kg})v_2$
Therefore, $v_2 = 0.556$ m/s is the final speed of the three combined masses.

(b) $\Delta KE = \frac{1}{2} m_1 v^2 - \frac{1}{2} (m_1 + m_2 + m_3)v_2^2$

$$= \frac{1}{2} [(1 \text{ kg})(5 \text{ m/s})^2 - (1 \text{ kg} + 6 \text{ kg} + 2 \text{ kg})(0.556 \text{ m/s})^2] = 11.1 \text{ J}$$

6.67 Let the puck which is initially at rest be m_2.
$$m_1v_{1i} = m_1v_{1f} \cos\theta + m_2v_{2f} \cos\phi$$
and $$0 = m_1v_{1f} \sin\theta - m_2v_{2f} \sin\phi$$
$$(0.2 \text{ kg})(2 \text{ m/s}) = (0.2 \text{ kg})(1 \text{ m/s}) \cos 53° + (0.3 \text{ kg})v_{2f}\cos\phi$$

CHAPTER SIX SOLUTIONS

and $0 = (0.2 \text{ kg})(1 \text{ m/s}) \sin 53° - (0.3 \text{ kg})v_{2f}\sin\phi$
From these equations, we find $\tan\phi = 0.571$ and $\phi = 29.7°$ and $v_{2f} = 1.07$ m/s
(b) $\Delta KE/KE = (KE_f - KE_i)/KE_i = -0.32$ or 32% lost

6.68 (a) The impulse is the area under the curve between 0 and 3 s. This is
Impulse = (4 N)(3 s) = 12.0 N s
(b) The area under the curve between 0 and 5 s is
Impulse = (4 N)(3 s) + (-2 N)(2 s) = 8.00 N s
(c) Use, $F\Delta t = \Delta p = mv_f - mv_i$.
at 3 s 12.0 N s = (1.5 kg)v - 0
  v = 8.00 m/s
at 5 s 8.00 N s = (1.5 kg)v - 0
  v = 5.33 m/s

6.69 Let us use conservation of momentum to find the velocity V of the can immediately after the bullet passes through.
(0.11 kg)(V) + (0.002 kg)(720 m/s) = (0.002)(900 m/s)
V = 3.27 m/s
Now, treat the projectile motion of the can. We have

$$y = v_{oy}t + \frac{1}{2}gt^2$$

$$-1.7 \text{ m} = 0 + \frac{1}{2}(-9.8 \text{ m/s}^2)t^2$$

which gives t = 0.589 s
and from its horizontal motion, we have
x = v_{ox}t = (3.27 m/s)(0.589 s) = 1.93 m

6.70 First, we will find the horizontal velocity, v_{0x}, of the block and embedded bullet just after impact. At this instant, the block-bullet combination has become a projectile, so we find the time to reach the floor by use of

$$y = v_{0y}t + \frac{1}{2}at^2$$

$$-1 \text{ m} = 0 + \frac{1}{2}(-9.8 \text{ m/s}^2)t^2.$$

t = 0.452 s
Thus, the initial horizontal velocity can be found as
v_{0x} = x/t = 2m/0.452 s = 4.43 m/s
Now use conservation of momentum for the collision, with v_b = velocity of incoming bullet.
(6 × 10^{-3} kg)v_b = (206 × 10^{-3} kg)(4.43 m/s)
v_b = 152 m/s (about 340 mph)

6.71 Let us apply conservation of energy to the block from the time just after the bullet has passed through until it reaches its maximum height in order to find its velocity V just after the collision.

$$\frac{1}{2}mv_i^2 + mgy_i = \frac{1}{2}mv_f^2 + mgy_f$$

$$\frac{1}{2}(1.5 \text{ kg})V^2 + 0 = 0 + (1.5 \text{ kg})g(0.12 \text{ m})$$

V = 1.53 m/s
Now use conservation of momentum from before until just after the collision in order to find the speed of the bullet, v_0.

CHAPTER SIX SOLUTIONS

$(7 \times 10^{-3} \text{ kg})(v_0 - 200 \text{ m/s}) = (1.5 \text{ kg})(1.534 \text{ m/s})$
$v_0 = 529 \text{ m/s}$

6.72 The mass of the third fragment is
$(17 \times 10^{-27} \text{ kg} - 13.4 \times 10^{-27} \text{ kg}) = 3.6 \times 10^{-27} \text{ kg}$.
First, we conserve momentum in the x direction.
$(8.4 \times 10^{-27} \text{ kg})(4 \times 10^6 \text{ m/s}) - (3.6 \times 10^{-27} \text{ kg})v_3\cos\theta = 0$
or, $v_3\cos\theta = 9.33 \times 10^6$ m/s   (1)
Now, conserve momentum in the y direction.
$(5 \times 10^{-27} \text{ kg})(6 \times 10^6 \text{ m/s}) - (3.6 \times 10^{-27} \text{ kg})v_3\sin\theta = 0$.
or, $v_3\sin\theta = 8.33 \times 10^6$ m/s   (2)
Dividing (2) by (1) gives $\tan\theta = 0.893$ or $\theta = 41.8°$
Then from either (1) or (2)
$v_3 = 1.25 \times 10^7$ m/s

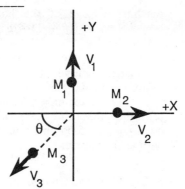

6.73 (a) The force exerted on the cart has the same magnitude as the force exerted on the bullets (third law). We find the force, F, on the bullets.
  $F = \Delta p/\Delta t$ = (number of bullets fired per second)(change in momentum of each bullet)
  $F = nmv$
(b) If 50 bullets are fired in 5 sec, n = 10 per second, and we have
  $F = (10 \text{ per sec})(5 \times 10^{-3} \text{ kg})(300 \text{ m/s}) = 15.0$ N
Apply Newton's second law to the cart.
  $F = Ma$ or $a = F/m = 15 \text{ N}/70 \text{ kg} = 2.14 \times 10^{-1}$ m/s²

6.74 First, we find the speed of the candy just before it lands:
  $v^2 = v_0^2 + 2ay$
  $v^2 = 0 + 2(-9.8 \text{ m/s}^2)(-0.2 \text{ m})$
  $v = 1.98$ m/s
The net force pushing down on the scale = weight of the candy already in the pan + force exerted by impacting candy. Thus,
  Scale Reading = $w + F$
At t = 10 s, the number of pieces in the pan is 40, each having a mass of 1 g. Thus,
  $w = (40 \times 10^{-3} \text{ kg})(9.8 \text{ m/s}^2) = 0.392$ N
To get F, we use $F\Delta t = \Delta p$ or $F = \Delta p/\Delta t$
  $\Delta p/\Delta t$ = (number of pieces hitting per second)(change in momentum of each piece)
$\Delta p/\Delta t$ = (4 per second)(mv) = (4 per second)(10^{-3} kg)(1.98 m/s) = 7.92×10^{-3} N
Thus, Scale Reading = $w + F = 0.392$ N + 7.92×10^{-3} N = 0.40 N

6.75 (a) Since the net force on the system (child and boat) is zero, the momentum of the system will remain equal to zero (its initial value).
Therefore, as the child walks to the right end of the boat, the boat will move to the left (toward the pier).
(b) Since there are no external forces acting on the system, the center of the mass will not move. The location of the center of mass (relative to the pier) is initially

89

CHAPTER SIX SOLUTIONS

$$x_{cm} = \frac{m_c x_c + m_b x_b}{m_c + m_b} = \frac{(3\text{ m})(40\text{ kg}) + (5\text{ m})(70\text{ kg})}{70\text{ kg} + 40\text{ kg}}$$

If we let d equal the distance which the boat moves to the left, when the child reaches the end of the boat

$$x_{cm} = \frac{(7 - d)(40\text{ kg}) + (5 - d)(70\text{ kg})}{40\text{ kg} + 70\text{ kg}}$$

Setting these two expressions for x_{cm} equal and solving for d, we find d = 1.45 m (distance moved to left by boat). Therefore, the position of the child relative to the pier will be

(7 - 1.45) m = 5.55 m

(c) Since the turtle is 7 m from the pier and the child at the right end of the boat will be 5.55 m from the pier, he will not be able to reach the turtle even with a 1 m reach.

6.76 The deceleration of the first block is $a = \frac{F}{m} = -\mu g$. Therefore, just before collision, M has a velocity $v = \sqrt{v_0^2 - 2\mu g d}$

Using conservation of momentum, we have $Mv = Mv_1 + 2Mv_2$ where v_1 and v_2 are the speeds of the two blocks after collision. Also, since this is an elastic collision, we have $\frac{1}{2}Mv^2 = \frac{1}{2}Mv_1^2 + \frac{1}{2}(2M)v_2^2$. Solving these two equations gives

$$v_2 = \frac{2}{3}v = \sqrt{\frac{4}{9}v_0^2 - \frac{8}{9}\mu g d} \quad (1)$$

The second block has the same deceleration as the first. It comes to rest when its velocity is 0, and therefore it will move a distance D given by $0 - v_2^2 = -2\mu g D$. Substituting Eq. (1) for v_2^2 into this expression, we find

$$D = \frac{2v_0^2}{9\mu g} - \frac{4d}{9}$$

6.77 (a) $\Delta p_x = m(-v\sin\theta) - mv\sin\theta = -2mv\sin\theta$

(b) $\Delta p_y = mv\cos\theta - mv\cos\theta = 0$

(c) The force on the ball is $F_B = \frac{-2mv\sin\theta}{t}$ Thus, the force on the wall from Newton's third law is $F_w = \frac{2mv\sin\theta}{t}$

CHAPTER SEVEN SOLUTIONS

7.1 The conversion factor between degrees and radians is π rad = 180°. Thus, to convert 30° to rad, we perform the following operation.

$$30°\left(\frac{\pi \text{ rad}}{180°}\right) = \frac{\pi}{6} \text{ rad} = 0.524 \text{ rad}$$

We proceed in a similar fashion to convert the remaining angles in degrees to radians.

$$45° = \frac{\pi}{4} \text{ rad} = 0.785 \text{ rad}$$

60° = 1.05 rad
90° = 1.57 rad
180° = 3.14 rad
270° = 4.71 rad, and
360° = 6.28 rad

7.2 The conversion relationship is π rad = 180°. To convert from rad to degrees, we use the conversion factor as follows.

$$\frac{\pi}{3} \text{ rad} = \frac{\pi}{3}\left(\frac{180°}{\pi \text{ rad}}\right) = 60°$$

Similarly, we find 1.2π rad = 216°, and 3π rad = 540°.

7.3 The circumference = $2\pi r$ = 25.76 m
$30° = \frac{1}{12}$ of full circle, so $s = \frac{1}{12}$ (circumference) = 2.15 m
$s = r\theta = (4.1 \text{ m})(30 \text{ rad}) = 123$ m
$s = 30$(circumference) = 773 m

7.4 The earth turns through 2π rad in one day (86,400 s). Thus,

$$\omega = \frac{2\pi \text{ rad}}{86400 \text{ s}} = 7.27 \times 10^{-5} \text{ rad/s}$$

7.5 The earth moves through 2π rad in one year (3.156 × 10^7 s). Thus,

$$\omega = \frac{2\pi \text{ rad}}{3.156 \times 10^7 \text{ s}} = 1.99 \times 10^{-7} \text{ rad/s}$$

Alternatively, the earth moves through 360° in one year (365.242 days). Thus,

$$\omega = \frac{360°}{365.2 \text{ days}} = 0.986 \text{ deg/day}$$

7.6 (a) ω = (33 rev/min)(2π rad/rev)(1 min/60 s) = 3.46 rad/s
(b) $\theta = \omega t$ = (3.46 rad/s)(1.5 s) = 5.19 rad

7.7 Use $\omega = \omega_0 + \alpha t$, with ω = 0.2 rev/s = 1.256 rad/s.
1.256 rad/s = 0 + α(30 s)
$\alpha = 4.19 \times 10^{-2}$ rad/s^2.

7.8 $\theta = \frac{s}{r} = \frac{2.5 \text{ m}}{1.5 \text{ m}} = 1.67$ rad = 95.7°

CHAPTER SEVEN SOLUTIONS

7.9 $\omega_f = 2.51 \times 10^4$ rev/min $= 2.63 \times 10^3$ rad/s

(a) $\alpha = \dfrac{\omega_f - \omega_0}{t} = \dfrac{2.63 \times 10^3 \text{ rad/s} - 0}{3.2 \text{ s}} = 8.22 \times 10^2$ rad/s^2

(b) $\theta = \omega_0 t + \dfrac{1}{2}\alpha t^2 = 0 + \dfrac{1}{2}(8.22 \times 10^2 \text{ rad/s}^2)(3.2 \text{ s})^2 = 4.21 \times 10^3$ rad

7.10 $\omega_0 = 100$ rev/min $= 10.47$ rad/s

(a) $t = \dfrac{\omega - \omega_0}{\alpha} = \dfrac{0 - 10.47 \text{ rad/s}}{-2 \text{ rad/s}^2} = 5.24$ s

(b) $\theta = \bar{\omega} t = \dfrac{\omega + \omega_0}{2} t = \dfrac{0 + 10.47 \text{ rad/s}}{2}(5.24 \text{ s}) = 27.4$ rad

7.11 (a) $33\dfrac{1}{3}$ rev/min is equivalent to 3.49 rad/s. Now, use

$\omega = \omega_0 + \alpha t$

3.49 rad/s $= 0 + \alpha(20 \text{ s})$

$\alpha = 1.75 \times 10^{-1}$ rad/s^2.

(b) $\theta = \omega_0 t + \dfrac{1}{2}\alpha t^2$.

$\theta = 0 + \dfrac{1}{2}(1.75 \times 10^{-1} \text{ rad/s}^2)(20 \text{ s})^2$.

$\theta = 35.0$ rad $= 5.57$ rev

7.12 $\theta = 4.7$ rev $= 29.5$ rad

$\theta = \omega_0 t + \dfrac{1}{2}\alpha t^2$.

29.5 rad $= 0 + \dfrac{1}{2}\alpha(1.2 \text{ s})^2$.

$\alpha = 41$ rad/s^2

7.13 $\omega = 5$ rev/s $= 10\pi$ rad/s

We will break the motion into two stages: (1) an acceleration period and (2) a deceleration period. While speeding up,

$\theta_1 = \bar{\omega} t = \dfrac{0 + 10\pi \text{ rad/s}}{2}(8 \text{ s}) = 125.7$ rad

While slowing down,

$\theta_2 = \bar{\omega} t = \dfrac{10\pi \text{ rad/s} + 0}{2}(12 \text{ s}) = 188.5$ rad

So, $\theta_{total} = \theta_1 + \theta_2 = 314.2$ rad $= 50$ rev

7.14 First find the speed attained before the wheel begins to slowing down. With $\omega_0 = 0$, we have

$\omega = \omega_0 + \alpha t = 0 + (5 \text{ rad/s}^2)(8 \text{ s}) = 40$ rad/s

Next, consider the wheel as it comes to rest. Now $\omega_0 = 40$ rad/s, and

$\omega^2 = \omega_0^2 + 2\alpha\theta$

becomes $0 = (40 \text{ rad/s})^2 + 2\alpha(20\pi \text{ rad})$

from which, $\alpha = -12.7$ rad/s^2

CHAPTER SEVEN SOLUTIONS

and $t = \dfrac{\theta}{\bar{\omega}} = \dfrac{20\pi \text{ rad}}{\dfrac{0 + 40 \text{ rad/s}}{2}} = 3.14$ s

---

7.15 $\omega_0 = 3600$ rev/min $= 3.77 \times 10^2$ rad/s, $\theta = 50$ rev $= 3.14 \times 10^2$ rad, and $\omega = 0$.
$\omega^2 = \omega_0^2 + 2\alpha\theta$.
$0 = (3.77 \times 10^2 \text{ rad/s})^2 + 2\alpha(3.14 \times 10^2 \text{ rad})$
$\alpha = -2.26 \times 10^2$ rad/s^2

7.16 $\omega^2 = \omega_0^2 + 2\alpha\theta$.
$(2.2 \text{ rad/s})^2 = (0.6 \text{ rad/s})^2 + 2(0.7 \text{ rad/s}^2)\theta$.
$\theta = 3.2$ rad

7.17 $\omega = \omega_0 + \alpha t = 2000$ rad/s $+ (-80$ rad/s$^2)(10$ s$) = 1200$ rad/s
(b) $t = \dfrac{\omega - \omega_0}{\alpha} = \dfrac{0 - 2000 \text{ rad/s}}{-80 \text{ rad/s}^2} = 25$ s

---

7.18 $\theta = \dfrac{\omega^2 - \omega_0^2}{2\alpha} = \dfrac{0 - (18 \text{ rad/s})^2}{2(-1.9 \text{ rad/s}^2)} = 85.3$ rad
$s = r\theta$ and $r = \dfrac{1}{2}$ diameter $= 1.2$ cm, so
$s = (1.2$ cm$)(85.3$ rad$) = 102$ cm $= 1.02$ m

7.19 (a) $v_t = r\omega$.
80 m/s $= (200$ m$)\omega$.
$\omega = 0.40$ rad/s
(b) $a_t = \Delta v_t/\Delta t = 0$ because the tangential velocity is a constant.

7.20 (a) $a_t = \Delta v_t/\Delta t = \dfrac{95 \text{ m/s} - 80 \text{ m/s}}{10 \text{s}} = 1.5$ m/s^2.
and $a_t = r\alpha$
1.5 m/s$^2 = (200$ m$)\alpha$
$\alpha = 7.5 \times 10^{-3}$ rad/s^2
(b) $s = \bar{v}_t t = \dfrac{80 \text{ m/s} + 95 \text{ m/s}}{2}(10$ s$) = 875$ m
and $\theta = \dfrac{s}{r} = 875$ m$/200$ m $= 4.38$ rad

7.21 $s = v_0 t + \dfrac{1}{2}at^2 = (17.0$ m/s$)(5$ s$) + \dfrac{1}{2}(2$ m/s$^2)(5$ s$)^2 = 110$ m
$s = r\theta$, and $r = 48$ cm $= 0.48$ m
Thus, $\theta = \dfrac{s}{r} = \dfrac{110 \text{ m}}{0.48 \text{ m}} = 229.2$ rad $= 36.5$ rev

---

7.22 $\omega_0 = \dfrac{v_0}{r} = \dfrac{24.0 \text{ m/s}}{0.48 \text{ m}} = 50.0$ rad/s
The time to stop is
$t = \dfrac{\omega - \omega_0}{\alpha} = \dfrac{0 - 50.0 \text{ rad/s}}{-1.35 \text{ rad/s}^2} = 37.0$ s

CHAPTER SEVEN SOLUTIONS

The angular displacement is
$$\theta = \bar{\omega}t = \frac{0 + 50 \text{ rad/s}}{2}(37.0 \text{ s}) = 925 \text{ rad} = 147 \text{ rev}$$

7.23 $\omega_1 = 8$ rev/s $= 16\pi$ rad/s, and $r_1 = 0.6$ m
Thus, $v_1 = (0.6 \text{ m})(16\pi \text{ rad/s}) = 30.2$ m/s
$\omega_2 = 6$ rev/s $= 12\pi$ rad/s, and $r_2 = 0.9$ m
Thus, $v_2 = (0.9 \text{ m})(12\pi \text{ rad/s}) = 33.9$ m/s
(a) He obtains greater linear speed with the larger sling.
(b) $a_c = \frac{v^2}{r} = \frac{(30.2 \text{ m/s})^2}{0.6 \text{ m}} = 1.52 \times 10^3$ m/s²
(c) $a_c = \frac{v^2}{r} = \frac{(33.9 \text{ m/s})^2}{0.9 \text{ m}} = 1.28 \times 10^3$ m/s²

7.24 $\omega = 200$ rev/min $= 20.9$ rad/s
$v = r\omega = (0.5 \text{ m})(20.9 \text{ rad/s}) = 10.5$ m/s
$a_c = \frac{v^2}{r} = r\omega^2 = (0.5 \text{ m})(20.9 \text{ rad/s})^2 = 218$ m/s²

7.25 The angular velocity of the earth is 7.27×10^{-5} rad/s. (See problem 7.4.)
The radius of the earth is 6.38×10^6 m.
(a) $a_r = r\omega^2 = (6.38 \times 10^6 \text{ m})(7.27 \times 10^{-5} \text{ rad/s})^2 = 3.37 \times 10^{-2}$ m/s².
(b) $a_r = r\omega^2 = 0$ since $r = 0$ (The north pole is on the rotation axis of the earth.)

7.26 The radius is 2.5 mi $= 4.02 \times 10^3$ m.
$a_r = r\omega^2$
9.8 m/s² $= (4.02 \times 10^3 \text{ m})\omega^2$.
$\omega = 4.94 \times 10^{-2}$ rad/s

7.27 (a) The final angular velocity is 78 rev/min $= 8.17$ rad/s, and the radius of the record is equal to 5 in $= 1.27 \times 10^{-1}$ m.
$\alpha = \Delta\omega/\Delta t = (8.17 \text{ rad/s})/(3 \text{ s}) = 2.72$ rad/s² .
and $a_t = r\alpha = 3.45 \times 10^{-1}$ m/s².
(b) $v_t = r\omega$.
$v_t = (1.27 \times 10^{-1} \text{ m})(8.17 \text{ rad/s}) = 1.04$ m/s
(c) At t= 1 s, $a_t = 3.45 \times 10^{-1}$ m/s². From $\omega = \omega_0 + \alpha t$, we find $\omega = 2.72$ rad/s, and from $v_t = r\omega$. we find $v_t = 0.345$ m/s. Thus,
$a_r = \frac{v_t^2}{r} = 0.937$ m/s².
$a = \sqrt{(a_t)^2 + (a_r)^2} = 0.998$ m/s².
$\tan\theta = a_t/a_r = .368 \quad \theta = 20.2°$

7.28 (a) $\omega = \omega_0 + \alpha t$
$\omega = 0 + (4 \text{ rad/s}^2)(2 \text{ s}) = 8$ rad/s
(b) $v_t = r\omega. = (0.3 \text{ m})(8 \text{ rad/s}) = 2.4$ m/s
$a_t = r\alpha = (0.3 \text{ m})(4 \text{ rad/s}^2) = 1.2$ m/s²
(c) $\theta = \theta_0 + \omega_0 t + \frac{1}{2}\alpha t^2$.

CHAPTER SEVEN SOLUTIONS

$\theta = 1 \text{ rad} + 0 + \frac{1}{2}$ (4 rad/s^2)(2s)^2 = 9 rad = 516°   (or at 156°counterclockwise from a horizontal reference line.)

7.29 Use $F = m\frac{v^2}{r}$

2.65 N = (0.08 kg)$\frac{v^2}{10 \text{ m}}$

Which gives,   v = 18.2 m/s = 40.7 mi/h

7.30 As he passes through the bottom of his swing, the force that the vine must supply is equal to the (1) his weight plus (2) the needed centripetal force. From Newton's second law, we have

$T - mg = m\frac{v^2}{r}$

or $T = mg + m\frac{v^2}{r}$ = (85 kg)(9.8 m/s^2) + (85 kg)$\frac{(8 \text{ m/s})^2}{10 \text{ m}}$ = 1377 N

Since 1377 N > 1000 N, the vine will break.

7.31 The friction force, μmg, must supply the required centripetal force. We have

$F = \frac{mv^2}{r}$

$\mu mg = \frac{mv^2}{r}$

The mass cancels, and we have $v^2 = \mu rg$ = 0.7(20 m)(9.8 m/s^2)
From which,   v = 11.7 m/s

7.32 The radius of the wheel is 45 ft = 13.7 m, its angular velocity is ω = 2π rad/6 s = 1.05 rad/s, and, the tangential velocity of a point on the rim of the wheel is v = rω = 14.4 m/s.

Consider, first, the equation $\Sigma F_y = 0$. We have

$F\cos\theta = mg$,   (1)

where F is the net force applied on the cars by the bolts. Now consider

$F_r = \frac{mv^2}{r}$

$F\sin\theta = \frac{mv^2}{r}$ (2)

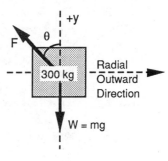

Divide (2) by (1) to find
   $\tan\theta = v^2/rg$ =(14.4 m/s)^2/(13.7 m)(9.8 m/s^2) =1.53
Which gives   θ = 56.9°
Then from (1), we find   F = (300 kg)(9.8 m/s^2)/cos56.9° = 5.38 X 10^3 N
But, one-half this force is supplied by each bolt. Thus,
   F_{bolt} = 2.69 X 10^3 N at 56.9° from the vertical.

7.33 (a) At A the forces on the car are the normal force, N, and its weight. We have

$F = \frac{mv^2}{r} = N - mg$, or $N = mg + \frac{mv^2}{r}$, which gives

CHAPTER SEVEN SOLUTIONS

$$N = (500 \text{ kg})(9.8 \text{ m/s}^2) + \frac{(500 \text{ kg})(20 \text{ m/s})^2}{10 \text{ m}} = 2.49 \times 10^4 \text{ N}$$

For the vehicle to remain on the track, we must have $N \geq 0$.

(b) At B we have, $F = \frac{mv^2}{r} = mg - N$, but N goes to zero when $v^2 = rg$. Thus,

$$v = \sqrt{(15 \text{ m})(9.8 \text{ m/s}^2)} = 12.1 \text{ m/s}$$

7.34 $F = \frac{mv^2}{r} = mr\omega^2$

so, $\omega^2 = \frac{F}{mr} = \frac{8.20 \times 10^{-8} \text{ N}}{(9.1 \times 10^{-31} \text{ kg})(5.3 \times 10^{-11} \text{ m})} = 1.7 \times 10^{33} \text{ rad}^2/\text{s}^2$, or

$\omega = 4.12 \times 10^{16} \text{ rad/s}$

The time for 1 rev = $\frac{2\pi}{\omega} = \frac{2\pi}{4.12 \times 10^{16} \text{ rad/s}} = 1.52 \times 10^{-16} \text{ s}$

and the number of revolutions in one sec = $\frac{1 \text{ s}}{1.52 \times 10^{-16} \text{ s}} = 6.56 \times 10^{15}$ rev/s

---

7.35 (a) The radial acceleration must not exceed $7g = 7(9.8 \text{ m/s}^2) = 68.6 \text{ m/s}^2$.

Thus, from $a_r = \frac{v_t^2}{r}$, we have $r = (100 \text{ m/s})^2/68.6 \text{ m/s}^2 = 1.46 \times 10^2 \text{ m}$.

(b) $F = ma_r = m(7g) = 7(mg) = 7(\text{pilot's weight}) = 7(80 \text{ kg})(9.8 \text{ m/s}^2) = 5488 \text{ N}$

7.36 (a) $a_r = r\omega^2 = (2 \text{ m})(3 \text{ rad/s})^2 = 18 \text{ m/s}^2$.

(b) $F = ma_r = (50 \text{ kg})(18 \text{ m/s}^2) = 900 \text{ N}$

(c) We know the centripetal force is equal to the force of friction. Thus, $f = 900$ N. Also, the normal force, N, is equal to 490 N. Thus,

$\mu = f/N = 900 \text{ N}/490 \text{ N} = 1.84$. A coefficient of friction greater than one is unreasonable. Thus, she is not going to be able to stay on the ride.

7.37 From $\Sigma F_y = 0$, we find the vertical component of the tension to equal the weight.

$T\cos 5° = mg$ (1) where T is the tension in the vine. Then, the inward component of T provides the centripetal force. We have,

$T\sin 5° = ma_r$ (2)

Dividing (2) by (1) gives

$\tan 5° = a_r/g$, or $a_r = g\tan 5° = 0.857 \text{ m/s}^2$.

$F = ma_r = (80 \text{ kg})(0.857 \text{ m/s}^2) = 68.6 \text{ N}$

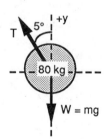

CHAPTER SEVEN SOLUTIONS

7.38 From $\Sigma F_y = 0$, we have
$N \cos\theta = mg$ (1)
where N is the normal force exerted on the car by the ramp.
Now, use $F = ma_r$
$N \sin\theta = mv^2/r$ (2)
Divide (2) by (1) $\tan\theta = v^2/rg$
(b) $\tan\theta = (13.4 \text{ m/s})^2/(50 \text{ m})(9.8 \text{ m/s}^2) = 0.366$
$\theta = 20.1°$

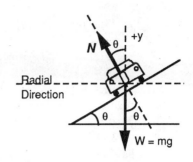

7.39 The velocity must be great enough so that the necessary centripetal force is equal to or greater than the weight mg.
$F_c \geq mg$ or $mv^2/r \geq mg$ or $v^2 \geq rg$
At the minimum speed, we have $v_{min}^2 = rg$ or $v_{min} = \sqrt{rg}$
Therefore, $v_{min} = \sqrt{(1 \text{ m})(9.8 \text{ m/s}^2)} = 3.13$ m/s

---

7.40 (a) At the beginning of the motion, $v = 0$, and from $F = \dfrac{mv^2}{r}$, we see that F, the required centripetal force is also zero. The centripetal force is T.
Thus, $T = 0$.
(b) Let us use conservation of mechanical energy to find an expression for v, the velocity of the stuntman at any point along the arc of his swing.
$\dfrac{1}{2} mv_i^2 + mgy_i = \dfrac{1}{2} mv_f^2 + mgy_f$
$0 + mgR = \dfrac{1}{2} mv^2 + mgy$
or $v^2 = 2g(R - y)$ (1)
where R is the radius of the path, and y is the vertical distance above the zero reference level at the bottom of the arc.

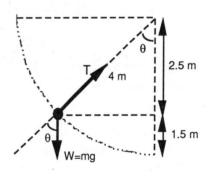

When $y = 1.5$ m,
$v^2 = 2g(4 \text{ m} - 1.5 \text{ m}) = 49$ m²/s².
The necessary centripetal force is
$F = \dfrac{mv^2}{R} = \dfrac{(70 \text{ kg})(49 \text{ m}^2/\text{s}^2)}{4 \text{ m}} = 858$ N
But the centipetal force F is given by
$T - mg\cos\theta = 858$ N
where $\cos\theta = 2.5 \text{ m}/4 \text{ m} = .625$ $\theta = 51.3°$
So, $T = 858$ N $+ (70$ kg$)g(\cos 51.3°) = 1287$ N
(c) At the bottom of the arc, $y = 0$ and equation (1) gives $v^2 = 78.4$ m²/s².
Thus, the necessary centripetal force is
$F = \dfrac{mv^2}{r} = (70 \text{ kg})(78.4 \text{ m}^2/\text{s}^2)/4\text{m} = 1372$ N.
Also, the centripetal force at the bottom of the arc is
$T - mg$ Thus, $F = T - mg$,
or $T = F + mg = 1372 + (70 \text{ kg})g = 2.06 \times 10^3$ N

CHAPTER SEVEN SOLUTIONS

7.41 $F = \dfrac{Gm_1m_2}{r^2} = \dfrac{(6.67 \times 10^{-11} \text{ Nm}^2/\text{kg}^2)(1.99 \times 10^{30} \text{ kg})(5.98 \times 10^{24} \text{ kg})}{(1.496 \times 10^{11} \text{ m})^2}$

$= 3.55 \times 10^{22}$ N

7.42 At the half-way point the spaceship is at a distance of 1.92×10^8 m from both bodies. The force exerted by the earth on the ship is

$F_1 = \dfrac{Gm_1m_2}{r^2} = \dfrac{(6.67 \times 10^{-11} \text{ Nm}^2/\text{kg}^2)(3 \times 10^4 \text{ kg})(5.98 \times 10^{24} \text{ kg})}{(1.92 \times 10^8 \text{ m})^2} = 325$ N

and the force exerted on the ship by the moon is

$F_2 = \dfrac{Gm_1m_2}{r^2} = \dfrac{(6.67 \times 10^{-11} \text{ Nm}^2/\text{kg}^2)(3 \times 10^4 \text{ kg})(7.36 \times 10^{22} \text{ kg})}{(1.92 \times 10^8 \text{ m})^2} = 4.00$ N

Thus, the resultant force is (325 N - 4.00 N) = 321 N directed toward the earth.

7.43 The mass of the 600 N student is 61.2 kg, and that of the 700 N student is 71.4 kg. Thus,

$F = \dfrac{Gm_1m_2}{r^2} = \dfrac{(6.67 \times 10^{-11} \text{ Nm}^2/\text{kg}^2)(61.2 \text{ kg})(71.4 \text{ kg})}{(0.5 \text{ m})^2} = 1.17 \times 10^{-6}$ N

7.44 The force exerted on the 2 kg mass by the 3 kg mass is in the positive y direction and is given by

$F_1 = \dfrac{Gm_1m_2}{r^2} = \dfrac{(6.67 \times 10^{-11} \text{ Nm}^2/\text{kg}^2)(3 \text{ kg})(2 \text{ kg})}{(2 \text{ m})^2} = 1.00 \times 10^{-10}$ N

and the force exerted on the 2 kg mass by the 4 kg mass is in the positive x direction, and is

$F_2 = \dfrac{Gm_1m_2}{r^2} = \dfrac{(6.67 \times 10^{-11} \text{ Nm}^2/\text{kg}^2)(4 \text{ kg})(2 \text{ kg})}{(4 \text{ m})^2} = 3.34 \times 10^{-11}$ N

The resultant F is found from the Pythagorean theorem to be 1.05×10^{-10} N, and $\tan\theta = \dfrac{F_1}{F_2}$ gives

$\theta = 71.5°$.

7.45 We find $g = \dfrac{GM}{R^2}$ where M is the mass of the planet and R is its radius. We write this equation down twice, once for Earth and once for Venus and divide the two to get,

$\dfrac{g_{venus}}{g_{earth}} = \dfrac{M_v}{M_e}\left(\dfrac{R_e}{R_v}\right)^2 = \dfrac{4.88 \times 10^{24} \text{ kg}}{5.98 \times 10^{24} \text{ kg}}\left(\dfrac{6.38 \times 10^6 \text{ m}}{6.06 \times 10^6 \text{ m}}\right)^2 = 0.905$

Thus, $g_v = 0.905 g_e = 8.87$ m/s^2.

7.46 Assume that the Moon moves around the center of the Earth, and

$F_c = F_G$

gives $\dfrac{mv^2}{r} = \dfrac{GmM_E}{r^2}$

or, $M_E = \dfrac{rv^2}{G}$

and $v = \dfrac{\text{distance traveled}}{\text{time required}} = \dfrac{2\pi r}{\text{period}} = \dfrac{2\pi(3.84 \times 10^8 \text{ m})}{27.32 \text{ days}(86,400 \text{ s/day})} = 1022$ m/s

CHAPTER SEVEN SOLUTIONS

Thus, $M_E = \dfrac{(3.84 \times 10^8 \text{ m})(1022 \text{ m/s})^2}{6.67 \times 10^{-11} \text{ Nm}^2/\text{kg}^2} = 6.01 \times 10^{24}$ kg

The estimate is slightly high because the Moon actually orbits the center of mass of the Earth-Moon system, not the center of the Earth.

7.47 (a) $F_c = ma_c = F_G$

so, $ma_c = \dfrac{GmM_{moon}}{r^2}$

and $a_c = \dfrac{GM_{moon}}{r^2} = \dfrac{(6.67 \times 10^{-11} \text{ Nm}^2/\text{kg}^2)(7.36 \times 10^{22} \text{ kg})}{(1.738 \times 10^6 \text{ m})^2} = 1.63$ m/s^2

(b) $a_c = \dfrac{v^2}{r}$, so $v^2 = a_c r = (1.63 \text{ m/s}^2)(1.738 \times 10^6 \text{ m})$ yields $v = 1.68 \times 10^3$ m/s

(c) The time to orbit $= \dfrac{\text{distance around}}{\text{speed}} = \dfrac{2\pi r}{v} = \dfrac{2\pi(1.738 \times 10^6 \text{ m})}{1.68 \times 10^3 \text{ m/s}} = 6.5 \times 10^3$ s $= 108.3$ min $= 1.81$ h

---

7.48 $g = \dfrac{GM}{R^2} = \dfrac{(6.67 \times 10^{-11} \text{ Nm}^2/\text{kg}^2)(5.98 \times 10^{24} \text{ kg})}{(4.66 \times 10^7 \text{ m})^2} = 0.184$ m/s^2.

7.49 Kepler's third law says $T^2 = Kr^3$, where K is a constant of proportionality. For Venus, $T = 1.94 \times 10^7$ s and $r = 1.08 \times 10^{11}$ m.
We find a value for K as

$\dfrac{T^2}{r^3} = \dfrac{(1.94 \times 10^7 \text{ s})^2}{(1.08 \times 10^{11} \text{ m})^3} = 2.98 \times 10^{-19}$ s^2/m^3.

This can be repeated for Mars and Neptune to show that the constant K is the same in each case.

7.50 (a) We have $F = \dfrac{mv^2}{r} = \dfrac{GM_E m}{r^2}$. With $r = 2R_E$

$v^2 = \dfrac{GM_E}{r} = \dfrac{(6.67 \times 10^{-11} \text{ Nm}^2/\text{kg}^2)(5.98 \times 10^{24} \text{ kg})}{(1.28 \times 10^7 \text{ m})}$. From which,

$v = 5.58 \times 10^3$ m/s

(b) The period is given by $T = \dfrac{2\pi r}{v} = \dfrac{2\pi(1.28 \times 10^7 \text{ m})}{5.58 \times 10^3 \text{ m/s}} = 1.44 \times 10^4$ s $= 240$ min $= 4$ h.

(c) $F_2 = \dfrac{Gm_1 m_2}{r^2} = \dfrac{(6.67 \times 10^{-11} \text{ Nm}^2/\text{kg}^2)(5.98 \times 10^{24} \text{ kg})(600 \text{ kg})}{(1.28 \times 10^7 \text{ m})^2}$
$= 1.46 \times 10^3$ N

7.51 At the equilibrium position, the magnitude of the force exerted by the earth on the object is equal to the magnitude of the force exerted by the sun on the object.

$\dfrac{GM_E m}{(1.50 \times 10^{11} - r)^2} = \dfrac{GM_S m}{r^2}$

This can be solved for r using the quadratic equation to find,

$r = 1.49 \times 10^{11}$ m (r is the distance from the sun.)

CHAPTER SEVEN SOLUTIONS

7.52 $F_c = F_G$

gives $\dfrac{mv^2}{R} = \dfrac{GmM_E}{R^2}$

or $M = \dfrac{Rv^2}{G}$ and

$v = \dfrac{\text{distance traveled}}{\text{time}} = \dfrac{2\pi R}{\text{period}} = \dfrac{2\pi(4.22 \times 10^8 \text{ m})}{(1.77 \text{ days})(86,400 \text{ s/day})} = 1.733 \times 10^4 \text{ m/s}$

Thus, $M = \dfrac{(4.22 \times 10^8 \text{ m})(1.733 \times 10^4 \text{ m/s})^2}{6.67 \times 10^{-11} \text{ Nm}^2/\text{kg}^2} = 1.90 \times 10^{27} \text{ kg}$

7.53 $F_c = F_G$ gives $\dfrac{mv^2}{r} = \dfrac{GmM_E}{r^2}$, which reduces to

$v = \sqrt{\dfrac{GM_E}{r}}$

and period $= \dfrac{2\pi r}{v} = 2\pi r\sqrt{\dfrac{r}{GM_E}}$

$r = R_E + 200 \text{ km} = 6380 \text{ km} + 200 \text{ km} = 6580 \text{ km}$

Thus, period $= 2\pi(6.58 \times 10^6 \text{ m})\sqrt{\dfrac{(6.58 \times 10^6 \text{ m})}{(6.67 \times 10^{-11} \text{ Nm}^2/\text{kg}^2)(5.98 \times 10^{24} \text{ kg})}} =$
$5310 \text{ s} = 88.5 \text{ min} = 1.48 \text{ h}$

(b) $v = \sqrt{\dfrac{GM_E}{r}} = \sqrt{\dfrac{(6.67 \times 10^{-11} \text{ Nm}^2/\text{kg}^2)(5.98 \times 10^{24} \text{ kg})}{(6.58 \times 10^6 \text{ m})}} = 7.79 \times 10^3 \text{ m/s} =$
7.79 km/s

(c) $KE_f + PE_f = KE_i + PE_i +$ energy input, gives

input $= \dfrac{1}{2}mv_f^2 - \dfrac{1}{2}mv_i^2 + \left(\dfrac{-GM_E m}{r_f}\right) - \left(\dfrac{-GM_E m}{r_i}\right)$ (1)

$r_i = R_E = 6.38 \times 10^6 \text{ m}$, $v_i = \dfrac{2\pi R_E}{24 \text{ h}} = 4.64 \times 10^2 \text{ m/s}$

Substituting the appropriate values into (1), yields
  minimum energy input $= 6.43 \times 10^9 \text{ J}$
This assumes that the launch is from the equator.

7.54 The graph should look something like that sketched below.

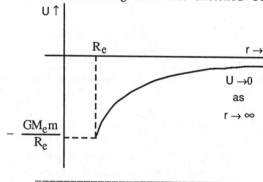

CHAPTER SEVEN SOLUTIONS

7.55 (a) $U = -\dfrac{GM_Em}{r} = \dfrac{(6.67 \times 10^{-11}\ Nm^2/kg^2)(5.98 \times 10^{24}\ kg)(100\ kg)}{(8.38 \times 10^6\ m)}$

$= -4.77 \times 10^9$ J

(b) $F = \dfrac{GM_Em}{r^2} = \dfrac{(6.67 \times 10^{-11}\ Nm^2/kg^2)(5.98 \times 10^{24}\ kg)(100\ kg)}{(8.38 \times 10^6\ m)^2} = 5.68 \times 10^2$ N

7.56 (a) The escape velocity from the moon is

$v = \sqrt{\dfrac{2GM}{R}}$ For the moon, $M = 7.36 \times 10^{22}$ kg and $R = 1.74 \times 10^6$ m.

$v = 2.38 \times 10^3$ m/s (5300 mph)

(b) The steps are the same for mercury, with $M = 3.18 \times 10^{23}$ kg and $R = 2.43 \times 10^6$ m.

$v = 4.18 \times 10^3$ m/s (9350 mph)

(c) For Jupiter, $M = 1.90 \times 10^{27}$ kg, $R = 6.99 \times 10^7$ m

and v is found to be 6.02×10^4 m/s (about 135,000 mph).

7.57 (a) We have the expression for the gravity on a body as $g = \dfrac{GM}{R^2}$. We write this equation down twice, once for the neutron star and once for the earth and divide the two. (See problem 45.) The result is

$\dfrac{g_{star}}{g_{earth}} = \dfrac{M_s}{M_e}\left(\dfrac{R_e}{R_s}\right)^2 = \dfrac{2.99 \times 10^{30}}{5.98 \times 10^{24}}\left(\dfrac{6.38 \times 10^6}{10^4}\right)^2$

which gives a value of g on the star of 1.99×10^{12} m/s^2.

(b) $w = mg = (0.12\ kg)(1.99 \times 10^{12}\ m/s^2) = 2.38 \times 10^{11}$ N

(c) $PE = mgh = (70\ kg)(1.99 \times 10^{12}\ m/s^2)(10^{-2}\ m) = 1.39 \times 10^{12}$ J

7.58 $\omega = 0.5$ rev/s $= \pi$ rad/s

(a) $v_t = r\omega = (0.8\ m)(\pi\ rad/s) = 2.51$ m/s

(b) $a_r = r\omega^2 = v_t^2/r = (2.51\ m/s)^2/0.8\ m = 7.90\ m/s^2$.

(c) $F = \dfrac{mv^2}{r}$ If the maximum value of T is 100 N, then we have

$100\ N = \dfrac{(5\ kg)v^2}{0.8\ m}$

From which, $v = 4.00$ m/s

7.59 Since the centripetal force on the satellite of mass m is the gravitational force exerted on it by Mars, mass M, we have

$\dfrac{mv^2}{r} = \dfrac{GMm}{r^2}$

From this, $v^2 = \dfrac{GM}{r}$.   (1)

But we know that the velocity of the satellite is found from

$v = \dfrac{2\pi r}{T}$.   (2)   Eliminating v by substituting from (2) into (1) gives

$r^3 = \dfrac{GMT^2}{4\pi^2}$ For $M = 6.42 \times 10^{23}$ kg and T = 459 min = 2.754×10^4 s, we find

$r = 9.37 \times 10^6$ m = 5823 mi

CHAPTER SEVEN SOLUTIONS

7.60 The distance of the satellite from the center of the earth is $r = R_E + h$, where R_E is the radius of the earth and h is the height of the satellite above the earth. We find $r = 6.62 \times 10^6$ m. Now use

$$v^2 = \frac{GM}{r}.$$ (See problem 59.) With $M = 5.98 \times 10^{24}$ kg, we have

(a) $v = 7.79 \times 10^3$ m/s (b) $T = \frac{2\pi r}{v} = \frac{2\pi(6.62 \times 10^6 \text{ m})}{7.77 \times 10^3 \text{ m/s}} = 5.36 \times 10^3$ s = 89.3 min

---

7.61 (a) Use $\theta = \omega_0 t + \frac{1}{2}\alpha t^2$.

$\theta = (3 \text{ rad/s})(2 \text{ s}) + \frac{1}{2}(1.5 \text{ rad/s}^2)(2 \text{ s})^2$

$\theta = 9$ rad $= 516°$ or $\theta = 516° - 360° = 156°$

(b) The angular velocity at 2 s is
$\omega = \omega_0 + \alpha t$
$\omega = 3$ rad/s $+ (1.5 \text{ rad/s}^2)(2 \text{ s}) = 6$ rad/s.

7.62 (a) We use the definition of angular velocity as
$\omega = \frac{\Delta\theta}{\Delta t} = \frac{2\pi \text{ rad}}{T} = \frac{2\pi \text{ rad}}{200.2 \times 10^{-3} \text{ s}} = 31.4$ rad/s

(b) The radius of the disk is 6.67×10^{-2} m.
$v_t = r\omega = (6.67 \times 10^{-2} \text{ m})(31.4 \text{ rad/s}) = 2.09$ m/s

---

7.63 (a) $\omega = (1200 \text{ rev/min})(1 \text{ min}/60 \text{ s})(2\pi \text{ rad/rev}) = 125.6$ rad/s
(b) $v_t = r\omega = (2 \times 10^{-2} \text{ m})(125.6 \text{ rad/s}) = 2.51$ m/s
(c) $a_r = r\omega^2 = 947$ m/s^2.
(d) $\theta = \omega t = (125.6 \text{ rad/s})(2 \text{ s}) = 251$ rad, and
$s = r\theta = (6 \times 10^{-2} \text{ m})(251 \text{ rad}) = 15.1$ m

---

7.64 (a) The force F_1 exerted on the 4 kg mass by the 2 kg mass at the origin is directed to the left and so is the force F_2 exerted on the 4 kg mass by the 3 kg mass. Thus, the resultant force is to the left and is equal to the sum of F_1 and F_2. (See Figure 1 below.)

$F_{net} = F_1 + F_2$

$= \frac{6.67 \times 10^{-11} \text{ Nm}^2/\text{kg}^2 (2 \text{ kg})(4 \text{kg})}{(4 \text{ m})^2} + \frac{6.67 \times 10^{-11} \text{ Nm}^2/\text{kg}^2 (3 \text{ kg})(4 \text{ kg})}{(2 \text{ m})^2}$

$= 2.34 \times 10^{-10}$ N   in -x direction

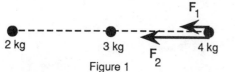
Figure 1

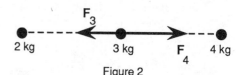
Figure 2

(b) The force exerted on the 3 kg mass by the 2 kg mass is to the left and the force exerted by the 4 kg mass is to the right. (See Figure 2 above.) The net force is

$F_{net} = \frac{6.67 \times 10^{-11} \text{ Nm}^2/\text{kg}^2 (3 \text{ kg})(4 \text{kg})}{(2 \text{ m})^2} - \frac{6.67 \times 10^{-11} \text{ Nm}^2/\text{kg}^2 (3 \text{ kg})(2 \text{ kg})}{(2 \text{ m})^2}$

$= 1.00 \times 10^{-10}$ N   (in +x direction)

CHAPTER SEVEN SOLUTIONS

7.65 From $v_t = r\omega$ we have
$$\omega = \frac{v_t}{r} = \frac{1.2 \text{ m/s}}{0.2 \text{ m}} = 6 \text{ rad/s}$$

7.66 The centripetal force is supplied by the force of friction between tires and road. When the car is on the verge of slipping,
$$F_c = f = \mu mg = \frac{mv^2}{r}$$
From this, $\mu = \frac{v^2}{rg} = \frac{(17.9 \text{ m/s})^2}{(150 \text{ m})(9.8 \text{ m/s}^2)} = 0.218$

7.67 From $\Sigma F_y = 0$, we have
$$N \cos\theta = mg \quad (1)$$ where N is the normal force exerted on the car by the road.
From $F = \frac{mv^2}{r}$, we have
$$N \sin\theta = \frac{mv^2}{r} \quad (2)$$
Divide (2) by (1) to get
$$\tan\theta = \frac{v^2}{rg} = \frac{(17.9 \text{ m/s})^2}{(150 \text{ m})(9.8 \text{ m/s}^2)} = 0.218 \text{ and } \theta = 12.3°$$

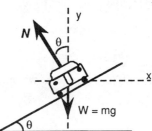

7.68 We have $r = 1.32$ m and $\theta = 41.4°$
The tensions in the strings are not equal. For the y direction, we have
$$T_1\cos\theta - T_2\cos\theta = mg \quad (1)$$
And for the x direction, $(T_1 + T_2)\sin\theta = \frac{mv^2}{r}$ (2)
Solving (1) and (2) we get $T_1 = m(v^2/r)/2\sin\theta + mg/2\cos\theta$
$= (4 \text{ kg})[(6 \text{ m/s})^2/1.32 \text{ m}]/2\sin 41.4° + (4 \text{ kg})(9.80 \text{ m/s}^2)/2\cos 41.4° = 108.77$ N
and $T_2 = 56.51$ N

7.69 r is the distance from the axis and θ is the angle between the strings and axis. From the given geometry, we see that $r = 1.32$ m and $\theta = 41.4°$
By symmetry, the tensions in the two strings are equal.

(a) At the lowest point, we have $-mg + 2T \sin\theta = \frac{mv^2}{r}$ which becomes
$$-(4 \text{ kg})(9.8 \text{ m/s}^2) + 2T\sin 41.4° = \frac{(4 \text{ kg})(4 \text{ m/s})^2}{1.32 \text{ m}}$$
From which, $T = 66.3$ N

(b) In the horizontal position $2T \sin\theta = \frac{mv^2}{r}$ From this, $T = 36.7$ N

7.70 (a) The forces acting on the car at the top of its circular arc are the normal force, N, and the weight. We have $F = \frac{mv^2}{r} = mg - N$, or
$$N = mg - \frac{mv^2}{r} \quad (1)$$
(b) When the normal force goes to zero, we have, from (1)
$$v^2 = rg = (30 \text{ m})g \quad \text{From which,}$$

103

CHAPTER SEVEN SOLUTIONS

$v = 17.1$ m/s

7.71 (a) Consider a person at the equator standing on a pair of scales. There are two forces on him W, his true weight, and the upward force exerted on him by the scale, W', which we shall call his apparent weight since it, by Newton's third law, is what the scale will read. We have

$$W - W' = \frac{mv^2}{r} \quad \text{or} \quad W' = W - \frac{mv^2}{r} \quad \text{so, } W > W'$$

(b) At the poles $v = 0$, and $W' = W = mg = (75 \text{ kg})(9.80 \text{ m/s}^2) = 735$ N
At the equator $W' = 735$ N $- (75 \text{ kg})(0.034 \text{ m/s}^2) = 732.5$ N

7.72 By conservation of energy: $\Delta PE + \Delta KE = 0$ or $(-mgh - 0) + (\frac{1}{2}mv^2 - 0) = 0$

Also, the centripetal force acting on the skier is $mg\cos\theta = mg(R - \frac{h}{R})$ and $a = \frac{v^2}{R}$

Solving these three equations gives $h = \frac{R}{3}$

7.73 (a) At A we must have $mg = \frac{mv_A^2}{R}$ in order for the coaster to remain on the track. By energy conservation we have $\Delta PE + \Delta KE = 0$

or $-\frac{mgh}{3} + \frac{m(v_A^2 - v_0^2)}{2} = 0$

Substituting for v_A and solving we find $v_0 = \sqrt{g(R - \frac{2h}{3})}$

(b) If the coaster just makes it to point B, then $v_B = 0$

By conservation of energy $[0 - \frac{mv_0^2}{2}] + mg(h' - h) = 0$

Substituting for v_0 gives $h' = \frac{R}{2} + \frac{2h}{3}$

7.74 If the car is about to slip down the plane,

$$-mg\sin\theta + f = \frac{mv^2}{r}\cos\theta \quad \text{where} \quad f = \mu[mg\cos\theta + \frac{mv^2}{r}\sin\theta]$$

From the above, we find $v_{min} = [Rg(\tan\theta - \mu)/(1 + \mu\tan\theta)]^{1/2}$

If the car slips up the plane, then $-mg\sin\theta - f = \frac{mv^2}{r}\cos\theta$

and $v_{max} = [Rg(\tan\theta + \mu)/(1 - \mu\tan\theta)]^{1/2}$

(b) $v_{min} = [(100)(9.80)(\tan 10° - 0.1)/(1 + 0.1 \tan 10°)]^{1/2} = 8.57$ m/s

$v_{max} = [(100)(9.80)(\tan 10° + 0.1)/(1 - 0.1 \tan 10°)]^{1/2} = 16.6$ m/s

---

7.75 We know $f = \mu_s N$ where f must equal mg to support the rider. The normal force is $N = m\omega^2 r$

Thus, $\mu_s = \frac{g}{\omega^2 r} = \frac{(9.8 \text{ m/s}^2)}{(5 \text{ rad/s})^2 (3 \text{ m})} = 0.131$

CHAPTER EIGHT SOLUTIONS

8.1 In order to exert the minimum force, the force must be applied perpendicular to the wrench. With the pivot at the nut, we have

$\tau = Fd$

or $\quad 40$ N m $= F(0.3$ m$)$

So, $\quad F = 133$ N

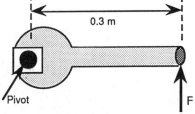

8.2 First, break the 100 N applied force into components along the pole and perpendicular to it, as shown.
$F_y = (100$ N$)\sin 20° = 34.2$ N
F_x has zero torque about the pivot point selected. Thus,

$\tau = (34.2$ N$)(2$ m$) = 68.4$ N m
(clockwise)

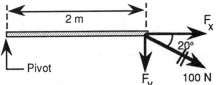

8.3 Resolve the 100 N force into components perpendicular to and parallel to the rod, as
$F_{par} = (100$ N$)\cos 57° = 54.5$ N
and $\quad F_{perp} = (100$ N$)\sin 57° = 83.9$ N
Torque of $F_{par} = 0$ since its line of action passes through the pivot point.
Torque of F_{perp} is

$\tau = (83.9$ N$)(2$ m$) = 168$ N m
(clockwise)

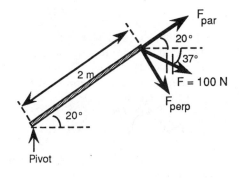

8.4 The weight of the diver is 470 N, and the torque about the pivot point through the center of the board is

$\tau = (470$ N$)(1.5$ m$) = 705$ N m (clockwise)

8.5 The lever arm is $d = (1.2 \times 10^{-2}$ m$)\cos 48° = 8.03 \times 10^{-3}$ m, and the torque is

$\tau = Fd = (80.0$ N$)(8.03 \times 10^{-3}$ m$) = 0.642$ N m counterclockwise

---

8.6 The components of F_1 are $F_{1x} = 94.0$ N and $F_{1y} = 34.2$ N, and the components of F_2 are $F_{2x} = 233$ N and $F_{2y} = 869$ N.

$\Sigma \tau_A = (-34.2$ N$)(0.6$ m$) - (233$ N$)(0.8$ m$) = -207$ N m (a clockwise torque)

$\Sigma \tau_B = (-800$ N$)(0.6$ m$) - (233$ N$)(0.8$ m$) + (869$ N$)(0.6$ m$) = -145$ N m (a clockwise torque)

$\Sigma \tau_C = (-94.0$ N$)(0.8$ m$) - (34.2$ N$)(0.6$ m$) = -95.7$ N m (a clockwise torque)

CHAPTER EIGHT SOLUTIONS

8.7 From $\Sigma\tau = 0$, we have

$(400 \text{ N})(2 \text{ m}) - (475 \text{ N})x = 0$

$x = 1.68 \text{ m}$

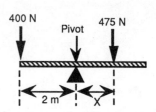

8.8 From $\Sigma\tau = 0$, we have

$(400 \text{ N})(1.5 \text{ m}) - W(0.5 \text{ m}) = 0$

$W = 1200 \text{ N}$

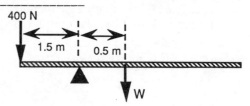

8.9 Taking torques about the left end of the scaffold, we have
$T_1(0) - (700 \text{ N})(1 \text{ m}) - (200 \text{ N})(1.5 \text{ m}) + T_2(3 \text{ m}) = 0$
From which, $T_2 = 333 \text{ N}$
Then from $\Sigma F_y = 0$, we have

$T_1 + T_2 - 700 \text{ N} - 200 \text{ N} = 0$

Since $T_2 = 333 \text{ N}$, we find
$T_1 = 567 \text{ N}$

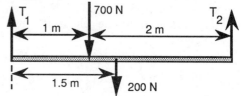

8.10 From $\Sigma\tau = 0$ about the pivot shown, we have
$(T\sin 30°)d - (200 \text{ N})d = 0$
where d is the length of the beam.
The forces H, V, and $T_x = T\cos 30°$ produce no torque about the pivot point. We find
$T = 400 \text{ N}$
(b) From $\Sigma F_x = 0$, we have
$H - T\cos 30° = 0$
$T = 400 \text{ N}$, so $H = 346 \text{ N}$ (to right)
From $\Sigma F_y = 0$,
$V + T\sin 30° - 200 \text{ N} = 0 \qquad V = 0$

8.11 (a) See the diagram below:

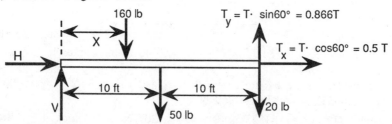

(b) If $x = 3$ ft, then
$\Sigma\tau_{\text{left end}} = (-160 \text{ lb})(3 \text{ ft}) - (50 \text{ lb})(10 \text{ ft}) - (20 \text{ lb})(20 \text{ ft}) + (0.866T)(20 \text{ ft}) = 0$
From which, $T = 79.7 \text{ lb}$
From $\Sigma F_x = 0$, $H = 0.5 T = 39.8 \text{ lb}$

CHAPTER EIGHT SOLUTIONS

From $\Sigma F_y = 0$, $V = 230 \text{ lb} - 0.866T = 161 \text{ lb}$
(c) If $T = 200 \text{ lb}$
$\Sigma \tau_{\text{left end}} = (-160 \text{ lb})(x) - (50 \text{ lb})(10 \text{ ft}) - (20 \text{ lb})(20 \text{ ft}) + (173.2 \text{ lb})(20 \text{ ft}) = 0$
From which, $x = 16.0 \text{ ft}$

8.12 We choose the pivot point at the left end, and use $\Sigma \tau = 0$.
$-(700 \text{ N})(0.5 \text{ m}) - (294 \text{ N})(1 \text{ m}) + (T_1 \sin 40°)(2 \text{ m}) = 0$
From which, $T_1 = 501 \text{ N}$
Now use $\Sigma F_x = 0$.
$T_1 \cos 40° - T_3 = 0$
With $T_1 = 501 \text{ N}$, this gives
$T_3 = 384 \text{ N}$
Finally, $\Sigma F_y = 0$ gives
$T_2 - 700 \text{ N} - 294 \text{ N} + T_1 \sin 40° = 0$
From which, $T_2 = 672 \text{ N}$

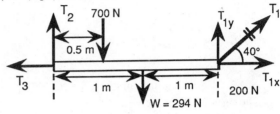

8.13

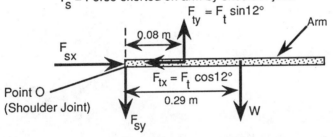 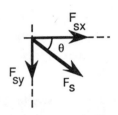

F_t = Tension force in Deltoid Muscle,
F_s = Force exerted on arm by shoulder joint

$\Sigma \tau_O = F_t \sin 12°(0.08 \text{ m}) - (41.5 \text{ N})(0.29 \text{ m}) = 0$
From which $F_t = 724 \text{ N}$ (Tension in deltoid muscle)
$\Sigma F_y = 0$ gives $-F_{sy} + F_t \sin 12° - 41.5 \text{ N} = 0$
yielding, $F_{sy} = 109 \text{ N}$
$\Sigma F_x = 0$ gives $F_{sx} = F_t \cos 12°$ and $F_{sx} = 709 \text{ N}$
Therefore, $F_s = \sqrt{(F_{sx})^2 + (F_{sy})^2} = 717 \text{ N}$
and $\tan \theta = \dfrac{F_{sy}}{F_{sx}} = 0.1539$ and $\theta = 8.75°$

CHAPTER EIGHT SOLUTIONS

8.14

(a) $\Sigma\tau_R = (m_1 g)(x_1) + (m_2 g)(x_2) + \circ \circ \circ$ (1)

(b) and $\Sigma\tau_R = W x_{cg}$ gives
$W x_{cg} = (Mg)(x_{cg}) = (m_1 g + m_2 g + \circ \circ \circ)(x_{cg})$ (2)

(c) Setting the right side of (1) equal to the right side of (2) and canceling g, we have

$$x_{cg} = \frac{\sum m_i x_i}{\sum m_i}$$

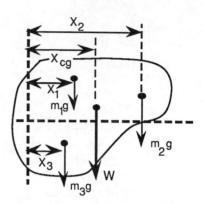

8.15 Choosing the pivot point at the point O shown, $\Sigma\tau = 0$ becomes
(50 N)(7.5 cm) + T(0) - R(3.5 cm) = 0
Thus, R = 107 N
Now, apply $\Sigma F_y = 0$.
  -50 N + T - 107 N = 0
and T = 157 N

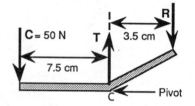

8.16 Let us first resolve all forces into components parallel to and perpendicular to the leg, as shown. Use $\Sigma\tau = 0$ about the pivot indicated.

$T_y(d/5) - C_y(d/2) - F_y(d) = 0$

d is the length of the lower leg.
$C_y = C \sin V = (30 \text{ N})\sin 40° = 19.3$ N, and
$F_y = F \sin V = (12.5 \text{ N})\sin 40° = 8.03$ N.
Thus, $T_y = 88.5$ N, but $T_y = T \sin 25°$
So, T = 209 N

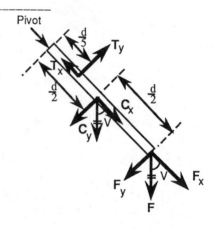

8.17

$\Sigma\tau_{\text{left end}} = 0$

$= (-1200 \text{ N})\frac{L}{2} \cos 65° + (T\cos 25°)(\frac{3L}{4} \sin 65°)$

$+ (T\sin 25°)(\frac{3L}{4} \cos 65°) - (2000 \text{ N})(L\cos 65°)$

From which, T = 1465 N
From $\Sigma F_x = 0$,

H = T cos 25° = 1328 N (toward right)
From $\Sigma F_y = 0$,

V = 3200 N - T sin 25° = 2581 N (upward)

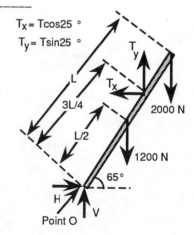

108

CHAPTER EIGHT SOLUTIONS

8.18 We call the tension in the cord at the left end of the sign, T_1 and the tension in the cord near the middle of the sign, T_2, and we choose our pivot point at the point where T_1 is attached.

$\Sigma\tau_{pivot} = 0 = (-w)(0.5\text{ m}) + T_2(0.75\text{ m}) = 0$

so, $T_2 = \frac{2}{3} w$

From $\Sigma F_y = 0$, $T_1 + T_2 - w = 0$
Substituting the expression for T_2 and solving, we find
$T_1 = \frac{1}{3} w$

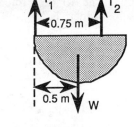

8.19 Use $\Sigma F_y = 0$.
$F_1 - 500\text{ N} - 800\text{ N} = 0$
$F_1 = 1300\text{ N}$ (1)

Now, apply $\Sigma F_x = 0$.
$f - F_2 = 0$
or $f = F_2$.   (2)

The lever arm for the 500 N force is
$(7.5\text{ m})\cos 60° = 3.75\text{ m}$
The lever arm for the 800 N force is
$d\cos 60° = d/2$
and for F_2, the lever arm is
$(15\text{ m})\sin 60° = 13\text{ m}$

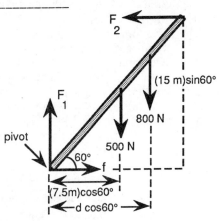

Using $\Sigma\tau = 0$ with pivot pt. at base of ladder,
$-(500\text{ N})(3.75\text{ m}) - (800\text{ N})(d/2) + F_2(13\text{ m}) = 0$ (3)

(a) When $d = 4$ m, Equation (3) gives a value for F_2 of
$F_2 = 267\text{ N}$
and we already know from (1) that $F_1 = 1300\text{ N}$.

(b) If $d = 9$ m, Equation (3) gives a value for F_2 as
$F_2 = 421\text{ N}$
Thus, from Equation (2), $f = 421\text{ N}$
and from eq (1) $F_1 = 1300\text{ N}$
If the ladder is on the verge of slipping, $f = \mu_s F_1$, and
$\mu_s = f/F_1 = 421\text{ N}/1300\text{ N} = 0.324$

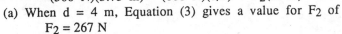

CHAPTER EIGHT SOLUTIONS

8.20 Use $\Sigma F_y = 0$.
$$F_1 - 200 \text{ N} - 800 \text{ N} = 0$$
Thus, $F_1 = 1000$ N
The friction force, f, at the base of the ladder is
$$f = \mu_s F_1$$
when the ladder is on the verge of slipping.
So, $f = 0.6(1000 \text{ N}) = 600$ N
Now use $\Sigma F_x = 0$.
$$f - F_2 = 0$$
So $F_2 = f = 600$ N

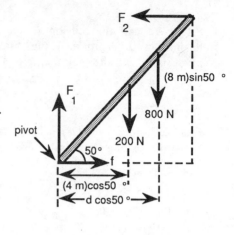

Finally, use $\Sigma \tau = 0$ with the pivot pt. at the base of the ladder.
The lever arm for the 200 N force is
$$(4m \cos 50°) = 2.57 \text{ m}$$
The lever arm for the 800 N force is $d\cos 50° = 0.643d$, where d is the distance from the base of the ladder up to the position of the person. Finally, the lever arm for the force F_2 is
$$(8 \text{ m}) \sin 50° = 6.13 \text{ m}$$
We have
$$-(200 \text{ N})(2.57 \text{ m}) - (800 \text{ N})(.643d) + (600 \text{ N})(6.13 \text{ m}) = 0$$
giving, $d = 6.15$ m

8.21 We are given that $V_1 = V_2 = w/2 = 125$ N.
Take $\Sigma \tau = 0$ about indicated pivot pt.
$$-(250 \text{ N})(0.5 \text{ m}) + H_2(2 \text{ m}) = 0$$
Giving, $H_2 = 62.5$ N
Now, use $\Sigma F_x = 0$.
$$H_1 - H_2 = 0$$
Therefore, $H_1 = 62.5$ N
The force on the lower hinge is a force of compression, and on the upper hinge the force is one of tension.

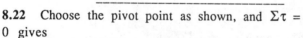

8.22 Choose the pivot point as shown, and $\Sigma \tau = 0$ gives
$$2F_R(0) - W(2 \text{ m}) + 2F_F(3 \text{ m}) = 0 \quad (1)$$
where F_R is the force on each rear tire and F_F is the force on each front tire. From (1),
$$F_F = W/3$$
Now, use $\Sigma F_y = 0$.
$$2F_R - W - 2F_F = 0$$
Since $F_F = W/3$, we find $F_R = W/6$ (where W = 15680 N)

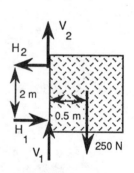

CHAPTER EIGHT SOLUTIONS

8.23 (a) When the door is horizontal, use $\Sigma\tau = 0$.

With the pivot point at the hinge, we have

$F(2 \text{ m}) - (360 \text{ N})(1 \text{ m}) = 0$

or $\quad F = 180 \text{ N}$

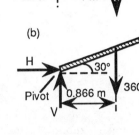

(b) At 30° to the horizontal, and the pivot point at the hinge, $\Sigma\tau = 0$ gives

$F(2 \text{ m}) - (360 \text{ N})(.866 \text{ m}) = 0$

where the lever arm for the 360 N force is

$(1 \text{ m}) \cos 30° = .866 \text{ m}$

We find, $\quad F = 156 \text{ N}$

8.24 $I = mr^2 = (4.5 \text{ kg})(2.5 \text{ m})^2 = 28.1 \text{ kg m}^2$.

8.25 $\tau = I\alpha = (28.1 \text{ kg m}^2)(2 \text{ rad/s}^2) = 56.3 \text{ N m}$

8.26 $I = \Sigma m_i r_i^2$

(a) First, apply the above equation about the x axis. We have,
$I_x = (3 \text{ kg})(9 \text{ m}^2) + (2 \text{ kg})(9 \text{ m}^2) + (2 \text{ kg})(9 \text{ m}^2) + (4 \text{kg})(9 \text{ m}^2) = 99.0 \text{ kg m}^2$.
(b) About the y axis, we have
$I_y = (3 \text{ kg})(4 \text{ m}^2) + (2 \text{ kg})(4 \text{ m}^2) + (2 \text{ kg})(4 \text{ m}^2) + (4 \text{kg})(4 \text{ m}^2) = 44.0 \text{ kg m}^2$.
(c) The distance, r, from an axis through O and perpendicular to the page out to each of the masses is found from the pythagorean theorem.

$r = \sqrt{(2 \text{ m})^2 + (3 \text{ m})^2} = \sqrt{13} \text{ m}^2$

and the moment of inertia is

$I_0 = (3 \text{ kg})(r^2) + (2 \text{ kg})(r^2) + (2 \text{ kg})(r^2) + (4 \text{kg})(r^2) = (11 \text{ kg})(13 \text{ m}^2)$
$I_0 = 143 \text{ kg m}^2$.

8.27 We use $\tau = I\alpha$.

$\tau_x = I_x \alpha = (99 \text{ kg m}^2)(1.5 \text{ rad/s}^2) = 149 \text{ N m}$

$\tau_y = I_y \alpha = (44 \text{ kg m}^2)(1.5 \text{ rad/s}^2) = 66.0 \text{ N m}$

$\tau_0 = I_0 \alpha = (143 \text{ kg m}^2)(1.5 \text{ rad/s}^2) = 215 \text{ N m}$

8.28 We have already found $I_y = 44.0 \text{ kg m}^2$, and we know $\tau = I\alpha$. Thus,

$$\alpha = \frac{\tau}{I} = \frac{(20 \text{ N m})}{(44 \text{ kg m}^2)} = 0.455 \text{ rad/s}^2.$$

and $\quad \omega = \omega_0 + \alpha t = 0 + (0.455 \text{ rad/s}^2)(3 \text{ s}) = 1.36 \text{ rad/s}$

8.29 (a) For a cylinder $I = \frac{1}{2} MR^2$, where M is the mass of the cylinder and R the radius.

$I = \frac{1}{2}(1.5 \text{ kg})(0.3 \text{ m})^2 = 6.75 \times 10^{-2} \text{ kg m}^2$.

(b) For a sphere, $I = \frac{2}{5} MR^2$, where M is the mass of the sphere and R the radius.

CHAPTER EIGHT SOLUTIONS

$$I = \frac{2}{5}(1.5 \text{ kg})(0.3 \text{ m}^2) = 5.40 \times 10^{-2} \text{ kg m}^2.$$

8.30 $\omega_0 = 2$ rev/s $= 4\pi$ rad/s. Let us now find the angular acceleration of the cylinder.
$$\omega = \omega_0 + \alpha t$$
$$0 = 4\pi \text{ rad/s} + \alpha(15 \text{ s})$$
$$\alpha = -0.838 \text{ rad/s}^2.$$
and $\tau = I\alpha$ gives $\tau = (6.75 \times 10^{-2} \text{ kg m}^2)(-0.838 \text{ rad/s}^2) = -5.65 \times 10^{-2}$ N m

---

8.31 (a) $I = \frac{1}{2} mr^2 = \frac{1}{2}(0.85 \text{ kg})(4 \times 10^{-2} \text{ m})^2 = 6.8 \times 10^{-4}$ kg m^2
$\tau_{net} = I\alpha = (6.8 \times 10^{-4} \text{ kg m}^2)(66 \text{ rad/s}^2) = 4.49 \times 10^{-2}$ kg m^2
The torque exerted by the fish = Fr, so the net torque is
$$\tau_{net} = Fr - 1.3 \text{ Nm} = 4.49 \times 10^{-2} \text{ kg m}^2$$
From which, $F = \dfrac{1.345 \text{ Nm}}{4 \times 10^{-2} \text{ m}} = 33.6$ N

(b) $\theta = \omega_0 t + \frac{1}{2}\alpha t^2 = 0 + \frac{1}{2}(66 \text{ rad/s}^2)(0.5 \text{ s})^2 = 8.25$ rad
and $s = r\theta = (4 \times 10^{-2} \text{ m})(8.25 \text{ rad}) = 0.33$ m = 33 cm

---

8.32 (a) $I = \frac{1}{2} mr^2 = \frac{1}{2}(100 \text{ kg})(0.5 \text{ m})^2 = 12.5$ kg m^2
$\omega_0 = 50$ rev/min $= 5.24$ rad/s
$$\alpha = \frac{\omega - \omega_0}{t} = \frac{0 - 5.24 \text{ rad/s}}{6 \text{ s}} = -0.873 \text{ rad/s}^2$$
$\tau = I\alpha = (12.5 \text{ kg m}^2)(-0.873 \text{ rad/s}^2) = -10.9$ Nm
Also, the magnitude of the torque is given by fr = 10.9 N m, where f is the force of friction.
So, $f = \dfrac{10.9 \text{ Nm}}{0.5 \text{ m}} = 21.8$ N
and $f = \mu_k N$
$$\mu_k = \frac{f}{N} = \frac{21.8 \text{ N}}{70 \text{ N}} = 0.312$$

8.33 We first calculate the moment of inertia as
$$I = \frac{1}{2} MR^2 = \frac{1}{2}(150 \text{ kg})(1.5 \text{ m})^2 = 168.8 \text{ kg m}^2.$$
Then we note that $\omega_f = 0.5$ rev/s $= \pi$ rad/s, and calculate α as,
$\omega = \omega_0 + \alpha t$ or π rad/s $= 0 + \alpha(2 \text{ s})$. Thus, $\alpha = \pi/2$ rad/s^2.
Then $\tau = I\alpha$ becomes $Fr = I\alpha$, or $F = \dfrac{I\alpha}{r} = \dfrac{(169 \text{ kg m}^2)(\pi/2 \text{ rad/s}^2)}{1.5 \text{ m}} = 177$ N

CHAPTER EIGHT SOLUTIONS

The moment of inertia of the pulley is
$$I = \tfrac{1}{2} MR^2 = \tfrac{1}{2}(5 \text{ kg})(0.6 \text{ m})^2 = 0.90 \text{ kg m}^2.$$

First we apply Newton's second law to the falling bucket of mass 3 kg (weight 29.4 N). We have
$$29.4 \text{ N} - T = (3 \text{ kg})a \quad (1)$$

Now apply $\tau = I\alpha$ to the pulley. We have
$$TR = I\alpha = I\left(\frac{a}{R}\right) \text{ Thus,} \quad T = \frac{Ia}{R^2} = \frac{0.90 \text{ kg m}^2}{(.6 \text{ m})^2} a \quad (2)$$

We can solve (1) and (2) simultaneously to find a. We find,

(a) $a = 5.35 \text{ m/s}^2$ downward.

(b) To find how far it drops, we use $y = v_0 t + \tfrac{1}{2} at^2$
$$y = 0 + \tfrac{1}{2}(5.35 \text{ m/s}^2)(4 \text{ s})^2 = 42.8 \text{ m}$$

(c) $\alpha = \dfrac{a}{R} = \dfrac{5.35 \text{ m/s}^2}{0.6 \text{ m}} = 8.91 \text{ rad/s}^2.$

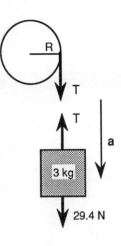

8.35 The resultant torque is given by
$$(120 \text{ N})(0.811 \text{ m}) - (100 \text{ N})(0.811 \text{ m}) = 16.2 \text{ N m}$$
$$I = \tfrac{1}{2} mr^2 = \tfrac{1}{2}(2.1 \text{ kg})(0.811 \text{ m})^2 = 0.691 \text{ kg m}^2$$

$\tau = I\alpha$ gives $\alpha = \dfrac{\tau}{I} = \dfrac{16.2 \text{ Nm}}{0.691 \text{ kg m}^2} = 23.5 \text{ rad/s}^2$

8.36 The free body diagram is shown at the right. The resultant torque about the center of the pulley is

$$(1.14 \times 10^4 \text{ N})r - T_1 r = (79.8 \text{ kg m}^4)\alpha \quad (1)$$

From Newton's second law, applied in the vertical direction, we have

$$T_1 - 9800 \text{ N} = (1000 \text{ kg})a \quad (2)$$

Since there is no slipping, we also have
$$a = r\alpha = (0.762 \text{ m})\alpha \quad (3)$$
Solving (1), (2), and (3) simultaneously, we find
$$a = 1.41 \text{ m/s}^2$$

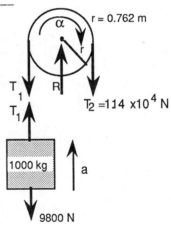

8.37 (a) Newton's second law for the cylinder is
$$mg - T = ma \quad (1)$$

(b) $\tau = I\alpha$ becomes
$$Tr = \tfrac{1}{2} mr^2 \alpha \quad (2)$$

(c) Solve (2) for T and substitute $\alpha = a/r$. We find
$$T = \frac{mr^2 a}{2r^2} = \frac{ma}{2} \quad (3)$$

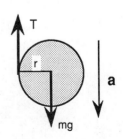

113

CHAPTER EIGHT SOLUTIONS

Finally, substitute (3) into (1) to find,
$$mg - \frac{ma}{2} = ma \quad \text{which gives } a = \frac{2}{3}g$$

8.38 (a) We have 4 rev = 8π rad. We first find the angular acceleration of the grindstone.
$$\omega^2 = \omega_0^2 + 2\alpha\theta$$
$$0 = (8 \text{ rad/s})^2 + 2a(8\pi \text{ rad})$$
$$\alpha = -4/\pi \text{ rad/s}^2.$$

We now use $\tau = I\alpha$
$$Fr = \frac{1}{2}mr^2\left(\frac{4 \text{ rad/s}^2}{\pi}\right)$$ Where F is the friction force on the grindstone and $I = \frac{1}{2}mr^2$ (m is the mass of the grindstone, and r is its radius). With m = 5 kg and r = 0.4 m, we find,
$$F = 1.27 \text{ N}$$
(b) We have $f = \mu N$. Thus,
$$N = f/\mu = 1.27 \text{ N}/0.4 = 3.18 \text{ N}$$

8.39 The initial angular velocity of the wheels is zero, and the final angular velocity is
$$\omega_f = \frac{v}{r} = \frac{50.0 \text{ m/s}}{1.25 \text{ m}} = 40.0 \text{ rad/s}$$
so, $\quad \alpha = \frac{\omega_f - \omega_0}{t} = \frac{40.0 \text{ rad/s} - 0}{0.48 \text{ s}} = 83.3 \text{ rad/s}^2$

$\tau_{\text{center of a wheel}} = I\alpha$ is $\quad fr = I\alpha$

or $\quad f = \frac{I\alpha}{r} = \frac{(110 \text{ kg m}^2)(83.3 \text{ rad/s}^2)}{1.25 \text{ m}} = 7.33 \times 10^3 \text{ N}$

and $\quad \mu_k = \frac{f}{N} = \frac{7.33 \times 10^3 \text{ N}}{1.4 \times 10^4 \text{ N}} = 0.524$

8.40 (Note that 2 rev/s = 4π rad/s.) Calculate the moment of inertia as
$$I = \frac{1}{2}mr^2 = \frac{1}{2}(6 \text{ kg})(0.35 \text{ m})^2 = 3.68 \times 10^{-1} \text{ kg m}^2.$$
and $\quad KE = \frac{1}{2}I\omega^2 = \frac{1}{2}(3.68 \times 10^{-1} \text{ kg m}^2)(4\pi \text{ rad/s})^2 = 29.0 \text{ J}$

8.41 (a) The moment of inertia of the rod when rotating about an axis through its center is
$$I = \frac{1}{12}mL^2 = \frac{1}{12}(0.1 \text{ kg})(5 \times 10^{-2} \text{ m})^2 = 2.08 \times 10^{-5} \text{ kg m}^2.$$
and its kinetic energy is
$$KE = \frac{1}{2}I\omega^2 = \frac{1}{2}(2.08 \times 10^{-5} \text{ kg m}^2)(3 \text{ rad/s})^2 = 9.38 \times 10^{-5} \text{ J}$$
(b) When the rotation axis is at one end, we have
$$I = \frac{1}{3}mL^2 = \frac{1}{3}(0.1 \text{ kg})(5 \times 10^{-2} \text{ m})^2 = 8.33 \times 10^{-5} \text{ kg m}^2.$$
and
$$KE = \frac{1}{2}I\omega^2 = \frac{1}{2}(8.33 \times 10^{-5} \text{ kg m}^2)(3 \text{ rad/s})^2 = 3.75 \times 10^{-4} \text{ J}$$

CHAPTER EIGHT SOLUTIONS

8.42 (a) $I = \Sigma mr^2 = (4 \text{ kg})(3 \text{ m})^2 + (2 \text{ kg})(2 \text{ m})^2 + (3 \text{ kg})(4 \text{ m})^2 = 92.0 \text{ kg m}^2$

$KE = \frac{1}{2} I\omega^2 = \frac{1}{2} (92.0 \text{ kg m}^2)(2 \text{ rad/s})^2 = 184 \text{ J}$

(b) $v_4 = r_4\omega = (3 \text{ m})(2 \text{ rad/s}) = 6 \text{ m/s}$
$v_2 = r_2\omega = (2 \text{ m})(2 \text{ rad/s}) = 4 \text{ m/s}$
$v_3 = r_3\omega = (4 \text{ m})(2 \text{ rad/s}) = 8 \text{ m/s}$
$KE = \Sigma \frac{1}{2} mv^2 = \frac{1}{2}(4 \text{ kg})(6 \text{ m/s})^2 + \frac{1}{2}(2 \text{ kg})(4 \text{ m/s})^2 + \frac{1}{2}(3 \text{ kg})(8 \text{ m/s})^2 = 184 \text{ J}$

8.43 (a) $I_x = 99.0 \text{ kg m}^2$ (See problem 26).

$KE = \frac{1}{2} I\omega^2 = \frac{1}{2}(99.0 \text{ kg m}^2)(2.5 \text{ rad/s})^2 = 309 \text{ J}$

(b) $I_y = 44.0 \text{ kg m}^2$ (See problem 26).

$KE = \frac{1}{2} I\omega^2 = \frac{1}{2}(44.0 \text{ kg m}^2)(2.5 \text{ rad/s})^2 = 138 \text{ J}$

8.44 The moment of inertia of the cylinder is

$I = \frac{1}{2} mr^2 = \frac{1}{2}(81.6 \text{ kg})(1.5 \text{ m})^2 = 91.8 \text{ kg m}^2.$

and the angular acceleration of the merry-go-round is found as

$\alpha = \tau/I = (F r)/I = (50 \text{ N})(1.5 \text{ m})/(91.8 \text{ kg m}^2) = 0.817 \text{ rad/s}^2.$

At $t = 3$ s, we find the angular velocity
$\omega = \omega_0 + \alpha t$
$\omega = 0 + (0.817 \text{ rad/s}^2)(3 \text{ s}) = 2.45 \text{ rad/s}$

and $KE = \frac{1}{2} I\omega^2 = \frac{1}{2}(91.8 \text{ kg m}^2)(2.45 \text{ rad/s})^2 = 276 \text{ J}$

8.45 If the kinetic energy of the cylinder is equal to the kinetic energy of the sphere, we have,

$\frac{1}{2}(\frac{1}{2} m_c r_c^2)\omega_c^2 = \frac{1}{2}(\frac{2}{5} m_s r_s^2)\omega_s^2.$ Note that $m_c = m_s$ and $r_c = r_s$

From which, we find $\omega_c^2 = \frac{4}{5}\omega_s^2 = \frac{4}{5}(9 \text{ rad/s})^2 = 64.8 \text{ rad}^2/s^2$, and

$\omega_c = 8.05 \text{ rad/s}$

8.46 (a) $KE_{trans} = \frac{1}{2} mv^2 = \frac{1}{2}(10 \text{ kg})(10 \text{ m/s})^2 = 500 \text{ J}$

(b) $KE_{rot} = \frac{1}{2} I\omega^2 = \frac{1}{2}(\frac{1}{2} mr^2)(\frac{v^2}{r^2}) = \frac{1}{4}(10 \text{ kg})(10 \text{ m/s})^2 = 250 \text{ J}$

(c) $KE_{total} = KE_{trans} + KE_{rot} = 750 \text{ J}$

8.47 $I = \frac{2}{5} mr^2 = \frac{2}{5}(150 \text{ kg})(0.2 \text{ m})^2 = 2.4 \text{ kg m}^2.$

The sphere rolls without slipping, so $v = r\omega = (0.2 \text{ m})(50 \text{ rad/s}) = 10 \text{ m/s}$

Work $= \Delta KE = KE_f - KE_i = \frac{1}{2} mv_f^2 + \frac{1}{2} I\omega_f^2 - 0$

$= \frac{1}{2}(150 \text{ kg})(10 \text{ m/s})^2 + \frac{1}{2}(2.4 \text{ kg m}^2)(50 \text{ rad/s})^2 = 10.5 \text{ kJ}$

CHAPTER EIGHT SOLUTIONS

8.48 The forces on a rolling object on an incline of slope angle θ are the force of friction, f, the normal force, N, and its weight, mg. We apply Newton's second law with the x axis along the plane, to find,
$$mg\sin\theta - f = ma \qquad (1)$$
Now use, $\tau = I\alpha$
$$fr = I\alpha = I(a/r)$$
Therefore, $\quad f = \dfrac{I}{r^2} a \qquad (2)$

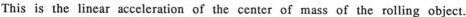

Substitute into (1) for the frictional force f from (2) and solve the resulting equation for a.
$$a = \frac{g\sin\theta}{(1 + \dfrac{I}{mr^2})}$$

This is the linear acceleration of the center of mass of the rolling object.

For a solid sphere, $I = \dfrac{2}{5} mr^2$ or $\dfrac{I}{mr^2} = \dfrac{2}{5}$

From this, we see that $a = g\sin\theta/1.4$.

Using an identical approach for a solid cylinder, $\dfrac{I}{mr^2} = \dfrac{1}{2}$
and $\quad a = g\sin\theta/1.5$

Finally, for a hollow ring, $\dfrac{I}{mr^2} = 1$
and $\quad a = g\sin\theta/2$

Thus, we find $a_{sphere} > a_{cylinder} > a_{ring}$. Thus, the sphere wins and the ring comes in last.

8.49 (a) The angular velocity of the flywheel is 5000 rev/min = 524 rad/s, and its moment of inertia is,
$$I = \frac{1}{2} mr^2 = \frac{1}{2} (500 \text{ kg})(2 \text{ m})^2 = 10^3 \text{ kg m}^2.$$
Therefore, the stored energy is
$$KE = \frac{1}{2} I\omega^2 = \frac{1}{2} (10^3 \text{ kg m}^2)(524 \text{ rad/s})^2 = 1.37 \times 10^8 \text{ J}$$

(b) A 10 hp motor is equivalent to 7460 W. Therefore, if energy is used at the rate of 7460 J/s, the stored energy will last for
$$t = E/P = 1.37 \times 10^8 \text{ J}/7.46 \times 10^3 \text{ J/s} = 1.84 \times 10^4 \text{ s} = 5.10 \text{ h}$$

8.50 Work done = $Fs = (5.57 \text{ N})(0.8 \text{ m}) = 4.46 \text{ J}$
and Work $= \Delta KE = \dfrac{1}{2} I\omega_f^2 - \dfrac{1}{2} I\omega_0^2$ (The last term is zero because the top starts from rest.)
Thus, $4.46 \text{ J} = \dfrac{1}{2} (4 \times 10^{-4} \text{ kg m}^2)\omega_f^2$
and from this, $\quad \omega_f = 149 \text{ rad/s}$

8.51 $I = \dfrac{2}{5} mr^2 = \dfrac{2}{5} (5.98 \times 10^{24} \text{ kg})(6.38 \times 10^6 \text{ m})^6 = 9.74 \times 10^{37} \text{ kg m}^2$

The initial period = 24 h = 86,400 s, and $\quad \omega_0 = \dfrac{2\pi \text{ rad}}{86400 \text{ s}} = 7.27 \times 10^{-5}$ rad/s

CHAPTER EIGHT SOLUTIONS

The final period = 24 h + 1 min = 86,460 s, $\quad \omega_f = \dfrac{2\pi \text{ rad}}{86460 \text{ s}} = 7.267 \times 10^{-5}$ rad/s

The energy available = $\Delta KE = \dfrac{1}{2} I\omega_0^2 - \dfrac{1}{2} I\omega_f^2$

Substituting the appropriate values gives, $\quad \Delta KE = 3.57 \times 10^{26}$ J

and $\quad$ time = $\dfrac{\text{energy available}}{\text{rate of energy use}} = \dfrac{3.57 \times 10^{26} \text{ J}}{2 \times 10^{20} \text{ J/y}} = 1.79 \times 10^6$ y

---

8.52 Let us first consider conservation of energy, with the zero level for potential energy at the base of the ramp. We have

$$\tfrac{1}{2} mv_i^2 + \tfrac{1}{2} I\omega_i^2 + mgy_i = \tfrac{1}{2} mv_f^2 + \tfrac{1}{2} I\omega_f^2 + mgy_f$$

$$\tfrac{1}{2} mv_i^2 + \tfrac{1}{2} I\omega_i^2 + 0 = 0 + 0 + (mg)(s)\sin 20°$$

where s is the distance the hoop rolls up the ramp.
For a hoop, $I = mr^2$. Thus,
$\quad I\omega_i^2 = mr^2\omega_i^2 = m(r\omega_i)^2 = mv_i^2$.
Therefore, our equation for conservation of energy becomes

$$\tfrac{1}{2} mv_i^2 + \tfrac{1}{2} mv_i^2 = (mg)(s)\sin 20°$$

or, solving for s, we have
$\quad s = v_i^2 / g\sin 20°$
We have $v_i = r\omega_i = (3 \text{ m})(3 \text{ rad/s}) = 9$ m/s, so
$\quad s = (9 \text{ m/s})^2 / (9.8 \text{ m/s}^2)\sin 20° = 24.2$ m

---

8.53 We shall first use conservation of energy with the zero level for potential energy chosen to be at the lowest point reached by the bucket, three meters below its starting position.

We have $I = \tfrac{1}{2} mr^2 = \tfrac{1}{2} (5 \text{ kg})(0.6 \text{ m})^2 = 0.90$ kg m^2.

$$\tfrac{1}{2} mv_i^2 + \tfrac{1}{2} I\omega_i^2 + mgy_i = \tfrac{1}{2} mv_f^2 + \tfrac{1}{2} I\omega_f^2 + mgy_f$$

$$0 + 0 + (3 \text{ kg})g(3 \text{ m}) = \tfrac{1}{2} (3 \text{ kg})(0.6 \text{ m})^2\omega^2 + \tfrac{1}{2} (0.90 \text{ kg m}^2)\omega^2 + 0$$

The substitution $v_f = r\omega$ has been used in the preceding equation. We find,
$\quad \omega^2 = 89.1$ rad^2/s^2
Thus, the kinetic energy of the pulley at this point is,
$\quad KE = \tfrac{1}{2} I\omega^2 = \tfrac{1}{2} (0.90 \text{ kg m}^2)(89.1 \text{ rad}^2/\text{s}^2) = 40.1$ J

8.54 (a) The angular velocity of the earth on its axis is
$\omega = 2\pi$ rad/day = 7.27×10^{-5} rad/s. We assume the earth is a uniform sphere of radius R and calculate its moment of inertia as
$\quad I = \tfrac{2}{5} mR^2 = \tfrac{2}{5} (5.98 \times 10^{24} \text{ kg})(6.38 \times 10^6 \text{ m})^2 = 9.74 \times 10^{37}$ kg m^2.
The angular momentum is found as
$\quad L = I\omega = (9.71 \times 10^{37} \text{ kg m}^2)(7.27 \times 10^{-5} \text{ rad/s}) = 7.08 \times 10^{33}$ J s
(b) For the orbiting earth, $\omega = 2\pi$ rad/yr = 1.99×10^{-7} rad/s, and the angular momentum is given by
$\quad L = I\omega = mr^2\omega = (5.98 \times 10^{24} \text{ kg})(1.49 \times 10^{11} \text{ m})^2 (1.99 \times 10^{-7} \text{ rad/s})$

CHAPTER EIGHT SOLUTIONS

$= 2.66 \times 10^{40}$ J s

8.55 (a) $\omega_{orbit} = \omega_{spin} = 2\pi$ rad/28 day $= 2.597 \times 10^{-6}$ rad/s. The orbital angular momentum is
$$L = I\omega = mr^2\omega = (7.36 \times 10^{22} \text{ kg})(3.84 \times 10^8 \text{ m})^2 (2.597 \times 10^{-6} \text{ rad/s})$$
$$= 2.82 \times 10^{34} \text{ J s}$$
(b) The spin angular momentum is
$$L = I\omega = \frac{2}{5}mR^2\omega = \frac{2}{5}(7.36 \times 10^{22} \text{ kg})(1.74 \times 10^6 \text{ m})^2 (2.597 \times 10^{-6} \text{ rad/s})$$
$$= 2.31 \times 10^{29} \text{ J s}$$

8.56 (a) We conserve angular momentum as
$I_f\omega_f = I_i\omega_i$ which gives
$$\omega_f = \frac{I_i}{I_f}\omega_i \quad (1)$$

However, we know $I_i = \frac{2}{5} mr_i^2$ and $I_f = \frac{2}{5} mr_f^2$. As a result, equation (1) reduces to

$$\omega_f = \frac{r_i^2}{r_f^2}\omega_i = \frac{\omega_i}{4}$$ (Where $r_f = 2r_i$) This result says the days would be 4 times longer.

8.57 We use conservation of angular momentum.
$I_f\omega_f = I_i\omega_i$
Thus, $\dfrac{I_f}{I_i} = \dfrac{\omega_i}{\omega_f} = \dfrac{0.5 \text{ rev/s}}{1.7 \text{ rev/s}} = 0.294$

8.58 We use conservation of angular momentum, with $I_i = I$, and $I_f = 0.9I$
$I_f\omega_f = I_i\omega_i$
$$\omega_f = \frac{I_i\omega_i}{\omega_f} = \frac{I}{0.9I}\omega_i = 1.111\omega_i$$

The % change in kinetic energy $= \left(\dfrac{KE_f - KE_i}{KE_i}\right)100\%$. Substituting appropriate values, gives
%change in KE $= 11.1\%$
The kinetic energy has increased because she has had to do work to pull her arms inward.

8.59 $I = I_{turntable} + I_{flies}$
where $I_{turntable} = \frac{1}{2} mr^2 = \frac{1}{2} (0.225 \text{ kg})r^2 = (0.1125 \text{ kg})r^2$
So $I_0 = (0.1125 \text{ kg})r^2 + 200(.002 \text{ kg})r^2 = (0.5125)r^2$
$I_f = (0.1125 \text{ kg})r^2 + 0$ (all flies are now on axis of rotation)
We now use conservation of angular momentum.
$I_f\omega_f = I_0\omega_0$
$$\omega_f = \frac{I_0\omega_0}{\omega_f} = \frac{(0.5125 \text{ kg})r^2\omega_0}{(0.1125 \text{ kg})r^2} = 4.56\omega_0 = 4.56(33\tfrac{1}{3}) = 152 \text{ rev/min}$$

8.60 The total angular momentum is given by

CHAPTER EIGHT SOLUTIONS

$I_{total} = I_{weights} + I_{student} = 2(mr^2) + (3 \text{ kg m}^2)$

Before: r = 1 m. Thus, $I_i = 2(3 \text{ kg})(1 \text{ m})^2 + (3 \text{ kg m}^2) = 9 \text{ kg m}^2$.
After: r = 0.3 m. Thus, $I_f = 2(3 \text{ kg})(0.3 \text{ m})^2 + (3 \text{ kg m}^2) = 3.54 \text{ kg m}^2$.
We now use conservation of angular momentum

$$I_f \omega_f = I_i \omega_i$$

or $\quad \omega_f = \dfrac{I_i}{I_f} \omega_i = \dfrac{9}{3.54} (0.75 \text{ rad/s}) = 1.91 \text{ rad/s}$

(b) $KE_i = \dfrac{1}{2} I_i \omega_i^2 = \dfrac{1}{2} (9 \text{ kg m}^2)(0.75 \text{ rad/s})^2 = 2.53 \text{ J}$

$KE_f = \dfrac{1}{2} I_f \omega_f^2 = \dfrac{1}{2} (3.54 \text{ kg m}^2)(1.91 \text{ rad/s})^2 = 6.44 \text{ J}$

8.61 The initial moment of inertia of the rotating disk is

$$I_i = \dfrac{1}{2} mr^2 = \dfrac{1}{2} (0.1 \text{ kg})(0.25 \text{ m})^2 = 3.125 \times 10^{-3} \text{ kg m}^2.$$

After the spider drops on the disk, the moment of inertia of the system becomes

$$I_f = I_i + m_{spider} r_s^2 = I_i + (0.015 \text{ kg})(0.02 \text{ m})^2 = 3.13 \times 10^{-3} \text{ kg m}^2$$

Thus, from $I_f \omega_f = I_i \omega_i$

we have $\quad \omega_f = \dfrac{I_i}{I_f} \omega_i = \dfrac{3.125 \times 10^{-3}}{3.13 \times 10^{-3}} (45 \text{ rev/min}) = 44.9 \text{ rev/min}$

8.62 (a) The moment of inertia of the system is given by

$$I = I_{man} + I_{wheel} = mr^2 + \dfrac{1}{2} MR^2 = (80 \text{ kg})r^2 + \dfrac{1}{2} (25 \text{ kg})(2 \text{ m})^2$$

When the man is at a distance of r = 2 m from the axis, the equation above gives

$$I_i = 370 \text{ kg m}^2$$

and when the man moves to a point 1 m from the center (r = 1 m), the moment of inertia becomes,

$$I_f = 130 \text{ kg m}^2$$

The initial angular velocity of the system is $\omega_i = 0.2 \text{ rev/s} = 1.26 \text{ rad/s}$, and we find the final angular velocity via conservation of angular momentum, as

$$\omega_f = \dfrac{I_i}{I_f} \omega_i = \dfrac{370}{130} (1.26 \text{ rad/s}) = 3.58 \text{ rad/s}$$

(b) The change in kinetic energy is

$$KE_f - KE_i = \dfrac{1}{2} I_f \omega_f^2 - \dfrac{1}{2} I_i \omega_i^2$$

$$= \dfrac{1}{2} (130 \text{ kg m}^2)(3.58 \text{ rad/s})^2 - \dfrac{1}{2}(370 \text{ kg m}^2)(1.26 \text{ rad/s})^2$$

$$= 831 \text{ J} - 292 \text{ J} = 539 \text{ J}$$

This difference results from work done by the man on the system as he walks inward.

8.63 (a) The initial moment of inertia of the system is just that of the wheel

$$I_i = I_{wheel} = 50 \text{ kg m}^2$$

and the moment of inertia after the man sits down on the edge is

$$I_f = I_{wheel} + I_{man} = 50 \text{ kg m}^2 + (75 \text{ kg})(2m)^2 = 350 \text{ kg m}^2$$

CHAPTER EIGHT SOLUTIONS

Also, the initial angular velocity of the wheel is $\omega_i = 0.2$ rev/s = 1.26 rad/s. The final angular velocity is found from conservation of angular momentum as
$$\omega_f = \frac{I_i}{I_f}\omega_i = \frac{50}{350}\,(1.26\ \text{rad/s}) = 0.180\ \text{rad/s}$$
(b) The change in kinetic energy of the system is
$$KE_f - KE_i = \frac{1}{2}I_f\omega_f^2 - \frac{1}{2}I_i\omega_i^2$$
$$= \frac{1}{2}(350\ \text{kg m}^2)(0.180\ \text{rad/s})^2 - \frac{1}{2}(50\ \text{kg m}^2)(1.26\ \text{rad/s})^2$$
$$= 5.64\ \text{J} - 39.5\ \text{J} = -33.8\ \text{J}$$

8.64 $I_0 = mr_0^2 = (0.12\ \text{kg})(0.4\ \text{m})^2 = 1.92 \times 10^{-2}\ \text{kg m}^2$
$I_f = mr_f^2 = (0.12\ \text{kg})(0.25\ \text{m})^2 = 7.5 \times 10^{-3}\ \text{kg m}^2$
and $\quad \omega_0 = \dfrac{v_0}{r_0} = \dfrac{0.8\ \text{m/s}}{0.4\ \text{m}} = 2$ rad/s
Now, use conservation of angular momentum
$$\omega_f = \frac{I_0}{I_f}\omega_0 = \frac{1.92 \times 10^{-2}\ \text{kg m}^2}{7.5 \times 10^{-3}\ \text{kg m}^2}\,(2\ \text{rad/s}) = 5.12\ \text{rad/s}$$
The work done = $\Delta KE = \dfrac{1}{2}I_f\omega_f^2 - \dfrac{1}{2}I_0\omega_0^2$
Substituting the appropriate values found earlier, we have
work done = 5.99×10^{-2} J

8.65 (a) The table turns opposite to the way the woman walks, so its angular momentum cancels that of the woman. From conservation of angular momentum, we have
$\quad L_f = L_i = 0$
so, $\quad L_f = I_w\omega_w + I_{table}\omega_{table} = 0$
and
$$\omega_{table} = -\frac{I_w}{I_{table}}\omega_{woman} = -\frac{m_w r^2}{I_{table}} \times \frac{v_{woman}}{r}$$
$$= -\frac{(60\ \text{kg})(2\ \text{m})^2}{500\ \text{kg m}^2} \times \frac{1.5\ \text{m/s}}{2\ \text{m}} = -0.36\ \text{rad/s}$$
or $\omega_{table} = 0.36$ rad/s (counterclockwise)
(b) work done = ΔKE
$$= KE_f - 0 = \frac{1}{2}m_{woman}v^2_{woman} + \frac{1}{2}I\omega^2_{table}$$
$$= \frac{1}{2}(60\ \text{kg})(1.5\ \text{m/s})^2 + \frac{1}{2}(500\ \text{kg m}^2)(0.36\ \text{rad/s})^2 = 99.9\ \text{J}$$

8.66 The moment of inertia of the sphere is
$$I = \frac{2}{5}mR^2 = 7.2 \times 10^{-2}\ \text{kg m}^2 \text{ when the mass is 2 kg and } R = 0.3\ \text{m}$$
Thus, $\quad \alpha = \tau/I = 100\ \text{N m}/7.2 \times 10^{-2}\ \text{kg m}^2 = 1.39 \times 10^3\ \text{rad/s}^2.$
The time to reach 2 rad/s is found as
$\quad \Delta t = \Delta\omega/\alpha = (2\ \text{rad/s})/1.39 \times 10^3\ \text{rad/s}^2 = 1.44 \times 10^{-3}$ s.
The steps are identical for the solid disk and the cylindrical shell. For the disk the moment of inertia is

CHAPTER EIGHT SOLUTIONS

$I = \frac{1}{2} mR^2 = 9.00 \times 10^{-2}$ kg m^2 and from this α is found to be 1.11×10^3 rad/s^2.
Finally,
$\Delta t = 1.80 \times 10^{-3}$ s
For the shell, the results are
$I = mR^2 = 1.80 \times 10^{-1}$ kg m^2, $\quad \alpha = 5.56 \times 10^2$ rad/s^2, and
$\Delta t = 3.60 \times 10^{-3}$ s.
Thus, we see that the sphere takes the shortest time to spin up to speed and therefore wins the race.

8.67 (a) The moment of inertia is
$$I = \frac{2}{5} mR^2 = \frac{2}{5} (2 \text{ kg})(0.5 \text{ m})^2 = 2 \times 10^{-1} \text{ kg m}^2.$$

(b) A point mass at a distance k from the axis would have a moment of inertia of mk^2. Thus,
$mk^2 = I$ or,
$k^2 = I/m = 2 \times 10^{-1}$ kg m^2/2 kg
From which, $k = 3.16 \times 10^{-1}$ m

8.68 $\omega_0 = (33 \frac{1}{3}$ rev/min$) = 3.49$ rad/s and $\omega_f = 0$

(a) $\alpha = \dfrac{\omega_f - \omega_0}{\Delta t} = \dfrac{0 - 3.49 \text{ rad/s}}{105 \text{ s}} = -3.32 \times 10^{-2}$ rad/s^2

(b) $\theta = \omega_0 t + \frac{1}{2} \alpha t^2 = (3.49 \text{ rad/s})(105 \text{ s}) + \frac{1}{2}(-3.32 \times 10^{-2} \frac{\text{rad}}{\text{s}^2})(105 \text{ s})^2 = 183$ rad = 29.2 rev

(c) $I = \frac{1}{2} mr^2 = \frac{1}{2}(3.5 \text{ kg})(0.15 \text{ m})^2 = 0.0394$ kg m^2

$\tau = I\alpha = (0.0394 \text{ kg m}^2)(-3.32 \times 10^{-2} \text{ rad/s}^2) = -1.31 \times 10^{-3}$ N m

---

8.69 (a) From conservation of angular momentum
$(I_1 + I_2)\omega = I_1 \omega_0$
$\omega = \dfrac{I_1}{I_1 + I_2} \omega_0$
$K_f = \frac{1}{2}(I_1 + I_2)\omega^2 \quad$ and $\quad K_i = \frac{1}{2} I_1 \omega_0^2$

so $\quad \dfrac{K_f}{K_i} =$ (after some algebra) $= \dfrac{I_1}{I_1 + I_2}$ which is less than 1.

Since $\dfrac{K_f}{K_i} < 1$ then $K_f < K_i$ so some energy has been lost.

8.70 We use conservation of angular momentum as $I_f \omega_f = I_i \omega_i$. From which,
$\omega_f = \dfrac{I_i}{I_f} \omega_i = \dfrac{1.5}{0.9} (0.1$ rev/s$) = 0.167$ rev/s

---

8.71 At a distance of 1.5 m from the axis of rotation, the moment of inertia of the two men is
$I_{men} = 2mr^2 = 2(80 \text{ kg})(1.5 \text{ m})^2 = 360$ kg m^2
and the total moment of inertia is the sum of the moment of inertia of the seesaw plus that of the men. This is

CHAPTER EIGHT SOLUTIONS

$I_0 = 3$ kg m^2 + 360 kg m^2 = 363 kg m^2.

The initial angular momentum is

$L_0 = I_0\omega_0 = (363$ kg m$^2)(2$ rad/s$) = 726$ kg m^2/s

We know from conservation of angular momentum that the final value of angular momentum equals the initial angular moment. We have

$(I'_{men} + 3$ kg m$^2)\omega_f = L_0$

where I'_{men} is the moment of inertia after the two men have scooted closer together.

$(I'_{men} + 3$ kg m$^2)(2.5$ rad/s$) = 726$ kg m^2/s

From this, we find

$I'_{men} = 287$ kg m$^2 = 2mr^2 = 2(80$ kg$)r^2$

This is solved for r to find, $\quad r = 1.34$ m

As a result, the men are 2.68 m apart at the end.

8.72 We will use conservation of energy in the form

$$\frac{1}{2} mv_i^2 + \frac{1}{2} I\omega_i^2 + mgy_i = \frac{1}{2} mv_f^2 + \frac{1}{2} I\omega_f^2 + mgy_f$$

$$0 + 0 + mg(6 \text{ m})\sin 37° = \frac{1}{2} m(0.2 \text{ m})^2\omega^2 + \left(\frac{1}{2}\right)\frac{2}{5} m(0.2)^2\omega^2 + 0$$

where we have used $v_f = r\omega$ and $I = \frac{2}{5} mR^2$

The mass cancels from the equation, and we find the angular velocity to be

$\omega = 35.6$ rad/s

8.73 The free body diagram is shown at the right.

$\Sigma\tau = 0$, yields

$-\frac{L}{2} (350$ N$) + T\sin 12° \left(\frac{2L}{3}\right) - (200$ N$)L = 0$

From which, $\quad T = 2705$ N

Compression force along the spine = R_x, and we find this from $\Sigma F_x = 0$, which gives

$R_x = T_x = T \cos 12° = 2646$ N

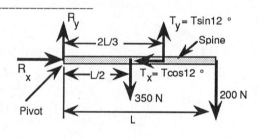

8.74 Use the lower left hand corner as the pivot point, and apply $\Sigma\tau = 0$.

$-(10$ N$)(.15$ m$) - (T_1\cos 50°)(.15$ m$) + (T_1\sin 50°)(.3$ m$) = 0$

Which gives, $\quad T_1 = 11.2$ N

Now use $\Sigma F_x = 0$.

$-F + (11.2)\cos 50° = 0$

and $\quad F = 7.23$ N.

Finally, $\Sigma F_y = 0$ yields

$T_2 - 10$ N $+ (11.2$ N$)\sin 50° = 0$

so, $\quad T_2 = 1.39$ N

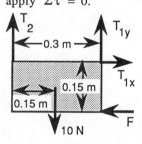

8.75 $\Sigma\tau = 0$ with pivot at left end yields

$-(500$ N$)(d/2) + T_y(d) = 0$

$T_y = T\sin 30° = .5T$, and d = the length of the flagpole.

122

CHAPTER EIGHT SOLUTIONS

Thus, we find   T = 500 N
Now use $\Sigma F_y = 0$
V - 500 N + 0.5(500 N) = 0
V = 250 N
Finally, use $\Sigma F_x = 0$
H - Tcos30° = 0
H = (500 N)cos30° = 433 N

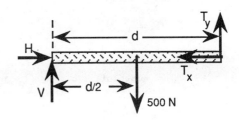

8.76 We have $T_y = T\sin30° = T/2$, and we use a pivot point at the left end to apply $\Sigma \tau = 0$.
$$-w\frac{d}{2} + \frac{T}{2}d = 0$$
where d is the length of the flagpole. This gives w = T
Thus, $w_{max} = T_{max} = 2000$ N.

8.77 The free body diagram is shown.
For a maximum value of x, the rod is on the verge of slipping, so $f = \mu_k N = 0.5\, N$.
Now, use $\Sigma F_x = 0$
$N = T_x = T\cos37°$, so $N = 0.799\, T$
Therefore,
    f = 0.5 N. = 0.399 T   (1)
From $\Sigma F_y = 0$, $f + T_y - 2w = 0$
or   0.399 T + 0.602 T = 2W which reduces to T = 2W
Taking torques about the left end, we have
    -Wx = W(2 m) + T_y (4 m) = 0
gives x = 2.81 m

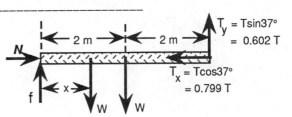

8.78 The free body diagram is shown at the right:
$\Sigma F_y = 0$ gives
    A_y - 29,400 N - 98,000 N = 0
so A_y = 127,400 N
Summing torques about point A, gives
B_x(1 m) - 29,400 N(2 m) - 98,000 N(6 m) = 0
and B_x = 646,800 N
Finally, $\Sigma F_x = 0$ gives
    $A_x = B_x$ = 646,800 N
The resultant force for A can be found from the Pythagorean theorem, and an appropriate trig function, to be
    6.59×10^5 N at an angle of 78.9° to the left of vertical.

8.79 We call the positive x direction down the plane and write down Newton's second law, as
    $mg\sin\theta - f = ma$ (1)
We now use $\tau = I\alpha$ with the axis of rotation at the center of the sphere.

CHAPTER EIGHT SOLUTIONS

$$fR = I\frac{a}{R}$$

From which, $\quad f = \dfrac{I}{R^2} a \quad (2)$

We now substitute (2) into (1) and solve for a, the linear acceleration of the center of mass,

$$a = \frac{g\sin\theta}{1 + \dfrac{I}{mR^2}}$$

For a solid cylinder, $I = \frac{1}{2} mR^2$, and thus, $\dfrac{I}{mR^2} = \dfrac{1}{2}$

From this, we find a for the solid cylinder to be, $\quad a = 3.93 \text{ m/s}^2$.

The time for the sphere to roll 6 m from rest is found as follows.

$$x = v_0 t + \frac{1}{2} at^2$$

$$6 \text{ m} = 0 + \frac{1}{2}(3.93 \text{ m/s}^2)t^2$$

So, $\quad t = 1.75$ s

For the cylindrical shell, $I = mR^2$. Thus, $\quad \dfrac{I}{mR^2} = 1$

and, from this, $\quad a = 2.95 \text{ m/s}^2$

At this acceleration, in a time of 1.75 s, the shell will roll a distance given by

$$x = v_0 t + \frac{1}{2} at^2$$

$$x = 0 + \frac{1}{2}(2.95 \text{ m/s}^2)(1.75 \text{ s})^2 = 4.50 \text{ m}$$

8.80 The initial moment of inertia of the system is

$$I_0 = \Sigma m_i r_i^2 = (4M)(1 \text{ m})^2$$

The moment of inertia of the system after the spokes are shortened is

$$I_f = \Sigma m_f r_f^2 = (4M)(0.5 \text{ m})^2$$

We conserve angular momentum as

$$\omega_f = \frac{I_i}{I_f}\omega_i = (4)(2 \text{ rev/s}) = 8 \text{ rev/s}$$

8.81 We have $L = I\omega = $ constant. Thus, with m the mass of the planet and r the distance of the planet from the sun, we have

$$\omega = \frac{L}{I} = \frac{L}{mr^2}$$

From this, we see that the angular velocity is a maximum when r is a minimum, or when the planet is closest to the sun.

8.82 The radius of the earth is $R_E = 6.38 \times 10^6$ m, and the final distance of the spaceship from the center of the earth is $r = 6.58 \times 10^6$ m. Apply conservation of angular momentum to the spaceship as,

$$\omega_f = \frac{I_i}{I_f}\omega_i = \frac{mR_E^2}{mr^2}\omega_E = \frac{(6.38)^2}{(6.58)^2}\omega_E = 0.94\,\omega_E$$

where ω_E is the angular velocity of the earth, which is the same angular velocity as that of the space ship initially. Thus, we see that the ship will orbit

CHAPTER EIGHT SOLUTIONS

the earth at a lower angular velocity that that at which the earth turns. As a result, the ship cannot stay directly above a fixed spot on the earth's surface.

8.83 From Newton's second law, we have
$mg\sin 37° - T = ma$
or $(12 \text{ kg})(9.8 \text{ m/s}^2)\sin 37° - T = (12 \text{ kg})a$
so, $T = 70.56 \text{ N} - (12 \text{ kg})a$ (1)

also $\alpha = \dfrac{a}{r} = \dfrac{2 \text{ m/s}^2}{0.1 \text{ m}} = 20 \text{ rad/s}^2$ (2)

(a) Thus, from (1), $T = 70.56 \text{ N} - (12 \text{ kg})(2 \text{ m/s}^2) = 46.6 \text{ N}$

(b) From $\tau = I\alpha$, we have $I = \dfrac{\tau}{\alpha} = \dfrac{Tr}{\alpha} = \dfrac{(46.6 \text{ N})(0.1 \text{ m})}{20 \text{ rad/s}} = 0.233 \text{ kg m}^2$.

(c) $\omega = \omega_0 + \alpha t$ becomes $\omega = 0 + (20 \text{ rad/s}^2)(2 \text{ s}) = 40 \text{ rad/s}$

8.84 (a) From conservation of energy, the work done is equal to the change in rotational kinetic energy. $W = \dfrac{1}{2}I\omega^2 - \dfrac{1}{2}I\omega_0^2$

and $I = \dfrac{1}{2}mR^2 = \dfrac{1}{2}(1 \text{ kg})(0.6 \text{ m})^2 = 0.18 \text{ kg m}^2$.

Thus, $W = \dfrac{1}{2}(0.18 \text{ kg m}^2)(6 \text{ rad/s})^2 = 3.24 \text{ J}$

(b) We have $\alpha = \dfrac{a}{r} = \dfrac{2.5 \text{ m/s}^2}{0.6 \text{ m}} = 4.17 \text{ rad/s}^2$. From the definition of angular

acceleration, we find the time as $t = \dfrac{\omega}{\alpha} = \dfrac{6 \text{ rad/s}}{4.17 \text{ rad/s}^2} = 1.44 \text{ s}$

(c) $\theta = \dfrac{1}{2}\alpha t^2 = \dfrac{1}{2}(4.17 \text{ rad/s}^2)(1.44 \text{ s}) = 4.32 \text{ rad}$

$s = r\theta = (0.6 \text{ m})(4.32 \text{ rad}) = 2.59 \text{ m}$. when the spool reaches 6 rad/s, 1.44 s will have elapsed and 2.59 m of cord will have been removed from the spool. YES!

8.85 We use conservation of energy with the zero level selected at the bottom of the bowl. At the initial position the gravitational potential energy is $mg(R - r)(1 - \cos\theta)$.

$$mg(R - r)(1 - \cos\theta) = \dfrac{1}{2}mv^2 + \dfrac{1}{2}\left(\dfrac{2}{5}mr^2\omega^2\right)$$

Since $v = r\omega$, we have $\omega = \sqrt{\dfrac{10}{7}\dfrac{g}{r^2}(1 - \cos\theta)(R - r)}$

8.86 (a) Although an external force (tension in rope) acts on the mass, no external torques act. Thus, angular momentum is conserved $mv_0r_0 = mvr$, so

$v = \dfrac{v_0 r_0}{r} = \dfrac{(1.5 \text{ m/s})(0.3 \text{ m})}{0.1 \text{ m}} = 4.5 \text{ m/s}$

(b) $T = \dfrac{mv^2}{r} = \dfrac{(0.050 \text{ kg})(4.5 \text{ m/s})^2}{0.1 \text{ m}} = 10.1 \text{ N}$

8.87 (a) If the center of mass of the cylinder is not to move in the vertical direction, we know that
$\Sigma F_y = 0$
Thus, $T - mg = 0$

CHAPTER EIGHT SOLUTIONS

or, $T = mg = (0.5 \text{ kg})(9.8 \text{ m/s}^2) = 4.9$ N

(b) We first find the angular acceleration of the cylinder from $\tau = I\alpha$.

$$\alpha = \frac{\tau}{I} = \frac{Tr}{I} = \frac{Tr}{\frac{1}{2}mr^2} = \frac{mgr}{\frac{1}{2}mr^2} = \frac{2g}{r} = \frac{19.6 \text{ m/s}^2}{0.1 \text{ m}} = 196 \text{ rad/s}^2.$$

We can now find ω from $\omega = \omega_0 + \alpha t$ as

$\omega = 0 + (196 \text{ rad/s}^2)(0.2 \text{ s}) = 39.2$ rad/s

---

8.88 We have $\quad \text{Work} = \bar{P}t = \Delta KE = \frac{1}{2}I\omega_f^2 - \frac{1}{2}I\omega_0^2 = \frac{1}{2}I\omega_f^2 - 0$

Therefore, $\quad \bar{P} = \dfrac{KE_f}{t} = \dfrac{\frac{1}{2}I\omega_f^2}{t}$

Then, $I = \frac{1}{2}mr^2 = \frac{1}{2}(184 \text{ kg})(0.7 \text{ m})^2 = 45.0 \text{ kg m}^2$,

and $\omega_f = 1400$ rev/min $= 147$ rad/s.

So, $\quad \bar{P} = \dfrac{\frac{1}{2}(45 \text{ kg m}^2)(147 \text{ rad/s})^2}{150 \text{ s}} = 3.22 \times 10^3$ W $= 4.32$ hp

---

8.89 Initially, the angular momentum of the system is zero because everything is at rest. As a result, the final angular momentum must also be zero. This means that

$L_{wheel} = - L_{child}$

The negative sign indicates that the wheel will turn in the opposite direction to that in which the child runs. Thus, the wheel must turn counterclockwise. The moment of inertia of the wheel is

$I_w = \frac{1}{2}mr^2 = \frac{1}{2}(150 \text{ kg})(2 \text{ m})^2 = 300 \text{ kg m}^2.$

Let us now equate the magnitude of the angular momentum of the child to that of the wheel, as

$I_w\omega = m_c r^2 \omega_c = m_c rv \quad$ where we have eliminated ω_c through use of $v = r\omega$.
Thus,

$\omega = \dfrac{m_c rv}{I_w} = \dfrac{(50 \text{ kg})(2 \text{ m})(2 \text{ m/s})}{300 \text{ kg m}^2} = 0.667$ rad/s

---

8.90 (a) and (b) First, we write down Newton's second law for the 3 kg mass.
$\quad T_2 = (3 \text{ kg})a \quad (1)$

The second law for the 4 kg mass is
$\quad 39.2 \text{ N} - T_1 = (4 \text{ kg})a \quad (2)$

Now apply $\tau = I\alpha$ to the pulley (axis of rotation at its center)
$\quad (T_1 - T_2)r = (0.5 \text{ kg m}^2)(a/r)$
or $\quad T_1 - T_2 = (5.56 \text{ kg})a \quad (3)$

Equations (1), (2) and (3) can be solved simultaneously to give
$\quad a = 3.12 \text{ m/s}^2 \qquad T_1 = 26.7 \text{ N, and} \qquad T_2 = 9.37 \text{ N}$

8.91 (a) and (b) First, we write down the second law for the 2 kg mass.
$\quad T_1 - 19.6 \text{ N} = (2 \text{ kg})a \quad (1)$

Now, we write the second law for the 5 kg mass.

CHAPTER EIGHT SOLUTIONS

$$49 \text{ N} - T_2 = (5 \text{ kg})a \quad (2)$$

Finally, $\tau = I\alpha$ for the pulley becomes

$$(T_2 - T_1)(0.5 \text{ m}) = (5 \text{ kg m}^2)(a/0.5 \text{ m})$$

or $\quad T_2 - T_1 = (20 \text{ kg})a \quad (3)$

Equations (1), (2), and (3) can be solved simultaneously to give

$$a = 1.09 \text{ m/s}^2 \qquad T_1 = 21.8 \text{ N and } T_2 = 43.6 \text{ N}$$

8.92 Note that initially the center of mass of the sphere is a distance $h + r$ above the bottom of the loop and as the mass reaches the top of the loop, this distance above the reference level is $2R - r$. The conservation of energy gives

$$mg(h + r) = mg(2R - r) + \frac{1}{2}mv^2 + \frac{1}{2}I\omega^2.$$

For a sphere $I = \frac{2}{5}mr^2$ and $v = r\omega$ so that the expression becomes

$$gh + 2gr = 2gR + \frac{7}{10}v^2 \quad (1)$$

Note that $h = h_{min}$ when the speed of the sphere at the top of the loop satisfies the condition

$$\Sigma F = mg = \frac{mv^2}{(R - r)} \quad \text{or} \quad v^2 = g(R - r).$$

Substituting this into Eq. (1) gives $\quad h_{min} = 2(R - r) + 0.7(R - r)$

or $\quad h_{min} = 2.7(R - r)$

8.93 (a) and (b) For the cylinder; $\Sigma\tau = (2T)R = \frac{1}{2}MR^2\frac{a}{R} \quad (1)$

where a is the acceleration of the falling masses. For each of the falling masses

$$\Sigma F = mg - T = ma \quad (2)$$

Combining Equations (1) and (2), we find $\quad T = \frac{Mmg}{M + 4m} \text{ and } a = \frac{4mg}{(M + 4m)}$

8.94 For the two masses we have $T_1 - m_1 g = m_1 a$

which becomes $\quad T_1 - (2 \text{ kg})(9.8 \text{ m/s}^2) = (2 \text{ kg})a \quad (1)$

and $\quad m_2 g - T_2 = m_2 a$, which becomes $(5 \text{ kg})(9.8 \text{ m/s}^2) - T_2 = (5 \text{ kg})a \quad (2)$

For the pulley $\quad \tau = I\alpha$ becomes $\quad (T_2 - T_1)R = I\frac{a}{R}$

or $(T_2 - T_1)(0.5 \text{ m}) = (5 \text{ kg m}^2)\frac{a}{0.5 \text{ m}} \quad (3)$

Solving (1), (2), and (3) together gives $a = 1.09 \text{ m/s}^2$, $T_1 = 21.8 \text{ N}$, and $T_2 = 43.6 \text{ N}$

CHAPTER NINE SOLUTIONS

9.1 Stress = $\frac{F}{A}$, where F = 0.3(weight) = 0.3(480 N) = 144 N,
and A = πr^2 = $\pi(0.5 \times 10^{-2}$ m$)^2$ = 7.85×10^{-5} m^2. Thus,
$$\text{Stress} = \frac{144 \text{ N}}{7.85 \times 10^{-5} \text{ m}^2} = 1.83 \times 10^6 \text{ Pa}$$

9.2 Y = $\frac{\text{stress}}{\text{strain}}$, or strain = $\frac{\text{stress}}{Y}$, and $Y_{al} = 7 \times 10^{10}$ Pa

so $\Delta L = \frac{FL}{AY} = \frac{(2 \times 10^3 \text{ kg})(9.8 \text{ m/s}^2)(7.0 \text{ m})}{\pi(0.3 \text{ m})^2(7.0 \times 10^{10} \text{ Pa})} = 6.93$ μm

---

9.3 (a) Stress = F/A = 19.6 N/ 1.26×10^{-5} m^2 = 1.56×10^6 Pa
(b) Strain = $\Delta L/L$ = stress/Y
Therefore, ΔL = L(stress)/Y = (4 m)(1.56×10^6 Pa)/11×10^{10} Pa = 5.67×10^{-5} m

---

9.4 The area of the beam is A = πr^2 = $\pi(1.5 \times 10^{-2}$ m$)^2$ = 7.07×10^{-4} m^2. The maximum allowable strain is
   Strain = $\Delta L/L$ = 5×10^{-5} m/6 m = 8.33×10^{-6}
Thus, the maximum stress is
   Stress = F/A = Y(Strain) = (20×10^{10} Pa)(8.33×10^{-6}) = 1.67×10^6 Pa
From which, we find the maximum force to be
   F = (1.67×10^6 Pa)(7.07×10^{-4} m^2) = 1.18×10^3 N

9.5 From the defining equation for the shear modulus, we find Δx as
Δx = h(F/A)/S = (5×10^{-3} m)(20 N)/(14×10^{-4} m^2)(3×10^6 Pa) = 2.38×10^{-2} mm

---

9.6 The maximum stress = F/A_{min} = 5×10^8 Pa
Therefore,
   A_{min} = (70 kg)(9.8 m/s^2)/5×10^8 Pa = 1.37×10^{-6} m^2
But, $A_{min} = \frac{\pi d^2}{4}$ from which, d_{min} = 1.32 mm

---

9.7 v_0 = 80 km/h = 22.2 m/s, and
a = $\frac{v - v_0}{t} = \frac{0 - 22.2 \text{ m/s}}{5 \times 10^{-3} \text{ s}}$ = $- 4.45 \times 10^3$ m/s^2.
The force on the bone can be found from F = ma, as
   F = (3.0 kg)(4.45×10^3 m/s^2) = 1.33×10^4 N
so the stress on the bone is
   stress = $\frac{F}{A} = \frac{1.33 \times 10^4 \text{ N}}{(2.5 \text{ cm}^2)(1 \text{ m}^2/10^4 \text{ cm}^2)}$ = 5.34×10^7 Pa
Since the stress on the bone is less than the breaking stress, the arm should survive, provided that it is not used to decelerate the rest of the body.

9.8 Let us find the force required to place the support at the elastic limit.

CHAPTER NINE SOLUTIONS

$F = A$ (Stress)

where $A = \pi r^2_{outer} - \pi r^2_{inner} = \pi((9 \times 10^{-3} \text{ m})^2 - (7 \times 10^{-3} \text{ m})^2) = 1.01 \times 10^{-4} \text{ m}^2$

Therefore, $F = (1.01 \times 10^{-4} \text{ m}^2)(29 \times 10^7 \text{ Pa}) = 2.91 \times 10^4 \text{ N}$

The total load which can be supported $= 2F = 5.82 \times 10^4 \text{ N}$

The total load $= 4000 \text{ N} + n(670 \text{ N})$

Where we have assumed an average weight of 670 N (150 lb) per person. Solving for n, the number of people, we find

$n = 80.9 = 80$ people.

9.9 From the definition of bulk modulus, we find the change in volume to be
$\Delta V = -(\Delta P)V/B = -(5 \times 10^4 \text{ Pa})(10^{-3} \text{ m}^3)/14 \times 10^{10} \text{ Pa} = -3.57 \times 10^{-10} \text{ m}^3$.

9.10 From the definition of bulk modulus, the change in pressure is

$\Delta P = -\dfrac{\Delta V}{V} B$

For our situation, $\Delta V/V = -1/100$

Then, $\Delta P = B/100 = 0.21 \times 10^{10} \text{ Pa}/100 = 2.1 \times 10^7 \text{ Pa}$

9.11 From the definition of Young's modulus, the area is given by

$A = (F)(L)/(\Delta L)(Y) = (0.025 \text{ N})(2 \times 10^{-1} \text{ m})/(7.6 \times 10^{10} \text{ Pa})(4 \times 10^{-4} \text{ m})$
$= 1.65 \times 10^{-10} \text{ m}^2$

Also, $A = \dfrac{\pi d^2}{4}$ Hence, $d = 1.45 \times 10^{-5}$ m

---

9.12 The area over which the shear must occur is equal to the circumference of the hole times its thickness. Thus,

$A = (2\pi r)t = 2\pi(5 \times 10^{-3} \text{ m})(5 \times 10^{-3} \text{ m}) = 1.57 \times 10^{-4} \text{ m}^2$.

So, $F = (A)\text{Stress} = (1.57 \times 10^{-4} \text{ m}^2)(4 \times 10^8 \text{ Pa}) = 6.28 \times 10^4$ N (about 14,100 lbs)

9.13 Let us find the tension to stretch the wire by 0.1 mm.

The area $= A = \dfrac{\pi d^2}{4} = 3.80 \times 10^{-8} \text{ m}^2$. Thus, the force is

$F = \dfrac{YA\Delta L}{L_0} = \dfrac{(18 \times 10^{10} \text{ Pa})(3.80 \times 10^{-8} \text{ m}^2)(10^{-4} \text{ m})}{3.1 \times 10^{-2} \text{ m}} = 22.1 \text{ N}$

We have an equilibrium situation, so

$\Sigma F_x = 0$ becomes $F\cos 30° - F\cos 30° = 0$

$\Sigma F_y = 0$ becomes $2F\sin 30° = 2(22.1 \text{ N})\sin 30° = 22.1$ N and will be directed down the page in the textbook figure.)

9.14 The cross-sectional area of the wire $A = \pi(2 \times 10^{-3} \text{ m})^2 = 1.256 \times 10^{-5} \text{ m}^2$, and the tension $= 5.8 \times 10^3$ N throughout both pieces. Let us compute the elongation of each part separately:

For aluminum

$\Delta L_{al} = \dfrac{L_0 F}{YA} = \dfrac{(1.3 \text{ m})(5.8 \times 10^3 \text{ N})}{(7 \times 10^{10} \text{ Pa})(1.256 \times 10^{-5} \text{ m}^2)} = 8.57 \times 10^{-3} \text{ m}$

Similarly, for the copper part

$\Delta L_{cu} = \dfrac{L_0 F}{YA} = \dfrac{(2.6 \text{ m})(5.8 \times 10^3 \text{ N})}{(11 \times 10^{10} \text{ Pa})(1.256 \times 10^{-5} \text{ m}^2)} = 10.9 \times 10^{-3} \text{ m}$

So, the total elongation $= 19.5 \times 10^{-3}$ m $= 1.95$ cm

CHAPTER NINE SOLUTIONS

9.15 The density of the cube is
$$\rho = 1.31 \text{ kg}/1.25 \times 10^{-4} \text{ m}^3 = 1.05 \times 10^4 \text{ kg/m}^3 \text{ (material is silver)}$$

9.16 $V = 185 \text{ cm}^3 = 1.85 \times 10^{-4} \text{ m}^3$
$$\rho = m/V = 5 \times 10^{-1} \text{ kg}/1.85 \times 10^{-4} \text{ m}^3 = 2.70 \times 10^3 \text{ kg/m}^3 \text{ (crown is aluminum)}$$

9.17 The area of one of the legs is $A = \pi r^2 = \pi(10^{-2} \text{ m})^2 = \pi \times 10^{-4} \text{ m}^2$
The force exerted by one leg on the floor is
$$F = \frac{1}{2} \text{ (weight of man + weight of chair)} = \frac{1}{2}(75 \text{ kg})(9.8 \text{ m/s}^2) = 368 \text{ N}$$
Thus, $P = F/A = 368 \text{ N}/(\pi \times 10^{-4}) \text{m}^2 = 1.17 \times 10^6 \text{ Pa}$

9.18 From the definition of pressure, we have
$$F = PA$$
The area of the hole is $A = \pi d^2/4 = \pi(4 \times 10^{-3} \text{ m})^2/4 = 1.26 \times 10^{-5} \text{ m}^2$
Therefore, $F = (5 \times 10^5 \text{ Pa})(1.26 \times 10^{-5} \text{ m}^2) = 6.28 \text{ N}$

9.19 $P = P_a + \rho g h$
$$P = (1.013 \times 10^5 \text{ Pa}) + (10^3 \text{ kg/m}^3)(9.8 \text{ m/s}^2)(30 \text{ m}) = 3.95 \times 10^5 \text{ Pa}$$

9.20 $P_g = P - P_a = \rho g h = (10^3 \text{ kg/m}^3)(9.8 \text{ m/s}^2)(1200 \text{ ft})(1\text{m}/3.28 \text{ ft})$
$$= 3.58 \times 10^6 \text{ Pa.}$$

9.21 (a) $7 \text{ mi} = 1.13 \times 10^4 \text{ m}$
$$P = P_a + \rho g h = 1.013 \times 10^5 \text{ Pa} + (10^3 \text{ kg/m}^3)(9.8 \text{ m/s}^2)(1.13 \times 10^4 \text{ m})$$
$$= 1.104 \times 10^8 \text{ Pa.}$$
(b) The gauge pressure at this depth is
$$P_g = P - P_a = 1.1037 \times 10^8 \text{ Pa}$$
and the inward force on the port hole is
$F = P_g A = (1.1037 \times 10^8 \text{ Pa})(1.77 \times 10^{-2} \text{ m}^2) = 1.95 \times 10^6 \text{ N}$ (about 4.38×10^5 lb)

9.22 We use $P = P_0 + \rho g h$, where $\rho = 806 \text{ km/m}^3$ for ethyl alcohol, and where $P_0 = 1.1 \text{ atm} = 1.114 \times 10^5 \text{ Pa}$.
Thus,
$P = 1.114 \times 10^5 \text{ Pa} + (806 \text{ kg/m}^3)(9.8 \text{ m/s}^2)(0.4 \text{ m}) = 1.146 \times 10^5 \text{ Pa} = 1.13 \text{ atm}$

9.23 The gauge pressure of the fluid at the level of the needle must equal the gauge pressure in the artery.
$\rho g h = 100 \text{ Torr} = 1.333 \times 10^4 \text{ Pa}$
so, $h = \dfrac{1.333 \times 10^4 \text{ Pa}}{(1.02 \times 10^3 \text{ kg/m}^3)(9.8 \text{ m/s}^2)} = 1.33 \text{ m}$

9.24 $P = P_t + \rho g h$ where P_t = the pressure at the top of the column of mercury, which is small enough to be taken to be zero. Thus,
$$P = (13.6 \times 10^3 \text{ kg/m}^3)(9.8 \text{ m/s}^2)(0.74 \text{ m}) = 9.86 \times 10^4 \text{ Pa}$$

CHAPTER NINE SOLUTIONS

9.25 Atmospheric pressure for a barometer is found as $P = \rho g h$ regardless of the substance in the barometer. As a result,

$$\rho_{water} g h_{water} = \rho_{mercury} g h_{mercury}$$

or $\quad h_{water} = \dfrac{13.6 \times 10^3}{1 \times 10^3} (0.77 \text{ m}) = 10.5 \text{ m}$

9.26 We have the gauge pressure at the height of the first floor as

$$P_{g1} = \rho g h_1$$

and the gauge pressure at the top floor is

$$P_{g2} = \rho g h_2$$

Therefore,

$$P_{g2} - P_{g1} = \Delta P_g = \rho g \Delta h$$

or $\quad \Delta h = \dfrac{\Delta P_g}{\rho g} = (1.5 \times 10^5 \text{ Pa} - 2.5 \times 10^5 \text{ Pa})/(10^3 \text{ kg/m}^3)(9.8 \text{ m/s}^2) = -10.2 \text{ m}$

Thus, the top floor is 10.2 m above the first floor (downward has been taken as the positive direction).

9.27 In this application, Pascal's law is written as

$$\dfrac{F_2}{A_2} = \dfrac{F_1}{A_1}$$

or $\quad F_2 = \dfrac{A_2}{A_1} F_1 = \dfrac{2 \text{ cm}^2}{400 \text{ cm}^2} (1.80 \times 10^5 \text{ lbs}) = 900 \text{ lb}$

9.28 $\dfrac{F_2}{A_2} = \dfrac{F_1}{A_1}$ Pascal's principle becomes

$$F_{brake} = \dfrac{A_{brake\ cylinder}}{A_{master\ cylinder}} (F_{pedal}) = \dfrac{1.75 \text{ cm}^2}{6.4 \text{ cm}^2} (44 \text{ N}) = 12.0 \text{ N}$$

This is the normal force exerted on the brake shoe. The frictional force is

$$f = \mu N = 0.5(12.0 \text{ N}) = 6.0 \text{ N}$$

and the torque is

$$\tau = fr = (6.0 \text{ N})(0.34 \text{ m}) = 2.04 \text{ N m}$$

9.29 $\dfrac{F_2}{A_2} = \dfrac{F_1}{A_1}$ Pascal's principle becomes

$$F_1 = \dfrac{A_1}{A_2} F_2 = \dfrac{\frac{\pi}{4}(0.25 \text{ in})^2}{\frac{\pi}{4}(1.5 \text{ in})^2} (500 \text{ lbs}) = 13.9 \text{ lb}$$

Now, consider the jack handle with the pivot point at the left end.

$$\Sigma \tau = 0 = (13.9 \text{ lb})(2 \text{ in}) - F(12 \text{ in}) = 0$$
$$F = 2.31 \text{ lb}$$

9.30 (a) $P = P_a + \rho g h = 1.013 \times 10^5 \text{ Pa} + (13.6 \times 10^3 \text{ kg/m}^3)(9.8 \text{ m/s}^2)(0.2 \text{ m})$
$\quad\quad\quad = 1.28 \times 10^5 \text{ Pa}.$
(b) $P_g = P - P_a = 2.67 \times 10^4 \text{ Pa}$

CHAPTER NINE SOLUTIONS

9.31 We first find the absolute pressure at the interface between oil and water. This is

$P = P_a + \rho gh = 1.013 \times 10^5$ Pa $+ (7 \times 10^2$ kg/m$^3)(9.8$ m/s$^2)(0.3$ m$) = 1.03 \times 10^5$ Pa.

This is the pressure at the top of the water. To find the absolute pressure at the bottom, we use

$P = P_t + \rho gh = 1.03 \times 10^5$ Pa $+ (1 \times 10^3$ kg/m$^3)(9.8$ m/s$^2)(0.2$ m$) = 1.05 \times 10^5$ Pa.

9.32 (a) B = (weight of water displaced) = $(\rho_w V_{displaced})g$
$= (10^3$ kg/m$^3)(0.2$ m$^3)(9.8$ m/s$^2) = 1.96 \times 10^3$ N

(b) The buoyant force remains the same because the same volume of water (and hence, weight) of water has been displaced.

9.33 Since the frog floats, the buoyant force = the weight of the frog. Also, the weight of the displaced water = weight of the frog, so

$\rho_{water} Vg = m_{frog} g$

or, $m_{frog} = \rho_{water} V = \rho_{water} \frac{1}{2}(\frac{4}{3}\pi r^3) = (1.35 \times 10^3$ kg/m$^3)\frac{2\pi}{3}(6 \times 10^{-2}$ m$)^3$

$m_{frog} = 0.611$ kg

9.34 The balloon is in equilibrium under the action of three forces, F_b, the buoyant force on the balloon, w, its weight, and T, the tension in the string. We have

$T = F_b - w$ (1)

F_b = weight of displaced air = $\rho_{air} Vg = (1.29$ kg/m$^3)(9.8$ m/s$^2)\frac{4}{3}\pi(0.5$ m$)^3 = 6.65$ N

w = weight of empty balloon + weight of enclosed helium
$= (0.012$ kg$)(9.8$ m/s$^2) + (0.181$ kg/m$^3)\frac{4}{3}\pi(0.5$ m$)^3 (9.8$ m/s$^2) = 1.05$ N

Then from equation (1), T = 6.65 N - 1.05 N = 5.60 N

9.35 The forces on the object when submerged are the tension in the string, the buoyant force of the water, and the weight of the object. We use $\Sigma F_y = 0$

B + T = w
B + 3.5 N = 5 N

Therefore, B = 1.5 N (This is also the weight of the water displaced.)
We also know,

$V_{displaced\ water} = V_{object}$

Thus, the buoyant force may be written as

$B = (\rho_w V_{object})g$

or, $V_{object} = B/\rho_w g$ (1)

The density of the object is found as

$\rho = m_{object}/V_{object}$ (2)

We eliminate the volume of the object from equation (2) by use of (1) to find,

$\rho = \frac{(w_{object})(\rho_w)}{B} = \frac{(5\ N)(10^3\ kg/m^3)}{1.5\ N} = 3.33 \times 10^3$ kg/m^3

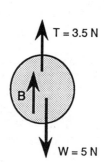

CHAPTER NINE SOLUTIONS

9.36 The buoyant force on the boat is equal to the weight of the displaced water, and also, because the boat floats, the buoyant force is equal to the weight of the boat. The volume of the water displaced is

$V = A(d - h)$ where d is the total height of the side of the boat and h is the height of the boat which is above water level. Thus, the buoyant force is

$$\rho_w A(d - h)g = 1000 \text{ N}$$

or

$$d - h = d - 5 \text{ cm} = \frac{10^3 \text{ N}}{(3 \text{ m}^2)(9.8 \text{ m/s}^2)(10^3 \text{ kg/m}^3)} = 3.4 \times 10^{-2} \text{ m} = 3.4 \text{ cm} \quad (1)$$

which gives, $d = 8.4$ cm

In salt water, the density is 1.03×10^3 kg/m^3, and as a result (1) gives

$$d - h = \frac{10^3 \text{ N}}{(3 \text{ m}^2)(9.8 \text{ m/s}^2)(1.03 \times 10^3 \text{ kg/m}^3)} = 3.30 \times 10^{-2} \text{ m} = 3.3 \text{ cm}$$

Therefore, $h = d - 3.3$ cm $= 8.4$ cm $- 3.3$ cm $= 5.10$ cm

---

9.37 Because the sea monster floats, its weight, w, is equal to the buoyant force (the weight of water displaced by the 52 m^3 volume of the monster). We have

$$w = mg = \rho_w V g$$

or, $m = \rho_w V = (1.03 \times 10^3 \text{ kg/m}^3)(52 \text{ m}^3) = 5.36 \times 10^4$ kg $= 53.6$ metric tons

---

9.38 The weight of the truck, W, is equal to the weight of the additional water displaced when it drives onto the boat. This is

$W = \rho_w (\Delta V)g = (10^3 \text{ kg/m}^3)(4 \text{m})(6 \text{ m})(0.04 \text{ m})(9.8 \text{ m/s}^2) = 9.41 \times 10^3$ N

9.39 (a) The forces on the object when suspended in water are the tension in the string, T_1, the weight of the object, and the buoyant force of the water, B_w. Since the object is in equilibrium, we have

$$T_1 + B_w = w$$

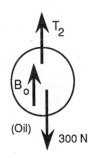

or,

$B_w = w - T_1 = 300 \text{ N} - 265 \text{ N} = 35 \text{ N} = \rho_w V g$

From which,

$$V = \frac{35 \text{ N}}{(10^3 \text{ kg/m}^3)(9.8 \text{ m/s}^2)}$$

$= 3.57 \times 10^{-3}$ m^3

The mass of the 300 N object is 30.6 kg, and the volume V found above is the volume of water displaced which is also the volume of the object. Thus, the density of the object is

$$\rho_{object} = m/V = 30.6 \text{ kg}/3.57 \times 10^{-3} \text{ m}^3 = 8.57 \times 10^3 \text{ kg/m}^3$$

(b) When submerged in the oil, the forces on the object are T_2, the tension in the string, the weight of the object, w, and the buoyant force of the oil, B_0. For equilibrium, we have

$$T_2 + B_0 = w$$

or, $B_0 = w - T_2 = 300$ N $- 275$ N $= 25$ N

However, the buoyant force exerted by the oil is also equal to the weight of the oil displaced. The volume of the oil displaced is equal to the volume of the object. Thus, the density of the oil is

133

CHAPTER NINE SOLUTIONS

$$\rho_{oil} = \frac{w_{oil}}{gV} = \frac{25 \text{ N}}{(9.8 \text{ m/s}^2)(3.57 \times 10^{-3} \text{ m}^3)} = 714 \text{ kg/m}^3$$

9.40 The buoyant force exerted by the water, F_w, is

$$F_w = \rho_w Vg = (10^3 \text{ kg/m}^3)(2 \text{ m})(0.5 \text{ m})(0.08 \text{ m})(9.8 \text{ m/s}^2) = 784 \text{ N}$$

Thus, the total weight that can be supported is $w = mg = 784 \text{ N}$
The mass to be supported is the sum of the mass of the mattress and the mass on the mattress,

$$m = \frac{784 \text{ N}}{9.8 \text{ m/s}^2} = 80 \text{ kg}$$

Therefore, the mass on top, m_s, is

$$m_s = 80 \text{ kg} - 2.3 \text{ kg} = 77.7 \text{ kg}$$

9.41 When the system floats,
$F_B = w$, or the weight of the displaced water equals the weight of the object,

$$\rho_w(\pi r^2 z)g = 1.96 \text{ N} \qquad \pi r^2 z = \text{volume of displaced water.}$$

$$z = \frac{1.96 \text{ N}}{(10^3 \text{ kg/m}^3)\pi(2 \times 10^{-2} \text{ m})^2(9.8 \text{ m/s}^2)} = 0.159 \text{ m} = 15.9 \text{ cm}$$

9.42 When the bar of soap, of cross-sectional area A, is in water only, the forces on it are the buoyant force of the water and its weight, w. Because it is floating, we know

$$B_w = w$$

The buoyant force is

$$B_w = \rho_w(A)(1.5 \times 10^{-2} \text{ m})g = w$$

From this equation, we find

$$\frac{w}{Ag} = \rho_w(1.5 \times 10^{-2} \text{ m}) \qquad (1)$$

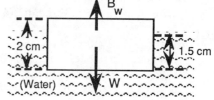

When the bar is floating in both water and oil, the forces on it are the buoyant force of the oil, the buoyant force of the water, and its weight. We have

$$B_o + B_w = w \qquad (2)$$

Let us call x the height of the bar that is in oil. Thus, the portion of the height of the bar which is in water is $(2 \times 10^{-3} \text{ m} - x)$.
Equation (2) becomes

$$\rho_o(A)(x)g + \rho_w A(2 \times 10^{-2} \text{ m} - x)g = w$$

or, $\quad \rho_o x + \rho_w(2 \times 10^{-2} \text{ m} - x) = \dfrac{w}{Ag}$

However, from (1) above, this is

$$\rho_o x + \rho_w(2 \times 10^{-2} \text{ m} - x) = \rho_w(1.5 \times 10^{-2} \text{ m})$$

When we substitute 1000 kg/m³ for the density of water and 600 kg/m³ for the density of oil, we find $\quad x = 1.25 \times 10^{-2} \text{ m} = 1.25 \text{ cm}$

9.43 When in water the buoyant force is

$$B_w = \rho_w Vg = (10^3 \text{ kg/m}^3)(450 \times 10^{-6} \text{ m}^3)(9.8 \text{ m/s}^2) = 4.41 \text{ N}$$

When the object is in air, this buoyant force will no longer be present and an additional weight of 4.41 N will have to be supported by the wire. This much

134

additional force stretches the wire by 0.035 mm. From the definition of Young's modulus, we find

$$L = \frac{(\Delta L)(A)(Y)}{F} = \frac{(3.5 \times 10^{-5} \text{ m})(1.5 \times 10^{-5} \text{ m}^2)(11 \times 10^{10} \text{ Pa})}{4.41 \text{ N}} = 13.1 \text{ m}$$

9.44 Because there are two edges (the inside and outside of the ring) we have,

$$\gamma = F/2L$$

We have $L = 2\pi r = 2\pi(1.75 \times 10^{-2} \text{ m}) = 1.01 \times 10^{-1} \text{ m}$

and $\gamma = (1.61 \times 10^{-2} \text{ N})/2(1.01 \times 10^{-1} \text{ m}) = 7.32 \times 10^{-2} \text{ N/m}$

9.45 The force on one of the legs of the insect is $F = \gamma L$ where L is the circumference of the circular depression in which the foot rests; $L = 2\pi r$. The vertical component of this force is

$F_v = F\cos 45° = \gamma L \cos 45°$

$= (7.3 \times 10^{-2} \text{ N/m})(2\pi)(2.5 \times 10^{-3} \text{ m}) \cos 45° = 8.11 \times 10^{-4} \text{ N}$

The total support force will be $6F_v$, which is equal to the weight of the insect, w. Thus, we see that the mass of the insect is

$$m = \frac{6F_v}{g} = \frac{6(8.11 \times 10^{-4} \text{ N})}{9.8 \text{ m/s}^2} = 4.96 \times 10^{-4} \text{ kg} = 0.496 \text{ g}$$

9.46 Because we have a two-sided surface, the force on the ring of radius r (diameter d) is

$$F = 2\gamma L = 2\gamma(2\pi r) = 4\pi\gamma r = 2\pi\gamma d$$

This force is the same for ethanol and the tissue fluid. Thus,

$2\pi\gamma_e d_e = 2\pi\gamma_f d_f$

or, $d_f = \frac{\gamma_e}{\gamma_f} d_e = \frac{0.0227}{0.050} (5 \text{ cm}) = 2.27 \text{ cm}$

9.47 The tension in the string attaching the sheet to the balance is equal to the sum of the vertical component of the surface force plus the weight of the sheet. This is a two sided surface, so the surface force is $F = \gamma 2L$. The vertical component of this force is

$F_v = \gamma 2L \cos\theta$.

When $\theta = 0°$, the tension is 0.4 N and we have

$T = F_v + w$

$0.4 \text{ N} = \gamma 2L + w$ (1)

When $\theta = 180°$, $T = 0.39$ N, and we have

$T = F_v + w$

$0.39 \text{ N} = -\gamma 2L + w$ (2)

We have $L = 5 \times 10^{-2}$ m, so equations (1) and (2) can be solved for γ. We find,

$\gamma = 5 \times 10^{-2}$ N/m

9.48 The vertical component of the surface force is equal to the weight of water inside the capillary tube.

$F_v = w$

CHAPTER NINE SOLUTIONS

$$\gamma L\cos\theta = \gamma(2\pi r)\cos\theta = w = \rho(\pi r^2)hg$$

Where h is the height of the water in the tube. We solve the above for the surface tension.

$$\gamma = \frac{\rho rhg}{2\cos\theta} = \frac{(1080 \text{ kg/m}^3)(5 \times 10^{-4} \text{ m})(2.1 \times 10^{-2} \text{ m})(9.8 \text{ m/s}^2)}{2}$$
$$= 5.56 \times 10^{-2} \text{ N/m}$$

9.49 The height the blood can rise is given by

$$h = \frac{2\gamma\cos\theta}{\rho g r} = \frac{2(5.8 \times 10^{-2} \text{ N/m})}{(1050 \text{ kg/m}^3)(9.8 \text{ m/s}^2)(2 \times 10^{-6} \text{ m})} = 5.64 \text{ m}$$

9.50 We have

$$h = \frac{2\gamma\cos\theta}{\rho g r} = \frac{2(7.3 \times 10^{-2} \text{ N/m})}{(1000 \text{ kg/m}^3)(9.8 \text{ m/s}^2)(5 \times 10^{-5} \text{ m})} = 2.98 \times 10^{-1} \text{ m} = 29.8 \text{ cm}$$

9.51 We have

$$h = 5 \times 10^{-2} \text{ m} = \frac{2\gamma\cos\theta}{\rho g r} = \frac{2(8.8 \times 10^{-2} \text{ N/m})}{(1035 \text{ kg/m}^3)(9.8 \text{ m/s}^2) r}$$

From which, we find $r = 3.47 \times 10^{-4}$ m, or a diameter of 0.694 mm.

9.52 Because the level of the fluid is below the level of the reservoir, h is a negative number. Thus, we find

$$h = -5.36 \times 10^{-3} \text{ m} = \frac{2\gamma\cos\theta}{\rho g r} = \frac{2(0.465 \text{ N/m})\cos\theta}{(13.6 \times 10^3 \text{ kg/m}^3)(9.8 \text{ m/s}^2)(10^{-3} \text{ m})}$$

This can be solved to find
$$\cos\theta = -0.768$$
and $\theta = 140°$

9.53 We assume a length for the femur of 0.5 m. The amount of compression ΔL is given by,

$$\Delta L = \frac{L(\text{stress})}{Y} = \frac{(5 \times 10^{-1} \text{ m})(160 \times 10^6 \text{ Pa})}{14.5 \times 10^9 \text{ Pa}} = 5.52 \times 10^{-3} \text{ m} = 5.5 \text{ mm}$$

9.54 The area of the tip of the needle is
$$A = \pi r^2 = \pi(1.25 \times 10^{-4} \text{ m})^2 = 4.91 \times 10^{-8} \text{ m}^2.$$
Thus, the pressure exerted is
$$P = \frac{F}{A} = \frac{10^{-1} \text{ N}}{4.91 \times 10^{-8} \text{ m}^2} = 2.04 \times 10^6 \text{ Pa}.$$

9.55 We use,
$$\frac{F_2}{A_2} = \frac{F_1}{A_1}$$
or $F_2 = \frac{A_2}{A_1} F_1 = \frac{(5 \text{ cm})^2}{(1 \text{ cm})^2} (80 \text{ N}) = 2000 \text{ N}$

The force found above is exerted on each wheel. Thus, the total braking force is 8000 N.

CHAPTER NINE SOLUTIONS

9.56 We use $\Delta P = \rho g h$, where h is the height above the arm. Thus,
$$\Delta P = (1050 \text{ kg/m}^3)(9.8 \text{ m/s}^2)(1 \text{ m}) = 1.03 \times 10^4 \text{ Pa}.$$

9.57 (a) $\Delta P = 0.012 \text{ bar} = 1200 \text{ Pa}$
(b) $F = \Delta P A = (1200 \text{ Pa})\pi(0.004)^2 = 0.06 \text{ N}$

9.58 Consider that portion of the tire in contact with the ground; there are two forces acting on this segment; a downward force due to the air in the tire, F_{down}, and an upward contact force of the ground on the tire, F_{up}.

$$F_{down} = P_{gauge} A \quad \text{and} \quad F_{up} = \frac{w}{4}$$

So, $P_{gauge} A = \frac{w}{4}$

From which, $A = \frac{w}{4(P_{gauge})} = \frac{1.2 \times 10^4 \text{ N}}{4(200 \times 10^3 \text{ Pa})} = 1.5 \times 10^{-2} \text{ m}^2 = 150 \text{ cm}^2$

9.59 The buoyant force, B, on the iceberg must be equal to its weight, w, in order for it to float.
$$B = w$$
But, the buoyant force is equal to the weight of the water displaced. Thus,
$$\rho_w V_{uw} g = \rho_{ice} V_{total} g$$
where V_{uw} is the volume of the iceberg under water and V_{total} is the total volume of the berg.

We have, $\frac{V_{uw}}{V_{total}} = \frac{\rho_{ice}}{\rho_w} = \frac{920}{1000} = 0.92$

Therefore, 0.92 of the volume is submerged and 0.08 of the volume is exposed.

9.60 (a) The forces on the object are the tension, T, in the string connecting it to a balance, The weight of the object, w, and the buoyant force on it, B. Because the object is in equilibrium when immersed in the alcohol, we have $\Sigma F_y = 0$, which gives,
$$T + B = w$$
or $200 \text{ N} + B = 300 \text{ N}$
Thus, $B = 100 \text{ N}$
We also know that the buoyant force is equal to the weight of the displaced alcohol. So,
$$B = \rho_{alcohol} V_{alcohol} g$$
But, $V_{alcohol}$ is also equal to the volume of the object because the object is completely submerged. Thus,
$$V_{object} = \frac{B}{\rho_{alcohol} g} = \frac{100 \text{ N}}{(700 \text{ kg/m}^3)(9.8 \text{ m/s}^2)} = 1.46 \times 10^{-2} \text{ m}^3$$
(b) The mass of the 300 N object is 30.6 kg, and now that we know its volume, its density can be found as
$$\rho = \frac{m}{V} = 2.10 \times 10^3 \text{ kg/m}^3$$

9.61 We know that the shear modulus is given by

CHAPTER NINE SOLUTIONS

$$S = \frac{\text{shear stress}}{\text{shear strain}} = \frac{\text{stress}}{\left(\frac{\Delta x}{h}\right)}$$

or, Stress = $(S)(\Delta x/h) = (1.5 \times 10^{10} \text{ N/m}^2)(5 \text{ m}/10^4 \text{ m}) = 7.50 \times 10^6$ Pa

---

9.62 (a) When immersed in water the forces on the metal are the tension, T_w, in the string which supports it, B_w, the buoyant force of the water, and its weight, w. The metal is in equilibrium, so, $T_w + B_w = w$
or,   36 N + B_w = 50 N
Which gives, B_w = 14 N
Thus, $B_w = \rho_w V g = 14$ N   (1)
An identical analysis as the above for the case when the metal is immersed in oil gives
$$B_{oil} = \rho_{oil} V g = 9 \text{ N} \quad (2)$$
In both (1) and (2) V is the total volume of the metal because it is completely submerged. From (1), we find
$$V = \frac{14 \text{ N}}{(10^3 \text{ kg/m}^3)(9.8 \text{ m/s}^2)} = 1.43 \times 10^{-3} \text{ m}^3.$$
The mass of the 50 N piece of metal is 5.10 kg. Thus, its density is
$$\rho = \frac{m}{V} = \frac{5.10 \text{ kg}}{1.43 \times 10^{-3} \text{ m}^3} = 3.57 \times 10^3 \text{ kg/m}^3.$$
(b) The density of the oil is now found from equation (2), as
$$\rho_{oil} = \frac{9 \text{ N}}{(1.43 \times 10^{-3} \text{ m}^3)(9.8 \text{ m/s}^2)} = 643 \text{ kg/m}^3.$$

9.63 Because the amount of water in the tube remains a constant, we know that the length of water in the side arms of the tube is 0.40 m. The situation after the alcohol has

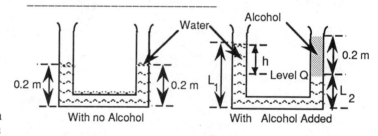

been added is shown in the sketch. We know that the added pressure at level Q must be F/A, where F equals the weight of a 20 cm column of alcohol. Therefore,
$$\rho_w g h = \frac{F}{A} = \frac{\rho_{alc} g A (0.2 \text{ m})}{A}$$
We solve for h to find,
$$h = \frac{\rho_{alc}(0.2 \text{ m})}{\rho_w} = \frac{(789.9)(0.2 \text{ m})}{1000} = 0.158 \text{ m}$$
But, $h + L_2 = L_1$
or, $L_1 = L_2 + .158$ m   (3)
We also recall that $L_1 + L_2 = 0.40$ m   (4)
From (3) and (4), we find
    $L_2 = 0.121$ m = 12.1 cm and $L_1 = 27.9$ cm

CHAPTER NINE SOLUTIONS

9.64 The forces on the balloon while in flight are B_a, the buoyant force of the air, w_{He}, the weight of the helium, w_B, the weight of the balloon, and w_L, the weight of the load. These quantities are found as follows,

$$B_a = \rho_a V_{balloon} g$$
$$w_{He} = \rho_{He} V_{balloon} g$$
$$w_B = (600 \text{ kg})g$$

and, $w_L = (4000 \text{ kg})g$

When floating in equilibrium, we have

$$B_a = w_{He} + w_B + w_L$$

or, $\rho_a V_{balloon} g = \rho_{He} V_{balloon} g + (600 \text{ kg})g + (4000 \text{ kg})g$

We are given the density of Helium as 0.178 kg/m^3 and the density of air as 1.29 kg/m^3. Thus, we can solve for the volume of the balloon to find

$$V_{balloon} = 4.14 \times 10^3 \text{ m}^3$$

9.65 The pressure at the surface of each liquid (in open containers) is given by

$$P = P_a + \rho_w g h_w = P_a + \rho g h. \quad \text{Therefore,} \quad \rho = \frac{h_w}{h} \rho_w$$

9.66 (a) $P = P_a + \rho_w g h = 1.013 \times 10^5 \text{ Pa} + (1.00 \text{ kg/m}^3)(9.8 \text{ m/s}^2)(1.50 \text{ m})$
$= 1.160 \times 10^5 \text{ Pa}$

(b) The tank bottom must now support both the weight of the water and the weight of the object, so $(P' - P)A = Mg = (P' - P)(\frac{\pi}{4})(6 \text{ m})^2 = (150 \text{ kg})(9.8 \text{ m/s}^2)$ gives

$$(P' - P) = 52.0 \text{ Pa}$$

9.67 The resultant force on the ball when it is beneath the surface of the water is B_w, the buoyant force of the water, minus the weight of the ball, w. This is, $B_w - w = \rho_w V_{ball} g - \rho_{ball} V_{ball} g$

We also know that the resultant force is equal to ma from Newton's second law. Thus, we find the acceleration as

$$a = \frac{(\rho_w V_{ball} g - \rho_{ball} V_{ball} g)}{m} = \frac{(\rho_w - \rho_{ball})g}{\rho_{ball}} = 88.2 \text{ m/s}^2. \quad \text{(Where the}$$

density of the ball is 100 kg/m^3.)

9.68 When the sinker alone is submerged, the forces on the system are T_1, the tension in the string attached to the block, w_B, the weight of the block, w_s, the weight of the sinker, and B_s, the buoyant force on the sinker. Since equilibrium exists, we have

$$T_1 = w_B + w_s - B_s \quad (1)$$

When both are submerged, the forces on the system are T_2, the tension in the string attached to the block, the weight of the block, the buoyant force on the block, B_B, and the buoyant force on the sinker. From the first condition for equilibrium, we have

$$T_2 = w_B + w_s - B_B - B_s \quad (2)$$

Subtract (2) from (1) to give

$$T_1 - T_2 = B_B$$

CHAPTER NINE SOLUTIONS

Thus, the buoyant force on the block is $B_B = 200\text{ N} - 140\text{ N} = 60\text{ N}$
However, the buoyant force on the block is equal to the weight of water displaced by the block.

$$B_B = 60\text{ N} = \rho_w V_{block} g$$

Thus, $V_{block} = \dfrac{60\text{ N}}{(1000\text{ kg/m}^3)(9.8\text{ m/s}^2)} = 6.12 \times 10^{-3}\text{ m}^3$.

The mass of the 50 N block is 5.10 kg. So, its density is

$$\rho = \dfrac{m}{V} = \dfrac{5.10\text{ kg}}{6.12 \times 10^{-3}\text{ m}^3} = 833\text{ kg/m}^3.$$

9.69 (a) When at rest, the tension in the cable is equal to the weight of the 800 kg mass, 7840 N. Thus, from the definition of Young's modulus, we find the amount the cable is stretched is

$$\Delta L = \dfrac{(F)(L)}{(Y)(A)} = \dfrac{(7840\text{ N})(25\text{ m})}{(20 \times 10^{10}\text{ Pa})(4 \times 10^{-4}\text{ m}^2)} = 2.45 \times 10^{-3}\text{ m} = 2.45\text{ mm}$$

(b) We write down Newton's second law for the block when it is accelerating upward.

$$T - mg = ma$$
or $\quad T = m(g + a) \quad (1)$

When $a = 3\text{ m/s}^2$, we find $T = 1.02 \times 10^4\text{ N}$

so, $\quad \Delta L_{new} = \dfrac{(F)(L)}{(Y)(A)} = \dfrac{(1.02 \times 10^4\text{ N})(25\text{ m})}{(20 \times 10^{10}\text{ Pa})(4 \times 10^{-4}\text{ m}^2)} = 3.20 \times 10^{-3}\text{ m} = 3.20\text{ mm}$

Therefore, the increase in elongation is 3.20 mm - 2.45 mm = 0.75 mm.

(c) If the stress (F/A) is not to exceed 2.2×10^8 Pa, the maximum force allowed is

$$F = T = (2.2 \times 10^8\text{ Pa})(4 \times 10^{-4}\text{ m}^2) = 8.8 \times 10^4\text{ N}$$

From (1) we find the largest mass to be

$$m = \dfrac{T}{a + g} = \dfrac{8.8 \times 10^4\text{ N}}{(3 + 9.8)\text{m/s}^2} = 6875\text{ kg}$$

9.70 We will first find the speed of the sphere just before impact by use of the free-fall equations.

$$v^2 = v_0^2 + 2ay$$
$$v^2 = 0 + 2(-9.8\text{ m/s}^2)(-10\text{ m})$$
or, $\quad v = -14\text{ m/s}$

This is the initial velocity of the sphere as it enters the water.
Once under water, the forces acting on the sphere are the buoyant force of the water, B, and its weight, w. Thus, the net force on the object is

$$F_{net} = B - w = \rho_w V g - \rho V g \quad (1)$$

But, we are given that the density of the sphere ρ is $\rho = 0.6\rho_w$. Thus, from (1) we see that the net force on the sphere is

$$F_{net} = 0.4\rho_w V g$$

From Newton's second law, the acceleration of the sphere is

$$a = \dfrac{F_{net}}{m} = \dfrac{F_{net}}{\rho V} = \dfrac{F_{net}}{0.6\rho_w V} = \dfrac{0.4\rho_w V g}{0.6\rho_w V} = \dfrac{2}{3} g$$

We can now find the stopping distance, h, as

$$v^2 = v_0^2 + 2ay$$

CHAPTER NINE SOLUTIONS

$0 = (-14 \text{ m/s})^2 + 2(\frac{2}{3}g)h$

$h = -15$ m The object sinks 15 m into the water.

9.71 (a) The pressure on the surface of the two hemispheres is constant at all points and the force on each element of surface area is directed along the radius of the hemispheres. The applied force along the axis must balance the force on the "effective" area which is the projection of the actual surface onto a plane perpendicular to the x axis, $A = \pi R^2$. Therefore,

$F = (P_a - P)\pi R^2$.

(b) For the values given

$F = (P_a - 0.1P_a)\pi(0.3 \text{ m})^2 = 0.254 P_a = 2.57 \times 10^4$ N

9.72 The Block floats, so

Total buoyant force = weight of block,

or

$F_{B(oil)} + F_{B(water)} =$ weight of block

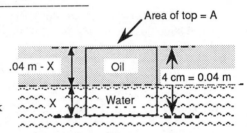

$\rho_{oil}(A(4 \times 10^{-2} \text{ m} - x))g + \rho_{water}(Ax)g = \rho_{block}(A(4 \times 10^{-2} \text{ m}))g$

where $A(4 \times 10^{-2} \text{ m} - x)$ = volume of oil displaced, and Ax = volume of water displaced.

Canceling A, and substituting into the equation above, we have

$(930 \text{ kg/m}^3)(4 \times 10^{-2} \text{ m} - x) + (1000 \text{ kg/m}^3)x = (960 \text{ kg/m}^3)(4 \times 10^{-2} \text{ m})$

Solving for x $x = 1.71 \times 10^{-2}$ m = 1.71 cm

9.73 We require that the buoyant force equal the total weight, or

$\rho_{air}gV = (Mg)_{balloon} + \rho_{He}gV$, and $V = \dfrac{M}{(\rho_a - \rho_{He})}$

9.74 Let us first find the density of the mixture. Let w = weight of mixture, w_e = weight of ethanol, and w_w = weight of water used.

Then, $\dfrac{w_e}{w} = \dfrac{m_e g}{m g} = \dfrac{m_e}{m} = 0.3$ (30% ethanol by weight)

and $\dfrac{w_w}{w} = \dfrac{m_w g}{m g} = \dfrac{m_w}{m} = 0.7$ (70% ethanol by weight)

Also, $V = V_e + V_w$ (Total volume = volume of ethanol plus volume of water)

or, $\dfrac{m}{\rho} = \dfrac{m_e}{\rho_e} + \dfrac{m_w}{\rho_w}$

So, $\dfrac{1}{\rho} = \dfrac{(m_e/m)}{\rho_e} + \dfrac{(m_w/m)}{\rho_w}$

or, $\dfrac{1}{\rho} = \dfrac{0.3}{\rho_e} + \dfrac{0.7}{\rho_w}$

Substituting the appropriate values for the densities gives

$\rho = 0.933 \times 10^3$ kg/m^3

Now, find the apparent weight of the olive.

Apparent weight = true weight - buoyant force = $mg - (\rho V g)$

CHAPTER NINE SOLUTIONS

$$= (5 \times 10^{-3} \text{ kg})(9.8 \text{ m/s}^2) - (0.933 \times 10^3 \text{ kg/m}^3)(\tfrac{4}{3}\pi\,(10^{-2})^3)(9.8 \text{ m/s}^2)$$

$$= 1.07 \times 10^{-2} \text{ N}$$

9.75 To have the object float fully submerged in the fluid, its density (average) must be the same as that of the fluid. Therefore, we must add ethanol to water until the density of the mixture is 900 kg/m^3.

$m = m_e + m_w$ (1)   (where m = mass of mixture, m_e = mass of ethanol, and m_w = mass of water)

and $\quad V = V_e + V_w$

(1) becomes $\quad \rho V = \rho_e V_e + \rho_w V_w$

which reduces to $\quad \dfrac{\rho}{\rho_w}(V_e + V_w) = \dfrac{\rho_e}{\rho_w}(V_e) + V_w$

or $\quad 0.9(V_e + V_w) = (0.806)(V_e) + V_w$

which becomes $\quad V_e = 1.064 V_w = (1.064)(500 \text{ cc}) = 532 \text{ cc}$

9.76 (a) $\bar{\rho} = \dfrac{M}{V} = \dfrac{(\rho_1 s^2 h_1 + \rho_2 s^2 h_2)}{s^2(h_1 + h_2)} = \dfrac{(\rho_1 h_1 + \rho_2 h_2)}{(h_1 + h_2)}$

(b) We need $\rho_w s^2 d = M$ so $d = \dfrac{M}{\rho_w s^2} = \dfrac{(\rho_1 h_1 + \rho_2 h_2)}{\rho_w}$

(c) Same as above. $d' = d$.

9.77 (a) Since the upward buoyant force is balanced by the weight of the sphere, we have

$$m_1 g = \rho V g = \rho(\tfrac{4}{3}\pi R^3) g$$

In this problem, $\rho = 0.78945$ g/cm^3 at 20°C, and $R = 1$ cm, so we find

$$m_1 = \rho(\tfrac{4}{3}\pi R^3) = (0.78945 \text{ g/cm}^3)(\tfrac{4}{3}\pi)(1 \text{ cm}^3)$$

$$m_1 = 3.307 \text{ g}$$

(b) Following the same procedure as in part (a), with $\rho' = 0.78097$ g/cm^3 at 30°C, we find

$$m_2 = \rho'(\tfrac{4}{3}\pi R^3) = (0.78097 \text{ g/cm}^3)(\tfrac{4}{3}\pi)(1 \text{ cm}^3)$$

$$m_2 = 3.271 \text{ g}$$

(c) When the first sphere is resting on the bottom of the tube, we see that

$$n + B = w_1 = m_1 g$$

where n is the normal force. But, $B = \rho' V g$, so

$$n = m_1 g - \rho' V g$$

$$= (3.307 \text{ g})(980 \text{ cm/s}^2) - (0.78097 \text{ g/cm}^3)(\tfrac{4}{3}\pi)(1 \text{ cm}^3)(980 \text{ cm/s}^2)$$

$$n = 35 \text{ dynes} = 3.5 \times 10^{-4} \text{ N}$$

CHAPTER TEN SOLUTIONS

10.1 (a) The flow rate is Av, which for this case is 0.25 liters/s = 250 cm^3/s. Thus, the velocity through the area of $\pi/4$ cm^2 is found as

$$v = \frac{\text{flow rate}}{A} = \frac{250 \text{ cm}^3/\text{s}}{\pi/4 \text{ cm}^2} = 318 \text{ cm/s} = 3.18 \text{ m/s}$$

(b) With the area of the smaller tube equal to 0.141 π cm^2, the continuity equation becomes

$$v_2 = \frac{A_1 v_1}{A_2} = \frac{250 \text{ cm}^3}{.141 \pi \text{ cm}^2} = 566 \text{ cm/s} = 5.66 \text{ m/s}$$

10.2 The cross-sectional area of the 2 cm diameter hose is A = $\pi \times 10^{-4}$ m^2, and the flow rate from the hose is

$$Av = (\pi \times 10^{-4} \text{ m}^2)(1.5 \text{ m/s}) = 1.5 \pi \times 10^{-4} \text{ m}^3/\text{s}.$$

The volume to be filled is (1.5 m)(0.6 m)(0.4 m) = 3.6 $\times$ 10^{-1} m^3. The time required to fill the trough is

$$\text{time} = \frac{\text{volume}}{\text{flow rate}} = \frac{3.6 \times 10^{-1} \text{ m}^3}{1.5 \pi \times 10^{-4} \text{ m}^3/\text{s}} = 764 \text{ s} = 12.7 \text{ min}$$

10.3 $A_1 = \frac{A_2 v_2}{v_1}$ Where A_2 = 1.96 $\times$ 10^{-5} m^2 = area of aorta. Then A_1 = total capillary cross-section needed.

$$A_1 = \frac{(1.96 \times 10^{-5} \text{ m}^2)(1 \text{ m/s})}{10^{-2} \text{ m/s}} = 1.96 \times 10^{-3} \text{ m}^2$$

But, A_1 = (number of capillaries)$A_{\text{single capillary}}$

And $A_{\text{single capillary}}$ = 7.85 $\times$ 10^{-11} m^2

So number of capillaries = $\frac{1.96 \times 10^{-3} \text{ m}^2}{7.85 \times 10^{-11} \text{ m}^2}$ = 2.5 $\times$ 10^7 = 25 million

---

10.4 (a) $v_2 = \frac{A_1 v_1}{A_2} = \frac{\pi(1 \text{ cm})^2(25 \text{ cm/s})}{\pi (0.15 \text{ cm})^2}$ = 1.11 $\times$ 10^3 cm/s = 11.1 m/s

(b) flow rate = Av = $A_1 V_1 = \pi(10^{-2} \text{ m})^2(0.25 \text{ m/s})$ = 7.85 $\times$ 10^{-5} m^3/s

The time for 4.0 $\times$ 10^{-3} m^3 to flow is

$$t = \frac{\text{quantity}}{\text{flow rate}} = \frac{4.0 \times 10^{-3} \text{ m}^3}{7.85 \times 10^{-5} \text{ m}^3/\text{s}} = 50.9 \text{ s}$$

10.5 The area of the water pipe is A = $\pi r^2 = \pi(1.5 \times 10^{-2} \text{ m})^2$ = 7.07 $\times$ 10^{-4} m^2. The flow rate is

$$Av = (7.07 \times 10^{-4} \text{ m}^2)(0.636 \text{ m/s}) = 4.50 \times 10^{-4} \text{ m}^3/\text{s} = 450 \text{ cm}^3/\text{s}.$$

10.6 Consider Bernoulli's equation

$$P_1 + \frac{1}{2}\rho v_1^2 + \rho g y_1 = P_2 + \frac{1}{2}\rho v_2^2 + \rho g y_2$$

We choose point 1 to be at the top of the fluid, and point 2 at the position of the puncture. At both of these locations, the pressure is atmospheric and cancels from Bernoulli's

CHAPTER TEN SOLUTIONS

equation. We also choose the zero level for the gravitational potential energy per unit volume to be at the level of the hole. Finally, the level of the water falls slowly enough that we set $v_2 = 0$. We have

$$\frac{1}{2}\rho v_1^2 = \rho g y_2$$

or, $v_1 = \sqrt{2(9.8 \text{ m/s}^2)(4 \text{ m})} = 8.85 \text{ m/s}$

10.7 (a) We find the flow velocity in the second section from the continuity equation.

$$v_2 = \frac{A_1 v_1}{A_2} = \frac{10}{2.5} v_1 = 4(2.75 \text{ m/s}) = 11 \text{ m/s}$$

(b) Choosing the zero level for y along the common center line of the pipes, we have

$$P_1 + \frac{1}{2}\rho v_1^2 = P_2 + \frac{1}{2}\rho v_2^2$$

or, $P_2 = P_1 + \frac{1}{2}\rho(v_1^2 - v_2^2)$

$= (1.2 \times 10^5 \text{ Pa}) + \frac{1}{2}(1650 \text{ kg/m}^3)[(2.75 \text{ m/s})^2 - (11 \text{ m/s})^2]$

$P_2 = 2.64 \times 10^4 \text{ Pa}$

10.8 $\Delta P = P_2 - P_1 = \frac{1}{2}\rho(v_1^2 - v_2^2)$ (Where we have ignored any differences in height.)

$\Delta P = \frac{1}{2}(1.29 \text{ kg/m}^3)((0.15 \text{ m/s})^2 - (0.30 \text{ m/s})^2) = -4.35 \times 10^{-2} \text{ Pa}$

10.9 $P_{1\text{gauge}} = P_1 - P_{atm} = \frac{F}{A_1} = \frac{2 \text{ N}}{2.5 \times 10^{-5} \text{ m}^2} = 8 \times 10^4 \text{ Pa}$

We write Bernoulli's equation as

$$\frac{1}{2}\rho v_2^2 = (P_1 - P_2) + \frac{1}{2}\rho v_1^2 + \rho g(y_1 - y_2)$$

The last term goes to zero because the syringe is in a horizontal position. Also, we realize that $P_1 - P_2 = P_1 - P_{atm} = P_{1\text{gauge}} = 8 \times 10^4 \text{ Pa}$

Finally, we assume $v_1 = 0$ in comparison to the speed inside the needle. Thus, with these substitutions, we find

$v_2 = 12.6 \text{ m/s}$

10.10 (a) We choose point 2 just beneath the roof where the velocity of the air is zero, and point 1 just above the roof. The difference in height between these points is negligible, and thus, Bernoulli's equation reduces to

$$P_2 - P_1 = \frac{1}{2}\rho v_1^2 = \frac{1}{2}(1.3 \text{ kg/m}^3)(30 \text{ m/s})^2 = 585 \text{ Pa}$$

Thus, we see that $P_2 > P_1$ and the net force is directed outward.

(b) This produces a net force on the house of

$F = (P_2 - P_1)A = (585 \text{ Pa})(175 \text{ m}^2) = 1.02 \times 10^5 \text{ N}$ upward.

10.11 We select point 1 just above the wing and point 2 just below it. As a result, the difference in vertical heights between these two points is negligible, and Bernoulli's equation reduces to

$$P_2 - P_1 = \frac{1}{2}\rho(v_1^2 - v_2^2) = \frac{1}{2}(1.3 \text{ kg/m}^3)[(300 \text{ m/s})^2 - (280 \text{ m/s})^2]$$

$= 7540 \text{ Pa}$

The net upward force is

$$\frac{250 \text{ cm}^3/\text{s}}{(.5)^2 \pi} = \frac{250}{.25\pi}$$

$$\frac{250}{\frac{\pi}{4}} = \frac{1000}{\pi}$$

The Luther Touch

381-0566
Purle

JAMES DIEMER -
7772 ARLINGTON AVE N.
BROOKLYN PARK, MN 55443

(612) 561-8946

(PAULINE SEGLEM)

Beautiful
appreciate

LutherHospital
· A Mayo Regional Hospital ·

CHAPTER TEN SOLUTIONS

$F = (P_2 - P_1)A = (7540 \text{ Pa})(20 \text{ m}^2) = 1.51 \times 10^5$ N upward.

10.12 The continuity equation tells us that Av = constant. The cross-sectional area of the circular pipe is a constant, and therefore, the velocity must also be a constant at all points. We will select point b at the top of the loop and point d at the point which is 0.5 m below the top of the loop. We also choose the reference level from which to measure y to be at the bottom of the loop. Bernoulli's equation becomes,

$$P_b + \rho g y_b = P_d + \rho g y_d$$

or, $P_b = P_d + \rho g(y_d - y_b)$
   $= 1.3 \times 10^4 \text{ Pa} + (1050 \text{ kg/m}^3)(9.8 \text{ m/s}^2)(1.3 \text{ m} - 1.8 \text{ m})$
   $= 7.86 \times 10^3$ Pa

(b) We parallel the approach used in part (a) with point c at the bottom of the loop to find

$P_c = P_d + \rho g(y_d - y_c) = 1.3 \times 10^4 \text{ Pa} + (1050 \text{ kg/m}^3)(9.8 \text{ m/s}^2)(1.3 \text{ m} - 0)$
   $= 2.64 \times 10^4$ Pa.

10.13 The velocity in the lower pipe, point 2, is given by the continuity equation.

$$v_2 = \frac{A_1 v_1}{A_2} = 4v_1 \quad (1)$$

We find v_1 the velocity in the upper pipe, point 1, from the flow rate.

$A_1 v_1 = 0.2 \text{ m}^3/\text{s}$

or $v_1 = \dfrac{0.2 \text{ m}^3/\text{s}}{\pi(0.3 \text{ m})^2} = 0.707$ m/s

Therefore, from (1) $v_2 = 2.83$ m/s

We now use Bernoulli's equation with the reference level for y at the level of the upper tube. We also note that the gauge pressure in the upper pipe is zero. Thus, we find

$$P_2 = \frac{1}{2}\rho(v_1^2 - v_2^2) + \rho g(y_1 - y_2)$$

$= \dfrac{1}{2}(10^3 \text{ kg/m}^3)[(0.707 \text{ m/s})^2 - (2.83 \text{ m/s})^2]$
   $+ (10^3 \text{ kg/m}^3)(9.8 \text{ m/s}^2)(0.6 \text{ m})$
$= 2125$ Pa which is the gauge pressure.

The absolute pressure is 1.03×10^5 Pa.

10.14 (a) Flow Rate = $Av = (2 \text{ cm}^2)(40 \text{ cm/s}) = 80 \text{ cm}^3/\text{s}$

However, since blood has a mass of 1 g/cm^3, this is equivalent to a flow rate of 80 g/s.

(b) From the equation of continuity, we have

$$v_2 = \frac{A_1 v_1}{A_2} = \frac{2}{3000}(40 \text{ cm/s}) = 2.67 \times 10^{-2} \text{ cm/s}$$

10.15 We choose the top surface of the water as point 1 and the location of the hole in the tank as point 2. Both of these points are at atmospheric pressure, and we take the velocity of the water at point 1 to be negligible. As a result, Bernoulli's equation reduces to

$$\rho g y_1 = \frac{1}{2}\rho v_2^2 + \rho g y_2$$

and $v_2^2 = 2g(y_1 - y_2) = 2g(1.5 \text{ m})$
   $v_2 = 5.42$ m/s

Now consider the water to behave as a projectile with an initial horizontal velocity of 5.42 m/s. The time to reach the floor is

145

CHAPTER TEN SOLUTIONS

$$y = v_{0y}t + \frac{1}{2}at^2$$

$$-1 \text{ m} = 0 + \frac{1}{2}(-9.8 \text{ m/s}^2)t^2$$

$$t = 0.452 \text{ s}$$

Now that we know the time of flight, we can find the range of the water droplets.

$$R = v_{0x}t = (5.42 \text{ m/s})(0.452 \text{ s}) = 2.45 \text{ m}$$

10.16 First, consider the path from the standpoint of projectile motion to find the speed at which the water emerges from the tank. The time to drop one meter with an intial vertical velocity of zero is

$$y = v_{0y}t + \frac{1}{2}at^2$$

$$1 \text{ m} = 0 + \frac{1}{2}(9.8 \text{ m/s}^2)t^2$$

$$t = 0.452 \text{ s}$$

and from the horizontal motion

$$v_x = v_0 = \frac{\Delta x}{t} = \frac{0.6 \text{ m}}{0.452 \text{ s}} = 1.33 \text{ m/s}$$

We now use Bernoulli's equation, with point 1 at the top of the tank and point 2 at the level of the hole. With $P_1 = P_2 = P_{atm}$ and v_1 approximately equal to zero, we have

$$\frac{1}{2}\rho v_2^2 = \rho g(y_1 - y_2) = \rho g h$$

$$h = \frac{v_0^2}{2g} = \frac{(1.33 \text{ m/s})^2}{2(9.8 \text{ m/s}^2)} = 9.0 \times 10^{-2} \text{ m} = 9 \text{ cm}$$

10.17 The gauge pressure in the large section (to left) = 10 cm of water
$$= \rho g h_1 = P_1 - P_{atm}$$

The gauge pressure in the constriction = 5 cm of water = $\rho g h_2 = P_2 - P_{atm}$

Now use Bernoulli's equation

$$\frac{1}{2}\rho v_2^2 = \frac{1}{2}\rho v_1^2 + P_1 - P_2$$

where the terms involving elevation drop out.
Also, $P_1 - P_2 = P_{atm} + \rho g h_1 - P_{atm} + \rho g h_2 = \rho g(h_2 - h_2)$
We need v_1, and to find this we turn to the equation of continuity.

flow rate = $A_1 v_1 = 1.8 \times 10^{-4}$ m^3/s   and $A_1 = 4.91 \times 10^{-4}$ m^2

from which $v_1 = 0.367$ m/s
We can now find v_2 from Bernoulli's equation,
  $v_2 = 1.06$ m/s
and from the flow rate, we find

$$A_2 = \frac{1.8 \times 10^{-4} \text{ m}^3/\text{s}}{1.06 \text{ m/s}} = 1.71 \times 10^{-4} \text{ m}^2 = \frac{\pi d_2^2}{4}$$

from which $d_2 = 1.47 \times 10^{-2}$ m = 1.47 cm

10.18 (a) The flow rate, Av, as given may be expressed as follows
  25 liters/30 s = 0.833 liters/s = 833 cm^3/s
The area of the faucet tap is π cm^2, so we can find the velocity as

$$v = \frac{\text{flow rate}}{A} = \frac{833 \text{ cm}^3/\text{s}}{\pi \text{ cm}^2} = 265 \text{ cm/s} = 2.65 \text{ m/s}$$

(b) We choose point 1 to be in the entrance pipe and point 2 to be at the faucet tap. Bernoulli's equation is

CHAPTER TEN SOLUTIONS

$$P_1 - P_2 = \tfrac{1}{2}\rho(v_2^2 - v_1^2) + \rho g(y_2 - y_1)$$

$$P_1 - P_2 = \tfrac{1}{2}(10^3 \text{ kg/m}^3)[(2.65 \text{ m/s})^2 - (0.295 \text{ m/s})^2]$$
$$+ (10^3 \text{ kg/m}^3)(9.8 \text{ m/s}^2)(2 \text{ m})$$
$$= 2.31 \times 10^4 \text{ Pa}$$

Since P_2 is atmospheric pressure, this is the gauge pressure.

10.19 We can find the velocity of the water as it exits the hose by use of the equations of motion with constant acceleration.

$$v^2 = v_0^2 + 2ay$$
$$0 = v_0^2 + 2(-9.8 \text{ m/s}^2)(4 \text{ m})$$
$$v_0 = 8.85 \text{ m/s}$$

We choose point 1 just inside the nozzle and point 2 just outside the nozzle. (The pressure at point 2 is atmospheric.) The difference in elevation between these points is negligible, and as a result, the potential energy per unit volume terms in Bernoulli's equation cancels. Also, we shall take the velocity inside the hose to be negligible in comparison to that of the water as it sprays out. Thus, we have

$$P_1 - P_a = \tfrac{1}{2}\rho v_2^2 = \tfrac{1}{2}(10^3 \text{ kg/m}^3)(8.85 \text{ m/s})^2 = 3.92 \times 10^4 \text{ Pa}$$

10.20 (a) We choose point 1 at the surface of the tank and point 2 at the hole. Both of these points are at atmospheric pressure, so the pressure cancels from Bernoulli's equation. We also assume that v_1 is negligibly small. Finally, we choose the zero level for y at the level of the hole. Under these conditions, we have

$$v_2 = \sqrt{2gy} = \sqrt{2g(16 \text{ m})} = 17.7 \text{ m/s}$$

(b) The area of the hole is found from the flow rate as

$$A = \frac{\text{flow rate}}{v_2} = \frac{4.17 \times 10^{-5} \text{ m}^3/\text{s}}{17.7 \text{ m/s}} = 2.35 \times 10^{-6} \text{ m}^2.$$

From which the diameter is easily found to be 1.73 mm.

10.21 We call the position of the lower hole point 1 and the position of the higher hole point 2. Both of these points are at atmospheric pressure.
From our earlier study of projectile motion, the range is given by

$$R = v_1 t_1 = v_2 t_2 \quad (1)$$

An expression for the time for the water to reach the floor can also be found through the projectile motion equations. We have

$$y = v_{0y}t + \tfrac{1}{2}at^2$$
$$y = 0 + \tfrac{1}{2}gt^2$$

or, $t = \sqrt{\tfrac{2}{g}h}$ where h is the height through which the projectile falls before it reaches ground level. Thus, we have the time for the water from the upper hole to reach the floor to be

$$t_2 = \sqrt{\tfrac{2}{g}(0.12 \text{ m})} \quad \text{and the time for the water from the lower hole as}$$

$$t_1 = \sqrt{\tfrac{2}{g}(0.05 \text{ m})}$$

We now substitute these two values for the times into (1) above to give

$$v_1^2 = 2.4 v_2^2 \quad (2)$$

CHAPTER TEN SOLUTIONS

Since the pressure cancels from Bernoulli's equation, it reduces to

$$(v_1^2 - v_2^2) = 2g(y_2 - y_1) \quad (3)$$

$y_2 = 0.12$ m, $y_1 = 0.05$ m and with the use of (2) and (3) we find

$$v_2 = 0.99 \text{ m/s}$$

Let us now consider Bernoulli's Equation once again, with a point 3 at the top of the tank and point 2 still at the position of the top hole. Both these points are at atmospheric pressure, and the velocity of fall of water at the top of the tank is negligibly small. We find

$$\rho g y_3 = \frac{1}{2}\rho v_2^2 + \rho g y_2$$

or $\rho g(y_3 - y_2) = \rho g h = \frac{1}{2}\rho v_2^2$

From which, $h = \dfrac{v_2^2}{2g} = \dfrac{(0.99 \text{ m/s})^2}{2(9.8 \text{ m/s}^2)} = 5 \times 10^{-2}$ m = 5 cm

Thus, the water surface is 5 cm above the top hole or 17 cm above the bottom of the tank.

10.22 With the zero level for y selected at the center line of the two pipes, Bernoulli's equation reduces to

$$P_1 + \frac{1}{2}\rho v_1^2 = P_2 + \frac{1}{2}\rho v_2^2$$

or, $P_1 - P_2 = \frac{1}{2}\rho(v_2^2 - v_1^2) \quad (1)$

Point 1 is an arbitrary point in the tube of large diameter, and point 2 a similar point in the smaller tube. From the continuity equation, we find

$$v_1 = \frac{A_2 v_2}{A_1} = \frac{(0.5)^2}{1} v_2 = 0.25 v_2 \quad (2)$$

We see from (2) that $v_2 > v_1$ and therefore, from (1) we find that $P_1 > P_2$. Thus, the pressure is higher in the larger pipe.

(b) Let us now substitute for v_1 from equation (2) into (1) and use the fact that $P_1 - P_2 = 6.66 \times 10^3$ Pa to find

$$6.66 \times 10^3 \text{ Pa} = \frac{1}{2}(1000 \text{ kg/m}^3)(v_2^2 - 0.0625 v_2^2)$$

From which,

$$v_2 = 3.77 \text{ m/s}$$

Then the flow rate is found from the equation of continuity.

$$A_2 v_2 = \pi(5 \times 10^{-3} \text{ m})^2(3.77 \text{ m/s}) = 2.96 \times 10^{-4} \text{ m}^3/\text{s} = 296 \text{ cm}^3/\text{s}$$

10.23 From the definition of the coefficient of viscosity, we have

$$F = \frac{\eta A v}{L} = \frac{(1.79 \times 10^{-3} \text{ Ns/m}^2)(0.8 \text{ m})(1.2 \text{ m})(0.5 \text{ m/s})}{10^{-4} \text{ m}} = 8.59 \text{ N}$$

10.24 $\eta = \dfrac{FL}{Av} = \dfrac{(1.9 \text{ N})(10^{-3} \text{ m})}{(4.8 \times 10^{-2} \text{ m}^2)(0.5 \text{ m/s})} = 7.92 \times 10^{-2}$ N s/m^2.

10.25 From the definition of the coefficient of viscosity, we have

$$F = \frac{\eta A v}{L} = \frac{(1500 \times 10^{-3} \text{ Ns/m}^2)(4 \times 10^{-4} \text{ m}^2)(0.3 \text{ m/s})}{1.5 \times 10^{-3} \text{ m}} = 0.120 \text{ N}$$

10.26 From Poiseuille's law

CHAPTER TEN SOLUTIONS

$$P_1 - P_2 = \frac{(\text{flow rate})8\eta L}{\pi R^4}$$

$$= \frac{(8.6 \times 10^{-5} \text{ m}^3/\text{s})8(0.12 \text{ Ns/m}^2)(50 \text{ m})}{\pi(5 \times 10^{-3} \text{ m})^4} = 2.1 \times 10^6 \text{ Pa} = 20.7 \text{ atm}.$$

Also, since $P_2 = 1$ atm, this is also the gauge pressure at the inlet point of the pipe.

10.27 Poiseuille's law gives the flow rate as

$$\text{Flow rate} = \frac{(\Delta P)\pi R^4}{8L\eta}$$

Therefore
$$\Delta P = \frac{(\text{flow rate})8L\eta}{\pi R^4}$$

$$= \frac{(10^{-6} \text{ m}^3/\text{s})8(0.03 \text{ m})(1 \times 10^{-3} \text{ N s/m}^2)}{\pi(1.5 \times 10^{-4} \text{ m})^4} = 1.51 \times 10^5 \text{ Pa}$$

10.28 (a) The pressure differential across the needle is

$$\Delta P = \rho g y = (1050 \text{ kg/m}^3)(9.8 \text{ m/s}^2)(1.0 \text{ m}) = 1.03 \times 10^4 \text{ Pa}.$$

Then we find the flow rate as

$$\text{Flow rate} = \frac{(\Delta P)\pi R^4}{8L\eta} = \frac{(1.03 \times 10^4 \text{ Pa})\pi(3 \times 10^{-4} \text{ m})^4}{8(3 \times 10^{-2} \text{ m})(4 \times 10^{-3} \text{ N s/m}^2)}$$

$$= 2.73 \times 10^{-7} \text{ m}^3/\text{s} = 2.73 \times 10^{-1} \text{ cm}^3/\text{s}$$

(b) At this flow rate, the time to inject 500 cm^3 is

$$\Delta t = \frac{500 \text{ cm}^3}{0.273 \text{ cm}^3/\text{s}} = 1.83 \times 10^3 \text{ s} = 30.6 \text{ min}$$

10.29 flow rate = 30.0 liter/min = 5×10^{-4} m^3/s
From Poiseuille's law, we have

$$R^4 = \frac{(\text{rate})8L\eta}{\pi(\Delta P)} = \frac{(5 \times 10^{-4} \text{ m}^3/\text{s})8(10^{-3} \text{ Pa s})(10^2 \text{ m})}{\pi(5.06 \times 10^4 \text{ Pa})}$$

Which gives $R = 7.08$ mm

10.30 The required flow rate is

$$\text{flow rate} = \frac{500 \text{ cm}^3}{1800 \text{ s}} = 2.78 \times 10^{-1} \text{ cm}^3/\text{s} = 2.78 \times 10^{-7} \text{ m}^3/\text{s}$$

If the solution is elevated 1 m, the pressure differential across the needle is

$$\Delta P = \rho g y = (1000 \text{ kg/m}^3)(9.8 \text{ m/s}^2)(1.0 \text{ m}) = 9800 \text{ Pa}$$

We find the radius via Poiseuille's law.

$$R^4 = \frac{8L\eta(\text{flow rate})}{\pi \Delta P}$$

$$= \frac{8(2.5 \times 10^{-2} \text{ m})(1 \times 10^{-3} \text{ N s/m}^2)(2.78 \times 10^{-7} \text{ m}^3/\text{s})}{\pi(9800 \text{ Pa})}$$

From which we find
$R = 2.06 \times 10^{-4}$ m $= 0.206$ mm, or diameter $= 0.412$ mm.

10.31 $\text{Flow rate} = \frac{(\Delta P)\pi R^4}{8L\eta} = \frac{(400 \text{ Pa})\pi(2.6 \times 10^{-3} \text{ m})^4}{8(4 \times 10^{-3} \text{ Pa s})(8.4 \times 10^{-2} \text{ m})}$

CHAPTER TEN SOLUTIONS

$$= 2.14 \times 10^{-5} \text{ m}^3/\text{s}$$

Then $v = \dfrac{\text{flow rate}}{\text{area}} = \dfrac{2.14 \times 10^{-5} \text{ m}^3/\text{s}}{\pi(2.6 \times 10^{-3} \text{ m})^2} = 1.01 \text{ m/s}$

10.32 From the definition of the Reynolds's number,

$$v_{max} = \frac{(RN)\eta}{\rho d} = \frac{(2000)(10^{-3} \text{ Ns/m}^2)}{(10^3 \text{ kg/m}^3)(2.5 \times 10^{-2} \text{ m})} = 8 \times 10^{-2} \text{ m/s} = 8 \text{ cm/s}$$

10.33 From the definition of the Reynolds's number,

$$RN = \frac{\rho v d}{\eta} = \frac{(850 \text{ kg/m}^3)(3 \text{ m/s})(1.2 \text{ m})}{(0.3 \text{ Ns/m}^2)} = 10,200$$

The onset of turbulence is at RN about 3000, so this is definitely turbulent.

10.34 The minimum value of the Reynold's Number for turbulent flow is 3000. From the definition of the Reynold's Number, we find the minimum velocity for turbulent flow as

$$v = \frac{\eta(RN)}{\rho d} = \frac{(10^{-3} \text{ Ns/m}^2)(3000)}{(10^3 \text{ kg/m}^3)(5 \times 10^{-3} \text{ m})} = 0.6 \text{ m/s}$$

10.35 The velocity of the blood is found from the definition of the Reynolds's number as,

$$v = \frac{\eta(RN)}{\rho d} = \frac{(4 \times 10^{-3} \text{ N s/m}^2)(980)}{(1050 \text{ kg/m}^3)(4.5 \times 10^{-3} \text{ m})} = 0.830 \text{ m/s}$$

10.36 The Reynold's number is

$$RN = \frac{\rho v d}{\eta} = \frac{(1050 \text{ kg/m}^3)(0.55 \text{ m/s})(2 \times 10^{-2} \text{ m})}{4 \times 10^{-3} \text{ N s/m}^2} = 2890$$

In this region the flow is unstable, but not necessarily turbulent.

10.37 Fick's law enables us to find the difference in concentration as

$$\Delta C = \frac{(\text{diffusion rate})L}{DA} = \frac{(5.33 \times 10^{-15} \text{ kg/s})(0.1 \text{ m})}{(5 \times 10^{-10} \text{ m}^2/\text{s})(6 \times 10^{-4} \text{ m}^2)} = 1.78 \times 10^{-3} \text{ kg/m}^3$$

10.38 From Fick's law, we find the concentration gradient by

$$\frac{\Delta C}{L} = \frac{(\text{diffusion rate})}{DA} = \frac{(5. \times 10^{-15} \text{ kg/s})}{(5 \times 10^{-10} \text{ m}^2/\text{s})(10^{-4} \text{ m}^2)} = 0.10 \text{ kg/m}^4.$$

10.39 We use Fick's law to find the diffusion coefficient.

$$D = \frac{\text{diffusion rate}}{A(\frac{\Delta C}{L})} = \frac{5.7 \times 10^{-15} \text{ kg/s}}{(2 \times 10^{-4} \text{ m}^2)(3 \times 10^{-2} \text{ kg/m}^4)} = 9.50 \times 10^{-10} \text{ m}^2/\text{s}$$

10.40 We have

$$\text{Diffusion rate} = DA\frac{\Delta C}{L} = (5 \times 10^{-10} \text{ m}^2/\text{s})(4 \times 10^{-4} \text{ m}^2)(0.2 \text{ kg/m}^4)$$
$$= 4 \times 10^{-14} \text{ kg/s}$$

In 10 s the amount diffused is

$$\Delta m = (\text{diffusion rate})\Delta t = 4.00 \times 10^{-13} \text{ kg}$$

CHAPTER TEN SOLUTIONS

10.41 We use
$$v_t = \frac{2r^2 g}{9\eta}(\rho - \rho_f)$$
or $(\rho - \rho_f) = \frac{9\eta v_t}{2r^2 g} = \frac{9(1 \times 10^{-3} \text{ N s/m}^2)(1.1 \times 10^{-2} \text{ m/s})}{2(5 \times 10^{-4})^2(9.8 \text{ m/s}^2)} = 20.2 \text{ kg/m}^3$.

Therefore, $\rho = \rho_f + 20.2 \text{ kg/m}^3 = 1000 \text{ kg/m}^3 + 20.2 \text{ kg/m}^3 = 1.02 \times 10^3 \text{ kg/m}^3$

10.42 The terminal velocity is
$$v_t = \frac{2r^2 g}{9\eta}(\rho - \rho_f) = \frac{2(2 \times 10^{-6} \text{ m})^2(9.8 \text{ m/s}^2)}{9(1 \times 10^{-3} \text{ N s/m}^2)}(1800 \text{ kg/m}^3 - 1000 \text{ kg/m}^3)$$
$$= 6.97 \times 10^{-6} \text{ m/s}$$
Thus, the time is
$$t = d/v_t = (1 \times 10^{-1} \text{ m})/(6.97 \times 10^{-6} \text{ m/s}) = 1.43 \times 10^4 \text{ s} = 3.99 \text{ h}$$

10.43 $F = 6\pi \eta r v = 6\pi(1.8 \times 10^{-5} \text{ N s/m}^2)(1 \times 10^{-7} \text{ m})(4 \times 10^{-4} \text{ m/s})$
$= 1.36 \times 10^{-14} \text{ N}$

10.44 From $F = 6\pi \eta r v$, we have
$$\eta = \frac{F}{6\pi r v} = \frac{3 \times 10^{-13} \text{ N}}{6\pi(2.5 \times 10^{-6} \text{ m})(4.5 \times 10^{-4} \text{ m/s})} = 1.41 \times 10^{-5} \text{ N s/m}^2$$

10.45 From $v_t = \frac{2r^2 g}{9\eta}(\rho - \rho_f)$, we find
$$r^2 = \frac{9\eta v_t}{2g(\rho - \rho_f)} = \frac{9(1.8 \times 10^{-5} \text{ N s/m}^2)(4 \times 10^{-5} \text{ m/s})}{2(9.8 \text{ m/s}^2)(800 - 1.29) \text{ kg/m}^3}$$
gives $r = 6.43 \times 10^{-7} \text{ m} = 0.643 \text{ microns}$

10.46 If at the end of one hour a particle is still in suspension, then its terminal velocity must be less than 5 cm/h = 1.39×10^{-5} m/s. Thus, we use
$v_t = \frac{2r^2 g}{9\eta}(\rho - \rho_f)$ to find
$$r^2 = \frac{9\eta v_t}{2g(\rho - \rho_f)} = \frac{9(10^{-3} \text{ N s/m}^2)(1.39 \times 10^{-5} \text{ m/s})}{2(9.8 \text{ m/s}^2)(800 \text{ kg/m}^3)}$$
and $r = 2.82 \times 10^{-6}$ m = 2.82 microns is the size of the largest particles that can still remain in suspension.

10.47 The equation of continuity enables us to find the velocity as
$$v_2 = \frac{A_1 v_1}{A_2} = \frac{d_1^2}{d_2^2} v_1 = \frac{(2)^2}{(0.25)^2}(0.5 \text{ m/s}) = 32.0 \text{ m/s}$$

10.48 (a) The speed at the narrow section is found from the equation of continuity.
$$v_2 = \frac{A_1 v_1}{A_2} = \frac{d_1^2}{d_2^2} v_1 = 16 v_1 = 16.0 \text{ m/s}$$

CHAPTER TEN SOLUTIONS

(b) We choose the zero level for y at the common center line of the horizontal pipes, and solve the resulting form of Bernoulli's equation for the pressure at the narrow section, point 2.

$$P_2 = P_1 + \frac{1}{2}\rho(v_1^2 - v_2^2) = (3 \times 10^5 \text{ Pa})$$

$$+ \frac{1}{2}(1000 \text{ kg/m}^3)[(1 \text{ m/s})^2 - (16 \text{ m/s})^2] = 1.73 \times 10^5 \text{ Pa}$$

10.49 The equation that we are to check is
Power = $(\Delta P)Av$

Power has SI units of J/s

$(\Delta P)Av$ has SI units of $\frac{(N/m^2)(m^2)(m)}{s} = \frac{N\ m}{s} = \frac{J}{s}$

All that dimensional analysis can tell us is that the units are consistent in the equation. Thus, the equation may be true. It can, however, be off by a constant dimensionless factor.

10.50 Let us first find the velocity with which the water leaves the jet of the fountain by use of the equations of motion with constant acceleration.

$$v^2 = v_0^2 + 2ay$$
$$0 = v_0^2 + 2(-9.8 \text{ m/s}^2)(10 \text{ m})$$
$$v_0 = 14 \text{ m/s}$$

Now, let us apply Bernoulli's equation with point 1 selected at some point inside the water reservoir where the velocity is negligible in comparison to its speed of exit at the jet. Point 2 is at the nozzle where the pressure is atmospheric, and we shall also assume that the difference in height between points 1 and 2 is negligble. The velocity of 14 m/s found above becomes the velocity of the water at point 2. Bernoulli's equation reduces to,

$$\Delta P = P_1 - P_2 = \frac{1}{2}\rho(v_2^2) = \frac{1}{2}(1000 \text{ kg/m}^3)(14 \text{ m/s})^2 = 9.8 \times 10^4 \text{ Pa}$$

The power equals the flow rate times the pressure difference across the tube. (See problem 49.)

Power = (flow rate)(ΔP) = $Av(\Delta P)$ = $(7.85 \times 10^{-3} \text{ m}^2)(14 \text{ m/s})(9.8 \times 10^4 \text{ Pa})$
= 1.08×10^4 W = 14.4 hp

10.51 $v_t = \frac{2r^2g}{9\eta}(\rho - \rho_f)$

$= \frac{2(5 \times 10^{-6} \text{ m})^2(9.8 \text{ m/s}^2)}{9(1500 \times 10^{-3} \text{ N s/m}^2)}(2700 \text{ kg/m}^3 - 900 \text{ kg/m}^3) = 6.53 \times 10^{-8}$ m/s

---

10.52 Let us first consider the glass sphere. Its terminal velocity is given by

$$v_t = \frac{2r^2g}{9\eta}(\rho - \rho_f)$$

We rewrite this as follows:

$$\frac{2r^2g}{9\eta} = \frac{v_t}{(\rho - \rho_f)} = \frac{0.3 \text{ m/s}}{(2500 - 1000)\text{kg/m}^3} = 2 \times 10^{-4} \text{ m}^4/\text{kg s}$$

Therefore, for the wooden sphere, we find

$$v_t = \frac{2r^2g}{9\eta}(\rho - \rho_f) = (2 \times 10^{-4} \text{ m}^4/\text{kg s})(850 - 1000) \text{ kg/m}^3 = -0.03 \text{ m/s}$$

Thus, the wooden ball rises at a rate of 3 cm/s.

CHAPTER TEN SOLUTIONS

10.53 We choose point 1 at the center-line of the ground-level pipe and point 2 at the center-line of the underground pipe. From the equation of continuity, we find the velocity of the fluid at point 1.

$$v_1 = \frac{A_2 v_2}{A_1} = \frac{0.21 \text{ m}^3/\text{s}}{7.07 \times 10^{-3} \text{ m}^2} = 2.97 \text{ m/s}$$

We now choose the zero level for y at the level of the underground pipe and write Bernoulli's equation.

$$P_2 + \frac{1}{2}\rho v_2^2 = P_1 + \frac{1}{2}\rho v_1^2 + \rho g y_1$$

$$P_2 + \frac{1}{2}\rho v_2^2$$
$$= 2.5 \times 10^4 \text{ Pa} + \frac{1}{2}(791 \text{ kg/m}^3)(2.97 \text{ m/s})^2 + (791 \text{ kg/m}^3)(9.8 \text{ m/s}^2)(2 \text{ m})$$
$$= 4.40 \times 10^4 \text{ Pa}$$

If P_2 cannot be greater than 3×10^4 Pa, then $\frac{1}{2}\rho v_2^2$ cannot be less than

$$\frac{1}{2}\rho v_2^2 = 4.40 \times 10^4 \text{ Pa} - P_2 \geq 4.40 \times 10^4 \text{ Pa} - 3 \times 10^4 \text{ Pa} = 1.40 \times 10^4 \text{ Pa}$$

or $\quad v_2^2 \geq \dfrac{2}{791 \text{ kg/m}^3}(1.40 \times 10 \text{ Pa})^4 = 35.4 \text{ m}^2/\text{s}^2$

$v_2 \geq 5.95$ m/s

But, $\quad A_2 v_2 = 0.21 \text{ m}^3/\text{s}$

Thus, $A_2 \leq \dfrac{0.21 \text{ m}^3/\text{s}}{5.95 \text{ m/s}} = 3.53 \times 10^{-2} \text{ m}^2$

Which leads to the necessity that the diameter of the pipe be less than 21.2 cm.

10.54 (a) Let us use the flow rate to find the required velocity of the fluid.

$$v = \frac{0.15 \text{ m}^3/\text{s}}{A} = \frac{0.15 \text{ m}^3/\text{s}}{\frac{\pi}{4}d^2} = \frac{0.191 \text{ m}^3/\text{s}}{d^2}$$

Now let us turn to the expression for the Reynold's number.

$$R_N = \frac{\rho v d}{\eta} = \frac{\rho}{\eta}\frac{0.191 \text{ m}^3/\text{s}}{d^2} d$$

From which,

$$d = \frac{(0.191 \text{ m}^3/\text{s})(750 \text{ kg/m}^3)}{(0.45 \text{ N s/m}^2)(R_N)} = \frac{318 \text{ m}}{R_N}$$

Therefore, if R_N cannot exceed 1000, then the diameter d cannot be less than 0.318 m.

(b) From Poiseuille's equation, we find the pressure difference as

$$\Delta P = \frac{(\text{flow rate})8L\eta}{\pi R^4} = \frac{(0.15 \text{ m}^3/\text{s})8(500 \text{ m})(0.45 \text{ N s/m}^2)}{\pi(0.159 \text{ m})^4}$$
$$= 1.34 \times 10^5 \text{ Pa}$$

10.55 (a) Using Bernoulli's equation, we have

$$P_1 + \frac{1}{2}\rho v_1^2 - \rho g y_1 = P_2 + \frac{1}{2}\rho v_2^2 + \rho g h_2$$

where $P_1 = P_2 = P_a$. For a large tank the velocity at the top, v_1, is approximately equal to zero, and $y_1 - y_2 = h$. This gives $\quad v_2 = \sqrt{2gh}$

(b) For $y = y_{max}$, the fluid will be at rest at the top of the system; v_2 becomes zero, and

153

CHAPTER TEN SOLUTIONS

$y_2 - y_2 = y_{max}$, so that $P_a = P + \rho g y$. The minimum value of P is 0. Therefore, $y_{max} = \dfrac{P_a}{\rho g}$.

CHAPTER ELEVEN SOLUTIONS

11.1 Since we have a linear graph, the pressure is related to the temperature as
P = A + BT, where A and B are constants. To find A and B, we use the given data
0.7 atm = A + (100°C)B   (1)
0.512 atm = A + 0   (2)
Thus, from (2)   A = 0.512 atm
Then from (1),   B = 1.88 X 10^{-3} atm/°C
Therefore,   P = 0.512 atm + (1.88 X 10^{-3} atm/°C)T
(a) When P = 0.0400 atm, we have
0.04 atm = 0.512 atm + (1.88 X 10^{-3} atm/°C)T
and   T = -251°C
(b) When T = 450°C,   P = 0.512 atm + (1.88 X 10^{-3} atm/°C)(450°C) = 1.358 atm

---

11.2 Since we have a linear graph, we know that the pressure is related to the temperature as P = A + BT, where A and B are constants. To find A and B, we use the given data
0.9 atm = A + (-80°C)B   (1)
1.635 atm = A + (78°C)B   (2)
Solving (1) and (2) simultaneously, we find
A = 1.272 atm
and   B = 4.652 X 10^{-3} atm/°C
Therefore,   P = 1.272 atm + (4.652 X 10^{-3} atm/°C)T
(a) At absolute zero
P = 0 = 1.272 atm + (4.652 X 10^{-3} atm/°C)T
which gives   T = -273.5°C
(b) At the freezing point of water
P = 1.272 atm + 0 = 1.272 atm
and at the boiling point
P = 1.272 atm + (4.652 X 10^{-3} atm/°C)(100°C) = 1.737 atm

11.3 (a) To convert from Fahrenheit to Celsius, we use
$T_C = \frac{5}{9}(T_F - 32) = \frac{5}{9}(98.6 - 32) = 37°$ C
and the Kelvin temperature is found as
$T_K = T_C + 273 = 37°$ C + 273 = 310 K
(b) In a fashion identical to that used in (a), we find
$T_C = -20.6°$ C and T_K = 252.6 K

11.4 (a) To convert from Celsius to Fahrenheit, we use
$T_F = \frac{9}{5} T_C + 32 = \frac{9}{5}(-252.87 °C) + 32 = -423°F$
and to convert to Kelvin, we use
$T_K = T_C + 273 = -253°C + 273 = 20$ K
(b) We use an approach here that is identical to that used in (a).
$T_F = 68°F$ and $T_K = 293$ K

CHAPTER ELEVEN SOLUTIONS

11.5 To illustrate the approach, we are going to assume here (incorrectly) that the temperature of the star is known to 8 significant figures. The Celsius temperature is
$$T_C = T_K - 273 = 19,999,727°C$$
and, from, $T_F = \frac{9}{5} T_C + 32$, we find $T_F = 35,999,954°F$

11.6 (a) We use
$$\Delta T_C = \frac{5}{9} (\Delta T_F)$$
which gives $\Delta T_C = \frac{5}{9} (54 °F) = 30°C$

(b) The relationship between a change on the kelvin scale and the Celsius scale is
$$\Delta T_K = \Delta T_C$$
Thus, $\Delta T_K = 30$ K

11.7 $T_C = \frac{5}{9} (T_F - 32) = \frac{5}{9} (136 - 32) = 57.8° C$

and $T_C = \frac{5}{9} (T_F - 32) = \frac{5}{9} (-127 - 32) = -88.3° C$

11.8 $T_F = \frac{9}{5} T_C + 32 = T_{boiling} = \frac{9}{5} (444.6) + 32 = 832.3° F$

Therefore, $T_{melting} = 832.3°F - 586.1°F = 246.2°F$

(a) $T_C = \frac{5}{9} (T_F - 32) = T_{melting} = \frac{5}{9} (246.2 - 32) = 119° C$

(b) See above

11.9 Let us use $T_C = \frac{5}{9} (T_F - 32)$ with $T_F = -40°C$. We find
$$T_C = \frac{5}{9} (-40 - 32) = -40°C$$

11.10 Let us apply $T_F = \frac{9}{5} T_C + 32$ to two different temperatures, which we will call 1 and 2. We have
$$T_{F1} = \frac{9}{5} T_{C1} + 32 \quad (1)$$
and $T_{F2} = \frac{9}{5} T_{C2} + 32 \quad (2)$

Let us subtract (1) from (2). We find
$$T_{F2} - T_{F1} = \frac{9}{5} (T_{C2} - T_{C1})$$
or, $\Delta T_F = \frac{9}{5} \Delta T_C$

11.11 Use $L = L_0(1 + \alpha(T - T_0))$
$L_{-20} = L_0 + \alpha L_0(-20°C) - L_0 T_0 \quad (1)$
$L_{35} = L_0 + \alpha L_0(35°C) - L_0 T_0 \quad (1)$

$\Delta L = L_{35} - L_{-20} = \alpha L_0(55°C)$
or $\Delta L = (11 \times 10^{-6} /°C)(518 \text{ m})(55°C) = 0.313 \text{ m} = 31.3 \text{ cm}$

CHAPTER ELEVEN SOLUTIONS

11.12 $\Delta L = \alpha L_0 \Delta T = (17 \times 10^{-6}\ °C^{-1})(2.0000\ m)(100°C) = 3.4\ mm$
Thus, $L = L_0 + \Delta L = 2.0000\ m + .0034\ m = 2.0034\ m$

11.13 The temperature range is $\Delta T = 30°C - (-6°C) = 36°C$, and the change in length is given by,
$\Delta L = \alpha L_0 \Delta T = (11 \times 10^{-6}\ °C^{-1})(20\ m)(36°C) = 7.92\ mm$

11.14 (a) The change in length is
$\Delta L = \alpha L_0 \Delta T = (19 \times 10^{-6}\ °C^{-1})(1.3\ m)(-20°C) = -4.94 \times 10^{-4}\ m$
Thus, the final length of the pendulum is
$L_f = 1.2995\ m$

(b) From the expression for the period, $T = 2\pi\sqrt{\dfrac{L}{g}}$, we see that as the length decreases the period decreases. Thus, the clock runs fast.

11.15 The gap should be equal to the amount of expansion of one section. Thus,
$\Delta L = \alpha L_0 \Delta T = (12 \times 10^{-6}\ °C^{-1})(30\ m)(35°C) = 1.26 \times 10^{-2}\ m = 1.26\ cm$

11.16 We shall choose the radius as our linear dimension.
$L = L_0(1 + \alpha(T - T_0))$
$2.21\ cm = 2.20\ cm(1 + (130 \times 10^{-6}\ C^{-1})\Delta T)$
$\Delta T = 35°C$
So, $T = 55°C$

11.17 (a) We must raise the diameter (a linear dimension) of the sleeve to match the diameter of the shaft, and it is necessary to increase the diameter of the sleeve by $\Delta L = 1.6 \times 10^{-2}\ cm$.
$\Delta L = \alpha L_0 \Delta T$
$1.6 \times 10^{-2}\ cm = (19 \times 10^{-6}\ °C^{-1})(3.196\ cm)(\Delta T)$
We find, $\Delta T = 263.5\ C°$
(b) We must cool the shaft to contract its diameter from 3.212 cm down to 3.196 cm, a change of $-1.6 \times 10^{-2}\ cm$.
$\Delta L = \alpha L_0 \Delta T$
$-1.6 \times 10^{-2}\ cm = (19 \times 10^{-6}\ °C^{-1})(3.212\ cm)(\Delta T)$
or, $\Delta T = -262.2\ C°$

11.18 (a) $L = L_0(1 + \alpha(T - T_0))$
or $5.050 = 5.000(1 + (24 \times 10^{-6}\ °C^{-1})\Delta T)$
From which, $\Delta T = 417\ C°$
and $T = 437°C$
(b) We must get $L_{al} = L_{brass}$ for some ΔT.
$L_{0al}(1 + \alpha_{al}\Delta T) = L_{0brass}(1 + \alpha_{brass}\Delta T)$
$5.0(1 + (24 \times 10^{-6}\ °C^{-1})\Delta T) = 5.05(1 + (19 \times 10^{-6}\ °C^{-1})\Delta T)$
Solving for ΔT gives $\Delta T = 2079\ C°$
So, $T = 2099°C$
This will not work because aluminum melts at 660°C

11.19 As the temperature of the pipe is raised, its length increases by
$\Delta L = \alpha L_0 \Delta T = (17 \times 10^{-6}\ °C^{-1})(2\ m)(80\ °C) = 2.72 \times 10^{-3}\ m$
Thus, the strain created is

CHAPTER ELEVEN SOLUTIONS

$$\text{Strain} = \frac{\Delta L}{L_0} = \frac{2.72 \times 10^{-3} \text{ m}}{2 \text{ m}} = 1.36 \times 10^{-3}$$

and the stress is found from the definition of Young's modulus.

$$\text{Stress} = Y(\text{strain}) = (11 \times 10^{10} \text{ Pa})(1.36 \times 10^{-3}) = 1.50 \times 10^8 \text{ Pa}$$

11.20 Let us find the elongation of the circumference when heated.

$$\Delta L = \alpha L_0 \Delta T = (17.3 \times 10^{-6} \text{ °C}^{-1})(3.142 \times 10^{-2} \text{ m})(103 \text{ °C}) = 5.6 \times 10^{-5} \text{ m}$$

When the band cools but is not allowed to contract, the tension in the band must be sufficient to stretch it by an amount ΔL.

$$\text{Stress} = Y \text{ strain} = Y\frac{\Delta L}{L_0} = (18 \times 10^{10} \text{ Pa})\frac{5.6 \times 10^{-5} \text{ m}}{3.142 \times 10^{-2} \text{ m}} = 3.21 \times 10^8 \text{ Pa}$$

But Stress $= \frac{F}{A}$ and A = width X thickness = $(4 \times 10^{-3} \text{ m})(5 \times 10^{-4} \text{ m}) = 20 \times 10^{-7} \text{ m}$

and $F = (20 \times 10^{-7} \text{ m}^2)(3.21 \times 10^8 \text{ Pa}) = 641 \text{ N}$

11.21 The section of the tape which matches the boy's height at 27 °C is 1.2 m long. We must find out how long was this section of the tape at 20 °C, when the scale markings were painted on.

$$\Delta L = L - L_0 = \alpha L_0 \Delta T$$

or, $L = (1 + \alpha \Delta T) L_0$

Which gives,

$$L_0 = \frac{L}{(1 + \alpha \Delta T)} = \frac{1.2 \text{ m}}{(1 + (11 \times 10^{-6} \text{ C}^{-1})(7 \text{ °C}))} = 1.11991 \text{ m}$$

The value for L_0 found above is the value that the tape would give for the height of the boy at a temperature of 27 °C.

As a result, the tape reads $(1.2 - 1.11991) = 9 \times 10^{-5}$ m too short.

11.22 Let us first find the length of the aluminum column at 29.4 °C.

$$\Delta L_c = L_c - L_{0c} = \alpha L_{0c} \Delta T$$

or, $L_c = (1 + \alpha \Delta T) L_{0c} = (1 + (24 \times 10^{-6} \text{ C}^{-1})(8.2 \text{ °C}))18.7 \text{ m} = 18.704 \text{ m}$

We must now find what length of tape (at 21.2 °C when the scale markings were painted on) will be 18.704 m long at 29.4 °C.

$$L_t = (1 + \alpha \Delta T) L_{0t}$$

or, $L_{0t} = \frac{L_t}{(1 + \alpha \Delta T)} = \frac{18.704 \text{ m}}{(1 + (11 \times 10^{-6} \text{ C}^{-1})(8.2 \text{ °C}))} = 18.702 \text{ m}$

Thus, the steel tape will read the length of the aluminum column to be 18.702 m at a temperature of 29.4 °C.

11.23 $V_0 = 45$ liters $= 45 \times 10^{-3} \text{ m}^3$

When the gasoline warms to 35°C, its new volume can be found from

$$\Delta V = \beta V_0 \Delta T$$

This is the volume which must overflow.

$$\Delta V = (9.6 \times 10^{-4} \text{ C}^{-1})(45 \times 10^{-3} \text{ m}^3)(25 \text{°C}) = 1.08 \times 10^{-3} \text{ m} = 1.08 \text{ liters}$$

(about 0.29 gallons)

11.24 The change in Celsius temperature of the gasoline as it leaves the truck and enters the storage tank is

$$\Delta T_F = \frac{9}{5} \Delta T_C$$

CHAPTER ELEVEN SOLUTIONS

$$-36\,°F = \frac{9}{5}\Delta T_C$$

Thus, $\Delta T_C = -20\,°C$

We are given that the final volume of gasoline, V_f, is 1000 gal, and we are to find the volume, V_0, when it was on the truck at 90 °F. We have,

$$\frac{\Delta V}{V_0} = \frac{V_f - V_0}{V_0} = \beta \Delta T = (9.6 \times 10^{-4}\,C^{-1})(-20\,°C) = -1.92 \times 10^{-2}$$

or, $V_f = (1 - 1.92 \times 10^{-2})V_0 = 0.9808 V_0$

From which, $V_0 = \dfrac{1000\,\text{gal}}{0.9808} = 1020$ gallons

---

11.25 As the temperature increases the volume will increase. Thus, From the definition of density, $\rho = \dfrac{m}{V}$, we see that the density will decrease. The change in volume of the aluminum is

$$\Delta V = \beta V_0 \Delta T = (7.2 \times 10^{-5}\,°C^{-1})(V_0)(80\,°C) = 5.76 \times 10^{-3}\,V_0$$

Thus, the final volume is

$$V = V_0 + 5.76 \times 10^{-3}\,V_0 = 1.00576 V_0$$

Therefore, the new value of the density is

$$\rho' = \frac{m}{V} = \frac{m}{1.00576 V_0} = \frac{1}{1.00576}\rho_0$$

where ρ_0 is the value of the density at 20 °C. Thus, we find

$$\rho' = \frac{2.70\,\text{g/cm}^3}{1.00576} = 2.685\,\text{g/cm}^3.$$

11.26 (a) From the density-temperature graph, we see that at 0°C, the density of water has a value of 0.9999 g/cm³. Thus,

$$V = \frac{m}{\rho} = \frac{1\,g}{0.9999\,\text{g/cm}^3} = 1.0001\,\text{cm}^3$$

(b) From the graph we see that the density of water at 4 °C is 1 g/cm³. Thus,

$$V = \frac{m}{\rho} = \frac{1\,g}{1\,\text{g/cm}^3} = 1.0000\,\text{cm}^3$$

11.27 (a) $n = \dfrac{PV}{RT} = \dfrac{(9\,\text{atm})(1.013 \times 10^5\,\text{Pa/atm})(8 \times 10^{-3}\,m^3)}{(8.31\,\text{Nm/mol K})(293\,K)} = 3.0$ moles

(b) $N = nN_a = (3.0\,\text{moles})(6.02 \times 10^{23}\,\text{molecules/mole}) = 1.80 \times 10^{24}$ molecules

---

11.28 (a) From $PV = nRT$, we have

$$n = \frac{PV}{RT} = \frac{(1.013 \times 10^5\,\text{Pa})(10^{-6}\,m^3)}{(8.31\,\text{J/mol K})(293\,K)} = 4.16 \times 10^{-5}\,\text{mol} \quad (1)$$

(b) Since V, R, and T are all constants, (1) above shows that the ratio of n to P is a constant. Thus,

$$\frac{n_f}{n_i} = \frac{P_f}{P_i}$$

or $n_f = (4.16 \times 10^{-5}\,\text{mol})\dfrac{10^{-11}\,\text{Pa}}{1.013 \times 10^5\,\text{Pa}} = 4.11 \times 10^{-21}$ mol

---

11.29 We may use of the ideal gas equation expressed as

CHAPTER ELEVEN SOLUTIONS

$$\frac{P_f V_f}{T_f} = \frac{P_i V_i}{T_i}$$

Because the container is kept at constant volume, we have

$$\frac{P_f}{T_f} = \frac{P_i}{T_i}$$

or, $\quad \dfrac{P_f}{323 \text{ K}} = \dfrac{3 \text{ atm}}{293 \text{ K}}$

and $\quad P_f = 3.31$ atm

11.30 $P_f = \dfrac{n_f R T_f}{V_f}$ and $\quad P_0 = \dfrac{n_0 R T_0}{V_0}$

so $\quad \dfrac{P_f}{P_0} = \dfrac{n_f T_f V_0}{n_0 T_0 V_f}$ But $V_0 = V_f$ and $n_f = \dfrac{1}{2} n_0$

so $\quad \dfrac{P_f}{P_0} = \dfrac{1}{2} \dfrac{T_f}{T_0} = \dfrac{1}{2} \dfrac{338}{288} = 0.587$

$P_f = 5.87$ atm, since $P_0 = 10$ atm

11.31 From $\dfrac{P_f V_f}{T_f} = \dfrac{P_i V_i}{T_i}$, we have

$$\frac{T_f}{T_i} = \frac{P_f V_f}{P_i V_i} = \frac{(3 P_i)(V_i/2)}{P_i V_i} = \frac{3}{2}$$

11.32 We first find the number of moles of gas required as

$$n = \frac{PV}{RT} = \frac{(5.07 \times 10^6 \text{ Pa})(10^{-2} \text{ m}^3)}{(8.31 \text{ J/mol K})(293 \text{ K})} = 20.8 \text{ mol}$$

But, $\quad n = \dfrac{m}{M}$, so $\quad m = nM$

(a) If the gas is oxygen, $M = 32$ g/mol (a diatomic gas)
Thus, $m = (20.8 \text{ mol})(32 \text{ g/mol}) = 666$ g
(b) If the gas is helium, $M = 4$ g/mol
and, $m = (20.8 \text{ mol})(4 \text{ g/mol}) = 83.2$ g

11.33 $\dfrac{P_f V_f}{T_f} = \dfrac{P_i V_i}{T_i}$

$$\frac{(0.8 \times 10^5 \text{ Pa})(0.7 \text{ m}^3)}{T_f} = \frac{(0.2 \times 10^5 \text{ Pa})(1.5 \text{ m}^3)}{300 \text{ K}}$$

From which, $T_f = 560$ K $= 287$ °C

11.34 $\dfrac{P_f}{T_f} = \dfrac{P_i}{T_i}$

$T_f = \dfrac{T_i P_f}{P_i} = \dfrac{(300 \text{ K}) 3 P_i}{P_i} = 900$ K $= 627$ °C

(b) $\dfrac{P_f V_f}{T_f} = \dfrac{P_i V_i}{T_i}$

$\dfrac{2 P_i 2 V_f}{T_f} = \dfrac{P_i V_i}{300 \text{ K}}$

$T_f = 1200$ K $= 927$°C

CHAPTER ELEVEN SOLUTIONS

11.35 The absolute pressure is the gauge pressure plus atmospheric. Thus, the initial pressure in the tire is

P_i = 50 lb/in^2 + 14.7 lb/in^2 = 64.7 lb/in^2

We use,
$$\frac{P_f V_f}{T_f} = \frac{P_i V_i}{T_i}$$

But with the volume constant, this becomes
$$\frac{P_f}{T_f} = \frac{P_i}{T_i}$$

or $\dfrac{P_f}{308 \text{ K}} = \dfrac{64.7 \text{ lb/in}^2}{293 \text{ K}}$

From which, P_f = 68.0 lb/in^2,
which is equivalent to a gauge pressure of 53.3 lb/in^2

11.36 P_i = (0.95)(1 atm) = 9.62 X 10^4 Pa

$P_h = P_0 + \rho g h$ = 1.013 X 10^5 Pa + (10^3 kg/m^3)(9.8 m/s^2)(10 m)
= 1.99 X 10^5 Pa (This is the absolute pressure at 10 m down in the water.)
Therefore $P_f = 0.95 P_h$ = 1.89 X 10^5 Pa
We assume that the temperature of the gas remains at body temperature throughout.

Thus, $V_f = V_i \dfrac{P_i}{P_f}$ = (0.82 liters) $\dfrac{9.62 \text{ X } 10^4 \text{ Pa}}{1.89 \text{ X } 10^5 \text{ Pa}}$ = 0.417 liters

---

11.37 We first find the pressure of the air in the bubble when at a depth of 100 m by use of

$P = P_a + \rho g h$ = 1.013 X 10^5 Pa + (1000 kg/m^3)(9.8 m/s^2)(100 m) = 1.08 X 10^6 Pa

We now use
$$\frac{P_f V_f}{T_f} = \frac{P_i V_i}{T_i}$$

However, at constant temperature, this becomes
$P_f V_f = P_i V_i$

or, (1.013 X 10^5 Pa)V_f = (1.08 X 10^6 Pa)(1.5 cm^3)
and V_f = 16.0 cm^3

11.38 In order to have a sufficient buoyancy force to lift the balloon, we must decrease the mass of the air inside by 100 kg (the mass of the balloon). The number of moles of air which must be allowed to escape (as the air expands) is thus,

$$\Delta n = \frac{100 \text{ kg}}{\text{molecular weight}} = \frac{10^5 \text{ gm}}{28.8 \text{ gm/mol}} = 3472 \text{ moles}$$

But $n = \dfrac{PV}{RT}$

so, $\Delta n = n_i - n_f = \dfrac{P_i V_i}{RT_i} - \dfrac{P_f V_f}{RT_f} = \dfrac{PV}{R}\left(\dfrac{1}{T_i} - \dfrac{1}{T_f}\right)$

Now use $P_f = P_i$ = 1 atm, and $V_f = V_i$ = 400 m^3, and T_i = 300 K
To find, T_f = 382 K = 109°C

11.39 Let us use $V = \dfrac{4}{3}\pi r^3$ as the volume of the balloon, and the ideal gas law in the form

161

CHAPTER ELEVEN SOLUTIONS

$$\frac{P_f V_f}{T_f} = \frac{P_i V_i}{T_i}$$

to give, $\quad r_i^3 = \frac{300 \text{ K}}{200 \text{ K}} \frac{.03 \text{ atm}}{1 \text{ atm}} (20 \text{ m})^3$

$r_i = 7.11$ m

11.40 We shall assume 1500 ft² of floor space and 8 ft high ceilings for a volume of 340 m³.
To find the mass of air, we first find the number of moles.

$$n = \frac{PV}{RT} = \frac{(1.013 \times 10^5 \text{ Pa})(340 \text{ m}^3)}{(8.31 \text{ J/mol K})(T)} = \frac{4.14 \times 10^6 \text{ mol K}}{T}$$

Also, $n = \frac{m}{M}$ So, $m = nM = n(28.8 \text{ g/mol})$

Therefore, $\quad m = \frac{4.14 \times 10^6 \text{ mol K}}{T}(28.8 \text{ g/mol}) = \frac{1.19 \times 10^8 \text{ g K}}{T}$

(a) At 0 °F (255 K), we find

$$m = \frac{1.19 \times 10^8 \text{ g K}}{T} = \frac{1.19 \times 10^8 \text{ g K}}{255 \text{ K}} = 4.68 \times 10^5 \text{ g} = 468 \text{ kg}$$

(b) At 100 °F (311 K), by the same approach as in (a) we find the mass of air in the house is

$m = 384$ kg

Thus, 84 kg must leave.

11.41 $V_i = (\pi r^2)h_i = \pi(1.5 \text{ m})^2(4 \text{ m}) = 28.3 \text{ m}^3$ (original volume of gas inside)
At 220 m down

$P = P_a + \rho gh = 1.013 \times 10^5 \text{ Pa} + (1025 \text{ kg/m}^3)(9.8 \text{ m/s}^2)(220 \text{ m})$

$\quad = 2.311 \times 10^6$ Pa

$V_f = V_i \frac{P_i}{P_f} \frac{T_f}{T_i} = 28.3 \text{ m}^3 \frac{(1.013 \times 10^5 \text{ Pa})(278)}{(2.311 \times 10^6 \text{ Pa})(298)} = 1.16 \text{ m}^3 = (\text{area})h'$

$\quad = \pi(1.5 \text{ m})^2 h'$

Which gives $h' = 0.164$ m $= 16.4$ cm $=$ height of the remaining air space
Thus, the water has risen a distance of $h - h' = 4$ m $- 0.164$ m $= 3.84$ m inside the hull.

11.42 $PV = nRT$, and $n = \frac{m}{M}$

Therefore, $\quad PV = \frac{mRT}{M}$

From which $\frac{m}{V} = \frac{PM}{RT} = \rho$

11.43 (a) To find the density, we use $\rho = \frac{PM}{RT}$ (See problem 42.)

$$\rho = \frac{(1.013 \times 10^5 \text{ Pa})(32 \times 10^{-3} \text{ kg/mol})}{(8.314 \text{ J/mol K})(273 \text{ K})} = 1.43 \text{ kg/m}^3.$$

(b) The approach is the same here as in part (a), with $M = 4 \times 10^{-3}$ kg/mol for helium.
We find,

$\rho = 1.79 \times 10^{-1}$ kg/m³.

CHAPTER ELEVEN SOLUTIONS

11.44 We use
$PV = NkT$

$$N = P\frac{1 \text{ m}^3}{(1.38 \times 10^{-23} \text{ N m/mol K})(300 \text{ K})} = P(2.42 \times 10^{20} \text{ mol m}^2/\text{N}) \quad (1)$$

(a) At atmospheric pressure, (1) gives
$N = (1.013 \times 10^5 \text{ Pa})(2.42 \times 10^{20} \text{ mol m}^2/\text{N}) = 2.45 \times 10^{25}$ molecules

(b) Using the same approach as in (a) for a pressure of 10^{-11} Pa gives
$N = 2.42 \times 10^9$ molecules

11.45 If N molecules are in a cubical volume of 1 m^3, then the number which must be fitted along each edge of length 1 m is $N^{1/3}$.
Therefore, the distance, d, between adjacent ones is $(1 \text{ m}/N^{1/3})$.
If $N = 2.45 \times 10^{25}$ molecules as found in (a) of 11.30, we find
$d = 1 \text{ m}/(2.45 \times 10^{25})^{1/3} = 3.44$ nm
With $N = 2.42 \times 10^9$ molecules we find
$d = 0.745$ nm

11.46 One mole of helium has Avogadro's number of molecules and contains a mass of 4 g. Let us call m the mass of one atom, and we have
$N_a m = 4$ g/mol

or $m = \dfrac{4 \text{ g/mol}}{6.02 \times 10^{23} \text{ molecules/mol}} = 6.64 \times 10^{-24}$ g/molecule
$= 6.64 \times 10^{-27}$ kg/molecule.

11.47 We first find the pressure exerted by the gas on the wall of the container.
$$P = \frac{NkT}{V} = \frac{3N_a kT}{V} = \frac{3RT}{V} = \frac{3(8.31 \text{ N m/mol K})(293 \text{ K})}{8 \times 10^{-3} \text{ m}^3} = 9.13 \times 10^5 \text{ Pa}$$

Thus, the force on one of the walls of the cubical container is
$F = PA = (9.13 \times 10^5 \text{ Pa})(4 \times 10^{-2} \text{ m}^2) = 3.65 \times 10^4$ N

11.48 For an elastic head-on collision, the change in momentum of one bullet of mass m is
$\Delta mv = mv_f - mv_i = mv - (-mv) = 2mv$
The total impulse delivered to the target in one minute is
$\bar{F}(60 \text{ s}) = 2mvN$
where N is the number of bullets hitting the target in the 1 minute interval.

Thus, $\bar{F} = \dfrac{(150)(2 \times 8 \times 10^{-3} \text{ kg} \times 400 \text{ m/s})}{60 \text{ s}} = 16.0$ N

The pressure is
$P = \dfrac{\bar{F}}{A} = \dfrac{16.0 \text{ N}}{5 \text{ m}^2} = 3.20$ Pa

---

11.49 (See problem 48 for an analysis of this problem. We use the approach and equations developed there.)

$$F = \frac{(5 \times 10^{23})(2 \times 4.68 \times 10^{-26} \text{ kg} \times 300 \text{ m/s})}{1 \text{ s}} = 14.0 \text{ N}$$

and $P = \dfrac{F}{A} = \dfrac{14.0 \text{ N}}{8 \times 10^{-4} \text{ m}^2} = 1.76 \times 10^4$ Pa

CHAPTER ELEVEN SOLUTIONS

11.50 When the impact with the target is inelastic, the change in momentum of a single bullet is

$$\Delta p = mv - 0 = mv$$

The average force on the wall during the 60 s interval is

$$\bar{F} = \frac{(mv)N}{60\text{ s}} = \frac{(8 \times 10^{-3}\text{ kg})(4 \times 10^2\text{ m/s})(150)}{60\text{ s}} = 8\text{ N}$$

and the pressure is

$$P = \frac{\bar{F}}{A} = \frac{8\text{ N}}{5\text{ m}^2} = 1.6\text{ Pa}$$

11.51 Let us set the y direction perpendicular to the wall. The x direction is parallel to the wall and in the plane in which the incoming velocity and outgoing velocity lie. For an elastic collision, there is no change in the x component of momentum, but the y component undergoes a change given by

$$\Delta P_y = P_y(\text{after}) - P_y(\text{before}) = mv\cos 30° - (-mv\cos 30°) = 2mv\cos 30°$$

The basic approach to this problem has been outlined in problem 48 for an elastic collision. We parallel those steps here.

$$F = \frac{(2mv\cos 30°)N}{1\text{ s}} = \frac{(2)(4.68 \times 10^{-26}\text{ kg})(300\text{ m/s})(\cos 30°)(5 \times 10^{23})}{1\text{ s}}$$
$$= 12.2\text{ N}$$

and the pressure is

$$P = \frac{F}{A} = \frac{12.2}{8 \times 10^{-4}\text{ m}^2} = 1.52 \times 10^4\text{ Pa}$$

11.52 $<KE> = \frac{3}{2}kT = \frac{3}{2}(1.38 \times 10^{-23}\text{ J/mol K})(300\text{ K}) = 6.21 \times 10^{-21}\text{ J}$

11.53 (a) The mass, m, of a molecule is given by $m = \text{molecular weight}/N_a$. For diatomic hydrogen, we have

$$m = (2 \times 10^{-3}\text{ kg/mol})/(6.02 \times 10^{23}\text{ molecules/mol}) = 3.32 \times 10^{-27}\text{ kg/mol}$$

Thus,

$$v_{rms} = \sqrt{\frac{3kT}{m}} = \sqrt{\frac{3(1.38 \times 10^{-23}\text{ J/mol K})(373\text{ K})}{3.32 \times 10^{-27}\text{ kg/mol}}} = 2.16 \times 10^3\text{ m/s}$$

(b) We use the same technique as in part (a). The mass of a nitrogen molecule is found to be 4.65×10^{-26} kg/molecule, and the rms speed is 576 m/s.

11.54 (a) $<KE> = \frac{3}{2}kT = \frac{3}{2}(1.38 \times 10^{-23}\text{ J/K})(423\text{ K}) = 8.76 \times 10^{-21}$ J/molecule

(b) $<KE> = \frac{1}{2}mv^2_{rms} = 8.76 \times 10^{-21}$ J

so $\quad v_{rms} = \sqrt{\frac{1.75 \times 10^{-20}\text{ J}}{m}} \qquad (1)$

For helium

$$m = \frac{4\text{ gm/mol}}{6.02 \times 10^{23}\text{ molecules/mol}} = 6.64 \times 10^{-24}\text{ gm/molecule}$$
$$= 6.64 \times 10^{-27}\text{ kg/molecule.}$$

Similarly for argon

CHAPTER ELEVEN SOLUTIONS

$$m = \frac{39.9 \text{ gm/mol}}{6.02 \times 10^{23} \text{ molecules/mol}} = 6.63 \times 10^{-23} \text{ gm/molecule}$$
$$= 6.63 \times 10^{-26} \text{ kg/molecule.}$$

Substituting in (1) above, we find
For helium $v_{rms} = 1620$ m/s
and for argon, $v_{rms} = 514$ m/s

11.55 $<KE> = \frac{1}{2} mv^2_{rms} = \frac{3}{2} kT$

so $T = \frac{m}{3k} v^2_{rms}$

For helium $m = 6.64 \times 10^{-27}$ kg (see preceding problem)
Thus, at 20°C,

$$v^2_{rms} = \frac{3kT}{m} = \frac{3(1.38 \times 10^{-23} \text{ J/mol K})(293 \text{ K})}{6.64 \times 10^{-27} \text{ kg/mole}} = 1.81 \times 10^6 \text{ m}^2/\text{s}^2$$

For nitrogen molecules,
$$m = \frac{28 \text{ g/mole}}{6.02 \times 10^{23} \text{ molecules/mol}} = 4.65 \times 10^{-23} \text{ g} = 4.65 \times 10^{-26} \text{ kg}$$

Therefore, when $v^2_{rms} = 1.81 \times 10^6$ m^2/s^2 (as it is for helium at 20°C)

$$T = \frac{(4.65 \times 10^{-26} \text{ kg/mol})}{3(1.38 \times 10^{-23} \text{ J/mol K})} (1.81 \times 10^6 \text{ m}^2/\text{s}^2) = 2040 \text{ K or } 1770°C$$

11.56 From the ideal gas equation,
$$P = \frac{nRT}{V} = \frac{(3 \text{ mol})(8.31 \text{ J/mol K})(300 \text{ K})}{2.24 \times 10^{-2} \text{ m}^3} = 3.34 \times 10^5 \text{ Pa}$$

11.57 (a) From $v_{rms} = \sqrt{\frac{3kT}{m}}$, we find the temperature as

$$T = \frac{(6.66 \times 10^{-27} \text{ kg})(1.12 \times 10^4 \text{ m/s})^2}{3(1.38 \times 10^{-23} \text{ J/mol K})} = 2.02 \times 10^4 \text{ K}$$

(b) $T = \frac{(6.66 \times 10^{-27} \text{ kg})(2.37 \times 10^3 \text{ m/s})^2}{3(1.38 \times 10^{-23} \text{ J/mol K})} = 9.04 \times 10^2 \text{ K}$

11.58 Note that the absolute pressure is 101 atm (1.02 $\times$ 10^7 Pa. From the ideal gas equation, we have

$$n = \frac{PV}{RT} = \frac{(1.02 \times 10^7 \text{ Pa})(0.75 \text{ m}^3)}{(8.31 \text{ J/mol K})(300 \text{ K})} = 3.08 \times 10^3 \text{ mol}$$

Nitrogen has a molecular weight of 28 g/mol. Thus, the mass of the gas is
$$m = nM = (3.08 \times 10^3 \text{ mol})(28 \times 10^{-3} \text{ kg/mol}) = 86.2 \text{ kg}$$

11.59 (a) The change in length of the rod is
$$\Delta L = \alpha L_0 \Delta T = (9 \times 10^{-6} \text{ °C}^{-1})(20 \text{ cm})(75 \text{ °C}) = 1.35 \times 10^{-2} \text{ cm}$$
(b) The change in diameter is
$$\Delta L = \alpha L_0 \Delta T = (9 \times 10^{-6} \text{ °C}^{-1})(1 \text{ cm})(75 \text{ °C}) = 6.75 \times 10^{-4} \text{ cm}$$
(c) The change in volume is
$$\Delta V = 3\alpha V_0 \Delta T = 3(9 \times 10^{-6} \text{ °C}^{-1})(15.7 \text{ cm}^3)(75 \text{ °C}) = 3.18 \times 10^{-2} \text{ cm}^3$$

CHAPTER ELEVEN SOLUTIONS

11.60 Let us consider a solid which has a volume given by $V_0 = A_0 L_0$ at T_0. At a temperature T, the volume is $V = AL$. Let us consider the area first. We have,
$$\Delta A = A - A_0 = 2\alpha A_0 \Delta T$$
Therefore, $A = (1 + 2\alpha \Delta T)A_0$
Similarly, the linear dimension is
$$L = (1 + \alpha \Delta T)L_0$$
So, $AL = (1 + 2\alpha \Delta T)A_0(1 + \alpha \Delta T)L_0 = A_0 L_0 (1 + \alpha \Delta T + 2\alpha \Delta T + 2\alpha^2 \Delta T)$

The term involving α^2 is negligibly small in comparison to the other terms in the parenthesis of the last equation. Thus,
$$V = AL = A_0 L_0 (1 + 3\alpha \Delta T)$$
or $V = V_0(1 + 3\alpha \Delta T) = V_0 + 3\alpha V_0 \Delta T$
Thus, $\Delta V = V - V_0 = \beta V_0 \Delta T$ (where $\beta = 3\alpha$)

11.61 For each gas alone, $P_1 = n_1 kT/V$ and $P_2 = n_2 kT/V$ and $P_3 = n_3 kT/V$, etc. For all the gases
$P_1 V_1 + P_2 V_2 + P_3 V_3 \ldots = (n_1 + n_2 + n_3 \ldots)kT$ and $(n_1 + n_2 + n_3 \ldots)kT = PV$.
Also, $V_1 + V_2 + V_3 \ldots = V$. Thus, $P = P_1 + P_2 + P_3 \ldots$

11.62 $P = P_O + P_N$, and $P_O = \dfrac{N_O kT}{V}$ while $P_N = \dfrac{N_N kT}{V}$

Therefore, $\dfrac{P_O}{P_N} = \dfrac{N_O}{N_N} = \dfrac{\dfrac{\text{mass of oxygen present}}{\text{mass of } O_2 \text{ molecule}}}{\dfrac{\text{mass of nitrogen present}}{\text{mass of } N_2 \text{ molecule}}} = \dfrac{\text{mass of } N_2 \text{ molecule}}{\text{mass of } O_2 \text{ molecule}}$

(Since we have equal masses of the 2 gases present.)

But, $\dfrac{\text{mass of } N_2 \text{ molecule}}{\text{mass of } O_2 \text{ molecule}} = \dfrac{\text{molecular weight of nitrogen}}{\text{molecular weight of oxygen}} = \dfrac{28}{32} = \dfrac{7}{8}$

Thus, $\dfrac{P_O}{P_N} = \dfrac{7}{8}$

So, $P = \dfrac{7}{8} P_N + P_N = 1.875 \, P_N = 1$ atm

Thus, $P_N = 0.533$ atm and $P_O = 0.467$ atm

11.63 From $PV = NkT$, we have
$$V_f = N_f \dfrac{kT}{P} \quad \text{and} \quad V_i = N_i \dfrac{kT}{P}$$
k, T, and P are the same for both cases, and $V_f = 1.4 \, V_i$, $N_f = 1.4 \, N_f$

But N_i = number of argon molecules present = $\dfrac{0.3 \text{ g}}{\text{mass of Ar molecule}}$

$= \dfrac{0.3 \text{ g}}{\dfrac{\text{molecular weight}}{\text{Avogadro's number}}}$

or $N_i = \dfrac{0.3 \text{ g}}{\dfrac{39.9}{6.02 \times 10^{23}}} = 4.53 \times 10^{21}$ molecules

and $N_f = 6.34 \times 10^{21}$ molecules

The number added = 6.34×10^{21} molecules $- 4.53 \times 10^{21}$ molecules
$= 1.81 \times 10^{21}$ molecules

CHAPTER ELEVEN SOLUTIONS

The mass of nitrogen added = (mass of a nitrogen molecule)(1.81×10^{21} molecules) = $\dfrac{\text{molecular weight}}{\text{Avogadro's number}}$ (1.81×10^{21} molecules) = $\dfrac{(28 \text{ g})(1.81 \times 10^{21} \text{ molecules})}{6.02 \times 10^{23}}$ = 0.084 g

---

11.64 We use the ideal gas equation as
$$\frac{P_f V_f}{T_f} = \frac{P_i V_i}{T_i}$$
The volume is held constant, so we have
$$T_f = \frac{P_f}{P_i} T_i = \frac{8 \times 10^4 \text{ Pa}}{1.07 \times 10^5 \text{ Pa}} (300 \text{ K}) = 225 \text{ K} = -48 \text{ °C}$$

11.65 We must first find the temperature of the gas from the ideal gas equation.
$$T = \frac{PV}{nR} = \frac{(8.10 \times 10^5 \text{ Pa})(5 \times 10^{-3} \text{ m}^3)}{(2 \text{ mol})(8.31 \text{ J/mol K})} = 244 \text{ K}$$
and $<KE> = \dfrac{3}{2} kT = \dfrac{3}{2} (1.38 \times 10^{-23} \text{ J/mol K})(244 \text{ K}) = 5.05 \times 10^{-21}$ J/molecule

---

11.66 We shall consider that we have a volume of 1 m^3, and find the number of moles of helium.
$$n = \frac{PV}{RT} = \frac{(1.013 \times 10^5 \text{ Pa})(1 \text{ m}^3)}{(8.31 \text{ J/mol K})(300 \text{ K})} = 40.6 \text{ moles}$$
For helium, one mole contains 4 g = 4×10^{-3} kg. Therefore, we find the mass as
$$m = nM = (40.6 \text{ mol})(4 \times 10^{-3} \text{ kg/mole}) = 1.63 \times 10^{-1} \text{ kg}$$
and the density is
$$\rho = \frac{m}{V} = \frac{1.63 \times 10^{-1} \text{ kg}}{1 \text{ m}^3} = 1.63 \times 10^{-1} \text{ kg/m}^3$$

11.67 (a) We assume that the change in the volume is negligible and find the final temperature of the air in the tire from the ideal gas law.
$$T_f = \frac{P_f}{P_i} T_i = \frac{3.2 \text{ atm}}{2.8 \text{ atm}} (300 \text{ K}) = 343 \text{ K}$$
(b) We use the ideal gas law to find the final volume in terms of the initial volume. The pressure remains constant, so
$$V_f = \frac{T_f}{T_i} V_i = \frac{343 \text{ K}}{300 \text{ K}} (V_i) = 1.143 V_i$$
Therefore, $\Delta V = V_f - V_i = 0.143 V_i$
This is the volume that would have to be released, about 14.3 % of the original volume.

11.68 From PV = NkT, we have
$$\frac{N}{V} = \frac{P}{kT} = \frac{3.04 \times 10^5 \text{ Pa}}{(1.38 \times 10^{-23} \text{ N m/mol K})(500 \text{ K})} = 4.04 \times 10^{25} \text{ molecules/m}^3$$
$$= 4.40 \times 10^{19} \text{ molecules/cm}^3$$

11.69 (a) The volume of the liquid increases as $\Delta V_{liq} = V\beta\Delta T$. The volume of the flask increases as $\Delta V_g = V 3\alpha \Delta T$. Therefore, the overflow in the capillary is

CHAPTER ELEVEN SOLUTIONS

$V_c = V\Delta T(\beta - 3\alpha)$; and in the capillary $V_c = A\Delta h$. Therefore,

$$\Delta h = \frac{V\Delta T(\beta - 3\alpha)}{A}$$

(b) For a mercury thermometer $\beta(Hg) = 1.82 \times 10^{-4}$ °C^{-1}
and for glass, $3\alpha = 3 \times 9 \times 10^{-6}$ °C^{-1}
Thus, $\beta - 3\alpha$ is approximately equal to β.

11.70 Using the results of the previous problem and neglecting the expansion of the glass,

$$\Delta h = \frac{V\Delta T\beta}{A} = \frac{\frac{4}{3}\pi(0.15 \text{ cm})^3(25 \text{ C°})(1.82 \times 10^{-4} \text{ °C}^{-1})}{\pi(0.0025 \text{ cm})^2} = 3.28 \text{ cm}$$

11.71 The rms speed is given by

$$v_{rms} = \sqrt{\frac{3kT}{m}} \quad (1)$$

and from the ideal gas equation, $PV = NkT$, we have

$$kT = \frac{PV}{N}$$

We use this expression to eliminate kT from (1) to find

$$v_{rms} = \sqrt{\frac{3PV}{mN}}$$

but N, the number of molecules, times m, the mass of one molecule, is equal to the total mass of the sample, m_s. So,

$$v_{rms} = \sqrt{\frac{3PV}{m_s}} = \sqrt{\frac{3P}{\rho}} \quad \text{(where } \rho \text{ is the density of the gas). Therefore,}$$

$$v_{rms} = \sqrt{\frac{3(1.013 \times 10^5 \text{ Pa})}{3.75 \text{ kg/m}^3}} = 285 \text{ m/s}$$

11.72 We will first find the pressure that existed at the bottom of the lake by finding the pressure that would be required to compress the trapped gas to the extent indicated by the marking indicator. The cross-sectional area of the container is A. We shall assume that the temperature of the gas remained constant as it was compressed.

$$P_f = P_i \frac{V_i}{V_f} = (1 \text{ atm}) \frac{(1.5 \text{ m})(A)}{(0.4 \text{ m})(A)} = 3.75 \text{ atm}$$

But, $P_f = P_a + \rho g h$, so $P_f - P_a = 3.75 \text{ atm} - 1 \text{ atm} = 2.75 \text{ atm}$

Therefore, $h = \dfrac{2.75 \text{ atm}}{\rho g} = \dfrac{2.75(1.013 \times 10^5 \text{ Pa})}{(1000 \text{ kg/m}^3)(9.8 \text{ m/s}^2)} = 28.4 \text{ m}$

11.73 We have
$d_{0c} = 0.99 d_{0p}$ (1)

The diameter is a linear dimension. Thus, we use $L = (1 + \alpha\Delta T)L_0$, with the ultimate goal of finding the temperature at which the final diameter of the piston and cylinder are equal.

$d_{fc} = d_{fp}$

Thus, $(1 + \alpha_c \Delta T)d_{0c} = (1 + \alpha_p \Delta T)d_{0p}$ (2)

Using (1) in (2), we have

CHAPTER ELEVEN SOLUTIONS

$(1 + \alpha_c \Delta T)(0.99) = (1 + \alpha_p \Delta T)$

or, $\Delta T = \dfrac{0.01}{0.99\alpha_c - \alpha_p} = \dfrac{0.01 \,°C}{(0.99)(24 \times 10^{-6}) - 11 \times 10^{-6}} = 784\,°C$

Therefore, $T_f = 804\,°C$

11.74 Because gas can flow from one container to the other, the pressure must be the same in the two containers at all times. Let us call container 1 the flask that is raised to 100 °C, and container 2 the flask that is maintained at 0 °C. We also know that the total number of moles in the two containers combined remains a constant. Thus,

$n_{1f} + n_{2f} = n_{1i} + n_{2i}$ (1)

But, $n = \dfrac{PV}{RT}$ The volume remains constant, so (1) becomes

$P_f\left(\dfrac{1}{T_{1f}} + \dfrac{1}{T_{2f}}\right) = P_i\left(\dfrac{1}{T_{1i}} + \dfrac{1}{T_{2i}}\right)$

We know that $T_{1f} = 373$ K, $T_{2f} = 273$ K, $T_{1i} = T_{2i} = 273$ K, and $P_i = 1$ atm. Thus, we find

$P_f = 1.15$ atm $= 1.17 \times 10^5$ Pa

11.75 $L_1 = L(1 + \alpha_{Al}\Delta T)$, and $L_2 = L(1 + \alpha_{inv}\Delta T)$. The angle θ between the two bars is

$\dfrac{L_2}{2} = L_1 \sin(\theta/2)$. Thus $\theta = 2\sin^{-1}\dfrac{L_2}{2L_1} = 2\sin^{-1}\dfrac{(1 + \alpha_{inv}\Delta T)}{2(1 + \alpha_{Al}\Delta T)}$

with $\alpha_{inv}\Delta T = 0.9 \times 10^{-6}(100\,°C) = 0.9 \times 10^{-4}$ and $\alpha_{Al}\Delta T = 24 \times 10^{-6}(100\,°C) = 2.4 \times 10^{-3}$

we find $\theta = 2\sin^{-1}(0.49885) = 59.9°$

11.76 (a) Start with $P\Delta V = nR\Delta T$ and multiply each side of the equation by VT. We have

$P\Delta V(VT) = nR\Delta T(VT)$ and using $PV = nRT$ this becomes

$T\Delta V = V\Delta T$. But $\beta = \dfrac{\Delta V}{V\Delta T}$ and therefore $\beta = \dfrac{1}{T}$

(b) At T = 0 °C, $\beta = \dfrac{1}{273.15} = 3.66 \times 10^{-3}\,°C^{-1}$

$\beta(He) = 3.66^{-3}\,°C^{-1}$ and $\beta(air) = 3.67^{-3}\,°C^{-1}$ Hence, both gases have thermal expansion coefficients very close to the values predicted by this approach.

11.77 In equilibrium $P_{gas} = \dfrac{mg}{A} + P_a$. Therefore

$\dfrac{nRT}{hA} = \dfrac{mg}{A} + P_a$ or $h = \dfrac{nRT}{(mg + P_a A)}$

$= \dfrac{(3\,mol)(8.314)(500\,K)}{(5\,kg)(9.8) + (1.01 \times 10^5\,Pa)(0.05\,m^2)} = 2.45$ m

---

11.78 (a) and (b) The pressure at any depth is $P = P_a + \rho gh$. Also, for an ideal gas at constant temperature $PV = P_a V_a$. This gives $P = \dfrac{V_a}{V} P_a = \left(\dfrac{r_a}{r}\right)^3 P_a$

Substituting this expression for P into the first equation and using the given numerical values, we have

169

CHAPTER ELEVEN SOLUTIONS

$h = \dfrac{P - P_a}{\rho g} = \dfrac{P_a}{\rho g}[(1.5)^3 - 1] = 23.9$ m ($P_a = 1.01 \times 10^5$ Pa and $\rho = 1025$ kg/m^3.)

CHAPTER TWELVE SOLUTIONS

12.1 From $Q = mc\Delta T$, we find
$$\Delta T = \frac{Q}{mc} = \frac{1200 \text{ J}}{(0.05 \text{ kg})(387 \text{ J/kg °C})} = 62 \text{ °C}$$
Thus, the final temperature is 87 °C

12.2 $Q = mc\Delta T = (0.1 \text{ kg})(129 \text{ J/kg °C})(80 \text{ °C}) = 1030 \text{ J}$

12.3 $Q = 400 \text{ cal} = 1674 \text{ J}$
So, $\Delta T = \dfrac{Q}{mc} = \dfrac{1674 \text{ J}}{(5 \times 10^{-2} \text{ kg})(230 \text{ J/kg °C})} = 146 \text{°C}$
and $T_f = 166 \text{ °C}$

12.4 The kinetic energy of the bullet is
$$KE = \frac{1}{2}mv^2 = \frac{1}{2}(5 \times 10^{-3} \text{ kg})(3 \times 10^2 \text{ m/s})^2 = 225 \text{ J}$$
If half of this goes into heat, $Q = (225 \text{ J}/2)$, and
$$\Delta T = \frac{Q}{mc} = \frac{(225 \text{ J}/2)}{(5 \times 10^{-3} \text{ kg})(128 \text{ J/kg °C})} = 176 \text{ °C}$$

12.5 $Q = 500 \text{ Cal} = 500(4184) \text{ J} = 2.09 \times 10^6 \text{ J}$
From $Q = \Delta PE = mgh$, we find
$$h = \frac{Q}{mg} = \frac{2.09 \times 10^6 \text{ J}}{(75 \text{ kg})(9.8 \text{ m/s}^2)} = 2.85 \times 10^3 \text{ m}$$

12.6 The amount of energy equivalent to a climb up a rope of length h is
$Q = \Delta PE = mgh$
This expression gives the energy expenditure in Joules. In units of "food calories" this is,
$Q = mgh/(4186 \text{ J/Cal}) = (2.34 \times 10^{-2} \text{ Cal/kg})m$, where m is your mass in kg.

12.7 $Q = \Delta PE = mgh = 2(1.5 \text{ kg})(9.8 \text{ m/s}^2)(3 \text{ m}) = 88.2 \text{ J}$
Also, $\Delta T = \dfrac{Q}{mc} = \dfrac{88.2 \text{ J}}{(0.2 \text{ kg})(4.186 \times 10^3 \text{ J/kg °C})} = 0.105 \text{ °C}$

12.8 The heat required to raise 1 kg of water by 0.10 °C is
$Q = mc\Delta T = (1 \text{ kg})(4186 \text{ J/kg °C})(0.1 \text{ °C}) = 418.6 \text{ J}$
Therefore, $mgh = 418.6 \text{ J} = (3 \text{ kg})(9.8 \text{ m/s}^2)h$
$h = 14.2 \text{ m}$

12.9 Consider a 1 kg mass of water
$Q = \Delta PE = mgh = (1.0 \text{ kg})(9.8 \text{ m/s}^2)(50 \text{ m}) = 490 \text{ J}$
Also, $\Delta T = \dfrac{Q}{mc} = \dfrac{490 \text{ J}}{(1.0 \text{ kg})(4.19 \times 10^3 \text{ J/kg °C})} = 0.117 \text{ °C}$
So, $T_f = 10.117 \text{ C°}$

CHAPTER TWELVE SOLUTIONS

12.10 One BTU is the heat required to raise the temperature of a lb mass (454 g) of water by 1 °F ($\frac{5}{9}$ °C). Therefore,

$$1 \text{ BTU} = mc\Delta T = (4.54 \times 10^{-1} \text{ kg})(4186 \text{ J/kg °C})(5/9 \text{ °C}) = 1056 \text{ J}$$

12.11 (a) We use $Q = mc\Delta T = (0.85)\frac{1}{2}mv^2$

or $\Delta T = \dfrac{0.85 v^2}{2c} = \dfrac{0.85(3 \text{ m/s})^2}{2(387 \text{ J/kg °C})} = 9.88 \times 10^{-3}$ °C

(b) The remaining energy is absorbed by the horizontal surface on which the block slides.

12.12 Let us find the heat extracted from the system in one minute.
$Q = m_{cup}c_{cup}\Delta T + m_{water}c_{water}\Delta T$
$= [(0.2 \text{ kg})(900 \text{ J/kg °C}) + (0.8 \text{ kg})(4186 \text{ J/kg °C})](1.5°C) = 5293 \text{ J}$

If this much heat is removed each minute, the rate of removal of heat is

$$P = \frac{Q}{\Delta t} = \frac{5293 \text{ J}}{60 \text{ s}} = 88.2 \text{ J/s} = 88.2 \text{ W}$$

12.13 Our heat loss = heat gain equation becomes
$m_{iron}c_{iron}\Delta T_{iron} = m_{water}c_w\Delta T_w$
or, $(0.4 \text{ kg})(448 \text{ J/kg °C})(500 \text{ °C} - T) = (20 \text{ kg})(4186 \text{ J/kg °C})(T - 22 \text{ °C})$
From which, we find $T = 23.0$ °C

12.14 Use heat loss = heat gain
$m_{unknown}c_u\Delta T_u = m_{water}c_w\Delta T_w$
$(0.400 \text{ kg})c_u(31 \text{ °C}) = (0.400 \text{ kg})(4186 \text{ J/kg °C})(9 \text{ °C})$
$c_u = 1215$ J/kg °C

12.15 heat gain = heat loss
$m_{water}c_w\Delta T_w = m_{gold}c_g\Delta T_g$
$m_{water}(4186 \text{ J/kg °C})(25 \text{ °C}) = (3 \text{ kg})(129 \text{ J/kg °C})(50 \text{ °C})$
$m_{water} = 0.185$ kg = 185 g

12.16 heat gain = heat loss
$(0.4 \text{ kg})(4186 \text{ J/kg °C})(T_f - 27 \text{ °C}) + (0.3 \text{ kg})(837 \text{ J/kg °C})(T_f - 27 \text{ °C})$
$= (0.2 \text{ kg})(387 \text{ J/kg °C})(90 \text{ °C} - T_f)$
From which, $T_f = 29.5$ °C

12.17 heat gain = heat loss
$m_{cup}c_c\Delta T_c + m_w c_w \Delta T_w + m_{stirrer}c_s\Delta T_s = m_{silver}c_{sil}\Delta T_{sil}$
$m_{cup}(900 \text{ J/kg °C})(5°C) + (0.225 \text{ kg})(4186 \text{ J/kg °C})(5°C)$
$+ (0.04 \text{ kg})(387 \text{ J/kg °C})(5 \text{ °C}) = (0.4 \text{ kg})(234 \text{ J/kg °C})(55 \text{ °C})$
We find $m_{cup} = 80.3 \times 10^{-3}$ kg = 80.3 g

12.18 heat loss = heat gain
$(0.09 \text{ kg})(2.45 \times 10^4 \text{ J/kg}) + (0.09 \text{ kg})(128 \text{ J/kg °C})(327.3 \text{ °C} - T_f)$
$= (0.3 \text{ kg})(448 \text{ J/kg °C})(T_f - 20 \text{ °C})$
From which, $T_f = 59.4$ °C

12.19 The heat needed to raise the temperature of the water to 25 °C is

CHAPTER TWELVE SOLUTIONS

$Q = (0.5 \text{ kg})(4186 \text{ J/kg °C})(5 \text{ °C}) = 1.05 \times 10^4$ J

The heat received from each pellet is

$Q = (10^{-3} \text{ kg})(128 \text{ J/kg °C})(175 \text{ °C}) = 22.4$ J/pellet

Thus, the number of pellets needed is

$$\text{number} = \frac{1.05 \times 10^4 \text{ J}}{22.4 \text{ J/pellet}} = 467 \text{ pellets}$$

12.20 $\Delta Q_{system} = 0$
$= (0.2 \text{ kg})(4.19 \times 10^3 \text{ J/kg°C})(T_f - 10\text{°C}) + (0.3 \text{ kg})(900 \text{ J/kg°C})(T_f - 10\text{°C})$
$\qquad + (0.1 \text{ kg})(4.19 \times 10^3 \text{ J/kg°C})(T_f - 100\text{°C})$

Gives $T_f = 34.7\text{°C}$

12.21 $\Delta Q_{system} = 0$
$= (0.1 \text{ kg})(900 \text{J/kg°C})(20\text{°C} - 10\text{°C}) + (0.25 \text{ kg})(4186 \text{ J/kg°C})(20\text{°C} - 10\text{°C})$
$\qquad + (0.05 \text{ kg})(387 \text{ J/kg°C})(20\text{°C} - 80\text{°C}) + (0.07 \text{ kg})(c_2)(20\text{°C} - 100\text{°C})$

Gives $c_2 = 1821$ J/kg °C = 0.435 cal/g °C

12.22 $\Delta Q_{system} = 0$
$= (0.2 \text{ kg})(138 \text{J/kg°C})(T_f - 0\text{°C}) + (0.05 \text{ kg})(2430 \text{ J/kg°C})(T_f - 50\text{°C})$
$\qquad + (0.1 \text{ kg})(4186 \text{J/kg°C})(T_f - 100\text{°C})$

Gives $T_f = 84.44\text{°C}$

(b) $\Delta Q_{mercury} = (0.2 \text{ kg})(138 \text{J/kg°C})(84.4\text{°C} - 0\text{°C}) = 2330$ J
$\Delta Q_{alc} = (0.05 \text{ kg})(2430 \text{J/kg°C})(84.4\text{°C} - 50\text{°C}) = 4184$ J
$\Delta Q_{water} = (0.1 \text{ kg})(4186 \text{J/kg°C})(84.4\text{°C} - 100\text{°C}) = -6514$ J

12.23 The heat needed is the sum of the following terms.
Q_{needed} = (heat to reach melting point) + (heat to melt) + (heat to reach boiling point) + (heat to vaporize) + (heat to reach 110 °C)

Thus, we have
$Q_{needed} = 0.04 \text{ kg}[(2090 \text{ J/kg °C})(10\text{°C}) + (3.33 \times 10^5 \text{ J/kg}) + (4186 \text{ J/kg °C})(100 \text{ °C})$
$\qquad + (2.26 \times 10^6 \text{ J/kg °C}) + (2010 \text{ J/kg °C})(10 \text{ °C})]$

$Q_{needed} = 1.22 \times 10^5$ J

12.24 The final temperature of the pellets will be 0 °C since not all of the ice can be melted.
heat gain = heat loss
$m_{melted\ ice} L_f = m_{pellets} c_p \Delta T_p$

$$m_{melted\ ice} = \frac{(.400 \text{ kg})(900 \text{ J/kg °C})(30 \text{ °C})}{3.33 \times 10^5 \text{ J/kg °C}} = 32.4 \times 10^{-3} \text{ kg} = 32.4 \text{ g}$$

12.25 Q = (heat to melt) + (heat to warm melted ice to 100 °C) + (heat to vaporize 5 g)
$Q = (5 \times 10^{-2} \text{ kg})(3.33 \times 10^5 \text{ J/kg}) + (5 \times 10^{-2} \text{ kg})(4186 \text{ J/kg °C})(100 \text{ °C})$
$\qquad + (5 \times 10^{-3} \text{ kg})(2.26 \times 10^6 \text{ J/kg}) = 4.89 \times 10^4$ J

12.26 heat loss = heat gain
(heat to cool steam to 100 °C) + (heat to condense steam) + (heat to cool condensed steam to 50°C) = (heat gain by cool water) + (heat gain by cup)
$m_s[(2010 \text{ J/kg})(20 \text{ °C}) + 2.26 \times 10^6 \text{ J/kg} + (4186 \text{ J/kg °C})(50 \text{ °C})] =$
$\qquad (.35 \text{ kg})(4186 \text{ J/kg °C})(30 \text{ °C}) + (.30 \text{ kg})(900 \text{ J/kg °C})(30 \text{ °C})$

CHAPTER TWELVE SOLUTIONS

From which, $m_s = 20.7$ g

12.27 $\Delta Q_{system} = 0$
$= (0.05 \text{ kg})(4186 \text{J/kg°C})(100°C - 80°C) + m_s(2.26 \times 10^6 \text{ J/kg})$
$\quad + (1.94 \text{ kg})(138 \text{J/kg°C})(100°C - 200°C)$
Gives $m_s = 10^{-2}$ kg = 10 g

12.28 Heat required to melt ice = $(0.020 \text{ kg})(3.34 \times 10^5 \text{ J/kg}) = 6680$ J
Heat steam can give up in condensing = $(0.010 \text{ kg})(2.26 \times 10^6 \text{ J/kg}) = 22{,}600$ J
Therefore, the steam can melt all the ice and still be able to yield as much as
22,600 J - 6680 J = 15,920 J to raise the temperature of the melted ice.
The heat required to raise the melted ice to 100°C
$\quad = (0.02 \text{ kg})(4186 \text{ J/kg°C})(100°C) = 8370$ J
Thus, the ice can absorb 6680 J + 8370 J = 15,050 J from the steam before it reaches the same temperature as the steam (i.e. thermal equilibrium).

So, we condense $\dfrac{15050 \text{ kg}}{2.26 \times 10^6 \text{ J/kg}} = 6.66$ g of steam, and we are left with a mixture of 3.34 g of steam and 26.66 g of liquid water all at 100°C.

12.29 heat loss = heat gain
(loss as water cools to T_f) = (heat to melt ice) + (heat to warm melted ice to T_f)
$(.65 \text{ kg})(4186 \text{ J/kg °C})(25 °C - T_f) = (0.1 \text{ kg})(3.33 \times 10^5 \text{ J/kg})$
$\quad\quad\quad\quad\quad\quad\quad\quad\quad\quad\quad\quad\quad\quad + (0.1 \text{ kg})(4186 \text{ J/kg °C})(T_f)$
$\quad T_f = 11.1$ °C

12.30 The final temperature is 0 °C since not all of the ice can be melted.
$\quad$ heat gain = heat loss
(heat to melt ice) = (heat loss as water cools to 0 °C) + (heat loss as cup cools to 0 °C)
$m(3.33 \times 10^5 \text{ J/kg}) = (.2 \text{ kg})(4186 \text{ J/kg °C})(10 °C) + (0.1 \text{ kg})(900 \text{ J/kg °C})(10 °C)$
$\quad m = 2.78 \times 10^{-2}$ kg = 27.8 g

12.31 (a) Q_1 = heat to melt all the ice = $(50 \times 10^{-3} \text{ kg})(3.33 \times 10^5 \text{ J/kg}) = 1.67 \times 10^4$ J
Q_2 = (heat to raise temp of ice to 100 °C)
$\quad = (50 \times 10^{-3} \text{ kg})(4186 \text{ J/kg °C})(100 °C) = 2.09 \times 10^4$ J
Thus, the total heat to melt ice and raise temp to 100 °C = 37,570 J
Q_3 = heat available as steam condenses = $(10 \times 10^{-3} \text{ kg})(2.26 \times 10^6 \text{ J/kg}) = 2.26 \times 10^4$ J
Thus, we see that $Q_3 > Q_1$, but $Q_3 < Q_1 + Q_2$.
Therefore, all the ice melts but $T_f < 100$ °C. Let us now find T_f.
$\quad$ heat gain = heat loss
$(50 \times 10^{-3} \text{ kg})(3.33 \times 10^5 \text{ J/kg}) + (50 \times 10^{-3} \text{ kg})(4186 \text{ J/kg °C})(T_f - 0 °C) =$
$\quad (10 \times 10^{-3} \text{ kg})(2.26 \times 10^6 \text{ J/kg}) + (10 \times 10^{-3} \text{ kg})(4186 \text{ J/kg °C})(100 °C - T_f)$
From which, $T_f = 40.4$ °C
(b) Q_1 = heat to melt all ice = 1.67×10^4 J (see part (a))
Q_2 = heat given up as steam condenses = $(10^{-3} \text{ kg})(2.26 \times 10^6 \text{ J/kg}) = 2.26 \times 10^3$ J
Q_3 = (heat given up as condensed steam cools to 0 °C)
$\quad = (10^{-3} \text{ kg})(4186 \text{ J/kg °C})(100 °C) = 418$ J
$\quad$ Note that $Q_2 + Q_3 < Q_1$ Therefore, the final temperature will be 0 °C with some ice remaining. Let us find the mass of ice which must melt to condense the steam and cool the condensate to 0 °C.

CHAPTER TWELVE SOLUTIONS

$mL_f = Q_2 + Q_3 = 2678$ J

Thus, $m = \dfrac{2678 \text{ J}}{3.33 \times 10^5 \text{ J/kg}} = 8.04 \times 10^{-3}$ kg $= 8.04$ g

Therefore, there are approximately 42 g of ice left over.

12.32 Heat to raise ice to 0°C $= \Delta Q_1 = m_{ice}c_{ice}(\Delta T)_{ice}$
    $= (40 \text{ g})(0.5 \text{ cal/g °C})(78 \text{ °C}) = 1560$ cal

The heat the cup and water can lose before they reach 0°C $= \Delta Q_2$
$= m_{cu}c_{cu}(\Delta T)_{cu} + m_w c_w(\Delta T)_w$
$= (0.08 \text{ kg})(387 \text{ J/kg°C})(25°C) + (0.56 \text{ kg})(4186 \text{ J/kg°C})(25°C) = 14{,}185$ cal

We see that $\Delta Q_2 > \Delta Q_1$ so the ice will reach 0°C and at least start to melt.

The heat to melt the ice is

$\Delta Q_3 = m_{ice}L_{f\,ice} = (0.04 \text{ kg})(3.34 \times 10^5 \text{ J/kg})(1\text{cal}/4.186 \text{ J}) = 3192$ cal

$\Delta Q_2 > \Delta Q_1 + \Delta Q_3$

Thus, the final temperature is greater than 0°C and all the ice melts. Let us find the final temperature T_f.

$\Delta Q_{system} = 0 = \Delta Q_{warm\ water} + \Delta Q_{cup} + \Delta Q_{to\ raise\ ice\ to\ 0°C} + \Delta Q_{to\ melt\ ice} + \Delta Q_{to\ warm\ melted\ ice} = 0$

$= (.56 \text{ kg})(4186 \text{ J/kg °C})(T_f - 25°C) + (.08 \text{ kg})(387 \text{ J/kg °C})(T_f - 25°C)$
    $+ 1560 \text{ cal}(4.186 \text{ J/cal}) + 3192 \text{ cal}(4.186 \text{ J/cal}) + (.04 \text{ kg})(4186 \text{ J/kg °C})(T_f - 0°C) = 0$

From which, $T_f = 15.5°C$

12.33 (a) Heat to cool system to 0 °C $= \Delta Q_1 = m_{cup}c_{al}(\Delta T)_{cup} + m_{water}c_{water}(\Delta T)_{water} =$
$(.1 \text{ kg})(900 \text{ J/kg °C})(30°C) + (0.18 \text{ kg})(4186 \text{ J/kg °C})(30°C) = 25{,}304$ J

The amount of ice which must melt to absorb this much heat $= \Delta Q_2 = m_{ice}L_f = 25{,}304$ J

$m_{ice} = \dfrac{25304 \text{ J}}{3.34 \times 10^5 \text{ J/kg}} = 7.58 \times 10^{-2}$ kg $= 75.8$ g

Therefore, if 100 g of ice is used, not all of the ice will melt,
    $T_f = 0°C$ with 24.2 g of ice left over

(b) if 50 g of ice is used, all the ice melts and $T_f > 0°C$. Let us compute T_f.

$\Delta Q_{system} = 0 = m_{ice}L_f + m_{ice}c_{water}(T_f - 0°C) + m_{water}c_{water}(T_f - 30°C)$
    $+ m_{cup}c_{al}(T_f - 30°C) = 0$

$= (0.05 \text{ kg})(3.34 \times 10^5 \text{ J/kg}) + (.05 \text{ kg})(4186 \text{ J/kg °C})T_f$
    $+ (0.18 \text{ kg})(4186 \text{ J/kg °C})(T_f - 30°C) + (0.1 \text{ kg})(900 \text{ J/kg°C})(T_f - 30°C) = 0$

$T_f = 8.2°C$

12.34 The energy required to melt 1 kg of snow is
    $Q = (1 \text{ kg})(3.33 \times 10^5 \text{ J}) = 3.33 \times 10^5$ J

The work done against friction is
    $f = \mu N = \mu mg = (0.2)(75 \text{ kg})(9.8 \text{ m/s}^2) = 147$ N

Therefore, the work done is
    $W = fs = (147 \text{ N})s$

Thus, $(147 \text{ N})s = 3.33 \times 10^5$ J
and $s = 2.27 \times 10^3$ m

12.35 (a) $H = \dfrac{\Delta Q}{\Delta t} = \dfrac{kA\Delta T}{L} = \dfrac{(397 \text{ J/s m°C})(15 \times 10^{-4} \text{ m}^2)(30 \text{ °C})}{0.08 \text{ m}} = 223$ J/s

(b) The steps are identical here as in (a) except $k = 0.0234$ J/s m °C. $H = 1.32 \times 10^{-2}$ J/s

CHAPTER TWELVE SOLUTIONS

(c) k = 0.08 J/s m °C , and H = 4.50 X 10^{-2} J/s

12.36 (a) $H = \frac{\Delta Q}{\Delta t} = \frac{kA\Delta T}{L} = \frac{(0.8 \text{ J/s m °C})(0.16 \text{ m}^2)(11.1°C)}{3 \times 10^{-3} \text{ m}}$ = 474 J/s (into house)

(b) $H = \frac{\Delta Q}{\Delta t} = \frac{kA\Delta T}{L} = \frac{(0.8 \text{ J/s m °C})(0.16 \text{ m}^2)(38.9°C)}{3 \times 10^{-3} \text{ m}}$ = 1.66 X 10^3 J/s (to outdoors)

12.37 $R_{total} = \Sigma R_i$ = (0.17 + 0.87 + 1.32 + 3(3.7) + 0.45 + 0.17)ft^2 F° h/BTU
= 14.1 ft^2 F° h/BTU

12.38 (a) $H = \frac{\Delta Q}{\Delta t} = \frac{kA\Delta T}{L} = \frac{(0.8 \text{ J/s m °C})(3 \text{ m}^2)(25 \text{ °C})}{6 \times 10^{-3} \text{ m}}$ = 10000 J/s

Thus, in one hour, $\Delta Q = Ht$ = (10000 J/s)(3600 s) = 3.6 X 10^7 J = 8.6 X 10^6 cal

12.39 A = $A_{end\ walls}$ + $A_{ends\ of\ attic}$ + $A_{side\ walls}$ + A_{roof}

A = 2(8 m X 5 m) + 2(2 X $\frac{1}{2}$ X (4 m) X (4 m) tan37°) + 2(10 m) X 5 m)

+ 2(10 m X $\frac{4 \text{ m}}{\cos 37°}$) = 304 m^2

$H = \frac{\Delta Q}{\Delta t} = \frac{kA\Delta T}{L} = \frac{(4.8 \times 10^{-4} \text{ kW/ m °C})(304 \text{ m}^2)(25 \text{ °C})}{0.21 \text{ m}}$ = 17.4 kW = 4.15 kcal/s

Thus, the heat lost per day = (4.15 kcal/s)(86,400 s) = 3.59 X 10^5 kcal/day

The gas needed to replace this loss = $\frac{3.59 \times 10^5 \text{ kcal/day}}{9300 \text{ kcal/m}^3}$ = 38.6 m^3/day

12.40 (a) Temperature gradient = $\frac{\Delta T}{L} = \frac{80 °C - 20°C}{0.6 \text{ m}}$ = 100 °C/m

(b) $H = \frac{\Delta Q}{\Delta t} = \frac{kA\Delta T}{L}$ = (79.5 J/s m °C)(2 X 10^{-4} m^2)(100 °C/m) = 1.59 J/s

(c) ΔT = temperature gradient (length)
or   80°C - T = (100 °C/m)(0.2 m)
gives T = 60°C

12.41 For the thermopane,

$H = \frac{\Delta Q}{\Delta t} = \frac{A\Delta T}{\Sigma \frac{L_i}{k_i}} = \frac{(1.0 \text{ m}^2)(23 \text{ °C} - 0°C)}{\frac{10^{-2} \text{ m}}{0.0234 \text{ J/sm°C}} + \frac{5 \times 10^{-3} \text{ m}}{0.8 \text{ J/sm°C}} + \frac{5 \times 10^{-3} \text{ m}}{0.8 \text{ J/sm°C}}}$ = 52.3 J/s

For the single pane,

$H = \frac{\Delta Q}{\Delta t} = \frac{kA\Delta T}{L} = \frac{(0.8 \text{ J/s m °C})(1 \text{ m}^2)(23 \text{ °C})}{1 \times 10^{-2} \text{ m}}$ = 1840 J/s

12.42 The answers will vary here depending on the details of construction. However, the approach to find the answer is like that used in problem 12.37.

12.43 The average of the inside and outside temperatures is equal to the temperature of the surface of the window. This is

$<T> = \frac{23 \text{ °C} + 5 \text{ °C}}{2}$ = 14 °C

CHAPTER TWELVE SOLUTIONS

Therefore, ΔT = temperature of the air - temperature of window surface = 9 °C. and the convection coefficient is

$$h = 1.77(\Delta T)^{1/4} \text{ J/s m}^2 \text{ °C} = 1.77(9)^{1/4} \text{ J/s m}^2 \text{ °C} = 3.07 \text{ J/s m}^2 \text{ °C}$$

Thus, the rate of heat transfer by convection is,

$$\Delta Q/\Delta t = hA(\Delta T) = (3.07 \text{ J/s m}^2 \text{ °C})(0.6 \text{ m}^2)(9 \text{ °C}) = 16.6 \text{ J/s}$$

12.44 We have
conduction flow rate through window = convection rate to window = 16.6 J/s

Therefore, from $\dfrac{\Delta Q}{\Delta t} = \dfrac{kA\Delta T}{L}$, or $\Delta T = \dfrac{(\Delta Q)L}{(\Delta t)kA}$

we have $\Delta T = \dfrac{(16.6 \text{ J/s})(1.25 \times 10^{-3} \text{ m})}{(0.8 \text{ J/s m °C})(0.6 \text{ m}^2)} = 4.32 \times 10^{-2}$ °C

Thus, $T_{inside} = <T> + \dfrac{\Delta T}{2}$ = 9 °C + .02 °C = 9.02 °C

and $T_{outside} = <T> - \dfrac{\Delta T}{2}$ = 9 °C - .02 °C = 8.98 °C

12.45 From $P_{net} = \sigma Ae(T^4 - T_0^4)$
For two identical objects, we have the ratio of the power emitted by the hotter to that from the colder is

$$\dfrac{P_h}{P_c} = \dfrac{(1200)^4 - (273)^4}{(1100)^4 - (273)^4}) = \dfrac{2.07 \times 10^{12}}{1.46 \times 10^{12}} = 1.42$$

12.46 The area of the sphere is $A = 4\pi r^2 = 4\pi(0.06 \text{ m})^2 = 4.52 \times 10^{-2}$ m^2. For a perfect radiator,
e = 1. Thus,

$$P = (5.67 \times 10^{-8} \text{ W/m}^2\text{K}^4)(4.52 \times 10^{-2} \text{ m}^2)((473 \text{ K})^4 - (295 \text{ K})^4) = 109 \text{ W}$$

12.47 The heat needed to melt 5 kg of ice = mL_f = (5 kg)(3.33 × 10^5 J/kg) = 1.67 × 10^6 J
Thus, the required flow rate during the 8 h (2.88 × 10^4 s) period is,

$$\dfrac{\Delta Q}{\Delta t} = 57.8 \text{ J/s}$$

Then, $\dfrac{\Delta Q}{\Delta t} = \dfrac{kA\Delta T}{L}$

becomes $57.8 \text{ J/s} = \dfrac{k(0.8 \text{ m}^2)(20 \text{ °C})}{2 \times 10^{-2} \text{ m}}$

and we find $k = 7.23 \times 10^{-2}$ J/s m °C

12.48 We use $P_{net} = \sigma Ae(T^4 - T_0^4)$

$$25 \text{ W} = (5.67 \times 10^{-8} \text{ W/m}^2\text{K}^4)(2.5 \times 10^{-5} \text{ m}^2)(0.25)(T^4 - (295 \text{ K})^4)$$

From which, T = 2.90 × 10^3 K = 2.63 × 10^3 °C

12.49 Let us call the copper rod object 1 and the aluminum rod object 2. At equilibrium, the flow rate through each must be the same. We have

$$\dfrac{k_2 A_2 \Delta T}{L_2} = \dfrac{k_1 A_1 \Delta T}{L_1}$$

Since the cross-sectional areas of the rods are the same and the temperature difference across the rods are also equal, we have

CHAPTER TWELVE SOLUTIONS

$$\frac{k_2}{L_2} = \frac{k_1}{L_1}$$

$$\frac{238 \text{ J/s m°C}}{L_2} = \frac{397 \text{ J/s m°C}}{0.15 \text{ m}}$$

From which, we find
$$L_2 = 9.00 \times 10^{-2} \text{ m} = 9 \text{ cm}$$

12.50 heat gain = heat loss
$$m_b c_b(T_f - 20 \text{ °C}) = m_w c_w(90 \text{ °C} - T_f) \quad (1)$$

The mass of the bullet, m_b, is 5 g, the mass of the water is 100 g, and the specific heat of water is 4186 J/kg °C. We substitute these into (1) and solve for T_f in terms of the specific heat of the bullet, c_b. We find

$$T_f = \frac{c_b(20 \text{ °C}) + 7.53 \times 10^6 \text{ J/kg}}{(c_b + 8.37 \times 10^4 \text{ J/kg °C})}$$

For the silver bullet, c_b = 234 J/kg °C, and T_f = 89.8 °C.
For the copper bullet, c_b = 387 J/kg °C, and T_f = 89.7 °C. The copper bullet wins.

12.51 To raise the temperature of water to 100 °C, we need an amount of heat equal to
$$Q = (0.25 \text{ kg})(4186 \text{ J/kg °C})(77 \text{ °C}) = 8.06 \times 10^4 \text{ J}$$
The collection rate is $(550 \text{ W/m}^2)(1 \text{ m}^2) = 550 \text{ W} = 550 \text{ J/s}$
Therefore, the time required is
$$t = 8.06 \times 10^4 \text{ J}/(550 \text{ J/s}) = 146 \text{ s} = 2.44 \text{ min}$$

12.52 (a) Heat capacity = $mc = (10^4 \text{ kg})(920 \text{ J/kg°C}) = 9.2 \times 10^6 \text{ J/°C}$
(b) Energy received = $(400 \text{ J/s m}^2)(20 \text{ m}^2)(6 \text{ h})(3600 \text{ s/h}) = 1.73 \times 10^8 \text{ J/day}$
Energy stored = $(0.3)(\text{energy received}) = 5.18 \times 10^7 \text{ J/day} = \Delta Q$
(c) $\Delta Q = mc\Delta T = (9.2 \times 10^6 \text{ J°C})(T_f - 15\text{°C}) = 5.18 \times 10^7 \text{ J}$
From which, $T_f = 20.6\text{°C}$

12.53 The rate energy is received = $(300 \text{ W/m}^2)(12.0 \text{ m}^2) = 3600 \text{ W} = 3600 \text{ J/s}$
Thus, it must be radiating away 3600 J/s.
$$P_{net} = \sigma A e (T^4 - T_0^4)$$
$$3600 \text{ W} = (5.67 \times 10^{-8} \text{ W/m}^2\text{K}^4)(12 \text{ m}^2)(1)(T^4 - (253 \text{ K})^4)$$
Which yields, T = 311 K = 38°C

12.54 The area of the bottom of the kettle is $\pi \times 10^{-2} \text{ m}^2$, and the temperature difference across the bottom is 2 °C. The rate of heat transfer is

$$H = \frac{\Delta Q}{\Delta t} = \frac{kA\Delta T}{L} = \frac{(397 \text{ J/s m°C})(\pi \times 10^{-2} \text{ m}^2)(2 \text{ °C})}{2 \times 10^{-3} \text{ m}} = 1.25 \times 10^4 \text{ J/s}$$

12.55 $Q = (50 \text{ kg})(380 \text{ J/kg °C})(20 \text{ °C}) = 3.8 \times 10^5 \text{ J} = 9.08 \times 10^4 \text{ cal}$

12.56 heat loss = heat gain
(heat lost by lead) = (heat to melt ice) + (heat to raise melted ice to final temp)
 + (heat to raise water to final temp) + (heat to raise temp of cup)
$m(128 \text{ J/kg °C})(86\text{°C}) = (0.04 \text{ kg})(3.33 \times 10^5 \text{ J/kg}) + (0.04 \text{ kg})(4186 \text{ J/kg °C})(12\text{°C})$
 $+ (.200 \text{ kg})(4186 \text{ J/kg°C})(12\text{°C}) + (0.1 \text{ kg})(387 \text{ J/kg °C})(12\text{°C})$
From which, m = 2.35 kg

CHAPTER TWELVE SOLUTIONS

12.57 (a) The heat required for the system to reach 0 °C with the ice still a solid is

Q_1 = (10^{-2} kg)(2090 J/kg °C)(20 °C) + (0.2 kg)(387 J/kg °C)(20 °C) = 1966 J

The heat to leave system at 0 °C with all ice melted is

Q_2 = 1966 J + (10^{-2} kg)(3.33 X 10^5 J/kg) = 5296 J

The heat to reach 100 °C with liquid water in the cup is

Q_3 = 5296 J + (10^{-2} kg)(4186 J/kg °C)(100 °C) + (.200 kg)(387J/kg °C)(100 °C)
= 17,220 J

Thus, when we add 500 cal (2092 J), the system reaches 0 °C and melts the following amount of ice.

$$\frac{2092 \text{ J} - 1966 \text{ J}}{3.33 \times 10^5 \text{ J/kg}} = 3.78 \times 10^{-4} \text{ kg}$$

(b) Adding 5000 cal (20,920 J) causes the system to reach 100 °C and vaporizes the following amount of water.

$$\frac{20920 \text{ J} - 17220 \text{ J}}{2.26 \times 10^6 \text{ J/kg}} = 1.64 \times 10^{-3} \text{ kg} = 1.64 \text{ g of water}$$

12.58 The heat added by the heater is

Q = Pt = (800 J/s)(300 s) = 2.4 X 10^5 J

and the temperature change is

$$\Delta T = \frac{Q}{mc} = \frac{2.4 \times 10^5 \text{ J}}{(2 \text{ kg})(4186 \text{ J/kg °C})} = 28.7 \text{ °C}$$

12.59 The heat added to the air in one hour is

Q = 10(200 J/s)(3600 s) = 7.2 X 10^6 J

The mass of the air in the room is

m = ρV = (6 m)(15 m)(3 m)(1.3 kg/m^3) = 351 kg

Thus, the change in temperature of the room is

$$\Delta T = \frac{Q}{mc} = \frac{7.2 \times 10^6 \text{ J}}{(351 \text{ kg})(837 \text{ J/kg °C})} = 24.5 \text{ °C}$$

Thus, the final temperature is 44.5 °C

12.60 The rate of heat loss by sweating = 0.9(300 W) = 270 W = 270 J/s

Thus, the heat dissipated in 1 h = (270 J/s) $(\frac{1 \text{ cal}}{4.186 \text{ J}} \frac{3600 \text{ s}}{\text{h}})$ = 2.32 X 10^5 cal/h

= 232 kcal/h

Therefore, the amount of water at about 37°C which must evaporate to absorb this much heat is, $\frac{232 \text{ kcal/h}}{575 \text{ kcal/kg}}$ = 0.404 kg = 404 g. (There is a loss of 404 cm^3 of water per hour.)

12.61 (a) P = Fv = (50 N)(40 m/s) = 2000 W

(b) The frictional work done on each object in 10 s is

W = Pt = (1000 J/s)(10 s) = 10,000 J.

and from Q = mcΔT, we have $\Delta T = \frac{Q}{mc} = \frac{10000 \text{ J}}{(5 \text{ kg})(448 \text{ J/kg °C})}$ = 4.47 °C

12.62 The kinetic energy of the car is

$\frac{1}{2}$ mv^2 = $\frac{1}{2}$ (1500 kg)(25 m/s)^2 = 4.69 X 10^5 J

The melting point of aluminum is 660°C. Thus, the heat required to raise the temperature of the brakes from 20°C to the melting point is

CHAPTER TWELVE SOLUTIONS

$Q = mc\Delta T = (60000 \text{ g})(0.9 \text{ J/g C°})(640 \text{ C°}) = 3.46 \times 10^7 \text{ J}$

Thus, the number of stop would be $\dfrac{3.46 \times 10^7 \text{ J}}{4.69 \times 10^5 \text{ J}} = 74$ stops

(b) This calculation assumes no heat loss to the surroundings, and that all the heat generated in one stops still remains with the brakes until the next application of the brakes.

12.63 During any time interval the heat energy lost by the rod equals the heat energy gained by the helium. Therefore,

$(mL)_{He} = (mc\Delta T)_{Al}$ or $(\rho V L)_{He} = (\rho V c \Delta T)_{Al}$ so that

$V_{He} = \dfrac{(\rho V c \Delta T)_{Al}}{(\rho L)_{He}} = \dfrac{(2.7 \text{ g/cm}^3)(100 \text{ cm}^3)(0.21 \text{ cal/g °C})(295.8°C)}{(0.125 \text{ g/cm}^3)(4.99 \text{ cal/g})}$

$= 2.69 \times 10^4 \text{ cm}^3 = 26.9$ liters

12.64 The power incident on the solar collector is

$P_i = IA = (600 \text{ W/m}^2)\pi(0.25 \text{ m})^2 = 117.8 \text{ W}$ or

$P_i = \dfrac{117.8 \text{ W}}{4.186 \text{ J/cal}} = 28.14$ cal/s

For a 50% reflector, the collected power is $P_c = 14.07$ cal/s. The total energy required to increase the temperature of the water to the boiling point and to evaporate it is

$Q = mc\Delta T + mL_v = (1000 \text{ g})[(1 \text{ cal/g°C})(80°C) + 540 \text{ cal/g}] = 6.2 \times 10^5$ cal

The time required is $\Delta t = \dfrac{Q}{P_c} = \dfrac{6.2 \times 10^5 \text{ cal}}{14.07 \text{ cal/s}} = 4.41 \times 10^4$ s $= 12.2$ h

---

12.65 The change in temperature of the rod is

$\Delta T = \dfrac{Q}{mc} = \dfrac{10^4 \text{ J}}{(0.35 \text{ kg})(900 \text{ J/kg °C})} = 31.7$ °C

and the new length of the rod is

$L = L_0(1 + \alpha \Delta T) = 20 \text{ cm}(1 + (24 \times 10^{-6} \text{ °C}^{-1})(31.7 \text{ °C})) = 20.02$ cm

---

12.66 The masses of all the liquids to be mixed are equal in this problem, so m will cancel from the calorimetry equations. Let us consider case 1 in which liquids 1 and 2 are mixed.

heat gain = heat loss   (liquids 1 and 2)

$c_1(7 \text{ °C}) = c_2(3 \text{ °C})$

Therefore, $c_2 = \dfrac{7}{3} c_1$ (1)

Now consider case 2 in which liquids 2 and 3 are mixed.

heat gain = heat loss

$c_2(8 \text{ °C}) = c_3(2 \text{ °C})$

Thus, $c_3 = 4c_2$ (2)

Eliminate c_2 between (1) and (2), and we find

$c_3 = \dfrac{28}{3} c_1$ (3)

Finally, consider case 3 in which liquids 1 and 3 are mixed.

heat gain = heat loss

$c_1(T_f - 10 \text{ °C}) = c_3(30 \text{ °C} - T_f)$ (4)

Use (3) to eliminate the specific heats from equation (4), and solve the resulting equation for T_f.

$T_f = 28.1$ °C

12.67 (a) With $Q = mc\Delta T$ and $m = \rho V$, the rate of addition of heat is found as

CHAPTER TWELVE SOLUTIONS

$Q = \rho V c \Delta T$, or the amount of heat ΔQ added to a volume ΔV is $\Delta Q = \rho \Delta V c \Delta T$, and from this the heat added in a time Δt is $\dfrac{\Delta Q}{\Delta t} = \rho c \Delta T \dfrac{\Delta V}{\Delta t}$

(b) From part (a) $c = \dfrac{\dfrac{\Delta Q}{\Delta t}}{\rho \Delta T \dfrac{\Delta V}{\Delta t}} = \dfrac{(40 \text{ J/s})(1 \text{ cal}/4.186 \text{ J})}{(0.72 \text{ g/cm}^3)(5.8 \text{ °C})(3.5 \text{ cm}^3/\text{s})} = 0.654 \text{ cal/g C°}$

12.68 (a) Let us call rod 1 the aluminum rod, and rod 2 is the iron rod. At equilibrium, the rate of heat transfer through the rods are equal. Thus, we have

$$\dfrac{k_2 A_2 \Delta T_2}{L_2} = \dfrac{k_1 A_1 \Delta T_1}{L_1}$$

But, since the rods have equal lengths and radii, this becomes

$$\Delta T_1 = \dfrac{k_2}{k_1} \Delta T_2,$$

Let us call the temperature at the interface T. Thus, we have

$$100 \text{ °C} - T = \dfrac{79.5}{238}(T - 0 \text{°C})$$

or $T = 75.0$ °C

(b) We find, $\dfrac{\Delta Q}{\Delta t} = \dfrac{k_2 A_2 \Delta T_2}{L_2} = \dfrac{(79.5 \text{ J/s m °C})(5 \times 10^{-4} \text{ m}^2)(75 \text{ °C})}{0.15 \text{ m}} = 19.9 \text{ J/s}$

Therefore, if $\Delta t = 30$ min $= 1800$ s, we have

$\Delta Q = 3.58 \times 10^4$ J

12.69 The heat required to vaporize 0.5 kg of water at 100 °C is

$Q = (0.5 \text{ kg})(2.26 \times 10^6 \text{ J/kg}) = 1.13 \times 10^6$ J

Thus, the rate of heat transfer is

$\dfrac{\Delta Q}{\Delta t} = \dfrac{1.13 \times 10^6 \text{ J}}{60 \text{ s}} = 1.88 \times 10^4$ J/s

From $\dfrac{\Delta Q}{\Delta t} = \dfrac{kA\Delta T}{L}$ we have

$\Delta T = \dfrac{\Delta Q}{\Delta t} \dfrac{L}{kA} = (1.88 \times 10^4 \text{ J/s})\dfrac{5 \times 10^{-3} \text{ m}}{(238 \text{ J/s m°C})\pi(.12 \text{ m})^2} = 8.75$ °C

Thus, $T = 108.75$ °C $= 109$ °C

CHAPTER THIRTEEN SOLUTIONS

13.1 $W = P\Delta V = PA\Delta L = (2 \times 10^5 \text{ Pa})\frac{\pi}{4}(16 \times 10^{-2} \text{ m})^2(0.2 \text{ m}) = 804 \text{ J}$

13.2 We call N the number of molecules, and the average kinetic energy per molecule is represented by <KE>. The total energy of our system is

$$U = N<KE> = (3N_a)\frac{3}{2}kT = \frac{9}{2}RT \text{ (where we have used } N_ak = R)$$

Thus, $U = \frac{9}{2}(8.31 \text{ J/mol K})(303 \text{ K}) = 1.13 \times 10^4 \text{ J}$

13.3 The sketches for (a) and (b) are shown below.

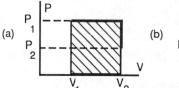 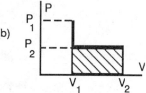

(c) There is more work done in process (a). We recognize this from the figures because there is more area under the PV curve in (a). Physically, more work is done because of the higher pressure during the expansion part of the process.

13.4 (a) W(IAF) = W(IA) + W(AF) (The notation IAF means the work done from I to A to F.) The term W(AF) = 0 because the volume is a constant during this part of the process. Thus,

$W(IAF) = W(IA) = P_I(V_A - V_I) = (4.05 \times 10^5 \text{ Pa})(2 \times 10^{-3} \text{ m}^3) = 810 \text{ J}$

(b) Along path IF, we find the work by finding the area under the PV curve. This consists of a triangular area plus a rectangular area. This is

$W(IF) = \frac{1}{2}(2 \times 10^{-3} \text{ m}^3)(3.04 \times 10^5 \text{ Pa}) + (1.013 \times 10^5 \text{ Pa})(2 \times 10^{-3} \text{ m}^3)$
$= 507 \text{ J}$

(c) W(IBF) = W(IB) + W(BF). The term W(IB) equals zero because the volume remains a constant during this part of the process. So,

$W(IBF) = W(BF) = (1.013 \times 10^5 \text{ Pa})(2 \times 10^{-3} \text{ m}^3) = 203 \text{ J}$

13.5 The sketch for the cycle is shown at the right.

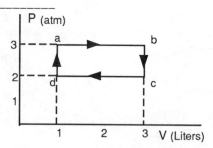

Let us now find the net work done during the process.
During the expansion from a to b we have
  $W(a \text{ to } b) = P_a(V_b - V_a)$
    $= 3(1.013 \times 10^5 \text{ Pa})(2 \times 10^{-3} \text{ m}^3) = 608 \text{ J}.$
  W(b to c) = 0 because volume does not change
  $W(c \text{ to } d) = P_c(V_d - V_c)$

CHAPTER THIRTEEN SOLUTIONS

$= 2(1.013 \times 10^5 \text{ Pa})(-2 \times 10^{-3} \text{ m}^3) = -405$ J.
W(d to a) = 0   volume remains constant
Thus, the net work is   W = 608 J - 405 J = 203 J
(Note: the net work could also be obtained by computing the area enclosed within one cycle on the PV diagram.)

13.6 P = 1.5 atm = 1.52×10^5 Pa
(a) W = PΔV, and ΔV = 4 m^3, so
  W = $(1.52 \times 10^5 \text{ Pa})(4 \text{ m}^3) = 6.08 \times 10^5$ J
(b) ΔV = -3 m^3
  W = $(1.52 \times 10^5 \text{ Pa})(-3 \text{ m}^3) = -4.56 \times 10^5$ J

13.7 For an ideal gas PV = nRT, and $V = \frac{nRT}{P}$, so $\Delta V = V_f - V_i = \frac{nR}{P}(T_f - T_i)$
This gives, W = PΔV = nRΔT = (0.2 mol)(8.31 J/K mol)(280 K) = 465 J

13.8 W = $-P(V_f - V_i)$ (W is negative because work is done on the gas.)
  W = $-P(V_f - 2V_f) = PV_f$
Thus, 100 J = $(1.25 \times 10^5 \text{ Pa})V_f$
We find, $V_f = 8 \times 10^{-4}$ m^3 = 800 cm^3.

13.9 The work is done on the gas, so
  W = $-P\Delta V$
Thus, $P = \frac{-W}{\Delta V} = \frac{-10^4 \text{ J}}{-35 \times 10^{-3} \text{ m}^3} = 2.86 \times 10^5$ Pa

---

13.10 $V_i = \frac{m}{\rho_i} = \frac{2 \text{ kg}}{999.17 \text{ kg/m}^3} = 2.00166 \times 10^{-3}$ m^3
$V_f = \frac{m}{\rho_f} = \frac{2 \text{ kg}}{999.02 \text{ kg/m}^3} = 2.00196 \times 10^{-3}$ m^3
Thus, $\Delta V = 3.01 \times 10^{-7}$ m^3
W = PΔV = $(1.013 \times 10^5 \text{ Pa})(3.01 \times 10^{-7} \text{ m}^3) = 3.04 \times 10^{-2}$ J

13.11 (a) W = PΔV = $[0.3(1.013 \times 10^5 \text{ Pa})(3 \times 10^{-3} \text{ m}^3 - 8 \times 10^{-3} \text{ m}^3)]$
  = -152 J.
(b) ΔU = Q - W We are given that Q = -400 J. Thus,
  ΔU = -400 J - (-152 J) = -248 J

13.12 (a) ΔU = Q - W = 418 J - 507 J = -89 J (It was shown in problem 4 that the work done along the diagonal path is 507 J.)
(b) Along IAF the work done is 810 J, (See problem 4.), and ΔU is the same as above. Thus,
  Q = ΔU + W = -89 J + 810 J = 721 J

13.13 ΔU = Q - W, or   Q = ΔU + W = -500 J - 220 J = -720 J

13.14 The initial and final pressures are 1 atm, and we can treat the system as an ideal gas, PV = nRT. We have
$\frac{V_f}{T_f} = \frac{V_i}{T_i}$, or $V_f = \frac{T_f}{T_i} V_i = \frac{310 \text{ K}}{273 \text{ K}}(0.6 \times 10^{-3} \text{ m}^3) = 6.81 \times 10^{-4}$ m^3

CHAPTER THIRTEEN SOLUTIONS

and $V_i = 6.00 \times 10^{-4}$ m^3
so $\Delta V = 8.1 \times 10^{-5}$ m^3
(a) $W = P\Delta V = (1.013 \times 10^5$ Pa$)(8.1 \times 10^{-5}$ m$^3) = 8.24$ J.
(b) $Q = mc\Delta T = (7.7 \times 10^{-4}$ kg$)(1010$ J/kg°C$)(37$°C$) = 28.8$ J
(c) $\Delta U = 28.8$ J $- 8.24$ J $= 20.5$ J

13.15 (a) The change in volume is
$\Delta V = A\Delta L = (0.15$ m$^2)(-0.2$ m$) = -3.0 \times 10^{-2}$ m^3.
Thus, $W = P\Delta V = (6 \times 10^3$ Pa$)(-3 \times 10^{-2}$ m$^3) = -180$ J.
(b) $Q = \Delta U + W = -8$ J $- 180$J $= -188$ J   (188 J of heat energy are removed from the gas.)

13.16 (a) $W = P\Delta V = (1.013 \times 10^5$ Pa$)(3.34 \times 10^{-3}$ m$^3) = 338$ J.
(b) The heat added is
$Q = mL_v = (2 \times 10^{-3}$ kg$)(2.26 \times 10^6$ J/kg$) = 4520$ J
(c) $\Delta U = Q - W = 4520$ J $- 338$ J $= 4182$ J

13.17 We are given $V_f = 3V_i$. For the work done during an isothermal process, we use
(a) $W = (nRT)\ln\dfrac{V_f}{V_i} = (2$ mol$)(8.31$ J/mol K$)(300$ K$)\ln(3) = 5.48 \times 10^3$ J

(b) $\Delta U = 0$ for an isothermal process of an ideal gas. Therefore, $\Delta U = Q - W = 0$, or
$Q = W = 5.48 \times 10^3$ J

13.18 $W_{BC} = 0$ (constant volume), $W_{CA} < 0$ ($\Delta V < 0$), $W_{AB} > 0$ ($\Delta V > 0$)
Now, $\Delta U = Q - W$ gives $\Delta U_{BC} = Q_{BC} - W_{BC} < 0$ (because $Q_{BC} < 0$, and $W_{BC} = 0$)
or, $U_C < U_B$
This is a cyclic process, so $\Delta U_{cycle} = 0$
That is, $\Delta U_{AB} + \Delta U_{BC} + \Delta U_{CA} = 0$
Thus, $\Delta U_{AB} > 0$ since both ΔU_{BC} and ΔU_{CA} are negative
Therefore, $\Delta U_{AB} > 0$, $\Delta U_{BC} < 0$, and $\Delta U_{CA} < 0$
$\Delta U_{CA} = Q_{CA} - W_{CA}$ becomes $Q_{CA} = \Delta U_{CA} + W_{CA} < 0$ since both ΔU_{CA} and W_{CA} are negative.
Also, $Q_{AB} = \Delta U_{AB} + W_{AB} > 0$ since both ΔU_{AB} and W_{AB} are positive

13.19 $\Delta U_{cycle} = Q_{cycle} - W_{cycle} = 0$ (for any complete cycle)

$Q_{cycle} = W_{cycle} =$ area enclosed in PV process diagram $= \dfrac{1}{2}(4$ m$^3)(6 \times 10^3$ Pa$)$
$= 12 \times 10^3$ J
or, $Q_{cycle} = 12$ kJ
If the cycle is reversed, then $Q_{cycle} = -12$ kJ

13.20 (a) $W = (nRT)\ln\dfrac{V_f}{V_i} = (2$ mol$)(8.31$ J/mol K$)(300$ K$)\ln\dfrac{3}{10} = -6000$ J

(b) The internal energy of an ideal gas is a function only of temperature. Therefore, if the temperature does not change,
$\Delta U = 0$
(c) $Q = \Delta U + W = 0 - 6000$ J $= -6000$ J, or 6000 J of heat must be removed from the gas.

184

CHAPTER THIRTEEN SOLUTIONS

13.21 (a) The change in length of the rod is
$$\Delta L = \alpha L_0 (\Delta T) = (11 \times 10^{-6} \, °C^{-1})(2 \, m)(20 \, °C) = 4.4 \times 10^{-4} \, m$$
The force exerted by the rod equals the weight of the load = 5.88×10^4 N
$$W = F\Delta L = (5.88 \times 10^4 \, N)(4.4 \times 10^{-4} \, m) = 25.9 \, J$$
(b) $Q = mc\Delta T = (10^2 \, kg)(448 \, J/kg \, °C)(20 \, °C) = 8.96 \times 10^5 \, J$
(c) $\Delta U = Q - W = 8.96 \times 10^5 \, J - 25.9 \, J = 8.96 \times 10^5 \, J$

13.22 (a) The change in volume of the aluminum is
$$\Delta V = \beta V_0 (\Delta T) = (3)(24 \times 10^{-6} \, °C^{-1})(3.7 \times 10^{-4} \, m^3)(18 \, °C) = 4.8 \times 10^{-7} \, m^3$$
Thus, $W = P\Delta V = (1.013 \times 10^5 \, Pa)(4.8 \times 10^{-7} \, m^3) = 4.86 \times 10^{-2} \, J$
(b) $Q = mc\Delta T = (1 \, kg)(900 \, J/kg \, °C)(18 \, °C) = 1.62 \times 10^4 \, J$
(c) $\Delta U = Q - W = 1.62 \times 10^4 \, J - 4.86 \times 10^{-2} \, J = 1.62 \times 10^4 \, J$

13.23 (a) $V_f = \dfrac{nRT}{P_f} = \dfrac{(1 \, mol)(8.31 \, J/mol \, K)(300 \, K)}{(0.8)(1.013 \times 10^5 \, Pa)} = 3.08 \times 10^{-2} \, m^3$.

(b) Also from the ideal gas law we find
$$\frac{V_f}{V_i} = \frac{P_i}{P_f} = \frac{1}{4}$$
Thus, $W = (nRT)\ln\dfrac{V_f}{V_i} = (1 \, mol)(8.31 \, J/mol \, K)(300 \, K)\ln(0.25) = -3.46 \times 10^3 \, J$

(c) $\Delta U = 0$ for an isothermal process of an ideal gas. Thus, $\Delta U = Q - W = 0$, or
$$Q = W = -3.46 \times 10^3 \, J$$

13.24 From $W = (nRT)\ln\dfrac{V_f}{V_i}$
we find, $\ln\dfrac{V_f}{V_i} = \dfrac{W}{nRT} = \dfrac{800 \, J}{(1 \, mol)(8.31 \, J/mol \, K)(300 \, K)} = 0.321$
Now, take the antilog to find
$$\frac{V_f}{V_i} = e^{0.321} = 1.38$$
Thus, $V_f = 1.38 V_i = 1.38(2 \, liters) = 2.76 \, liters$

13.25 (a) $W = (nRT)\ln\dfrac{V_f}{V_i} = (1 \, mol)(8.31 \, J/mol \, K)(400 \, K)\ln(2) = 2300 \, J$

(b) $\Delta U_{ab} = Q_{ab} - W_{ab} = 0$ ($\Delta U = $ in isothermal process with ideal gas.)
$Q_{ab} = W_{ab} = 2300 \, J$

(c) From the ideal gas equation, $\dfrac{P_b}{P_a} = \dfrac{V_a}{V_b} = \dfrac{1}{2}$

13.26 (b) From the ideal gas equation,
$$T = \frac{P_f V_f}{nR} = \frac{(1.013 \times 10^5 \, Pa)(25 \times 10^{-3} \, m^3)}{(1 \, mol)(8.31 \, J/mol \, K)} = 305 \, K = 32 \, °C$$
(a) $\ln\dfrac{V_f}{V_i} = \dfrac{W}{nRT} = \dfrac{3000 \, J}{(1 \, mol)(8.31 \, J/mol \, K)(305 \, K)} = 1.185$
$\dfrac{V_f}{V_i} = 3.27$, and $V_i = \dfrac{V_f}{3.27} = \dfrac{25 \, L}{3.27} = 7.65 \, liters$

13.27 (a) $W = 0$ because the volume equals a constant.

CHAPTER THIRTEEN SOLUTIONS

The system gives off heat. Thus,
$Q < 0$
$\Delta U = Q - W$ and since $W = 0$, we have $\Delta U = Q$
Therefore, $\Delta U < 0$
(b) Again, $W = 0$ because the volume remains constant.
$Q > 0$ because the water receives heat.
Again, $\Delta U = Q$
So $\Delta U > 0$

13.28 (a) $W = P\Delta V$ = area under PV curve.
$W(IAF) = (1.5 \text{ atm})(0.8 - 0.3)\text{liters} = 1.5(1.013 \times 10^5 \text{ Pa})(0.5 \times 10^{-3} \text{ m}^3) = 76$ J
$W(IBF) = (2 \text{ atm})(0.8 - 0.3)\text{liters} = 2(1.013 \times 10^5 \text{ Pa})(0.5 \times 10^{-3} \text{ m}^3) = 101$ J
$W(IF) = W(IAF) + \frac{1}{4}(1.013 \times 10^5 \text{ Pa})(0.5 \times 10^{-3} \text{ m}^3) = 88.7$ J

(b) We are given that $\Delta U = (182 \text{ J} - 91 \text{ J}) = 91$ J
Thus, $Q = \Delta U + W = 91 \text{ J} + W$
$Q(IAF) = 91 \text{ J} + 76 \text{ J} = 167$ J
$Q(IBF) = 91 \text{ J} + 101 \text{ J} = 192$ J
$Q(IF) = 91 \text{ J} + 88.7 \text{ J} = 180$ J

13.29 (a) $\text{Eff} = \dfrac{W}{Q_h} = \dfrac{25 \text{ J}}{375 \text{ J}} = 0.067$
(b) $Q_c = Q_h - W = 375 \text{ J} - 25 \text{ J} = 350$ J

13.30 (a) $W = Q_h - Q_c = 1700 \text{ J} - 1200 \text{ J} = 500$ J
$\text{Eff} = \dfrac{W}{Q_h} = \dfrac{500 \text{ J}}{1700 \text{ J}} = 0.294$
(b) The work done in each cycle has been found to be 500 J in part (a).
(c) $P = \dfrac{W}{\Delta t} = \dfrac{500 \text{ J}}{0.3 \text{ s}} = 1667$ W

13.31 $W = Q_h - Q_c = 200$ J (1)
$\text{Eff} = \dfrac{W}{Q_h} = 1 - \dfrac{Q_c}{Q_h} = 0.3$ (2)
From (2) $Q_c = 0.7 Q_h$ (3)
Solving (3) and (1) simultaneously, we have
$Q_h = 667$ J, and $Q_c = 467$ J

13.32 (a) $\text{Eff} = \dfrac{W}{Q_h} = \dfrac{W}{3W} = \dfrac{1}{3} = 0.333$
(b) $Q_c = Q_h - W = 3W - W = 2W$
Therefore, $\dfrac{Q_c}{Q_h} = \dfrac{2W}{3W} = \dfrac{2}{3}$

13.33 (a) We have $\text{Eff} = \dfrac{W}{Q_h} = \dfrac{Q_h - Q_c}{Q_h} = 1 - \dfrac{Q_c}{Q_h} = 0.25$, With $Q_c = 8000$ J, we have
$Q_h = 10667$ J
(b) $W = Q_h - Q_c = 2667$ J
and from $P = \dfrac{W}{\Delta t}$, we have $\Delta t = \dfrac{W}{P} = \dfrac{2667 \text{ J}}{5000 \text{ J/s}} = 0.533$ s

CHAPTER THIRTEEN SOLUTIONS

13.34 eff = $1 - \frac{Q_c}{Q_h} = 0.3$, which gives, $Q_c = 0.7Q_h$

(a) Therefore, $Q_c = 0.7(800 \text{ J}) = 560 \text{ J}$

(b) For a Carnot Cycle, eff = $1 - \frac{T_c}{T_h} = 0.3$, from which $T_c = 0.7T_h = 0.7(500 \text{ K}) = 350 \text{ K}$

13.35 The maximum efficiency equals the Carnot efficiency.
$$\text{Eff}_c = \frac{T_h - T_c}{T_h} = 1 - \frac{T_c}{T_h} = 1 - \frac{293}{573} = 0.488 \text{ (or 48.8 \%)}$$

13.36 $\text{Eff}_c = 1 - \frac{T_c}{T_h} = 1 - \frac{339}{422} = 0.197$ (or 19.7 %)

13.37 We use $\text{Eff}_c = 1 - \frac{T_c}{T_h}$

as, $0.3 = 1 - \frac{573 \text{ K}}{T_h}$

From which, $T_h = 819 \text{ K} = 546 \text{ °C}$

13.38 We have $W = Q_h - Q_c = 200 \text{ J}$

and $\text{Eff} = \frac{W}{Q_h} = \frac{200 \text{ J}}{500 \text{ J}} = 0.4$

If $\text{Eff} = 0.6\text{Eff}_c$, then $\text{Eff}_c = \frac{0.4}{0.6} = 0.667$.

But, $\text{Eff}_c = 1 - \frac{T_c}{T_h} = 0.667$

Thus, $\frac{T_c}{T_h} = 0.333 = \frac{1}{3}$

13.39 We have, $\text{Eff}_c = 1 - \frac{T_c}{T_h} = 1 - \frac{353}{623} = 0.433$

and $\text{Eff} = \frac{W}{Q_h}$

Thus, $W = Q_h(\text{Eff}) = 21{,}000 \text{ J}(0.433) = 9.10 \times 10^3 \text{ J}$

Therefore, $P = \frac{W}{t} = \frac{9.10 \times 10^3 \text{ J}}{1 \text{ s}} = 9.10 \times 10^3 \text{ W} = 9.10 \text{ kW}$

(b) $Q_c = Q_h - W = 21000 \text{ J} - 9100 \text{ J} = 1.19 \times 10^4 \text{ J}$

13.40 The heat expelled in each cycle (which lasts for one second) is 1.19×10^4 J.
If $\Delta T = 2$ °C, then
$Q_c = mc\Delta T$
$1.19 \times 10^4 \text{ J} = m(4184 \text{ J/kg °C})(2 \text{ °C})$
From which, $m = 1.42$ kg. This is the mass of water which must flow through the engine each second. The mass flow needed per hour is
$M = (1.42 \text{ kg/s})(3600 \text{ s/h}) = 5.12 \times 10^3$ kg/h, or
$V = \frac{M}{\rho} = \frac{5.12 \times 10^3 \text{ kg/m}}{10^3 \text{ kg/m}^3} = 5.12 \text{ m}^3/\text{hr}$

CHAPTER THIRTEEN SOLUTIONS

13.41 (a) $\text{Eff}_{max} = 1 - \frac{T_c}{T_h} = 1 - \frac{278}{293} = 5.12 \times 10^{-2}$ (or 5.12 %)

(b) $P = \frac{W}{\Delta t} = 75 \times 10^6$ J/s

Therefore, $W = (75 \times 10^6 \text{ J/s})(3600 \text{ s/h}) = 2.70 \times 10^{11}$ J/h

From, $\text{Eff} = \frac{W}{Q_h}$ we find

$$Q_h = \frac{W}{\text{Eff}} = \frac{2.70 \times 10^{11} \text{ J/h}}{5.12 \times 10^{-2}} = 5.27 \times 10^{12} \text{ J/h}$$

13.42 $\text{COP} = 0.1 \text{COP}_{\text{Carnot Cycle}}$ or

$$\frac{Q_h}{W} = 0.1 \left(\frac{Q_h}{W}\right)_{\text{Carnot Cycle}} = 0.1 \left(\frac{1}{\text{Carnot efficiency}}\right)$$

$$= 0.1 \left(\frac{T_h}{T_h - T_c}\right) = 0.1 \frac{293 \text{ K}}{293 \text{ K} - 268 \text{ K}} = 1.17$$

Thus, 1.17 Joules of heat are delivered for each joule of work done.

13.43 Q_c = heat to melt 15 g of Hg = mL_f = $(0.015 \text{ kg})(1.18 \times 10^4 \text{ J/kg}) = 177$ J

Q_h = heat absorbed to freeze 1 g of aluminum = mL_f = $(10^{-3} \text{ kg})(3.97 \times 10^5 \text{ J/kg}) = 397$ J

and the work output = $W = Q_h - Q_c = 220$ J

(a) $\text{Eff} = \frac{W}{Q_h} = \frac{220 \text{ J}}{397 \text{ J}} = 0.554$, or 55.4 %

(b) $\text{Eff (Carnot)} = \left(\frac{T_h}{T_h - T_c}\right) = \frac{933 \text{ K} - 234.1 \text{ K}}{933 \text{ K}} = 0.749 = 74.9\%$

13.44 $Q = mL_v = (1 \text{ kg})(2.26 \times 10^6 \text{ J/kg}) = 2.26 \times 10^6$ J

$\Delta S = \frac{Q}{T} = \frac{2.26 \times 10^6 \text{ J}}{373 \text{ K}} = 6.06 \times 10^3$ J/K

13.45 The heat generated is equal to the potential energy given up by the log.

$Q = mgh = (70 \text{ kg})(9.8 \text{ m/s}^2)(25 \text{ m}) = 1.72 \times 10^4$ J

Thus, $\Delta S = \frac{Q}{T} = \frac{1.72 \times 10^4 \text{ J}}{300 \text{ K}} = 57.2$ J/K

13.46 The heat generated is equal to the kinetic energy lost.

$Q = (2)(\frac{1}{2} mv^2) = (2000 \text{ kg})(20 \text{ m/s})^2 = 8 \times 10^5$ J

So, $\Delta S = \frac{Q}{T} = \frac{8 \times 10^5 \text{ J}}{296 \text{ K}} = 2.70 \times 10^3$ J/K

13.47 The heat lost by the water is

$Q = -mL_f = -(0.5 \text{ kg})(3.33 \times 10^5 \text{ J/kg}) = -1.67 \times 10^5$ J

Therefore, $\Delta S = \frac{Q}{T} = \frac{-1.67 \times 10^5 \text{ J}}{273 \text{ K}} = -612$ J/K

CHAPTER THIRTEEN SOLUTIONS

13.48 (a) The table is shown below. On the basis of the table, the most probably result of a toss is 2 heads and 2 tails.
(b) The most ordered state is the least likely state. Thus, on the basis of the table this is either all heads or all tails.
(c) The most disordered is the most likely state. Thus, this is 2 heads and 2 tails.

Result	Possible Combinations	Total
all heads	HHHH	1
3H,1T	THHH,HTHH,HHTH,HHHT	4
2H,2T	TTHH,THTH,THHT,HTTH, HTHT,HHTT	6
1H,3T	HTTT,THTT,TTHT,TTTH	4
all tails	TTTT	1

13.49 (a)

Result	Possible Combinations	Total
all red	RRR	1
2R,1G	RRG,RGR,GRR	3
1R,2G	RGG,GRG,GGR	3
all green	GGG	1

(b)

Result	Possible Combinations	Total
all red	RRRRR	1
4R,1G	RRRRG,RRRGR,RRGRR,RGRRR,GRRRR	5
3R,2G	RRRGG,RRGRG,RGRRG,GRRRG,RRGGR,RGRGR, GRRGR,RGGRR,GRGRR,GGRRR	10
2R,3G	GGGRR,GGRGR,GRGGR,RGGGR,GGRRG,GRGRG, RGGRG,GRRGG,RGRGG,RRGGG	10
1R,4G	RGGGG,GRGGG,GGRGG,GGGRG,GGGGR	5
all green	GGGGG	1

13.50 (a) A 12 can only be obtained one way 6 + 6
(b) A 7 can be obtained six ways: 6 + 1, 5 + 2, 4 + 3, 3 + 4, 2 + 5, 1 + 6

13.51 Work done each second = 1000 MJ = 10^9 J

Eff = $\dfrac{W}{Q_h}$ = 0.33, so Q_h = 3 W = $3(10^9 \text{ J})$ = 3 X 10^9 J

But, also   W = $Q_h - Q_c$, so $Q_c = Q_h - W$ = 3 X 10^9 J - 10^9 J = 2 X 10^9 J

Therefore, 2 X 10^9 J of heat must be absorbed each second by 10^6 kg of river water.

$$\Delta T = \dfrac{Q}{mc} = \dfrac{2 \times 10^9 \text{ J}}{(10^6 \text{ J})(4190 \text{ J/kg °C})} = 0.477°C$$

13.52 W = $(nRT)\ln\dfrac{V_f}{V_i}$ = (1 mol)(8.31 J/mol K)(300 K)$\ln\dfrac{1}{2}$ = -1730 J (work done by gas)
Therefore, done on the gas is +1730 J

13.53 The first law requires that $\Delta U = Q - W = (Q_h - Q_c) - W$. With the numbers claimed by the inventor, this becomes.

ΔU = 7 X 10^4 J - 2 X 10^4 J - 1 X 10^4 J = 4 X 10^4 J

CHAPTER THIRTEEN SOLUTIONS

Either the engine violates the first law or it does not operate in a cyclic process (which would make it incapable of continuous operation). You should not invest.

13.54 (a) $\Delta U(123) = Q - W = 418 \text{ J} - 167 \text{ J} = 251 \text{ J}$
(b) We use $\Delta U(143) = Q - W$, with $\Delta U = 251$ J.
Thus, $251 \text{ J} = Q - 63 \text{ J}$
or $Q = 314$ J
(c) $W(12341) = W(123) - W(143) = 167 \text{ J} - 63 \text{ J} = 104 \text{ J}$
(d) $W(14321) = W(143) - W(123) = 63 \text{ J} - 167 \text{ J} = -104 \text{ J}$
(e) The internal energy change is zero in both cases because both are cyclic processes.

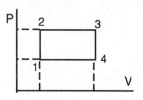

13.55 (a) $Q = mc\Delta T = (1 \text{ mol})(28.74 \text{ J/mol K})(120 \text{ K}) = 3.45 \times 10^3$ J
(b) $\Delta U = N_a \langle KE \rangle_f - N_a \langle KE \rangle_i = N_a[\frac{3}{2} kT_f - \frac{3}{2} kT_i] = \frac{3}{2}(N_a k)T_f - \frac{3}{2}(N_a k)T_i$

$= \frac{3}{2} R(T_f - T_i) = \frac{3}{2}(8.31 \text{ J/mol K})(120 \text{ K}) = 1.50 \times 10^3$ J

(c) $W = Q - \Delta U = 3.45 \times 10^3 \text{ J} - 1.50 \times 10^3 \text{ J} = 1.95 \times 10^3$ J

13.56 (a) The original volume of the aluminum is
$$V = \frac{m}{\rho} = \frac{5 \text{ kg}}{2.7 \times 10^3 \text{ kg/m}^3} = 1.85 \times 10^{-3} \text{ m}^3.$$
The change in volume is
$\Delta V = \beta V_0 (\Delta T) = (3)(24 \times 10^{-6} \text{ °C}^{-1})(1.85 \times 10^{-3} \text{ m}^3)(70 \text{ °C}) = 9.32 \times 10^{-6} \text{ m}^3$
The work done is $W = P\Delta V = (1.013 \times 10^5 \text{ Pa})(9.32 \times 10^{-6} \text{ m}^3) = 0.945$ J
(b) $Q = mc\Delta T = (5 \text{ kg})(900 \text{ J/kg °C})(70 \text{ °C}) = 3.15 \times 10^5$ J
(c) $\Delta U = Q - W \cong Q = 3.15 \times 10^5$ J

13.57 From the ideal gas law, the final pressure of the gas is
$$P_f = P_i \frac{V_i}{V_f} = \frac{1}{3} P_i = \frac{1}{3} \text{ atm}$$
and the initial volume is
$$V_i = \frac{nRT_i}{P_i} = \frac{(1 \text{mol})(8.31 \text{ J/mol K})(573 \text{ K})}{1.013 \times 10^5 \text{ Pa}}$$
$= 4.7 \times 10^{-2} \text{ m}^3.$
The sketch for this process is shown at the right.

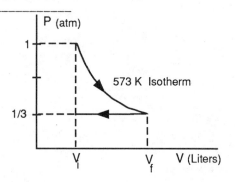

13.58 (a) $W = (nRT) \ln \frac{V_f}{V_i} = (1 \text{ mol})(8.31 \text{ J/mol K})(573 \text{ K}) \ln(3) = 5.23 \times 10^3$ J

(b) $W = -P\Delta V = -\frac{1}{3}(1.013 \times 10^5 \text{ Pa})(4.7 \times 10^{-2} - 0.141)\text{m}^3 = 3.17 \times 10^3$ J

CHAPTER THIRTEEN SOLUTIONS

13.59 The maximum COP for the refrigerator = $COP_{Carnot} = \left(\dfrac{T_c}{T_h - T_c}\right) = \dfrac{273 \text{ K}}{20 \text{ K}}$
$= 13.7$

For a given Q_c, $COP_{max} = \dfrac{Q_c}{W_{min}}$, so $W_{min} = \dfrac{Q_c}{COP_{max}} = \dfrac{400 \text{ J}}{13.7} = 29.3 \text{ J}$

13.60 $COP_{max} = COP_{Carnot} = \left(\dfrac{Q_c}{W}\right)_{Carnot} = \left(\dfrac{T_c}{T_h - T_c}\right) = \dfrac{273}{297 - 273} = 11.4$

$COP_{actual} = 0.075 \, COP_{max} = (0.075)(11.4) = 0.853 = \dfrac{Q_c}{W}$

or $W = 1.172 \, Q_c$

The rate heat is conducted into the refrigerator =
$\dfrac{kA(T_h - T_c)}{L} = (2.1 \times 10^{-2} \text{ J/ms°C}) \dfrac{(4.0 \text{ m}^2)(24°C)}{3 \times 10^{-2} \text{ m}} = 67.2 \text{ J/s}$

Therefore, each second, the amount of heat conducted is $Q_c = 67.2$ J in order to maintain a temperature of 0°C inside. The work per second, or power, is
$P = 1.172(67.2 \text{ J/s}) = 78.8 \text{ W} = 0.106 \text{ hP}$

13.61 $COP = 3 = \dfrac{Q_c}{W}$; Therefore, $W = \dfrac{Q_c}{3}$ (1)

The heat removed each minute is
$H = (0.03 \text{ kg})(4184 \text{ J/kg °C})(22 \text{ °C}) + (0.03 \text{ kg})(3.33 \times 10^5 \text{ J/kg})$
$+ (0.03 \text{ kg})(2090 \text{ J/kg °C})(20 \text{ °C}) = 1.40 \times 10^4 \text{ J/min}$

or, $H = \dfrac{Q_c}{\Delta t} = 233 \text{ J/s}$

Thus, the work done per sec = $P = \dfrac{233 \text{ J/s}}{3} = 77.8 \text{ W}$

13.62 (a) In a cyclic process, $\Delta U = 0$. Thus,
$\Delta U = Q - W = (Q_c - Q_h) - W = 0$

Thus the work output, or the work done by the system, is
$W = -Q_h + Q_c$

and the work done on the system, the work input, is
$W = -Q_c + Q_h = -25 \text{ cal} + 32 \text{ cal} = +7 \text{ cal/cycle} = 29.3 \text{ J/cycle}$

Power = (work/cycle)(cycles/s) = (29.3 J/cycle)(60 cycles/s) = 1.76×10^3 J/s
= 1.76×10^3 W

(b) $COP = \dfrac{Q_c}{W} = \dfrac{25 \text{ cal}}{7 \text{ cal}} = 3.57$

13.63 We have, $Eff = \dfrac{W}{Q_h} = 0.25$. Thus,

$Q_h = \dfrac{W}{0.25} = \dfrac{1500 \times 10^3 \text{ J/s}}{0.25} = 6.00 \times 10^6 \text{ J/s}$

and, $Q_c = Q_h - W = 6.0 \times 10^6 \text{ J/s} - 1.5 \times 10^6 \text{ J/s} = 4.5 \times 10^6$ J/s (This is the heat to be absorbed by the coolant water.)

The coolant flow is 60 liters/s = 6×10^{-2} m^3/s

Thus, the mass flow is mass/s = $(6 \times 10^{-2} \text{ m}^3\text{/s})(1000 \text{ kg/m}^3) = 60.0$ kg/s

Therefore, we see that the 60 kg of water must absorb 4.5×10^6 J.

CHAPTER THIRTEEN SOLUTIONS

From, $Q = mc\Delta T$

$$\Delta T = \frac{Q}{mc} = \frac{4.5 \times 10^6 \text{ J}}{(60 \text{ kg})(4184 \text{ J/kg°C})} = 17.9 \text{ °C}$$

13.64 For the isothermal process AB,

$$W_{AB} = P_A V_A \ln\left(\frac{V_B}{V_A}\right) = (5)(1.013 \times 10^5 \text{ Pa})(10 \times 10^{-3} \text{ m}^3)\ln\left(\frac{50}{10}\right) = 8.15 \times 10^3 \text{ J}$$

$W_{BC} = P_B \Delta V = (1.01 \times 10^5 \text{ Pa})[(10 - 50) \times 10^{-3}] \text{ m}^3. = -4.05 \times 10^3 \text{ J}$

$W_{CA} = 0$ and $W = W_{AB} + W_{BC} = 4.11 \times 10^3 \text{ J}$

13.65 (a) For a complete cycle, $\Delta U = 0$ and $W = Q_h - Q_c = Q_c[\frac{Q_h}{Q_c} - 1]$

We have already shown that for a Carnot cycle (and only for a Carnot cycle) that

$$\frac{Q_h}{Q_c} = \frac{T_h}{T_c}$$

Therefore, $W = Q_c\left(\frac{T_h - T_c}{T_c}\right)$

13.66 Let us first find the temperature at point C by use of the ideal gas law.

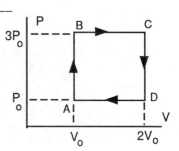

$$\frac{P_c V_c}{T_c} = \frac{P_0 V_0}{T_0}$$

Thus, $T_c = T_0 \frac{P_c V_c}{P_0 V_0} = 6 T_0$.

Now, find the work done along path ABC. This work is equal to the area under the PV curve.

$W(ABC) = (3P_0)(2V_0 - V_0) = 3P_0V_0$

But, for one mole of gas, the ideal gas law says

$P_0V_0 = RT_0$

Thus, $W(ABC) = 3 RT_0$

The internal energy of one mole of an ideal gas is found as

$$U = N\langle KE\rangle = N_a(\frac{3}{2} kT) = \frac{3}{2} RT$$

Thus, $\Delta U(ABC) = \frac{3}{2} RT_c - \frac{3}{2} RT_0 = \frac{3}{2} R(6T_0) - \frac{3}{2} RT_0 = \frac{15}{2} RT_0$

Now, from the first law, we find the heat entering the system as

$$Q(ABC) = \Delta U + W = \frac{15}{2} RT_0 + 3RT_0 = \frac{21}{2} RT_0$$

(b) Using an approach similar to that used in (a), we find

$W(CDA) = P_0(V_0 - 2V_0) = -P_0V_0 = -RT_0$

and $Q(CDA) = \Delta U + W = U_A - U_C + W(CDA) = \frac{3}{2} RT_0 - \frac{3}{2} R(6T_0) - RT_0 = -\frac{17}{2} RT_0$

Thus, the heat leaving the system is $\frac{17}{2} RT_0$

(c) $\text{Eff} = \frac{W_{net}}{Q_{input}} = \frac{3RT_0 - RT_0}{\frac{21}{2}RT_0} = \frac{4}{21}$ (approximately 19%)

(d) $\text{Eff}_{max} = 1 - \frac{T_A}{T_C} = 1 - \frac{T_0}{6T_0} = \frac{5}{6}$ (about 83.3 %)

CHAPTER THIRTEEN SOLUTIONS

13.67 (a) The net work done in the entire cycle = area enclosed within the cycle on a PV diagram. Thus,

$$W_{net} = (3P_0 - P_0)(3V_0 - V_0) = 4P_0V_0$$

(b) $\Delta U = 0$ for complete cycle. Thus, from the first law,

$$Q_{net} = W_{net} = 4P_0V_0$$

(c) From the ideal gas law, we have
$P_0V_0 = RT_0 = (1 \text{ mol})(8.31 \text{ J/mol K})(273 \text{ K}) = 2.27 \times 10^3$ J

Thus, $W_{net} = 4P_0V_0 = 4(2.27 \times 10^3 \text{ J}) = 9.07 \times 10^3$ J

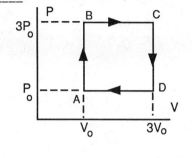

13.68 (a) A PV diagram of the process is shown at the right where $P_0 = 4$ atm $= 4.04 \times 10^5$ Pa,

$T_0 = 300$ K, and $V_0 = \dfrac{nRT_0}{P_0}$

$= \dfrac{(1)(8.314)(300)}{4.04 \times 10^5 \text{ Pa}} = 6.17 \times 10^{-3}$ m^3

(b) Over the isothermal paths,

$W_1 = P_0V_0 \ln\dfrac{2V_0}{V_0}$ and $W_3 = \dfrac{P_0}{2}V_0 \ln\dfrac{V_0}{2V_0}$

Over the constant pressure paths:

$W_2 = \dfrac{P_0}{2}(V_0 - 2V_0) = -\dfrac{P_0V_0}{2}$ and

$W_4 = P_0(V_0 - \dfrac{V_0}{2}) = \dfrac{P_0V_0}{2}$

The net work is $W = W_1 + W_2 + W_3 + W_4$ or

$W = \dfrac{P_0}{2}V_0\ln(2) = 864$ J

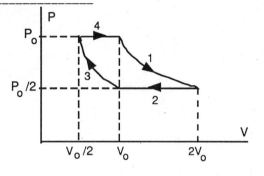

13.69 (a) $\dfrac{W}{t} = 1.5 \times 10^7$ W$_{(electrical)}$, $Q = mL = \dfrac{W}{0.15t}\Delta t$ and $L = 7.8 \times 10^6$ cal/kg.

$m = \dfrac{W}{0.15t}\dfrac{\Delta t}{L} = \dfrac{(1.5 \times 10^8 \text{ W})(86,400 \text{ s/day})}{(0.15)(7.8 \times 10^6 \text{ cal/kg})(4.184 \text{ J/cal})}$

$= 2.65 \times 10^6$ kg/day $= 2.65 \times 10^3 \dfrac{\text{metric tons}}{\text{day}}$

(b) Cost = (\$8/metric ton)(2.65 $\times 10^3 \dfrac{\text{metric tons}}{\text{day}}$)(365 days/y)

= \$7.74 $\times 10^6$ per year

CHAPTER THIRTEEN SOLUTIONS

(c) First find the rate at which heat energy must be discharged to the water.

$$e = \frac{W}{Q_h} = \frac{W}{(W + Q_c)} = \frac{\frac{W}{t}}{\frac{W}{t} + \frac{Q_c}{t}}, \text{ and } \frac{Q_c}{t} = \frac{W}{t}\left(\frac{1}{e} - 1\right) = (1.5 \times 10^8 \text{ W})\left(\frac{1}{0.15} - 1\right)$$

$$= 8.5 \times 10^8 \text{ W}$$

$\frac{Q_c}{t} = 2.03 \times 10^8$ cal/s. Now require $\frac{Q_c}{t} = \frac{mc\Delta T}{t}$ and find

$$\frac{m}{t} = \frac{1}{c\Delta T}\frac{Q_c}{t} = \frac{2.03 \times 10^8 \text{ cal/s}}{(1 \text{ cal/g C°})5 \text{ C°}} = 4 \times 10^7 \text{ g/s}$$

CHAPTER FOURTEEN SOLUTIONS

14.1 (a) At equilibrium, $mg = kx$. Thus,
$$k = \frac{mg}{x} = \frac{(0.4 \text{ kg})(9.8 \text{ m/s}^2)}{(8 \times 10^{-2} \text{ m})} = 49 \text{ N/m}$$
(b) A 0.575 kg mass (weight 5.64 N) stretches the spring by
$$x = \frac{F}{k} = \frac{5.64 \text{ N}}{49 \text{ N/m}} = 1.15 \times 10^{-1} \text{ m} = 11.5 \text{ cm}$$

14.2 (a) The force on the mass is
$$F = kx = (160 \text{ N/m})(0.15 \text{ m}) = 24 \text{ N}$$
From Newton's second law, the acceleration is
$$a = \frac{F}{m} = \frac{24 \text{ N}}{0.4 \text{ kg}} = 60 \text{ m/s}^2.$$
(b) The maximum value of both the force and acceleration occur at the amplitude of the motion. Thus, the maximum force is 24 N, and the maximum acceleration is 60 m/s^2.

14.3 (a) The spring constant is given by
$$k = \frac{mg}{x} = \frac{50 \text{ N}}{5 \times 10^{-2} \text{ m}} = 1000 \text{ N/m}$$
and the force required to stretch the spring 11 cm (0.11 m) is
$$F = kx = (1000 \text{ N/m})(0.11 \text{ m}) = 110 \text{ N}$$
(b) The graph will be a straight line passing through the origin and having a slope of 1000 N/m.

14.4 From Newton's second law,
$$a = \frac{kx}{m}, \text{ or } 9.8 \text{ m/s}^2 = k\frac{0.5 \text{ m}}{3 \text{ kg}}$$
From which, $k = 58.8$ N/m

14.5 We have $PE)_{\text{elastic at end}} = PE)_{\text{gravitational initially}}$
so, $32(\frac{1}{2} kx^2) = mgh_i$
$$32\left(\frac{1}{2}(5000 \text{ N/m})x^2\right) = (40 \text{ kg})(9.8 \text{ m/s}^2)(1.93 \text{ m})$$
gives $x = 9.72 \times 10^{-2}$ m = 9.72 cm

14.6 (a) $F = kx$, so $k = \frac{F}{x} = \frac{230 \text{ N}}{0.4 \text{ m}} = 575$ N/m
(b) $PE = \frac{1}{2} kx^2 = \frac{1}{2}(575 \text{ N/m})(0.4 \text{ m})^2 = 46$ J

14.7 From conservation of mechanical energy, we have
$$\frac{1}{2} mv_f^2 = \frac{1}{2} kx_i^2$$
or $\frac{1}{2}(0.1 \text{ kg})v_f^2 = \frac{1}{2}(200 \text{ N/m})(6 \times 10^{-2} \text{ m})^2$

CHAPTER FOURTEEN SOLUTIONS

From which, $v_f = 2.68$ m/s

14.8 (a) We use conservation of mechanical energy with the zero level for gravitational potential energy at the base of the incline. At the end of its trip up the plane the final velocity is zero. Let us call h the vertical height the disk has risen when it reaches its highest point, and s the distance of travel up the incline. The gravitational potential energy is

$$mgh = mgs(\sin 10°)$$

Thus, from conservation of energy, we have

$$mgs(\sin 10°) = \frac{1}{2}mv^2$$

$$(9.8 \text{ m/s}^2)s(\sin 10°) = \frac{1}{2}(2.68 \text{ m/s})^2$$

From which, $s = 2.12$ m

(b) At half the maximum distance up the incline, s = 1.06 m, and conservation of energy is

$$(9.8 \text{ m/s}^2)(1.06 \text{ m})\sin 10° = \frac{1}{2}v^2$$

$$v = 1.90 \text{ m/s}$$

14.9 The elastic potential energy is given by $\frac{1}{2}kx^2$. Thus, the ratio requested is

$$\frac{PE(\text{at } 0.4)}{PE(\text{at } 0.2)} = \frac{(0.4)^2}{(0.2)^2} = 4$$

14.10 $k = \dfrac{F}{x} = \dfrac{(4 \text{ kg})(9.8 \text{ m/s}^2)}{2.5 \times 10^{-2} \text{ m}} = 1568$ N/m

(a) $x = \dfrac{F}{k} = \dfrac{(1.5 \text{ kg})(9.8 \text{ m/s}^2)}{1568 \text{ N/m}} = 9.38 \times 10^{-3}$ m = 0.938 cm

(b) $W = PE = \frac{1}{2}kx^2 = \frac{1}{2}(1568 \text{ N/m})(4 \times 10^{-2} \text{ m})^2 = 1.25$ J

14.11 (a) At $x = \dfrac{A}{2}$, $PE = \frac{1}{2}k\left(\dfrac{A}{2}\right)^2 = \frac{1}{4}E$, where $E = \frac{1}{2}kA^2$

Thus, $KE + \frac{1}{4}E = E$, so $KE = \frac{3}{4}E$

(b) If KE = PE, then

$$PE + PE = E, \text{ or } PE = \frac{1}{2}E$$

Thus, $\frac{1}{2}kx^2 = \frac{1}{2}\left(\frac{1}{2}kA^2\right)$

and $x = \dfrac{A}{\sqrt{2}}$

14.12 (a) In the absence of friction, conservation of mechanical energy becomes

$$\frac{1}{2}mv_f^2 = \frac{1}{2}kx_i^2$$

$$\frac{1}{2}(1.5 \text{ kg})v_f^2 = \frac{1}{2}(2000 \text{ N/m})(3 \times 10^{-3} \text{ m})^2$$

CHAPTER FOURTEEN SOLUTIONS

$v_f = 1.10 \times 10^{-1}$ m/s = 11 cm/s

(b) In the presence of non-conservative forces, we use

$$W_{nc} = \tfrac{1}{2} mv_f^2 - \tfrac{1}{2} mv_i^2 + mgy_f - mgy_i + \tfrac{1}{2} kx_f^2 - \tfrac{1}{2} kx_i^2$$

If f = 2.0 N, then $W_{nc} = -fs = -(2\text{ N})(3 \times 10^{-3}\text{ m}) = -6 \times 10^{-3}$ J

We have,

$$-6 \times 10^{-3} \text{ J} = \tfrac{1}{2}(1.5 \text{ kg})v_f^2 - 0 + 0 - 0 + 0 - \tfrac{1}{2}(2000 \text{ N/m})(3 \times 10^{-3} \text{ m})^2$$

and $v_f = 6.32 \times 10^{-2}$ m/s = 6.32 cm/s

(c) If $v_f = 0$, then

$$W_{nc} = \tfrac{1}{2} mv_f^2 - \tfrac{1}{2} mv_i^2 + mgy_f - mgy_i + \tfrac{1}{2} kx_f^2 - \tfrac{1}{2} kx_i^2$$

becomes $-f(3 \times 10^{-3} \text{ m}) = 0 - 0 + 0 - 0 + 0 - \tfrac{1}{2}(2000 \text{ N/m})(3 \times 10^{-3} \text{ m})^2$

f = 3 N

14.13 First, let us apply conservation of momentum to the collision. We call V, the velocity of the block and embedded bullet just after the collision.

$$(10 \times 10^{-3} \text{ kg})(300 \text{ m/s}) = (2 \text{ kg} + 10 \times 10^{-3} \text{ kg})V$$

Thus, V = 1.49 m/s

We now apply conservation of mechanical energy from just after the end of the collision until the spring is fully compressed.

$$\tfrac{1}{2} kx_f^2 = \tfrac{1}{2}(m + M)V^2$$

$$\tfrac{1}{2}(19.6 \text{ N/m})x_f^2 = \tfrac{1}{2}(2.01 \text{ kg})(1.49 \text{ m/s})^2$$

$x_f = 4.78 \times 10^{-1}$ m = 47.8 cm

14.14 In the presence of non-conservative forces, we use

$$W_{nc} = \tfrac{1}{2} mv_f^2 - \tfrac{1}{2} mv_i^2 + mgy_f - mgy_i + \tfrac{1}{2} kx_f^2 - \tfrac{1}{2} kx_i^2$$

$$(20 \text{ N})(0.3 \text{ m}) = \tfrac{1}{2}(1.5 \text{ kg})v_f^2 - 0 + 0 - 0 + \tfrac{1}{2}(19.6 \text{ N/m})(0.3 \text{ m})^2 - 0$$

$v_f = 2.61$ m/s

14.15 $v = \sqrt{\dfrac{k}{m}(A^2 - x^2)} = \sqrt{\dfrac{10 \text{ N/m}}{0.05 \text{ kg}}((0.25)^2 - (0.125)^2)} = 3.06$ m/s

---

14.16 We use $v = \sqrt{\dfrac{k}{m}(A^2 - x^2)}$ which becomes, when squared,

$v^2 = \dfrac{k}{m}(A^2 - x^2)$. This becomes,

$$v^2 = \dfrac{19.6 \text{ N/m}}{0.4 \text{ kg}}((4 \times 10^{-2} \text{ m})^2 - x^2) = 49 \text{ s}^{-2}(1.6 \times 10^{-3} \text{ m}^2 - x^2) \qquad (1)$$

(a) If x = 0, (1) gives v = 0.28 m/s = 28 cm/s (This is the maximum velocity.)

(b) If $x = -1.5 \times 10^{-2}$ m, (1) gives v = 0.26 m/s = 26 cm/s

(c) if $x = 1.5 \times 10^{-2}$ m, (1) gives v = 0.26 m/s = 26 cm/s

(d) One-half the maximum velocity is 0.14 m/s. (See part (a).) We use this for v in (1) and solve for x to find

x = 3.46 cm

CHAPTER FOURTEEN SOLUTIONS

14.17 (a) The maximum elastic potential energy is stored when the spring is fully compressed. Thus, $PE(max) = \frac{1}{2}kx^2 = \frac{1}{2}(19.6 \text{ N/m})(5 \times 10^{-2} \text{ m})^2$
$= 2.45 \times 10^{-2}$ J

(b) We use $v = \sqrt{\frac{k}{m}(A^2 - x^2)}$ and note that the maximum velocity occurs when $x = 0$. Thus,

$v(max) = \sqrt{\frac{19.6 \text{ N/m}}{0.3 \text{ kg}}}(5 \times 10^{-2} \text{ m}) = 4.04 \times 10^{-1}$ m/s = 40.4 cm/s

14.18 The maximum velocity occurs when $x = 0$ and is given by

$v = \sqrt{\frac{k}{m}}A$

$(0.4 \text{ m/s}) = \sqrt{\frac{16 \text{ N/m}}{m}}(0.2 \text{ m})$

From which, $m = 4.0$ kg
and the weight is $w = mg = (4.0 \text{ kg})(9.8 \text{ m/s}^2) = 39.2$ N

14.19 (a) $E = \frac{1}{2}kA^2 = \frac{1}{2}(250 \text{ N/m})(3.5 \times 10^{-2} \text{ m})^2 = 0.153$ J

(b) $v = \sqrt{\frac{k}{m}(A^2 - x^2)}$ becomes, with $v = v_{max}$ at $x = 0$,

$v_{max} = \sqrt{\frac{k}{m}}A = \sqrt{\frac{250 \text{ N/m}}{0.5 \text{ kg}}}(3.5 \times 10^{-2} \text{ m}) = 0.783$ m/s

(c) From $F = kx = ma$, we have $a = \frac{k}{m}x$. Thus, $a = a_{max}$ at $x = x_{max} = A$

or, $a_{max} = \frac{k}{m}A = \frac{250 \text{ N/m}}{0.5 \text{ kg}}(3.5 \times 10^{-2} \text{ m}) = 17.5$ m/s^2

14.20 (a) From, $T = 2\pi\sqrt{\frac{m}{k}}$, we have $k = \frac{4\pi^2 m}{T^2} = \frac{4\pi^2(0.2 \text{ kg})}{(0.25 \text{ s})^2} = 126.3$ N/m

(b) From $E = \frac{1}{2}kA^2$, we have $A = \sqrt{\frac{2E}{k}} = \sqrt{\frac{2(2 \text{ J})}{126.3 \text{ N/m}}} = 0.178$ m = 17.8 cm

14.21 We know that $v_x = -v\sin\theta$.
(a) at $\theta = 0$; $v_x = 0$
(b) at $\theta = 60°$; $v_x = -4.33$ m/s
(c) at $\theta = 90°$; $v_x = -5.0$ m/s
(d) at $\theta = 180°$; $v_x = 0$
(e) at $\theta = 270°$; $v_x = +5.0$ m

14.22 (a) and (b) The distance around the circle is $s = 2\pi r = 2\pi(0.4 \text{ m})$, and the time to make one revolution is the period T. We have

$T = \frac{s}{v} = \frac{2\pi(0.4 \text{ m})}{5 \text{ m/s}} = 0.503$ s

The period and the frequency are related as

$f = \frac{1}{T} = \frac{1}{0.503 \text{ s}} = 1.99$ Hz

CHAPTER FOURTEEN SOLUTIONS

14.23 $T = 2\pi\sqrt{\dfrac{m}{k}} = \dfrac{1}{f} = 0.2$ s

$0.2\text{ s} = 2\pi\sqrt{\dfrac{4 \times 10^{-3}\text{ kg}}{k}}$ From which, $k = 3.95$ N/m

14.24 (a) $v = \dfrac{2\pi r}{T} = \dfrac{2\pi(0.2\text{ m})}{2\text{ s}} = 0.628$ m/s

(b) $f = \dfrac{1}{T} = \dfrac{1}{2\text{ s}} = 0.5$ Hz

(c) $\omega = 2\pi f = 2\pi(0.5\text{ s}^{-1}) = 3.14$ rad/s

14.25 (a) $E = \dfrac{1}{2}kA^2$, so if $A' = 2A$, $E' = \dfrac{1}{2}kA'^2 = \dfrac{1}{2}k(2A)^2 = 4E$ (The total energy is quadrupled.)

(b) $v_{max} = \sqrt{\dfrac{k}{m}}\,A$, so if A is doubled, v_{max} is also doubled.

(c) $a_{max} = \dfrac{k}{m}A$, so if A is doubled, a_{max} also doubles.

(d) $T = 2\pi\sqrt{\dfrac{m}{k}}$ is independent of A, so the period is unchanged.

14.26 $k = \dfrac{F}{x} = \dfrac{(0.01\text{ kg})(9.8\text{ m/s}^2)}{3.9 \times 10^{-2}\text{ m}} = 2.51$ N/m

and $T = 2\pi\sqrt{\dfrac{m}{k}} = 2\pi\sqrt{\dfrac{0.025\text{ kg}}{2.51\text{ N/m}}} = 0.627$ s

14.27 From $f = \dfrac{1}{T} = \dfrac{1}{2\pi}\sqrt{\dfrac{k}{m}}$, we find $k = 4\pi^2 f^2 m = 4\pi^2(5\text{ Hz})^2(4 \times 10^{-3}\text{ kg}) = 3.95$ N/m

14.28 We find the spring constant from
$mg = kx$
$(320\text{ kg})(9.8\text{ m/s}^2) = k(8 \times 10^{-3}\text{ m})$
Thus, $k = 3.92 \times 10^5$ N/m

Then, $f = \dfrac{1}{2\pi}\sqrt{\dfrac{k}{m}} = \dfrac{1}{2\pi}\sqrt{\dfrac{3.92 \times 10^5\text{ N/m}}{2 \times 10^3\text{ kg}}} = 2.23$ Hz

14.29 We use $x = A\cos(2\pi ft)$
(a) at $t = 0$, $x = A = 20$ cm
(b) at $t = T/8$, $x = (20\text{ cm})\cos\left(\dfrac{2\pi}{8}\right) = (20\text{ cm})(0.707) = 14.1$ cm

We now continue as outlined in part (b) to find
(c) at $t = T/4$, $x = 0$
(d) at $t = 3T/8$, $x = -14.1$ cm
(e) at $t = T/2$, $x = -20$ cm
(f) at $t = 5T/8$, $x = -14.1$ cm
(g) at $t = 3T/4$, $x = 0$
(h) at $t = 7T/8$, $x = 14.1$ cm
(i) at $t = T$, $x = 20$ cm
When plotted the graph is a cosine curve with amplitude 20 cm.

CHAPTER FOURTEEN SOLUTIONS

14.30 (a) We have $\qquad x = (0.3 \text{ m}) \cos \pi t/3 \quad (1)$
At $t = 0$, we have $\qquad x = (0.3 \text{ m}) \cos 0 = 0.3$ m
At $t = 0.2$ s, we have $\qquad x = (0.3 \text{ m}) \cos(0.2\pi/3) = (0.3 \text{ m})\cos(0.209 \text{ rad}) = 0.293$ m
(b) The general form for oscillatory motion is
$\qquad x = A \cos 2\pi f t \quad (2)$
Thus, by comparing (1) to (2), we see that $A = 0.3$ m
(c) Using comparison as in (b), we see that
$\qquad 2\pi f = \pi/3$
and $\quad f = 1/6$ Hz
(d) $T = \dfrac{1}{f} = 6$ s

14.31 The period of the motion is

$$T = 2\pi\sqrt{\dfrac{m}{k}} = 2\pi\sqrt{\dfrac{0.5 \text{ kg}}{30 \text{ N/m}}} = 0.258\pi \text{ s}$$

and the frequency is $\quad f = \dfrac{1}{T} = 1.23$ Hz

We also know the amplitude of motion is 0.25 m. Thus, the equation of motion becomes
$\qquad x = A\cos(2\pi f t) = (0.25 \text{ m})\cos 2.47\pi t$

14.32 The equation of motion of the object is
$\qquad x = A\cos(2\pi f t) = A\cos(2\pi t/3)$
(a) When $x = A/2$, we have
$\qquad A/2 = A\cos(2\pi t/3)$
or, $\quad \cos(2\pi t/3) = 0.5$
and, $\quad 2\pi t/3 = \cos^{-1}(0.5)$
Thus, $2\pi t/3 = 1.05$ rad
and $\quad t = 0.5$ s
(b) The steps for solving this portion are outlined in (a). We find
$\qquad t = 1$ s
(c) (See part (a) for the details of how to solve this.)
$\qquad t = 0.75$ s

14.33 (a) $k = \dfrac{F}{x} = \dfrac{7.5 \text{ N}}{3 \times 10^{-2} \text{ m}} = 250$ N/m

(b) $\qquad T = 2\pi\sqrt{\dfrac{m}{k}} = 2\pi\sqrt{\dfrac{0.5 \text{ kg}}{250 \text{ N/m}}} = 0.281$ s

and $\quad f = \dfrac{1}{T} = 3.56$ Hz

$\qquad \omega = 2\pi f = 2\pi(3.56 \text{ Hz}) = 22.4$ rad/s
(c) $E = \dfrac{1}{2} kA^2 = \dfrac{1}{2}(250 \text{ N/m})(5 \times 10^{-2} \text{ m})^2 = 0.313$ J
(d) $E = \text{const} = \dfrac{1}{2} mv_0^2 + \dfrac{1}{2} kx_0^2 = \dfrac{1}{2} kA^2$ When $v_0 = 0$, $x_0 = A = 5$ cm
(e) $v_{max} = \omega A = (22.4 \text{ rad/s})(0.05 \text{ m}) = 1.12$ m/s
$a_{max} = \omega^2 A = (22.5 \text{ rad/s})^2(0.05 \text{ m}) = 25$ m/s^2
(f) $x = A\cos\omega t = (5 \text{ cm})\cos((22.4 \text{ rad/s})(0.5 \text{ s})) = 0.919$ cm

CHAPTER FOURTEEN SOLUTIONS

14.34 $T = 2\pi\sqrt{\dfrac{L}{g}} = 2\pi\sqrt{\dfrac{2.0 \text{ m}}{9.8 \text{ m/s}^2}} = 2.84 \text{ s}$

$f = \dfrac{1}{T} = 0.352 \text{ Hz}$

The number of oscillations in 5 min = (f)(t) = (0.352 Hz)(5 min)(60 s/min) = 105.7
or, 105 complete oscillations in 5 min.

14.35 $T = 2\pi\sqrt{\dfrac{L}{g}}$

$1.4 \text{ s} = 2\pi\sqrt{\dfrac{0.5 \text{ m}}{g}}$ From which, $g = 10.1 \text{ m/s}^2$.

14.36 $T = 2\pi\sqrt{\dfrac{L}{g}}$

$1 \text{ s} = 2\pi\sqrt{\dfrac{L}{9.8 \text{ m/s}^2}}$ From which, $L = 0.248 \text{ m}$

14.37 $\dfrac{T_{earth}}{T_{moon}} = \dfrac{2\pi\sqrt{\dfrac{L}{g_e}}}{2\pi\sqrt{\dfrac{L}{g_m}}} = \sqrt{\dfrac{g_m}{g_e}} = \dfrac{1}{\sqrt{6}}$ If $T_{earth} = 2.5$ s, then $T_{moon} = \sqrt{6}\,(2.5 \text{ s})$

$= 6.12 \text{ s}$

14.38 The period in Tokyo is $T_t = 2\pi\sqrt{\dfrac{L_t}{g_t}}$

and the period in Cambridge is $T_c = 2\pi\sqrt{\dfrac{L_c}{g_c}}$

We know $T_t = T_c = 2$ s

From which, we see $\dfrac{L_t}{g_t} = \dfrac{L_c}{g_c}$ or, $\dfrac{g_c}{g_t} = \dfrac{L_c}{L_t} = \dfrac{0.9942}{0.9927} = 1.0015$

14.39 $T = 2\pi\sqrt{\dfrac{L}{g}}$

Thus, if $L' = 4L$, $T' = 2\pi\sqrt{\dfrac{4L}{g}} = 2T$ (The period is doubled.)

Since $f = \dfrac{1}{T}$, if the period is doubled, the frequency is cut in half.

14.40 $v = \lambda f = (0.085 \text{ m})(2 \text{ Hz}) = 0.17 \text{ m/s}$, and $t = \dfrac{d}{v} = \dfrac{10.0 \text{ m}}{0.17 \text{ m/s}} = 58.8 \text{ s}$

14.41 (a) $T = \dfrac{1}{f} = \dfrac{1}{88 \times 10^6 \text{ Hz}} = 1.14 \times 10^{-8} \text{ s}$

(b) $\lambda = \dfrac{v}{f} = \dfrac{3 \times 10^8 \text{ m/s}}{88 \times 10^6 \text{ Hz}} = 3.41 \text{ m}$

14.42 $v = \lambda f = (0.6 \text{ m})(4 \text{ Hz}) = 2.4 \text{ m/s}$

CHAPTER FOURTEEN SOLUTIONS

14.43 $\lambda_{long} = \dfrac{v}{f} = \dfrac{343 \text{ m/s}}{28 \text{ Hz}} = 12.3 \text{ m}$

$\lambda_{short} = \dfrac{v}{f} = \dfrac{343 \text{ m/s}}{4200 \text{ Hz}} = 0.082 \text{ m}$

14.44 $f = \dfrac{40 \text{ vib}}{30 \text{ s}} = 1.333 \text{ Hz}$

$v = \dfrac{425 \text{ cm}}{10 \text{ s}} = 42.5 \text{ cm/s} = 0.425 \text{ m/s}$

$\lambda = \dfrac{v}{f} = \dfrac{0.425 \text{ m/s}}{1.333 \text{ Hz}} = 0.319 \text{ m} = 31.9 \text{ cm}$

14.45 (a) The vertical distance from the center line to the top of a crest or the bottom of a trough is the amplitude. Thus, by inspection, we see
$A = 9 \text{ cm}$
(b) The wavelength is the horizontal distance along the wave between two points behaving identically. Thus, by inspection, we see
$\lambda = 20 \text{ cm}$
(c) $T = \dfrac{1}{25 \text{ Hz}} = 0.04 \text{ s}$
(d) $v = \lambda f = (20 \text{ cm})(25 \text{ Hz}) = 500 \text{ cm/s} = 5 \text{ m/s}$

14.46 $f = \dfrac{8}{12 \text{ s}} = 0.667 \text{ Hz}$, and $v = \lambda f = (1.2 \text{ m})(0.667 \text{ Hz}) = 0.80 \text{ m/s}$

14.47 (a) The distance to the sun is 93×10^6 miles $= 1.50 \times 10^{11}$ m.
Thus, $t = \dfrac{d}{v} = \dfrac{1.50 \times 10^{11} \text{ m}}{3 \times 10^8 \text{ s}} = 500 \text{ s} = 8.31 \text{ min}$
(b) distance to moon $= 3.84 \times 10^8$ m
Therefore, $t = \dfrac{d}{v} = \dfrac{3.84 \times 10^8 \text{ m}}{3 \times 10^8 \text{ s}} = 1.28 \text{ s}$

14.48 $d = vt = (3 \times 10^8 \text{ m/s})(3.156 \times 10^7 \text{ s/y}) = 9.47 \times 10^{15}$ m

14.49 If d is the distance to the reflecting object, the total distance traveled by the wave is 2d.
Thus, $v = \dfrac{2d}{t}$, or $d = \dfrac{vt}{2} = \dfrac{(343 \text{ m/s})(2.6 \text{ s})}{2} = 446 \text{ m}$

14.50 The speed of the wave is
$v = d/t = 20 \text{ m}/0.8 \text{ s} = 25 \text{ m/s}$
We now use, $v = \sqrt{\dfrac{F}{\mu}}$. We have $\mu = \dfrac{0.35 \text{ kg}}{1 \text{ m}} = 0.35 \text{ kg/m}$
Thus, $F = v^2 \mu = (25 \text{ m/s})^2(0.35 \text{ kg/m}) = 219 \text{ N}$

14.51 $v = \dfrac{4 \text{ m}}{0.1 \text{ s}} = 40 \text{ m/s}$ and $\mu = \dfrac{0.2 \text{ kg}}{4 \text{ m}} = 5 \times 10^{-2}$ kg/m, so
$F = \mu v^2 = (5 \times 10^{-2} \text{ kg/m})(40 \text{ m/s})^2 = 80.0 \text{ N}$

CHAPTER FOURTEEN SOLUTIONS

14.52 The hanging block is in equilibrium under the action of the tension in the string and its weight. From the first law, we find
$$F = mg = (2 \text{ kg})(9.8 \text{ m/s}^2) = 19.6 \text{ N}$$
So, the mass per unit length of the string is
$$\mu = \frac{F}{v^2} = \frac{19.6 \text{ N}}{(15 \text{ m/s})^2} = 8.71 \times 10^{-2} \text{ kg/m}$$
Therefore, the mass of the string is $\mu L = (8.71 \times 10^{-2} \text{ kg/m})(3 \text{ m}) = 2.61 \times 10^{-1}$ kg

---

14.53 (a) The mass per unit length is $\mu = \frac{0.6 \text{ kg}}{5 \text{ m}} = 1.2 \times 10^{-2}$ kg/m

The required tension is $F = v^2\mu = (50 \text{ m/s})^2(1.2 \times 10^{-2} \text{ kg/m}) = 30$ N

(b) $v = \sqrt{\frac{F}{\mu}} = \sqrt{\frac{8 \text{ N}}{1.2 \times 10^{-2} \text{ kg/m}}} = 25.8$ m/s

---

14.54 $F = A(\text{Stress})$. Thus, $v = \sqrt{\frac{F}{\mu}} = \sqrt{\frac{A(\text{Stress})}{\frac{m}{L}}} = \sqrt{\frac{\text{Stress}}{\frac{m}{AL}}} = \sqrt{\frac{\text{Stress}}{\frac{m}{\text{Volume}}}}$

$= \sqrt{\frac{\text{Stress}}{\rho}}$

where ρ is the density. The maximum velocity occurs when the stress is a maximum.

$$v_{max} = \sqrt{\frac{2.7 \times 10^9 \text{ Pa}}{7860 \text{ kg/m}^3}} = 586 \text{ m/s}$$

14.55 We know that the mass per unit length of the string is the same in both instances. Thus,
$$\frac{F_1}{v_1^2} = \frac{F_2}{v_2^2}$$
$$\frac{6 \text{ N}}{(20 \text{ m/s})^2} = \frac{F_2}{(30 \text{ m/s})^2} \quad \text{Which gives,} \quad F_2 = 13.5 \text{ N}$$

14.56 The mass of an object is given by $m = \rho V = \rho(AL)$. Thus, the mass per unit length is given by $\mu = \frac{m}{L} = \rho A$. Which for a wire of circular cross-section becomes
$$\mu = \rho \pi r^2$$
Thus, the ratio of the mass per unit length for the first and the second wire is
$$\frac{\mu_2}{\mu_1} = \frac{r_2^2}{r_1^2}$$
Also, we have $v = \sqrt{\frac{F}{\mu}}$ Thus, the ratio of the speeds on wires 1 and 2 is
$$\frac{v_2^2}{v_1^2} = \frac{\mu_1}{\mu_2} = \frac{r_1^2}{r_2^2} = \frac{1}{4}$$
Thus, $v_2 = \frac{v_1}{2} = \frac{80 \text{ m/s}}{2} = 40$ m/s

CHAPTER FOURTEEN SOLUTIONS

14.57 (a) If the end is fixed, there is inversion of the pulse upon reflection. Thus, when they meet, they cancel and the amplitude is zero.
(b) If the end is free there is no inversion on reflection. When they meet the amplitude is 2A = 2(0.15 m) = 0.3 m.

14.58 (a) The maximum occurs when the two waves interfere constructively. The individual amplitudes add for a resultant amplitude of A = 0.5 m.
(b) The minimum occurs when the waves interfere destructively. In this case, the individual amplitudes subtract, and we have a resultant of A = 0.1 m.

14.59 From $v = \sqrt{\dfrac{F}{\mu}}$. The ratio of the wave speeds for the two cases (with μ constant) is

$$\frac{v_2^2}{v_1^2} = \frac{F_2}{F_1}$$

Therefore, if $F_2 = 2F_1$, we have

$$v_2 = \sqrt{2}\, v_1 = (90 \text{ m/s})\sqrt{2} = 127.3 \text{ m/s}$$

14.60 (a) We compare $x = (0.25 \text{ m})\cos(0.4\pi t)$ to the general equation of motion, $x = A\cos(2\pi f t)$, to find $A = 0.25$ m
(b) By comparison of the two equations as in (a), we find $2\pi f = 0.4\pi$. Also, we know that

$$2\pi f = \sqrt{\frac{k}{m}} \quad \text{so,} \quad \frac{k}{m} = (0.4\pi)^2. \quad \text{From which}$$

$$k = (0.4\pi)^2 (0.3 \text{ kg}) = 0.474 \text{ N/m}$$

(c) At $t = 0.3$ s, the position is

$$x = (0.25 \text{ m})\cos(0.4\pi \cdot 0.3) = (0.25)\cos(0.377 \text{ rad}) = (0.25 \text{ m})(0.93) = 0.232 \text{ m}$$

(d) $v = -\sqrt{\dfrac{k}{m}(A^2 - x^2)} = -(0.4\pi)\sqrt{(0.25 \text{ m})^2 - (0.232)^2} = -0.116 \text{ m/s}$

The negative sign is used above because the object is moving toward the equilibrium position (in the negative direction).

14.61 The force is found from $F = -kx = -(1000 \text{ N/m})x$, and the acceleration is found from $a = \dfrac{F}{m}$

(a) At $x = 0.5$ m, $F = -500$ N, and $a = -333$ m/s^2
(b) At $x = 0.1$ m, $F = -100$ N, and $a = -66.7$ m/s^2
(c) At $x = 0$ m, $F = 0$, and $a = 0$
(d) At $x = -0.2$ m, $F = 200$ N, and $a = 133$ m/s^2
(e) At $x = -0.5$ m, $F = 500$ N, and $a = 333$ m/s^2

14.62 We have $\mu_1 = \dfrac{m_1}{L_1} = \dfrac{2m_2}{L_2} = 2\mu_2$, and $F_1 = F_2$.

$$v = \sqrt{\frac{2F_1}{\mu_1}} = \sqrt{2}\, v_1 = \sqrt{2}\,(5 \text{ m/s}) = 7.07 \text{ m/s}$$

14.63 (a) The tension in the string is $F = (3 \text{ kg})(9.8 \text{ m/s}^2) = 29.4$ N.

$$\mu = \frac{F}{v^2} = \frac{29.4 \text{ N}}{(24 \text{ m/s})^2} = 5.1 \times 10^{-2} \text{ kg/m}$$

CHAPTER FOURTEEN SOLUTIONS

(b) If m = 2 kg, then F = 19.6 N, and

$$v = \sqrt{\frac{F}{\mu}} = \sqrt{\frac{19.6 \text{ N}}{5.10 \times 10^{-2} \text{ kg/m}}} = 19.6 \text{ m/s}$$

14.64 $W = \Delta PE = \frac{1}{2} k(x^2 - x_0^2) = \frac{1}{2} (30 \text{ N/m})((0.3 \text{ m})^2 - (0.2 \text{ m})^2) = 0.75 \text{ J}$

14.65 From conservation of mechanical energy, we have

$$mgh = \frac{1}{2} kx^2$$

$$(0.5 \text{ kg})(9.8 \text{ m/s}^2)(2 \text{ m}) = \frac{1}{2} (20 \text{ N/m})x^2$$

From which, $x = 0.99$ m

14.66 We have $2\pi f = \sqrt{\frac{k}{m}}$

Therefore, $\frac{k}{m} = 4\pi^2 (5 \text{ Hz})^2 = 100\pi^2 \text{ s}^{-2}$

Now use, $v = \sqrt{\frac{k}{m}(A^2 - x^2)} = (10\pi \text{ s}^{-1})\sqrt{((0.4 \text{ cm})^2 - (0.1 \text{ cm})^2)}$
$= 12.2$ cm/s

14.67 (a) The muzzle velocity is found from conservation of energy.

$$\frac{1}{2} mv^2 = \frac{1}{2} kx^2$$

$$\frac{1}{2} (10^{-3} \text{ kg})v^2 = \frac{1}{2} (9.8 \text{ N/m})(0.2 \text{ m})^2$$

$v = 19.8$ m/s

(b) From the equations for projectile motion, we find the time to reach the floor.

$$y = v_{0y}t + \frac{1}{2} at^2$$

$$-1 \text{ m} = 0 + \frac{1}{2} (9.8 \text{ m/s}^2)t^2$$

$t = 0.452$ s

and the range is
$R = v_{0x}t = (19.8 \text{ m/s})(0.452 \text{ s}) = 8.94$ m

14.68 From the definition of Young's modulus,

$$F = \frac{YA}{L} \Delta L$$

This is of Hooke's law form, $F = kx = k\Delta L$ if we identify k as $\frac{YA}{L}$.

14.69 From $f = \frac{1}{2\pi}\sqrt{\frac{k}{m}}$, $k = 4\pi^2 f^2 m = 4\pi^2 (36 \text{ Hz})^2 (5.5 \times 10^{-3} \text{ kg})$
$= 281.4$ N/m (the effective force constant)

Also, (see problem 68) $Y = \frac{kL}{A} = \frac{(281.4 \text{ N/m})(3.5 \times 10^{-2} \text{ m})}{\pi(10^{-3} \text{ m})^2} = 3.14 \times 10^6$ Pa

CHAPTER FOURTEEN SOLUTIONS

14.70 $\omega = \sqrt{\dfrac{k}{m}} = \sqrt{\dfrac{4.7 \times 10^{-4} \text{ N/m}}{3 \times 10^{-4} \text{ kg}}} = 1.252$ rad/s

and $v_{max} = \omega A = (1.252 \text{ rad/s})(2 \times 10^{-3} \text{ m}) = 2.50 \times 10^{-3}$ m/s = 250 mm/s
This is the maximum velocity of the wingtip 3 mm from the fulcrum. To get the maximum velocity of the outer tip of the wing, treat the wing as a rigid body rotating about the fulcrum. All parts have the same angular velocity, so

$\dfrac{v}{r} = \omega$ leads to $\dfrac{v_{\text{far tip}}}{r_{\text{far tip}}} = \dfrac{v_{\text{near tip}}}{r_{\text{near tip}}}$

or $\quad v_{\text{far tip}} = 15 \text{ mm} \dfrac{2.50 \text{ mm/s}}{3.0 \text{ mm}} = 12.5$ mm/s = 1.25 cm/s

---

14.71 (a) Since no external forces act on the block-bullet system, momentum is conserved

$(0.005 \text{ kg})(400 \text{ m/s}) = (0.005 \text{ kg})v_1 + (1 \text{ kg})v_2 \quad$ where v_1 and v_2 are the speeds of the bullet and block after collision. Also, since no nonconservative forces act on the block after the bullet leaves it, we can use conservation of mechanical energy.

$\dfrac{1}{2}(1 \text{ kg})v_2^2 = \dfrac{1}{2}(900 \text{ N/m})(0.05 \text{ m})^2$.

Solving these two equations gives $\quad v_1 = 100$ m/s and $\quad v_2 = 1.5$ m/s
(b) The energy lost in the collision is the bullet's initial energy minus the sum of its final energy and the energy imparted to the block. Therefore

$E_{\text{loss}} = \dfrac{1}{2}(0.005 \text{ kg})(400 \text{ m/s})^2 - [\dfrac{1}{2}(0.005 \text{ kg})(100 \text{ m/s})^2$

$\qquad + \dfrac{1}{2}(1 \text{ kg})(1.5 \text{ m/s})^2] = 374$ J

14.72 Referring to the sketch, we have
$F = -mg \sin\theta$ and $\tan\theta = \dfrac{x}{R}$.

For small displacements, $\tan\theta$ is approximately equal to $\sin\theta$ and $F = -\dfrac{mg}{R}x = -kx$.

Since the ball obeys a Hooke's law-type equation, the motion is simple harmonic.

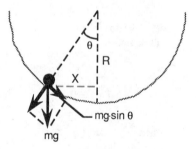

14.73 (a) From the work-energy theorem, we have

$W = \dfrac{1}{2}mv^2 - \dfrac{1}{2}mv_0^2$

$W = \dfrac{1}{2}(8 \text{ kg})[(3 \text{ m/s})^2 - (4 \text{ m/s})^2] = -28$ J

(b) The distance moved against friction is s, and $f = \mu N = \mu mg$. Thus, $W = -\mu mgs$
So, $\quad s = \dfrac{28 \text{ J}}{(0.4)(8 \text{ kg})(9.8 \text{ m/s}^2)} = 0.893$ m

But s is twice the distance the spring was compressed during the in-and-out motion of the block. Therefore, the distance of compression = 0.446 m

---

CHAPTER FOURTEEN SOLUTIONS

14.74 Newton's law of gravity is

$$F = -\frac{GMm}{r^2} = -\frac{Gm}{r^2}\left(\frac{4}{3}\pi r^3\right)\rho$$

Thus, $F = -\left(\frac{4}{3}\pi\rho G m\right)r$ Which is of Hooke's law form with $k = \frac{4}{3}\pi\rho Gm$

14.75 Let us choose the zero level for gravitational potential energy at the initial height of the mass and use conservation of mechanical energy.

$$\frac{1}{2}mv_f^2 + mgy_f + \frac{1}{2}kx_f^2 = \frac{1}{2}mv_i^2 + mgy_i + \frac{1}{2}kx_i^2$$

$$0 - mgh + \frac{1}{2}kh^2 = 0 + 0 + 0 \qquad \text{(h is the distance the mass has dropped,)}$$

Thus, $k = \frac{2mg}{h} = \frac{2(3 \text{ kg})(9.8 \text{ m/s}^2)}{0.1 \text{ m}} = 588$ N/m

(b) We use conservation of mechanical energy again, as

$$\frac{1}{2}mv_f^2 + mgy_f + \frac{1}{2}kx_f^2 = \frac{1}{2}mv_i^2 + mgy_i + \frac{1}{2}kx_i^2$$

$$\frac{1}{2}(3 \text{ kg})v_f^2 - (3 \text{ kg})(9.8 \text{ m/s}^2)(0.05 \text{ m}) + \frac{1}{2}(588 \text{ N/m})(0.05 \text{ m})^2 = 0$$

and $v_f = 0.7$ m/s

14.76 Consider the motion of the firefighter during the three intervals: (1) before, (2) during, and (3) after collision with the platform.
(a) While falling a height of 5 m, his speed changes from v_0 to v_1. We can use conservation of energy to find the velocity at the initial position of the platform, which also is taken to be the zero level for gravitational potential energy. We have

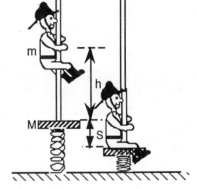

$fh = \frac{1}{2}mv_1^2 - mgh$, or $v_1 = \sqrt{2\frac{(-fh + mgh)}{m}}$

$= \sqrt{2\frac{(-300)(5)}{60} + 2(9.8)(5)} = 6.93$ m/s

(b) During the inelastic collision, momentum is conserved and if v_2 is the speed of the firefighter and platform just after collision, we have $mv_1 = = (m + M)v_2$ or

$v_2 = \frac{60}{80}(6.93 \text{ m/s}) = 5.2$ m/s . Following the collision and again using conservation of energy and using the distances labeled in the figure, (zero level of PE remains unchanged)

$$fs = 0 + (m + M)g(-s) + \frac{1}{2}ks^2 - \frac{1}{2}(m + M)v_2^2$$

This results in a quadratic equation in s:

$$s^2 - \frac{2}{k}[-f + (m + M)g]s - [\frac{(m + M)}{k}]v^2 = 0.$$

Inserting the given values, we find $s = 1.14$ m.

CHAPTER FOURTEEN SOLUTIONS

14.77 The nonconservative work (due to friction) must equal the change in the kinetic energy plus the change in potential energy. Therefore,

$$-\mu mgx\cos\theta = \frac{1}{2}kx^2 - mgx\sin\theta.$$

Thus, $-m(2)(9.80)(\cos 37°)(0.2) = \frac{1}{2}(100)(0.2)^2 - 2(9.80)(\sin 37°)(0.2)$

From which, $\mu = 0.115$

14.78 Since no nonconservative forces act, mechanical energy is conserved. Therefore,

$$\frac{1}{2}(m_1 + m_2)v^2 + m_2 gx\sin\theta - m_1 gx - \frac{1}{2}kx^2 = 0$$

$$\frac{1}{2}(25 + 30)v^2 + 25(9.80)(0.2)\sin 40° - 30(9.80)0.2 - \frac{1}{2}(200)(0.2)^2 = 0$$

Solving for v, we find $v = 1.07$ m/s

CHAPTER FIFTEEN SOLUTIONS

15.1 We use $v = (331 \text{ m/s})\sqrt{1 + \frac{T_C}{273}}$

(a) For $T_C = 27$ °C, $v = (331 \text{ m/s})\sqrt{1 + \frac{27}{273}} = 347$ m/s

(b) Using the same approach as in (a), we find $v = 387$ m/s at 100 °C, and

(c) $v = 436$ m/s at 200 °C.

15.2 The speed of sound at 27 °C is 347 m/s. (See problem 1.) Thus, the wavelength is

$$\lambda = \frac{v}{f} = \frac{347 \text{ m/s}}{20 \text{ Hz}} = 17.3 \text{ m}$$

$$\lambda = \frac{v}{f} = \frac{347 \text{ m/s}}{2 \times 10^4 \text{ Hz}} = 1.73 \times 10^{-2} \text{ m}$$

15.3 $v = (331 \text{ m/s})\sqrt{1 + \frac{22}{273}} = 344$ m/s

If d is the distance to the object, the total distance traveled by the sound (there and back) is 2d.

Thus, $d = v(t/2) = (344 \text{ m/s})(3 \text{ s}/2) = 516$ m

---

15.4 The speed of sound in seawater at 25 °C is 1530 m/s. Therefore, the time for the sound to reach the sea floor and return is
$$t = 2d/v = 2(150 \text{ m})/1533 \text{ m/s} = 0.196 \text{ s}$$

15.5 At 27 °C, the speed of sound is 347 m/s (see problem 1), and at 10 °C, the speed is

$$v = (331 \text{ m/s})\sqrt{1 + \frac{10}{273}} = 337 \text{ m/s}$$

The wavelength at 27 °C is $\lambda_1 = \frac{v}{f} = \frac{347 \text{ m/s}}{4000 \text{ Hz}} = 8.68 \times 10^{-2}$ m

At 10 °C, the wavelength is $\lambda_2 = \frac{v}{f} = \frac{337 \text{ m/s}}{4000 \text{ Hz}} = 8.43 \times 10^{-2}$ m

Thus, the change in wavelength is $\Delta\lambda = \lambda_1 - \lambda_2 = 2.50 \times 10^{-3}$ m = 2.5 mm, and the

percent change is $\frac{\Delta\lambda}{\lambda_1} (100 \%) = \frac{2.50 \times 10^{-3} \text{ m}}{8.68 \times 10^{-2} \text{ m}} (100 \%) = 2.88 \%$

15.6 $\lambda = \frac{v}{f} = \frac{10^4 \text{ m/s}}{2 \times 10^{10} \text{ Hz}} = 5 \times 10^{-7}$ m

---

15.7 $v = \sqrt{\frac{Y}{\rho}} = \sqrt{\frac{11 \times 10^{10} \text{ Pa}}{8.92 \times 10^3 \text{ kg/m}^3}} = 3.51 \times 10^3$ m/s

CHAPTER FIFTEEN SOLUTIONS

15.8 $v = \sqrt{\dfrac{B}{\rho}}$ Therefore, $B = v^2\rho = (1140 \text{ m/s})^2(0.8 \times 10^3 \text{ kg/m}^3) = 1.04 \times 10^9$ Pa.

15.9 (a) $\beta = 10 \log\left(\dfrac{I}{I_0}\right) = 10 \log\left(\dfrac{10^{-6}}{10^{-12}}\right) = 10 \log(10^6) = 60$ dB

(b) $\beta = 10 \log\left(\dfrac{I}{I_0}\right) = 10 \log\left(\dfrac{10^{-5}}{10^{-12}}\right) = 10 \log(10^7) = 70$ dB

15.10 (a) We use $\beta = 10 \log\left(\dfrac{I}{I_0}\right)$

$$40 = 10 \log\left(\dfrac{I}{10^{-12} \text{ W/m}^2}\right)$$

or, $4 = \log\left(\dfrac{I}{10^{-12} \text{ W/m}^2}\right)$ Taking the antilog of both sides gives

$$\dfrac{I}{10^{-12} \text{ W/m}^2} = 10^4$$

From which, $I = 10^{-8}$ W/m^2.

(b) Following the approach used in (a), we find $I = 10^{-2}$ W/m^2.

15.11 (a) $P = IA = I(5 \times 10^{-5} \text{ m}^2)$
At the threshold of hearing, $I = 10^{-12}$ W/m^2. Thus,
$P = (10^{-12} \text{ W/m}^2)(5 \times 10^{-5} \text{ m}^2) = 5 \times 10^{-17}$ W

(b) At the threshold of pain, $I = 1$ W/m^2.
$P = (1 \text{ W/m}^2)(5 \times 10^{-5} \text{ m}^2) = 5 \times 10^{-5}$ W

15.12 At 80 dB, the intensity is $I = 10^{-4}$ W/m^2. At 90 dB, $I = 10^{-3}$ W/m^2. Therefore, to reach 90 dB from 80 dB would require a sound intensity 10 times the level of one machine. Thus, 10 machines would be required to produce this level.

15.13 $I = \dfrac{P}{A}$

Therefore, $\beta = 10 \log\left(\dfrac{I}{I_0}\right) = 10 \log\left(\dfrac{P/A}{P_0/A}\right) = 10 \log\left(\dfrac{P}{P_0}\right)$

15.14 On a work day $\beta = 70$ dB $= 10 \log\left(\dfrac{I}{I_0}\right)$. With $I_0 = 10^{-12}$ W/m^2, we solve for I (see problem 10 for the mathematical details) to find $I = 10^{-5}$ W/m^2. If the number of cars is reduced from 100 to 25 (to one-fourth the initial value) the intensity will be reduced to one-fourth its original value. Therefore,

$I' = 0.25 \times 10^{-5}$ W/m^2

Thus, $\beta = 10 \log\left(\dfrac{2.5 \times 10^{-6}}{10^{-12}}\right) = 10 \log(2.5 \times 10^6) = 10[\log(2.5) + \log(10^6)]$
$= 10[0.398 + 6] = 64$ dB

CHAPTER FIFTEEN SOLUTIONS

15.15 At r = 5 m from door, I = I_5, and at the desired distance r = R. Let I = I_0, the threshold of hearing. We have, $I = \dfrac{const}{r^2}$ or $Ir^2 = const$

Therefore, $I_0 R^2 = I_5 (5\ m)^2$

But, $80 = 10 \log \dfrac{I_5}{I_0}$, or $8 = \log \dfrac{I_5}{I_0} = \log \left(\dfrac{R}{5\ m}\right)^2 = 2 \log \dfrac{R}{5\ m}$

Which gives $R = 5 \times 10^4$ m

15.16 At A, B, and C,

$I_A = \dfrac{const}{r_A^2}$, $I_B = \dfrac{const}{r_B^2}$, $I_C = \dfrac{const}{r_C^2}$

Thus, $\dfrac{I_A}{I_B} = \dfrac{r_B^2}{r_A^2} = \dfrac{2 \times 10^4\ m^2}{10^4\ m^2} = 2$

and $\dfrac{I_A}{I_C} = \dfrac{r_C^2}{r_A^2} = \dfrac{5 \times 10^4\ m^2}{10^4\ m^2} = 5$

So, $I_A = 2 I_B$ and $I_A = 5 I_C$

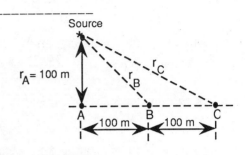

15.17 (a) $I = \dfrac{P}{4\pi r^2} = \dfrac{100\ W}{4\pi (10\ m)^2} = 7.96 \times 10^{-2}$ W/m^2.

(b) $\beta = 10 \log \left(\dfrac{7.96 \times 10^{-2}}{10^{-12}}\right) = 10 \log(7.96 \times 10^{10}) = 109$ dB

(c) If β = 120 dB (the threshold of pain), I = 1 W/m^2, and from $I = \dfrac{P}{4\pi r^2}$, we find

$r^2 = \dfrac{P}{4\pi I} = \dfrac{100\ W}{4\pi (1\ W/m^2)} = 7.96$ m^2.

and r = 2.82 m

15.18 $\beta = 10 \log \left(\dfrac{I}{I_0}\right) = 70$ dB $= 10 \log \left(\dfrac{I}{10^{-12}\ W/m^2}\right)$

From which, I = 10^{-5} W/m^2. (See problem 10 for mathematical details.)
Now use, $P = IA = I(4\pi r^2) = (10^{-5}\ W/m^2)(4\pi (5\ m)^2) = 3.14 \times 10^{-3}$ W

15.19 If β = 120 dB at 3 m, then I = 1 W/m^2 at 3 m. Therefore,
$P = IA = I(4\pi r^2) = (1\ W/m^2)(4\pi (3\ m)^2) = 113$ W

At β = 100 dB, I = 10^{-2} W/m^2. Thus,

$r^2 = \dfrac{P}{4\pi I} = \dfrac{113\ W}{4\pi (10^{-2}\ W/m^2)} = 900$ m^2.

and r = 30 m

15.20 At r = 2m, β = 100 dB. Thus,

$100 = 10 \log \left(\dfrac{I}{10^{-12}\ W/m^2}\right)$

From which, we find I = 10^{-2} W/m^2. The power emitted by the source is
$P = IA = I(4\pi r^2) = (10^{-2}\ W/m^2)(4\pi (2\ m)^2) = 0.503$ W
The intensity is given by

CHAPTER FIFTEEN SOLUTIONS

$$I = \frac{P}{A} = \frac{P}{4\pi r^2} = \frac{0.503 \text{ W}}{4\pi r^2} \quad (1)$$

(a) at $r = 4$ m, (1) gives $I = 2.5 \times 10^{-3}$ W/m^2. And $\beta = 10 \log\left(\frac{2.5 \times 10^{-3}}{10^{-12}}\right) = 94$ dB

(b) and (c) Using the approach outlined in (a), we find $\beta = 90.5$ dB at 6 m, and $\beta = 88$ dB at 8 m.

15.21 We use $\quad f' = f\dfrac{v \pm v_0}{v \pm v_s}$

with $v_0 = -30$ m/s (away) and $v_s = 0$

$$f' = (1000 \text{ Hz})\frac{345 \text{ m/s} - 30 \text{ m/s}}{345 \text{ m/s}} = 913 \text{ Hz}$$

15.22 (a) We have $v_0 = 0$ and $v_s = -30$ m/s (toward observer)

$$f' = f\frac{v \pm v_0}{v \pm v_s} = (1000 \text{ Hz})\frac{345}{345 - 30} = 1095 \text{ Hz}$$

(b) $v_s = +30$ m/s (away from observer)

$$f' = f\frac{v \pm v_0}{v \pm v_s} = (1000 \text{ Hz})\frac{345}{345 + 30} = 920 \text{ Hz}$$

15.23 Since the observer hears a reduced frequency, the source must be moving away from the observer. The cyclist must be behind the car. Since both source and observer are in motion, we use

$$f' = f\frac{v \pm v_0}{v \pm v_s}$$

Since the observer moves toward the source, use upper sign in numerator. Since source moves away from observer, use plus sign in denominator. Then with $v_0 = \frac{1}{3} v_s$, we have

$$415 \text{ Hz} = 440 \text{ Hz}\frac{345 \text{ m/s} + \frac{1}{3}v_s}{345 \text{ m/s} + v_s}$$

which can be solved to find $v_s = 32.1$ m/s $= 72$ mph with the bicyclist behind the car.

15.24 As observer approaches stationary source, $f' = f\dfrac{v + v_0}{v}$

or $\quad 520 \text{ Hz} = 500 \text{ Hz}\dfrac{345 \text{ m/s} + v_0}{345 \text{ m/s}}$

which gives $v_0 = 13.8$ m/s

Thus, as observer moves away from stationary source, the detected frequency is

$$f' = f\frac{v - v_0}{v} = 500 \text{ Hz}\frac{345 \text{ m/s} - 13.8 \text{ m/s}}{345 \text{ m/s}} = 480 \text{ Hz}$$

15.25 (a) $f' = f\dfrac{v \pm v_0}{v \pm v_s} = (500 \text{ Hz})\dfrac{345 + 15}{345 - 10} = 537$ Hz

(b) $f' = f\dfrac{v \pm v_0}{v \pm v_s} = (500 \text{ Hz})\dfrac{345 - 15}{345 + 10} = 465$ Hz

---

15.26 When the train is moving toward the observer at a speed v, we have

CHAPTER FIFTEEN SOLUTIONS

$$442 \text{ Hz} = f\frac{345}{345 - v} \quad (1)$$

When the train is moving away from the observer, at v, we have

$$441 \text{ Hz} = f\frac{345}{345 + v} \quad (2)$$

Divide equation (1) by (2), f cancels, and the resulting equation can be solved for v. We find,

$$v = 0.391 \text{ m/s}$$

15.27 Both observer and source are moving, so we use

$$f' = f\frac{v \pm v_0}{v \pm v_s} = 500 \text{ Hz}\frac{345 \text{ m/s} + 38.4 \text{ m/s}}{345 \text{ m/s} - 25.0 \text{ m/s}} = 599 \text{ Hz}$$

15.28 In a reference frame in which the air is at rest, both source and observer are moving, $v_s = 14$ m/s, and $v_0 = 14$ m/s

$$f' = f\frac{v \pm v_0}{v \pm v_s} = 1000 \text{ Hz}\frac{v + 14 \text{ m/s}}{v + 14 \text{ m/s}}$$

Conclusion: The velocity of the medium has no effect. Only the velocities of the source and observer relative to the medium are important.

15.29 (a) $\sin \theta = \frac{v}{v_s}$

Thus, $\sin 40° = \frac{345 \text{ m/s}}{v_s}$ From which, $v_s = 537$ m/s

(b) $\sin 30° = \frac{345 \text{ m/s}}{v_s}$ From which, $v_s = 690$ m/s

---

15.30 $\sin \theta = \frac{v}{v_s} = \frac{1}{1.5} = 0.667$ From which, $\theta = 41.8°$

15.31 (a) The wavelength emitted by the speakers has a length of

$$\lambda = \frac{v}{f} = \frac{345 \text{ m/s}}{500 \text{ Hz}} = 0.69 \text{ m}$$

To produce destructive interference, the path difference should be $\lambda/2$. Therefore, the top speaker must be moved back 34.5 cm.

(b) If the path difference is λ, constructive interference will occur.

15.32 If constructive interference is to occur for the first time, the path difference must be equal to one wavelength. Thus, $\lambda = 0.5$ m, and

$$f = \frac{v}{\lambda} = \frac{345 \text{ m/s}}{0.5 \text{ m}} = 690 \text{ Hz}$$

15.33 (a) The wavelength emitted by the speaker is

$$\lambda = \frac{v}{f} = \frac{345 \text{ m/s}}{400 \text{ Hz}} = 0.863 \text{ m}$$

If destructive interference is presently taking place, one must increase the path length by $\lambda/2$ in order to change this to constructive interference. Thus, the increase is 0.431 m.

(b) To hear destructive interference once again, one needs a total increase (from the initial position) of one wavelength. Thus, the increase is 0.863 m.

CHAPTER FIFTEEN SOLUTIONS

15.34 The wavelength emitted by the speakers has a wavelength of
$$\lambda = \frac{v}{f} = \frac{345 \text{ m/s}}{1000 \text{ Hz}} = 0.345 \text{ m}$$
For destructive interference, we need the path difference between the two speakers to equal one-half wavelength. We call d_2 the distance from the top speaker, and d_1 the distance from the lower speaker. He will hear destructive interference when
$$d_2 - d_1 = 0.173 \text{ m}$$
or, $\quad d_2 = 5\text{m} + 0.173 \text{ m} = 5.173 \text{ m}$

15.35 For minima, the difference in pathlengths must be odd half wavelengths. (λ = 0.429 m/s)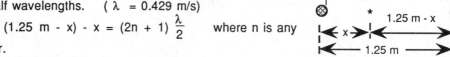
So, $\quad (1.25 \text{ m} - x) - x = (2n + 1)\frac{\lambda}{2} \quad$ where n is any integer.
Solving, gives the following results:
n = 0, x = 0.518 m
n = 1, x = 0.3035
n = 2, x = 0.089 m
n = -1, x = 0.732 m
n = -2, x = 0.947 m
n = -3, x = 1.161 m
Thus, at distances of 0.089 m, 0.304 m, 0.518 m, 0.732 m, 0.947 m, and 1.161 m from either speaker.

15.36 The length of the diagonal of the triangle formed by the two sides 800 m and 600 m is found from the Pythagorean theorem to be d_1 = 1000 m. Thus, the extra distance of travel of the wave from A over that from B is 400 m. For constructive interference, we must have
$$\frac{\lambda}{2} = 400 \text{ m, or } \lambda = 800 \text{ m}$$

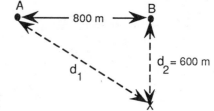

15.37 (a) For the fundamental mode of vibration, the length of the string L = λ/2,
or, $\quad \lambda = 2L = 100$ cm
If f = 20 Hz, then $\quad v = \lambda f = (1 \text{ m})(20 \text{ Hz}) = 20$ m/s
and $\quad v = \sqrt{\frac{F}{\mu}}$

$$20 \text{ m/s} = \sqrt{\frac{F}{20 \times 10^{-5} \text{ kg/m}}}$$
From which, $F = 8 \times 10^{-2}$ N
(b) If f = 4500 Hz, then the wave speed is 4500 m/s, and using the approach of (a) we find,
$\quad F = 4.05 \times 10^3$ N

15.38 $\mu = \frac{m}{L} = \frac{4.3 \times 10^{-3} \text{ kg}}{0.7 \text{ m}} = 6.14 \times 10^{-3}$ kg/m
For the fundamental, $\lambda = 2L = 2(0.7 \text{ m}) = 1.4$ m, and if f = 261.6 Hz,
$\quad v = \lambda f = 1.4 \text{ m}(261.6 \text{ Hz}) = 366.2$ m/s
and $\quad F = \mu v^2 = (6.14 \times 10^{-3} \text{ kg/m})(366.2 \text{ m/s})^2 = 824$ N

CHAPTER FIFTEEN SOLUTIONS

15.39 In the second harmonic, $\lambda = L = 1.6$ m. (Also, note that $\mu = 0.015$ g/cm $= 1.5 \times 10^{-3}$ kg/m.) Thus, if $f = 460$ Hz,
$$v = \lambda f = (1.6 \text{ m})(460 \text{ m/s}) = 736 \text{ m/s}$$
and $\quad F = \mu v^2 = (1.5 \times 10^{-3} \text{ kg/m})(736 \text{ m/s})^2 = 813$ N

15.40 The mass per unit length of the string is
$$\mu = \frac{m}{L} = \frac{3 \times 10^{-4} \text{ kg}}{0.7 \text{ m}} = 4.29 \times 10^{-4} \text{ kg/m}$$
and the wave speed is
$$v = \sqrt{\frac{F}{\mu}} = \sqrt{\frac{600 \text{ N}}{4.29 \times 10^{-4} \text{ kg/m}}} = 1183 \text{ m/s}$$
The frequency of the first harmonic, the fundamental, has a wavelength of $\lambda = 2L = 1.4$ m.
Thus,
$$f_1 = \frac{v}{\lambda} = \frac{1183 \text{ m/s}}{1.4 \text{ m}} = 845 \text{ Hz}$$
The second harmonic is
$$f_2 = 2f_1 = 1690 \text{ Hz}$$
The third harmonic is
$$f_3 = 2f_1 = 2535 \text{ Hz}$$

15.41 (a) The wavelength of the fundamental is $\lambda = 2L = 1.4$ m, and the wave speed for a frequency of 20,000 Hz is
$$v = \lambda f = (1.4 \text{ m})(20000 \text{ Hz}) = 2.8 \times 10^4 \text{ m/s}$$
Thus, $\quad F = \mu v^2 = (4.29 \times 10^{-4} \text{ kg/m}^2)(2.8 \times 10^4 \text{ m/s})^2 = 3.36 \times 10^5$ N
(b) A wire this small would surely break under this tension.

15.42 $\mu = \frac{m}{L} = \frac{0.04 \text{ kg}}{8 \text{ m}} = 5 \times 10^{-3}$ kg/m
and $\quad v = \frac{F}{\mu} = \frac{49 \text{ N}}{5 \times 10^{-3} \text{ kg/m}} = 99$ m/s
For the third harmonic, $L = \frac{3\lambda}{2} = 8$ m, giving $\lambda = 5.33$ m
(a) Therefore nodes occur at 0, 2.67 m, 5.33 m, and 8.0 m from one end, and antinodes occur at 1.33 m, 4.0 m, and 6.67 m from the end.
(b) $f = \frac{v}{\lambda} = \frac{99 \text{ m/s}}{5.33 \text{ m}} = 18.6$ Hz

---

15.43 μ is the same for both wires. We have $F_{\text{long wire}} = 4F_{\text{short wire}}$.
Since $v = \sqrt{\frac{F}{\mu}}$ then
$$v_{\text{long wire}} = 2v_{\text{short wire}}$$
$$v_{\text{short wire}} = \lambda f = (2L_{\text{short}})(60 \text{ Hz}) = (120 \text{ Hz})L_{\text{short}}$$
So, $\quad v_{\text{long wire}} = 2v_{\text{short wire}} = (240 \text{ Hz})L_{\text{short}}$
$$f_{\text{long wire}} = \frac{v_{\text{long wire}}}{\lambda_{\text{long wire}}} = \frac{(240 \text{ Hz})L_{\text{short}}}{2L_{\text{short}}} = 120 \text{ Hz} = \text{frequency of 2nd}$$
harmonic in long wire.

15.44 If the string vibrates in five segments at a frequency of 630 Hz, the wavelength is

CHAPTER FIFTEEN SOLUTIONS

$$\lambda = \frac{2}{5} L.$$

and the wave speed is

$$v = \lambda f = \frac{2}{5} L(630 \text{ Hz}) = (252)L \text{ s}^{-1}$$

For three segments, we must have $\lambda = \frac{2}{3} L$. Thus, the required frequency is

$$f = \frac{v}{\lambda} = \frac{252L \text{ s}^{-1}}{\frac{2}{3} L} = 378 \text{ Hz}$$

15.45 The natural frequency of vibration of the swing is

$$f_0 = \frac{1}{2\pi} \sqrt{\frac{g}{L}} = \frac{1}{2\pi} \sqrt{\frac{(9.8 \text{ m/s}^2)}{2 \text{ m}}} = 0.352 \text{ Hz}$$

Thus, for resonance, we can push the swing at frequencies of f_0, $f_0/2$, $f_0/3$, $f_0/4$, these correspond to frequencies of 0.352 Hz, 0.176 Hz, 0.117 Hz, and so forth.

15.46 Resonance at the fundamental mode occurs when the circumference and the wavelength are the same. Thus,

$$f = \frac{v}{\lambda} = \frac{345 \text{ m/s}}{0.1 \text{ m}} = 3450 \text{ Hz}.$$

15.47 For the open pipe (and the fundamental of it), $\lambda = \frac{v}{f} = \frac{345 \text{ m/s}}{261.6 \text{ Hz}} = 1.319 \text{ m}$

and $L = \frac{\lambda}{2} = 0.659 \text{ m} = 65.9 \text{ cm}$

For the closed pipe (and the third harmonic of it), $\lambda = \frac{v}{f} = \frac{345 \text{ m/s}}{261.6 \text{ Hz}} = 1.319 \text{ m}$

and $L = \frac{3}{4}(1.319 \text{ m}) = 0.989 \text{ m} = 98.9 \text{ cm}$

15.48 We know that $\lambda = \frac{2}{3} L = \frac{2}{3}(0.7 \text{ m}) = 0.467 \text{ m}$

Thus, $v = \lambda f = (0.467 \text{ m})(748 \text{ Hz}) = 349 \text{ m/s}$

15.49 The wavelength is $\lambda = \frac{v}{f} = \frac{345 \text{ m/s}}{261.6 \text{ Hz}} = 1.32 \text{ m}$

(a) The length of the shortest open pipe that will produce resonance is
$L = \lambda/2 = 1.32 \text{ m}/2 = 0.66 \text{ m}$,
(b) The length of the shortest closed pipe is
$L = \lambda/4 = 1.32 \text{ m}/4 = 0.33 \text{ m}$

15.50 The frequency of the fundamental mode is

$$f = \frac{v}{\lambda} = \frac{345 \text{ m/s}}{(4)(3 \text{ m})} = 28.8 \text{ Hz}.$$

Thus, resonance occurs in the interval between 20 Hz and 20,000 Hz for frequencies given by
$f_n = (28.8 \text{ Hz})n$ where n = any odd integer between 1 and 694

15.51 (a) At 0°C, the speed of sound is 331 m/s. The wavelength of the wave is

CHAPTER FIFTEEN SOLUTIONS

$$\lambda = \frac{v}{f} = \frac{331 \text{ m/s}}{300 \text{ Hz}} = 1.10 \text{ m}$$

The wavelength and the length of the tube (for the fundamental frequency) are related as

$$L = \frac{\lambda}{2} = 0.55 \text{ m}$$

(b) At T = 30 °C, the speed of sound is

$$v = (331 \text{ m/s})\sqrt{1 + \frac{T_C}{273}} = (331 \text{ m/s})\sqrt{1 + \frac{30}{273}} = 349 \text{ m/s}$$

Thus, $f = \frac{v}{\lambda} = \frac{v}{2L} = \frac{349 \text{ m/s}}{2(0.552 \text{ m})} = 317 \text{ Hz}$

15.52 (a) The space between successive resonance points is $\lambda/2$. Therefore,

$\lambda/2 = (0.24 \text{ m} - 0.08 \text{ m}) = 0.16 \text{ m}$

or, $\lambda = 0.32 \text{ m}$

The third resonance point will be one-half wavelength further down the tube. This location is at $0.24 \text{ m} + 0.16 \text{ m} = 0.40 \text{ m}$

(b) $f = \frac{v}{\lambda} = \frac{345 \text{ m/s}}{0.32 \text{ m}} = 1080 \text{ Hz}$

15.53 For the open pipe, the frequency of the nth harmonic is, $f_n = n\frac{v}{2L}$

Thus, $f_{n+1} - f_n = (n+1)\frac{v}{2L} - n\frac{v}{2L} = \frac{v}{2L}$

Thus, if $f_n = 410$ Hz, and $f_{n+1} = 492$ Hz, when L = 2 m,

$$492 \text{ Hz} - 410 \text{ Hz} = \frac{v}{2(2 \text{ m})}$$

giving $v = 328$ m/s

15.54 (a) For an open pipe, $L = \frac{\lambda}{2}$ for the fundamental

For f = 8 Hz, $\lambda = \frac{v}{f} = \frac{343 \text{ m/s}}{8 \text{ Hz}} = 42.8$ m, and L = 21.4 m

For f = 30000 Hz $\lambda = \frac{v}{f} = \frac{343 \text{ m/s}}{30000 \text{ Hz}} = 1.14 \times 10^{-2}$ m, and L = 5.72×10^{-3} m

(b) For the closed pipe, $L = \frac{3}{4}\lambda$ for the fundamental

For f = 8 Hz, $L = \frac{3}{4}(42.8 \text{ m}) = 32.1$ m

For f = 30000 Hz, $L = \frac{3}{4}(1.14 \times 10^{-2} \text{ m}) = 8.58 \times 10^{-3}$ m = 8.58 mm

The range of lengths for the open pipe: 5.72 mm to 21.4 m
and the range for the closed pipe: 8.58 mm to 32.1 m

15.55 The beat frequency is equal to the difference between the two frequencies beating together. Thus,

$\Delta f = f_1 - f_2 = 2$ Hz

Thus, the two possible frequencies for the piano key are 442 Hz and 438 Hz.

15.56 (a) The speed of sound in air at 20 °C is

CHAPTER FIFTEEN SOLUTIONS

$$v = (331 \text{ m/s})\sqrt{1 + \frac{T_C}{273}} = (331 \text{ m/s})\sqrt{1 + \frac{20}{273}} = 343 \text{ m/s}$$

and the wavelength of the sound is

$$\lambda = \frac{v}{f} = \frac{343 \text{ m/s}}{261.6 \text{ Hz}} = 1.31 \text{ m}$$

For an open pipe the length of the fundamental wavelength and the length of the tube are related as

$$L = \frac{\lambda}{2} = \frac{1.31 \text{ m}}{2} = 0.655 \text{ m}$$

(b) From $f = \frac{v}{\lambda}$, we see that the frequency of the player in the colder room is lower than that of the person in the warmer room because the speed of sound is lower in the cold room. Thus, if the beat frequency is 3 Hz, the frequency played in the colder room is

f' = 258.6 Hz.

Thus, $v' = \lambda f' = (1.31 \text{ m})(258.6 \text{ Hz}) = 338$ m/s and the temperature is found from

$$v = (331 \text{ m/s})\sqrt{1 + \frac{T_C}{273}}$$

$$338 \text{ m/s} = (331 \text{ m/s})\sqrt{1 + \frac{T_C}{273}} \quad \text{Which gives, } T_C = 11.7 \text{ °C}$$

15.57 The wavelength of the sound emitted by the violin of the first player is

$\lambda = 2L = 2(30 \text{ cm}) = 60 \text{ cm}$

As the second player shortens the length of her string, the frequency it emits will increase. Thus, the frequency her instrument emits is f' = f + 2 Hz = 198 Hz The velocity of the waves are the same for both instruments. Therefore,

$\lambda f = \lambda' f'$

and, $\lambda' = \frac{\lambda f}{f'} = \frac{(60 \text{ cm})(196 \text{ Hz})}{198 \text{ Hz}} = 59.4 \text{ cm}$

Finally, the length of the shortened string is $L' = \frac{0.594 \text{ m}}{2} = 0.297 \text{ m}$

15.58 The speed of sound at T = 37 °C is 353 m/s, and the wavelength in the ear canal is,

$$\lambda = \frac{v}{f} = \frac{353 \text{ m/s}}{3000 \text{ Hz}} = 0.118 \text{ m}$$

At resonance, we have $L = \frac{\lambda}{4} = 2.94 \text{ cm}$

15.59 At normal body temperature of 37 °C, the speed of sound is 353 m/s, and the wavelength of a 20,000 Hz sound is

$$\lambda = \frac{v}{f} = \frac{353 \text{ m/s}}{20000 \text{ Hz}} = 1.76 \times 10^{-2} \text{ m}$$

Thus, the diameter of the eardrum is 1.76 cm.

15.60 At a temperature of 22 °C, the speed of sound is 344 m/s. The lowest frequency corresponds to the longest wavelength. For the fundamental mode, λ and L are related as,

$\lambda = 2L = 2(8 \text{ m}) = 16 \text{ m}$

Thus, $f = \frac{v}{\lambda} = \frac{344 \text{ m/s}}{16 \text{ m}} = 21.5 \text{ Hz}$

CHAPTER FIFTEEN SOLUTIONS

15.61 We use $f' = f\dfrac{v \pm v_0}{v \pm v_s}$

(a) As the train approaches, we have
$$f' = f\dfrac{v \pm v_0}{v \pm v_s} = (320\text{ Hz})\dfrac{345}{345 - 40} = 362\text{ Hz}$$

(b) as the train recedes, we have
$$f' = f\dfrac{v \pm v_0}{v \pm v_s} = (320\text{ Hz})\dfrac{345}{345 + 40} = 287\text{ Hz}$$

(c) The wavelength when the train is approaching is
$$\lambda = \dfrac{v}{f} = \dfrac{345\text{ m/s}}{362\text{ Hz}} = 0.953\text{ m}$$
and for the receding train, it is
$$\lambda = \dfrac{v}{f} = \dfrac{345\text{ m/s}}{287\text{ Hz}} = 1.20\text{ m}$$

15.62 The beat frequency is 2 Hz and f = 180 Hz. Thus, f' = 182 Hz if the moving train is coming toward the station, or f' = 178 Hz if the moving train is going away from the station.

(Source moving toward stationary observer) $\quad f' = f\dfrac{v}{v - v_s}\quad$ becomes

$$182\text{ Hz} = 180\text{ Hz}\dfrac{345\text{ m/s}}{345\text{ m/s} - v_s}$$
from which, we find v_s = 3.79 m/s

(Source moving away from stationary observer) $\quad f' = f\dfrac{v}{v + v_s}$

becomes $\quad 178\text{ Hz} = 180\text{ Hz}\dfrac{345\text{m/s}}{345\text{ m/s} + v_s}$

from which, v_s = 3.88 m/s
Thus, the second train is either moving at 3.78 m/s toward the station or at 3.88 m/s away from the station.

15.63 $\lambda = \dfrac{v}{f} = \dfrac{333\text{ m/s}}{300\text{ Hz}} = 1.11\text{ m}$

Thus, $L = \dfrac{\lambda}{2}$ for the fundamental = 0.555 m

For the second harmonic
$$f = \dfrac{v}{\lambda} = \dfrac{344\text{ m/s}}{0.555\text{ m}} = 619.8\text{ Hz}$$

15.64 (a) For a string vibrating in four segments, the wavelength of the wave is
λ = L/2 = 240 cm/2 = 120 cm

(b) The speed of the wave on the string is
$v = \lambda f$ = (120 cm)(120 Hz) = $(120)^2$ cm/s

If the string vibrates in its fundamental mode, the wavelength of the wave is
λ = 2L = 2(240 cm) = 480 cm

Therefore, the frequency of the fundamental is
$$f = \dfrac{v}{\lambda} = \dfrac{(120)^2\text{ cm/s}}{480\text{ cm}} = 30\text{ Hz}$$

15.65 At its fundamental frequency, we have

CHAPTER FIFTEEN SOLUTIONS

$$f = \frac{v}{\lambda} = \frac{1}{2L}\sqrt{\frac{F}{\mu}}$$

(a) If L is doubled, we see that f decreases by a factor of 2.
(b) If μ is doubled, we see that f decreases by a factor of $\sqrt{2}$.
(c) If F is doubled, f is increased by a factor of $\sqrt{2}$.

15.66 We have $P_1 = I_1 A$, and $P_2 = I_2 A$. Let us divide the former by the latter and solve for I_1.

$$I_1 = \frac{P_1}{P_2}(I_2) = \frac{100}{200}(I_2), \quad \text{or, } I_2 = 2I_1$$

Therefore, $\beta_2 = 10 \log\left(\frac{I_2}{I_0}\right) = 10 \log\left(\frac{2I_1}{I_0}\right) = 10 \log\left[(2)\left(\frac{I_1}{I_0}\right)\right] = 10\left[\log 2 + \log\left(\frac{I_1}{I_0}\right)\right]$

or, $\beta_2 = 10 \log 2 + \beta_1$

Thus, $\beta_2 - \beta_1 = 10 \log 2 = 3.01$ dB

15.67 The tension in the string is $F = (4 \text{ kg})(9.8 \text{ m/s}^2) = 39.2$ N, and the wave speed on the string is
$$v = \sqrt{\frac{F}{\mu}} = \sqrt{\frac{39.2 \text{ N}}{1.6 \times 10^{-3} \text{ kg/m}}} = 156.5 \text{ m/s}$$

In its fundamental mode of vibration, we have $\lambda = 2L = 2(5 \text{ m}) = 10$ m.

Thus, $f = \frac{v}{\lambda} = \frac{156.5 \text{ m/s}}{10 \text{ m}} = 15.7$ Hz

---

15.68 Let us first find the intensity of a 40 dB sound from $\beta = 10 \log\left(\frac{I}{I_0}\right)$.

We have, $40 = 10 \log\left(\frac{I}{I_0}\right)$, or $\log\left(\frac{I}{I_0}\right) = 4$, and $I = 10^4 I_0$

A similar procedure for a 50 dB sound gives $I' = 10^5 I_0$

Therefore to increase the dB level from 40 to 50 dB, we must increase the intensity by a factor of 10. This means we need 10 identical sources, or 10 mosquitoes.

15.69 The half angle of the shock wave cone is given by $\sin\theta = \frac{v}{v_s}$. Thus,

$$v_s = \frac{v}{\sin\theta} = \frac{2.25 \times 10^8 \text{ m/s}}{\sin 53°} = 2.82 \times 10^8 \text{ m/s}$$

15.70 From $\beta = 10 \log\frac{I}{I_0}$, we have $I = (10^{-12} \text{ W/m}^2) 10^{\beta/10}$

Thus, $I_{vac} = (10^{-12} \text{ W/m}^2) 10^7 = 10^{-5}$ W/m^2
$I_{mower} = (10^{-12} \text{ W/m}^2) 10^{10} = 10^{-2}$ W/m^2
$I_{traffic} = (10^{-12} \text{ W/m}^2) 10^8 = 10^{-4}$ W/m^2
$I_{total} = (10^{-5} + 10^{-2} + 10^{-4})$ W/m^2 which is approximately equal to 10^{-2} W/m^2

or $\beta_{total} = 10 \log\frac{10^{-2}}{10^{-12}} = 100$ dB

15.71 The first resonance will occur for $L = \frac{\lambda}{4} = 0.34$ m.

CHAPTER FIFTEEN SOLUTIONS

Therefore, λ_{air} = 1.36 m and f = $\frac{v}{\lambda}$ = 250 Hz

Since the wire is vibrating in its third harmonic, L = $\frac{3\lambda}{2}$ and λ_{wire} = 0.8 m. The wave velocity in the wire is then v = fλ_{wire} = (250 Hz)(0.8 m) = 200 m/s

---

15.72 The velocity of sound in the pipes is v_1 = 347 m/s and v_2 = 350 m/s. In the fundamental mode $\frac{\lambda_1}{4} = \frac{\lambda_2}{4}$ or $f_2 = f_1(\frac{v_2}{v_1})$ = (480 Hz)$\frac{350}{347}$ = 484 Hz.
Therefore, the beat frequency is 4 Hz.

15.73 We use f' = f$\frac{v \pm v_0}{v \pm v_s}$ Both sources move toward the observer. We have for train 1,

f_1' = (300 Hz)$\frac{345}{345 - 30}$ = 328.6 Hz

and for the second train,

f_2' = (300 Hz)$\frac{345}{345 - v_{s2}}$

We are told that the second train travels faster than the first.
Therefore, $f_2' > f_1'$
We also know that $f_2' - f_1'$ = 3 Hz.
Therefore,

(300 Hz)$\frac{345}{345 - v_{s2}}$ - 328.6 Hz = 3 Hz.

From which, v_{s2} = 32.9 m/s

15.74 We have

$v_{long} = \sqrt{\frac{Y}{\rho}}$ and $v_{trans} = \sqrt{\frac{F}{\mu}}$

If $\frac{v_l}{v_t}$ = 8, then $\sqrt{\frac{Y/\rho}{F/\mu}}$ = 8, or $\frac{Y\mu}{\rho F}$ = 64. But, $\frac{\mu}{\rho} = \frac{V}{L}$ = A. Thus, $\frac{(Y)(A)}{F}$ = 64

We solve the above for F, to find

$F = \frac{YA}{64} = \frac{(6.8 \times 10^{11} \text{ dynes/cm}^2)(4 \times 10^{-2} \pi \text{ cm}^2)}{64}$ = 1.34 × 10^9 dynes

= 1.34 × 10^4 N

15.75 We use f' = f$\frac{v \pm v_0}{v \pm v_s}$ with f_1' = frequency of the speaker in front of the student and f_1' = frequency of the speaker behind the student.

f_1' = (456 Hz)$\frac{(331 \text{ m/s} + 1.5 \text{ m/s})}{(331 \text{ m/s} - 0)}$ = 458 Hz and

f_2' = (456 Hz)$\frac{(331 \text{ m/s} - 1.5 \text{ m/s})}{(331 \text{ m/s} - 0)}$ = 454 Hz

Therefore, the beat frequency is 4 Hz.

15.76 The moving student hears two frequencies-- that due to receding from the source $f_1' = f\frac{(v - v_0)}{v}$

CHAPTER FIFTEEN SOLUTIONS

and that due to approaching the reflected wave $f_2' = f\dfrac{(v + v_0)}{v}$

The number of beats per second $= f_2' - f_1' = f\dfrac{2v_0}{v}$

Therefore, $v_0 = \dfrac{v}{2f}$ (#beats/s), but $f = \dfrac{v}{\lambda}$ where

$v = \sqrt{\dfrac{T}{m/L}}$ and $\lambda = \dfrac{2L}{3}$

so $f = \dfrac{3}{2}\sqrt{\dfrac{T}{mL}} = \dfrac{3}{2}\sqrt{\dfrac{400\ N}{(2.25 \times 10^{-3}\ kg)(0.75\ m)}} = 7.3 \times 10^2$ Hz

and $v_0 = \dfrac{v}{2f}$ (#beats/s) $= \dfrac{(340\ m/s)(8.3\ beats/s)}{(2)(7.3 \times 10^2\ Hz)} = 1.93$ m/s

---

15.77 When observer is moving in front of and in the same direction as the source

$f_0 = f_s \dfrac{v - v_0}{v - v_s}$ where v_0 and are v_s measured relative to the medium in which

the sound is propagated. In this case the ocean current is opposite the direction of travel of the ships and

$v_0 = 45$ km/h $-(-10$ km/h$) = 55$ km/h $= 15.3$ m/s

and $v_s = 64$ km/h $-(-10$ km/h$) = 74$ km/h $= 20.55$ m/s

Therefore, $f_0 = (1200\ Hz)\dfrac{1520\ m/s - 15.3\ m/s}{1520\ m/s - 20.55\ m/s} = 1204.2$ Hz.

15.78 The time required for a sound pulse to travel a distance L at a speed v is given by

$t = \dfrac{L}{v} = \dfrac{L}{\sqrt{\dfrac{Y}{\rho}}}$ Using this expression, we find

$t_1 = \dfrac{L_1}{\sqrt{\dfrac{Y_1}{\rho_1}}} = \dfrac{L_1}{\sqrt{\dfrac{7 \times 10^{10}\ Pa}{2.7 \times 10^3\ kg/m^3}}} = 1.96 \times 10^{-4}\ L_1$

$t_2 = \dfrac{(1.5 - L_1)}{\sqrt{\dfrac{1.6 \times 10^{10}\ Pa}{11.3 \times 10^3\ kg/m^3}}} = 1.26 \times 10^{-3} - 8.40 \times 10^{-4}\ L_1$

$t_3 = \dfrac{1.5\ m}{\sqrt{\dfrac{11 \times 10^{10}\ Pa}{8.8 \times 10^3\ kg/m^3}}} = 4.24 \times 10^{-4}$ We required $t_1 + t_2 = t_3$ Therefore,

$1.96 \times 10^{-4}\ L_1 + 1.26 \times 10^{-3} - 8.40 \times 10^{-4}\ L_1 = 4.24 \times 10^{-4}$

Which yields $L_1 = 1.30$ m   and $\dfrac{L_1}{L_2} = 6.5$

15.79 The wavelength behind the duck, λ_2, is related to the wavelength in front of the duck, λ_1, as $\lambda_2 = 1.5\lambda_1$.

Thus, $\dfrac{v}{f_2} = 1.5\dfrac{v}{f_1}$ or $f_1 = 1.5\ f_2$

But $f_1 = f\dfrac{v}{v - v_s}$ and $f_2 = f\dfrac{v}{v + v_s}$

CHAPTER FIFTEEN SOLUTIONS

Therefore, $\quad f\dfrac{v}{v - v_s} = 1.5\, f\dfrac{v}{v + v_s}$

From which, $v_s = \dfrac{v}{5}$

Thus, if $v = 0.5$ m/s, $v_s = 0.1$ m/s or 10 cm/s

CHAPTER SIXTEEN SOLUTIONS

16.1 (a) $Q = n(-e) = 5 \times 10^{14}(-1.6 \times 10^{-19} \text{ C}) = -8 \times 10^{-5} \text{ C} = -80 \text{ μC}$
(b) $Q = n_p(+e) + n_e(-e) = (n_p - n_e)e$
 $= (7 \times 10^{13} - 4 \times 10^{13})1.6 \times 10^{-19} \text{ C} = 4.8 \times 10^{-6} \text{ C} = 4.8 \text{ μC}$

16.2 $F = \dfrac{kq_1q_2}{r^2} = (9 \times 10^9 \text{ Nm}^2/\text{C}^2)\dfrac{(3 \times 10^{-9} \text{ C})(6 \times 10^{-9} \text{ C})}{(3 \times 10^{-1} \text{ m})^2} = 1.8 \times 10^{-6} \text{ N}$

16.3 The force is one of attraction. Its magnitude is
$F = \dfrac{kq_1q_2}{r^2} = (9 \times 10^9 \text{ Nm}^2/\text{C}^2)\dfrac{(4.5 \times 10^{-9} \text{ C})(2.8 \times 10^{-9} \text{ C})}{(3.2 \text{ m})^2} = 1.11 \times 10^{-8} \text{ N}$

16.4 (a) The force is one of attraction. The distance r in Coulomb's law is the distance between centers. The magnitude of the force is
$F = \dfrac{kq_1q_2}{r^2} = (9 \times 10^9 \text{ Nm}^2/\text{C}^2)\dfrac{(12 \times 10^{-9} \text{ C})(18 \times 10^{-9} \text{ C})}{(0.3 \text{ m})^2} = 2.16 \times 10^{-5} \text{ N}$
(b) The net charge of -6×10^{-9} C will be equally split between the two spheres, or -3×10^{-9} C on each. The force is one of repulsion, and its magnitude is
$F = \dfrac{kq_1q_2}{r^2} = (9 \times 10^9 \text{ Nm}^2/\text{C}^2)\dfrac{(3 \times 10^{-9} \text{ C})(3 \times 10^{-9} \text{ C})}{(0.3 \text{ m})^2} = 9 \times 10^{-7} \text{ N}$

16.5 1 g of hydrogen contains $N_A = 6.02 \times 10^{23}$ atoms = number of protons = number of electrons. Thus,
$q = N_A e = (6.02 \times 10^{23})(1.6 \times 10^{-19} \text{ C}) = 9.63 \times 10^4 \text{ C}$
The distance of separation of these charges = $2R_e = 1.276 \times 10^7$ m
Thus, $F = \dfrac{kq_1q_2}{r^2} = (9 \times 10^9 \text{ Nm}^2/\text{C}^2)\dfrac{(9.63 \times 10^4 \text{ C})^2}{(1.276 \times 10^7 \text{ m})^2} = 5.13 \times 10^5 \text{ N}$

16.6 We set $F_g = F_e$,
$G\dfrac{m^2}{r^2} = k\dfrac{e^2}{r^2}$, which gives

$m = e\sqrt{\dfrac{k}{G}} = (1.6 \times 10^{-19} \text{ C})\sqrt{\dfrac{9 \times 10^9 \text{ Nm}^2/\text{C}^2}{6.67 \times 10^{-11} \text{ Nm}^2/\text{kg}^2}} = 1.86 \times 10^{-9}$ kg
This is the mass of a single proton, and the mass for a pair would be just twice this = 3.72×10^{-9} kg.

16.7 If $F_e = F_g$, we have
$\dfrac{kQ^2}{r^2} = \dfrac{GMm}{r^2}$
where M is the mass of the earth, m the mass of the moon, and Q the charge that would have to be on each. We solve for Q to find,

$Q = \sqrt{\dfrac{GMm}{k}} = \sqrt{\dfrac{(6.67 \times 10^{-11} \text{ N m}^2/\text{kg}^2)(5.98 \times 10^{24} \text{ kg})(7.36 \times 10^{22} \text{ kg})}{(9 \times 10^9 \text{ N m}^2/\text{C}^2)}}$

CHAPTER SIXTEEN SOLUTIONS

$= 5.71 \times 10^{13}$ C

16.8 We have $F_e = mg$, or

$\dfrac{ke^2}{r^2} = mg$, from which

$r = \sqrt{\dfrac{(9 \times 10^9 \ Nm^2/C^2)(1.6 \times 10^{-19} \ C)^2}{(9.11 \times 10^{-31} \ kg)(9.8 \ m/s^2)}} = 5.08$ m

16.9 The force exerted on the 3.5×10^{-9} C charge by the 2.2×10^{-9} C charge is in the positive x direction (a force of repulsion) and has a magnitude of

$F_1 = \dfrac{kq_1q_3}{r^2} = (9 \times 10^9 \ Nm^2/C^2)\dfrac{(2.2 \times 10^{-9} \ C)(3.5 \times 10^{-9} \ C)}{(1.5 \ m)^2} = 3.08 \times 10^{-8}$ N

and the force on the 3.5×10^{-9} C charge which is exerted by the 5.4×10^{-9} C charge is in the negative x direction (also a force of repulsion). Its magnitude is

$F_2 = \dfrac{kq_2q_3}{r^2} = (9 \times 10^9 \ Nm^2/C^2)\dfrac{(5.4 \times 10^{-9} \ C)(3.5 \times 10^{-9} \ C)}{(2 \ m)^2} = 4.25 \times 10^{-8}$ N

The net force is

$F_{net} = -4.25 \times 10^{-8}$ N $+ 3.08 \times 10^{-8}$ N $= -1.17 \times 10^{-8}$ N (in negative x direction)

16.10 The forces are as shown in the sketch below.

$F_1 = \dfrac{kq_1q_2}{r_{12}^2} = (9 \times 10^9 \ Nm^2/C^2)\dfrac{(6 \times 10^{-6} \ C)(1.5 \times 10^{-6} \ C)}{(3 \times 10^{-2} \ m)^2} = 90$ N

$F_2 = \dfrac{kq_1q_3}{r_{13}^2} = (9 \times 10^9 \ Nm^2/C^2)\dfrac{(6 \times 10^{-6} \ C)(2 \times 10^{-6} \ C)}{(5 \times 10^{-2} \ m)^2} = 43.2$ N

$F_3 = \dfrac{kq_2q_3}{r_{23}^2} = (9 \times 10^9 \ Nm^2/C^2)\dfrac{(1.5 \times 10^{-6} \ C)(2 \times 10^{-6} \ C)}{(2 \times 10^{-2} \ m)^2} = 67.5$ N

The net force on the 6 μC charge $= F_1 - F_2 = 46.8$ N (toward left).
The net force on the 1.5 μC charge $= F_1 + F_3 = 157.5$ N (toward right).
The net force on the -2 μC charge $= F_2 + F_3 = 110.7$ N (toward left).

16.11 $F_{compression} = \dfrac{kq_1q_2}{r^2} = (9 \times 10^9 \ Nm^2/C^2)\dfrac{(1.6 \times 10^{-19} \ C)^2}{(2.17 \times 10^{-6} \ m)^2}$
$= 4.89 \times 10^{-17}$ N

This force causes the molecule to compress by a distance of $0.01(2.17 \times 10^{-6}$ m$) = 2.17 \times 10^{-8}$ m. Thus, the effective force constant is

$k = \dfrac{F}{x} = \dfrac{4.89 \times 10^{-17} \ N}{2.17 \times 10^{-8} \ m} = 2.25 \times 10^{-9}$ N/m

CHAPTER SIXTEEN SOLUTIONS

16.12 The force exerted on the charge at the origin by the charge on the x axis is in the negative x direction (repulsive). Its magnitude is

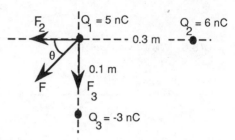

$$F_2 = (9 \times 10^9 \text{ Nm}^2/\text{C}^2)\frac{(5 \times 10^{-9} \text{ C})(6 \times 10^{-9} \text{ C})}{(0.3 \text{ m})^2} = 3 \times 10^{-6} \text{ N}$$

The force exerted on the charge at the origin by the charge on the y axis is in the negative y direction (attractive), with magnitude,

$$F_3 = (9 \times 10^9 \text{ Nm}^2/\text{C}^2)\frac{(3 \times 10^{-9} \text{ C})(5 \times 10^{-9} \text{ C})}{(0.1 \text{ m})^2} = 1.35 \times 10^{-5} \text{ N}$$

The resultant is found from the Pythagorean theorem, as

$$F = \sqrt{(F_2)^2 + (F_3)^2} = 1.38 \times 10^{-5} \text{ N}$$

The angle θ is $\tan\theta = \frac{F_3}{F_2} = 4.5 \qquad \theta = 77.5°$

16.13 The force F_1 exerted on the 6×10^{-9} C charge by the 2×10^{-9} C is repulsive and in the direction indicated in the sketch. Its magnitude is

$$F_1 = (9 \times 10^9 \text{ Nm}^2/\text{C}^2)\frac{(2 \times 10^{-9} \text{ C})(6 \times 10^{-9} \text{ C})}{2(0.5 \text{ m})^2} = 2.16 \times 10^{-7} \text{ N}$$

The force F_2 on the 6×10^{-9} C charge by the 3×10^{-9} C is repulsive and in the direction shown. The magnitude is

$$F_2 = (9 \times 10^9 \text{ Nm}^2/\text{C}^2)\frac{(3 \times 10^{-9} \text{ C})(6 \times 10^{-9} \text{ C})}{2(0.5 \text{ m})^2} = 3.24 \times 10^{-7} \text{ N}$$

Now, resolve the forces into their x and y components as

Force	x-comp	y-comp
F_1	$+1.53 \times 10^{-7}$ N	$+1.53 \times 10^{-7}$ N
F_2	$+2.29 \times 10^{-7}$ N	-2.29×10^{-7} N
F_R	3.82×10^{-7} N	-7.61×10^{-8} N

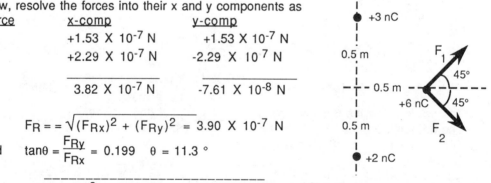

$$F_R = \sqrt{(F_{Rx})^2 + (F_{Ry})^2} = 3.90 \times 10^{-7} \text{ N}$$

and $\tan\theta = \frac{F_{Ry}}{F_{Rx}} = 0.199 \qquad \theta = 11.3°$

16.14 (a) $F = \frac{ke^2}{r^2} = (9 \times 10^9 \text{ Nm}^2/\text{C}^2)\frac{(1.6 \times 10^{-19} \text{ C})^2}{(0.51 \times 10^{-10} \text{ m})^2} = 8.86 \times 10^{-8}$ N

(b) We have $F = \frac{mv^2}{r}$ From which

$$v = \sqrt{\frac{Fr}{m}} = \sqrt{\frac{(8.86 \times 10^{-8} \text{ N})(0.51 \times 10^{-10} \text{ m})}{(9.11 \times 10^{-31} \text{ kg})}} = 2.23 \times 10^6 \text{ m/s}$$

16.15 Let us call Q the charge on each. The force of repulsion between them is

226

CHAPTER SIXTEEN SOLUTIONS

$$F = \frac{kQ^2}{r^2} \quad \text{From which,}$$

$$Q = \sqrt{\frac{Fr^2}{k}} = \sqrt{\frac{(2 \text{ N})(5 \times 10^{-2} \text{ m})^2}{(9 \times 10^9 \text{ Nm}^2/\text{C}^2)}} = 7.45 \times 10^{-7} \text{ C}$$

16.16 (a) One gram of copper = $\frac{1}{63.5}$ mol = 1.58×10^{-2} mol
and $(1.58 \times 10^{-2} \text{ mol})(6.02 \times 10^{23} \text{ atoms/mol}) = 9.48 \times 10^{21}$ atoms
In copper, there are 29 electrons per atom. Therefore, the total number of electrons is
   $N = 29(9.48 \times 10^{21} \text{ atoms}) = 2.75 \times 10^{23}$ electrons
(b) The total charge is
   $Q = -Ne = (2.75 \times 10^{23})(-1.6 \times 10^{-19} \text{ C}) = -4.40 \times 10^4$ C

16.17 If the net force is zero, $F_1 = F_2$. (See the sketch.) We see that
$$\frac{k(2 \times 10^{-9} \text{ C})q}{x^2} = \frac{k(4 \times 10^{-9} \text{ C})q}{(1.5 \text{ m} - x)^2}$$
which reduces to $(1.5 \text{ m} - x)^2 = 2x^2$
Now take square root of both sides of the
equation to yield
   $(1.5 \text{ m} - x) = \sqrt{2}\, x$
Note that the square root of 2 can be either plus or minus. We select, the plus sign so that the position of equilibrium will be between the two charges.
We find, $x = 0.621$ m

16.18 The distance of separation between the two charges at equilibrium, r, is
   $r = 2(0.3 \text{ m})\sin 5° = (0.6 \text{ m})\sin 5°$
Therefore, the Coulomb force is
$$F = \frac{kQ^2}{r^2} = \frac{kQ^2}{((0.6 \text{ m})\sin 5°)^2}$$
We now isolate one sphere and use $\Sigma F_y = 0$, and $\Sigma F_x = 0$
$(\Sigma F_y = 0) \quad T\cos 5° = mg \quad (1)$
$(\Sigma F_x = 0) \quad T\sin 5° = F \quad (2)$
Divide (2) by (1),
$$\tan 5° = \frac{F}{mg} = \frac{kQ^2}{mg((0.6 \text{ m})\sin 5°)^2} = \frac{(9 \times 10^9 \text{ Nm}^2/\text{C}^2)Q^2}{(2 \times 10^{-4} \text{ kg})(9.8 \text{ m/s}^2)((0.6 \text{ m})\sin 5°)^2}$$
From which, $Q = 7.22 \times 10^{-9}$ C

16.19 The required position is indicated in the sketch. We find
$$\frac{k(6 \times 10^{-9} \text{ C})q}{(x + .6 \text{ m})^2} = \frac{k(3 \times 10^{-9} \text{ C})q}{(x)^2}$$
or $2x^2 = (x + .6 \text{ m})^2$
Take square root of both sides, to yield
   $\sqrt{2}\, x = x + .6$ m
We must choose the signs as shown in order to have $x > 0$ so that the forces on the charge q will be in opposite directions. We solve for x.
   $x = 1.45$ m (1.45 m beyond the -3×10^{-9} C charge)

227

CHAPTER SIXTEEN SOLUTIONS

16.20 We are given that $Q_1 + Q_2 = 6 \times 10^{-4}$ C   (1)

Also, we know $F = 30$ N $= \dfrac{kQ_1Q_2}{r^2}$ From which, we find

$$Q_1Q_2 = \dfrac{(30\ N)(0.9\ m)^2}{(9 \times 10^9\ Nm^2/C^2)} = 2.7 \times 10^{-9}\ C^2. \quad (2)$$

We solve (1) and (2) simultaneously to find
$Q_1 = 4.53 \times 10^{-6}$ C and $Q_2 = 5.95 \times 10^{-4}$ C

16.21 $E = \dfrac{F}{q} = \dfrac{3.8 \times 10^{-3}\ N}{5 \times 10^{-9}\ C} = 7.6 \times 10^5$ N/C (in +x direction)

16.22 (a) $F = qE = (1.6 \times 10^{-19}\ C)(500\ N/C) = 8 \times 10^{-17}$ N (force is in westward direction)

(b) The magnitude of the force is the same, but the direction is eastward for the electron.

16.23 $E = \dfrac{kq}{r^2} = \dfrac{(9 \times 10^9\ Nm^2/C^2)(1.6 \times 10^{-19}\ C)}{(0.1\ m)^2} = 1.44 \times 10^{-7}$ N/C (toward electron)

16.24 $E = \dfrac{kq}{r^2}$

$200\ N/C = \dfrac{(9 \times 10^9\ Nm^2/C^2)q}{(0.8\ m)^2}$

$q = 1.42 \times 10^{-8}$ C

16.25 $E = \dfrac{kq}{r^2} = \dfrac{(9 \times 10^9\ Nm^2/C^2)(1.6 \times 10^{-19}\ C)}{(0.51 \times 10^{-10}\ m)^2} = 5.54 \times 10^{11}$ N/C (directed away from proton)

16.26 (a) E_1 = field due to 30×10^{-9} C charge.

$E_1 = \dfrac{kq}{r^2} = \dfrac{(9 \times 10^9\ Nm^2/C^2)(30 \times 10^{-9}\ C)}{(.15\ m)^2}$
$= 1.20 \times 10^4$ N/C

Similarly E_2 (the field due to the 60×10^{-9} C charge is
$E_2 = 2.40 \times 10^4$ N/C

Directions of E_1 and E_2 are shown in the sketch.

The resultant electric field is $E_2 - E_1 = 1.20 \times 10^4$ N/C (directed toward 30×10^{-9} C charge.)

(b) If $q_2 = -60 \times 10^{-9}$ C, then E_2 is of the same magnitude, but opposite in direction. Thus, the net electric field is $E_2 + E_1 = 3.60 \times 10^4$ N/C (toward the -60×10^{-9} C charge)

CHAPTER SIXTEEN SOLUTIONS

16.27

```
    +6 μC              E₁  +1.5 μC              -2 μC
  ----●--------------- ←●------●---------------●----
                   E₂  → ←
                       E₃
        |←─ 0.02 m ─→|←──  0.02 m ──→|
                   |←→|
                   0.01 m
```

(a) See sketch.
$E = E_1 - E_2 + E_3$

$= (9 \times 10^9 \text{ Nm}^2/\text{C}^2)\left(\frac{6 \times 10^{-6} \text{ C}}{(2 \times 10^{-2} \text{ m})^2} - \frac{1.5 \times 10^{-6} \text{ C}}{(1 \times 10^{-2} \text{ m})^2} + \frac{2 \times 10^{-6} \text{ C}}{(3 \times 10^{-2} \text{ m})^2}\right)$

$= 2.0 \times 10^7$ N/C
Directed toward the right.
(b) $F = |q|E = (2.0 \times 10^{-6} \text{ N})(2.0 \times 10^7 \text{ N/C}) = 40$ N (to left)

16.28 $eE = mg$, or, for an electron,

$E = \frac{mg}{e} = \frac{(9.11 \times 10^{-31} \text{ kg})(9.8 \text{ m/s}^2)}{1.6 \times 10^{-19} \text{ C}} = 5.58 \times 10^{-11}$ N/C (directed downward)

(b) For a proton,

$E = \frac{mg}{e} = \frac{(1.67 \times 10^{-27} \text{ kg})(9.8 \text{ m/s}^2)}{1.6 \times 10^{-19} \text{ C}} = 1.02 \times 10^{-7}$ N/C (directed upward)

16.29 If there is zero tension in the string,
$qE = mg$
Thus, $(3 \times 10^{-6} \text{ C})E = 0.49$ N
and $E = 1.63 \times 10^5$ N/C

16.30 The electrical force on the floating object = $F = (24 \times 10^{-6} \text{ C})(610 \text{ N/C}) = 1.464 \times 10^{-2}$ N, and we know $F = mg$, so

$m = \frac{F}{g} = \frac{1.464 \times 10^{-2} \text{ N}}{9.8 \text{ m/s}^2} = 1.49 \times 10^{-3}$ kg = 1.49 g

---

16.31 (a) The magnitude of the force on the electron is
$F = qE = (1.6 \times 10^{-19})(300 \text{ N/C}) = 4.8 \times 10^{-17}$ N
Hence, from Newton's second law, the acceleration is

$a = \frac{F}{m} = \frac{4.8 \times 10^{-17} \text{ N}}{9.11 \times 10^{-31} \text{ kg}} = 5.27 \times 10^{13}$ m/s².

(b) We use $v = v_0 + at$
$v = 0 + (5.27 \times 10^{13} \text{ m/s}^2)(10^{-8} \text{ s}) = 5.27 \times 10^5$ m/s

---

16.32 (a) The force on the proton is
$F = qE = (1.6 \times 10^{-19})(2 \times 10^3 \text{ N/C}) = 3.2 \times 10^{-16}$ N (in the +x direction)
(b) The acceleration is found from Newton's second law.

$a = \frac{F}{m} = \frac{3.2 \times 10^{-16} \text{ N}}{1.67 \times 10^{-27} \text{ kg}} = 1.91 \times 10^{11}$ m/s².

(c) $v = v_0 + at$
10^6 m/s = $0 + (1.91 \times 10^{11}$ m/s$^2)t$

CHAPTER SIXTEEN SOLUTIONS

$t = 5.23 \times 10^{-6}$ s

16.33 (a) The force on the proton is
$F = qE = (1.6 \times 10^{-19}$ C$)(640$ N/C$) = 1.02 \times 10^{-16}$ N
(a) The acceleration is found from Newton's second law.

$$a = \frac{F}{m} = \frac{1.02 \times 10^{-16} \text{ N}}{1.67 \times 10^{-27} \text{ kg}} = 6.13 \times 10^{10} \text{ m/s}^2.$$

(b) $t = \dfrac{v}{a} = \dfrac{1.2 \times 10^6 \text{ m/s}}{6.13 \times 10^{10} \text{ m/s}^2} = 1.96 \times 10^{-5}$ s $= 19.6$ μs

(c) $s = \dfrac{v^2}{2a} = \dfrac{(1.2 \times 10^6 \text{ m/s})^2}{2(6.13 \times 10^{10} \text{ m/s}^2)} = 11.7$ m

(d) KE $= \dfrac{1}{2} mv^2 = \dfrac{1}{2}(1.67 \times 10^{-27}$ s $)(1.2 \times 10^6$ m/s$)^2 = 1.20 \times 10^{-15}$ J

16.34 Work done $= \Delta$KE $= $ KE$_f$ - KE$_i$ = KE$_f$ - 0

so, $F(2 \times 10^{-3}$ m$) = \dfrac{1}{2}(9.11 \times 10^{-31}$ kg$)(3 \times 10^6$ m/s$)^2$

From which $F = 2.05 \times 10^{-15}$ N

But, $F = qE$, so $E = \dfrac{F}{q} = \dfrac{2.05 \times 10^{-15} \text{ N}}{1.6 \times 10^{-19} \text{ C}} = 1.28 \times 10^4$ N/C

(b) Using the reasoning of part (a),

$\dfrac{1}{2} mv^2 = qEs$, or

$$v = \sqrt{\frac{2qEs}{m}} = \sqrt{\frac{2(1.6 \times 10^{-19} \text{ C})(1.28 \times 10^4 \text{ N/C})(4 \times 10^{-3} \text{ m})}{9.11 \times 10^{-31} \text{ kg}}}$$

$= 4.24 \times 10^6$ m/s

16.35 The required electric field will be in the direction of motion. We know,
Work done $= \Delta$KE

so, $-Fs = -\dfrac{1}{2} mv_0^2$ (since the final velocity $= 0$)

which becomes $eEs = \dfrac{1}{2} mv_0^2$

and $E = \dfrac{\frac{1}{2} mv_0^2}{es} = \dfrac{1.6 \times 10^{-17} \text{ J}}{(1.6 \times 10^{-19} \text{ C})(0.1 \text{ m})} = 10^3$ N/C (in direction of electron's motion)

16.36 From problem 16.12, the resultant force exerted on the 5×10^{-9} C charge located at the origin is 1.38×10^{-5} N at 77.5° below the -x axis. Therefore, the electric field at the origin is given by

$E = \dfrac{F}{q} = \dfrac{1.38 \times 10^{-5} \text{ N}}{5 \times 10^{-9} \text{ C}} = 2.76 \times 10^3$ N/C at 77.5 ° below -x axis.

CHAPTER SIXTEEN SOLUTIONS

16.37 The length of the diagonal can be shown, from the Pythagorean theorem, to be 0.63 m. Also, the angle ϕ (see figure) is found from the tangent function to be 18.4°. The field, E_1, due to the 3 X 10^{-9} C charge is (direction shown on sketch).

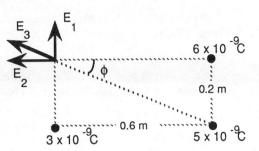

$$E_1 = \frac{kq}{r^2} = \frac{(9 \times 10^9 \text{ Nm}^2/\text{C}^2)(3 \times 10^{-9} \text{ C})}{(0.2 \text{ m})^2} = 675 \text{ N/C}$$

Likewise, E_2, due to the 6 X 10^{-9} C charge is

$$E_2 = \frac{kq}{r^2} = \frac{(9 \times 10^9 \text{ Nm}^2/\text{C}^2)(6 \times 10^{-9} \text{ C})}{(0.6 \text{ m})^2} = 150 \text{ N/C}$$

and E_3, due to the 5 X 10^{-9} C charge is

$$E_3 = \frac{kq}{r^2} = \frac{(9 \times 10^9 \text{ Nm}^2/\text{C}^2)(5 \times 10^{-9} \text{ C})}{(0.63 \text{ m})^2} = 112.5 \text{ N/C}$$

Now, resolve the fields into their x and y components as

Field	x-comp	y-comp
E_1	0	675 N/C
E_2	-150 N/C	0
E_3	-107 N/C	35.6 N/C
E_R	-257 N/C	711 N/C

$E_R = \sqrt{(E_{Rx})^2 + (E_{Ry})^2} = 756$ N/C

and $\tan\theta = \frac{E_{Ry}}{E_{Rx}} = 2.76 \quad \theta = 70.1°$ (above - x axis)

16.38 The magnitude of the fields remain the same, but the direction of E_3 is changed. (See sketch.) We resolve the fields into their components.
Now, resolve the fields into their x and y components as

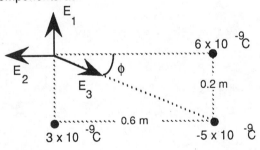

Field	x-comp	y-comp
E_1	0	675 N/C
E_2	-150 N/C	0
E_3	107 N/C	-36 N/C
E_R	-43 N/C	639 N/C

$E_R = \sqrt{(E_{Rx})^2 + (E_{Ry})^2} = 641$ N/C

and $\tan\theta = \frac{E_{Ry}}{E_{Rx}}$ gives

$\theta = 86.1°$ (above -x axis)

CHAPTER SIXTEEN SOLUTIONS

16.39 Students should recognize that $E_{net} = 0$ at the center from the symmetry of the arrangement. If not,
$$E_1 = E_2 = E_3$$
where E_1 is the field due to the $-5\ \mu C$ charge in the upper right quadrant, E_2 is the field due to the $-5\ \mu C$ in the upper left quadrant, and E_3 is the field due to the $-5\ \mu C$ charge directly below the point.
$$E_x = E_{1x} - E_{2x} = E_1\cos 30° - E_2\cos 30° = 0$$
and $\quad E_y = E_{1y} + E_{2y} - E_{3y} = 2E_1\sin 30° - E_2 = E_1 - E_2 = 0$
Thus, $E = 0$

16.40 (a) By symmetry, $E = 0$ at center of triangle
(b) See sketch (b).
$$E_1 = E_2 = \frac{kq}{a^2}$$
By symmetry
$$E_{1x} = -E_{2x}$$
so the resultant x component = 0

$$E_y = E_{1y} + E_{2y} = 2E_1\sin\theta = 2E_1\frac{b}{a} \quad \text{where } b = \frac{\sqrt{3}}{2}a$$

Thus, $E = E_y = \sqrt{3}E_1 = \sqrt{3}\frac{kq}{a^2}$

$$= \sqrt{3}\frac{(9 \times 10^9\ Nm^2/C^2)(2.7 \times 10^{-6}\ C)}{(0.35\ m)^2}$$

$$= 3.44 \times 10^5\ N/C \text{ (directed straight upward)}$$

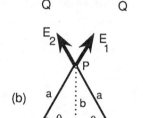

16.41 The field, E_1, due to the 8×10^{-9} C charge is (direction shown on sketch).
$$E_1 = \frac{kq}{r^2} = \frac{(9 \times 10^9\ Nm^2/C^2)(8 \times 10^{-9}\ C)}{(0..25\ m)^2}$$
$$= 1152\ N/C$$
Likewise, E_2, due to the -5×10^{-9} C charge is
$$E_2 = \frac{kq}{r^2} = \frac{(9 \times 10^9\ Nm^2/C^2)(5 \times 10^{-9}\ C)}{(0.25\ m)^2} = 720\ N/C$$
and E_3, due to the 3×10^{-9} C charge is
$$E_3 = \frac{kq}{r^2} = \frac{(9 \times 10^9\ Nm^2/C^2)(3 \times 10^{-9}\ C)}{(0.433\ m)^2} = 144\ N/C$$
$E_{Rx} = \Sigma E_x = E_1 + E_2 = 1872\ N/C$ and $E_{Ry} = \Sigma E_y = -E_3 = -144\ N/C$
$$E_R = = \sqrt{(E_{Rx})^2 + (E_{Ry})^2} = 1880\ N/C$$
and from $\quad \tan\theta = \frac{E_{Ry}}{E_{Rx}} \quad \theta = -4.4°$ (below x axis)

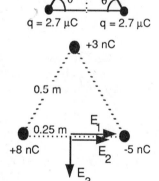

16.42 The force needed would be
$$F = ma = (1.67 \times 10^{-27}\ kg)(9.8\ m/s^2) = 1.64 \times 10^{-26}\ N$$

CHAPTER SIXTEEN SOLUTIONS

Thus, $E = \dfrac{F}{q} = \dfrac{1.64 \times 10^{-26} \text{ N}}{1.6 \times 10^{-19} \text{ C}} = 1.02 \times 10^{-7}$ N/C

16.43 The point is designated in the sketch.
The electric field, E_1, due to the -2.5×10^{-6} C
charge is

$E_1 = \dfrac{kq}{r^2} = \dfrac{(9 \times 10^9 \text{ Nm}^2/\text{C}^2)(2.5 \times 10^{-6} \text{ C})}{d^2}$ (1)

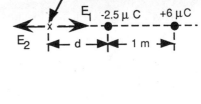

and E_2 due to the 6×10^{-6} C charge is

$E_2 = \dfrac{kq}{r^2} = \dfrac{(9 \times 10^9 \text{ Nm}^2/\text{C}^2)(6 \times 10^{-6} \text{ C})}{(d + 1 \text{ m})^2}$ (2)

Equate the right sides of (1) and (2), gives
$(d + 1 \text{ m})^2 = 2.4 d^2$
or, $d + 1 \text{ m} = \pm 1.55 d$
which yields d = either 1.82 m or d = -0.392 m
The negative value for d is unsatisfactory because that locates a point between the charges where both fields are in the same direction. Thus,

d = 1.82 m to the left of the -2.5×10^{-6} C charge.

16.44 The fields must be in opposite
direction to cancel. Therefore, the point must
be between the two charges. We equate the
magnitudes of the fields as

$\dfrac{k(2.5 \times 10^{-6} \text{ C})}{d^2} = \dfrac{k(6 \times 10^{-6} \text{ C})}{(1 \text{ m} - d)^2}$

Which can be reduced to $1 \text{ m} - d = \pm 1.55 d$.
We must choose the + sign to make d > 0 so the point is between the two charges.
Thus, d = 0.392 m

16.45 Note in the sketches below that electric field lines originate on positive charges and terminate on negative charges. The density of lines is twice as great for the -2 μC charge in (b) as they are for the 1 μC charge in (a).

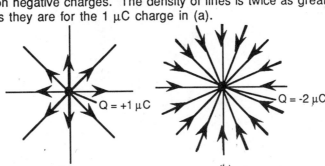

(a)   (b)

233

CHAPTER SIXTEEN SOLUTIONS

16.46 Rough sketches for these charge configurations are shown below.

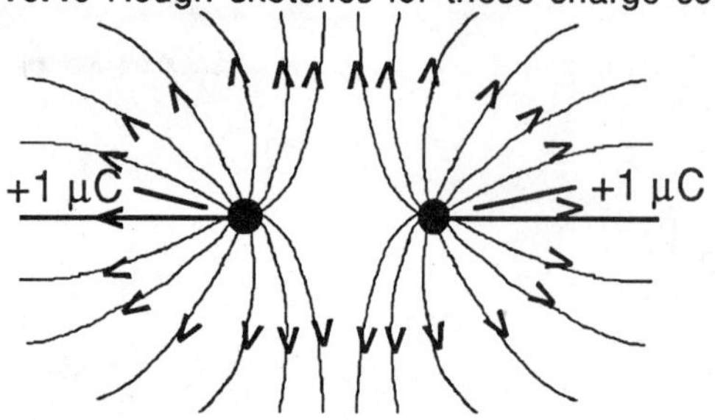

(a) +1 µC, +1 µC

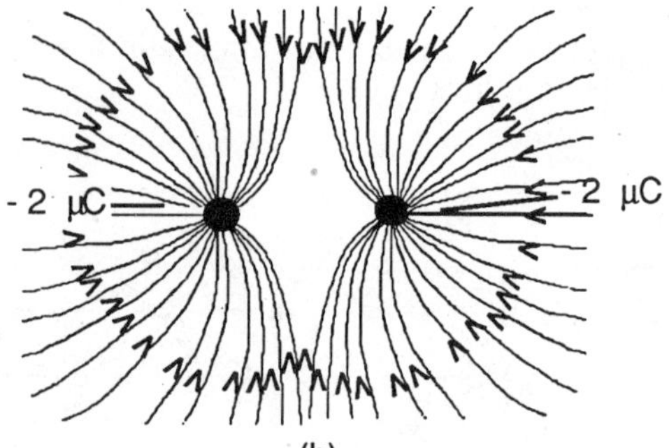

(b) −2 µC, −2 µC

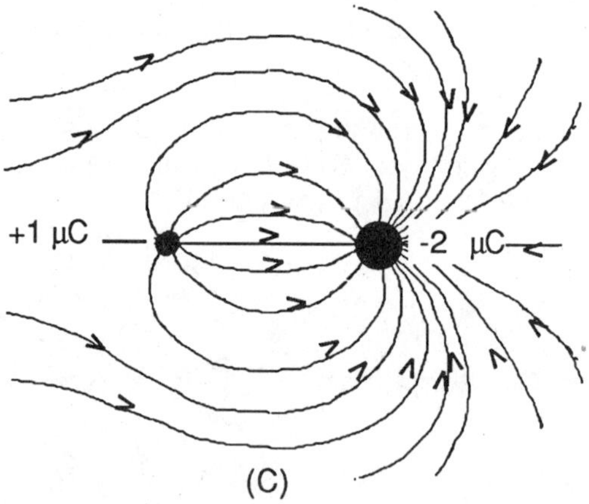

(c) +1 µC, −2 µC

16.47 (a) The magnitude of q_2 is three times the magnitude of q_1 because 3 times as many lines emerge from q_2 as enter q_1. ($q_2 = 3q_1$)
(b) $q_2 > 0$ because lines emerge from it, and $q_1 < 0$ because lines terminate on it.

16.48 (a) In the sketch for (a) below note that there are no lines inside the sphere. On the outside of the sphere, the field lines are uniformly spaced and radially outward.

CHAPTER SIXTEEN SOLUTIONS

(b) In the sketch for (b) below, not that the lines are perpendicular to the surface and symmetrical about the symmetry axes of the cube.

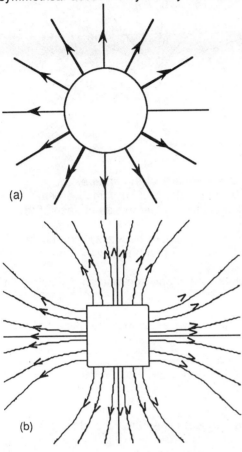

16.49 Note in the sketch at the right that the lines strike the surface at right angles. The pattern is symmetrical about the axis of the rod and symmetrical about a line perpendicular to the rod and passing through the center of the rod.

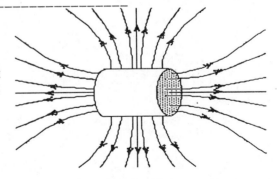

CHAPTER SIXTEEN SOLUTIONS

16.50 (a) The sketch for (a) is shown at the right. Note that approximately four times as many lines leave q_1 as emerge from q_2.

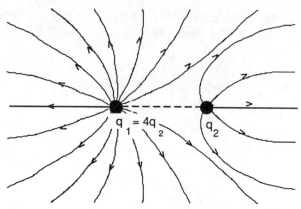

(b) The field pattern looks the same here as that shown for (a) if the arrowheads are reversed on the lines of force.

16.51 (a) Zero net charge on inside and outside surface of the sphere.
(b) The positive charge lowered into the sphere attracts -5 μC of negative charge to the inside surface of the sphere, and +5 μC is left behind on the outside surface of the sphere.
(c) The negative charge on the inside surface of the sphere neutralizes the positive charge lowered inside the sphere. However, +5 μC is left on the outside surface of the sphere.
(d) When the object is removed from the sphere, the +5 μC remains on the outside of the sphere.

16.52 The explanation for the charge distribution here is almost the same as that given for problem 51 above.
(a) zero net charge on either surface
(b) +5 μC, -5 μC
(c) zero, -5 μC
(d) zero, 5 μC

16.53 (a) The dome is a closed conducting surface. Therefore, the electric field is zero inside.

(b) $E = \dfrac{kq}{R^2} = \dfrac{(9 \times 10^9 \text{ Nm}^2/\text{C}^2)(2 \times 10^{-4} \text{ C})}{(1 \text{ m})^2} = 1.8 \times 10^6$ N/C

(c) $E = \dfrac{kq}{r^2} = \dfrac{(9 \times 10^9 \text{ Nm}^2/\text{C}^2)(2 \times 10^{-4} \text{ C})}{(4 \text{ m})^2} = 1.13 \times 10^5$ N/C

16.54 The field is strongest just outside the surface of a sphere. Thus,

$$E = \dfrac{kq}{R^2}$$

and $q = \dfrac{Er^2}{k} = \dfrac{3 \times 10^6 \text{ N/C}(2 \text{ m})^2}{9 \times 10^9 \text{ Nm}^2/\text{C}^2} = 1.33 \times 10^{-3}$ C

16.55 (a) $a = \dfrac{F}{m} = \dfrac{qE}{m} = \dfrac{(1.6 \times 10^{-19} \text{ C})(3 \times 10^6 \text{ N/C})}{9.11 \times 10^{-31} \text{ kg}} = 5.27 \times 10^{17}$ m/s²

(b) Anticipating that this distance is very small, we assume the field is uniform over this short distance

work done = $\Delta KE = KE_f - KE_i = KE_f - 0$

So, $Fs = \dfrac{1}{2} mv^2$

CHAPTER SIXTEEN SOLUTIONS

and $\quad s = \dfrac{mv^2}{2qE} = \dfrac{(9.11 \times 10^{-31} \text{ kg})(3 \times 10^7)^2}{2(1.6 \times 10^{-19} \text{ C})(3 \times 10^6 \text{ N/C})} = 8.54 \times 10^{-4}$ m = 0.854 mm

---

16.56 (a) $F = qE = (1.6 \times 10^{-19} \text{ C})(3 \times 10^4 \text{ N/C}) = 4.8 \times 10^{-15}$ N

(b) $a = \dfrac{F}{m} = \dfrac{4.8 \times 10^{-15} \text{ N}}{1.67 \times 10^{-27} \text{ kg}} = 2.87 \times 10^{12}$ m/s^2.

16.57 The force exerted on q_3 by the charge q_1 on the y axis is directed toward q_1 (attractive). It makes an angle of 45° with the x axis. (See sketch.) Its magnitude is

$F_1 = (9 \times 10^9 \text{ Nm}^2/\text{C}^2)\dfrac{(3.5 \times 10^{-9} \text{ C})(2.6 \times 10^{-9} \text{ C})}{(2.88 \times 10^{-2} \text{ m}^2)} = 2.84 \times 10^{-6}$ N

The force exerted on q_3 by the -1.8×10^{-9} C charge at the origin is in the positive x direction (attractive), with magnitude,

$F_2 = (9 \times 10^9 \text{ Nm}^2/\text{C}^2)\dfrac{(1.8 \times 10^{-9} \text{ C})(2.6 \times 10^{-9} \text{ C})}{(0.12 \text{ m})^2} = 2.93 \times 10^{-6}$ N

The resultant force in the x direction is
$F_x = F_{1x} + F_2 = 4.94 \times 10^{-6}$ N
and the resultant force in the y direction is
$F_y = F_{1y} = 2.01 \times 10^{-6}$ N
The resultant is found from the Pythagorean theorem,
as $\quad F = \sqrt{(F_x)^2 + (F_y)^2} = 5.33 \times 10^{-6}$ N
The angle θ is found from
$\tan\theta = \dfrac{F_y}{F_x} \quad \theta = 22.2°$

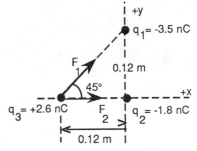

16.58 (a) $F = qE = (1.6 \times 10^{-19} \text{ C})(400 \text{ N/C}) = 6.4 \times 10^{-17}$ N

$a = \dfrac{F}{m} = \dfrac{6.4 \times 10^{-17} \text{ N}}{1.67 \times 10^{-27} \text{ kg}} = 3.83 \times 10^{10}$ m/s^2.

(b) $v = v_0 + at$
$v = 0 + (3.83 \times 10^{10} \text{ m/s}^2)(10^{-8} \text{ s}) = 383$ m/s

(c) $KE = \dfrac{1}{2}mv^2 = \dfrac{1}{2}(1.67 \times 10^{-27} \text{ kg})(383 \text{ m/s})^2 = 1.23 \times 10^{-22}$ J

16.59 (a) The field, E_1, due to the 4×10^{-9} C charge is in the -x direction (see sketch).

$E_1 = \dfrac{kq}{r^2} = \dfrac{(9 \times 10^9 \text{ Nm}^2/\text{C}^2)(4 \times 10^{-9} \text{ C})}{(2.5 \text{ m})^2} = 5.76$ N/C

Likewise, E_2, due to the 5×10^{-9} C charge is

$E_2 = \dfrac{kq}{r^2}$

$= \dfrac{(9 \times 10^9 \text{ Nm}^2/\text{C}^2)(5 \times 10^{-9} \text{ C})}{(2 \text{ m})^2}$

= 11.3 N/C

and E_3, due to the 3×10^{-9} C charge is

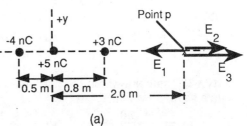

(a)

237

CHAPTER SIXTEEN SOLUTIONS

$$E_3 = \frac{(9 \times 10^9 \text{ Nm}^2/\text{C}^2)(3 \times 10^{-9} \text{ C})}{(1.2 \text{ m})^2} = 18.8 \text{ N/C}$$

$E_R = -E_1 + E_2 + E_3 = 24.2$ N/C in + x direction.
(b) The directions of the various fields are shown in the sketch, and the magnitudes are:
$E_1 = 8.47$ N/C
$E_2 = 11.3$ N/C
$E_3 = 5.82$ N/C
The resultant x component is
$E_x = -E_{1x} - E_{3x} = -4.21$ N/C
and the y component is
$E_y = -E_{1y} + E_{2y} + E_{3y} = 8.44$ N/C
From the Pythagorean theorem,
$E_R = 9.43$ N/C
and $\theta = 63.5°$ above -x axis

(b)

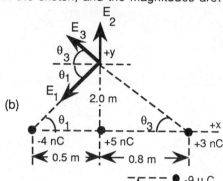

16.60 We know that $E_1 = E_2$ so,
$$\frac{k(9 \text{ μc})q_3}{(6-y)^2} = \frac{k(8 \text{ μc})q_3}{(4+y)^2}$$
(y = distance from origin to point where field is zero)
This reduces to $9(4+y)^2 = 8(6-y)^2$
Which can be solved as a quadratic equation to yield
y = 0.853 m or y = -168.9 m The last answer must be ruled out since the two fields would have the same direction at this location.

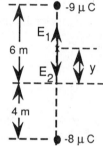

16.61 From the free body diagram shown,
$\Sigma F_y = 0 = T \cos 15° = 1.96 \times 10^{-2}$ N
and $T = 2.03 \times 10^{-2}$ N
From
$\Sigma F_x = 0$, we have
$qE = T \sin 15°$
or $q = \frac{T \sin 15°}{E} = \frac{(2.03 \times 10^{-2} \text{ N}) \sin 15°}{10^3 \text{ N/C}}$
$q = 5.25 \times 10^{-6}$ C = 5.25 μC

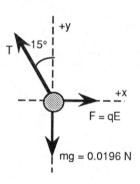

16.62 (a) The magnitude of the force on each charge is
$F = qE = (2 \times 10^{-9} \text{ C})(200 \text{ N/C}) = 4 \times 10^{-7}$ N
We use a line through -q and perpendicular to the page as the axis of rotation.
$\tau = FL = (4 \times 10^{-7} \text{ N})(3 \times 10^{-8} \text{ m}) = 1.2 \times 10^{-14}$ N m (CCW)
(b) Use same axis of rotation as in (a).
$\tau = F(\text{lever arm}) = F(L \cos 30°)$
$= (4 \times 10^{-7} \text{ N})(3 \times 10^{-8} \text{ m}) \cos 30° = 1.04 \times 10^{-14}$ N m (CCW)

16.63 The point where the net force is zero is on the y axis at a point y > 5 m. We have,
$$\frac{k(3 \times 10^{-6} \text{ C})q_3}{(y-5 \text{ m})^2} = \frac{k(8 \times 10^{-6} \text{ C})q_3}{(y-2 \text{ m})^2} \quad (1)$$
(y is the distance from the origin to the point where the net force = 0.)

238

CHAPTER SIXTEEN SOLUTIONS

(1) reduces to
   $y - 2\text{ m} = \pm 1.63(y - 5\text{ m})$
We must choose the + sign to have $y > 5$ m.
Thus, $y = 9.79$ m

16.64 The force on the proton is downward, and its magnitude is
   $F = qE = (1.6 \times 10^{-19}\text{ C})(500\text{ N/C}) = 8 \times 10^{-17}$ N
and the acceleration is also downward of magnitude
   $a = \dfrac{F}{m} = \dfrac{8 \times 10^{-17}\text{ N}}{1.67 \times 10^{-27}\text{ kg}} = 4.79 \times 10^{10}\text{ m/s}^2$.
To find the maximum height, H, reached, we use
   $v_y^2 = v_{0y}^2 + 2ay$
   $0 = (2.0 \times 10^5\text{ m/s})^2 - 2(4.79 \times 10^{10}\text{ m/s}^2)H$
   $H = 4.18 \times 10^{-1}$ m = 41.8 cm

16.65 The distance between the two charges at equilibrium is
   $r = 2(0.1\text{ m})\sin 10° = 3.47 \times 10^{-2}$ m
Now consider the forces on the sphere with charge +q [see sketch (a)], and use $\Sigma F_y = 0$

   $T\cos 10° = mg$, or $T = \dfrac{mg}{\cos 10°}$ (1)

Now use $\Sigma F_x = 0$
   $F_{net} = T\sin 10°$ (2)
F_{net} is the net electrical force on the charged sphere.
Eliminate T from (2) by use of (1).
$F_{net} = \dfrac{mg\sin 10°}{\cos 10°} = mg\tan 10° = (2 \times 10^{-3}\text{ kg})(9.8\text{ m/s}^2)\tan 10°$
   $= 3.46 \times 10^{-3}$ N
F_{net} is the resultant of two forces, F_1 and F_2 [see sketch (b)].
F_1 is the attractive force on +q exerted by -q, and
F_2 is the force exerted on +q by the external electric field.
   $F_{net} = F_2 - F_1$ or, $F_2 = F_{net} + F_1$
$F_1 = (9 \times 10^9\text{ Nm}^2/\text{C}^2)\dfrac{(5 \times 10^{-8}\text{ C})(5 \times 10^{-8}\text{ C})}{(3.47 \times 10^{-3}\text{ m})^2} = 1.87 \times 10^{-2}$ N
Thus, $F_2 = F_{net} + F_1$ yields
   $F_2 = 3.46 \times 10^{-3}$ N $+ 1.87 \times 10^{-2}$ N $= 2.21 \times 10^{-2}$ N
and, $F_2 = qE$, or $E = \dfrac{F_2}{q} = \dfrac{2.21 \times 10^{-2}\text{ N}}{5 \times 10^{-8}\text{ C}}$
   $= 4.43 \times 10^5$ N/C

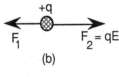
(b)

CHAPTER SIXTEEN SOLUTIONS

16.66 The forces on q are shown in the sketch. Their magnitudes are

$F_1 = \dfrac{kq^2}{a^2}$ $F_2 = \dfrac{kq^2}{a^2}$ $F_3 = \dfrac{kq^2}{2a^2}$

Now, resolve the forces into their x and y components as

Force	x-comp	y-comp
F_1	F_1	0
F_2	0	F_2
F_3	$F_3\cos 45°$	$F_3\sin 45°$
F_R	$F_1 + F_3\cos 45°$	$F_2 + F_3\sin 45°$

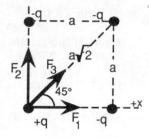

Or, $F_x = \dfrac{kq^2}{a^2} + \dfrac{kq^2}{2a^2}(.707) = 1.35 \dfrac{kq^2}{a^2}$ and $F_y = \dfrac{kq^2}{a^2} + \dfrac{kq^2}{2a^2}(.707) = 1.35 \dfrac{kq^2}{a^2}$

$F = \sqrt{(F_x)^2 + (F_y)^2} = 1.91 \dfrac{kq^2}{a^2}$ N $\tan\theta = \dfrac{F_y}{F_x} = 1$ $\theta = 45°$ (at 45° above x axis)

16.67 (a) The magnitude of the acceleration is given by $a = \dfrac{qE}{m}$ as

$a = \dfrac{(1.6 \times 10^{-19}\text{ C})(2.5 \times 10^4\text{ N/C})}{9.1 \times 10^{-31}\text{ kg}} = 4.4 \times 10^{15}$ m/s², or calling the direction of initial motion the positive direction, the acceleration is -4.4×10^{15} m/s².

(b) $v = v_0 + at$ becomes $0 = 4.0 \times 10^6$ m/s $+ (-4.4 \times 10^{15}$ m/s²$)t$
giving $t = 9.09 \times 10^{-10}$ s

(c) $x = v_0 t + \dfrac{1}{2}at^2$, becomes

$x = (4.0 \times 10^6$ m/s$)(9.09 \times 10^{-10}$ s$) + \dfrac{1}{2}(-4.4 \times 10^{15}$ m/s²$)(9.09 \times 10^{-10}$ s$)^2$

$= 1.82 \times 10^{-3}$ m

16.68 The electric field at any point x is

$E = \dfrac{kq}{(x-a)^2} - \dfrac{kq}{(x-(-a))^2} = \dfrac{kq(4ax)}{(x^2-a^2)^2}$ When x is much, much greater than a, we find

$E = \dfrac{(4a)(kq)}{x^3}$

16.69 (a) Let us sum force components to find
$\Sigma F_x = E_x q - T\sin\theta = 0$, and $\Sigma F_y = E_y q + T\cos\theta - mg = 0$ Combining these two equations, we get $q = \dfrac{mg}{(E_x \cot\theta + E_y)} = \dfrac{(0.001)(9.8)}{(3\cot 37° + 5)} = 1.09 \times 10^{-8}$ C

(b) From the two equations for ΣF_x and ΣF_y we also find

$T = \dfrac{E_x q}{\sin 37°} = 5.43 \times 10^{-3}$ N

16.70 By symmetry, we see that there is no component of electric field in the x direction. Thus, $E_y = 2\dfrac{kq}{r^2}\sin\theta$ where $r^2 = a^2 + y^2$

240

CHAPTER SIXTEEN SOLUTIONS

and $\sin\theta = \dfrac{y}{(a^2 + y^2)^{1/2}}$

From this, we have $E_y = \dfrac{2kqy}{(a^2 + y^2)^{3/2}}$

16.71 (a) $\phi = EA\cos\theta = EA = (6.24 \times 10^5 \text{ N/C})(3.2 \text{ m}^2) = 1.98 \times 10^6 \text{ N m}^2/\text{C}$
(b) $\phi = EA\cos\theta = EA\cos 90° = 0$

16.72 $E = \dfrac{\phi}{A} = \dfrac{5.2 \times 10^5 \text{ N m}^2/\text{C}}{\pi(0.2 \text{ m})^2} = 4.14 \times 10^6 \text{ N/C}$

16.73 $\phi = EA = (9 \times 10^9 \text{ N m}^2/\text{C}^2)\dfrac{5 \times 10^{-6} \text{ C}}{(0.12 \text{ m})^2} 4\pi(0.12 \text{ m})^2 = 5.65 \times 10^5 \text{ N m}^2/\text{C}$

16.74 (a) Outside the shell, the net charge enclosed in the gaussian surface is zero, $Q = q - q = 0$. Thus, $E = 0$. (b) $\Sigma EA \cos\theta = EA$ and $\dfrac{Q}{\varepsilon_0} = \dfrac{q}{\varepsilon_0}$, so $E = k\dfrac{q}{r^2}$

16.75 We use a gaussian surface in the shape of a cylinder with A as the area of each end cap. $\cos\theta = \cos 90° = 0$ on the cylindrical surface. $E = 0$ inside the conductor, so ϕ through the end cap inside the conductor is zero. For the endcap outside the conductor, $\Sigma EA \cos\theta = EA$ and $\dfrac{Q}{\varepsilon_0} = \dfrac{\sigma A}{\varepsilon_0}$ Thus, $E = \dfrac{\sigma}{\varepsilon_0}$

16.76 Construct a gaussian surface inside the conductor, where $E = 0$, and arbitrarily close to the surface. Since $E = 0$ inside, Gauss' law says $\dfrac{Q}{\varepsilon_0} = 0$ inside.

16.77 E inside the conductor $= 0$, and $\cos\theta = \cos 90° = 0$ on the cylindrical surface. On the endcap outside, $\Sigma EA \cos\theta = EA$ and $\dfrac{Q}{\varepsilon_0} = \dfrac{\sigma A}{\varepsilon_0}$. Thus, $E = \dfrac{\sigma}{\varepsilon_0}$

CHAPTER SEVENTEEN SOLUTIONS

17.1 (a) The force on the proton is
$$F = qE = (1.6 \times 10^{-19} \text{ C})(200 \text{ N/C}) = 3.2 \times 10^{-17} \text{ N}.$$
The work done is
$$W = Fs = (3.2 \times 10^{-17} \text{ N})(2 \times 10^{-2} \text{ m}) = 6.40 \times 10^{-19} \text{ J}$$
(b) $\Delta PE = -W = -6.40 \times 10^{-19}$ J

17.2 $1\dfrac{\text{volt}}{\text{meter}} = \dfrac{1 \text{ joule/coul}}{1 \text{ meter}} = \dfrac{1 \text{ joule}}{1 \text{ coul meter}} = \dfrac{1 \text{ N meter}}{1 \text{ coul meter}} = 1 \dfrac{\text{N}}{\text{C}}$

17.3 (a) We follow the path from (0,0) to (20 cm,0) to (20 cm,50 cm).
$\Delta PE = -$ (work done)
$= -$(work from origin to (20 cm,0)) $-$ (work from (20 cm,0) to (20 cm,50 cm))
The last term = 0 because the force is perpendicular to the displacement.
So, $\Delta PE = -(qE_x)(\Delta x) = -(12 \times 10^{-6} \text{ C})(250 \text{ V/m})(0.2 \text{ m}) = -6 \times 10^{-4}$ J

(b) $\Delta V = \dfrac{\Delta PE}{q} = -\dfrac{6 \times 10^{-4} \text{ J}}{12 \times 10^{-6} \text{ C}} = -50$ J/C $= -50$ V

17.4 $E = \dfrac{V}{d} = \dfrac{25 \times 10^3 \text{ J/C}}{1.5 \times 10^{-2} \text{ m}} = 1.67 \times 10^6$ N/C

17.5 $\Delta V = -Ed = -E(3 \times 10^{-4} \text{ m}) = -20$ V
so $E = 6.67 \times 10^4$ V/m

17.6 (a) $E = \dfrac{\Delta V}{d} = \dfrac{600 \text{ V}}{5.33 \times 10^{-3} \text{ m}} = 1.13 \times 10^5$ V/m
(b) $F = qE = (1.6 \times 10^{-19} \text{ C})(1.13 \times 10^5 \text{ V/m}) = 1.80 \times 10^{-14}$ N
(c) Work $= Fs = qEs = (1.80 \times 10^{-14} \text{ N})(5.33 \times 10^{-3} \text{ m} - 2.9 \times 10^{-3} \text{ m})$
$= 4.38 \times 10^{-17}$ J

17.7 $\Delta V = -16$ V and $\quad Q = -N_a e = -(6.02 \times 10^{23})(1.6 \times 10^{-19} \text{ C})$
$= -9.63 \times 10^4$ C
$\Delta V = W/Q$, so $\quad W = Q(\Delta V) = (-9.63 \times 10^4 \text{ C})(-16 \text{ J/C}) = 1.54 \times 10^6$ J

17.8 $W = q\Delta V = (1.6 \times 10^{-19} \text{ C})(90 \times 10^{-3} \text{ J/C}) = 1.44 \times 10^{-20}$ J

17.9 From the definition of potential difference,
$q = \dfrac{W}{\Delta V} = \dfrac{1.92 \times 10^{-17} \text{ J}}{60 \text{ V}} = 3.2 \times 10^{-19}$ C

17.10 (a) The strength of the electric field is
$E = \dfrac{V}{d} = \dfrac{5 \times 10^3 \text{ V}}{2 \times 10^{-2} \text{ m}} = 2.5 \times 10^5$ N/C

CHAPTER SEVENTEEN SOLUTIONS

and $F = qE = (1.6 \times 10^{-19} \text{ C})(2.5 \times 10^5 \text{ N/C}) = 4.0 \times 10^{-14}$ N

(b) $a = \dfrac{F}{m} = \dfrac{4 \times 10^{-14} \text{ N}}{1.67 \times 10^{-27} \text{ kg}} = 2.40 \times 10^{13}$ m/s^2.

17.11 $W = \Delta KE = q\Delta V$

$\dfrac{1}{2} mv^2 = e(120 \text{ V}) = 1.92 \times 10^{-17}$ J

Thus, $v = \sqrt{\dfrac{3.84 \times 10^{-17} \text{ J}}{m}}$

(a) For a proton, this becomes

$v = \sqrt{\dfrac{3.84 \times 10^{-17} \text{ J}}{1.67 \times 10^{-27} \text{ kg}}} = 1.52 \times 10^5$ m/s

(b) If an electron,

$v = \sqrt{\dfrac{3.84 \times 10^{-17} \text{ J}}{9.11 \times 10^{-31} \text{ kg}}} = 6.49 \times 10^6$ m/s

---

17.12 $W = \Delta KE = -q\Delta V$

$0 - \dfrac{1}{2} (9.11 \times 10^{-31} \text{ kg})(4.2 \times 10^5 \text{ m/s})^2 = (-1.6 \times 10^{-19} \text{ C})\Delta V$

From which, $\Delta V = -0.502$ V

17.13 Sixty percent of the speed of light (3×10^8 m/s) is 1.8×10^8 m/s. From conservation of energy, we have

$\dfrac{1}{2} mv^2 = qV$, or $V = \dfrac{mv^2}{2q}$

(a) For an electron, $|V| = \dfrac{mv^2}{2q} = \dfrac{(9.11 \times 10^{-31})(1.8 \times 10^8 \text{ m/s})^2}{2(1.6 \times 10^{-19} \text{ C})}$

$= 9.22 \times 10^4$ V

(b) For a proton, $|V| = \dfrac{mv^2}{2q} = \dfrac{(1.67 \times 10^{-27})(1.8 \times 10^8 \text{ m/s})^2}{2(1.6 \times 10^{-19} \text{ C})}$

$= 1.69 \times 10^8$ V

17.14 Conservation of energy yields $\dfrac{1}{2} mv^2 = qV$, or

$v = \sqrt{\dfrac{2qV}{m}}$

(a) For an electron, $v = \sqrt{\dfrac{2qV}{m}} = \sqrt{\dfrac{2(1.6 \times 10^{-19} \text{ C})(2000 \text{ V})}{9.11 \times 10^{-31} \text{ kg}}} = 2.65 \times 10^7$ m/s

(b) For a proton, $v = \sqrt{\dfrac{2qV}{m}} = \sqrt{\dfrac{2(1.6 \times 10^{-19} \text{ C})(2000 \text{ V})}{1.67 \times 10^{-27} \text{ kg}}} = 6.19 \times 10^5$ m/s

---

17.15 $V = k\dfrac{q}{r}$, or $r = \dfrac{kq}{V} = \dfrac{(9 \times 10^9 \text{ N m}^2/\text{C}^2)(6 \times 10^{-6} \text{ C})}{2.7 \times 10^4 \text{ N m/C}} = 2.0$ m

---

17.16 (a) The potential at 1 cm is

CHAPTER SEVENTEEN SOLUTIONS

$$V_1 = k\frac{q}{r} = \frac{(9 \times 10^9 \text{ N m}^2/\text{C}^2)(1.6 \times 10^{-19} \text{ C})}{10^{-2} \text{ m}} = 1.44 \times 10^{-7} \text{ V}$$

(b) The potential at 2 cm is

$$V_2 = k\frac{q}{r} = \frac{(9 \times 10^9 \text{ N m}^2/\text{C}^2)(1.6 \times 10^{-19} \text{ C})}{2 \times 10^{-2} \text{ m}} = 0.72 \times 10^{-7} \text{ V}$$

Thus, the difference in potential between the two points is

$$\Delta V = V_2 - V_1 = -7.2 \times 10^{-8} \text{ V}$$

(c) The approach is the same as above except the charge is -1.6×10^{-19} C. This changes the sign of all the answers, with the magnitudes remaining the same. That is, the potential at 1 cm is -1.44×10^{-7} V and the potential at 2 cm is -0.72×10^{-7} V, so $\Delta V = V_2 - V_1 = 7.2 \times 10^{-8}$ V

17.17 $V = k\dfrac{q}{r} = \dfrac{(9 \times 10^9 \text{ N m}^2/\text{C}^2)(1.6 \times 10^{-19} \text{ C})}{0.51 \times 10^{-10} \text{ m}} = 28.2$ V

17.18 The net potential is the algebraic sum of the potentials set up by the individual charges.

Thus, $V = k\left(\dfrac{q_1}{r_1} + \dfrac{q_1}{r_1}\right)$

(a) At $y = 0.6$ m, $r_1 = 0.6$ m and $r_2 = 0.3$ m. We have

$$V = (9 \times 10^9 \text{ N m}^2/\text{C}^2)\left(\frac{3 \times 10^{-9} \text{ C}}{0.6 \text{ m}} + \frac{6 \times 10^{-9} \text{ C}}{0.3 \text{ m}}\right) = 225 \text{ V}$$

(b) At $y = -0.6$ m, $r_1 = 0.6$ m and $r_2 = 0.9$ m, and

$$V = (9 \times 10^9 \text{ N m}^2/\text{C}^2)\left(\frac{3 \times 10^{-9} \text{ C}}{0.6 \text{ m}} + \frac{6 \times 10^{-9} \text{ C}}{0.9 \text{ m}}\right) = 105 \text{ V}$$

(c) If $q_2 = -6 \times 10^{-9}$ C and $y = 0.6$ m, we have

$$V = (9 \times 10^9 \text{ N m}^2/\text{C}^2)\left(\frac{3 \times 10^{-9} \text{ C}}{0.6 \text{ m}} - \frac{6 \times 10^{-9} \text{ C}}{0.3 \text{ m}}\right) = -135 \text{ V}$$

If $q_2 = -6 \times 10^{-9}$ C and $y = -0.6$ m, we have

$$V = (9 \times 10^9 \text{ N m}^2/\text{C}^2)\left(\frac{3 \times 10^{-9} \text{ C}}{0.6 \text{ m}} - \frac{6 \times 10^{-9} \text{ C}}{0.9 \text{ m}}\right) = -15 \text{ V}$$

17.19 (a) The distance, r_1, from the -4×10^{-9} C charge to $(0,-0.45)$ is found from the Pythagorean theorem as

$$r_1 = \sqrt{(0.45)^2 + (0.45)^2} = 0.636 \text{ m}$$

Thus, at the first point,

$$V = (9 \times 10^9 \text{ N m}^2/\text{C}^2)\left(\frac{6 \times 10^{-9} \text{ C}}{0.6 \text{ m}} - \frac{4 \times 10^{-9} \text{ C}}{0.636 \text{ m}}\right) = 33.4 \text{ V}$$

(b) The distance, r_2, from the 6×10^{-9} C charge to $(0.15, 0)$ is found as

$$r_2 = \sqrt{(0.15)^2 + (0.15)^2} = 0.212 \text{ m}$$

and,

$$V = (9 \times 10^9 \text{ N m}^2/\text{C}^2)\left(\frac{6 \times 10^{-9} \text{ C}}{0.212 \text{ m}} - \frac{4 \times 10^{-9} \text{ C}}{0.6 \text{ m}}\right) = 195 \text{ V}$$

17.20 First, consider an arbitrary point P on the x axis.

CHAPTER SEVENTEEN SOLUTIONS

$$V = k\left(\frac{-3 \times 10^{-9} \text{ C}}{|x|} + \frac{8 \times 10^{-9} \text{ C}}{|2 \text{ m} - x|}\right) = 0 \quad (1)$$

By inspection of this equation, we see that the desired point must be either between the two charges or it must be located to the left of the origin ($x < 0$).
(1) reduces to

$$\frac{3}{|x|} = \frac{8}{|2 \text{ m} - x|}, \text{ or } \quad 3|2 \text{ m} - x| = 8|x| \quad (2)$$

First, assume $x > 0$. Then, (2) becomes
$$3(2 \text{ m} - x) = 8x$$
From which, $x = 0.546$ m
If there is a solution for $x < 0$, then $|x| = -x$, and $|2 \text{ m} - x| = 2 \text{ m} - x$
Thus, $3(2m - x) = -8x$. From which, $x = -1.2$ m

17.21 $W = q'\Delta V = q'(V_f - V_i) = q'\left(k\frac{q}{r_f} - k\frac{q}{r_i}\right)$, but $k\frac{q}{r_i} = 0$ (r_i = infinity).

Thus,

$$W = k\frac{q'q}{r_f} = (9 \times 10^9 \text{ N m}^2/\text{C}^2)\frac{(3 \times 10^{-9} \text{ C})(9 \times 10^{-9} \text{ C})}{0.3 \text{ m}} = 8.10 \times 10^{-7} \text{ J}$$

17.22 (a) $V = (9 \times 10^9 \text{ N m}^2/\text{C}^2)\left(\frac{5 \times 10^{-9} \text{ C}}{0.175 \text{ m}} + \frac{-3 \times 10^{-9} \text{ C}}{0.175 \text{ m}}\right) = 103$ V

(b) $PE = k\frac{q_1 q_2}{r_{12}} = \frac{(9 \times 10^9 \text{ N m}^2/\text{C}^2)(5 \times 10^{-9} \text{ C})(-3 \times 10^{-9} \text{ C})}{0.35 \text{ m}}$

$= -3.86 \times 10^{-7}$ J

Positive work must be done on them to separate the charges.

17.23 Work done on 8 µC charge = $\Delta PE = PE_f - PE_i = 0 - PE_i$
(The final potential energy = 0 because at the end, the charge is at an infinite distance from all other charges.)
$= -(PE_{\text{due to presence of 2 µC charge}} + PE_{\text{due to presence of 4 µC charge}})$

$$= -\left(\frac{(9 \times 10^9 \text{ N m}^2/\text{C}^2)(8 \times 10^{-6} \text{ C})(2 \times 10^{-6} \text{ C})}{3 \times 10^{-2} \text{ m}}\right.$$

$$\left. + \frac{(9 \times 10^9 \text{ N m}^2/\text{C}^2)(8 \times 10^{-6} \text{ C})(4 \times 10^{-6} \text{ C})}{\sqrt{(0.03 \text{ m})^2 + (0.06 \text{ m})^2}}\right) = -9.09 \text{ J}$$

17.24 $W = q\Delta V = q(V_f - V_i)$

$= (-1.6 \times 10^{-19} \text{ C})(9 \times 10^9 \text{ N m}^2/\text{C}^2)\left(\frac{1.6 \times 10^{-19} \text{ C}}{1 \text{ m}} - \frac{1.6 \times 10^{-19} \text{ C}}{0.5 \text{ m}}\right)$

$= 2.30 \times 10^{-28}$ J

17.25 (a) 1 eV = 1.6×10^{-19} J. Thus, this is the kinetic energy of the electron. We have,

$$KE = \frac{1}{2}mv^2 = \frac{1}{2}(9.11 \times 10^{-31} \text{ kg})v^2 = 1.6 \times 10^{-19} \text{ J}.$$

From which, $v = 5.93 \times 10^5$ m/s
(b) If the particle is a proton, we have

$$KE = \frac{1}{2}mv^2 = \frac{1}{2}(1.67 \times 10^{-27} \text{ kg})v^2 = 1.6 \times 10^{-19} \text{ J}.$$

CHAPTER SEVENTEEN SOLUTIONS

and $v = 1.38 \times 10^4$ m/s

17.26 (a) From conservation of energy, we have
$$KE = qV = (1.6 \times 10^{-19} \text{ C})(2.5 \times 10^4 \text{ V}) = 4 \times 10^{-15} \text{ J}$$
The conversion factor, 1 eV = 1.6×10^{-19} J, gives us
$$KE = 4 \times 10^{-15} \text{ J} = 2.5 \times 10^4 \text{ eV (an answer that can be found by inspection)}$$
(b) $\frac{1}{2} mv^2 = \frac{1}{2}(1.67 \times 10^{-27} \text{ kg})v^2 = 4 \times 10^{-15}$ J.
From which, $v = 2.19 \times 10^6$ m/s

17.27 $\Delta PE = qV = (75 \times 10^{-6} \text{ C})(90 \text{ J/C}) = 6.75 \times 10^{-3}$ J $= 4.22 \times 10^{16}$ eV

17.28 $C = \frac{\varepsilon_0 A}{d}$ gives $A = \frac{Cd}{\varepsilon_0} = \frac{(1 \text{ F})(10^{-3} \text{ m})}{(8.85 \times 10^{-11} \text{ F/m})} = 1.13 \times 10^8$ m^2 = 43.6 sq miles

---

17.29 (a) $Q = CV = (4 \times 10^{-6} \text{ F})(12 \text{ V}) = 4.8 \times 10^{-5}$ C = 48 µC
(b) $Q = CV = (4 \times 10^{-6} \text{ F})(1.5 \text{ V}) = 6 \times 10^{-6}$ C = 6 µC

17.30 $V = \frac{Q}{C} = \frac{98 \text{ µC}}{2 \text{µF}} = 49$ V

17.31 $C = \frac{\varepsilon_0 A}{d}$ gives $A = \frac{Cd}{\varepsilon_0} = \frac{(2 \times 10^{-12} \text{ F})(1 \times 10^{-4} \text{ m})}{8.85 \times 10^{-12} \text{ F/m}}$
$= 2.26 \times 10^{-5}$ m^2.

17.32 (a) If d is doubled while A and Q remain constant, then when d doubles, C is reduced by a factor of 2. Thus, if Q remains constant, then
$$V_2 = \frac{Q_2}{C_2} = \frac{Q_1}{\frac{1}{2}C_1} = 2\frac{Q_1}{C_1}$$
or $V_2 = 2 V_1 = 800$ V
(b) If d is doubled while V remains constant, then, doubling d cuts the capacitance in half. Thus,
$$Q_2 = C_2 V_2 = \frac{1}{2} C_1 (V_1) = \frac{1}{2} C_1 V_1 = \frac{1}{2} Q_1$$
or, the charge must be halved.

17.33 $C = \frac{\varepsilon_0 A}{d} = \frac{(8.85 \times 10^{-12} \text{ F/m})(2 \times 10^{-4} \text{ m}^2)}{2 \times 10^{-3} \text{ m}} = 8.85 \times 10^{-13}$ F
and $Q = CV = (8.85 \times 10^{-13} \text{ F})(6 \text{ V}) = 5.31 \times 10^{-12}$ C

17.34 $C = \frac{\varepsilon_0 A}{d}$, so if $A_2 = 2A_1$ and $d_2 = \frac{1}{2} d_1$, then
$$C_2 = \frac{\varepsilon_0 A_2}{d_2} = \frac{\varepsilon_0 2A_1}{\frac{1}{2}d_1} = 4\frac{\varepsilon_0 A_1}{d_1}$$

CHAPTER SEVENTEEN SOLUTIONS

or $C_2 = 4C_1$ (The capacitance is quadrupled.)

17.35 (a) $C = \dfrac{\varepsilon_0 A}{d} = \dfrac{(8.85 \times 10^{-12} \text{ F/m})(5 \times 10^{-4} \text{ m}^2)}{1 \times 10^{-3} \text{ m}} = 4.43 \times 10^{-12}$ F

$V = \dfrac{Q}{C} = \dfrac{400 \times 10^{-12} \text{ C}}{4.43 \times 10^{-12} \text{ F}} = 90.4$ V

(b) $E = \dfrac{V}{d} = \dfrac{90.4 \text{ V}}{10^{-3} \text{ m}} = 9.04 \times 10^4$ V/m

---

17.36 (a) $C = \dfrac{\varepsilon_0 A}{d} = \dfrac{(8.85 \times 10^{-12} \text{ F/m})(10^6 \text{ m}^2)}{800 \text{ m}} = 1.11 \times 10^{-8}$ F $= 111$ μF

(b) $\Delta V_{max} = E_{max} d = (2 \times 10^6 \text{ V/m})(800 \text{ m}) = 1.6 \times 10^9$ V

and $Q_{max} = CV_{max} = (1.11 \times 10^{-8} \text{ F})(1.6 \times 10^9 \text{ V}) = 17.8$ C

17.37 Capacitors in parallel add. Thus, the equivalent capacitor has a value of

$C_{eq} = C_1 + C_2 = 5$ μF $+ 12$ μF $= 17$ μF

17.38 (a) The potential difference across each branch is the same and equal to the voltage of the battery.
$V = 9$ V

(b) $Q_5 = CV = (5$ μF$)(9$ V$) = 45$ μC, and $Q_{12} = CV = (12$ μF$)(9$ V$) = 108$ μC

17.39 (a) In series Capacitors add as

$\dfrac{1}{C_{eq}} = \dfrac{1}{C_1} + \dfrac{1}{C_2} = \dfrac{1}{5 \text{ μF}} + \dfrac{1}{12 \text{ μF}}$

and $C_{eq} = 3.53$ μF

The charge on the equivalent capacitor is,

$Q_{eq} = C_{eq} V = (3.53$ μF$)(9$ V$) = 31.8$ μC

Each of the series capacitors have this same charge on them.

$Q_1 = Q_2 = 31.8$ μC

The voltage across each is

$V_1 = \dfrac{Q_1}{C_1} = \dfrac{31.8 \text{ μC}}{5 \text{ μF}} = 6.35$ V, and $V_2 = \dfrac{Q_2}{C_2} = \dfrac{31.8 \text{ μC}}{12 \text{ μF}} = 2.65$ V

17.40 (a) $\dfrac{1}{C_{eq}} = \dfrac{1}{0.05 \text{ μF}} + \dfrac{1}{0.01 \text{ μF}}$

and $C_{eq} = 0.033$ μF

Thus, $Q = C_{eq} V = (0.033$ μF$)(400$ V$) = 13.3$ μC on each capacitor.

(b) The voltage is the same across both capacitors and equal to 400 V

Thus, $Q_1 = C_1 V = (0.05$ μF$)(400$ V$) = 20$ μC

$Q_2 = C_2 V = (0.1$ μF$)(400$ V$) = 40$ μC

17.41 (a) $C_{eq} = C_1 + C_2 + C_3 = 5$ μF $+ 4$ μF $+ 9$ μF $= 18$ μF

(b) $\dfrac{1}{C_{eq}} = \dfrac{1}{C_1} + \dfrac{1}{C_2} + \dfrac{1}{C_3} = \dfrac{1}{5 \text{ μF}} + \dfrac{1}{4 \text{ μF}} + \dfrac{1}{9 \text{ μF}}$

CHAPTER SEVENTEEN SOLUTIONS

Which gives, $C_{eq} = 1.78 \; \mu F$

17.42 (a) We let $C_1 = 5 \; \mu F$, $C_2 = 4 \; \mu F$, and $C_3 = 9 \; \mu F$.
The potential difference across each capacitor in parallel is the same, and in this case, equal to the voltage of the battery. Thus,

$$V_1 = V_2 = V_3 = 12 \; V$$

$$Q_1 = C_1 V_1 = (5 \; \mu F)(12 \; V) = 60 \; \mu C$$

$$Q_2 = C_2 V_2 = (4 \; \mu F)(12 \; V) = 48 \; \mu C, \text{ and } Q_3 = C_3 V_3 = (9 \; \mu F)(12 \; V) = 108 \; \mu C$$

(b) When connected in series, the charge is the same on each and equal to the charge on the equivalent capacitance. Thus,

$$Q_1 = Q_2 = Q_3 = C_{eq}V = (1.78 \; \mu F)(12 \; V) = 21.4 \; \mu C$$

The voltage across each is found as

$$V_1 = \frac{Q_1}{C_1} = \frac{21.4 \; \mu C}{5 \; \mu F} = 4.28 \; V, \qquad V_2 = \frac{Q_2}{C_2} = \frac{21.4 \; \mu C}{4 \; \mu F} = 5.35 \; V,$$

$$V_3 = \frac{Q_3}{C_3} = \frac{21.4 \; \mu C}{9 \; \mu F} = 2.38 \; V$$

17.43 We reduce the circuit in steps as shown at the right. Using the equivalent circuit, we find

$Q_{total} = CV = (4 \; \mu F)(24 \; V)$
$\qquad = 96 \; \mu C$

Then in the second circuit,

$$V_{AB} = \frac{Q_{total}}{C_{AB}} = \frac{96 \; \mu C}{6 \; \mu F} = 16 \; V$$

$$V_{BC} = \frac{Q_{total}}{C_{BC}} = \frac{96 \; \mu C}{12 \; \mu F} = 8 \; V$$

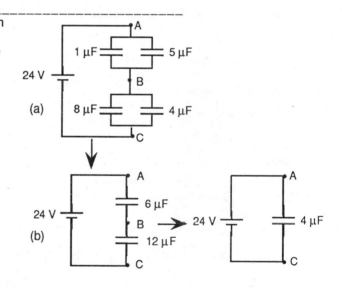

Finally, using the first circuit above,

$Q_1 = C_1 V_{AB} = (1 \; \mu F)(16 \; V) =$
$16 \; \mu C$

$Q_5 = C_5 V_{AB} = (5 \; \mu F)(16 \; V) = 80 \; \mu C$

$Q_8 = C_8 V_{BC} = (8 \; \mu F)(8 \; V) = 64 \; \mu C$

$Q_4 = C_4 V_{BC} = (4 \; \mu F)(8 \; V) = 32 \; \mu C$

17.44 (a) Using the rules for combining capacitors in series and in parallel, the circuit is reduced in steps as shown below. The equivalent capacitor is shown to be a 2 μF capacitor.

248

CHAPTER SEVENTEEN SOLUTIONS

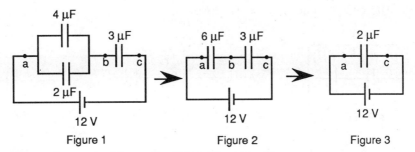

Figure 1   Figure 2   Figure 3

(b) From Fig. 3, $Q_{ac} = C_{ac}V_{ac} = (2\ \mu F)(12\ V) = 24\ \mu C$
From Fig. 2,
$$Q_{ab} = Q_{bc} = Q_{ac} = 24\ \mu C$$
Thus, the charge on the 3 μF capacitor is 24 μC.
We now find the potential differences across the points indicated in Fig. 2.

$$V_{ab} = \frac{Q_{ab}}{C_{ab}} = \frac{24\ \mu C}{6\ \mu F} = 4\ V,\ \text{and}\qquad V_{bc} = \frac{Q_{bc}}{C_{bc}} = \frac{24\ \mu C}{3\ \mu F} = 8\ V$$

Thus, the potential difference across the 2 μF and the 4 μF is 4 V, and the potential difference across the 3 μF = 8V.

Finally, using Fig. 1, the charge on the 4 μF and the 2 μF is found as follows.

$$Q_4 = C_4 V_{ab} = (4\ \mu F)(4\ V) = 16\ \mu C,\ \text{and}\ Q_2 = C_2 V_{ab} = (2\ \mu F)(4\ V) = 8\ \mu C$$

17.45 The circuit reduces as shown below.

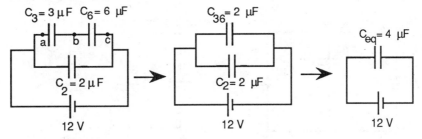

(b) $Q_2 = C_2 V_2 = (2\ \mu F)(12\ V) = 24\ \mu C$,
$Q_{36} = C_{36}V_2 = (2\ \mu F)(12\ V) = 24\ \mu C = Q_3 = Q_6$

Thus, $V_3 = \dfrac{Q_3}{C_3} = \dfrac{24\ \mu C}{3\ \mu F} = 8\ V$,

$V_6 = \dfrac{Q_6}{C_6} = \dfrac{24\ \mu C}{6\ \mu F} = 4\ V$

CHAPTER SEVENTEEN SOLUTIONS

17.46 The circuit reduces as shown below.

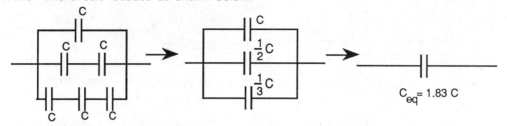

17.47 There are 7 possible distinct values.
(1) using 1 capacitor   resultant = C
(2) using 2 capacitors, the resultant = $\frac{1}{2}$C when they are in series, and 2C when in parallel.
(3) Using 3 capacitors, the resultant = $\frac{C}{3}$ when all three are in series, or 3C when all three are in parallel. Two in parallel with one in series yields $\frac{2}{3}$C, or two in series which is in turn in parallel with one capacitor yield $\frac{3}{2}$C.

17.48 (a) All four should be connected in parallel.
(b) Two in parallel followed by another group of two in parallel.
or, two in series which are in parallel with another group of two in series.
(c) One in series with a group of three in parallel.
(d) All four in series.

17.49 (a) The capacitors combine following the steps of the figures below. $C_{eq} = 9$ μF

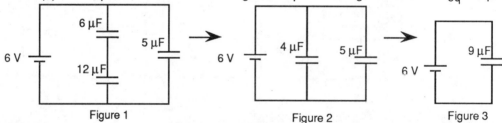

Figure 1   Figure 2   Figure 3

(b) In Fig. 2, $V_4 = V_5 = 6$ V
Thus, $Q_4 = (4\ \mu F)(6\ V) = 24\ \mu C$, and $Q_5 = (5\ \mu F)(6\ V) = 30\ \mu C$
From Fig. 1, $Q_6 = Q_{12} = Q_4 = 24\ \mu C$

Therefore, $V_6 = \dfrac{Q_6}{C_6} = \dfrac{24\ \mu C}{6\ \mu F} = 4$ V, and $V_{12} = \dfrac{Q_{12}}{C_{12}} = \dfrac{24\ \mu C}{12\ \mu F} = 2$ V

CHAPTER SEVENTEEN SOLUTIONS

17.50 The combination reduces to an equivalent capacitance of 12 µF in stages shown below.

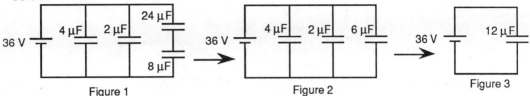

Figure 1   Figure 2   Figure 3

(b) From Fig. 2, Q_4 = (4 µF)(36 V) = 144 µC, Q_2 = (2 µF)(36 V) = 72 µC, and Q_6 = (6 µF)(36 V) = 216 µC

Then, from Fig. 1 $Q_{24} = Q_8 = Q_6$ = 216 µC

Finally, $V_{24} = \dfrac{Q_{24}}{C_{24}} = \dfrac{216 \text{ µC}}{24 \text{ µF}}$ = 9 V, and $V_8 = \dfrac{Q_8}{C_8} = \dfrac{216 \text{ µC}}{8 \text{ µF}}$ = 27 V

17.51 When the capacitor is connected across the battery, it receives a charge of
$Q = CV$ = (1 µF)(10 V) = 10 µC

When it is connected across the 2 µF, the charges move about until Q_1 is on the 1 µF and Q_2 is on the 2 µF. But, we know that

$Q_1 + Q_2$ = 10 µC   (1)

Also, since they are in parallel, the voltage across each must be the same, so

$V_1 = V_2$

or $\dfrac{Q_1}{C_1} = \dfrac{Q_2}{C_2} = \dfrac{Q_1}{1 \text{ µF}} = \dfrac{Q_2}{2 \text{ µF}}$

yielding $Q_2 = 2Q_1$ (2)

Solving (1) and (2) simultaneously, we find $Q_1 = \dfrac{10}{3}$ µC   and $Q_2 = \dfrac{20}{3}$ µC.

17.52 (a) $W = \dfrac{1}{2}CV^2 = \dfrac{1}{2}$ (3 µF)(12 V)^2 = 2.16 × 10^{-4} J

(b) $W = \dfrac{1}{2}CV^2 = \dfrac{1}{2}$ (3 µF)(6 V)^2 = 5.4 × 10^{-5} J

17.53 (a) $C = \dfrac{Q}{V} = \dfrac{36 \text{ µC}}{120 \text{ V}}$ = 0.3 µF

(b) $W = \dfrac{1}{2}QV = \dfrac{1}{2}$ (36 µC)(120 V) = 2.16 × 10^{-3} J

or, $W = \dfrac{1}{2}CV^2 = \dfrac{1}{2}$ (0.3 µF)(120 V)^2 = 2.16 × 10^{-3} J

or, $W = \dfrac{1}{2}\dfrac{Q^2}{C} = \dfrac{1}{2}\dfrac{(36 \text{ µC})^2}{(0.3 \text{ µF})}$ = 2.16 × 10^{-3} J

17.54 Use $W = \dfrac{1}{2}\dfrac{Q^2}{C}$ and $C = \dfrac{\varepsilon_0 A}{d}$

If $d_2 = 2d_1$, $C_2 = \dfrac{1}{2}C_1$

251

CHAPTER SEVENTEEN SOLUTIONS

so, $W_2 = \dfrac{Q^2}{2(\frac{1}{2}C_1)} = 2W_1$ (The stored energy doubles.)

17.55 $C = \dfrac{\varepsilon_0 A}{d} = \dfrac{(8.85 \times 10^{-12} \text{ F/m})(2 \times 10^{-4} \text{ m}^2)}{5 \times 10^{-3} \text{ m}} = 3.54 \times 10^{-13}$ F

and, $W = \dfrac{1}{2}CV^2 = \dfrac{1}{2}(3.54 \times 10^{-13} \text{ F})(12 \text{ V})^2 = 2.55 \times 10^{-11}$ J

17.56

$V_4 = 4$ V, $C_4 = 4$ μF; $W = \dfrac{1}{2}CV^2 = \dfrac{1}{2}(4 \times 10^{-6} \text{ F})(4 \text{ V})^2 = 3.2 \times 10^{-5}$ J

$V_2 = 4$ V, $C_2 = 2$ μF; $W = \dfrac{1}{2}CV^2 = \dfrac{1}{2}(2 \times 10^{-6} \text{ F})(4 \text{ V})^2 = 1.6 \times 10^{-5}$ J

$V_3 = 8$ V, $C_3 = 3$ μF; $W = \dfrac{1}{2}CV^2 = \dfrac{1}{2}(3 \times 10^{-6} \text{ F})(8 \text{ V})^2 = 9.6 \times 10^{-5}$ J

17.57 $C = \dfrac{\kappa \varepsilon_0 A}{d} = \dfrac{4.9(8.85 \times 10^{-12} \text{ F/m})(5 \times 10^{-4} \text{ m}^2)}{2 \times 10^{-3} \text{ m}} = 1.08 \times 10^{-11}$ F = 10.8 pF

17.58 (a) With air between the plates, we find
$C_0 = \dfrac{Q}{V} = \dfrac{48 \text{ μC}}{12 \text{ V}} = 4$ μF

(b) When teflon is inserted, the charge remains the same (48 μC) because the plates are isolated. However, the capacitance, and hence, the voltage changes. The new capacitance is

$C' = \kappa C_0 = 2.1(4 \text{ μF}) = 8.4$ μF

The voltage on the capacitor now is $V' = \dfrac{Q}{C'} = \dfrac{48 \text{ μC}}{8.4 \text{ μF}} = 5.71$ V

(c) The new value of the capacitance has been found in part (b).

17.59 The energy stored is given by $W = \dfrac{1}{2}CV^2$. The potential difference across the capacitor does not change (It remains connected to the 12 V battery.). Thus, the energy increases by the same factor as does the capacitance, a factor equal to the dielectric constant of the material between the plates. For nylon, this is $\kappa = 3.4$.

17.60 Initially, the potential difference across the capacitor is $V_0 = \dfrac{Q}{C_0}$

After removal of the capacitor from the circuit and insertion of the glass,
$V' = \dfrac{Q}{C'}$

The charge remains unchanged. Thus,

$\dfrac{V_0}{V'} = \dfrac{C'}{C_0} = \kappa$

Thus, $\kappa = \dfrac{100 \text{ V}}{25 \text{ V}} = 4.00$

CHAPTER SEVENTEEN SOLUTIONS

17.61 (a) $C = \dfrac{\kappa \varepsilon_0 A}{d} = \dfrac{2.1(8.85 \times 10^{-12} \text{ F/m})(1.75 \times 10^{-2} \text{ m}^2)}{4 \times 10^{-5} \text{ m}}$

$= 8.13 \times 10^{-9}$ F $= 8.13$ nF

(b) $V_{max} = E_{max} d = (60 \times 10^6 \text{ V/m})(4 \times 10^{-5} \text{ m}) = 2.4$ kV

17.62 $Q_{max} = C V_{max}$, but $V_{max} = E_{max} d$. Also, $C = \dfrac{\kappa \varepsilon_0 A}{d}$. Thus,

$Q_{max} = \dfrac{\kappa \varepsilon_0 A}{d}(E_{max} d) = \kappa \varepsilon_0 A E_{max}$

(a) With air between the plates, $\kappa = 1$ and $E_{max} = 3 \times 10^6$ V/m.
Therefore,
$Q_{max} = \kappa \varepsilon_0 A E_{max} = (8.85 \times 10^{-12} \text{ F/m})(5 \times 10^{-4} \text{ m}^2)(3 \times 10^6 \text{ V/m})$
$= 1.33 \times 10^{-8}$ C.

(b) With polystyrene between the plates $\kappa = 2.56$ and $E_{max} = 24 \times 10^6$ V/m.
$Q_{max} = \kappa \varepsilon_0 A E_{max} = 2.56(8.85 \times 10^{-12} \text{ F/m})(5 \times 10^{-4} \text{ m}^2)(24 \times 10^6 \text{ V/m})$
$= 2.72 \times 10^{-7}$ C.

17.63 $W = \Delta KE = Fs = F(10^{-2} \text{ m}) = 9 \times 10^{-18}$ J
Thus, $F = 9 \times 10^{-16}$ N

and $E = \dfrac{F}{q} = \dfrac{9 \times 10^{-16} \text{ N}}{1.6 \times 10^{-19} \text{ C}} = 5.63 \times 10^3$ N/C

17.64 (a) We have $E = \dfrac{kq}{r^2}$ and $V = \dfrac{kq}{r}$. Thus, $\dfrac{V}{E} = r = \dfrac{600 \text{ V}}{200 \text{ V/m}} = 3$ m

(b) Then from $V = \dfrac{kq}{r}$, we have $q = \dfrac{Vr}{k} = \dfrac{600 \text{ V}(3\text{ m})}{9 \times 10^9 \text{ N m}^2/\text{C}^2} = 2 \times 10^{-7}$ C

17.65 The distance across the diagonal (from the Pythagorean theorem) is 0.403 m.

$V = k\Sigma\dfrac{q}{r} = (9 \times 10^9 \text{ N m}^2/\text{C}^2)\left(\dfrac{8 \times 10^{-6} \text{ C}}{0.2 \text{ m}} + \dfrac{12 \times 10^{-6} \text{ C}}{0.403 \text{ m}} - \dfrac{8 \times 10^{-6} \text{ C}}{0.35 \text{ m}}\right)$

$= 4.22 \times 10^5$ V

17.66 The distance from the charge at the apex of the triangle to the mid-point of the base is found from the Pythagorean theorem to be 3.87 cm. We have

$V = k\Sigma\dfrac{q}{r} = (9 \times 10^9 \text{ N m}^2/\text{C}^2)\left(\dfrac{-5 \times 10^{-9} \text{ C}}{0.01 \text{ m}} - \dfrac{5 \times 10^{-6} \text{ C}}{0.01 \text{ m}} + \dfrac{5 \times 10^{-9} \text{ C}}{3.87 \times 10^{-2} \text{ m}}\right)$

$= -7.84 \times 10^3$ V

CHAPTER SEVENTEEN SOLUTIONS

17.67 The stages for the reduction of this circuit are shown below.

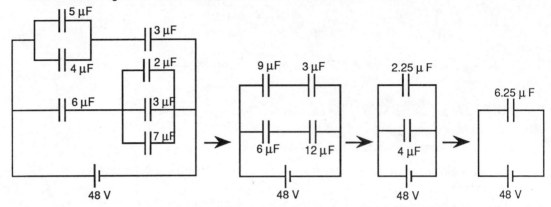

17.68 The stages for the reduction of this circuit are shown below.

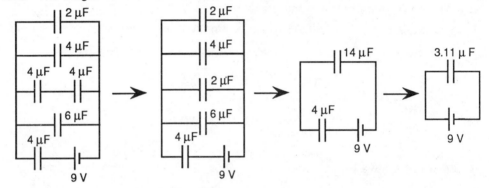

17.69 The capacitors reduce according to the stages below to a resultant of $\frac{4}{3}$ μF.

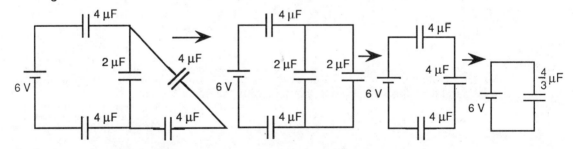

17.70 The charge stored on the equivalent capacitance is

$$Q = C_{eq}V = (\tfrac{4}{3} \times 10^{-6} \text{ F})(6 \text{ V}) = 8 \times 10^{-6} \text{ C}$$

Thus, $W_{total} = \tfrac{1}{2}QV = \tfrac{1}{2}(8 \times 10^{-6} \text{ C})(6 \text{ V}) = 2.4 \times 10^{-5}$ J

CHAPTER SEVENTEEN SOLUTIONS

17.71 We use the Nernst potential,
$$V_N = \frac{kT}{q} \ln\frac{c_i}{c_0} = \frac{(1.38 \times 10^{-23} \text{ J/K})(310 \text{ K})}{1.6 \times 10^{-19} \text{ C}} \ln\frac{0.138}{0.0045} = 9.15 \times 10^{-2} \text{ V}$$
$$= 91.5 \text{ mV}$$

17.72 We use $C = \frac{\kappa\varepsilon_0 A}{d}$ with $A = L(0.05 \text{ m})$. This gives
$$L = \frac{Cd}{\kappa\varepsilon_0(0.05)} = \frac{(2 \times 10^{-8})(2 \times 10^{-5})}{3\varepsilon_0(0.05)} = 0.301 \text{ m}$$

17.73 $\frac{Q_1}{Q_2} = \frac{C_1 V}{C_2 V} = \frac{\frac{\varepsilon_0 A}{d}}{\frac{\kappa\varepsilon_0 A}{d}} = \frac{1}{\kappa}$ so, $\kappa = \frac{Q_2}{Q_1} = \frac{150 \text{ }\mu\text{C} + 200 \text{ }\mu\text{C}}{150 \text{ }\mu\text{C}} = 2.33$

17.74 (a) From $V = Ed$, we find $d = \frac{V}{E} = \frac{6000 \text{ V}}{E}$

If $E_{max} = 14 \times 10^6$ V/m for pyrex glass, the minimum plate separation is
$$d_{min} = \frac{6000 \text{ V}}{E_{max}} = \frac{6000 \text{ V}}{14 \times 10^6 \text{ V/m}} = 4.29 \times 10^{-4} \text{ m}$$

Thus, for a given capacitance, the minimum plate area is given by
$$A_{min} = \frac{Cd_{min}}{\kappa\varepsilon_0} = \frac{(2 \times 10^{-7} \text{ F})(4.29 \times 10^{-4} \text{ m})}{(5.6)(8.85 \times 10^{-12} \text{ F/m})} = 1.73 \text{ m}^2.$$

(b) $W = \frac{1}{2} CV^2 = \frac{1}{2}(2 \times 10^{-7} \text{ F})(6000 \text{ V})^2 = 3.6 \text{ J}$

17.75 Let C = the capacitance of an individual capacitor, and C_p and C_s represent the equivalent capacitance of the group in parallel and series. While being charged in parallel, the charge on each capacitor is
$$Q = CV_{chg} = (10)(C)(V_{chg})$$
This charge is unaltered while switches are thrown, and the capacitors are connected in series. Thus, $V_{total} = NV_1 = N\frac{Q}{C} = N\frac{CV_{chg}}{C} = NV_{chg} = V_{discharge}$

In our case, $V_{discharge} = 10(800 \text{ V}) = 8000 \text{ V}$

17.76 $q_1 = C_1 V_1 = (2 \text{ }\mu\text{F})(200 \text{ V}) = 400 \text{ }\mu\text{C}$ and $q_2 = C_2 V_2 = (4 \text{ }\mu\text{F})(400 \text{ V})$
$= 1600 \text{ }\mu\text{C}$

(b) After reconnection (positive plate to negative plate)
$q' = q_2 - q_1 = 1200 \text{ }\mu\text{C},$ $C'_{eq} = C_1 + C_2 = 6 \text{ }\mu\text{F}$

so that $V'_1 = V'_2 = \frac{q'}{C'} = \frac{1200 \text{ }\mu\text{C}}{6 \text{ }\mu\text{F}} = 200 \text{ V}$

(a) Therefore, $q'_1 = C_1 V' = (2 \text{ }\mu\text{F})(200 \text{ V}) = 400 \text{ }\mu\text{C}$
and $q'_2 = C_2 V' = (4 \text{ }\mu\text{F})(200 \text{ V}) = 800 \text{ }\mu\text{C}$

17.77 Initially with the capacitors in series, $q_1 = q_2 = C_{eq}V.$
$$q_1 = q_2 = [\frac{1}{C_1} + \frac{1}{C_2}]^{-1} V = [\frac{1}{4 \text{ }\mu\text{F}} + \frac{1}{2 \text{ }\mu\text{F}}]^{-1}(100 \text{ V}) = 133 \text{ }\mu\text{C}$$
After reconnection (in parallel), $q_{total} = q_1 + q_2 = 266 \text{ }\mu\text{C}$ and $C'_{eq} = C_1 + C_2 = 6 \text{ }\mu\text{F}$

CHAPTER SEVENTEEN SOLUTIONS

so that $V' = \dfrac{q_{total}}{C'_{eq}} = \dfrac{266\ \mu C}{6\ \mu F} = 44.3$ V

Therefore, $q'_1 = C_1V' = (4\ \mu F)(44.3\ V) = 177\ \mu C$ and $q'_2 = C_2V' = (2\ \mu F)(44.3\ V) = 88.6\ \mu C$

17.78 $q_1 = C_1V_1 = (4\ \mu V)(800\ V) = 3200\ \mu C$. The equivalent capacitance of the two capacitors is

$C' = C_1 + C_2 = 10\ \mu F$ and the voltage across the combination is

$V' = \dfrac{q'}{C'} = \dfrac{q_1}{C'} = \dfrac{3200\ \mu C}{10\ \mu F} = 320$ V.

Therefore, $q'_1 = C_1V' = (4\ \mu F)(320\ V) = 1280\ \mu C$ and $q'_2 = C_2V' = (6\ \mu F)(320\ V) = 1920\ \mu C$

17.79 Initially (capacitors charged in parallel),

$q_1 = C_1V = (6\ \mu V)(250\ V) = 1500\ \mu C$. and $q_2 = C_2V = (2\ \mu V)(250\ V) = 500\ \mu C$

After reconnecting (positive plate to negative plate)

$q_{total} = q_1 - q_2 = 1000\ \mu C$

and $V' = \dfrac{q_{total}}{C_{total}} = \dfrac{1000\ \mu C}{8\ \mu F} = 125$ V. Therefore,

$q'_1 = C_1V' = (6\ \mu F)(125\ V) = 750\ \mu C$ and $q'_2 = C_2V' = (2\ \mu F)(125\ V) = 250\ \mu C$

17.80 Let x = the separation between the top plate and center section and y = the separation between the bottom plate and the center section. Then

$C_{top} = \dfrac{\varepsilon_0 A}{x}$ and $C_{bottom} = \dfrac{\varepsilon_0 A}{y}$

$C_{eq} = [\dfrac{1}{C_{top}} + \dfrac{1}{C_{bottom}}]^{-1} = [\dfrac{y + x}{\varepsilon_0 A}]^{-1} = \dfrac{\varepsilon_0 A}{x + y}$ But, if you refer to the

figure in the text, we see that $y + x = a - b$. Therefore $C_{eq} = \dfrac{\varepsilon_0 A}{a - b}$

CHAPTER EIGHTEEN SOLUTIONS

18.1 $\Delta Q = I \Delta T = (90 \text{ A})(0.5 \text{ s}) = 45 \text{ C}$

18.2 $\Delta Q = ne$. Thus, $n = \dfrac{\Delta Q}{e} = \dfrac{45 \text{ C}}{1.6 \times 10^{-19} \text{ C}} = 2.81 \times 10^{20}$ electrons

18.3 $\Delta Q = N_a e = (6.02 \times 10^{23} \text{ electrons})(1.6 \times 10^{-19} \text{ C/electron}) = 9.63 \times 10^4 \text{ C}$
From which, $I = \dfrac{\Delta Q}{\Delta t} = \dfrac{9.63 \times 10^4 \text{ C}}{3600 \text{ s}} = 26.8 \text{ A}$

18.4 $\omega = \dfrac{v}{r} = \dfrac{80 \times 10^3 \text{ m/s}}{2.15 \text{ m}} = 3.72 \times 10^4 \text{ rad/s}$
and $f = \dfrac{\omega}{2\pi} = \dfrac{3.72 \times 10^4 \text{ rad/s}}{2\pi} = 5.92 \times 10^3 \text{ Hz}$
so, $I = \dfrac{\Delta Q}{\Delta t} = qf = (1.67 \times 10^{-6} \text{ C})(5.92 \times 10^3 \text{ Hz}) = 9.89 \times 10^{-3} \text{ A} = 9.89 \text{ mA}$

18.5 $I = ne = 60 \times 10^{-6} \text{ A}$
Thus, $n = \dfrac{I}{e} = \dfrac{60 \times 10^{-6} \text{ A}}{1.6 \times 10^{-19} \text{ C}} = 3.75 \times 10^{14}$ electrons/s

18.6 $\Delta Q = I \Delta T = (80 \times 10^{-3} \text{ C/s})(600 \text{ s}) = 48 \text{ C}$
number of electrons $= \dfrac{\Delta Q}{e} = \dfrac{48 \text{ C}}{1.6 \times 10^{-19} \text{ C/electron}} = 3.0 \times 10^{20}$ electrons

The direction of the current is opposite to the direction of the electron's velocity.

18.7 The atomic weight of gold = 197, so the mass of one atom is
mass of atom $= \dfrac{197 \text{ g}}{6.02 \times 10^{23} \text{ atoms}} = 3.27 \times 10^{-22} \text{ g} = 3.27 \times 10^{-25} \text{ kg}$
Thus, the number of gold atoms deposited $= \dfrac{3.25 \times 10^{-3} \text{ kg}}{3.27 \times 10^{-25} \text{ kg/atom}}$
$= 9.93 \times 10^{21}$ ions
and the charge deposited $= (9.93 \times 10^{21} \text{ ions})(1.6 \times 10^{-19} \text{ C/ion}) = 1.59 \times 10^3 \text{ C}$
The elapsed time $= 2.78 \text{ h} = 1.0 \times 10^4 \text{ s}$, so
$I = \dfrac{\Delta Q}{\Delta t} = \dfrac{1.59 \times 10^3 \text{ C}}{1.0 \times 10^4 \text{ s}} = 1.59 \times 10^{-1} \text{ C/s} = 0.159 \text{ A}$

CHAPTER EIGHTEEN SOLUTIONS

18.8 The atomic weight of silver = 107.9, and
Volume = (area)(thickness) = $(700 \times 10^{-4} \text{ m}^2)(0.133 \times 10^{-3} \text{ m})$
= $9.31 \times 10^{-6} \text{ m}^3$
The mass of silver deposited = (density)(volume)
= $(10.5 \times 10^{-3} \text{ kg/m}^3)(9.31 \times 10^{-6} \text{ m}^3) = 9.78 \times 10^{-2}$ kg

and the number of silver atoms deposited = n = 9.78×10^{-2} kg $\dfrac{6.02 \times 10^{26} \text{ atoms}}{107.9 \text{ kg}}$
= 5.454×10^{23}

$$I = \frac{V}{R} = \frac{12 \text{ V}}{1.8 \text{ }\Omega} = 6.667 \text{ A} = 6.667 \text{ C/s}$$

and $\Delta t = \dfrac{\Delta Q}{I} = \dfrac{ne}{I} = \dfrac{(5.454 \times 10^{23})(1.6 \times 10^{-19} \text{ C})}{6.667 \text{ C/s}} = 1.31 \times 10^4 \text{ s} = 3.64 \text{ h}$

18.9 The atomic weight of gold = 197, and its density = $19.3 \times 10^3 \text{ kg/m}^3$
Thus, the mass of 1 m^3 = 19.3×10^3 kg.

N = the number of gold atoms in 1 m^3 = 19.3×10^3 kg $\dfrac{6.02 \times 10^{26} \text{ atoms/kg-mol}}{197 \text{ kg/kg-mol}}$
= 5.90×10^{28} atoms/m^3

The number of free electrons per m^3 = n = (number atoms/m^3)(number of free electrons/atom) = $(5.90 \times 10^{28} \text{ atoms/m}^3)(1 \text{ electron/atom})$
= 5.90×10^{28} electrons/m^3

18.10 We use, $I = nqAv_d$
Where n = (number of charge carriers per unit volume) = (number of atoms per unit volume). We assume a contribution of 1 free electron per atom in the relationship above. For aluminum, which has a molecular weight of 27, we know that Avogadro's number of atoms, N_a, has a mass of 27 g. Thus, the mass per atom is

$$\frac{27 \text{ g}}{N_a} = \frac{27 \text{ g}}{6.02 \times 10^{23}} = 4.49 \times 10^{-23} \text{ g/atom}.$$

Thus, n = $\dfrac{\text{density of aluminum}}{\text{mass per atom}} = \dfrac{2.7 \text{ g/cm}^3}{4.49 \times 10^{-23} \text{ g/atom}}$

= $6.02 \times 10^{22} \dfrac{\text{atoms}}{\text{cm}^3} = 6.02 \times 10^{28} \dfrac{\text{atoms}}{\text{m}^3}$

Therefore,
$$v_d = \frac{I}{nqA} = \frac{5 \text{ A}}{(6.02 \times 10^{28} \text{ m}^{-3})(1.6 \times 10^{-19} \text{ C})(4 \times 10^{-6} \text{ m}^2)}$$
= 1.3×10^{-4} m/s
or, v_d = 0.13 mm/s

18.11 $I = nqAv_d = (2.5 \times 10^{28} \text{ 1/m}^3)(1.6 \times 10^{-19} \text{ C})(5 \times 10^{-4} \text{ m/s})(7.85 \times 10^{-7} \text{ m}^2)$
= 1.57 A

18.12 (a) In one second, all the electrons in a 0.03 cm length of the wire must pass the cross-section. Therefore,
n = $(1.5 \times 10^{20} \text{ electrons/cm})(0.03 \text{ cm/s}) = 4.5 \times 10^{18}$ electrons/s
(b) $I = \dfrac{\Delta Q}{\Delta t}$ = ne = $(4.5 \times 10^{18} \text{ electrons/s})(1.6 \times 10^{-19} \text{ C/electron}) = 0.72$ C/s
= 0.72 A

CHAPTER EIGHTEEN SOLUTIONS

18.13 $R = \dfrac{\rho L}{A} = \dfrac{(1.59 \times 10^{-8} \;\Omega\text{ m})(1\text{ m})}{1 \times 10^{-5}\text{ m}^2} = 1.59 \times 10^{-3}\;\Omega.$

18.14 $R = (0.2\;\Omega/\text{m})(120\text{ m}) = 24\;\Omega.$

18.15 $A = \dfrac{\rho L}{R} = \dfrac{(5.6 \times 10^{-8}\;\Omega\text{ m})(2 \times 10^{-2}\text{ m})}{5 \times 10^{-2}\;\Omega} = 2.24 \times 10^{-8}\text{ m}^2.$
But $A = \pi d^2/4.$ From which, $d = 1.69 \times 10^{-4}\text{ m}$

18.16 $R = \dfrac{\rho L}{A} = \dfrac{(1.7 \times 10^{-8}\;\Omega\text{ m})(15\text{ m})}{8.24 \times 10^{-7}\text{ m}^2} = 0.310\;\Omega.$

18.17 If $R = 4 \times 10^5\;\Omega,$ $V_{max} = (4 \times 10^5\;\Omega)(8 \times 10^{-5}\text{ A}) = 32.0\text{ V}$
If $R = 2000\;\Omega,$ $V_{max} = (2 \times 10^3\;\Omega)(8 \times 10^{-5}\text{ A}) = 0.16\text{ V}$

18.18 $I = \dfrac{V}{R} = \dfrac{120\text{ V}}{240\;\Omega} = 0.5\text{ A}$

18.19 $R = \dfrac{V}{I} = \dfrac{120\text{ V}}{2.5\text{ A}} = 48\;\Omega.$

18.20 (a) The resistance of the device is $R = \dfrac{V}{I} = \dfrac{120\text{ V}}{0.5\text{ A}} = 240\;\Omega.$
If the voltage is lowered to 90 V, we have $I = \dfrac{V}{R} = \dfrac{90\text{ V}}{240\;\Omega} = 0.375\text{ A}.$
(b) If the voltage is raised to 130 V, we have $I = \dfrac{V}{R} = \dfrac{130\text{ V}}{240\;\Omega} = 0.542\text{ A}.$

18.21 If the voltage is 120 V and $R = 20\;\Omega,$ then $I = \dfrac{V}{R} = \dfrac{120\text{ V}}{20\;\Omega} = 6\text{ A}.$ (His claim violates Ohm's law.)

18.22 (a) $R = \dfrac{\rho L}{A} = \dfrac{(150 \times 10^{-8}\;\Omega\text{ m})(2\text{ m})}{2 \times 10^{-6}\text{ m}^2} = 1.50\;\Omega.$
(b) $I = \dfrac{V}{R} = \dfrac{3\text{ V}}{1.5\;\Omega} = 2.0\text{ A}$

18.23 Volume = AL = const
Thus, $A_2 L_2 = A_1 L_1$, and if $L_2 = 3L_1$, then $A_2 = \dfrac{1}{3} A_1$
So, $R_2 = \dfrac{\rho L_2}{A_2} = \dfrac{\rho 3 L_1}{\frac{1}{3} A_1} = 9 \dfrac{\rho L_1}{A_1}$
Thus, $R_2 = 9 R_1 = 9 R$

18.24 $L = \dfrac{RA}{\rho} = \dfrac{(8.5\;\Omega)(\frac{\pi}{4} 10^{-6}\text{ m}^2)}{150 \times 10^{-8}\;\Omega\text{ m}} = 4.45\text{ m}$

CHAPTER EIGHTEEN SOLUTIONS

18.25 (a) The resistance of the wire is $R = \dfrac{V}{I} = \dfrac{12\ V}{0.4\ A} = 30\ \Omega$.

and $\rho = \dfrac{RA}{L} = \dfrac{30\ \Omega\ \pi(4 \times 10^{-3}\ m)^2}{3.2\ m} = 4.71 \times 10^{-4}\ \Omega m$

(b) Resistance has been found in part (a).

18.26 The gold wire must have the same resistance as the iron wire. Thus,

$\dfrac{\rho_{gold} L_{gold}}{A_{gold}} = \dfrac{\rho_{iron} L_{iron}}{A_{iron}}$ But, $A_{gold} = A_{iron}$

Therefore, $L_{gold} = \dfrac{\rho_{iron}}{\rho_{gold}} L_{iron} = \dfrac{10 \times 10^{-8}\ \Omega\ m}{2.44 \times 10^{-8}\ \Omega\ m}(2\ m) = 8.20\ m$

18.27 The possible cross-sectional areas are:
$A_1 = (20\ cm)(40\ cm) = 800\ cm^2$, $A_2 = (10\ cm)(40\ cm) = 400\ cm^2$,
and $A_3 = (20\ cm)(10\ cm) = 200\ cm^2$.

From $R = \dfrac{\rho L}{A}$ we see that the minimum resistance occurs for the minimum L and the maximum A. Thus, $R_{min} = (1.78 \times 10^{-8}\ \Omega\ m)\dfrac{0.10\ m}{800 \times 10^{-4}\ m^2}$

$= 2.13 \times 10^{-8}\ \Omega$

and $R_{max} = (1.78 \times 10^{-8}\ \Omega\ m)\dfrac{0.40\ m}{200 \times 10^{-4}\ m^2} = 3.40 \times 10^{-7}\ \Omega$

(a) Using the information above, $I_{max} = \dfrac{V}{R_{min}} = \dfrac{6\ V}{2.13 \times 10^{-8}\ \Omega}$

$= 2.82 \times 10^8\ A$

and, (b) $I_{min} = \dfrac{V}{R_{max}} = \dfrac{6\ V}{3.4 \times 10^{-7}\ \Omega} = 1.76 \times 10^7\ A$

18.28 Consider a length of wire 1 meter long. The potential difference across this length is seen to be 100 V.
Therefore, $R = \dfrac{V}{I} = \dfrac{100\ V}{5\ A} = 20\ \Omega$.

and, $\rho = \dfrac{RA}{L} = \dfrac{20\ \Omega\ \pi(5 \times 10^{-3}\ m)^2}{1.0\ m} = 1.57 \times 10^{-3}\ \Omega m$

18.29 $R = R_0(1 + \alpha(\Delta T)) = 10\ \Omega(1 + (3.8 \times 10^{-3}\ °C^{-1})(20\ °C)) = 10.8\ \Omega$

18.30 $A = 3\ mm^2 = 3 \times 10^{-6}\ m^2$. At 20 °C the resistance of the wire is

$R = \dfrac{\rho L}{A} = \dfrac{(1.7 \times 10^{-8}\ \Omega\ m)(10\ m)}{3 \times 10^{-6}\ m^2} = 5.67 \times 10^{-2}\ \Omega$.

(a) At T = 30 °C, we have
$R = R_0(1 + \alpha(\Delta T)) = 5.67 \times 10^{-2}\ \Omega(1 + (3.9 \times 10^{-3}\ °C^{-1})(10\ °C))$
$= 5.89 \times 10^{-2}\ \Omega$

(b) At T = 10 °C,
$R = R_0(1 + \alpha(\Delta T)) = 5.67 \times 10^{-2}\ \Omega(1 + (3.9 \times 10^{-3}\ °C^{-1})(-10\ °C))$
$= 5.45 \times 10^{-2}\ \Omega$

CHAPTER EIGHTEEN SOLUTIONS

18.31 We use, $R = R_0(1 + \alpha(\Delta T))$, as
$$100 \; \Omega = R_0(1 + (3.4 \times 10^{-3} \; °C^{-1})(20 \; °C)) = R_0(1.068)$$
From which, $R_0 = 93.63 \; \Omega$
At the unknown temperature; $R = 97 \; \Omega$. Thus,
$$97 \; \Omega = (93.63 \; \Omega)(1 + (3.4 \times 10^{-3} \; °C^{-1})(\Delta T))$$
From which, we find $\Delta T = 10.6 \; °C$
and, $T = 30.6 \; °C$

18.32 At 80 °C the resistance has decreased to
$$R = R_0(1 + \alpha(\Delta T)) = 200 \; \Omega(1 + (-0.5 \times 10^{-3} \; °C^{-1})(60 \; °C)) = 194 \; \Omega$$
Thus, $I = \dfrac{V}{R} = \dfrac{5 \; V}{194 \; \Omega} = 2.58 \times 10^{-2} \; A = 25.8 \; mA$

18.33 $R = R_0(1 + \alpha(\Delta T))$
$$41.4 \; \Omega = 41 \; \Omega(1 + \alpha(9°C))$$
gives $\alpha = 1.084 \times 10^{-3} \; °C^{-1}$

18.34 % change $= \dfrac{R - R_0}{R_0} 100\% = (\dfrac{R}{R_0} - 1)100$, and $\dfrac{R}{R_0} = 1 + \alpha(\Delta T)$
Therefore, % change $= \alpha(\Delta T)100 = 100 \; (-0.5 \times 10^{-3} \; °C^{-1})(140°C) = -7 \; \%$

18.35 We are given $R = 1.2 R_0$. Thus,
$R = R_0(1 + \alpha(\Delta T))$ becomes
$$1.2 R_0 = R_0(1 + \alpha(\Delta T))$$
or, $\alpha \Delta T = 0.2$
Thus, $\Delta T = \dfrac{2 \times 10^{-1}}{3.9 \times 10^{-3} \; °C^{-1}} = 51.3 \; °C$
Therefore, the final temperature is 71.3 °C.

18.36 $R = R_0(1 + \alpha(\Delta T))$
$$140 \; \Omega = 19 \; \Omega(1 + (4.5 \times 10^{-3} \; °C^{-1})(\Delta T))$$
Which gives, $\Delta T = 1.415 \times 10^3 \; °C$
And, the final temperature is $T = 1435 \; °C$

18.37 (a) The resistance of the wire is
$$R = \dfrac{\rho L}{A} = \dfrac{(1.7 \times 10^{-8} \; \Omega \; m)(1 \; m)}{\pi(0.5 \times 10^{-2} \; m)^2} = 2.165 \times 10^{-4} \; \Omega.$$
Then, $V = IR = (3 \; A)(2.165 \times 10^{-4} \; \Omega) = 6.49 \times 10^{-4} \; V = 0.649 \; mV$
(b) At $T = 200 \; °C$, the resistance is
$$R = R_0(1 + \alpha(\Delta T)) = 2.165 \times 10^{-4} \; \Omega(1 + (3.9 \times 10^{-3} \; °C^{-1})(180 \; °C))$$
$= 3.69 \times 10^{-4} \; \Omega$
Thus, $V = IR = (3A)(3.69 \times 10^{-4} \; \Omega) = 1.11 \times 10^{-3} \; V = 1.11 \; mV$

18.38 We use, $\rho = \rho_0(1 + \alpha(\Delta T))$
For tungsten, $\rho_0 = 5.6 \times 10^{-8} \; \Omega \; m$ and $\alpha = 4.5 \times 10^{-3} \; °C^{-1}$
If $\rho = 4(\rho_0)_{copper} = 4(1.7 \times 10^{-8} \; \Omega \; m) = 6.8 \times 10^{-8} \; \Omega \; m$, then
$$6.8 \times 10^{-8} \; \Omega \; m = 5.6 \times 10^{-8} \; \Omega \; m(1 + 4.5 \times 10^{-3} \; °C^{-1} \Delta T)$$

CHAPTER EIGHTEEN SOLUTIONS

We find, $\Delta T = 47.6 \, °C$
and $T = 67.6 \, °C$

18.39 $I = \dfrac{P}{V} = \dfrac{600 \text{ W}}{120 \text{ V}} = 5 \text{ A}$

and, $R = \dfrac{V}{I} = \dfrac{120 \text{ V}}{5 \text{ A}} = 24 \, \Omega$.

18.40 (a) The energy W = power times the time used. Thus,
$W = Pt = (90 \text{ J/s})(3600 \text{ s}) = 3.24 \times 10^5 \text{ J}$
(b) The power consumed by the color set is,
$P = VI = (120 \text{ V})(2.5 \text{ A}) = 300 \text{ W}$
Thus, $t = \dfrac{W}{P} = \dfrac{3.24 \times 10^5 \text{ J}}{300 \text{ W}} = 1080 \text{ s} = 18 \text{ min}$

18.41 $I = \dfrac{P}{V} = \dfrac{120 \times 10^3 \text{ W}}{240 \text{ V}} = 500 \text{ A}$

18.42 $P = VI = (9 \text{ V})(0.300 \text{ A}) = 2.70 \text{ W}$

18.43 $R = \dfrac{V^2}{P} = \dfrac{(120 \text{ V})^2}{1500 \text{ W}} = 9.6 \, \Omega$.

and, $A = \dfrac{\rho L}{R} = \dfrac{(5.6 \times 10^{-8} \, \Omega \text{ m})(3 \text{ m})}{9.6 \, \Omega} = 1.75 \times 10^{-8} \text{ m}^2$.

18.44 $R = \dfrac{\rho L}{A} = \dfrac{(1.7 \times 10^{-8} \, \Omega \text{ m})(1 \text{ m})}{\pi (10^{-3} \text{ m})^2} = 5.41 \times 10^{-3} \, \Omega$.

and $P = I^2 R = (40 \text{ A})^2 (5.41 \times 10^{-3} \, \Omega) = 8.67 \text{ W}$

18.45 % change $= \dfrac{P_f - P_i}{P_i} 100\% = \dfrac{\dfrac{V_f^2}{R} - \dfrac{V_i^2}{R}}{\dfrac{V_i^2}{R}} 100 = \dfrac{V_f^2 - V_i^2}{V_i^2} 100$

$= \dfrac{(140 \text{ V})^2 - (120 \text{ V})^2}{(120 \text{ V})^2} 100 = 36.1\%$

18.46 The maximum power that can be dissipated in the circuit is,
$P = VI = (120 \text{ V})(15 \text{ A}) = 1800 \text{ W}$
Thus, one can operate at most 18 bulbs rated at 100 W per bulb.

18.47 $P = 0.25 \text{ hp} = 186.5 \text{ W}$. Thus, $I = \dfrac{P}{V} = \dfrac{186.5 \text{ W}}{120 \text{ V}} = 1.55 \text{ A}$

18.48 The heat that must be added to the water is
$Q = mc\Delta T = (1.5 \text{ kg})(4184 \text{ J/kg °C})(40 \, °C) = 2.51 \times 10^5 \text{ J}$
Thus, the power supplied by the heater is
$P = \dfrac{W}{t} = \dfrac{Q}{t} = \dfrac{2.51 \times 10^5 \text{ J}}{600 \text{ s}} = 418 \text{ W}$

CHAPTER EIGHTEEN SOLUTIONS

and the resistance is $\quad R = \dfrac{V^2}{P} = \dfrac{(120 \text{ V})^2}{418 \text{ W}} = 34.4 \ \Omega.$

18.49 The power appearing in electrical form is
$P = 0.90 \ (2000 \text{ hp})(746 \text{ W/hp}) = 1.34 \times 10^6 \text{ W}$
Thus, $I = \dfrac{P}{V} = \dfrac{1.34 \times 10^6 \text{ W}}{3000 \text{ V}} = 448 \text{ A}$

18.50 The diameters of conductors A and B and their lengths are related, respectively, as
$d_a = 2d_b$, and $L_a = 2L_b$.

From $R = \dfrac{\rho L}{A}$, which becomes $R = \dfrac{4\rho}{\pi} \dfrac{L}{d^2}$ where we have used, $A = \dfrac{\pi}{4} d^2$. Thus, the resistance of A can be expressed in terms of the resistance of B as

$R_a = \dfrac{4\rho}{\pi} \dfrac{2L_b}{4d_b^2} = \dfrac{R_b}{2}$, or $\quad R_b = 2R_a$

Now consider the power delivered by the two conductors.
$P_b = \dfrac{V^2}{R_b} = \dfrac{V^2}{2R_a} = \dfrac{1}{2} \dfrac{V^2}{R_a} = \dfrac{P_a}{2}$
Thus, $P_a = 2P_b$

18.51 $W = Pt = (90 \text{ W})(21 \text{ h}) = 1890 \text{ Wh} = 1.89 \text{ kWh}.$
Thus, the cost is $\quad$ cost $= (7 \text{ cents/kWh}) (1.89 \text{ kWh}) = 13.2$ cents

18.52 The energy requirement for the house expressed in joules is
$W = 2500 \text{ kWh} \dfrac{3.6 \times 10^6 \text{ J}}{1 \text{ kWh}} = 9 \times 10^9 \text{ J}$
The energy available from the coal is
$7 \times 10^6 \text{ cal/kg} = (7 \times 10^6 \dfrac{\text{cal}}{\text{kg}})(\dfrac{4.184 \text{ J}}{1 \text{ cal}}) = 2.93 \times 10^7 \text{ J/kg}$
The energy attainable from the coal at 40% efficiency is
$(2.93 \times 10^7 \text{ J/kg})(0.4) = 1.17 \times 10^7 \text{ J/kg}$
Thus, the mass of coal, m, needed is
$m = \dfrac{9 \times 10^9 \text{ J}}{1.17 \times 10^7 \text{ J/kg}} = 768 \text{ kg}$

18.53 We shall assume five minutes of use per day. The energy required is
$W = Pt = (1.5 \text{ kW})\left(\dfrac{5 \text{ min}}{\text{day}}\right)\left(\dfrac{1 \text{ h}}{60 \text{ min}}\right)\left(\dfrac{365 \text{ days}}{\text{yr}}\right) = 45.7 \text{ kWh (per year)}$
Thus, cost $= (8 \text{ cents/kWh})(45.7 \text{ kWh/yr}) = 365$ cents per year $= \$3.65$ /year

18.54 (a) The power input to the motor is
$P_{input} = VI = (120 \text{ V})(1.75 \text{ A}) = 210 \text{ W} = 0.282 \text{ hp}$
The energy used in four hours is
$W = Pt = (0.21 \text{ kW})(4 \text{ h}) = 0.84 \text{ kWh}$
and the cost is $\quad$ cost $= (\$0.06/\text{kWh})(0.84 \text{ kWh}) = 5.04 \times 10^{-2}$ dollars
$= 5.04$ cents.
(b) $\quad \text{Eff} = \dfrac{P_{output}}{P_{input}} = \dfrac{0.2 \text{ hp}}{0.282 \text{ hp}} = 0.71$, or 71 %

CHAPTER EIGHTEEN SOLUTIONS

18.55 $P = 1500 \text{ kcal/h} = 1744 \text{ W}$
and $I = \dfrac{P}{V} = \dfrac{1744 \text{ W}}{110 \text{ V}} = 15.9 \text{ A}$ (A 20 A fuse is required.)

18.56 400 cm^3 of water = 400 g of water = 0.4 kg
Assume the water is already at the boiling point. Then,
$$P = \dfrac{W}{t} = \dfrac{mL_v}{t} = \dfrac{(0.4 \text{ kg})(2.26 \times 10^6 \text{ J/kg})}{3600 \text{ s}} = 251 \text{ W},$$
and $R = \dfrac{V^2}{P} = \dfrac{(110 \text{ V})^2}{251 \text{ W}} = 48.2 \text{ }\Omega$

18.57 (power output) = (efficiency)(power input)
$(2.5 \text{ hp})\dfrac{746 \text{ W}}{\text{h p}} = (0.9) P_{input}$
so $P_{input} = 2072 \text{ W}$
(a) $I = \dfrac{P}{V} = \dfrac{2072 \text{ W}}{110 \text{ V}} = 18.8 \text{ A}$
(b) $W = Pt = (2072 \text{ W})(1 \text{ h}) = 2.072 \text{ kWh}$
(c) cost = (energy used)(rate) = (2.072 kWh)(8 cents/kWh) = 16.6 cents

18.58 500 cm^3 of water = 500 g of water = 0.5 kg
energy output = $mc\Delta T$ = (500 g)(1 cal/g °C)(80°C) = 4×10^4 cal
= 1.674×10^5 J
energy out = (eff)(energy in)
so, energy in = $\dfrac{\text{energy out}}{\text{eff}} = \dfrac{1.674 \times 10^5 \text{ J}}{0.5} = 3.35 \times 10^5 \text{ J}$
and $P = \dfrac{\text{energy input}}{\text{time}} = \dfrac{3.35 \times 10^5 \text{ J}}{600 \text{ s}} = 558 \text{ W}$

18.59 $Q = IA = (90 \text{ A})(1 \text{ h}) = (90 \text{ C/s})(3600 \text{ s}) = 3.24 \times 10^5 \text{ C}$

18.60 $W = Pt = (VI)t = (120 \text{ V})(6 \text{ A})(1200 \text{ s}) = 8.64 \times 10^5 \text{ J}$

18.61 $W = 8.64 \times 10^5 \text{ J} = 0.24 \text{ kWh}$ (See problem 42)
Thus, cost = (8 cents/kWh)(0.24 kWh) = 1.92 cents

18.62 The energy available from the battery is
$W = Pt = (VI)t = (12 \text{ V})(90 \text{ A})(3600 \text{ s}) = 3.89 \times 10^6 \text{ J}$
The rate of energy dissipation by the two lights is, 80 J/s
Thus, the time to consume the available energy is
$$t = \dfrac{3.89 \times 10^6 \text{ J}}{80 \text{ J/s}} = 4.86 \times 10^4 \text{ s} = 13.5 \text{ h}$$

18.63 The current drawn by the wire is
$I = \dfrac{P}{V} = \dfrac{48 \text{ W}}{20 \text{ V}} = 2.4 \text{ A}$
and the resistance of the wire is, $R = \dfrac{V}{I} = \dfrac{20 \text{ V}}{2.4 \text{ A}} = 8.33 \text{ }\Omega$.

CHAPTER EIGHTEEN SOLUTIONS

Thus, from $R = \frac{\rho L}{A}$ we have $L = \frac{RA}{\rho} = \frac{(8.33 \ \Omega) \ 4 \times 10^{-6} \ m^2}{3 \times 10^{-8} \ \Omega \ m} = 1.11 \times 10^3 \ m$

18.64 The resistance per unit length of the wire is

$\frac{R}{L} = \frac{\rho}{A} = \frac{1.7 \times 10^{-8} \ \Omega \ m}{\pi(2.2 \times 10^{-2} \ m)^2} = 1.12 \times 10^{-5} \ \Omega/m$

Thus, the resistance of the wire that is between the bird's feet is
R = (resistance per unit length of wire)(distance between feet)
= $(1.12 \times 10^{-5} \ \Omega/m)(4 \times 10^{-2} \ m) = 4.47 \times 10^{-7} \ \Omega$.

And the voltage between the feet is
$V = IR = (50 \ A)(4.47 \times 10^{-7} \ \Omega) = 2.24 \times 10^{-5} \ V$

18.65 The area of the wire is $A = 7.85 \times 10^{-5} \ m^2$.

The current in the wire is $I = \frac{V}{R} = \frac{15 \ V}{0.1 \ \Omega} = 150 \ A$

We now can find the density of free electrons from $I = nqAv_d$.

$n = \frac{I}{qv_dA} = \frac{150 \ A}{(1.6 \times 10^{-19} \ C)(3.17 \times 10^{-4} \ m/s)(7.85 \times 10^{-5} \ m^2)}$
$= 3.77 \times 10^{28}$ electrons/m^3.

18.66 The power dissipated in a wire is $P = I^2R = \frac{V^2}{R}$ and $R = \frac{\rho L}{A}$. Thus, from the latter, we see that the smaller wire has the largest resistance.
(a) When I is the same in both wires, consider the $P = I^2R$ form to see that P is proportional to R. Thus, the smaller wire dissipates the most power.
(b) When the wires are in parallel, they both have the same potential difference across them. Thus, consider the equation $P = \frac{V^2}{R}$ to see that P is inversely proportional to the resistance. Thus, the wire with the smallest resistance (the larger diameter) will dissipate the most heat.

18.67 $\omega = 2\pi f = 100\pi$ rad/s, so f = 50 Hz

$I = \frac{\Delta Q}{\Delta t} = \frac{q}{\frac{1}{f}} = qf = (8 \times 10^{-9} \ C)(50 \ Hz) = 400 \times 10^{-9} \ C/s = 0.4 \ \mu A$

18.68 $R = \frac{V}{I} = \frac{12 \ V}{I} = \frac{6 \ V}{(I - 3 \ A)}$ thus $12I - 36 = 6I$ and I = 6 A.

Therefore, $R = \frac{12 \ V}{6 \ A} = 2 \ \Omega$

18.69 (a) $R = \frac{\rho L}{A}$ and since the radius equals L/2, we have $R = \frac{\rho L}{\pi r^2} = \frac{4\rho}{\pi L}$

Density, $D = \frac{m}{V} = \frac{m}{\pi \left(\frac{L}{2}\right)^2 L} = \frac{4m}{\pi L^3}$ so that $L = \left(\frac{4m}{\pi D}\right)^{1/3} = \left(\frac{4(0.115 \ kg)}{\pi(2.7 \times 10^3 \ kg/m^3)}\right)^{1/3}$

$= 3.79 \times 10^{-2}$ m

Using this value and the resistivity of Al in the equation for R, we get

CHAPTER EIGHTEEN SOLUTIONS

$$R = \frac{4(2.8 \times 10^{-8} \, \Omega \, m)}{\pi (3.79 \times 10^{-2} \, m)} = 9.4 \times 10^{-7} \, \Omega$$

(b) In this case, $D = \frac{m}{V} = \frac{m}{L^3}$ so $L = \left(\frac{m}{D}\right)^{1/3} = 3.49 \times 10^{-2}$ m

and $R = \frac{\rho L}{A} = \frac{\rho L}{L^2} = \frac{\rho}{L} = \frac{2.8 \times 10^{-8} \, \Omega \, m}{3.49 \times 10^{-2} \, m} = 8.02 \times 10^{-7} \, \Omega$

18.70 $R = \frac{\rho L}{A} = \frac{\rho L}{\pi(r_b^2 - r_a^2)} = \frac{(3.5 \times 10^5 \, \Omega \, m)(0.04 \, m)}{\pi((0.012 \, m)^2 - (0.005 \, m)^2)} = 37.4 \, M\Omega$

18.71 (a) From the definition of current, we see that $\Delta Q = I \Delta T$.
And from this, we see that the charge can be found by finding the total area under a curve of I versus t. Thus,
Q = (area of rectangle A_1) + 2(area of triangle A_2) + (rectangular area A_3)

$Q = (2 \, A)(5 \, s) + 2(\frac{1}{2} (1 \, s)(4 \, A)) + (1 \, s)(4 \, A) = 10 \, C + 4 \, C + 4 \, C = 18 \, C$

(b) The constant current would be $I = \frac{\Delta Q}{\Delta t} = \frac{18 \, C}{5 \, s} = 3.6 \, A$

18.72 The volume of the wire remains a constant during the stretching process. Thus,
V = LA = constant

If $r = \frac{r_0}{4}$ then $A = \pi r^2 = \pi \frac{r_0^2}{16} = \frac{A_0}{16}$

Thus, $LA = L(\frac{A_0}{16}) = L_0 A_0$

or, $L = 16 L_0$

and, $R = \frac{\rho L}{A} = \frac{\rho(16 L_0)}{\frac{A_0}{16}} = 256 \frac{\rho L_0}{A_0} = 256 \, R_0$

If $R_0 = 1 \, \Omega$ then $R = 256 \, \Omega$.

18.73 (a) The volume of the wire is $Vol = \frac{50 \, g}{7.86 \, g/cm^3} = 6.36 \, cm^3$.

We also know that Vol = AL, or, $A = \frac{Vol}{L} = \frac{6.36 \, cm^3}{L} = \frac{6.36 \times 10^{-6} \, m^3}{L}$ (1)

From the definition of resistance, we find

$R = \frac{\rho L}{A}$

becomes, $1.5 \, \Omega = (11 \times 10^{-8} \, \Omega \, m) \frac{L}{A}$, or, $\frac{L}{A} = 1.36 \times 10^7 \, m^{-1}$ (2)

Let us solve (1) and (2) simultaneously to find
L = 9.31 m and A = 6.83 × 10^{-7} m².

(b) From the area of the wire, we can find its diameter to be
d = 9.33 × 10^{-4} m

CHAPTER EIGHTEEN SOLUTIONS

18.74 (a) $R = \dfrac{\rho L}{A} = \dfrac{(1.7 \times 10^{-8})(0.24)}{(0.08)(0.002)} = 2.55 \times 10^{-5}\ \Omega$

(b) m = (volume)(density) but $R = \dfrac{\rho L}{A}$ and $A = \dfrac{\text{volume}}{L}$. Therefore,

$m = \dfrac{\rho L^2}{R}$ (density) $= \dfrac{1.72 \times 10^{-8}(1.5 \times 10^3)^2}{(4.5)}\ (8.89 \times 10^3) = 76.5$ kg

---

18.75 From $R = R_0(1 + \alpha(T - T_0))$, we have $\dfrac{R - R_0}{R_0} = \alpha(T - T_0)$

or $\alpha = \dfrac{1}{R}\dfrac{\Delta R}{\Delta T}$

In this case, we require for the composite device that

$\dfrac{\Delta R}{\Delta T} = 0$ where $\Delta T_1 = \Delta T_2$ or $\Delta R_1 + \Delta R_2 = 0$

Therefore, $\alpha_1 R_1 + \alpha_2 R_2 = 0$ and $\alpha_1 \dfrac{\rho_1 L_1}{A_1} + \alpha_2 \dfrac{\rho_2 L_2}{A_2} = 0$ But $A_1 = A_2$

so $\dfrac{\alpha_1}{\alpha_2} = -\dfrac{\rho_2 L_2}{\rho_1 L_1} = -\dfrac{1}{(3.2)(2.6)} = -0.136$ (One of the materials must have a negative temperature coefficient of resistance.)

CHAPTER NINETEEN SOLUTIONS

19.1 (a) Resistors in series add. Thus, the equivalent resistance is
$R_{eq} = R_1 + R_2 + R_3 = 4\,\Omega + 8\,\Omega + 12\,\Omega = 24\,\Omega$.
(b) $I = \dfrac{V}{R_{eq}} = \dfrac{24\text{ V}}{24\,\Omega} = 1\text{ A}$

19.2 The equivalent resistance is,
$R_{eq} = R_1 + R_2 = 9\,\Omega + 6\,\Omega = 15\,\Omega$.
Therefore, the voltage setting of the power supply is
$V = IR_{eq} = \left(\dfrac{1}{3}\text{ A}\right)(15\,\Omega) = 5\text{ V}$

19.3 (a) The equivalent resistance for a parallel combination of resistors is given by
$\dfrac{1}{R_{eq}} = \dfrac{1}{R_1} + \dfrac{1}{R_2} + \dfrac{1}{R_3} = \dfrac{1}{4\,\Omega} + \dfrac{1}{8\,\Omega} + \dfrac{1}{12\,\Omega}$. From which, $R_{eq} = \dfrac{24}{11}\,\Omega$.
(b) The potential difference across each parallel element is the same. Thus,
$I_4 = \dfrac{V}{R_4} = \dfrac{24\text{ V}}{4\,\Omega} = 6\text{ A}$, $I_8 = \dfrac{V}{R_8} = \dfrac{24\text{ V}}{8\,\Omega} = 3\text{ A}$, and $I_{12} = \dfrac{V}{R_{12}} = \dfrac{24\text{ V}}{12\,\Omega} = 2\text{ A}$.

19.4 Consider them all in parallel.
$\dfrac{1}{R_{eq}} = \dfrac{1}{R_1} + \dfrac{1}{R_2} + \dfrac{1}{R_3} = \dfrac{1}{500\,\Omega} + \dfrac{1}{250\,\Omega} + \dfrac{1}{250\,\Omega}$ From which, $R_{eq} = 100\,\Omega$.

19.5 There are 7 distinct values possible:
(1) use one alone, value = R
(2) use two in series, value = 2R
(3) three in series, value = 3R
(4) two in parallel, value = $\dfrac{R}{2}$
(5) three in parallel, value = $\dfrac{R}{3}$
(6) two in series with one in parallel, value = $\dfrac{2}{3}R$
(7) two in parallel with one in series, value = $\dfrac{3}{2}R$

19.6 The equivalent resistance of the two resistors in parallel is
$\dfrac{1}{R_{eq}} = \dfrac{1}{R_1} + \dfrac{1}{R_2} = \dfrac{1}{9\,\Omega} + \dfrac{1}{6\,\Omega}$ Which yields, $R_{eq} = \dfrac{18}{5}\,\Omega$.
Therefore, $V = IR_{eq} = \left(\dfrac{1}{3}\text{ A}\right)\left(\dfrac{18\,\Omega}{5}\right) = \dfrac{6}{5}\text{ V} = 1.20\text{ V}$

19.7 $\dfrac{1}{R_{eq}} = \dfrac{1}{R_1} + \dfrac{1}{R_2} + \dfrac{1}{R_3} = \dfrac{1}{10\,\Omega} + \dfrac{1}{20\,\Omega} + \dfrac{1}{30\,\Omega}$ From which, $R_{eq} = \dfrac{60}{11}\,\Omega$.

CHAPTER NINETEEN SOLUTIONS

(a) The voltage drop across the parallel branch is
$$V = IR_{eq} = (5 \text{ A})(\frac{60}{11} \Omega) = 27.3 \text{ V}$$

(b) $I_{10} = \frac{V}{10 \Omega} = \frac{27.3 \text{ V}}{10 \Omega} = 2.73 \text{ A}$

$I_{20} = \frac{V}{20 \Omega} = \frac{27.3 \text{ V}}{20 \Omega} = 1.36 \text{ A}$

$I_{30} = \frac{V}{30 \Omega} = \frac{27.3 \text{ V}}{30 \Omega} = 0.910 \text{ A}$

19.8 With R the value of the load resistor, the current in a series circuit composed of a 12 V battery, a resistor of 10 Ω, and a load resistor is
$$I = \frac{12 \text{ V}}{R + 10 \text{ }\Omega}$$

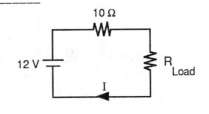

and the power delivered to the load resistor is
$$P = I^2 R = \frac{(144 \text{ V}^2) R}{(R + 10 \text{ }\Omega)^2}$$

Some typical values are

R	P_{load}
1 Ω	1.19 W
5 Ω	3.20 W
10 Ω	3.60 W
15 Ω	3.46 W
20 Ω	3.20 W
25 Ω	2.94 W
30 Ω	2.70 W

The curve peaks at $P_{load} = 3.6$ W at a load resistance of R = 10 Ω.

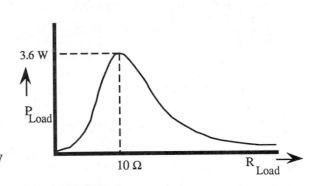

19.9 The three resistors in parallel are replaced by an equivalent resistance of
$$\frac{1}{R_{eqp}} = \frac{1}{R_1} + \frac{1}{R_2} + \frac{1}{R_3} = \frac{1}{18 \text{ }\Omega} + \frac{1}{9 \text{ }\Omega} + \frac{1}{6 \text{ }\Omega} \quad \text{From which, } R_{eqp} = 3 \text{ }\Omega.$$
This resistor is now in series with the 12 Ω resistor. The two add for an end result of
$$R_{eq} = R_1 + R_2 = 12 \text{ }\Omega + 3 \text{ }\Omega = 15 \text{ }\Omega.$$

19.10 The equivalent resistor for the two in series is
$$R_{eq} = R_1 + R_2 = 18 \text{ }\Omega + 6 \text{ }\Omega = 24 \text{ }\Omega.$$
From this, we can find the current through the circuit as,
$$I = \frac{V}{R_{eq}} = \frac{18 \text{ V}}{24 \text{ }\Omega} = 0.75 \text{ A}.$$
All elements in a series circuit carry the same current. Thus, this is also the current through the 18 Ω and 6 Ω resistors.
The voltage drop across each resistor can now be found as follows.
$V_{18} = IR_{18} = (0.75 \text{ A})(18 \text{ }\Omega) = 13.5 \text{ V},$ and $V_6 = IR_6 = (0.75 \text{ A})(6 \text{ }\Omega) = 4.5 \text{ V}.$
(b) For the parallel combination, the potential difference is the same across both resistors and equal to that of the battery. Thus,
$$V_{18} = V_6 = 18 \text{ V}$$
The currents are given by

CHAPTER NINETEEN SOLUTIONS

$$I_{18} = \frac{V}{R_{18}} = \frac{18 \text{ V}}{18 \text{ }\Omega} = 1 \text{ A, and} \qquad I_6 = \frac{V}{R_6} = \frac{18 \text{ V}}{6 \text{ }\Omega} = 3 \text{ A.}$$

19.11 The rules for combining resistors in series and parallel are used to reduce the circuit to an equivalent resistor according to the stages indicated below. The resultant is 9.83 Ω.

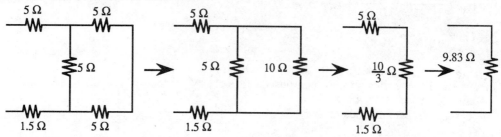

19.12 (a) We are given that, $V_{bc} = 12$ V (See the sketch.)
Therefore, $I = \frac{V_{bc}}{R_{bc}} = \frac{12 \text{ V}}{6 \text{ }\Omega} = 2 \text{ A}$

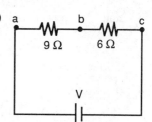

From this, we can find the voltage drop across the 9 Ω resistor as
$$V_{ab} = IR_{ab} = (2 \text{ A})(9 \text{ }\Omega) = 18 \text{ V}$$
Thus, the voltage of the battery is
$V = V_{bc} + V_{ab} = 18 \text{ V} + 12 \text{ V} = 30 \text{ V}$
(b) The voltage drop across the 9 Ω resistor is
$$V = IR = \left(\frac{1}{4} \text{ A}\right)(9 \text{ }\Omega) = 2.25 \text{ V}$$

The voltage drop across both resistors is the same and equal to that of the battery. Thus, the voltage of the battery is also 2.25 V.

CHAPTER NINETEEN SOLUTIONS

19.13 The rules for combining resistors in series and parallel are used to reduce the circuit to an equivalent resistor according to the stages indicated below. The resultant is 5.13 Ω.

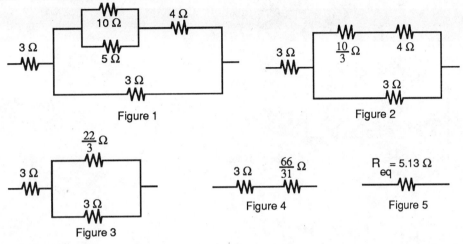

19.14 We are given that the total power supplied to the circuit by the battery is 4 W. Thus,

$$P = I^2 R_{eq}, \quad \text{From which,} \quad I = \sqrt{\frac{P}{R_{eq}}} = \sqrt{\frac{4 \text{ W}}{5.13 \text{ Ω}}} = 0.883 \text{ A}$$

Thus, the voltage across the equivalent resistor, which is the voltage of the battery, is

$$V = IR = (0.883 \text{ A})(5.13 \text{ Ω}) = 4.53 \text{ V}$$

CHAPTER NINETEEN SOLUTIONS

19.15 The resistors in the circuit can be combined in the stages shown below to yield an equivalent resistance of $\frac{63}{11} \Omega$.

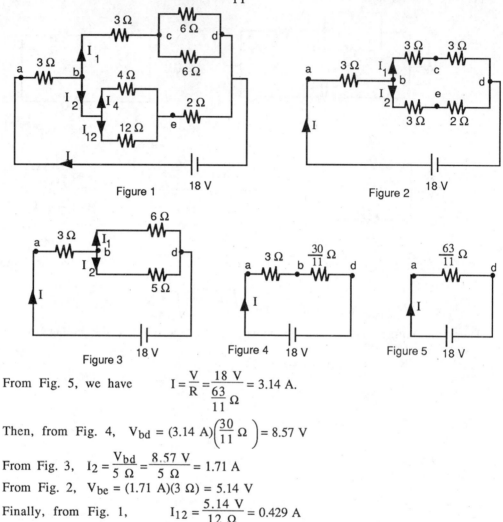

From Fig. 5, we have $I = \frac{V}{R} = \frac{18 \text{ V}}{\frac{63}{11} \Omega} = 3.14$ A.

Then, from Fig. 4, $V_{bd} = (3.14 \text{ A})\left(\frac{30}{11} \Omega\right) = 8.57$ V

From Fig. 3, $I_2 = \frac{V_{bd}}{5 \Omega} = \frac{8.57 \text{ V}}{5 \Omega} = 1.71$ A

From Fig. 2, $V_{be} = (1.71 \text{ A})(3 \Omega) = 5.14$ V

Finally, from Fig. 1, $I_{12} = \frac{5.14 \text{ V}}{12 \Omega} = 0.429$ A

19.16 First, consider the parallel case. We know the voltage drop across B is 6 V, and we are given that the current through it is 2 A. Thus, we can find the resistance of B as

$$R_B = \frac{6 \text{ V}}{2 \text{ A}} = 3 \Omega.$$

Now, consider the series connection. We are given that the voltage across A is 4 V, but we also know that the voltage across B must be 2 V. Thus, let us apply Ohm's law to B to find the current in the circuit. We find

$$I = \frac{V}{R_B} = \frac{2 \text{ V}}{3 \Omega} = \frac{2}{3} \text{ A}$$

Thus, the resistance of A can now be found as

CHAPTER NINETEEN SOLUTIONS

$$R_A = \frac{V}{I} = \frac{4 \text{ V}}{\frac{2}{3} \text{ A}} = 6 \text{ }\Omega.$$

19.17 (a) The resistors can be combined as shown below to yield an equivalent of 6.6 Ω.

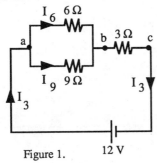

Figure 1.

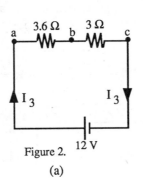

Figure 2.   Figure 3.

(a)

From Fig. 3, $I_3 = \frac{12 \text{ V}}{6.6 \text{ }\Omega} = 1.82 \text{ A}$

Then from Fig. 2, the voltage drop across the 3.6 Ω resistor is
$V_{3.6} = (1.82 \text{ A})(3.6 \text{ }\Omega) = 6.55 \text{ V}.$

From Fig. 1, $I_6 = \frac{6.55 \text{ V}}{6 \text{ }\Omega} = 1.09 \text{ A}$, and

$I_9 = \frac{6.55 \text{ V}}{9 \text{ }\Omega} = 0.727 \text{ A}$,

(b) KIRCHOFF'S RULES
First, we will apply the junction rule at a. (See sketch.)
$I_3 = I_6 + I_9$ (1)
Now, we apply the loop rule to acba
$-6I_6 + 9I_9 = 0$
From which, we find $I_6 = 1.5 I_9$ (2)
Next, we apply the loop rule to abcdea to find
$-9I_9 - 3I_3 + 12 = 0$, or, $3I_9 + I_3 = 4$ (3)
(1), (2), and (3) can be solved simultaneously to find
$I_3 = 1.82 \text{ A}$, $I_6 = 1.09 \text{ A}$,   and $I_9 = 0.727 \text{ A}$

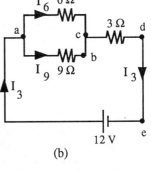

(b)

19.18 (a) If A is grounded, $V_A = 0$, $V_B = 10 \text{ V}$, $V_C = 5 \text{ V}$,
(b) If B is grounded, $V_A = -10$, $V_B = 0 \text{ V}$, $V_C = -5 \text{ V}$,
(c) If C is grounded, $V_A = -5 \text{ V}$, $V_B = 5 \text{ V}$, $V_C = 0 \text{ V}$.

19.19 The current in the circuit is
$I = \frac{3 \text{ V}}{300 \text{ }\Omega} = 0.01 \text{ A} = 0.01 \text{ C/s}$

Thus, if the available charge = 180 C, the batteries will last
$t = \frac{180 \text{ C}}{0.01 \text{ C/s}} = 18{,}000 \text{ s} = 5 \text{ h}$

CHAPTER NINETEEN SOLUTIONS

19.20   From the loop rule
$-(0.255\ \Omega)(0.6\ A) + 1.5\ V - (0.153\ \Omega)(0.6\ A)$
$\qquad + 1.5\ V - (0.6\ A)R = 0$
From which, $\qquad R = 4.59\ \Omega$.

The total power dissipated = I^2R
$\qquad = (0.6\ A)^2(0.255\ \Omega + 0.153\ \Omega + 4.59\ \Omega) = 1.8\ W$
and the total power dissipated within the batteries
$\qquad = (0.6\ A)^2(0.255\ \Omega) + (0.6\ A)^2(0.153\ \Omega) = 0.147\ W$

Thus, the fraction of the power dissipated internally = $\dfrac{0.147}{1.8\ W} = 0.082$

19.21

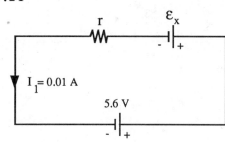
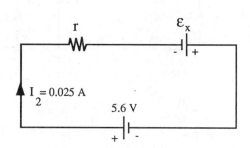

For the situation shown first,
$-\mathcal{E}_x - 0.01r + 5.6 = 0$ (1)
For the second situation,
$5.6 - 0.025\ r + \mathcal{E}_x = 0$
These two equations can be solved simultaneously to find
$\qquad r = 320\ \Omega$
and $\qquad \mathcal{E}_x = 2.4\ V$

19.22   The loop rule becomes
$\qquad 10\ V - (600\ \Omega)I + 20\ V - (400\ \Omega)I = 0$
and $\qquad I = 0.03\ A = 30\ mA$
also $\qquad V = IR = (0.03\ A)(400\ \Omega) = 12\ V$

19.23   (a) Moving clockwise around the loop, we find
$\qquad 20 - 2000\ I - 30 - 2500\ I + 25 - 500\ I = 0$
from which $I = 0.003\ A = 3\ mA$
(b) Start at the grounded point and move up the left side.
$\qquad V_A = 20\ V - 2000\ \Omega(0.003\ A) - 30\ V - 1000\ \Omega(0.003\ A)$
$\qquad V_A = -19.0\ V$
(c) $V_{1500} = (1500\ \Omega)(0.003\ A) = 4.5\ V$ (The upper end is at the higher potential.)

19.24   See sketch for the designation of
currents and their directions.
We apply the junction rule at point a.
$\qquad I_3 = I_1 + I_2$ (1)

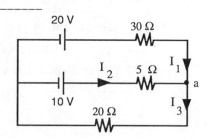

Now, we use the loop rule on the upper loop.
$+20 - 30I_1 + 5I_2 - 10 = 0$,
or $\qquad 6I_1 - I_2 = 2$ (2)
Finally, apply the loop rule to the lower loop.
$10 - 5I_2 - 20I_3 = 0$, or $\qquad I_2 + 4I_3 = 2$ (3)

274

CHAPTER NINETEEN SOLUTIONS

Equations (1), (2), and (3) can be solved together to find,
$I_1 = 0.353$ A, $I_2 = 0.118$ A, and $I_3 = 0.471$ A

19.25 See sketch for current designations and directions.
Apply the junction rule at point a.
$I_2 = I_1 + 2$ (1)

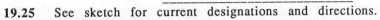

Use loop rule on outer loop, cdabc:
$-2I_1 + 6 + 2(2) - 12 = 0$
From which, $I_1 = -1$ A (2)
From (1), we can now find I_2.
$I_2 = 1$ A.
Finally, use the loop rule on the loop on the right.
$-\varepsilon + 3I_2 + 2(2) - 12 = 0$
Which yield, $\varepsilon = -5$ V (Thus, the polarity of the battery is reversed from that shown.)

19.26 Since the center branch is not a continuous conducting path, no current flows in that branch. We apply the loop rule to the outer loop of the circuit.
$12 \text{ V} - 10I - 5I - 8\text{V} - 2I - 3I = 0$
so $I = \dfrac{4 \text{ V}}{20 \text{ }\Omega} = 0.2$ A

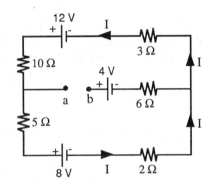

Now, start at b and go around the upper loop
$V_{ab} + 10(0.2 \text{ A}) - 12 \text{ V} + (3 \text{ }\Omega)(0.2 \text{ A}) + 0 + 4 \text{ V} = 0$
from which, $V_{ab} = 5.4$ V (with point a at a higher potential than b)

19.27 Apply the junction rule at point a,
$I_3 = I_1 + I_2$
so, $I_3 = 0.833$ A from the known values of the other currents.

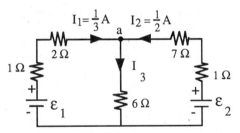

Now, start at point a and apply the loop rule, going clockwise.

$(-6 \text{ }\Omega)(0.833 \text{ A}) + \varepsilon_1 - (1 \text{ }\Omega)(0.333 \text{ A}) - (2 \text{ })(0.333 \text{ A}) = 0$
so $\varepsilon_1 = 6$ V
Next, start at point a and go around the right loop clockwise, applying the loop rule.
$(7 \text{ }\Omega)(0.5 \text{ A}) + (1 \text{ }\Omega)(0.5 \text{ A}) - \varepsilon_2 + (6 \text{ }\Omega)(0.833 \text{ A}) = 0$
which gives $\varepsilon_2 = 9$ V

CHAPTER NINETEEN SOLUTIONS

19.28 Junction rule at point a:
$I_1 = I_2 + I_3$ (1)
loop rule on upper loop,
$-3(I_3) + 24 - 2(I_2 + I_3) - 4(I_2 + I_3) = 0$
which reduces to $2I_2 + 3I_3 = 8$ (2)
and the loop rule for the lower loop
$-(I_2) - 5(I_2) + 12 + 3(I_3) = 0$
or $2I_2 - I_3 = 4$ (3)
Solve (1), (2), and (3) simultaneously, to find
$I_1 = 3.5$ A, $I_2 = 2.5$ A, and $I_3 = 1$ A.

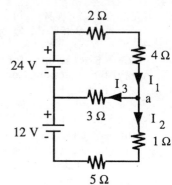

19.29
Apply the junction rule at point a.
$I_1 + I_2 = 2$
or,
$I_2 = 2 - I_1$.

Apply the loop rule to the right hand loop moving clockwise from a.

$-8 V + 6I_1 - 12(2 - I_1) = 0.$ From which, $I_1 = 1.78$ A.
Finally, use the loop rule on the left hand loop moving clockwise from a.
$+ \varepsilon_1 - (4 \Omega)(2 A) - (6 \Omega)(1.78 A) + 8 V = 0$
gives $\varepsilon_1 = 10.67$ V with the polarity indicated.

19.30 First, apply the junction rule at a.
$I_3 = I_1 + I_2$ (1)
Now apply the loop rule to the loop on the left.
$-3 - 4I_3 - 5I_1 + 12 = 0$,   or
$4I_3 + 5I_1 = 9$ (2)
Apply loop rule to right loop
$-18 + 2I_2 + 3I_2 + 4I_3 + 3 = 0,$
or $5I_2 + 4I_3 = 15$ (3)
Solving (1), (2), and (3) simultaneously yields
$I_1 = 0.323$ A, $I_2 = 1.523$ A, and $I_3 = 1.846$ A
Therefore,
$V_2 = 2I_2 = 3.05$ V, $V_3 = 3I_2 = 4.57$ V, $V_4 = 4I_3 = 7.38$ V, and $V_5 = 5I_1 = 1.62$ V.

19.31 (a) $\tau = RC = (2 \times 10^6 \, \Omega)(6 \times 10^{-6} \, F) = 12$ s
(b) $Q_{max} = CV = (6 \times 10^{-6} \, F)(20 \, V) = 120 \, \mu C$
(c) $I_0 = \dfrac{V}{R} = \dfrac{20 \, V}{2 \times 10^6 \, \Omega} = 1.0 \times 10^{-5}$ A

19.32 (a) $\tau = RC = (100 \, \Omega)(20 \times 10^{-6} \, F) = 2 \times 10^{-3}$ s

CHAPTER NINETEEN SOLUTIONS

(b) $Q_{max} = CV = (20 \times 10^{-6} F)(9 V) = 180 \mu C$

(c) $I_{max} = \frac{V}{R} = \frac{9 V}{100 \Omega} = 0.09 A$

(d) After one time constant, $Q = 0.63 Q_{max} = 113 \mu C$

(e) After one time constant $I = 0.37 I_{max} = 3.33 \times 10^{-2} A$

19.33 In an interval equal to a time constant, the capacitor gains an amount of charge equal to 0.63 X (the difference between the charge at the beginning of the interval and the eventual final charge).

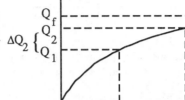

Thus, during $\Delta t = \tau$ (starting at $t = 0$),
 charge gained $= Q_1 = 0.63(Q_f)$
and during $\Delta t = \tau$ (starting at $t = \tau$),
$\Delta Q_2 = (Q_2 - Q_1) = 0.63(Q_f - Q_1)$, or
$Q_2 = Q_1 + 0.63 Q_f - 0.63 Q_1 = 0.37 Q_1 + 0.63 Q_f = 0.37(0.63)Q_f + 0.63 Q_f = 0.863 Q_f$

19.34 $\tau = RC = (10^6 \Omega)(5 \times 10^{-6} F) = 5 s$
Thus, at $t = 10 s$, $t = 2\tau$.

$$I_0 = \frac{\varepsilon}{R} = \frac{30 V}{10^6 \Omega} = 30 \mu A$$

at $t = \tau$, $I_1 = 0.37 I_0 = 0.37(30 \mu A) = 11.1 \mu A$
at $t = 2\tau$, $I_2 = 0.37 I_1 = 0.37(11.1 \mu A) = 4.1 \mu A$

19.35 (a) $R = \frac{\tau}{C} = \frac{5 s}{10^{-5} F} = 5 \times 10^5 \Omega$.

(b) After one time constant, the current will be

$$I = 0.37 I_{max} = 0.37 \left(\frac{9 V}{5 \times 10^5 \Omega.}\right) = 6.66 \times 10^{-6} A$$

After a second time constant, $I = (0.37)(6.66 \times 10^{-6} A) = 2.46 \times 10^{-6} A$

19.36 (a) We use $Q = Q_{max}(1 - e^{-t/\tau})$
with, $Q_{max} = CV = (2 \times 10^{-6} F)(9 V) = 18 \mu C$ and $\tau = RC = 6 s$
Thus, $Q = 18 \mu C(1 - e^{-10/6}) = 18 \mu C(1 - e^{-1.67}) = 18 \mu C(0.811) = 14.6 \mu C$

(b) $I_{max} = \frac{V}{R} = \frac{9 V}{3 \times 10^6 \Omega} = 3 \times 10^{-6} A$

Then, $I = I_{max} e^{-t/\tau} = (3 \times 10^{-6} A)e^{-1.67} = 0.567 \mu A$

19.37 (a) The resistors in the circuit can be replaced by an equivalent resistance which has a magnitude of 5 Ω. (See sketch.) This means that the current drawn from the battery is 1 A. The voltage across the parallel branch is then easily found to be 2 V.

CHAPTER NINETEEN SOLUTIONS

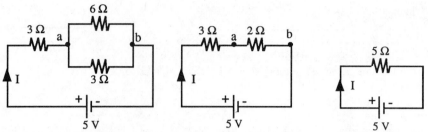

(b) If a 1 Ω resistance is then connected between a and b (in parallel with the elements already present), then $R_{ab} = 0.667$ Ω. The total resistance of the circuit is then 3.667 Ω and the current will be,

$$I = \frac{5 \text{ V}}{3.667 \text{ Ω}} = 1.364 \text{ A}$$

From this the voltage drop between points a and b (which will be the voltage read by the meter) is,

$$V_{ab} = (0.667 \text{ Ω})(1.364 \text{ A}) = 0.909 \text{ V}$$

(c) If a 1 MΩ resistor is connected between points a and b, then $R_{ab} = 1.999996$ Ω, and the current in the circuit is

$$I = \frac{5 \text{ V}}{4.999996 \text{ Ω}} = 1.0000008 \text{ A}$$

and the voltage drop between points a and b which will be read by the meter is

$$V_{ab} = 1.9999976 \text{ V}$$

(Note that we have carried out the answers here to several more decimal points than one could hope to read with a conventional meter. The purpose, however, is to show that the high resistance voltmeter is necessary in order to approach a reading equal to the voltage that would be present in the absence of a meter.)

19.38 (a) The loop rule gives
$$(30 \times 10^3 \text{ Ω})I = 275 \text{ V}$$
$$I = 9.17 \text{ mA}$$
Thus, $V_1 = IR_1 = (9.17 \times 10^{-3} \text{ A})(10 \times 10^3 \text{ Ω}) = 91.7$ V

(b) (See diagram b.)
$$\frac{1}{R'_1} = \frac{1}{(10 \times 10^3 \text{ Ω})} + \frac{1}{30.8 \times 10^3 \text{ Ω}}$$
From which, $R'_1 = 7.55 \times 10^3$ Ω.
and $R_{eq} = R'_1 + R_2 = 27.55$ kΩ

(c) Thus, from the loop rule
$$(27.55 \times 10^3 \text{ Ω})I' = 275 \text{ V}$$
giving $I' = 9.98$ mA
Finally,
$V'_1 = V_{ab} = I'R'_1$
$= (9.98 \times 10^{-3} \text{ A})(7.55 \times 10^3 \text{ Ω}.)$
$= 75.4$ V

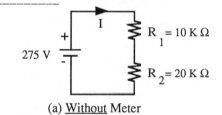

(a) <u>Without</u> Meter

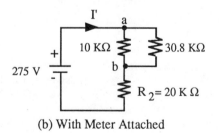

(b) With Meter Attached

19.39 In the absence of the meter,
$$I = \frac{V}{R} = \frac{100 \text{ V}}{25 \text{ Ω}} = 4.00 \text{ A}$$

CHAPTER NINETEEN SOLUTIONS

with the meter in place $I' = \dfrac{V}{R'} = \dfrac{100 \text{ V}}{30 \text{ }\Omega} = 3.33$ A

percentage error $= \dfrac{4.00 \text{ A} - 3.33 \text{ A}}{4.00 \text{ A}} 100\% = 16.7\%$

19.40 (a) $I = \dfrac{12 \text{ V}}{6 \text{ }\Omega} = 2$ A

(b) $I = \dfrac{12 \text{ V}}{106 \text{ }\Omega} = 0.113$ A

(c) $I = \dfrac{12 \text{ V}}{6.1 \text{ }\Omega} = 1.97$ A

19.41 $R_x = R_2 \dfrac{R_3}{R_1} = 25 \text{ }\Omega \left(\dfrac{40}{15} \right) = 66.7 \text{ }\Omega$.

19.42 We use $R_x = R_3 \dfrac{R_1}{R_2}$, and $R = \dfrac{\rho L}{A}$.

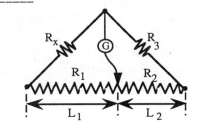

$R_x = R_3 \dfrac{\frac{\rho L_1}{A}}{\frac{\rho L_2}{A}} = R_3 \dfrac{L_1}{L_2}$

19.43 $R_x = R_3 \dfrac{R_2}{R_1} = \dfrac{1000 \text{ }\Omega}{2.5} = 400 \text{ }\Omega$.

19.44 (a) The current drawn by the heater is $I = \dfrac{P}{V} = \dfrac{1300 \text{ W}}{120 \text{ V}} = 10.8$ A

Toaster: $I = \dfrac{P}{V} = \dfrac{1000 \text{ W}}{120 \text{ V}} = 8.33$ A

Grill: $I = \dfrac{P}{V} = \dfrac{1500 \text{ W}}{120 \text{ V}} = 12.5$ A

(b) The total current in the wire leading directly to the power source is the sum of the currents drawn by each element. This is 31.6 A. Thus, a 30 A circuit is insufficient.

19.45 The current drawn by a single bulb is
$I = \dfrac{P}{V} = \dfrac{100 \text{ W}}{120 \text{ V}} = 0.833$ A

Thus, the number of bulbs that a 20 A circuit can supply is
$n = \dfrac{20 \text{ A}}{0.833 \text{ A}} = 24$ bulbs

19.46 Let us first find the current that would be drawn by the heating element when connected to 240 V. This is
$I = \dfrac{P}{V} = \dfrac{3000 \text{ W}}{240 \text{ V}} = 12.5$ A

Ohm's law gives us the resistance of the element.
$R = \dfrac{V}{I} = \dfrac{240 \text{ V}}{12.5 \text{ A}} = 19.2 \text{ }\Omega$.

CHAPTER NINETEEN SOLUTIONS

If the element is connected to a 120 V source, the current drawn is
$$I = \frac{V}{R} = \frac{120 \text{ V}}{19.2 \text{ }\Omega} = 6.25 \text{ A}$$
and the power dissipated is
$$P = VI = (120 \text{ V})(6.25 \text{ A}) = 750 \text{ W}$$

19.47 Total length of wire = 16 ft = 4.88 m,
and the cross-sectional area A = 8.24 X 10^{-7} m^2

Thus, $R = \frac{\rho L}{A} = \frac{(1.7 \times 10^{-8} \text{ }\Omega \text{ m})(4.88 \text{ m})}{8.24 \times 10^{-7} \text{ m}^2} = 0.101 \text{ }\Omega$

(a) $P = I^2 R = (1 \text{ A})(0.101 \text{ }\Omega) = 0.101 \text{ W}$
(b) $P = I^2 R = (10 \text{ A})(0.101 \text{ }\Omega) = 10.1 \text{ W}$

19.48 The cross-sectional area and lengths are the same for both wire, and the power is to be the same in both wires. Thus,
$$I_{al}^2 R_{al} = I_{cu}^2 R_{cu}$$
which reduces to

$$I_{al} = I_{cu} \sqrt{\frac{\rho_{cu}}{\rho_{al}}} = (20 \text{ A}) \sqrt{\frac{1.7 \times 10^{-8} \text{ }\Omega \text{ m}}{2.82 \times 10^{-8} \text{ }\Omega \text{ m}}} = 15.5 \text{ A (max safe current)}$$

19.49 (a) The current drawn by each individual element is found below.

lamp: $I = \frac{V}{R} = \frac{120 \text{ V}}{150 \text{ }\Omega} = 0.8 \text{ A}$

heater: $I = \frac{V}{R} = \frac{120 \text{ V}}{25 \text{ }\Omega} = 4.8 \text{ A}$

fan: $I = \frac{V}{R} = \frac{120 \text{ V}}{50 \text{ }\Omega} = 2.4 \text{ A}$

The total current drawn by the combination is 8.0 A
(b) The elements are connected in parallel to the source. Thus, the voltage across each is also 120 V.
(c) The current in the lamp has been found above to be 0.8 A.
(d) The power expended in the heater is
$$P = IV = (120 \text{ V})(4.8 \text{ A}) = 576 \text{ W}$$

19.50 Using one resistor, the following values are possible.
2 Ω, 4 Ω, and 6 Ω (3 distinct values)
Using two resistors, either in series or in parallel, yields the following possible values.
6 Ω, 8 Ω, 10 Ω, 1.33 Ω, 1.5 Ω, and 2.4 Ω (5 distinct values, and one repeat)
Using all three resistors in all the possible combinations, yields
12 Ω, 4.4 Ω, 5.5 Ω, 7.33 Ω, 1.09 Ω, (5 distinct values)
Total number of distinct values = 13

19.51 (a) The total resistance of the circuit in this case is
$$R_{total} = R_1 + 100 \text{ }\Omega.$$
But, from Ohm's law, $R_{total} = \frac{2 \text{ V}}{10^{-3} \text{ A}} = 2000 \text{ }\Omega.$
Therefore, $R_1 = 1900 \text{ }\Omega.$
(b) In this case, $R_{total} = R_2 + 2000 \text{ }\Omega.$

CHAPTER NINETEEN SOLUTIONS

and, $R_{total} = \dfrac{20 \text{ V}}{10^{-3} \text{ A}} = 20000 \text{ }\Omega$.

Therefore, $R_2 = 18{,}000 \text{ }\Omega$.

(c) $R_{total} = R_3 + 20000 \text{ }\Omega$.

$R_{total} = \dfrac{200 \text{ V}}{10^{-3} \text{ A}} = 200{,}000 \text{ }\Omega$

and, $R_3 = 180{,}000 \text{ }\Omega$.

19.52 (a) When a 5 V power supply is connected between points a and b, the circuit looks like that shown below, and reduces as shown to an equivalent resistance of 0.0999 Ω.

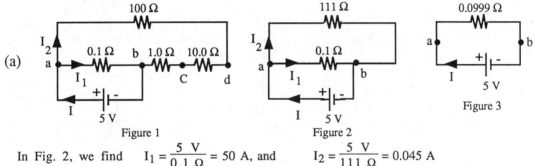

In Fig. 2, we find $I_1 = \dfrac{5 \text{ V}}{0.1 \text{ }\Omega} = 50 \text{ A}$, and $I_2 = \dfrac{5 \text{ V}}{111 \text{ }\Omega} = 0.045 \text{ A}$

(b) When the 5 V source is connected between points a and c, the circuit reduces to an equivalent resistance as shown.

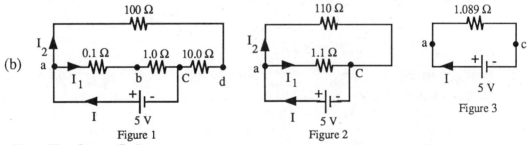

From Fig. 2, we find

$I_1 = \dfrac{5 \text{ V}}{1.1 \text{ }\Omega} = 4.55 \text{ A}$, and $I_2 = \dfrac{5 \text{ V}}{110 \text{ }\Omega} = 0.045 \text{ A}$

(c) Finally, when the 5 V source is connected between points a and d, we have an equivalent resistance of 9.99 Ω.

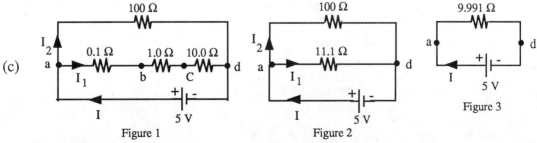

and from Fig. 2,

CHAPTER NINETEEN SOLUTIONS

$I_1 = \dfrac{5 \text{ V}}{11.1 \text{ }\Omega} = 0.45$ A, and $I_2 = \dfrac{5 \text{ V}}{100 \text{ }\Omega} = 0.05$ A

19.53 The resistors combine to an equivalent resistance of 15 Ω as shown.

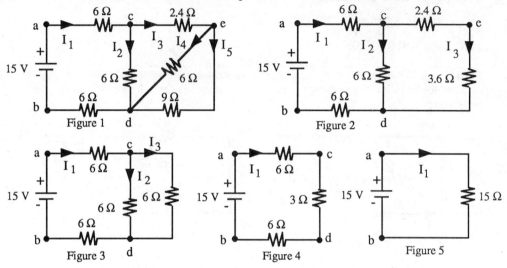

Figure 1　　　Figure 2　　　Figure 3　　　Figure 4　　　Figure 5

19.54 (a) From Fig. 4 of the solution to problem 53, we find

$I_1 = \dfrac{V_{ab}}{R_{ab}} = \dfrac{15 \text{ V}}{15 \text{ }\Omega} = 1$ A

Then $V_{cd} = I_1 R_{cd} = (1 \text{ A})(3 \text{ }\Omega) = 3$ V

From Fig. 3: $I_2 = \dfrac{V_{cd}}{6 \text{ }\Omega} = \dfrac{3 \text{ V}}{6 \text{ }\Omega} = 0.5$ A, and $I_3 = \dfrac{V_{cd}}{6 \text{ }\Omega} = \dfrac{3 \text{ V}}{6 \text{ }\Omega} = 0.5$ A

From Fig. 2: $V_{ed} = I_3 R_{ed} = (0.5 \text{ A})(3.6 \text{ }\Omega) = 1.8$ V

Finally, from Fig. 1: $I_4 = \dfrac{V_{ed}}{6 \text{ }\Omega} = \dfrac{1.8 \text{ V}}{6 \text{ }\Omega} = 0.3$ A, and $I_5 = \dfrac{V_{ed}}{9 \text{ }\Omega} = \dfrac{1.8 \text{ V}}{9 \text{ }\Omega} = 0.2$ A

(b) $V_{ac} = I_1(6 \text{ }\Omega) = 6$ V; $V_{ce} = I_3(2.4 \text{ }\Omega) = 1.2$ V; $V_{ed} = I_4(6 \text{ }\Omega) = 1.8$ V; $V_{fd} = I_5(9 \text{ }\Omega) = 1.8$ V; $V_{cd} = I_2(6 \text{ }\Omega) = 3$ V; and $V_{db} = I_1(6 \text{ }\Omega) = 6$ V

(c) $P_{ac} = I_1^2(6 \text{ }\Omega) = 6$ W; $P_{ce} = I_3^2(2.4 \text{ }\Omega) = 0.6$ W; $P_{ed} = I_4^2(6 \text{ }\Omega) = 0.54$ W; $P_{fd} = I_5^2(9 \text{ }\Omega) = 0.36$ W; $P_{cd} = I_2^2(6 \text{ }\Omega) = 1.5$ W; and $P_{db} = I_1^2(6 \text{ }\Omega) = 6$ W;

19.55 We first find the voltage drop across the 5 Ω resistor.

$V_{cd} = (3 \text{ A})(5 \text{ }\Omega) = 15$ V

This is also the voltage drop across the 10 Ω resistor. Hence,

$I_{10} = \dfrac{15 \text{ V}}{10 \text{ }\Omega} = 1.5$ A

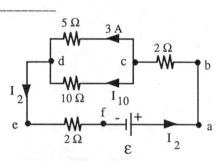

We also see that the current through the two 2 Ω resistors must be equal to the sum of the currents through the 5 Ω and 10 Ω resistors. Thus,

$I_2 = 4.5$ A

We now apply Kirchoff's loop rule to loop abcdefa.

$-2I_2 - 10 I_{10} - 2I_2 + \varepsilon = 0$

282

CHAPTER NINETEEN SOLUTIONS

From which, $\varepsilon = 33$ V

19.56 (a) The time constant is given by
$$\tau = RC = 0.96 \text{ s} \quad (1)$$
and the resistance can be found from the maximum current as
$$R = \frac{V}{I_{max}} = \frac{48 \text{ V}}{0.5 \text{ A}} = 96 \text{ }\Omega. \quad (2)$$
Thus, from (1), $C = 10^{-2}$ F
(b) After one time constant, the charge on the capacitor is
$$Q_1 = Q_{max} - 0.37 Q_{max} ,$$
and after two time constants, (two time constants equal 1.92 s), the charge is
$$Q_2 = Q_{max} - (0.37)^2 Q_{max} = 0.863 Q_{max}$$
Thus, $Q_2 = (0.863)CV = (0.863)(10^{-2} \text{ F})(48 \text{ V}) = 0.414$ C.

19.57 The current in the circuit is
$$I = \frac{V}{R} = \frac{6 \text{ V}}{12 \text{ }\Omega} = 0.5 \text{ A} \quad \text{(clockwise)}$$
Thus, we have
power delivered by 12 V battery = (12 V)(0.5 A) = 6 W
power absorbed by 6 V battery = (6 V)(0.5 A) = 3 W
power dissipated by 7 Ω resistor = $(0.5 \text{ A})^2 (7 \text{ }\Omega)$ = 1.75 W
power dissipated by 3 Ω resistor = $(0.5 \text{ A})^2 (3 \text{ }\Omega)$ = 0.75 W
power dissipated by 2 Ω resistor = $(0.5 \text{ A})^2 (2 \text{ }\Omega)$ = 0.5 W

19.58 Apply the junction rule at a. (See sketch.)
$$I_2 = I_1 + I_3 \quad (1)$$
Now use the loop rule on the left loop.
$-4 - 10I_2 + 9 - 5I_1 = 0$, or $I_1 + 2I_2 = 1 \quad (2)$
Now apply the loop rule to the right loop.
$-4 - 10I_2 + 14 - 10I_3 = 0$, or $I_2 + I_3 = 1 \quad (3)$
Solving (1), (2), and (3) simultaneously yields
$$I_1 = 0; \quad I_2 = I_3 = 0.5 \text{ A}$$

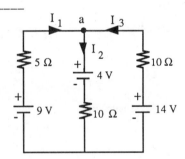

19.59 Apply the junction rule at point a.
$$I = I_1 + I_2$$
or, $1 \text{ A} = 0.1 \text{ A} + I_2$
gives, $I_2 = 0.9$ A
Now use the loop rule around loop abcda.
$$-RI_2 + (41 \text{ }\Omega)I_1 = 0$$
Knowing I_1 and I_2 we find, $R = 4.56$ Ω

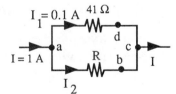

19.60 (a) Around the outside loop we have
$9 \text{ V} - (2 \text{ A})R + 5 \text{ V} = 0,$ or $R = 7$ Ω
(b) For the loop containing the 8 Ω resistor and the 3 V emf we have
$$3 \text{ V} - (8 \text{ }\Omega)I_8 = 0 \quad \text{or} \quad I_8 = \frac{3}{8} \text{ A}$$
And for the loop containing the 8 Ω resistor, the 3 V emf and the 5 V emf, we find

CHAPTER NINETEEN SOLUTIONS

$$5 \text{ V} + 3 \text{ V} - (6 \text{ }\Omega)I_6 = 0 \qquad \text{or} \qquad I_6 = \frac{4}{3} \text{ A}$$

19.61 The 5.1 Ω and 3.5 Ω resistors are in series and reduce to an equivalent of 8.6 Ω.

This 8.6 Ω resistor is in parallel with the 1.8 Ω resistor, and these two reduce to an equivalent of 1.50 Ω. Finally, the 1.5 Ω resistor is in series with the 2.4 Ω and 3.6 Ω resistors. These three add to give a resultant resistance of 7.5 Ω.

19.62 (a) The circuit is reduced as shown below to an equivalent resistance of 14 Ω.

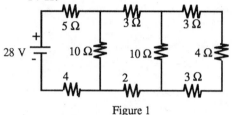
Figure 1

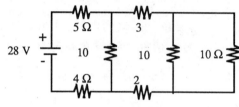
Figure 2

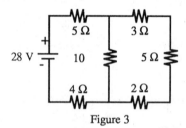
Figure 3

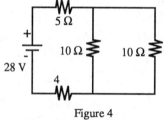
Figure 4

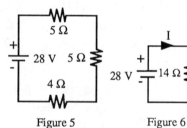
Figure 5   Figure 6

(b) The power dissipated in the entire circuit is:
$$P = I^2 R_{eq} = (2 \text{ A})^2 (14 \text{ }\Omega) = 56 \text{ W}$$

(c) From Fig. 6, we have $I = \dfrac{28 \text{ V}}{14 \text{ }\Omega} = 2 \text{ A}$

19.63 The time constant is $\tau = RC = (2 \times 10^6 \text{ }\Omega)(3 \times 10^{-6} \text{ F}) = 6 \text{ s}$
$$Q = Q_{max}(1 - e^{-t/\tau})$$

From which, $e^{-t/\tau} = 1 - \dfrac{Q}{Q_{max}}$

If $Q = 0.9 Q_{max}$, then $e^{-t/\tau} = 1 - 0.9 = 0.1$.

Now take antilog of above. $\ln(e^{-t/\tau}) = \ln(0.1)$

or, $-\dfrac{t}{\tau} = \ln(0.1)$, or $t = -(6 \text{ s})(\ln(0.1)) = 13.8 \text{ s}$

19.64 (a) The current in the circuit is
$$I = \frac{V}{R} = \frac{6 \text{ V}}{12 \text{ }\Omega} = 0.5 \text{ A}$$

(b) The total resistance of the circuit is 12 Ω. Thus, the power dissipated as heat is,
$$P = I^2 R = (0.5 \text{ A})^2 (12 \text{ }\Omega) = 3 \text{ W}$$

(c) The power absorbed in the 2 V battery through chemical changes is

CHAPTER NINETEEN SOLUTIONS

$P = IV = (0.5 \text{ A})(2 \text{ V}) = 1 \text{ W}$

The power dissipated as heat in the internal resistance of the 2 V battery is

$P = I^2R = (0.5 \text{ A})^2(0.25 \text{ }\Omega) = 0.063 \text{ W}$

Thus, the total power absorbed is $P = 1.06 \text{ W}$

19.65 Using the loop rule in case 1, we have

$\mathcal{E} - (0.3 \text{ A})r - (0.3 \text{ A})(5 \text{ }\Omega) = 0$ (1)

Once again, use the loop rule for case 2.

$\mathcal{E} - (0.2 \text{ A})r - (0.2 \text{ A})(8 \text{ }\Omega) = 0$ (2)

(2) and (1) are solved simultaneously to find

$r = 1 \text{ }\Omega$, and $\mathcal{E} = 1.8 \text{ V}$

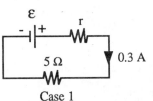
Case 1

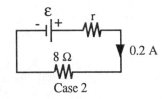
Case 2

19.66 From case 1, (see sketch)

$V = R_1(2 \text{ A})$ (1)

and from case 2,

$V = (R_1 + 3 \text{ }\Omega)(1.6 \text{ A})$ (2)

(1) and (2) are solved simultaneously, and the result is

$R_1 = 12 \text{ }\Omega$.

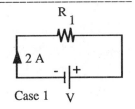
Case 1

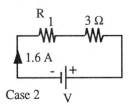
Case 2

19.67 Choosing directions and labeling currents as shown in the figure, we have the following equations:

$6 - I_1 - 2I_2 = 0$
$2 - 3I_3 + I_1 = 0$
$4 - 4I_4 - 2I_2 = 0$
$I_1 - I_2 + I_3 + I_4 = 0$

Solving for the currents, we find

$I_1 = \frac{34}{25} \text{ A}$,

$I_2 = 2.32 \text{ A}$, $I_3 = \frac{28}{25} \text{ A}$, and $I_4 = -0.16 \text{ A}$

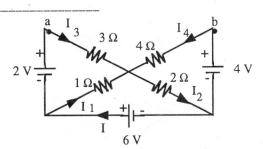

The current through the 6 V battery is $I = I_1 + I_3 = \frac{34 + 28}{25} \text{ A} = 2.48 \text{ A}$

CHAPTER NINETEEN SOLUTIONS

19.68 Labeling currents and directions as shown in the sketch, we have
$I_1 + I_4 = I_2 + I_3$
$6I_1 - 6I_4 = 0$
$6I_1 + 5I_3 - 4.5 = 0$
$6I_1 + 10I_2 - 6 = 0$

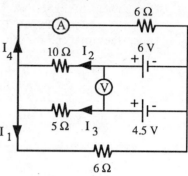

Solution of these equations yield ammeter reading
$$I_4 = 0.39 \text{ A}$$

Going around the loop containing the 6 V and 4.5 V emfs and the voltmeter, we have
$-6 \text{ V} + 4.5 \text{ V} + V_m = 0$.
Thus, $V_m = 1.5$ V

19.69 In series $R_{eq} = R_1 + R_2 + R_3 + \ldots + R_n$ and if each resistor equals R then
$R_{eq} = nR$

$$I_s = \frac{V}{R_{eq}} = \frac{V}{nR} \quad \text{and} \quad P_s = I_s^2 R_{eq} = \frac{V^2}{n^2 R^2} nR = \frac{V^2}{nR}$$

In parallel $\frac{1}{R_{eq}} = \frac{1}{R_1} + \frac{1}{R_2} + \frac{1}{R_3} + \ldots$ and if each resistor equals R then
$\frac{1}{R_{eq}} = \frac{n}{R}$ or $R_{eq} = \frac{R}{n}$

$$I_p = \frac{V}{R_{eq}} = \frac{nV}{R} \quad \text{and} \quad P_p = I_p^2 R_{eq} = \frac{n^2 V^2}{R^2} \frac{R}{n} = \frac{nV^2}{R}$$

Therefore, $\frac{P_s}{P_p} = \frac{1}{n^2}$ or $P_s = \frac{P_p}{n^2}$

CHAPTER TWENTY SOLUTIONS

20.1 From, $F = qvB\sin\theta$

we find, $v = \dfrac{F}{qB} = \dfrac{6 \times 10^{-11} \text{ N}}{(1.6 \times 10^{-19} \text{ C})(2 \text{ T})} = 1.88 \times 10^8$ m/s

---

20.2 $F = qvB\sin\theta = (1.6 \times 10^{-19} \text{ C})(2 \times 10^7 \text{ m/s})(1 \times 10^{-1} \text{ T}) = 3.2 \times 10^{-13}$ N
and from Newton's second law,

$a = \dfrac{F}{m} = \dfrac{3.2 \times 10^{-13} \text{ N}}{1.67 \times 10^{-27} \text{ kg}} = 1.92 \times 10^{14}$ m/s^2.

20.3 (a) $F = qvB\sin\theta = (1.6 \times 10^{-19} \text{ C})(3 \times 10^6 \text{ m/s})(3 \times 10^{-1} \text{ T})\sin 37°$
 $= 8.67 \times 10^{-14}$ N

(b) and $a = \dfrac{F}{m} = \dfrac{8.67 \times 10^{-14} \text{ N}}{1.67 \times 10^{-27} \text{ kg}} = 5.19 \times 10^{13}$ m/s^2.

20.4 To balance the weight of the proton, the magnitude of the magnetic force must equal the weight.
 $F = qvB = mg$

and, $B = \dfrac{mg}{qv} = \dfrac{(1.67 \times 10^{-27} \text{ kg})(9.8 \text{ m/s}^2)}{(1.6 \times 10^{-19} \text{ C})(2.5 \times 10^6 \text{ m/s})} = 4.09 \times 10^{-14}$ T

(b) B should be in the horizontal plane if F is to be in the vertical direction.

20.5 (a) These are all applications of the right hand rule.
(a) to left, (b) into page (c) out of page
(d) toward top of page (e) into page (f) out of page
(b) If the charge is negative, the right hand rule is still used, but the direction for the force is changed by 180°. Thus, the answers here are reversed in direction from those given above.

20.6 Use the right hand rule, realizing that, at the magnetic equator, B is parallel to the surface of Earth and directed northward and that the electron has a negative charge to find the following answers:
(a) westward
(b) no deflection, v is parallel to B
(c) upward
(d) downward

20.7 (a) into page, (b) toward the right (c) toward bottom of page

---

20.8 n = number of Na$^+$ ions present = 3×10^{20} 1/cm^3 (100 cm^3) = 3×10^{22} ions.
The force on a single ion = $F = qvB\sin\theta$
 $= (1.6 \times 10^{-19} \text{ C})(0.851 \text{ m/s})(0.254 \text{ T})\sin 51° = 2.69 \times 10^{-20}$ N
Thus, assuming all ions move in the same direction through the field, the total force is

CHAPTER TWENTY SOLUTIONS

$F_{total} = nF = (3 \times 10^{22} \text{ ions})(2.69 \times 10^{-20} \text{ N}) = 806 \text{ N}$

20.9 From $F = qvB\sin\theta$, we have

$$\sin\theta = \frac{F}{qvB} = \frac{3 \times 10^{-12} \text{ N}}{(1.6 \times 10^{-19} \text{ C})(5 \times 10^7 \text{ m/s})(2 \text{ T})} = 0.188$$

and, $\theta = 10.8°$

20.10 First find the speed of the electron:

$$\Delta KE = \frac{1}{2} mv^2 = eV = \Delta PE$$

$$v = \sqrt{\frac{2eV}{m}} = \sqrt{\frac{2(1.6 \times 10^{-19} \text{ C})(2400 \text{ J/C})}{(9.11 \times 10^{-31} \text{ kg})}} = 2.9 \times 10^7 \text{ m/s}$$

(a) $F_{max} = qvB = (1.6 \times 10^{-19} \text{ C})(2.9 \times 10^7 \text{ C})(1.7 \text{ T}) = 7.90 \times 10^{-12} \text{ N}$

(b) $F_{min} = 0$ occurs when v is either parallel to or anti-parallel to B

20.11 $F = ma = (1.67 \times 10^{-27} \text{ kg})(2 \times 10^{13} \text{ m/s}^2)$
$= 3.34 \times 10^{-14} \text{ N}$

$B = \frac{F}{qv} = \frac{3.34 \times 10^{-14} \text{ N}}{(1.6 \times 10^{-19} \text{ C})(1 \times 10^7 \text{ m/s})} = 2.09 \times 10^{-2} \text{ T}$

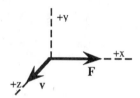

The right hand rule shows that B must be in the -y direction to yield a force in the +x direction when v is in the z direction.

20.12 The work done by a force is $W = Fs\cos\theta$ where θ is the angle between the direction of the force and the direction of the displacement s. But, if F is a magnetic force acting on a charged particle, then F is always perpendicular to the direction of the velocity and hence the displacement. Therefore, $\theta = 90°$, and $\cos\theta = 0$. Thus, $W = 0$.

20.13 $F = BIL = (1.2 \text{ T})(10 \text{ A})(2 \text{ m}) = 24 \text{ N}$

20.14 $F = BIL\sin\theta = (0.3 \text{ T})(10 \text{ A})(5 \text{ m})\sin 30° = 7.5 \text{ N}$

20.15 $F = BIL = (5 \times 10^{-5} \text{ T})(10 \text{ A})(20 \text{ m}) = 1.0 \times 10^{-2} \text{ N}$ The force is westward (use right-hand rule)

20.16 $\frac{F}{L} = BI$

Thus, $(0.12 \text{ N/m}) = B(15 \text{ A})$, and $B = 8 \times 10^{-3}$ T
The direction of B must be out of the page (in +z direction) to have F directed as specified.

20.17 (a) to the left,   (b) into page   (c) out of page
(d) toward top of page   (e) into page   (f) out of page

20.18 (a) into page   (b) toward right   (c) toward bottom of page

---

20.19 To find the minimum B field, B must be perpendicular to I, and from the right hand rule, it must be directed downward.
the friction force per unit length = μN

CHAPTER TWENTY SOLUTIONS

$= \mu$(mass per unit length)(g) $= 0.2\left(\dfrac{10^{-3} \text{ kg}}{10^{-2} \text{ m}}\right)$(9.8 m/s^2) = 0.196 N/m

Thus, the magnetic force per unit length must be 0.196 N/m to overcome the friction force.

$$\dfrac{F}{L} = BI = 0.196 \text{ N/m}$$

or $B = \dfrac{0.196 \text{ N/m}}{1.5 \text{ A}} = 0.131 \text{ T}$

20.20 For minimum B, B should be perpendicular to the current, and the right hand rule shows B should be directed eastward to have the magnetic force directed upward when I flows to the south.
To lift wire $F_{mag} = mg = BIL$

so, $B = \dfrac{m}{L}\dfrac{g}{I}$, and $\dfrac{m}{L} = 5 \times 10^{-2}$ kg/m

Thus, $B = (5 \times 10^{-2} \text{ kg/m})\dfrac{9.8 \text{ m/s}^2}{2 \text{ A}} = 0.245$ T

20.21 In order for the rod to float, the magnetic force must have the same magnitude as the weight of the rod. Thus,

$F = BIL\sin\theta = mg$

or, $I = \dfrac{mg}{BL\sin\theta}$. The minimum value of I occurs when $\sin\theta = 1$

Thus, $I_{min} = \dfrac{(5 \times 10^{-2} \text{ kg})(9.8 \text{ m/s}^2)}{(2 \text{ T})(1 \text{ m})} = 0.245$ A

20.22 For the wire to move upward at constant speed, the net force acting on it must be zero. Therefore,

$F_{magnetic} = mg$

or, $BIL\sin\theta = mg$

at $\theta = 90°$, we have $B = \dfrac{mg}{IL} = \dfrac{(0.15 \text{ kg})(9.8 \text{ m/s}^2)}{(5 \text{ A})(0.15 \text{ m})} = 1.96$ T

The magnetic force must be directed upward. Therefore, B must be directed out of page.

20.23 (a) $F = BIL\sin\theta = (0.6 \times 10^{-4} \text{ T})(15 \text{ A})(10 \text{ m})\sin 90° = 9.0 \times 10^{-3}$ N
F is perpendicular to B and by the right hand rule is directed at 15° above the horizontal in the northward direction.
(b) $F = BIL\sin\theta = (0.6 \times 10^{-4} \text{ T})(15 \text{ A})(10 \text{ m})\sin 165° = 2.33 \times 10^{-3}$ N
right hand rule shows F is horizontal and directed due west

20.24 $\tau = NBIA\sin\theta = 100(0.8 \text{ T})(1.2 \text{ A})(0.12 \text{ m}^2)\sin 60° = 9.98$ N m
The right hand rule shows the torque will tend to rotate the coil clockwise as viewed from above.

20.25 $\tau_{max} = NBIA$, and $A = (4 \times 10^{-2} \text{ m})(5 \times 10^{-2} \text{ m}) = 2 \times 10^{-3}$ m^2.

Thus, $I = \dfrac{\tau_{max}}{NBA} = \dfrac{(0.18 \text{ N m})}{(500)(0.65 \text{ T})(2 \times 10^{-3} \text{ m}^2)} = 0.277$ A

20.26 The area of the circular loop is 0.785 m^2, and $\tau = NBIA\sin\theta$.

CHAPTER TWENTY SOLUTIONS

(a) For τ_{max}, $\theta = 90°$, and we have
$$\tau_{max} = BIA = (0.4 \text{ T})(2 \text{ A})(0.785 \text{ m}^2) = 0.628 \text{ N m}$$
(b) If $\tau = \frac{1}{2}\tau_{max}$, then $\sin\theta = \frac{1}{2}$ Thus, $\theta = 30°$

20.27 (See sketch.) The current through the parallel resistor is
$$I = 3 \text{ A} - 10 \text{ mA} = 2.99 \text{ A}$$
Use the loop rule on the closed loop, and
$$-(0.01 \text{ A})(50 \text{ }\Omega) + R_p(2.99 \text{ A}) = 0$$
Which yields, $R_p = 0.167 \text{ }\Omega$.

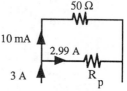

20.28 The current through the coil at full scale deflection is
$$i_c = \frac{50 \times 10^{-3} \text{ V}}{50 \text{ }\Omega} = 10^{-3} \text{ A}$$
Hence, the current that must exist in the parallel branch is
$$I_p = 10 \text{ A} - i_c = 10 \text{ A} - 10^{-3} \text{ A} = 9.999 \text{ A}$$
Therefore, the parallel resistor has a value of
$$R_p = \frac{50 \times 10^{-3} \text{ V}}{9.999 \text{ A}} = 5.00 \times 10^{-3} \text{ }\Omega.$$

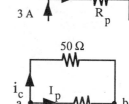

20.29 The voltage drop across the terminals is given by
$$V_{AB} = iR + ir = i(R + r)$$ where $r = 60 \text{ }\Omega$, and R is the unknown resistance connected in series with the meter movement.
$$1 \text{ V} = (5 \times 10^{-4} \text{ A})(R + 60 \text{ }\Omega)$$
gives $R = 1940 \text{ }\Omega$

20.30 The series resistor that must be used has a value found as
$$V = 150 \text{ V} = (R_s + 40 \text{ }\Omega)(2 \times 10^{-3} \text{ A})$$
Which yields, $R_s = 74,960 \text{ }\Omega$.

20.31 See sketch. If $I = 0.3$ A, and A and D are the terminals,
$$I_p(R_1 + R_2 + R_3) = i_c(100 \text{ }\Omega)$$
In this case, $I_p = 0.2999$ A and $i_c = 100 \times 10^{-6}$ A
Thus, $R_1 + R_2 + R_3 = 3.33 \times 10^{-2} \text{ }\Omega$ (1)
If $I = 3$ A, and A and C are the terminals,
$$I_p(R_1 + R_2) = i_c(100 \text{ }\Omega + R_3)$$
Here, $I_p = 2.9999$ A and $i_c = 100 \times 10^{-6}$ A
Thus, we have $R_1 + R_2 - 3.33 \times 10^{-5}R_3 = 3.33 \times 10^{-3} \text{ }\Omega$ (2)
Finally, if $I = 30$ A and A and B are the terminals,
$$I_pR_1 = i_c(100 \text{ }\Omega + R_2 + R_3)$$
With $I_p = 29.9999$ A and $i_c = 100 \times 10^{-6}$ A, this reduces to
$$R_1 - 3.33 \times 10^{-6}R_2 - 3.33 \times 10^{-6}R_3 = 3.33 \times 10^{-4} \text{ }\Omega \quad (3)$$
Solving (1), (2), and (3) simultaneously, we find
$R_1 = 3.33 \times 10^{-4} \text{ }\Omega$, $R_2 = 2.998 \times 10^{-3} \text{ }\Omega$, and $R_3 = 2.997 \times 10^{-2} \text{ }\Omega$,

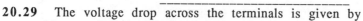

CHAPTER TWENTY SOLUTIONS

20.32 (See sketch.) We have
$$V_{ab} = i_c(R_1 + 100 \, \Omega)$$
or,
$$3 \, V = (100 \times 10^{-6} \, A)(R_1 + 100 \, \Omega)$$
From which, $R_1 = 2.99 \times 10^4 \, \Omega$.
Likewise,
$$V_{ac} = i_c(R_1 + R_2 + 100 \, \Omega)$$
and,
$$30 \, V = (100 \times 10^{-6} \, A)(R_2 + 2.99 \times 10^4 \, \Omega + 100 \, \Omega)$$
Which yields, $R_2 = 2.7 \times 10^5 \, \Omega$.
Finally, we see that
$$V_{ad} = i_c(R_1 + R_2 + R_3 + 100 \, \Omega)$$
or, $300 \, V = (100 \times 10^{-6} \, A)(R_3 + 2.99 \times 10^4 \, \Omega + 2.7 \times 10^5 \, \Omega + 100 \, \Omega)$
and $R_3 = 2.70 \times 10^6 \, \Omega$

20.33 $B = \dfrac{mv}{rq} = \dfrac{(6.68 \times 10^{-27} \, kg)(10^4 \, m/s)}{(1.6 \times 10^{-19} \, C)(3 \times 10^{-2} \, m)} = 1.39 \times 10^{-2} \, T$ (out of page)

20.34 $r = \dfrac{mv}{qB} = \dfrac{(1.67 \times 10^{-27} \, kg)(2 \times 10^5 \, m/s)}{(1.6 \times 10^{-19} \, C)(0.2 \, T)} = 1.04 \times 10^{-2} \, m = 1.04 \, cm$

20.35 $r = \dfrac{mv}{qB}$ gives $v = \dfrac{qrB}{m}$ and

the time to complete one orbit $= \dfrac{\text{circumference}}{\text{speed}}$

$= \dfrac{2\pi r}{v} = \dfrac{2\pi r}{\frac{qrB}{m}} = \dfrac{2\pi m}{qB} = \dfrac{2\pi(1.67 \times 10^{-27} \, kg)}{(1.6 \times 10^{-19} \, C)(0.758 \, T)} = 8.65 \times 10^{-8} \, s$

20.36 $r = \dfrac{mv}{qB}$

Thus, if B doubles, r will be reduced to $\dfrac{1}{2}$ its original value. So,

$r = \dfrac{1}{2} (1.77 \, m) = 0.885 \, m$

20.37 We have $KE = \dfrac{1}{2} mv^2 = 1.6 \times 10^{-19} \, J$

From which, $v = \sqrt{\dfrac{2(1.6 \times 10^{-19} \, J)}{9.11 \times 10^{-31} \, kg}} = 5.93 \times 10^5 \, m/s$

and, $r = \dfrac{mv}{qB} = \dfrac{(9.11 \times 10^{-31} \, kg)(5.93 \times 10^5 \, m/s)}{(1.6 \times 10^{-19} \, C)(0.2 \, T)} = 1.69 \times 10^{-5} \, m$

20.38 To pass through undeflected, the magnitude of the magnetic force must equal the magnitude of the electric force, and the two forces must be opposite in direction so they will cancel. Thus,
$$qvB = qE$$

CHAPTER TWENTY SOLUTIONS

and $v = \dfrac{E}{B}$

20.39 $v = \dfrac{E}{B}$, or

$$B = \dfrac{E}{v} = \dfrac{5 \times 10^5 \text{ N/C}}{10^5 \text{ m/s}} = 5 \text{ T}$$

20.40 We use conservation of energy to find the velocity of the ion upon entering the field.

$$\tfrac{1}{2} mv^2 = qV$$

or, $v = \sqrt{\dfrac{2(1.6 \times 10^{-19} \text{ C})(250 \text{ V})}{2.5 \times 10^{-26} \text{ kg}}} = 5.66 \times 10^4 \text{ m/s}$

and $r = \dfrac{mv}{qB} = \dfrac{(2.5 \times 10^{-26} \text{ kg})(5.66 \times 10^4 \text{ m/s})}{(1.6 \times 10^{-19} \text{ C})(0.5 \text{ T})} = 1.77 \times 10^{-2} \text{ m} = 1.77 \text{ cm}$

20.41 $v = \dfrac{E}{B} = \dfrac{950 \text{ V/m}}{0.93 \text{ T}} = 1.02 \times 10^3 \text{ m/s}$

and $r = \dfrac{mv}{qB} = \dfrac{(2.18 \times 10^{-26} \text{ kg})(1.02 \times 10^3 \text{ m/s})}{(1.6 \times 10^{-19} \text{ C})(0.93 \text{ T})} = 1.50 \times 10^{-4} \text{ m} = 0.15 \text{ mm}$

20.42 (a) The radius of the path as a function of charge is given by

$$r = \dfrac{mv}{qB} = \dfrac{(6.64 \times 10^{-27} \text{ kg})(10^6 \text{ m/s})}{q(0.2 \text{ T})} = \dfrac{3.32 \times 10^{-20} \text{ C m}}{q} \quad (1)$$

If the ion is singly ionized ($q = 1.6 \times 10^{-19}$ C) then (1) gives the radius of the orbit as

$r_1 = 2.08 \times 10^{-1}$ m $= 20.8$ cm

and for the doubly charged ion ($q = 3.2 \times 10^{-19}$ C), we have

$r_2 = 1.04 \times 10^{-1}$ m $= 10.4$ cm

(b) Their distance of separation after one-half their circular path is

$\Delta = 2(20.8 \text{ cm} - 10.4 \text{ cm}) = 20.8$ cm

20.43 $r = \dfrac{mv}{qB} = m\dfrac{v}{qB} = m\dfrac{(3 \times 10^5 \text{ m/s})}{(1.6 \times 10^{-19} \text{ C})(0.6 \text{ T})} = m(3.125 \times 10^{24} \text{ m/kg})$

For U^{235}, $m = 235(1.67 \times 10^{-27} \text{ kg}) = 3.925 \times 10^{-25}$ kg

Thus, $r = (3.125 \times 10^{24} \text{ m/kg})(3.925 \times 10^{-25} \text{ kg}) = 1.226$ m

for U^{238}, $m = 238(1.67 \times 10^{-27} \text{ kg}) = 3.975 \times 10^{-25}$ kg

and $r = (3.125 \times 10^{24} \text{ m/kg})(3.975 \times 10^{-25} \text{ kg}) = 1.242$ m

and $\Delta d = 2\Delta r = 2(1.242 - 1.226) = 3.2 \times 10^{-2}$ m $= 3.2$ cm

20.44 $m = \dfrac{qBr}{v} = \dfrac{(1.6 \times 10^{-19} \text{ C})(1.8 \text{ T})(7.94 \times 10^{-3} \text{ m})}{(4.6 \times 10^5 \text{ m/s})} = 4.97 \times 10^{-27}$ kg

20.45 (a) $B = \dfrac{\mu_0 I}{2\pi r} = \dfrac{(4\pi \times 10^{-7} \text{ T m/A})(5 \text{ A})}{2\pi(10^{-1} \text{ m})} = 10^{-5}$ T

(b) and (c) The method is the same as shown in (a). The answers are

(b) 2×10^{-6} T, and (c) 5×10^{-7} T

CHAPTER TWENTY SOLUTIONS

20.46 We use $B = \dfrac{\mu_0 I}{2\pi r}$, to find $r = \dfrac{\mu_0 I}{2\pi B} = \dfrac{(4\pi \times 10^{-7} \text{ T m/A})(5 \text{ A})}{2\pi(5 \times 10^{-5} \text{ T})}$

$= 2 \times 10^{-2}$ m $= 2$ cm

20.47 We use $B = \dfrac{\mu_0 I}{2\pi r}$,

At the surface of the wire, this becomes

$$0.1 \text{ T} = \dfrac{(4\pi \times 10^{-7} \text{ T m/A})I}{2\pi(10^{-3} \text{ m})}$$

from which, $I = 500$ A

20.48 (a) right to left, (b) out of page, (c) lower left to upper right

20.49 Let us call the wire on the left wire 1 and the wire on the right wire 2. B as a function of r reduces to

$$B = \dfrac{\mu_0 I}{2\pi r} = \dfrac{(4\pi \times 10^{-7} \text{ T m/A})(5 \text{ A})}{2\pi r} = \dfrac{10^{-6} \text{ T m}}{r}$$

(a) At a point midway between the wires, $B_1 = B_2 = \dfrac{10^{-6} \text{ T m}}{0.05 \text{ m}} = 2 \times 10^{-5}$ T

Using the right hand rule, we see that both fields are in the same direction (out of the page), which gives us a resultant field of

$B_{net} = B_1 + B_2 = 4 \times 10^{-5}$ T (directed into page)

(b) At the point P_1, $B_1 = \dfrac{10^{-6} \text{ T m}}{0.2 \text{ m}} = 5 \times 10^{-6}$ T (into the page)

and, $B_2 = \dfrac{10^{-6} \text{ T m}}{0.1 \text{ m}} = 10 \times 10^{-6}$ T (out of the page)

Thus, $B_{net} = 5 \times 10^{-6}$ T (out of page)

(c) At P_2, $B_1 = \dfrac{10^{-6} \text{ T m}}{0.2 \text{ m}} = 5 \times 10^{-6}$ T (out of the page)

and $B_2 = \dfrac{10^{-6} \text{ T m}}{0.3 \text{ m}} = 3.33 \times 10^{-6}$ T (into the page)

for $B_{net} = 1.67 \times 10^{-6}$ T (out of page)

20.50 (a) Let us call the wire on the left 1 and the wire on the right 2.

$B_1 = \dfrac{\mu_0 I}{2\pi r} = \dfrac{(4\pi \times 10^{-7} \text{ T m/A})(3 \text{ A})}{2\pi(10^{-1} \text{ m})} = 6 \times 10^{-6}$ T (upward)

and $B_2 = \dfrac{\mu_0 I}{2\pi r} = \dfrac{(4\pi \times 10^{-7} \text{ T m/A})(5 \text{ A})}{2\pi(10^{-1} \text{ m})} = 1 \times 10^{-5}$ T (downward)

$B_{net} = 4 \times 10^{-6}$ T (downward)

(b) (See the sketch.)
The distance R is
R = $\sqrt{(0.2 \text{ m})^2 + (0.2 \text{ m})^2}$ = 0.283 m
The wire on the right is wire 2 and the field it produces at P_2 is
$B_2 = \frac{\mu_0 I}{2\pi r} = \frac{(4\pi \times 10^{-7} \text{ T m/A})(5 \text{ A})}{2\pi(0.2 \text{ m})} = 5 \times 10^{-6}$ T
and the field from wire 1 on the left is
$B_1 = \frac{\mu_0 I}{2\pi R} = \frac{(4\pi \times 10^{-7} \text{ T m/A})(3 \text{ A})}{2\pi(0.283 \text{ m})} = 2.12 \times 10^{-6}$ T
The angle θ shown in the sketch is 45°.
Thus, the fields can be broken into their x and y components. These resultant components are
$B_x = -6.5 \times 10^{-6}$ T, $B_y = 1.5 \times 10^{-6}$ T,
The Pythagorean theorem gives a resultant field of
B = 6.67×10^{-6} T at 77° to the left of the vertical.

20.51 From $\frac{F}{L} = \frac{\mu_0 I_1 I_2}{2\pi d}$, we have
$d = \frac{\mu_0 I_1 I_2 L}{2\pi F} = \frac{(4\pi \times 10^{-7} \text{ T m/A})(10 \text{ A})(10 \text{ A})(0.5 \text{ m})}{2\pi(1 \text{ N})} = 10^{-5}$ m = 10 μm
It is highly unlikely that a wire of this radius could carry 10 A of current without melting.

20.52 Let us find the force on the wire carrying a current of 5 A. We shall designate this wire as wire 1. The magnitude of the magnetic field set up by wire 2 at the position of wire 1 is.
$B_2 = \frac{\mu_0 I_2}{2\pi r} = \frac{(4\pi \times 10^{-7} \text{ T m/A})(8 \text{ A})}{2\pi(0.3 \text{ m})} = 5.33 \times 10^{-6}$ T
The force on wire 1 is
$F_1 = B_2 I_1 L = (5.33 \times 10^{-6} \text{ T})(5 \text{ A})(2 \text{ m}) = 5.33 \times 10^{-5}$ N (directed toward the other wire)

20.53 (a) We use, $\frac{F_1}{L} = \frac{\mu_0 I_1 I_2}{2\pi d} = (2 \times 10^{-7} \text{ N/A}^2)\frac{(10 \text{ A})^2}{0.1 \text{ m}} = 2 \times 10^{-4}$ N/m (attracted)
(b) The magnitude remains the same as calculated in (a), but the wires are repelled.

20.54 In order for the system to be in equilibrium, the magnetic force per unit length on the top wire must be equal to its weight per unit length.
Thus, $\frac{F}{L}$ = weight per unit length, where $\frac{F}{L} = (2 \times 10^{-7} \text{ T m/A})\frac{I_1 I_2}{d}$
We have, $(2 \times 10^{-7} \text{ T m/A})\frac{(30 \text{ A})(60 \text{ A})}{d} = 0.08$ N/m
We find, $d = 4.5 \times 10^{-3}$ m = 4.5 mm

20.55 The number of turns per unit length is
$n = \frac{N}{L} = \frac{100}{0.15 \text{ m}} = 6.67 \times 10^2$ m^{-1}.

CHAPTER TWENTY SOLUTIONS

and, $B = \mu_0 n I$. From which,

$$I = \frac{B}{\mu_0 n} = \frac{5 \times 10^{-5} \text{ T}}{(4\pi \times 10^{-7} \text{ T m/A})(667/\text{m})} = 5.97 \times 10^{-2} \text{ A} = 59.7 \text{ mA}$$

20.56 We have $B = \mu_0 n I$. Therefore, to increase the field B by a factor of 3, we must increase n by a factor of 3. But, $n = \frac{N}{L}$, or $N = nL$. Thus, if n increases by a factor of 3, N must increase by a factor of 3. If N(initial) = 100 then N(final) = 300 turns.

20.57 The magnetic field at the center of the solenoid is

$$B = \mu_0 n I = (4\pi \times 10^{-7} \text{ T m/A})(3000 \text{ turns/m})(15 \text{ A}) = 5.66 \times 10^{-2} \text{ T}$$

The force on one of the sides of the loop is

$$F = BIL = (5.66 \times 10^{-4} \text{ T})(0.2 \text{ A})(0.02 \text{ m}) = 2.26 \times 10^{-4} \text{ N}$$

With the loop aligned as in the problem, all the forces are directed such that they tend to stretch the loop. There is no tendency for the forces to cause rotation. Thus,

$$\tau = 0.$$

20.58 Ampere's rule is, $\Sigma B_{||} \Delta L = \mu_0 I_{enclosed}$

We apply this rule along the path indicated in the textbook.

$(B_{||})_1 (\Delta L)_1 + (B_{||})_2 (\Delta L)_2 + (B_{||})_3 (\Delta L)_3 + (B_{||})_4 (\Delta L)_4 = \mu_0 I$

$BL + 0 + 0 + 0 = \mu_0 N I$

where N is the number of loops enclosed by the path selected over which to apply Ampere's rule.

Thus, $B = \mu_0 \frac{N}{L} I = \mu_0 n I$

20.59 $n = \frac{N}{L} = \frac{500 \text{ turns}}{0.2 \text{ m}} = 2500 \text{ turns/m}$

and from $B = \mu_0 n I$,

$$I = \frac{B}{\mu_0 n} = \frac{1.2 \times 10^{-4} \text{ T}}{(4\pi \times 10^{-7} \text{ T m/A})(2500 \text{ turns/m})} = 3.82 \times 10^{-2} \text{ A} = 38.2 \text{ mA}$$

20.60 $F = qvB = (1.6 \times 10^{-19} \text{ C})(1.8 \times 10^8 \text{ m/s})(0.6 \text{ T}) = 1.73 \times 10^{-11} \text{ N}$

The force on the negatively charged particle is directed in the +z direction.

20.61 A kinetic energy of 400 eV is equivalent to 6.4×10^{-17} J. Thus, we find the velocity v as

$$v = \sqrt{\frac{2KE}{m}} = \sqrt{\frac{2(6.4 \times 10^{-17} \text{ J})}{9.11 \times 10^{-31} \text{ kg}}} = 1.19 \times 10^7 \text{ m/s}$$

Thus, $B = \frac{mv}{qr} = \frac{(9.11 \times 10^{-31} \text{ kg})(1.19 \times 10^7 \text{ m/s})}{(1.6 \times 10^{-19} \text{ C})(0.8 \text{ m})} = 8.44 \times 10^{-5} \text{ T}$

20.62 $\tau = NBIA\sin\theta = (1)(.3 \text{ T})(25 \text{ A})\pi(0.3 \text{ m})^2 \sin 55° = 1.74 \text{ N m}$

20.63 The radius of the orbit as a function of the mass M of the particle is

CHAPTER TWENTY SOLUTIONS

$$r = \frac{Mv}{qB} = \frac{M(1 \times 10^5 \text{ m/s})}{(1.6 \times 10^{-19} \text{ C})(0.2 \text{ T})} = (3.13 \times 10^{24} \text{ m/kg})M$$

The radius of the orbit of the first particle is
$$r_1 = (3.13 \times 10^{24})(20 \times 10^{-27} \text{ kg}) = 6.25 \times 10^{-2} \text{ m},$$
and the radius followed by the second is
$$r_2 = (3.13 \times 10^{24})(23.4 \times 10^{-27} \text{ kg}) = 7.31 \times 10^{-2} \text{ m}.$$
Thus, the separation after one-half their circular path is
$$\Delta = 2(r_2 - r_1) = 2.12 \text{ cm}$$

20.64 The weight per unit length of the wire $= (0.04 \text{ kg/m})(9.8 \text{ m/s}^2)$
$= 0.392 \text{ N/m}$.
For equilibrium, the magnetic force per unit length must equal the weight per unit length. Thus, $\frac{F}{L} = BI = 0.392 \text{ N/m}$

And, we find $I = \frac{0.392 \text{ N/m}}{3.6 \text{ T}} = 0.109 \text{ A}$. The current must be directed toward the right in order to produce an upward force to counteract the gravitational force.

20.65 Let us call the wire carrying the 3 A current wire 1 and the wire with the 5 A current wire 2. The field at P produced by wire 1 is
$$B_1 = \frac{\mu_0 I}{2\pi r} = \frac{(4\pi \times 10^{-7} \text{ T m/A})(3 \text{ A})}{2\pi(0.3 \text{ m})} = 2 \times 10^{-6} \text{ T} \quad \text{(into page)}$$
and the field due to wire 2 is
$$B_2 = \frac{\mu_0 I}{2\pi r} = \frac{(4\pi \times 10^{-7} \text{ T m/A})(5 \text{ A})}{2\pi(0.4 \text{ m})} = 2.5 \times 10^{-6} \text{ T} \quad \text{(out of page)}$$
Thus, the net field is $B_{net} = 5 \times 10^{-7} \text{ T}$ (out of page)
(b) We use the same designation for discussing the wires as above. The field due to wire 1 is
$$B_1 = \frac{\mu_0 I}{2\pi r} = \frac{(4\pi \times 10^{-7} \text{ T m/A})(3 \text{ A})}{2\pi(0.3 \text{ m})} = 2 \times 10^{-6} \text{ T}$$

Sketch indicates direction of field.
The field due to wire 2 is
$$B_2 = \frac{\mu_0 I}{2\pi r} = \frac{(4\pi \times 10^{-7} \text{ T m/A})(5 \text{ A})}{2\pi(0.3 \text{ m})} = 3.33 \times 10^{-6} \text{ T}$$

For direction, see sketch.

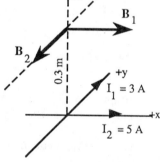

The net field is found from the Pythagorean theorem to be
$$B_{net} = 3.89 \times 10^{-6} \text{ T}$$
The angle, as seen from above the wires (second sketch) is
$\tan\theta = \frac{B_1}{B_2} = 0.6$ and $\theta = 31°$

20.66 (a) $\tau_{max} = NIAB$ so $I = \frac{\tau_{max}}{NAB} = \frac{6 \times 10^{-4}}{(150)(0.12)(0.15)} = 2.22 \times 10^{-4} \text{ A}$

CHAPTER TWENTY SOLUTIONS

(b) Let A', r', and N' represent the area, radius, and number of turns in the single turn coil. Then $\frac{A'}{A} = \left(\frac{r'}{r}\right)^2 = \left(\frac{N}{N'}\right)^2 = \left(\frac{150}{1}\right)^2$ and since the torque is proportional to NA, we have

$$\left(\frac{\tau'}{\tau}\right) = \left(\frac{N'}{N}\right)\left(\frac{A'}{A}\right) = \left(\frac{1}{150}\right)\left(\frac{150}{1}\right)^2 = 150$$

Therefore, in order to maintain the same torque in the larger single turn loop the required current would be smaller by a factor of 150.

20.67 The net magnetic force on the loop is due only to the force on the horizontal segment of the loop in the field. We require that the magnetic force on this segment balance the weight of the added mass, so

$$ILB = mg, \quad \text{or } B = \frac{mg}{IL} = \frac{(13.5 \times 10^{-3} \text{ kg})(9.80 \text{ m/s}^2)}{(2 \text{ A})(20 \times 10^{-2} \text{ m})} = 0.331 \text{ T}$$

20.68 We have $ILB = mg$, which is equivalent to $JALB = \rho(AL)g$ where J is the current density, ρ is the mass density, and A is the cross sectional area. Using the given values, we find $B = \frac{\rho g}{J} = \frac{(2.7 \times 10^3)(9.80)}{2.4 \times 10^6} = 1.10 \times 10^{-2}$ T

20.69 At point C the two currents give rise to oppositely directed magnetic fields. Therefore, to have $B = 0$ at C, we require $B_1 = B_2$ where B_1 is due to 1 and B_2 is due to the 10 A current. Using $B = \frac{\mu_0 I}{2\pi r}$ we have

$$\frac{B_1}{B_1} = \frac{I_1 r_2}{I_2 r_1} = \frac{I \frac{d}{2}}{(10\text{A})\frac{3d}{2}} = 1 \quad \text{or } I = 30 \text{ A}$$

(b) at point A, between the wires,

$$B_1 = \frac{\mu_0 I}{2\pi r} = \frac{(4\pi \times 10^{-7} \text{ T m/A})(30 \text{ A})}{2\pi(5 \times 10^{-2} \text{ m})} = 1.2 \times 10^{-4} \text{ T} \quad \text{(out of page)}$$

and

$$B_2 = \frac{\mu_0 I}{2\pi r} = \frac{(4\pi \times 10^{-7} \text{ T m/A})(10 \text{ A})}{2\pi(5 \times 10^{-2} \text{ m})} = 4 \times 10^{-5} \text{ T} \quad \text{(also out of page)}$$

Thus, $B_{total} = B_1 + B_2 = 1.6 \times 10^{-4}$ T   (out of page)

20.70 The distance to the center of the square is found from the Pythagorean theorem to be $r = 0.141$ m
The magnitude of the magnetic field of all four wires is the same and equal to

$$B_1 = B_2 = B_3 = B_4 = \frac{\mu_0 I}{2\pi r} = \frac{(4\pi \times 10^{-7} \text{ T m/A})(4 \text{ A})}{2\pi(0.141 \text{ m})} = 5.66 \times 10^{-6} \text{ T}$$

CHAPTER TWENTY SOLUTIONS

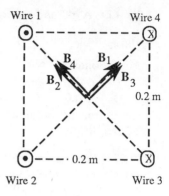

The directions of these fields are indicated in the sketch.
The net field pointing toward wire 4, B_{IV} is
$B_{IV} = B_1 + B_3 = 1.13 \times 10^{-5}$ T

The net field toward wire 1, B_I is
$B_I = B_2 + B_4 = 1.13 \times 10^{-5}$ T
These two fields are perpendicular. Thus, from the Pythagorean theorem, we find
$B_{net} = 1.6 \times 10^{-5}$ T
Direction is toward top of page.

20.71 (a) $B = \dfrac{mv}{qr} = \dfrac{(9.11 \times 10^{-31}\text{ kg})(10^4 \text{ m/s})}{(1.6 \times 10^{-19}\text{ C})(2 \times 10^{-2}\text{ m})} = 2.85 \times 10^{-6}$ T

(b) From $B = \mu_0 n I$,

$I = \dfrac{B}{\mu_0 n} = \dfrac{2.85 \times 10^{-6}\text{ T}}{(4\pi \times 10^{-7}\text{ T m/A})(2500\text{ turns/m})} = 9.06 \times 10^{-4}$ A $= 0.906$ mA

20.72 We have $v = \dfrac{qB}{m} r$

Thus, $KE = \dfrac{1}{2}mv^2 = \dfrac{1}{2} m \dfrac{q^2 B^2}{m^2} r^2$

or, $KE = (\text{constant}) r^2$

Thus, the kinetic energy is proportional to the square of the radius of the orbit.

20.73 The magnetic forces on the top and bottom segments of the rectangle cancel. The force on the left vertical segment is $F_L = \dfrac{\mu_0 I_1 I_2 l}{2\pi c}$ (to the left)
and the force on the right vertical segment is

$F_R = \dfrac{\mu_0 I_1 I_2 l}{2\pi(c+a)}$ (to the right)

Substituting numerical values and finding the resultant of the two ($F_L - F_R$) gives
$F = -2.70 \times 10^{-5}$ N (Force to the left)

20.74 Let Δx_1 be the elongation due to the weight of the wire and let Δx_2 be the additional elongation of the springs when the magnetic field is turned on. Then

$F_{magnetic} = 2k\Delta x_2$ where k is the force constant of the spring and can be determined from $k = \dfrac{mg}{2\Delta x_1}$ (The factor 2 is included in the two previous equations since there are 2 springs supporting the wire.) Combining these two equations, we find

$F_{magnetic} = 2 \dfrac{mg}{2\Delta x_1} \Delta x_2 = \dfrac{mg\Delta x_2}{\Delta x_1}$ But $F = ILB$. Therefore, $B = \dfrac{mg\Delta x_2}{IL\Delta x_1}$

$I = \dfrac{24\text{ V}}{12\ \Omega} = 2$ A so $B = \dfrac{(0.01)(9.8)(0.003)}{(2)(0.05)(0.005)} = 0.588$ T

CHAPTER TWENTY SOLUTIONS

20.75 We require that the torque on the current loop due to the magnetic field balance the elastic restoring torque due to the spring.
$\tau_{el} = \tau_{magnetic}$ or $K\theta = NIAB$ Thus, the full scale deflection current is

$$I = \frac{K\theta}{NAB} = \frac{2 \times 10^{-7} \frac{\pi}{4}}{\frac{0.7}{2\pi(0.015)}\pi(0.015)^2(0.5)} = 60 \ \mu A \text{ (Note that we have used } N = \frac{L}{2\pi r} \text{ for the number of turns.)}$$

CHAPTER TWENTY-ONE SOLUTIONS

21.1 $\Phi = BA = (.25 \text{ T})\pi(0.3 \text{ m})^2 = 7.07 \times 10^{-2} \text{ T m}^2$.

21.2 (a) We have $\Phi = BA\cos\theta$, but when B is perpendicular to the plane of the loop, $\theta = 0°$. Thus,
$\Phi = (5 \times 10^{-5} \text{ T})(20 \times 10^{-4} \text{ m}^2) = 1 \times 10^{-7} \text{ T m}^2$.
(b) $\Phi = (5 \times 10^{-5} \text{ T})(20 \times 10^{-4} \text{ m}^2)\cos 30° = 8.66 \times 10^{-8} \text{ T m}^2$.
(c) $\Phi = (5 \times 10^{-5} \text{ T})(20 \times 10^{-4} \text{ m}^2)\cos 90° = 0$

21.3 $\Phi = BA\cos\theta = (0.3 \text{ T})(4 \text{ m}^2)\cos 50° = 7.71 \times 10^{-1} \text{ T m}^2$.

21.4 (a) There is not a net number of flux lines threading through the cross-sectional area of the coil. Thus, $\Phi = 0$.
(b) The magnetic field lines are in the plane of the coil and do not thread through it. $\Phi = 0$.

21.5 We have $B = \mu_0 nI$, where $n = \dfrac{230 \text{ turns}}{0.2 \text{ m}} = 1250$ turns/m.
Thus,
$\Phi = BA\cos\theta = \mu_0 nIA = (4\pi \times 10^{-7} \text{ T m/A})(1250 \text{ turns/m})(15 \text{ A})\pi(2 \times 10^{-2} \text{ m})^2$
$= 2.96 \times 10^{-5} \text{ T m}^2$

21.6 The field outside the solenoid can be ignored because $B_{out} \ll B_{in}$. Thus, this problem is identical to problem 21.5.

21.7 $\Phi = (B\cos\theta)A =$ (component of B perpendicular to the surface)A
(a) $\Phi_{\text{shaded side}} = (5 \text{ T})(2.5 \times 10^{-2} \text{ m})^2 = 3.13 \times 10^{-3} \text{ T m}^2$
(b) $\Phi_{\text{total}} = 0$ (Since magnetic flux lines have no beginning or end, all flux lines that emerge at one point on a surface enter into the box at some other point.

21.8 (a) $\Phi = BA\cos\theta = B(\pi r^2)\cos\omega t$
(b) $\Phi = BA\cos\theta$ but $\theta = 90°$ at all times, thus $\cos\theta = 0$ at all time, so
$\Phi = 0$ at all times

21.9 $\varepsilon = N\dfrac{\Delta\Phi}{\Delta t} = (2)(3 \text{ T m}^2/\text{s}) = 6 \text{ V}$

21.10 The initial flux through the coil is
$\Phi_i = B_i A = (0.2 \text{ T})\pi(0.2 \text{ m})^2 = 2.51 \times 10^{-2} \text{ T m}^2$

CHAPTER TWENTY-ONE SOLUTIONS

and the final flux linkage is
$$\Phi_f = 0$$
Thus, $\Delta\Phi = \Phi_f - \Phi_i = -2.51 \times 10^{-2}$ T m^2
and the magnitude of the induced emf is
$$\mathcal{E} = N\frac{\Delta\Phi}{\Delta t} = \frac{2.51 \times 10^{-2} \text{ T m}^2}{0.3 \text{ s}} = 8.38 \times 10^{-2} \text{ V} = 83.8 \text{ mV}$$

21.11 At a distance r from the long straight current
$$B_i = \frac{\mu_0 I_i}{2\pi r} = \frac{2 \times 10^{-7} \text{ Tm/A}(6.02 \times 10^6 \text{ A})}{200 \text{ m}} = 6.02 \times 10^{-3} \text{ T}$$
and $B_f = 0$ since $I_f = 0$
so, $\Phi_i = B_i A = (6.02 \times 10^{-3} \text{ T})\pi(0.8 \text{ m})^2 = 1.21 \times 10^{-2}$ T m^2
giving $\mathcal{E} = N\frac{\Delta\Phi}{\Delta t} = (100)\frac{1.21 \times 10^{-2} \text{ T m}^2 - 0}{10.5 \times 10^{-6} \text{ s}} = 1.15 \times 10^5$ V

21.12 $\Phi_i = B_i A = (1.6 \text{ T})(0.2 \text{ m}^2) = 0.32$ T m^2
and $\Phi_f = 0$
so $<\mathcal{E}> = N\frac{\Delta\Phi}{\Delta t} = (200)\frac{0.32 \text{ T m}^2 - 0}{0.02 \text{ s}} = 3200$ V
and $I = \frac{\mathcal{E}}{R} = \frac{3200 \text{ V}}{20 \text{ }\Omega} = 160$ A

21.13 (a) We have, $\Delta\Phi = (\Delta B)A = (0.5 \text{ T} - 0.25 \text{ T})\pi(0.15 \text{ m})^2 = 1.77 \times 10^{-2}$ T m^2
Thus, the average value of the emf is
$$<\mathcal{E}> = \frac{\Delta\Phi}{\Delta t} = \frac{1.77 \times 10^{-2} \text{ T m}^2}{0.7 \text{ s}} = 2.5 \times 10^{-2} \text{ V} = 25.2 \text{ mV}$$
(b) $R = \frac{\mathcal{E}}{i} = \frac{2.5 \times 10^{-2} \text{ V}}{0.8 \text{ A}} = 3.16 \times 10^{-2}$ Ω

21.14 $\Delta\Phi = (B_f - B_i)A = (-0.2 \text{ T} - 0.3 \text{ T})\pi(0.3 \text{ m})^2 = -1.41 \times 10^{-1}$ T m^2.
Thus, the magnitude of the average value of the induced emf is
$$<\mathcal{E}> = \frac{\Delta\Phi}{\Delta t} = \frac{1.41 \times 10^{-1} \text{ T m}^2}{1.5 \text{ s}} = 9.42 \times 10^{-2} \text{ V} = 94.2 \text{ mV}$$

21.15 $\Delta\Phi = \Phi_i - \Phi_f = B\frac{\pi d^2}{4} - 0 = B\frac{\pi d^2}{4}$
and $<\mathcal{E}> = N\frac{\Delta\Phi}{\Delta t} = N\frac{B}{\Delta t}\frac{\pi d^2}{4}$
or $B = \frac{4 \Delta t <\mathcal{E}>}{N\pi d^2} = \frac{4(2.77 \times 10^{-3} \text{ s})(0.166 \text{ V})}{(500)\pi(15 \times 10^{-2} \text{ m})^2} = 5.2 \times 10^{-5}$ T

21.16 $\mathcal{E} = N\frac{\Delta\Phi}{\Delta t}$ but, $\frac{\Delta\Phi}{\Delta t} = \frac{\Delta(BA)}{\Delta t} = B\frac{\Delta A}{\Delta t}$
so, $<\mathcal{E}> = NB\frac{\Delta A}{\Delta t}$
gives 18×10^{-3} V $= (1) B(0.1$ m^2/s$)$
and $B = 0.18$ T

CHAPTER TWENTY-ONE SOLUTIONS

21.17 $\Delta\Phi = (A_f - A_i)B = (0 - A_i)B = -(0.15\ T)\ \pi(0.12\ m)^2 = -6.79 \times 10^{-3}\ T\ m^2$. The magnitude of the average value of the induced emf is

$$<\varepsilon> = \frac{\Delta\Phi}{\Delta t} = \frac{6.79 \times 10^{-3}\ T\ m^2}{0.2\ s} = 3.39 \times 10^{-2}\ V = 33.9\ mV$$

21.18 To produce 0.1 A in an 8 Ω coil, the induced emf must be
$\varepsilon = IR = (0.1\ A)(8\ \Omega) = 0.8\ V$

$\varepsilon = N\frac{\Delta\Phi}{\Delta t}$ but, $\frac{\Delta\Phi}{\Delta t} = \frac{\Delta(BA)}{\Delta t} = A\frac{\Delta B}{\Delta t}$

so $0.8\ V = (75)\frac{\Delta B}{\Delta t}(0.05\ m)(0.08\ m)$

and $\frac{\Delta B}{\Delta t} = 2.67\ T/s$

21.19 $\varepsilon = N\frac{\Delta\Phi}{\Delta t}$ and $\Delta\Phi = \Delta(BA) = (\Delta B)A$

$(\Delta B) = (B_i - B_f) = 0.2\ T - 0 = 0.2\ T$

we have $10 \times 10^3\ V = (500)\frac{(0.2\ T)(\pi(0.05\ m)^2)}{\Delta t}$

gives $\Delta t = 78.5\ \mu s$

21.20 (a) $B_i = \mu_0 n I_i = (4\pi \times 10^{-7}\ Tm/A)\frac{300\ turns}{0.2\ m}(2\ A) = 3.77 \times 10^{-3}\ T$

and, $B_f = \mu_0 n I_f = (4\pi \times 10^{-7}\ Tm/A)\frac{300\ turns}{0.2\ m}(5\ A) = 9.42 \times 10^{-3}\ T$

(a) $\Delta\Phi = (B_f - B_i)A = (9.42 \times 10^{-3}\ T - 3.77 \times 10^{-3}\ T)\pi(0.015\ m)^2$
$= 4.0 \times 10^{-6}\ T\ m^2$.

(b) $<\varepsilon> = N\frac{\Delta\Phi}{\Delta t} = 4\left(\frac{4 \times 10^{-6}\ T\ m^2}{0.9\ s}\right) = 1.78 \times 10^{-5}\ V$

21.21 (a) The magnetic field set up inside the solenoid is
$B = \mu_0 n I = (4\pi \times 10^{-7}\ Tm/A)\frac{100\ turns}{0.2\ m}(3\ A) = 1.88 \times 10^{-3}\ T$

$\Phi_i = BA = (1.88 \times 10^{-3}\ T)(10^{-2}\ m)^2 = 1.88 \times 10^{-7}\ T\ m^2$.

(b) When the current is reduced to zero, $\Phi_f = 0$.

Therefore, $\Delta\Phi = 1.88 \times 10^{-7}\ T\ m^2$.

and $<\varepsilon> = N\frac{\Delta\Phi}{\Delta t} = \frac{1.88 \times 10^{-7}\ T\ m^2}{3\ s} = 6.28 \times 10^{-8}\ V$

21.22 We have $<\varepsilon> = \frac{\Delta\Phi}{\Delta t} = 80 \times 10^{-3}\ V$

Therefore, $\Delta\Phi = \varepsilon\Delta t = (80 \times 10^{-3}\ V)(0.4\ s) = 3.2 \times 10^{-2}\ T\ m^2$.
But, $\Phi = NBA\cos\theta$, where θ is the angle between B and a line drawn perpendicular to the plane of the coil. Thus,
$\Delta\Phi = N\Delta BA\cos\theta$

or, $A = \frac{\Delta\Phi}{N\Delta B\cos\theta} = \frac{3.2 \times 10^{-2}\ T\ m^2}{(50)((600 - 200) \times 10^{-6}\ T)\cos 30°} = 1.85\ m^2$.

CHAPTER TWENTY-ONE SOLUTIONS

But, $A = L^2$, where L is the length of the sides of the coil.
Thus, $L = \sqrt{A} = \sqrt{1.85 \text{ m}^2} = 1.36$ m
Therefore, the total length of wire is
Length = (number of turns)(total length of wire per turn)
     = 50(4L) = 50(4)(1.36 m) = 272 m.

21.23 $\mathcal{E} = BLv = (0.2 \text{ T})(0.3 \text{ m})(0.2 \text{ m/s}) = 1.2 \times 10^{-2}$ V = 12.0 mV

21.24 From, $\mathcal{E} = BLv$ we have

$$v = \frac{\mathcal{E}}{BL} = \frac{1.5 \text{ V}}{(6 \times 10^{-5} \text{ T})(40 \text{ m})} = 625 \text{ m/s} = 1398 \text{ mi/h}$$

This speed exceeds that of a typical airplane. Thus, the induced voltage will not reach 1.5 V.

21.25 $\mathcal{E} = BLv = (40 \times 10^{-6} \text{ T})(5 \text{ m})(10 \text{ m/s}) = 2 \times 10^{-3}$ V = 2 mV
Using the right hand rule shows that the direction of the magnetic force on a positive charge in the wire is directed toward the west. Thus, a charge will drift to the western end of the wire, so the western end is positive relative to the eastern end.

21.26 To produce a 0.5 A current through 6 Ω of resistance, the induced emf in the bar must be
$$\mathcal{E} = IR = (0.5 \text{ A})(6 \text{ }\Omega) = 3.0 \text{ V}$$
But, $v = \dfrac{\mathcal{E}}{BL} = \dfrac{3.0 \text{ V}}{(2.5 \text{ T})(1.2 \text{ m})} = 1.00$ m/s

---

21.27 (a) $\mathcal{E} = BLv = (40 \times 10^{-6} \text{ T})(0.5 \text{ m})(5 \text{ m/s}) = 1 \times 10^{-4}$ V = 100 μV

(b) $i = \dfrac{\mathcal{E}}{R} = \dfrac{1 \times 10^{-4} \text{ V}}{5 \text{ }\Omega} = 2 \times 10^{-5}$ A = 20 μA

21.28 (a) $P_{supplied} = \mathcal{E}i = (1 \times 10^{-4} \text{ V})(2 \times 10^{-5} \text{ A}) = 2 \times 10^{-9}$ W = 2 nanowatts
(b) $P_{dissipated} = i^2R = (2 \times 10^{-5} \text{ A})^2(5 \text{ }\Omega) = 2 \times 10^{-9}$ W = 2 nanowatts
(c) $F_{applied} = F_{retarding}$ force exerted by magnetic field = BIL
     = $(40 \times 10^{-6} \text{ T})(2 \times 10^{-5} \text{ A})(0.5 \text{ m}) = 4 \times 10^{-10}$ N
(d) $P = Fv = (4 \times 10^{-10} \text{ N})(5 \text{ m/s}) = 2 \times 10^{-9}$ W = 2 nanowatts

---

21.29 We must first find the speed of the beam just before impact. We use conservation of mechanical energy.
$$\frac{1}{2} mv^2 = mgh$$
or, $v = \sqrt{2gh} = \sqrt{2(9.8 \text{ m/s}^2)(9 \text{ m})} = 13.3$ m/s
The magnitude of the induced field is
$$\mathcal{E} = BLv = (18 \times 10^{-6} \text{ T})(12 \text{ m})(13.3 \text{ m/s}) = 2.87 \times 10^{-3} \text{ V} = 2.87 \text{ mV}$$

21.30 (a) The top of the loop must behave as a south pole in order to oppose the approaching south pole of the bar magnet. Thus, the current must be clockwise as viewed from above.

CHAPTER TWENTY-ONE SOLUTIONS

(b) After the magnet falls through the loop, the lower side of the loop must act as a south pole to oppose the movement of the north pole of the falling magnet. Thus, the current is counterclockwise as viewed from above.

21.31 (a) The current is left to right
(b) The current is right to left.

21.32 (a) The current is left to right.
(b) No current is present since B is constant.
(c) The current is right to left.

21.33 (a) The current is right to left.
(b) The current is right to left.
(c) The current is left to right.
(d) The current is left to right.

21.34 The current is left to right.

21.35 The current is left to right.

21.36 The current flows from top toward the bottom.

21.37 (a) $A_i = \pi r_i^2 = \pi(10^{-2}\text{ m})^2 = \pi 10^{-4}\text{ m}^2$ and $A_f = 0$
Therefore, $\Delta\Phi = BA_i - BA_f = BA_i = (25 \times 10^{-3}\text{ T})(\pi 10^{-4}\text{ m}^2) = 7.85 \times 10^{-6}\text{ T m}^2$
so $<\mathcal{E}> = N\dfrac{\Delta\Phi}{\Delta t} = \dfrac{7.85 \times 10^{-6}\text{ T m}^2}{50 \times 10^{-3}\text{ s}} = 1.57 \times 10^{-4}\text{ V} = 0.157\text{ mV}$
and Lenz' law shows that the induced current will flow from A to B. (End B will be positive.)
(b) $\Delta\Phi = B_f A - B_i A = (100\text{ mT})A - (25\text{ mT})A = (75 \times 10^{-3}\text{ T})(\pi 10^{-4}\text{ m}^2)$
$= 2.36 \times 10^{-5}\text{ T m}^2$
and $<\mathcal{E}> = N\dfrac{\Delta\Phi}{\Delta t} = \dfrac{2.36 \times 10^{-5}\text{ T m}^2}{4 \times 10^{-3}\text{ s}} = 5.89 \times 10^{-3}\text{ V} = 5.89\text{ mV}$
In this case, the magnetic force on a + charge in the wire causes it to drift toward end A, so end A is positive. (Current flow is from B to A.)

21.38 I is proportional to $<\mathcal{E}> = N\dfrac{\Delta\Phi}{\Delta t}$, so I is proportional to the rate of change of the flux, or slope of the Φ versus time graph. (The maximum flux occurs when the magnet is perpendicular to the plane of the coil.) The curve will be somewhat like the sketch at the right.

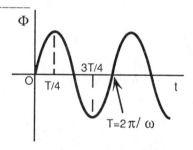

CHAPTER TWENTY-ONE SOLUTIONS

The induced current versus time curve is somewhat like the sketch at the right. Notice the phase difference between this curve and the Φ versus time curve sketched above.

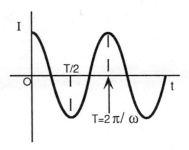

21.39 We compare the given equation $\mathcal{E} = 240\sin 500t$ to the general form for such an equation,
$$\mathcal{E} = \mathcal{E}_{max}\sin\omega t$$
By comparison, we see that $\mathcal{E}_{max} = 240$ V.
Also by comparison, we see that
$$\omega = 2\pi f = 500 \text{ s}^{-1}$$
Therefore, $f = \dfrac{500 \text{ s}^{-1}}{2\pi} = 79.6$ Hz

21.40 We have, $\omega = 2\pi f = 2\pi(20 \text{ Hz}) = 126 \text{ s}^{-1}$
and $\mathcal{E}_{max} = NAB\omega$
Thus, $3 \text{ V} = (50)\pi(10^{-2} \text{ m})^2 B(126 \text{ s}^{-1})$
Which yields, $B = 1.52$ T

21.41 $\mathcal{E}_{max} = NBA\omega = (100)(2 \times 10^{-4} \text{ T})(4 \times 10^{-2} \text{ m}^2)(157 \text{ rad/s}) = 0.126$ V

21.42 $\mathcal{E}_{max} = NAB\omega$
$2 \times 10^{-3} \text{ V} = (500)\pi(0.2 \text{ m})^2(0.1 \text{ T})\omega$
and, $\omega = 3.18 \times 10^{-4}$ rad/s

21.43 (a) $\omega = 2\pi f = 2\pi(60 \text{ Hz}) = 377 \text{ s}^{-1}$
and $\mathcal{E}_{max} = NAB\omega = (40)(0.1 \text{ T})(0.12 \text{ m}^2)(377 \text{ s}^{-1}) = 181$ V
(b) $I_{max} = \dfrac{\mathcal{E}_{max}}{R} = \dfrac{181 \text{ V}}{30 \text{ }\Omega} = 6.03$ A
(c) $\mathcal{E} = \mathcal{E}_{max}\sin\omega t = (181 \text{ V})\sin(377 \text{ s}^{-1}t)$
(d) $i = \dfrac{\mathcal{E}}{R} = (6.03 \text{ A})\sin(377 \text{ s}^{-1}t)$

21.44 (a) $I_0 = \dfrac{\mathcal{E}}{R} = \dfrac{240 \text{ V}}{30 \text{ }\Omega} = 8$ A
(b) $I = \dfrac{\mathcal{E} - \mathcal{E}_{back}}{R} = \dfrac{240 \text{ V} - 145 \text{ V}}{30 \text{ }\Omega} = 3.17$ A
(c) $\mathcal{E}_{back} = \mathcal{E} - IR = 240 \text{ V} - (6 \text{ A})(30 \text{ }\Omega) = 60$ V

21.45 When $\omega = 0$, $I = 11$ A. The resistance of the windings is found as
$$R = \dfrac{\mathcal{E}}{I} = \dfrac{120 \text{ V}}{11 \text{ A}} = 10.9 \text{ }\Omega.$$

CHAPTER TWENTY-ONE SOLUTIONS

When ω has reached operating speed, the current is 4 A, and
$$\mathcal{E} - \mathcal{E}_{back} = IR$$
Thus, $\mathcal{E}_{back} = \mathcal{E} - IR = 120 \text{ V} - (4 \text{ A})(10.9 \text{ }\Omega) = 76.4 \text{ V}$

21.46 $\tau_{max} = NBIA = (80)(0.8 \text{ T})(10 \text{ A})(2.5 \times 10^{-2} \text{ m})(4 \times 10^{-2} \text{ m}) = 0.64 \text{ Nm}$

21.47 (a) $\omega = 120 \text{ rev/min} = 12.6 \text{ rad/s}$, and
$\mathcal{E}_{max} = NAB\omega = (500)(0.6 \text{ T})(0.08 \text{ m})(0.2 \text{ m})(12.6 \text{ rad/s}) = 60.3 \text{ V}$
(b) $\mathcal{E} = \mathcal{E}_{max}\sin\omega t = (60.3 \text{ V})\sin(12.6 \text{ s}^{-1}(\pi/32)) = 56.9 \text{ V}$
(c) The emf will be a maximum at $t = T/4 = \dfrac{2\pi/\omega}{4} = \dfrac{\pi}{2\omega} = \dfrac{\pi}{2(12.6 \text{ rad/s})} = 0.125 \text{ s}$

21.48 $\mathcal{E} = L\dfrac{\Delta I}{\Delta t} = (3 \times 10^{-3} \text{ H})\dfrac{1.5 \text{ A} - 0.2 \text{ A}}{0.2 \text{ s}} = 1.95 \times 10^{-2} \text{ V} = 19.5 \text{ mV}$

21.49 From $L = \dfrac{\mu_0 N^2 A}{L}$, we find

$$N = \sqrt{\dfrac{(275 \times 10^{-6} \text{ H})(1.5 \text{ m})}{(4\pi \times 10^{-7} \text{ N/A}^2)\pi(0.04 \text{ m})^2}} = 256 \text{ turns}$$

21.50 (a) $L = \dfrac{\mu_0 N^2 A}{L} = \dfrac{(4\pi \times 10^{-7} \text{ N/A}^2)(400)^2 \pi(2.5 \times 10^{-2} \text{ m})^2}{(0.2 \text{ m})}$
$= 1.97 \times 10^{-3} \text{ H} = 1.97 \text{ mH}$
(b) From $\mathcal{E} = L\dfrac{\Delta I}{\Delta t}$

$\dfrac{\Delta I}{\Delta t} = \dfrac{\mathcal{E}}{L} = \dfrac{75 \times 10^{-3} \text{ V}}{1.97 \times 10^{-3} \text{ H}} = 38.1 \text{ A/s}$

21.51 From $\mathcal{E} = N\dfrac{\Delta \Phi}{\Delta t}$, we have

$\dfrac{\Delta \Phi}{\Delta t} = \dfrac{\mathcal{E}}{N} = \dfrac{75 \times 10^{-3} \text{ V}}{400} = 1.88 \times 10^{-4} \text{ T m}^2/\text{s}$

21.52 From $\mathcal{E} = L\dfrac{\Delta I}{\Delta t}$, we have

$L = \dfrac{\mathcal{E}}{\dfrac{\Delta I}{\Delta t}} = \dfrac{24 \times 10^{-3} \text{ V}}{10 \text{ A/s}} = 2.4 \times 10^{-3} \text{ H}$

and from $L = \dfrac{N\Phi}{I}$, we have

$\Phi = \dfrac{LI}{N} = \dfrac{(2.4 \times 10^{-3} \text{ H})(4 \text{ A})}{500} = 1.92 \times 10^{-5} \text{ T m}^2$.

21.53 $L = \dfrac{\Phi_{total}}{I} = \dfrac{N\Phi_{single\ loop}}{I}$

so $\Phi_{single\ loop} = \dfrac{LI}{N} = \dfrac{(7.2 \times 10^{-3} \text{ H})(10^{-2} \text{ A})}{300} = 2.4 \times 10^{-7} \text{ T m}^2$.

CHAPTER TWENTY-ONE SOLUTIONS

21.54 Let us use $\varepsilon = L\frac{\Delta I}{\Delta t}$ to find the fundamental units of L.

$$L = \frac{\varepsilon}{\frac{\Delta I}{\Delta t}} = \frac{V \cdot s}{A} = \Omega \cdot s$$

Thus, $\tau = \frac{L}{R} = \frac{\Omega \cdot s}{\Omega} = s$

21.55 $\tau = \frac{L}{R} = 600 \times 10^{-6}$ s $= 6 \times 10^{-4}$ s

$I_{max} = \frac{V}{R} = \frac{6 \text{ V}}{R} = 0.3$ A, which yields $R = 20 \, \Omega$

Therefore, $L = \tau R = (6 \times 10^{-4} \text{ s})(20 \, \Omega) = 1.2 \times 10^{-2} \, \Omega \cdot s = 12.0$ mH

21.56 (a) $V_R = iR$ At $t = 0$, $i = 0$. thus, $V_R = 0$

(b) At $t = \tau$, $i = 0.37 I_{max}$. Also, $I_{max} = \frac{V}{R} = \frac{6 \text{ V}}{8 \, \Omega} = 0.75$ A

Thus, $i = 0.37(0.75 \text{ A}) = 0.278$ A, and
$V_R = iR = (0.278 \text{ A})(8 \, \Omega) = 2.22$ V

(c) At $t = 0$, the voltage drop across the resistor is zero. Thus, the total voltage of the battery is dropped across the inductor.
$V_L = 6$ V

(d) At $t = \tau$, the voltage drop across the resistor is 2.22 V. Thus, the voltage drop across the inductor is 6 V - 2.22 V = 3.78 V

21.57 (a) $I_{max} = \frac{\varepsilon}{R} = \frac{24 \text{ V}}{6 \, \Omega} = 4$ A

(b) The time constant of the circuit is $\tau = \frac{L}{R} = \frac{3 \text{ H}}{6 \, \Omega} = 0.5$ s

Thus, at $t = 0.5$ s the current is
$i = I_{max}(1 - 0.37) = (0.63)(4 \text{ A}) = 2.52$ A

21.58 We use $i = I_{max}(1 - e^{-t/\tau})$ with $I_{max} = \frac{\varepsilon}{R} = \frac{24 \text{ V}}{6 \, \Omega} = 4$ A,

and $\tau = \frac{L}{R} = \frac{3 \text{ H}}{6 \, \Omega} = 0.5$ s

Thus, at $t = 0.7$ s, $i = (4 \text{ A})(1 - e^{-1.4}) = 3.01$ A

21.59 $W = \frac{1}{2}LI^2 = \frac{1}{2}(70 \times 10^{-3} \text{ H})(2 \text{ A})^2 = 0.140$ J

21.60 (a) $I_{max} = \frac{\varepsilon}{R} = \frac{24 \text{ V}}{8 \, \Omega} = 3$ A, and $W = \frac{1}{2}LI^2 = \frac{1}{2}(4 \text{ H})(3 \text{ A})^2 = 18$ J

(b) At $t = \tau$, $i = 0.63 I_{max} = (0.63)(3 \text{ A}) = 1.89$ A, and
$W = \frac{1}{2}LI^2 = \frac{1}{2}(4 \text{ H})(1.89 \text{ A})^2 = 7.14$ J

21.61 From $\varepsilon_{max} = NAB\omega$ we find

CHAPTER TWENTY-ONE SOLUTIONS

$$B = \frac{\varepsilon_{max}}{NA\omega} = \frac{0.5 \text{ V}}{(50)(0.2 \text{ m})(0.3 \text{ m})(90 \text{ rad/s})} = 1.85 \times 10^{-3} \text{ T}$$

21.62 $L = \frac{N\Phi}{I} = \frac{(200)(3.7 \times 10^{-4} \text{ T m}^2)}{1.75 \text{ A}} = 4.23 \times 10^{-2}$ H.

When I = 1.75 A in the coil, the energy stored in the magnetic field is

$$W = \frac{1}{2}LI^2 = \frac{1}{2}(4.23 \times 10^{-2} \text{ H})(1.75 \text{ A})^2 = 6.48 \times 10^{-2} \text{ J}$$

21.63 We use $\varepsilon = N\frac{\Delta\Phi}{\Delta t}$, with $\Delta\Phi = \Delta BA = (0.3 \text{ T} - 0)\pi(0.2 \text{ m})^2 = 3.77 \times 10^{-2}$ T m^2

Thus, $\varepsilon = \frac{50(3.77 \times 10^{-2} \text{ T m}^2)}{0.4 \text{ s}} = 4.71$ V

21.64 From $\varepsilon = L\frac{\Delta I}{\Delta t}$, we find

$$L = \frac{\varepsilon}{\frac{\Delta I}{\Delta t}} = \frac{100 \times 10^{-3} \text{ V}}{\frac{1.5 \text{ A}}{0.3 \text{ s}}} = \frac{0.1 \text{ V}}{5 \text{ A/s}} = 2 \times 10^{-2} \text{ H} = 20 \text{ mH}$$

21.65 When not rotating, $\varepsilon = IR$, and from this,

$$R = \frac{\varepsilon}{I} = \frac{12 \text{ V}}{18 \text{ A}} = 0.667 \text{ }\Omega$$

When rotating, $\varepsilon - \varepsilon_{back} = IR$

or, $\varepsilon_{back} = \varepsilon - IR = 12 \text{ V} - (3.5 \text{ A})(0.667 \text{ }\Omega) = 9.67$ V

21.66 The flux passing through the coil when B = 0.15 T is,

$$\Phi_i = NB_iA = (5)(0.15 \text{ T})\pi(0.15 \text{ m})^2 = 5.30 \times 10^{-2} \text{ T m}^2.$$

and when B = 0.2 T, the flux is

$$\Phi_f = NB_fA = (5)(0.2 \text{ T})\pi(0.15 \text{ m})^2 = 7.07 \times 10^{-2} \text{ T m}^2.$$

Thus, $\Delta\Phi = 1.77 \times 10^{-2}$ T m^2.

and $\varepsilon = \frac{\Delta\Phi}{\Delta t} = \frac{1.77 \times 10^{-2} \text{ T m}^2}{3 \text{ s}} = 5.89 \times 10^{-3}$ V

With the constant induced emf found above during the 3 seconds, we find the steady induced current to be

$$I = \frac{\varepsilon}{R} = \frac{5.89 \times 10^{-3} \text{ V}}{8 \text{ }\Omega} = 7.36 \times 10^{-4} \text{ A}$$

But, $I = \frac{\Delta Q}{\Delta t}$ Thus, $\Delta Q = I\Delta t = (7.36 \times 10^{-4} \text{ A})(3 \text{ s}) = 2.21 \times 10^{-3}$ C

21.67 The curve is a sine function with amplitude 300 V and period $\pi/10$ s.

21.68 The inductance is

$$L = \frac{\mu_0 N^2 A}{L} = \frac{(4\pi \times 10^{-7} \text{ N/A}^2)(300)^2 \pi(0.05 \text{ m})^2}{0.2 \text{ m}} = 4.44 \times 10^{-3} \text{ H}$$
= 4.44 mH

and, $W = \frac{1}{2}LI^2 = \frac{1}{2}(4.44 \times 10^{-3} \text{ H})(0.5 \text{ A})^2 = 5.55 \times 10^{-4}$ J

CHAPTER TWENTY-ONE SOLUTIONS

21.69 $\Phi = BA = (0.15\text{ T})\pi(0.1\text{ m})^2 = 4.71 \times 10^{-3}\text{ T m}^2$.

21.70 The maximum current in the circuit is
$$I_{max} = \frac{V}{R} = \frac{12\text{ V}}{1000\text{ }\Omega} = 1.2 \times 10^{-2}\text{ A}$$
and $L = 4.44$ mH (see problem 21.68)

(a) After one time constant, $i = 0.63 I_{max} = 7.56 \times 10^{-3}$ A

and $W = \frac{1}{2}LI^2 = \frac{1}{2}(4.44 \times 10^{-3}\text{ H})(7.56 \times 10^{-3}\text{ A})^2 = 1.27 \times 10^{-7}$ J

(b) After two time constants, $i = I_{max}(1 - (0.37)^2) = 1.04 \times 10^{-2}$ A

and $W = \frac{1}{2}LI^2 = \frac{1}{2}(4.44 \times 10^{-3}\text{ H})(1.04 \times 10^{-3}\text{ A})^2 = 2.38 \times 10^{-7}$ J

21.71

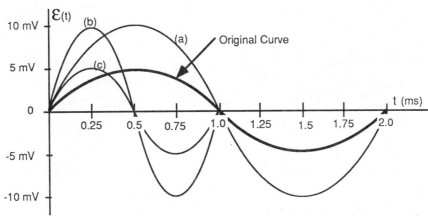

(a) Doubling the number of turns doubles the amplitude but does not alter the period.
(b) Doubling the angular velocity doubles the amplitude and also cuts the period in half.
(c) Doubling the angular velocity while reducing the number of turns to one half the original value leaves the amplitude unchanged but does cut the period in half.

21.72 Use $L = \frac{\mu_0 N^2 A}{L}$ For the two coils $N_A = 2N_B$ and since the two coils are wound with the same spacing between turns of equal length of wire $r_A = \frac{r_B}{2}$

or $A_A = \frac{A_B}{4}$

Also, the length of each coil is the product of the diameter of the wire (plus spacing between turns) and the number of turns. Therefore, $L_A = 2L_B$

So the ratio of the inductances is $\frac{L_A}{L_B} = \frac{1}{2}$

21.73 The flux through the solenoid is given by
$\Phi = BA = \mu_0 n I A$
Thus, the change in flux is $\Delta\Phi = \mu_0 n A \Delta I$

CHAPTER TWENTY-ONE SOLUTIONS

and $\varepsilon = N\dfrac{\Delta\Phi}{\Delta t} = \mu_0 n A \dfrac{\Delta I}{\Delta t} = (4\pi \times 10^{-7} \text{ N/A}^2)\dfrac{1600}{0.8 \text{ m}}\pi(0.05 \text{ m})^2\left(\dfrac{6 \text{ A} - 1.5 \text{ A}}{0.2 \text{ s}}\right)$

$= 4.44 \times 10^{-4} \text{ V} = 444 \text{ }\mu\text{V}$

21.74 The time constant is $\tau = \dfrac{L}{R} = \dfrac{3 \times 10^{-3} \text{ H}}{2000 \text{ }\Omega} = 1.5 \times 10^{-6}$ s

From $i = I_{max}(1 - e^{-t/\tau})$

we find $e^{-t/\tau} = 1 - \dfrac{i}{I_{max}}$

when $i = 0.9 I_{max}$ $\quad e^{-t/\tau} = 1 - 0.9 = 0.1$

Thus, $-\dfrac{t}{\tau} = \ln(0.1)$ and we find

$t = -\tau \ln(0.1) = -(1.5 \times 10^{-6} \text{ s})\ln(0.1) = 3.45 \times 10^{-6}$ s $= 3.45$ μs

---

21.75 (a) $\varepsilon_{max} = NAB\omega = (50)((0.15 \text{ m})(0.25 \text{ m})(1.4 \text{ T})(20 \text{ rad/s}) = 52.5$ V
(b) $\varepsilon = \varepsilon_{max}\sin\omega t = (52.5 \text{ V})\sin(20(0.05)) = 52.5 \text{ V}\sin(1 \text{ rad}) = 44.2$ V
(c) At the maximum emf, the current is

$i = \dfrac{52.5 \text{ V}}{12 \text{ }\Omega} = 4.38$ A

and the torque is

$\tau = NBIA\sin\theta = (50)(1.4 \text{ T})(4.38 \text{ A})(3.75 \times 10^{-2} \text{ m})\sin 90° = 11.5$ N m

---

21.76 At constant velocity $\Sigma F = 0$, thus $ILB = mg$, where $I = \dfrac{\varepsilon}{R} = \dfrac{BLv}{R}$

so $v_t = \dfrac{mgR}{L^2B^2}$

21.77 Refer to the sketch. Taking $\theta = 30°$ and $B = 2.6$ T, the net force along the direction of the incline is

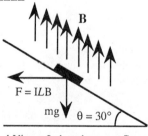

$\Sigma F = mg\sin\theta - ILB\cos\theta \quad$ where $I = \dfrac{\varepsilon}{R} = \dfrac{BLv\cos\theta}{R}$

$\Sigma F = 0$ when $v = v_t$ so $\quad mg\sin\theta - \dfrac{BLv_t\cos\theta}{R}LB\cos\theta$
$= 0$

and $v_t = \dfrac{Rmg\sin\theta}{B^2L^2\cos^2\theta}$.

(End View - Induced current flows out toward the viewer in the strip)

For the values given $v_t = \dfrac{(30)(0.04)(9.80)\sin 30°}{(2.6)^2(0.2)^2\cos^2 30°} = 29$ m/s

CHAPTER TWENTY-TWO SOLUTIONS

22.1 $V = \dfrac{V_m}{\sqrt{2}} = 100$ V

(a) $V_m = 100\sqrt{2} = 141$ V

(b) $I_m = \dfrac{V_m}{R} = \dfrac{141 \text{ V}}{5 \text{ }\Omega} = 28.3$ A

(c) $I = \dfrac{V}{R} = \dfrac{100 \text{ V}}{5 \text{ }\Omega} = 20$ A

22.2 $I_m = I\sqrt{2} = 6\sqrt{2} = 8.49$ A, and $V_m = V\sqrt{2} = 240\sqrt{2} = 339$ V

22.3 (a) $R = \dfrac{V}{I} = \dfrac{110 \text{ V}}{7.5 \text{ A}} = 14.7$ Ω

(b) $P = I^2 R = (7.5 \text{ A})^2 (14.7 \text{ }\Omega) = 825$ W

22.4 $R = \dfrac{V^2}{P} = \dfrac{(120 \text{ V})^2}{P}$ (1)

(a) if $P = 75$ W, (1) gives $R = 192$ Ω

(b) if $P = 100$ W, (1) gives $R = 144$ Ω

22.5 $P = IV$ and $V = 110$ V for each bulb (parallel circuit), so

$I_1 = I_2 = \dfrac{P_1}{V} = \dfrac{150 \text{ W}}{110 \text{ V}} = 1.36$ A, and $R_1 = \dfrac{V}{I_1} = 80.7$ $\Omega = R_2$

$I_3 = \dfrac{P_3}{V} = \dfrac{100 \text{ W}}{110 \text{ V}} = 0.909$ A, and $R_3 = \dfrac{V}{I_3} = 121$ Ω

$I_{total} = I_1 + I_2 + I_3 = 3.64$ A

22.6 $R_{total} = 8.2$ Ω + 10.4 Ω = 18.6 Ω

$I = \dfrac{V}{R} = \dfrac{15 \text{ V}}{18.6 \text{ }\Omega} = 0.806$ A

$P_{speaker} = I^2{}_{speaker} R = (0.806 \text{ A})^2 (10.4 \text{ }\Omega) = 6.76$ W

22.7 We must compare the expression given for the voltage, $v = 150\sin 377t$, with the general expression for an ac voltage, $v = V_m \sin 2\pi f t$. By comparison, we see

(a) $V_m = 150$ V, and from this, $V = \dfrac{V_m}{\sqrt{2}} = \dfrac{150 \text{ V}}{\sqrt{2}} = 106$ V

(b) We also see that $\omega = 2\pi f = 377$ rad/s. Thus, $f = \dfrac{377 \text{ rad/s}}{2\pi} = 60$ Hz

(c) At $t = \dfrac{1}{120}$ s, $v = 150\sin\left(377\left(\dfrac{1}{120}\right)\right) = 150\sin\pi = 0$

(d) $I_m = \dfrac{V_m}{R} = \dfrac{150 \text{ V}}{50 \text{ }\Omega} = 3$ A

CHAPTER TWENTY-TWO SOLUTIONS

22.8 We know that $v = V_m \sin 2\pi ft$. Thus,
$$\sin 2\pi ft = \frac{v}{V_m} = 0.25$$
From this, we find $2\pi ft = 0.253$ rad.
At $t = 0.002$ s, $\quad f = \frac{0.253}{2\pi(0.002 \text{ s})} = 20.1$ Hz

---

22.9 We have $X_C = \frac{1}{2\pi fC}$

The units of C are $F = \frac{C}{V}$ and the units of f are $\frac{1}{s}$. Thus, $X_C = \frac{1}{\frac{1}{s}F} = \frac{1}{\frac{C}{sV}} = \frac{C}{\frac{C}{s}} = \frac{V}{A} = \Omega$

22.10 The ratio of the capacitive reactance at the higher frequency to that at the lower is
$$\frac{X_{C(high)}}{X_{C(low)}} = \frac{2\pi f_{low} C}{2\pi f_{high} C} = \frac{f_{low}}{f_{high}} = \frac{120}{10000}$$
Thus, $X_{C(high)} = (30 \text{ }\Omega)(\frac{120}{10000}) = 0.36 \text{ }\Omega$.

22.11 The capacitive reactance is
$$X_C = \frac{1}{2\pi fC} = \frac{1}{2\pi(90 \text{ Hz})(3.7 \times 10^{-6} \text{ F})} = 478 \text{ }\Omega.$$
Therefore, $I = \frac{V}{X_C} = \frac{48 \text{ V}}{478 \text{ }\Omega} = 0.100$ A

22.12 $X_C = \frac{V}{I} = \frac{30 \text{ V}}{0.3 \text{ A}} = 100 \text{ }\Omega.$

Therefore, $\frac{1}{2\pi fC} = 100 \text{ }\Omega$, or $\quad f = \frac{1}{2\pi(100 \text{ }\Omega)(4 \times 10^{-6} \text{ F})} = 398$ Hz

---

22.13 $X_C = \frac{1}{\omega C}$, where $\omega = 2\pi f$ = the angular frequency.

Thus, $X_C = \frac{1}{(120\pi \text{ rad/s})(6 \times 10^{-6} \text{ F})} = 442 \text{ }\Omega.$

Also, $V_m = 140$ V $= \sqrt{2}$ V, and $V = \frac{140 \text{ V}}{\sqrt{2}} = 99$ V

Thus, $I = \frac{V}{X_C} = \frac{99 \text{ V}}{442 \text{ }\Omega} = 0.224$ A

22.14 $V = \frac{V_m}{\sqrt{2}} = \frac{170 \text{ V}}{\sqrt{2}} = 120$ V, and $\quad X_C = \frac{V}{I} = \frac{120 \text{ V}}{0.75 \text{ A}} = 160 \text{ }\Omega.$

So, $C = \frac{1}{2\pi f X_C} = \frac{1}{2\pi(60 \text{ Hz})(160 \text{ }\Omega)} = 1.66 \times 10^{-5}$ F $= 16.6$ µF

22.15 (a) $X_L = 2\pi fL = 2\pi(10 \text{ Hz})(3 \times 10^{-3} \text{ H}) = 0.188 \text{ }\Omega.$
(b) $X_L = 2\pi fL = 2\pi(10^6 \text{ Hz})(3 \times 10^{-3} \text{ H}) = 1.88 \times 10^4 \text{ }\Omega.$

22.16 The basic units of L are $\quad H = \frac{V \text{ s}}{A} = \Omega \text{ s}.$

CHAPTER TWENTY-TWO SOLUTIONS

Thus, $2\pi fL$ has units of $\left(\dfrac{1}{s}\right)(\Omega\ s) = \Omega$.

22.17 $X_C = \dfrac{V}{I} = \dfrac{9\ V}{25 \times 10^{-3}\ A} = 360\ \Omega$

$\omega = \dfrac{1}{CX_C} = \dfrac{1}{(360\ \Omega)(2.4 \times 10^{-6}\ F)} = 1157$ rad/s

and $f = \dfrac{\omega}{2\pi} = \dfrac{1157\ rad/s}{2\pi} = 184.2$ Hz

(b) $X_L = \omega L = (1157\ rad/s)(0.16\ H) = 185\ \Omega$

$I = \dfrac{V}{X_L} = \dfrac{9\ V}{185\ \Omega} = 4.86 \times 10^{-2}\ A = 48.6$ mA

22.18 $I = \dfrac{V}{X_L} = \dfrac{50\ V}{X_L} < \dfrac{80 \times 10^{-3}\ A}{\sqrt{2}}$

so $X_L > \dfrac{50\sqrt{2}\ V}{80 \times 10^{-3}\ A} = 884\ \Omega$

$X_L = \omega L = 2\pi(20\ Hz)L > 884\ \Omega$

$L > \dfrac{884\ \Omega}{40\pi\ rad/s}$

so $L > 7.03$ H

22.19 (a) $X_L = \dfrac{V}{I} = \dfrac{120\ V}{10\ A} = 12\ \Omega$.

Thus, from $X_L = 2\pi fL$, we have $L = \dfrac{X_L}{2\pi f} = \dfrac{12\ \Omega}{2\pi(60\ Hz)} = 3.18 \times 10^{-2}\ H = 31.8$ mH

(b) The ratio of the current at the unknown higher frequency to that at 60 Hz is found from,

$I = \dfrac{V}{X_L}$ as

$\dfrac{I_h}{I_{60}} = \dfrac{VX_{L60}}{VX_{Lh}}$, or $f_h = \dfrac{I_{60}}{I_h}(60\ Hz) = \dfrac{10\ A}{5\ A}(60\ Hz) = 120$ Hz.

22.20 The current as a function of frequency can be expressed as,

$I = \dfrac{V}{X_L} = \dfrac{V_{max}}{\sqrt{2}(2\pi fL)} = \dfrac{200\ V}{\sqrt{2}(2\pi(30 \times 10^{-3}\ H)f)} = \dfrac{750\ V/\Omega\ s}{f}$

(a) If $f = 60$ Hz, $I = \dfrac{750\ V/\Omega\ s}{(60\ Hz)} = 12.5$ A

(b) If $f = 6000$ Hz, $I = \dfrac{750\ V/\Omega\ s}{(6000\ Hz)} = 0.125$ A

22.21 From $X_L = 2\pi fL = \omega L$, where ω is the angular frequency, we have

$L = \dfrac{X_L}{\omega} = \dfrac{40\ \Omega}{754\ rad/s} = 5.31 \times 10^{-2}\ H = 53.1$ mH

22.22 $X_L = 2\pi fL = 120\pi\ rad/s\ L = 54\ \Omega$

so $L = \dfrac{54\ \Omega}{120\pi\ rad/s} = 0.143$ H

Thus, when $f = 50$ Hz and $V = 100$ V

CHAPTER TWENTY-TWO SOLUTIONS

$$I = \frac{V}{X_L} = \frac{100 \text{ V}}{(100\pi \text{ rad/s})(0.143 \text{ H})} = 2.22 \text{ A}$$

and $I_{max} = \sqrt{2} \, I = \sqrt{2} \, (2.22 \text{ A}) = 3.14 \text{ A}$

22.23 $X_L = 2\pi fL = (120\pi \text{ rad/s})(0.460 \text{ H}) = 173.4 \, \Omega$

$$X_C = \frac{1}{2\pi fC} = \frac{1}{(120\pi \text{ rad/s})(21 \times 10^{-6} \text{ F})} = 126.3 \, \Omega$$

$$\tan\phi = \frac{X_L - X_C}{R} = \frac{173.4 \, \Omega - 126.3 \, \Omega}{150 \, \Omega} = 0.314, \text{ so} \quad \phi = 17.4°$$

(b) $X_L > X_C$, so voltage leads current

22.24 We have, $X_C = \frac{1}{2\pi fC} = \frac{1}{2\pi(60 \text{ Hz})(40 \times 10^{-6} \text{ F})} = 66.3 \, \Omega.$

Thus, $Z = \sqrt{R^2 + (X_L - X_C)^2}$ becomes,

$Z = \sqrt{R^2 + (X_C)^2} = \sqrt{(150 \, \Omega)^2 + (66.3 \, \Omega)^2}$

$= 83.1 \, \Omega$

(a) $I = \frac{V}{Z} = \frac{30 \text{ V}}{83.1 \, \Omega} = 0.361 \text{ A}$

(b) $V_R = IR = (0.361 \text{ A})(50 \, \Omega) = 18.1 \text{ V}$

(c) $V_C = IX_C = (0.361 \text{ A})(66.3 \, \Omega) = 23.9 \text{ V}$

(d) $\tan\phi = \frac{X_L - X_C}{R} = \frac{-X_C}{R} = \frac{-66.3 \, \Omega}{50 \, \Omega} = -1.33,$

and $\phi = -53°$

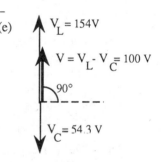

22.25 We have,

$X_C = \frac{1}{2\pi fC} = \frac{1}{2\pi(60 \text{ Hz})(10^{-5} \text{ F})} = 265 \, \Omega.$

and $X_L = 2\pi fL = 2\pi(60 \text{ Hz})(2 \text{ H}) = 754 \, \Omega.$

So, $Z = \sqrt{R^2 + (X_L - X_C)^2} = \sqrt{(X_L - X_C)^2} = 489 \, \Omega.$

(a) $I = \frac{V}{Z} = \frac{100 \text{ V}}{489 \, \Omega} = 0.205 \text{ A}$

(b) $V_L = IX_L = (0.205 \text{ A})(754 \, \Omega) = 154 \text{ V}$

(c) $V_c = IX_C = (0.205 \text{ A})(265 \, \Omega) = 54.3 \text{ V}$

(d) $\tan\phi = \frac{X_L - X_C}{R} = \frac{-X_C}{R} = \frac{489 \, \Omega}{0 \, \Omega} =$ infinity, and $\phi = 90°$

22.26 $X_L = X_C$ leads to $\omega = \frac{1}{\sqrt{LC}} = \frac{1}{\sqrt{(57 \times 10^{-6} \text{ H})(57 \times 10^{-6} \text{ F})}}$

$= 1.75 \times 10^4 \text{ rad/s}$

from which $f = \frac{\omega}{2\pi} = \frac{1.75 \times 10^4 \text{ rad/s}}{2\pi} = 2792 \text{ Hz} = 2.79 \text{ kHz}$

22.27 $Z = \sqrt{R^2 + (X_L)^2} = \sqrt{(100 \, \Omega)^2 + (X_L)^2} = 141 \, \Omega$

gives $X_L = 99.4 \, \Omega$

and $\omega = \frac{X_L}{L} = \frac{99.4 \, \Omega}{0.320 \text{ H}} = 3.11 \times 10^2 \text{ rad/s} = 2\pi f.$ Therefore $f = 49.4 \text{ Hz}.$

CHAPTER TWENTY-TWO SOLUTIONS

22.28 We have,
$X_L = 2\pi f L = 2\pi(60\text{ Hz})(2 \times 10^{-2}\text{ H}) = 7.54\ \Omega$.
Thus, $Z = \sqrt{R^2 + (X_L - X_C)^2} = \sqrt{R^2 + (X_L)^2} = 21.4\ \Omega$.

(a) $I = \dfrac{V}{Z} = \dfrac{100\text{ V}}{21.4\ \Omega} = 4.68\text{ A}$

(b) $V_L = IX_L = (4.68\text{ A})(7.54\ \Omega) = 35.3\text{ V}$

(c) $V_R = IR = (4.68\text{ A})(20\ \Omega) = 93.6\text{ V}$

(d) $\tan\phi = \dfrac{X_L - X_C}{R} = \dfrac{X_L}{R} = \dfrac{7.54\ \Omega}{20\ \Omega} = 0.377$, and $\phi = 20.7°$

(e)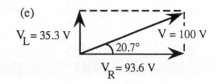

22.29 $X_C = \dfrac{1}{2\pi f C}$
$= \dfrac{1}{2\pi(60\text{ Hz})(3 \times 10^{-6}\text{ F})} = 884\ \Omega$,
and $X_L = 2\pi f L = 2\pi(60\text{ Hz})(2\text{ H}) = 754\ \Omega$.

(a) $Z = \sqrt{R^2 + (X_L - X_C)^2}$
$= \sqrt{(30\ \Omega)^2 + (754\ \Omega - 884\ \Omega)^2} = 134\ \Omega$.

(b) $I = \dfrac{V}{Z} = \dfrac{100\text{ V}}{134\ \Omega} = 0.748\text{ A}$

(c) $V_R = IR = (0.748\text{ A})(30\ \Omega) = 22.5\text{ V}$,
$V_L = IX_L = (0.748\text{ A})(754\ \Omega) = 564\text{ V}$, and
$V_c = IX_C = (0.748\text{ A})(884\ \Omega) = 662\text{ V}$

(d) $\tan\phi = \dfrac{X_L - X_C}{R} = \dfrac{754 - 884}{30} = -4.33$, and $\phi = -77°$

(e)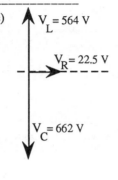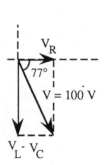

22.30 $X_C = \dfrac{1}{2\pi f C}$
$= \dfrac{1}{2\pi(60\text{ Hz})(10 \times 10^{-6}\text{ F})} = 265\ \Omega$.
and $X_L = 2\pi f L = 2\pi(60\text{ Hz})(0.1\text{ H}) = 37.7\ \Omega$.

(a) $V_R = IR = (2.75\text{ A})(50\ \Omega) = 138\text{ V}$

(b) $V_L = IX_L = (2.75\text{ A})(37.7\ \Omega) = 104\text{ V}$

(c) $V_c = IX_C = (2.75\text{ A})(265\ \Omega) = 729\text{ V}$

(d) $V = \sqrt{V_R^2 + (V_L - V_C)^2}$
$= \sqrt{(138\text{ V})^2 + (104\text{ V} - 729\text{ V})^2} = 640\text{ V}$

(e) The phasor diagram is sketched at the right.

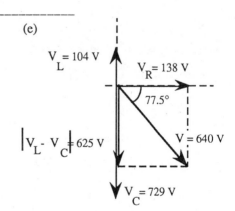

22.31 $X_C = \dfrac{1}{2\pi f C} = \dfrac{1}{2\pi(60\text{ Hz})(3 \times 10^{-6}\text{ F})} = 884\ \Omega$
and $X_L = 2\pi f L = 2\pi(60\text{ Hz})(0.4\text{ H}) = 151\ \Omega$.
$Z = \sqrt{(60\ \Omega)^2 + (151\ \Omega - 884\ \Omega)^2} = 736\ \Omega$.
Thus, $I = \dfrac{V}{Z} = \dfrac{90\text{ V}}{736\ \Omega} = 0.122\text{ A}$

CHAPTER TWENTY-TWO SOLUTIONS

(a) The voltage drop across the capacitor-inductor combination is
$$V_{LC} = IZ_{LC} = I\sqrt{(X_L - X_C)^2} = (0.122 \text{ A})(|151 \text{ }\Omega - 884 \text{ }\Omega|) = 89.7 \text{ V}$$
(b) The voltage drop across the RC combination is
$$V_{RC} = IZ_{RC} = I\sqrt{R^2 + (X_C)^2} = (0.122 \text{ A})\sqrt{(60 \text{ }\Omega)^2 + (884 \text{ }\Omega)^2}$$
$$= 108 \text{ V}$$

22.32 $X_C = \dfrac{1}{2\pi fC} = \dfrac{1}{2\pi(60 \text{ Hz})(20 \times 10^{-12} \text{ F})} = 1.33 \times 10^8 \text{ }\Omega$

$Z = \sqrt{(50 \times 10^3 \text{ }\Omega)^2 + (1.33 \times 10^8 \text{ }\Omega)^2}$ which is approximately equal to $1.33 \times 10^8 \text{ }\Omega$

so, $I = \dfrac{V}{Z} = \dfrac{5000 \text{ V}}{1.33 \times 10^8 \text{ }\Omega} = 3.77 \times 10^{-5} \text{ A}$

and $V_{body} = IR_{body} = (3.77 \times 10^{-5} \text{ A})(50 \times 10^3 \text{ }\Omega) = 1.88 \text{ V}$

22.33 We find $X_C = 49.0 \text{ }\Omega$, $X_L = 58.1 \text{ }\Omega$, and $Z = 41.0 \text{ }\Omega$, so

$I_{max} = \dfrac{150 \text{ V}}{41.0 \text{ }\Omega} = 3.66 \text{ A}$

(a) $V_{ab} = IR = (3.66 \text{ A})(40.0 \text{ }\Omega) = 146 \text{ V}$
(b) $V_{bc} = IX_L = (3.66 \text{ A})(58.1 \text{ }\Omega) = 213 \text{ V}$
(c) $V_{cd} = IX_C = (3.66 \text{ A})(49 \text{ }\Omega) = 179 \text{ V}$
(d) $V_{bd} = I\sqrt{(X_L - X_C)^2} = (3.66 \text{ A})\sqrt{(58.1 \text{ }\Omega - 49.0 \text{ }\Omega)^2} = 33.3 \text{ V}$

22.34 (a) $X_C = \dfrac{1}{2\pi fC} = \dfrac{1}{2\pi(60 \text{ Hz})(15 \times 10^{-6} \text{ F})} = 177 \text{ }\Omega$

$Z = \sqrt{R^2 + (X_L - X_C)^2} = \sqrt{R^2 + (X_C)^2} = 184 \text{ }\Omega$.

$I = \dfrac{V}{Z} = \dfrac{120 \text{ V}}{184 \text{ }\Omega} = 0.653 \text{ A} = 653 \text{ mA}$

(b) If $I_2 = \dfrac{1}{2} I_1$, then $Z_2 = 2Z_1$

Therefore, $Z_2^2 = 4Z_1^2$

or, $R^2 + (X_L - X_C)^2 = 4(R^2 + (X_C)^2)$ which reduces to
$X_L = X_C \pm 365 \text{ }\Omega = 177 \text{ }\Omega \pm 365 \text{ }\Omega$. We must use the + sign so that the inductive reactance will be greater than zero.
Thus, $X_L = 542 \text{ }\Omega = 2\pi fL$, and

$L = \dfrac{542 \text{ }\Omega}{2\pi(60 \text{ Hz})} = 1.44 \text{ H}$

22.35 (a) $X_C = \dfrac{1}{2\pi fC}$. For a dc source, the frequency is zero, and X_C approaches infinity. Thus,
$I = 0$.

(b) If we have a 10 V, 60 Hz source, then $X_C = \dfrac{1}{2\pi fC} = \dfrac{1}{2\pi(60 \text{ Hz})(3 \times 10^{-6} \text{ F})} = 884 \text{ }\Omega$

and $Z = 890 \text{ }\Omega$.

Thus, $I = \dfrac{V}{Z} = \dfrac{10 \text{ V}}{890 \text{ }\Omega} = 1.12 \times 10^{-2} \text{ A} = 11.2 \text{ mA}$

CHAPTER TWENTY-TWO SOLUTIONS

22.36 (a) The total resistance in the circuit is the sum of the resistive element plus the resistance of the wire of the inductor. Thus,
$R = 20\ \Omega + 15\ \Omega = 35\ \Omega$.
Also, $X_C = \dfrac{1}{2\pi f C} = \dfrac{1}{2\pi(60\ \text{Hz})(40 \times 10^{-6}\ \text{F})} = 66.3\ \Omega$
and $X_L = 2\pi f L = 2\pi(60\ \text{Hz})(0.2\ \text{H}) = 75.4\ \Omega$.
Thus, $Z = \sqrt{(35\ \Omega)^2 + (75.4\ \Omega - 66.3\ \Omega)^2} = 36.2\ \Omega$.
and $I = \dfrac{V}{Z} = \dfrac{70\ \text{V}}{36.2\ \Omega} = 1.94\ \text{A}$

(b) $V_{inductor} = IZ_{inductor} = I\sqrt{R_L^2 + (X_L)^2} = (1.94\ \text{A})\sqrt{(15\ \Omega)^2 + (75.4\ \Omega)^2} = 149\ \text{V}$

22.37 $X_L = 2\pi f L = (1.51 \times 10^3\ \text{rad/s})(2.5\ \text{H}) = 3770\ \Omega$.
$X_C = \dfrac{1}{2\pi f C} = \dfrac{1}{(1.51 \times 10^3\ \text{rad/s})(0.25 \times 10^{-6}\ \text{F})} = 2653\ \Omega$

(a) $Z = \sqrt{(900\ \Omega)^2 + (3770\ \Omega - 2653\ \Omega)^2} = 1435\ \Omega$.

(b) $I_{max} = \dfrac{V_{max}}{Z} = \dfrac{140\ \text{V}}{1435\ \Omega} = 9.76 \times 10^{-2}\ \text{A} = 97.6\ \text{mA}$

(c) $\tan\phi = \dfrac{X_L - X_C}{R} = \dfrac{3770\ \Omega - 2653\ \Omega}{900\ \Omega} = 1.241$, and $\phi = 51.1°$

(d) $X_L > X_C$ ϕ is greater than zero, so the voltage leads the current.

22.38 (a) Power factor = $\cos\phi$ = $\cos 51.1° = 0.627$ (see solution to problem 37)

(b) $I = \dfrac{V}{Z} = \dfrac{100\ \text{V}}{1435\ \Omega} = 0.68\ \text{A}$

(c) $P = VI\cos\phi = \dfrac{V_{max}}{\sqrt{2}} \dfrac{I_{max}}{\sqrt{2}} \cos\phi = \dfrac{(140\ \text{V})(0.0976\ \text{A})}{2} \cos 51.1° = 4.22\ \text{W}$

22.39 We find $X_L = 75.4\ \Omega$, $X_C = 177\ \Omega$, and $Z = 113\ \Omega$.
Thus, $I = \dfrac{V}{Z} = \dfrac{90\ \text{V}}{113\ \Omega} = 0.795\ \text{A}$
and $\tan\phi = \dfrac{X_L - X_C}{R} = \dfrac{75.4 - 177}{50} = -2.04$, and $\phi = -63.9°$
The power is found as $P = IV\cos\phi = (0.795\ \text{A})(90\ \text{V})\cos(-63.9°) = 31.6\ \text{W}$
or $P = I^2 R = (0.795)^2(50\ \Omega) = 31.6\ \text{W}$

22.40 (a) We find, $X_C = 88.4\ \Omega$, $Z = 102\ \Omega$, and $I = \dfrac{V}{Z} = \dfrac{100\ \text{V}}{102\ \Omega} = 0.980\ \text{A}$
Then, $\tan\phi = \dfrac{X_L - X_C}{R} = \dfrac{-X_C}{R} = \dfrac{-88.4}{50} = -1.77$, and $\phi = -60.5°$
Thus, the power factor = $\cos\phi = \cos(-60.5°) = 0.492$
and $P = IV\cos\phi = (0.980\ \text{A})(100\ \text{V})(0.492) = 48.3\ \text{W}$

(b) For this case, $X_L = 113\ \Omega$, $Z = 124\ \Omega$, and $I = \dfrac{V}{Z} = \dfrac{100\ \text{V}}{124\ \Omega} = 0.806\ \text{A}$
$\tan\phi = \dfrac{X_L - X_C}{R} = \dfrac{X_L}{R} = \dfrac{113}{50} = 2.26$, and $\phi = 66.1°$

CHAPTER TWENTY-TWO SOLUTIONS

Thus, the power factor $= \cos\phi = \cos(66.1°) = 0.404$
and $P = IV\cos\phi = (0.806\text{ A})(100\text{ V})(0.404) = 32.6\text{ W}$

22.41 (a) From the data, we find, $\cos\phi = \dfrac{R}{Z} = \dfrac{80\text{ }\Omega}{180\text{ }\Omega} = 0.444$

$I = \dfrac{V}{Z} = \dfrac{120\text{ V}}{180\text{ }\Omega} = 0.667\text{ A}$

Thus, $P = IV\cos\phi = (0.667\text{ A})(120\text{ V})(0.444) = 35.6\text{ W}$

22.42 (a) $Z = \dfrac{V}{I} = \dfrac{240\text{ V}}{6\text{ A}} = 40\text{ }\Omega$.

And, $R = Z\cos\phi = (40\text{ }\Omega)\cos(-53°) = 24.1\text{ }\Omega$

(b) From $\tan\phi = \dfrac{X_L - X_C}{R}$, we find

$X_L - X_C = R\tan\phi = (24.1\text{ }\Omega)\tan(-53°) = -31.9\text{ }\Omega$.

(c) $P = IV\cos\phi = (6\text{ A})(240\text{ V})\cos(-53°) = 867\text{ W}$

22.43 (a) $I = \dfrac{P}{V\cos\phi} = \dfrac{108\text{ W}}{(110\text{ V})\cos 37°} = 1.23\text{ A}$

(b) $Z = \dfrac{V}{I} = \dfrac{110\text{ V}}{1.23\text{ A}} = 89.5\text{ }\Omega$.

22.44 (a) We know, $\dfrac{V_R}{V} = \cos\phi = \dfrac{50\text{ V}}{90\text{ V}} = 0.555$

Thus, $I = \dfrac{P}{V\cos\phi} = \dfrac{14\text{ W}}{(90\text{ V})0.555} = 0.28\text{ A}$

From which, $R = \dfrac{V_R}{I} = \dfrac{50\text{ }\Omega}{0.28\text{ A}} = 179\text{ }\Omega$.

(b) From $\cos\phi = 0.555$, we find $\phi = 56.3°$

For the circuit, $\tan\phi = \dfrac{X_L - X_C}{R} = \dfrac{X_L}{R}$

Which gives, $X_L = R\tan\phi = (179\text{ }\Omega)\tan(56.3°) = 267\text{ }\Omega$

And, $L = \dfrac{X_L}{\omega} = \dfrac{267\text{ }\Omega}{2\pi(60\text{ Hz})} = 0.709\text{ H}$

22.45 $f_{\text{resonance}} = 99.7 \times 10^6\text{ Hz}$, and $\omega = 2\pi(99.7 \times 10^6\text{ Hz}) = 6.26 \times 10^8\text{ rad/s}$
at resonance,

$C = \dfrac{1}{\omega^2 L} = \dfrac{1}{(6.26 \times 10^8\text{ rad/s})(1.4 \times 10^{-6}\text{ H})} = 1.82 \times 10^{-12}\text{ F} = 1.82\text{ pF}$

22.46 at $f = 10$ Hz, $\omega = 62.8$ rad/s, $X_L = 15.7\text{ }\Omega$, $X_C = 7074\text{ }\Omega$, $Z = 7060\text{ }\Omega$
at $f = 100$ Hz, $\omega = 628$ rad/s, $X_L = 157\text{ }\Omega$, $X_C = 707\text{ }\Omega$, $Z = 559\text{ }\Omega$
at $f = 1000$ Hz, $\omega = 6280$ rad/s, $X_L = 1570\text{ }\Omega$, $X_C = 70.7\text{ }\Omega$, $Z = 1503\text{ }\Omega$
at $f = 10000$ Hz, $\omega = 62{,}800$ rad/s, $X_L = 15{,}700\text{ }\Omega$, $X_C = 7.07\text{ }\Omega$, $Z = 15{,}700\text{ }\Omega$

(a) $I = \dfrac{V}{Z} = \dfrac{3.54\text{ V}}{Z}$ gives

at 10 Hz, $I = 0.5$ mA   at 100 Hz, $I = 6.32$ mA, at 1000 Hz, $I = 2.35$ mA

CHAPTER TWENTY-TWO SOLUTIONS

at 10,000 Hz, I = 0.225 mA

(b) $\phi = \tan^{-1}\dfrac{X_L - X_C}{R}$ gives

at 10 Hz, $\phi = -89.2°$, at 100 Hz, $\phi = -79.7°$, at 1000 Hz, $\phi = 86.2°$
at 10,000 Hz, $\phi = 89.6°$

(c) I is a maximum at the resonance frequency. So,

$$f = \frac{1}{2\pi\sqrt{LC}} = \frac{1}{2\pi\sqrt{(0.25\ H)(2.25 \times 10^{-6}\ F)}} = 212\ Hz$$

22.47 (a) $f_0 = \dfrac{1}{2\pi\sqrt{LC}} = \dfrac{1}{2\pi\sqrt{(2 \times 10^{-3}\ H)(2 \times 10^{-6}\ F)}} = 2.52 \times 10^3\ Hz$

(b) At resonance, $Z = R = 40\ \Omega$, and $I = \dfrac{V}{Z} = \dfrac{60\ V}{40\ \Omega} = 1.5\ A$

(c) At $f = \dfrac{f_0}{2} = 1.26 \times 10^3\ Hz$, we find $X_L = 15.8\ \Omega$, and $X_C = 63.2\ \Omega$,
Thus, $Z = 62.0\ \Omega$,
so, $I = \dfrac{V}{Z} = \dfrac{60\ V}{62\ \Omega} = 0.967\ A$

22.48 The frequency of the station is equal to the resonant frequency of the tuning circuit.

$$f_0 = \frac{1}{2\pi\sqrt{LC}} = \frac{1}{2\pi\sqrt{(2 \times 10^{-4}\ H)(30 \times 10^{-12}\ F)}}$$

$= 2.05 \times 10^6\ Hz = 2.05$ megahertz

and, $\lambda = \dfrac{c}{f} = \dfrac{3 \times 10^8\ m/s}{2.05 \times 10^6\ Hz} = 146\ m$

22.49 For f_{min}: $C_{max} = \dfrac{1}{(2\pi(5 \times 10^5\ Hz))^2(2 \times 10^{-6}\ H)} = 5.07 \times 10^{-8}\ F$

for f_{max}: $C_{min} = \dfrac{1}{(2\pi(1.6 \times 10^6\ Hz))^2(2 \times 10^{-6}\ H)} = 4.95 \times 10^{-9}\ F$

22.50 At the low end $f_{min} = 500\ kHz$, so $\omega_{min} = 3.14 \times 10^6\ rad/s$

$$C_{max} = \frac{1}{(3.14 \times 10^6\ rad/s)^2(1 \times 10^{-4}\ H)} = 1.01 \times 10^{-9}\ F$$

At the high end $f_{max} = 1600\ kHz$, so $\omega_{max} = 1.01 \times 10^7\ rad/s$

$$C_{min} = \frac{1}{(1.01 \times 10^7\ rad/s)^2(1 \times 10^{-4}\ H)} = 9.89 \times 10^{-11}\ F$$

But, $C = C_1 + C_2$ (capacitors in parallel) gives $C_2 = C - C_1$
Therefore, $C_{2max} = C_{max} - C_1 = 1010\ pF - 80\ pF = 930\ pF$
and $C_{2min} = C_{min} - C_1 = 98.9\ pF - 80\ pF = 18.9\ pF$

22.51 From $f_0 = \dfrac{1}{2\pi\sqrt{LC}}$, we find

$$L = \frac{1}{(2\pi f_0)^2 C} = \frac{1}{(2.64 \times 10^5\ rad/s)^2(3 \times 10^{-10}\ F)} = 4.79 \times 10^{-2}\ H$$

22.52 (a) At the resonant frequency, $Z = R = 30\ \Omega$. The current in the circuit is

CHAPTER TWENTY-TWO SOLUTIONS

$$I = \frac{V}{Z} = \frac{120 \text{ V}}{30 \text{ }\Omega} = 4 \text{ A}$$

and the power is $P = I^2R = (4 \text{ A})^2(30 \text{ }\Omega) = 480 \text{ W}$

(b) At one-half the resonant frequency, the following can easily be calculated,
$X_C = 2000 \text{ }\Omega$, $X_L = 500 \text{ }\Omega$, and $Z = 1500 \text{ }\Omega$

Thus, $I = \frac{V}{Z} = \frac{120 \text{ V}}{1500 \text{ }\Omega} = 0.08 \text{ A}$

and $P = I^2R = 0.192 \text{ W}$

(c) At one-fourth the resonant frequency, we find
$X_C = 4000 \text{ }\Omega$, $X_L = 250 \text{ }\Omega$, and $Z = 3750 \text{ }\Omega$
$I = 0.03 \text{ A}$, and $P = 0.03 \text{ W}$

(d) At twice the resonant frequency,
$X_C = 500 \text{ }\Omega$, $X_L = 2000 \text{ }\Omega$, and $Z = 1500 \text{ }\Omega$
$I = 0.08 \text{ A}$, and $P = 0.192 \text{ W}$

(e) At four times the resonant frequency,
$X_C = 250 \text{ }\Omega$, $X_L = 4000 \text{ }\Omega$, and $Z = 3750 \text{ }\Omega$
$I = 0.03 \text{ A}$, and $P = 0.03 \text{ W}$

The power delivered to the circuit is maximum when the frequency of the source is equal to the resonant frequency of the circuit.

22.53 (a) $Q = \frac{V_L}{V_R}$ (at the resonant frequency)

Thus, $Q = \frac{V_L}{V_R} = \frac{2\pi f_0 L I_{max}}{R I_{max}} = \frac{2\pi f_0 L}{R}$

(b) In problem 52, $f_0 = 53.1 \text{ Hz}$

Thus, $Q = \frac{2\pi (53.1 \text{ Hz})(3 \text{ H})}{30 \text{ }\Omega} = 33.3$

22.54 (a) For the circuit of problems 52 and 53, the resonant frequency was 53.1 Hz and Q was 33.3.

Thus, $\Delta f = \frac{53.1 \text{ Hz}}{33.3} = 1.59 \text{ Hz}$

(b) If $R = 300 \text{ }\Omega$, $Q = (2\pi f_0)\frac{L}{R} = (333.3 \text{ rad/s})\frac{3 \text{ H}}{300 \text{ }\Omega} = 3.33$

So, $\Delta f = \frac{53.1 \text{ Hz}}{3.3} = 15.9 \text{ Hz}$

22.55 We use $V_s = \frac{N_s}{N_p} V_p = \frac{600}{150} 110 \text{ V} = 440 \text{ V}$

22.56 (a) From $V_s = \frac{N_s}{N_p} V_p$, we have

$N_s = \frac{V_s}{V_p} N_p = \frac{3600}{360000} (10^4) = 100 \text{ turns}$

(b) From $P_{in} = P_{out}$, or $V_p I_p = V_s I_s$, we find

$I_p = \frac{V_s}{V_p} I_s = \frac{3600}{360000} (600 \text{ A}) = 6 \text{ A}$

22.57 We have, $V_s = \frac{I_p}{I_s} V_p = \frac{6.5 \text{ A}}{0.8 \text{ A}} 96 \text{ V} = 780 \text{ V}$

CHAPTER TWENTY-TWO SOLUTIONS

22.58 (a) We are given that $P_{out} = 0.9 P_{in}$. Thus, $P_{in} = \dfrac{1000 \text{ kW}}{0.9} = 1110 \text{ kW}$

(b) We have $I_p = \dfrac{P_{in}}{V_p} = \dfrac{1110 \times 10^3 \text{ W}}{3600 \text{ V}} = 309 \text{ A}$

(c) $I_s = \dfrac{P_{out}}{V_s} = \dfrac{1000 \times 10^3 \text{ W}}{120 \text{ V}} = 8.33 \times 10^3 \text{ A}$

22.59 (a) The total power required by the city is
$P = (2 \times 10^4)(100 \text{ W}) = 2 \times 10^6 \text{ W}$

Thus, $I = \dfrac{P}{V} = \dfrac{2 \times 10^6 \text{ W}}{120 \text{ V}} = 1.67 \times 10^4 \text{ A}$

(b) $I = \dfrac{P}{V} = \dfrac{2 \times 10^6 \text{ W}}{200000 \text{ V}} = 10 \text{ A}$

(c) If $I = 1.67 \times 10^4$ A, then
$P_{loss} = I^2 R = (1.67 \times 10^4 \text{ A})^2 (5 \times 10^{-4} \text{ }\Omega/\text{m}) = 1.39 \times 10^5 \text{ W/m}$
If $I = 10$ A, then $P_{loss} = I^2 R = (10 \text{ A})^2 (5 \times 10^{-4} \text{ }\Omega/\text{m}) = 5 \times 10^{-2} \text{ W/m}$.

(d) case a: number of lines needed $= \dfrac{I}{100 \text{ A/line}} = \dfrac{1.67 \times 10^4 \text{ A}}{100 \text{ A/line}} = 167$ lines

case b: number of lines needed $= \dfrac{I}{100 \text{ A/line}} = \dfrac{10 \text{ A}}{100 \text{ A/line}} = 0.1$ line (1 line)

22.60 $X_C = \dfrac{1}{2\pi f C} = \dfrac{1}{2\pi (50 \text{ Hz})(15 \times 10^{-6} \text{ F})} = 212 \text{ }\Omega$,
$X_L = 2\pi f L = 2\pi (50 \text{ Hz})(0.3 \text{ H}) = 94.2 \text{ }\Omega$, and
$Z = \sqrt{R^2 + (X_L - X_C)^2} = \sqrt{(300 \text{ }\Omega)^2 + (94.2 \text{ }\Omega - 212 \text{ }\Omega)^2} = 322 \text{ }\Omega$.

(a) $I = \dfrac{V}{Z} = \dfrac{50 \text{ V}}{322 \text{ }\Omega} = 0.155 \text{ A}$

(b) $V_{RC} = I\sqrt{R^2 + (X_C)^2} = (0.155 \text{ A})\sqrt{(300 \text{ }\Omega)^2 + (212 \text{ }\Omega)^2} = 57 \text{ V}$,

(c) $P = I^2 R = (0.155 \text{ A})^2 (300 \text{ }\Omega) = 7.22 \text{ W}$

22.61 (a) From $X_L = 2\pi f L$, we have $f = \dfrac{20 \text{ }\Omega}{2\pi (0.25 \text{ H})} = 12.7 \text{ Hz}$

(b) $V = \dfrac{V_m}{\sqrt{2}} = \dfrac{90 \text{ V}}{\sqrt{2}}$, and $I = \dfrac{V}{X_L} = \dfrac{90 \text{ V}}{\sqrt{2}(20 \text{ }\Omega)} = 3.18 \text{ A}$

22.62 (a) From $f_0 = \dfrac{1}{2\pi \sqrt{LC}}$, we find

$C = \dfrac{1}{(2\pi f)^2 L} = \dfrac{1}{(2\pi (60 \text{ Hz}))^2 (0.25 \text{ H})} = 2.81 \times 10^{-5} \text{ F} = 28.1 \text{ }\mu\text{F}$

(b) At resonance, $X_L = X_C$
Therefore, Z across the capacitor-inductor combination is zero, and the voltage drop across the combination of elements is also zero.

22.63 $\tan\phi = \dfrac{X_L - X_C}{R} = \dfrac{X_L}{R} = \dfrac{2\pi f L}{R} = \dfrac{2\pi f (15.3 \times 10^{-3} \text{ H})}{5 \text{ }\Omega}$

From which, $f = (52 \text{ Hz})(\tan\phi) = (52 \text{ Hz})(\tan 60°) = 90.1 \text{ Hz}$

CHAPTER TWENTY-TWO SOLUTIONS

22.64 (a) $X_L = 942 \, \Omega$, and $Z = 946 \, \Omega$
so $I = \dfrac{V}{Z} = \dfrac{110 \text{ V}}{946 \, \Omega} = 0.116 \text{ A}$

(b) $I = \dfrac{V}{R} = \dfrac{110 \text{ V}}{80 \, \Omega} = 1.38 \text{ A}$

(c) $P = I^2 R = (0.116 \text{ A})^2 (80 \, \Omega) = 1.08 \text{ W}$ for case (a)
$P = I^2 R = (1.38 \text{ A})^2 (80 \, \Omega) = 151 \text{ W}$ for case (b)

22.65 (a) $V = \sqrt{V_R^2 + (V_L - V_C)^2} = \sqrt{(75 \text{ V})^2 + (75 \text{ V} - 150 \text{ V})^2} = 106 \text{ V}$

(b) $\tan\phi = \dfrac{V_L - V_C}{V_R} = \dfrac{75 - 150}{75} = -1$, and $\phi = -45°$

22.66 (a) We use $V_s = \dfrac{N_s}{N_p} V_p = \dfrac{(120 \text{V}) N_s}{200} = (0.6 \text{ V}) N_s$ \hfill (1)

The total number of turns on the secondary is determined by the maximum secondary voltage required. Thus, if the maximum voltage output is to be 9 V, we have
$9 \text{ V} = (0.6 \text{ V}) N_s$
or, $N_s = 15$ turns

(b) If the voltage output is to be 9 V, all 15 turns are used.
If the output voltage is to be 6 V, from (1) we find $N_s = 10$ turns, and in a similar fashion, when the output voltage is to be 3 V, $N_s = 5$ turns.

22.67 When connected to the battery, the only impedance to the current is the resistance of the coil. Thus, the coil resistance is
$R = \dfrac{V}{I} = \dfrac{12 \text{ V}}{3 \text{ A}} = 4 \, \Omega.$

The impedance when connected to the ac source is
$Z = \dfrac{V}{I} = \dfrac{12 \text{ V}}{2 \text{ A}} = 6 \, \Omega.$

We now find X_L as $Z^2 = R^2 + X_L^2$
$(6 \, \Omega)^2 = (4 \, \Omega)^2 + X_L^2$
Which yields, $X_L = 4.47 \, \Omega$
and, $L = \dfrac{X_L}{2\pi f} = \dfrac{4.47 \, \Omega}{2\pi (60 \text{ Hz})} = 1.19 \times 10^{-2} \text{ H} = 11.9 \text{ mH}$

22.68 $X_L = 2\pi f L = 2\pi (60 \text{ Hz})(0.7 \text{ H}) = 264 \, \Omega.$
Now use $V^2 = V_R^2 + V_L^2$
$(120 \text{ V})^2 = (40 \text{ V})^2 + V_L^2$
From which, $V_L = 113 \text{ V}$
But, $V_L = I X_L$
and $I = \dfrac{V_L}{X_L} = \dfrac{113 \text{ V}}{264 \, \Omega} = 0.429 \text{ A}$

22.69 (a) $N_s = \dfrac{V_s}{V_p} N_p = \dfrac{6 \text{ V}}{110 \text{ V}} (220) = 12$ turns

(b) The impedance of the primary $= X_L = 2\pi f L = 2\pi (60 \text{ Hz})(0.15 \text{ H}) = 56.6 \, \Omega.$

CHAPTER TWENTY-TWO SOLUTIONS

Thus, $I = \dfrac{V}{Z} = \dfrac{110 \text{ V}}{56.6 \text{ }\Omega} = 1.95$ A

(c) $P = IV\cos\phi$, but $\cos\phi = \dfrac{R}{Z} = 0$. Therefore, $P = 0$.

22.70 $X_C = 531$ Ω, and $Z = 567$ Ω. Thus, $I = \dfrac{V}{Z} = \dfrac{120 \text{ V}}{567 \text{ }\Omega} = 0.217$ A
and the power dissipated is
$P = I^2R = (0.217 \text{ A})^2(200 \text{ }\Omega) = 8.96$ W $= 8.96 \times 10^{-3}$ kW
The energy used in 24 h is $E = Pt = (8.96 \times 10^{-3} \text{ kW})(24 \text{ h}) = 0.215$ kWh
and the cost is cost $= (0.215 \text{ kWh})(8 \text{ cents/kWh}) = 1.72$ cents

---

22.71 The maximum voltage delivered to the circuit is 80 V, and the rms voltage is

$$V = \dfrac{V_m}{\sqrt{2}} = \dfrac{80 \text{ V}}{\sqrt{2}} = 56.6 \text{ V}$$

and the angular frequency, $\omega = 2\pi f$, is $1000/\pi$ rad/s.
We also find, $X_L = 191$ Ω, $X_C = 126$ Ω, and $Z = 119$ Ω. Thus,

(a) $I = \dfrac{V}{Z} = \dfrac{56.6 \text{ V}}{119 \text{ }\Omega} = 0.474$ A, and $I_{max} = \sqrt{2}\, I = 0.670$ A

(b) $\tan\phi = \dfrac{X_L - X_C}{R} = \dfrac{191 - 126}{100} = 0.65$, and $\phi = 33.2°$

(c) $\cos\phi = \cos 33.2° = 0.837$

(d) $V_L = IX_L = (0.474 \text{ A})(191 \text{ }\Omega) = 90.5$ V

(e) $P = I^2R = (0.474 \text{ A})^2(100 \text{ }\Omega) = 22.5$ W

22.72 $V = V_{max}\sin 2\pi ft$ and when $V = 0.7 V_m$, we have

$$f = \dfrac{1}{2\pi t}\sin^{-1}(0.7) = \dfrac{1}{2\pi(0.003 \text{ s})}\sin^{-1}(0.7) = 41.1 \text{ Hz}$$

22.73 Combining $X_L = \omega L = 12$ Ω and $X_C = \dfrac{1}{\omega C} = 8$ Ω we find $L = 96C$
Substitute this into the expression $\omega_0 = \dfrac{1}{\sqrt{LC}} = 4000$ s^{-1} to find

$C = 25.5$ μF and $L = 2.44$ mH

22.74 (a) The circumference of the solenoid $= 2\pi r = 2\pi(0.03 \text{ m}) = 0.188$ m. The number of turns that 5 m of wire can make is $N = \dfrac{5 \text{ m}}{0.188 \text{ m/turn}} = 26.5$ turns.
The length of the solenoid is
Length = (number of turns)(wire diameter) $= (26.5)(10^{-3} \text{ m}) = 2.65 \times 10^{-2}$ m $= 2.65$ cm

(b) From chapter 21)
$L = \dfrac{\mu_0 N^2 A}{\text{Length}} = \dfrac{(4\pi \times 10^{-7} \text{ N/A}^2)(26.5)^2 \pi(0.03 \text{ m})^2}{2.65 \times 10^{-2} \text{ m}} = 9.41 \times 10^{-5}$ H

(c) $R = \dfrac{\rho L}{A} = \dfrac{(1.7 \times 10^{-8} \text{ }\Omega\text{ m})(5 \text{ m})}{\pi(5 \times 10^{-4} \text{ m})^2} = 0.108$ Ω.

(d) We find $X_L = 3.55 \times 10^{-2}$ Ω, and $Z = 0.114$ Ω.

CHAPTER TWENTY-TWO SOLUTIONS

From which, $I = \dfrac{V}{Z} = \dfrac{20\ V}{0.114\ \Omega} = 176\ A$

22.75 (a) and (b) When a dc source is connected, there is a current in the circuit. Thus, neither of the two elements in the box can be a capacitor for if they were, no current could exist. Also, because the direct current is finite, one of the elements must be a resistor. The value of this resistance is

$$R = \dfrac{V}{I} = \dfrac{3\ V}{0.3\ A} = 10\ \Omega.$$

When an ac source is used, we find that the alternating current is less than the direct current. Thus, one of the elements must be an inductor. The impedance is

$$Z = \dfrac{V}{I} = \dfrac{3\ V}{0.2\ A} = 15\ \Omega.$$

Therefore, we use $Z^2 = (15\ \Omega)^2 = (10\ \Omega)^2 + X_L^2$
to find $X_L = 11.2\ \Omega = 2\pi f L$
and $L = \dfrac{11.2\ \Omega}{2\pi(60\ Hz)} = 2.97 \times 10^{-2}\ H = 29.7\ mH$

22.76 (a) $R = (4.5 \times 10^{-4}\ \Omega/m)(6.44 \times 10^5\ m) = 290\ \Omega$
and $I = \dfrac{P}{V} = \dfrac{5 \times 10^6\ W}{5 \times 10^5\ V} = 10\ A$
$P_{loss} = I^2 R = (10\ A)^2(290\ \Omega) = 2.9 \times 10^4\ W$

(b) $\dfrac{P_{loss}}{P} = \dfrac{2.9 \times 10^4\ W}{5 \times 10^6\ W} = 0.0058$

(c) In this case, we expect $I = \dfrac{5 \times 10^6\ W}{4500\ V} = 1.1 \times 10^3\ A$ and

the ratio $\dfrac{P_{loss}}{P} = \dfrac{(1.1 \times 10^3\ A)^2 (290\ \Omega)}{5 \times 10^6\ W} = 71.6$

This means that 71.6 times as much power would be used to offset line loss as would be transmitted.

22.77 We know $\dfrac{N_1}{N_2} = \dfrac{V_1}{V_2}$ Let the output impedance be $Z_1 = \dfrac{V_1}{I_1}$ and the input impedance $Z_2 = \dfrac{V_2}{I_2}$

so that $\dfrac{N_1}{N_2} = \dfrac{Z_1 I_1}{Z_2 I_2}$ but we also know $\dfrac{I_1}{I_2} = \dfrac{V_2}{V_1} = \dfrac{N_2}{N_1}$

so combining with the previous result, we have

$\dfrac{N_1}{N_2} = \left(\dfrac{Z_1}{Z_2}\right)^{1/2} = \left(\dfrac{8000}{8}\right)^{1/2} = 31.6$

22.78 $P_{loss} = I^2 R = I^2\left(\dfrac{\rho 2L}{A}\right),\quad P_L = VI\cos\phi,\ \text{or}\ I = \dfrac{P_L}{V\cos\phi}.$

$P_{loss} = \left(\dfrac{P_L}{V\cos\phi}\right)^2 \left(\dfrac{\rho 2L}{A}\right) = \dfrac{2\rho L P_L^2}{A V^2 \cos^2\phi}$

CHAPTER TWENTY-THREE SOLUTIONS

23.1 For the FM band, the frequencies range from a low of 88 MHz to a high of 108 MHz. Thus, from $f_0 = \dfrac{1}{2\pi\sqrt{LC}}$, we have

$C = \dfrac{1}{4\pi^2 L f_0^2}$ which, with $L = 2 \times 10^{-6}$ H, becomes $C = \dfrac{1}{(7.90 \times 10^{-5} \, \Omega \, s) f_0^2}$

(a) for a frequency of 108 MHz, $C = 1.09 \times 10^{-12}$ F = 1.09 pF
(b) for a frequency of 88 MHz, $C = 1.64 \times 10^{-12}$ F = 1.64 pF

23.2 In the fundamental mode there is a single loop in the standing wave between the plates. Thus, the distance between the plates is equal to half a wavelength.
$\lambda = 2L = 2(2 \text{ m}) = 4 \text{ m}$
Thus, $f = \dfrac{c}{\lambda} = \dfrac{3 \times 10^8 \text{ m/s}}{4 \text{ m}} = 7.5 \times 10^7$ Hz = 75 MHz

---

23.3 $c = \dfrac{1}{\sqrt{\mu_0 \varepsilon_0}}$ If $\mu_0 = 4\pi \times 10^{-7}$ Ns^2/C^2 and $\varepsilon_0 = 8.854 \times 10^{-12}$ C^2/Nm^2, then

$c = \dfrac{1}{\sqrt{1.1126 \times 10^{-17} \text{ s}^2/\text{m}^2}} = 2.99796 \times 10^8$ m/s, or rounding to four places,

$c = 2.998 \times 10^8$ m/s

23.4 The units of μ_0 are $\dfrac{N}{A^2} = \dfrac{\text{kg m/s}^2}{(C/s)^2} = \dfrac{\text{kg m}}{C^2} = \dfrac{ML}{Q^2}$

The units of ε_0 are $\dfrac{F}{m} = \dfrac{C/V}{m} = \dfrac{C}{\frac{J}{C} m} = \dfrac{C^2}{(N \, m) m} = \dfrac{C^2}{(\text{kg m/s}^2) m^2} = \dfrac{C^2 s^2}{\text{kg m}^3} = \dfrac{Q^2 T^2}{ML^3}$

Thus, the units of $\dfrac{1}{\sqrt{\mu_0 \varepsilon_0}}$ are $\dfrac{1}{\sqrt{\dfrac{MLQ^2T^2}{Q^2 ML^3}}} = \dfrac{L}{T}$

23.5 $B = \dfrac{E}{c} = \dfrac{150 \text{ V/m}}{3 \times 10^8 \text{ m/s}} = 5 \times 10^{-7}$ T

23.6 Average Power per unit area $= \dfrac{E_{max}^2}{2\mu_0 c}$

$= \dfrac{(150 \text{ V/m})^2}{2(4\pi \times 10^{-7} \text{ Ns}^2/\text{C}^2)(3 \times 10^8 \text{ m/s})} = 29.8$ W/m^2.

23.7 The intensity as a function of distance from the source is
$I = \dfrac{P}{A} = \dfrac{15 \text{ W}}{4\pi r^2} = \dfrac{1.19 \text{ W}}{r^2}$ (1)

CHAPTER TWENTY-THREE SOLUTIONS

(a) At $r = 1$ m, (1) becomes $\quad I = 1.19$ W/m^2 = $\dfrac{cB_{max}^2}{2\mu_0}$

From which, $\quad B_{max} = 10^{-7}$ T

and $\quad E_{max} = cB_{max} = (3 \times 10^8$ m/s$)(10^{-7}$ T$) = 30$ N/C

(b) Repeating the procedure outlined above with $r = 5$ m, $I = 4.78 \times 10^{-2}$ W/m^2, and

$\quad\quad B_{max} = 2 \times 10^{-8}$ T $\quad E_{max} = 6$ N/C

23.8 $I = \dfrac{P}{A} = \dfrac{P}{4\pi r^2}$

so $\quad P = (4\pi r^2)I = 4\pi(1.49 \times 10^{11}$ m$)^2(1340$ W/m$^2) = 3.74 \times 10^{26}$ W

23.9 From, Average Power per unit area $= \dfrac{E_{max}^2}{2\mu_0 c} = 1340$ W/m^2 we have

$E_{max} = \sqrt{(2\mu_0 c)(1340 \text{ W/m}^2)}$
$= \sqrt{2(4\pi \times 10^{-7} \text{ Ns}^2/\text{C}^2)(3 \times 10^8 \text{ m/s})(1340 \text{ W/m}^2)}$
$= 1.01 \times 10^3$ N/C

and $\quad B_{max} = \dfrac{E_{max}}{c} = \dfrac{1.01 \times 10^3 \text{ N/C}}{3 \times 10^8 \text{ m/s}} = 3.35 \times 10^{-6}$ T

23.10 $I = \dfrac{P}{A} = \dfrac{P}{4\pi r^2}$ and $I = \dfrac{E_{max}^2}{2\mu_0 c}$

Thus, $E_{max}^2 = \dfrac{\mu_0 c}{2\pi r^2} P = \dfrac{(4\pi \times 10^{-7} \text{ Ns}^2/\text{C}^2)(3 \times 10^8 \text{ m/s})(15 \times 10^3 \text{ J/s})}{2\pi r^2}$

$= \dfrac{(9 \times 10^5 \text{ N}^2 \text{ m}^2/\text{C}^2)}{r^2}$

(a) at $r = 1$ km, the above gives $E_{max} = 0.949$ N/C
(b) at $r = 10$ km, $\quad E_{max} = 0.0949$ N/C
(c) at $r = 100$ km, $\quad E_{max} = 9.49 \times 10^{-3}$ N/C

23.11 (a) $E = cB = (3 \times 10^8$ m/s$)(1.5 \times 10^{-7}$ T$) = 45.0$ N/C

(b) Average Power per unit area $= \dfrac{cB_{max}^2}{2\mu_0}$

$= \dfrac{(3 \times 10^8 \text{ m/s})(1.5 \times 10^{-7} \text{ T})^2}{2(4\pi \times 10^{-7} \text{ Ns}^2/\text{C}^2)} = 2.69$ W/m^2.

23.12 (a) The frequency of the magnetic field is equal to the frequency of the electric field.
$\quad\quad f = 10^{14}$ Hz

(b) The vibration is perpendicular to that of the electric field. It is north and south in a horizontal plane.

23.13 $I = \dfrac{P}{A} = \dfrac{5 \times 10^{-3} \text{ W}}{\dfrac{\pi (5 \times 10^{-3} \text{ m})^2}{4}} = 255$ W/m^2.

CHAPTER TWENTY-THREE SOLUTIONS

and from $I = \dfrac{cB_{max}^2}{2\mu_0}$, we have

$$B_{max} = \sqrt{\dfrac{2\mu_0}{c}(I)} = \sqrt{\dfrac{2(4\pi \times 10^{-7}\ Ns^2/C^2)(255\ W/m^2)}{(3 \times 10^8\ m/s)}} = 1.46 \times 10^{-6}\ T$$

and, $E_{max} = cB_{max} = (3 \times 10^8\ m/s)(1.46 \times 10^{-6}\ T) = 438\ N/C$

23.14 We have, $P = IA = (1340\ W/m^2)(20\ m)(30\ m) = 8.04 \times 10^5\ W$
and Energy $= Pt = (8.04 \times 10^5\ W)(8\ h)(3600\ s/h) = 2.32 \times 10^{10}\ J$

23.15 Power output = (power input)(efficiency)

Thus, Power input $= \dfrac{\text{power out}}{eff} = \dfrac{10^6\ W}{0.3} = 3.33 \times 10^6\ W$

and $A = \dfrac{P}{I} = \dfrac{3.33 \times 10^6\ W}{10^3\ W/m^2} = 3.33 \times 10^3\ m^2$

23.16 $P_{incident} = IA = (300\ W/m^2)(0.2\ m)(0.4\ m) = 24\ W$

and $P_{absorbed} = \dfrac{1}{2} P_{incident} = 12\ W$

Therefore, $E_{absorbed} = (P_{absorbed})t = (12\ J/s)(60\ s) = 720\ J$

23.17 (a) $f = \dfrac{c}{\lambda} = \dfrac{3 \times 10^8\ m/s}{2\ m} = 1.5 \times 10^8\ Hz = 150\ MHz$

(b) $f = \dfrac{c}{\lambda} = \dfrac{3 \times 10^8\ m/s}{20\ m} = 1.5 \times 10^7\ Hz = 15\ MHz$

(c) $f = \dfrac{c}{\lambda} = \dfrac{3 \times 10^8\ m/s}{200\ m} = 1.5 \times 10^6\ Hz = 1.5\ MHz$

23.18 (a) $\lambda = \dfrac{c}{f} = \dfrac{3 \times 10^8\ m/s}{10^6\ Hz} = 300\ m$ (AM radio)

(b) $\lambda = \dfrac{c}{f} = \dfrac{3 \times 10^8\ m/s}{10^8\ Hz} = 3\ m$ (TV, FM radio)

(c) $\lambda = \dfrac{c}{f} = \dfrac{3 \times 10^8\ m/s}{10^{10}\ Hz} = 3 \times 10^{-2}\ m = 3\ cm$ (microwave)

(d) $\lambda = \dfrac{c}{f} = \dfrac{3 \times 10^8\ m/s}{10^{13}\ Hz} = 30 \times 10^{-6}\ m = 30\ microns$ (infrared)

(e) $\lambda = \dfrac{c}{f} = \dfrac{3 \times 10^8\ m/s}{10^{15}\ Hz} = 3 \times 10^{-7}\ m = 300\ nm$ (ultraviolet)

(f) $\lambda = \dfrac{c}{f} = \dfrac{3 \times 10^8\ m/s}{10^{17}\ Hz} = 3 \times 10^{-9}\ m = 3\ nm$ (x-rays)

(g) $\lambda = \dfrac{c}{f} = \dfrac{3 \times 10^8\ m/s}{10^{21}\ Hz} = 3 \times 10^{-13}\ m = 0.003\ Å$ (γ rays)

23.19 (a) $\lambda_{max} = \dfrac{c}{f_{min}} = \dfrac{3 \times 10^8\ m/s}{540 \times 10^3\ Hz} = 556\ m$

$\lambda_{min} = \dfrac{c}{f_{max}} = \dfrac{3 \times 10^8\ m/s}{1600 \times 10^3\ Hz} = 186\ m$

CHAPTER TWENTY-THREE SOLUTIONS

(b) for the fm band,

$$\lambda_{max} = \frac{c}{f_{min}} = \frac{3 \times 10^8 \text{ m/s}}{88 \times 10^6 \text{ Hz}} = 3.41 \text{ m}$$

$$\lambda_{min} = \frac{c}{f_{max}} = \frac{3 \times 10^8 \text{ m/s}}{108 \times 10^6 \text{ Hz}} = 2.78 \text{ m}$$

23.20 $f = \frac{c}{\lambda} = \frac{3 \times 10^8 \text{ m/s}}{5.5 \times 10^{-7} \text{ m}} = 5.45 \times 10^{14}$ Hz

---

23.21 $\lambda = \frac{c}{f} = \frac{3 \times 10^8 \text{ m/s}}{63 \times 10^6 \text{ Hz}} = 4.76$ m

$L = \frac{\lambda}{4} = 1.19$ m (length of each rod)

23.22 (a) $f = \frac{v}{\lambda} = \frac{2.5 \times 10^8 \text{ m/s}}{4 \times 10^{-7} \text{ m}} = 6.25 \times 10^{14}$ Hz

(b) $\lambda = \frac{c}{f} = \frac{3 \times 10^8 \text{ m/s}}{6.25 \times 10^{14} \text{ Hz}} = 4.8 \times 10^{-7}$ m $= 480$ nm

---

23.23 The diagonal distance from the point of transmission to a point located directly above the midway point between transmitter and receiver is

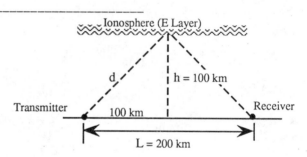

$d = \sqrt{(100 \text{ km})^2 + (100 \text{ km})^2}$
$= \sqrt{2} (100 \text{ km}) = 141.4$ km

The time to travel the straight line path is

$t_1 = \frac{L}{c} = \frac{200 \times 10^3 \text{ m}}{3 \times 10^8 \text{ m/s}} = 6.67 \times 10^{-4}$ s

and the time to travel the reflected path $= t_2$

$= \frac{2d}{c} = \frac{2(141.4 \times 10^3 \text{ m})}{3 \times 10^8 \text{ m/s}} = 9.43 \times 10^{-4}$ s

for a time difference of $t_2 - t_1 = 0.276$ ms.

23.24 (a) $t = \frac{\text{distance}}{\text{speed}} = \frac{100 \times 10^3 \text{ m}}{3 \times 10^8 \text{ m/s}} = 3.33 \times 10^{-4}$ s

(b) $d = vt = (345 \text{ m/s})(3.33 \times 10^{-4} \text{ s}) = 1.15 \times 10^{-1}$ m $= 11.5$ cm

---

23.25 $c = 3 \times 10^8$ m/s $= 1.86 \times 10^5$ mi/s

(a) $t = \frac{d}{v} = \frac{93 \times 10^6 \text{ mi}}{1.86 \times 10^5 \text{ mi/s}} = 500$ s $= 8.33$ min

(b) $t = \frac{d}{v} = \frac{2.5 \times 10^5 \text{ mi}}{1.86 \times 10^5 \text{ mi/s}} = 1.34$ s

---

23.26 The frequency being detected is

CHAPTER TWENTY-THREE SOLUTIONS

$$f_0 = \frac{1}{2\pi\sqrt{LC}} = \frac{1}{2\pi\sqrt{(5.4 \times 10^{-6} \text{ H})(3.1 \times 10^{-12} \text{ F})}} = 3.89 \times 10^7 \text{ Hz}$$

and from this, $\lambda = \frac{c}{f} = \frac{3 \times 10^8 \text{ m/s}}{3.89 \times 10^7 \text{ Hz}} = 7.71 \text{ m}$

23.27 time to reach object = $\frac{1}{2}$ total time of flight = $\frac{1}{2}(4 \times 10^{-4} \text{ s}) = 2 \times 10^{-4}$ s

Thus, $d = vt = (3 \times 10^8 \text{ m/s})(2 \times 10^{-4} \text{ s}) = 6 \times 10^4 \text{ m} = 60 \text{ km}$

23.28 The total distance the wave must travel is 12000 m.
(a) $t = \frac{d}{v} = \frac{12000 \text{ m}}{3 \times 10^8 \text{ m/s}} = 4.0 \times 10^{-5}$ s
(b) $d = vt = (50 \text{ m/s})(4 \times 10^{-5} \text{ s}) = 2 \times 10^{-3} \text{ m} = 2 \text{ mm}$

23.29 $E = cB = (3 \times 10^8 \text{ m/s})(2.4 \times 10^{-6} \text{ T}) = 720 \text{ N/C}$

23.30 (a) $f = \frac{c}{\lambda} = \frac{3 \times 10^8 \text{ m/s}}{10^{-6} \text{ m}} = 3 \times 10^{14} \text{ Hz}$

(b) $f = \frac{c}{\lambda} = \frac{3 \times 10^8 \text{ m/s}}{10^{-9} \text{ m}} = 3 \times 10^{17} \text{ Hz}$

(c) $f = \frac{c}{\lambda} = \frac{3 \times 10^8 \text{ m/s}}{10^{-10} \text{ m}} = 3 \times 10^{18} \text{ Hz}$

23.31 From $f_0 = \frac{1}{2\pi\sqrt{LC}}$ we have

$L = \frac{1}{4\pi^2 C f_0^2} = \frac{1}{4\pi^2(1.5 \times 10^{-12} \text{ F})(5.71 \times 10^7 \text{ Hz})^2} = 5.17 \times 10^{-6} \text{ H} = 5.17 \text{ μH}$

23.32 $\frac{E}{B} = \frac{\text{N/C}}{\text{T}} = \frac{\text{N/C}}{\text{N/A m}} = \frac{\text{A m}}{\text{C}} = \frac{\text{m C/s}}{\text{C}} = \text{m/s}$

23.33 $k = \frac{2\pi}{\lambda} = \frac{2\pi f}{c} = \frac{2\pi(4.8 \times 10^{14} \text{ Hz})}{3 \times 10^8 \text{ m/s}} = 10^7 \text{ m}^{-1}$.

23.34 $L = 4$ light years $= c(4 \text{ years})$
$= 4(3 \times 10^8 \text{ m/s})(3.156 \times 10^7 \text{ s/yr}) = 3.79 \times 10^{16}$ m

23.35 The time for a one-way trip to Jupiter for the radio signal is
$t = \frac{d}{v} = \frac{630 \times 10^9 \text{ m}}{3 \times 10^8 \text{ m/s}} = 2.1 \times 10^3 \text{ s} = 35 \text{ min}$

Thus, the total time for the command to reach Jupiter and for the signal to return is
70 min.

23.36 $I = \frac{E_{max}^2}{2\mu_0 c} = \frac{(15 \text{ V/m})^2}{2(4\pi \times 10^{-7} \text{ Ns}^2/\text{C}^2)(3 \times 10^8 \text{ m/s})} = 0.298 \text{ W/m}^2$.

CHAPTER TWENTY-THREE SOLUTIONS

and $I = \dfrac{P}{A} = \dfrac{P}{4\pi r^2}$ leads to $r = \sqrt{\dfrac{P}{4\pi I}} = \sqrt{\dfrac{100 \text{ W}}{4\pi(0.298 \text{ W/m}^2)}} = 5.16$ m

23.37 The intensity is
$I = \dfrac{cB_{max}^2}{2\mu_0} = \dfrac{(3 \times 10^8 \text{ m/s})(7 \times 10^{-8} \text{ T})^2}{2(4\pi \times 10^{-7} \text{ Ns}^2/\text{C}^2)} = 0.585$ W/m^2.

At $r = 2$m, $P = IA = (0.585 \text{ W/m}^2)(4\pi(2 \text{ m})^2) = 29.4$ W

23.38 (a) $B = \dfrac{E}{c} = \dfrac{6 \text{ V/m}}{3 \times 10^8 \text{ m/s}} = 2 \times 10^{-8}$ T

(b) $I = \dfrac{E_{max}B_{max}}{2\mu_0} = \dfrac{(6 \text{ V/m})(2 \times 10^{-8} \text{ T})}{2(4\pi \times 10^{-7} \text{ Ns}^2/\text{C}^2)} = 4.77 \times 10^{-2}$ W/m^2.

and $P = IA = 4\pi r^2 I = 4\pi(10^3 \text{ m})^2(4.77 \times 10^{-2} \text{ W/m}^2) = 6 \times 10^5$ W $= 600$ kW

23.39 (a) $B_{max} = \dfrac{E_{max}}{c} = \dfrac{2 \times 10^{-7} \text{ V/m}}{3 \times 10^8 \text{ m/s}} = 6.67 \times 10^{-16}$ T

(b) $I = \dfrac{E_{max}^2}{2\mu_0 c} = \dfrac{(2 \times 10^{-7} \text{ V/m})^2}{2(4\pi \times 10^{-7} \text{ Ns}^2/\text{C}^2)(3 \times 10^8 \text{ m/s})} = 5.31 \times 10^{-17}$ W/m^2.

(c) $P = IA = (5.31 \times 10^{-17} \text{ W/m}^2)(100\pi \text{ m}^2) = 1.67 \times 10^{-14}$ W

23.40 First determine the energy density in this sunlight (at the top of the atmosphere) by considering a shaft of sunlight one square meter in cross-sectional area and 1 m long (i.e one cubic meter of sunlight). All the energy in this volume strikes one square meter of the atmosphere in a time of

$$t = \dfrac{d}{v} = \dfrac{1 \text{ m}}{3 \times 10^8 \text{ m/s}} = 3.33 \times 10^{-9} \text{ s}.$$

The energy striking a unit area in time t is given by

$$E = It = \left(\dfrac{\text{power}}{\text{area}}\right)t = \dfrac{\text{energy}}{\text{area}}$$

Thus, if $I = 1340$ W/m^2, the energy striking one square meter in 3.33×10^{-9} s is
$U = (1340 \text{ J/s/m}^2)(3.33 \times 10^{-9} \text{ s/m}) = 4.47 \times 10^{-6}$ J/m^3

Thus, the energy in 1 liter of sunlight = (energy density)(volume)
$= (4.47 \times 10^{-6} \text{ J/m}^3)(10^{-3} \text{ m}^3) = 4.47 \times 10^{-9}$ J

CHAPTER TWENTY-FOUR SOLUTIONS

24.1 The additional time required for the light to cross the diameter of the earth's orbit is 22 min (1320 s). Therefore,
$$v = \frac{d}{t} = \frac{3 \times 10^{11} \text{ m}}{1320 \text{ s}} = 2.27 \times 10^8 \text{ m/s}$$

24.2 The time for the light to cross the diameter of the earth's orbit is 1320 s. Thus,
$$\text{diameter} = vt = (2.998 \times 10^8 \text{ m/s})(1320 \text{ s}) = 3.96 \times 10^{11} \text{ m}$$
and the radius is 1.98×10^{11} m

24.3 (a) The time for the light to travel 40 m is
$$t = \frac{40 \text{ m}}{3 \times 10^8 \text{ m/s}} = 1.33 \times 10^{-7} \text{ s}$$
At the lowest speed, the wheel will have turned through 1/360 rev in the time t. Thus,
$$\omega = \frac{\Delta \theta}{\Delta t} = \frac{1/360 \text{ rev}}{1.33 \times 10^{-7} \text{ s}} = 2.08 \times 10^4 \text{ rev/s}$$
The next lowest speed occurs when the wheel turns through 2/360 rev in the time t.
$$\omega = \frac{\Delta \theta}{\Delta t} = \frac{2/360 \text{ rev}}{1.33 \times 10^{-7} \text{ s}} = 4.17 \times 10^4 \text{ rev/s}$$
(b) The steps are identical to those used in part a.
The flight time of the light is 1.33×10^{-5} s, and the lowest speed is 2.08×10^2 rev/s. The next lowest speed is 4.17×10^2 rev/s.

24.4 (a) The time for the light to travel to the mirror and back is
$$t = \frac{2(35 \times 10^3 \text{ m})}{3 \times 10^8 \text{ m/s}} = 2.33 \times 10^{-4} \text{ s}$$
At the lowest speed, the mirror will have made 1/8 rev in time t. Thus,
$$\omega = \frac{\Delta \theta}{\Delta t} = \frac{1/8 \text{ rev}}{2.33 \times 10^{-4} \text{ s}} = 536 \text{ rev/s}$$
(b) The next higher speed will occur when the wheel makes 2/8 rev in time t. The angular velocity for this case is
$$\omega = \frac{\Delta \theta}{\Delta t} = \frac{2/8 \text{ rev}}{2.33 \times 10^{-4} \text{ s}} = 1070 \text{ rev/s}$$

24.5 $t = \dfrac{d}{c} = \dfrac{1.8 \times 10^4 \text{ m}}{3 \times 10^8 \text{ m/s}} = 6 \times 10^{-5}$ s = 60 μs

CHAPTER TWENTY-FOUR SOLUTIONS

24.6 We have $\alpha = i + r = 2i$. (See sketch.)

$$\alpha = \frac{5}{12} \text{ of a full circle} = \frac{5}{12}(360°) = 150°$$

Thus, $i = \frac{150°}{2} = 75°$

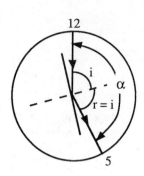

24.7 (a) From geometry, $1.25 \text{ m} = d \sin 40°$

so $d = 1.94$ m

(b) at 50° above horizontal, or parallel to the incident ray.

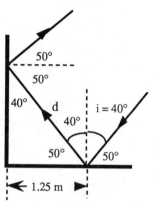

24.8 Direct graphical solution of this problem shows that the light ray reflects from mirror 1 six times and from mirror 2 five times befor escaping from between the two mirrors.

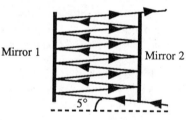

24.9 $v = \frac{c}{n} = \frac{3 \times 10^8 \text{ m/s}}{1.501} = 2.00 \times 10^8$ m/s

24.10 (a) $v = \frac{c}{n} = \frac{3 \times 10^8 \text{ m/s}}{1.66} = 1.81 \times 10^8$ m/s

(b) $v = \frac{c}{n} = \frac{3 \times 10^8 \text{ m/s}}{1.333} = 2.25 \times 10^8$ m/s

(c) $v = \frac{c}{n} = \frac{3 \times 10^8 \text{ m/s}}{1.923} = 1.56 \times 10^8$ m/s

24.11 $\lambda_{medium} = \frac{\lambda_{vac}}{n_{medium}} = \frac{589 \text{ nm}}{1.361} = 433$ nm

CHAPTER TWENTY-FOUR SOLUTIONS

24.12 (a) $\lambda_{water} = \dfrac{\lambda_{vac}}{n_{water}} = \dfrac{436 \text{ nm}}{1.333} = 327$ nm

(b) $\lambda_{glass} = \dfrac{\lambda_{vac}}{n_{glass}} = \dfrac{436 \text{ nm}}{1.52} = 287$ nm

---

24.13 We call medium 1 the air, and medium 2 is the water. Snell's law becomes
$$n_2 \sin\theta_2 = n_1 \sin\theta_1$$
$$\tfrac{4}{3} \sin\theta_2 = \sin 30°$$
From which, $\sin\theta_2 = 0.375$, and $\theta = 22.0°$

24.14 $n_{oil} = \dfrac{c}{v_{oil}} = \dfrac{3 \times 10^8 \text{ m/s}}{2.17 \times 10^8 \text{ m/s}} = 1.382$

Thus, from Snell's law $n_{oil} \sin\theta_2 = n_{air} \sin\theta_1$
$$1.382 \sin\theta_2 = (1.0) \sin 23.1°$$
$$\theta_2 = 16.5°$$

24.15 $n_2 \sin\theta_2 = n_1 \sin\theta_1$
$$\tfrac{4}{3} \sin\theta_2 = (1.0) \sin 35°$$
$$\theta_2 = 25.5°$$
$$\lambda_2 = \dfrac{\lambda_1 n_1}{n_2} = \dfrac{589 \text{ nm}(1)}{\tfrac{4}{3}} = 442 \text{ nm}$$

24.16 $\sin\theta_1 = \dfrac{n_2 \sin\theta_2}{n_1} = \dfrac{\tfrac{4}{3}\sin 31°}{1} = 0.687$
$$\theta_1 = 43.4°$$

24.17 Using the law of reflection, we find that $d = 6$ cm. From Snell's law (assuming ultrasonic waves obey it)
$$n_m \sin 50° = n_L \sin\theta$$
$$\sin\theta = \dfrac{n_m}{n_L} \sin 50° \quad (1)$$
but, $\dfrac{n_m}{n_L} = \dfrac{v_L}{v_m} = 0.9 \quad (2)$

solve (1) and (2) simultaneously to find
$\theta = 43.6°$
then $h = \dfrac{d}{\tan 43.6°} = 6.3$ cm

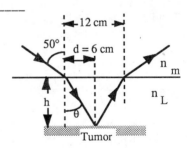

333

CHAPTER TWENTY-FOUR SOLUTIONS

24.18 The air is medium 1, and the ice is medium 2. Snell's law gives the angle of refraction as
$$n_2 \sin\theta_2 = n_1 \sin\theta_1$$
$$1.309 \sin\theta_2 = \sin 40°$$
and $\sin\theta_2 = 0.491$, and $\theta_2 = 29.4°$
Also, from the law of reflection, $\Phi = 40°$
The angle between the reflected and refracted ray (see sketch) is found as $\theta_2 + \alpha + \Phi = 180°$.
Thus, $\alpha = 110.6°$.

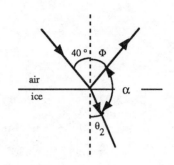

24.19 At the first surface, we call the air medium 1 and the glass medium 2. We find the angle of refraction from Snell's law.
$$n_2 \sin\theta_2 = n_1 \sin\theta_1$$
$$1.50 \sin\theta_2 = \sin 30°$$
and $\sin\theta_2 = 0.333$, and $\theta_2 = 19.5°$
At the second surface, we call medium 2 the glass and medium 3 the air into which the light exits. We find,
$$n_3 \sin\theta_3 = n_2 \sin\theta_2$$
$$1.00 \sin\theta_3 = 1.50 \sin 19.5°$$
and $\sin\theta_3 = 0.50$, and $\theta_3 = 30°$

24.20 The angle of refraction, θ_2, equals 19.5°. See problem 19. The distance h is found by use of triangle abc as.
$$\frac{2 \text{ cm}}{h} = \cos 19.5°.$$
From which, $h = 2.12$ cm.
We also see that
$$\theta_2 + \alpha = \theta_1$$
or, $\alpha = \theta_1 - \theta_2 = 30° - 19.5° = 10.5°$
Finally, $d = h \sin\alpha = (2.12 \text{ cm}) \sin 10.5° = 0.387$ cm

---

24.21 We find the index of refraction from Snell's law.
$$n_2 \sin\theta_2 = n_1 \sin\theta_1$$
$$n_2 \sin 25° = 1.33 \sin 37°$$
$$n_2 = 1.90$$
and $v = \frac{c}{n} = \frac{3 \times 10^8 \text{ m/s}}{1.90} = 1.58 \times 10^8$ m/s

---

24.22 The distance, d, traveled by the light (see sketch) is
$$d = \frac{2 \text{ cm}}{\cos 19.5°} = 2.12 \text{ cm}$$
The speed of light in the material is
$$v = \frac{c}{n} = \frac{3 \times 10^8 \text{ m/s}}{1.50} = 2 \times 10^8 \text{ m/s}$$
Therefore, $t = \frac{d}{v} = \frac{2.12 \times 10^{-2} \text{ m}}{2 \times 10^8 \text{ m/s}} = 1.06 \times 10^{-10}$ s

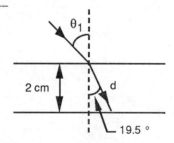

CHAPTER TWENTY-FOUR SOLUTIONS

24.23 At the first surface, the angle of incidence is 30° and the angle of refraction is found as

$n_2 \sin\theta_2 = n_1 \sin\theta_1$
$1.5 \sin\theta_2 = (1) \sin 30°$
$\theta_2 = 19.5°$

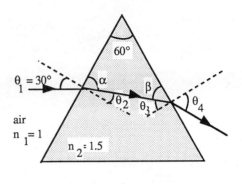

(See the sketch.) At the first surface, we see

$\alpha + \theta_2 = 90°$, or $\alpha = 90° - 19.5° = 70.5°$

We also see that

$\alpha + \beta + 60° = 180°$. Thus, $\beta = 49.5°$

Therefore, $\theta_3 = 90° - \beta = 90° - 49.5° = 40.5°$.

θ_3 is the angle of incidence at the second surface, and it is also the angle of reflection at the second surface.
Snell's law gives us the angle of refraction at the second surface.

$n_4 \sin\theta_4 = n_3 \sin\theta_3$
$(1)\sin\theta_4 = 1.5 \sin 40.5°$
$\theta_4 = 77.1°$ (angle of refraction at second surface)

24.24 Applying Snell's law at the air-oil interface,

$n_{air} \sin\theta = n_{oil} \sin 20°$

yields $\theta = 30.4°$

Applying Snell's law at the oil-water interface

$n_w \sin\theta' = n_{oil} \sin 20°$

yields $\theta' = 23.3°$

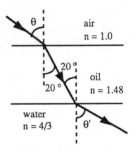

24.25 For the first placement, Snell's law becomes, with 1 referring to the upper sheet and 2 to the lower,

$n_2 = n_1 \dfrac{\sin 26.5°}{\sin 31.7°}$ (1)

Given Conditions and Observed Results

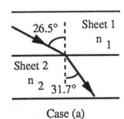 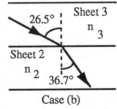 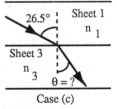

Case (a)   Case (b)   Case (c)

For the second placement, with 3 referring to the remaining sheet,

$n_3 = n_2 \dfrac{\sin 36.7°}{\sin 26.5°}$ (2)

Combining (1) and (2), we have

$n_3 = n_1 \dfrac{\sin 36.7°}{\sin 31.7°}$ (3)

Applying Snell's law to the final situation, we have

$\sin\theta = n_1 \dfrac{\sin 26.5°}{n_3} = \dfrac{\sin 31.7°}{\sin 36.7°} \sin 26.5°$ [utilizing (3)]

yields $\theta = 23.1°$

CHAPTER TWENTY-FOUR SOLUTIONS

24.26 (See sketch.)
$$\tan\theta_1 = \frac{90}{100} = 0.9,$$
and $\theta_1 = 42°$
From Snell's law, we find
$$n_2\sin\theta_2 = n_1\sin\theta_1$$
$(1)\sin\theta_2 = (1.33)\sin 42°$
$\theta_2 = 63.1°$
Finally,
$$h = \frac{210 \text{ m}}{\tan 63.1°} = 106 \text{ m}$$

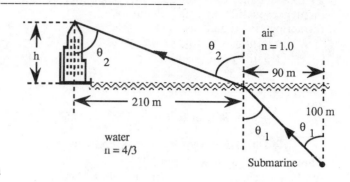

24.27 (a) From the dispersion curve for crown glass, $n = 1.515$ when $\lambda = 589$ nm. Thus,
$$v = \frac{c}{n} = \frac{3 \times 10^8 \text{ m/s}}{1.515} = 1.980 \times 10^8 \text{ m/s}$$
(b) We have, $\lambda_{glass} = \frac{\lambda_{vac}}{n_{glass}} = \frac{589 \text{ nm}}{1.515} = 389 \text{ nm}$
Thus, $t = 100\lambda_{glass} = (389 \text{ nm})(100) = 3.89 \times 10^{-5} \text{ m}$

24.28 time difference = (time for light to travel 6.2 m in ice) - (time to travel 6.2 m in air)
$$\Delta t = \frac{6.2 \text{ m}}{v_{ice}} - \frac{6.2 \text{ m}}{c}$$
but $v = \frac{c}{n}$, so $\Delta t = (6.2 \text{ m})\left(\frac{1.309}{c} - \frac{1}{c}\right) = \frac{6.2 \text{ m}}{c}(0.309) = 6.39 \times 10^{-9}$ s
= 6.39 ns

24.29 In a time t, the light travels a distance d_w in water, where
$$d_w = v_w t = \frac{c}{n_w} t$$
In the same time, the distance traveled in glass is
$$d_g = v_g t = \frac{c}{n_g} t$$
Thus, $\frac{d_w}{d_g} = \frac{n_g}{n_w}$
Or, $d_w = \frac{1.50}{1.33} (10 \text{ m}) = 11.25 \text{ m}$

CHAPTER TWENTY-FOUR SOLUTIONS

24.30 From Snell's law we see (see sketch)
$$n_2 \sin\theta_2 = n_1 \sin\theta_1$$
$$\sin\theta_2 = \frac{\sin\theta_1}{\frac{4}{3}} = \frac{3}{4}\sin\theta_1 \quad (1)$$

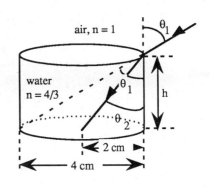

Also, $\tan\theta_1 = \frac{4\text{ cm}}{h}$, and $\tan\theta_2 = \frac{2\text{ cm}}{h}$

Thus, $\tan\theta_1 = 2\tan\theta_2$

or, $\frac{\sin\theta_1}{\cos\theta_1} = 2\frac{\sin\theta_2}{\cos\theta_2}$

Which, using Eq. (1), gives, $\cos\theta_2 = 2\frac{\sin\theta_2}{\sin\theta_1}$

$\cos\theta_2 = 2\left(\frac{3}{4}\right)\cos\theta_1$

And $\cos\theta_2 = \frac{3}{2}\cos\theta_1 \quad (2)$

Then, $\sin^2\theta_2 + \cos^2\theta_2 = 1$, or making use of equations (1) and (2):

$\frac{9}{16}\sin^2\theta_1 + \frac{9}{4}\cos^2\theta_1 = 1$, which reduces to $\sin^2\theta_1 + 4\cos^2\theta_1 = \frac{16}{9}$

or $(1 - \cos^2\theta_1) + 4\cos^2\theta_1 = \frac{16}{9}$. This can be solved for θ_1 to find, $\theta_1 = 59.4°$

Finally, $\tan\theta_1 = \frac{4\text{ cm}}{h}$ becomes $\tan 59.4° = \frac{4\text{ cm}}{h}$, which yields

$h = 2.37$ cm

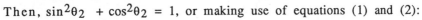

24.31 When the Sun is 28° from horizon, incident rays are 62° from normal at air-water interface. (See the sketch at the right.)

Thus, $n_w \sin\theta_w = (1.0)\sin 62°$

$\sin\theta_w = \frac{\sin 62°}{\frac{4}{3}} = 0.662$ and $\theta_w = 41.5°$

and $\frac{3\text{ m}}{h} = \tan\theta_w$, so $h = \frac{3\text{ m}}{\tan 41.5°} = 3.39$ m

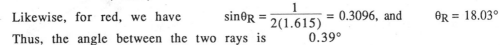

24.32 The angle of refraction for the blue light is found from Snell's law as

$\sin\theta_B = \frac{1}{2(1.650)} = 0.303$, and $\theta_B = 17.64°$

Likewise, for red, we have $\sin\theta_R = \frac{1}{2(1.615)} = 0.3096$, and $\theta_R = 18.03°$

Thus, the angle between the two rays is $0.39°$

24.33 For acrylic with $\lambda = 400$ nm, we find $n = 1.507$. (See dispersion graph in text.)
To be refracted at the same angle in the fused quartz, the wavelength used must have the same index of refraction as the 400 nm light had in the acrylic. From Fig. 24.16, it is seen that the wavelength for which $n = 1.507$ in fused quartz is

337

CHAPTER TWENTY-FOUR SOLUTIONS

$\lambda = 245$ nm

24.34 (a) From Fig. 24.16, we find the index of refraction is n = 1.455 for $\lambda =$ 700 nm in fused quartz. We find the angle of refraction at the first surface from Snell's law.

$$\sin\theta_2 = \frac{\sin 75°}{1.455} \text{ and } \theta_2 = 41.6°$$

(b) (See sketch.) $\alpha = 90° - \theta_2 = 48.4°$
and $\theta_2 + \gamma = 75°$
or, $\gamma = 75° - \theta_2 = 33.4°$
Then $\alpha + \beta + 60° = 180°$,
or $\beta = 120° - \alpha = 71.6°$
Thus, $\theta_3 = 90° - \beta = 18.4°$

(c) The angle of refraction at the second surface is found from Snell's Law as

(1)$\sin\theta_4 = 1.455\sin 18.4°$
and $\theta_4 = 27.3°$

(d) $\varepsilon = \theta_4 - \theta_3 = 8.94°$
Finally, $\delta = \gamma + \varepsilon = 33.4° + 8.94° = 42.3°$

24.35 This problem parallels the approach used for problem 24.34. See that solution.

$$\sin\theta_2 = \frac{\sin 75°}{n}$$

For blue light ($\lambda = 430$ nm), n = 1.650. We find
$\theta_2 = 35.83°$, $\alpha = 54.17°$, $\gamma = 39.17°$, $\beta = 65.83°$, $\theta_3 = 24.17°$,
$\theta_4 = 42.50°$,
and $\varepsilon = 18.33°$
So $\delta_B = 57.5°$

For red light ($\lambda = 680$ nm), n = 1.615. We find
$\theta_2 = 36.73°$, $\alpha = 53.27°$, $\gamma = 38.27°$, $\beta = 66.73°$, $\theta_3 = 23.27°$,
$\theta_4 = 39.64°$, and $\varepsilon = 16.37°$
and $\delta_R = 54.64°$

Therefore, the dispersion is $\delta_B - \delta_R = 2.86°$

24.36 (a) From geometry $\theta_1 = 60°$ and the law of reflection gives $\theta_2 = 60°$
Thus, angle $\alpha = 90° - \theta_2 = 30°$
and
$(\theta_3 + 90°) + 30° + 30° = 180°$ (sum of angles in triangle)
so $\theta_3 = 30°$
From Snell's law

$$\sin\theta_4 = (1.66)\frac{\sin 30°}{\frac{4}{3}} \text{ and}$$

$\theta_4 = 38.5°$

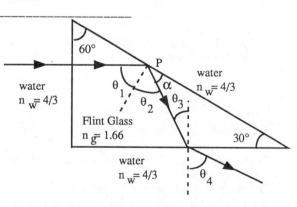

338

CHAPTER TWENTY-FOUR SOLUTIONS

(b) Total reflection will be on the verge of ceasing when $\theta_1 = 60° =$ critical angle

$$\sin 60° = \frac{n_w}{n_g} \quad \text{gives} \quad n_w = 1.66 \sin 60° = 1.44 \quad \text{as the largest value of } n_w$$

which allows total internal reflection to occur at point P.

24.37 We use $\sin\theta_c = \frac{n_2}{n_1}$

(a) For zircon: $\sin\theta_c = \frac{1}{1.923} = 0.5200$, and $\theta_c = 31.3°$

(b) for fluorite: $\sin\theta_c = \frac{1}{1.434} = 0.6973$, and $\theta_c = 44.2°$

(c) for ice: $\sin\theta_c = \frac{1}{1.309} = 0.7639$, and $\theta_c = 49.8°$

24.38 We use $\sin\theta_c = \frac{n_2}{n_1}$

(a) For zircon: $\sin\theta_c = \frac{1.333}{1.923} = 0.6934$, and $\theta_c = 43.9°$

(b) For fluorite: $\sin\theta_c = \frac{1.333}{1.434} = 0.9298$, and $\theta_c = 68.4°$

(c) For ice: $\sin\theta_c = \frac{1.333}{1.309} > 0$ Therefore, impossible. Total internal reflection cannot occur in this case.

24.39 From Snell's law, we find the index of refraction of the fluid.
$n_2 \sin\theta_2 = n_1 \sin\theta_1$
$n_2 \sin 22° = (1)\sin 30°$
$n_2 = 1.335$

Then, from $\sin\theta_c = \frac{n_2}{n_1}$

$\sin\theta_c = \frac{1}{1.335} = 0.749$, and $\theta_c = 48.5°$

24.40 The critical angle for a water-air boundary is
$\sin\theta_c = \frac{n_2}{n_1} = \frac{1}{\frac{4}{3}} = \frac{3}{4}$, and $\theta_c = 48.6°$

The circular disk must cover the area of the surface through which light from the diamond could emerge. Thus, it must form the base of a cone (with apex at the diamond) whose half angle is θ, where θ is greater than or equal to the critical angle. We have

$\frac{r_{min}}{h} = \tan\theta_c$, or $r_{min} = (2 \text{ m})\tan 48.6° = 2.27$ m
and the diameter is   4.54 m

24.41 $\sin\theta_c = \frac{n_2}{n_1} = \frac{1}{2.419} = 0.4134$, and $\theta_c = 24.4°$

CHAPTER TWENTY-FOUR SOLUTIONS

24.42 At Surface 2, $\sin\theta_c = \dfrac{n_m}{n_g} = \sin 42°$
= 0.669, and $n_m = 0.669 n_g$

$\alpha = 90° - \theta_c = 48°$
$\beta = 180° - 60° - \alpha = 180° - 60° - 48° = 72°$

Thus, $\theta_r = 90° - \beta = 90° - 72° = 18°$

Then, Snell's law applied at the first surface gives

$n_m \sin\theta_i = n_g \sin\theta_r$

$\sin\theta_i = \dfrac{n_g}{0.669 n_g} \sin 18° = 0.462$

or $\theta_i = 27.5°$.

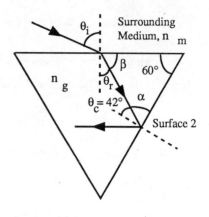

24.43 If the beam follows the path shown in Figure 24.26b, the angle of incidence when it strikes the short face of the prism is 45°. The index of refraction of the prism must be such that the angle 45° is greater than or equal to the critical angle. We find

$\sin\theta_c = \dfrac{n_2}{n_1} = \dfrac{1}{n}$ where n is the index of refraction of the prism material.

We solve for n,

$n = \dfrac{1}{\sin\theta_c}$

However, if the critical angle is less than or equal to $\sin 45°$, then $\sin\theta_c$ is less than or equal to $\sin 45°$, and $\sin 45° = \dfrac{\sqrt{2}}{2}$. Thus, $n \geq \dfrac{2}{\sqrt{2}} = \sqrt{2}$

24.44 If the prism is surrounded by water, then we find

$\sin\theta_c = \dfrac{1.333}{n}$, or

$n = \dfrac{1.333}{\sin\theta_c}$

Therefore, if the critical angle is less than or equal to 45°, as it must be...see problem 43, then $\sin\theta_c$ is less than or equal to $\dfrac{\sqrt{2}}{2}$, and we have

$n \geq \dfrac{1.333}{\dfrac{\sqrt{2}}{2}} = 1.333\sqrt{2}$ or, $n \geq 1.89$.

24.45 The critical angle for light trying to exit the cube is

$\sin\theta_c = \dfrac{1}{1.59}$, from which $\theta_c = 38.97°$

and (see sketch)

$\dfrac{r}{h} = \tan\theta_c$, or $h = \dfrac{r}{\tan\theta_c} = \dfrac{r}{0.8089} = 1.236 r$

r is the radius of a circular area of surface through which light from s could emerge. The penny obscures s if $r < r_{penny}$, or

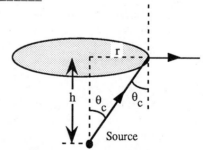

340

CHAPTER TWENTY-FOUR SOLUTIONS

$h < 1.236(0.95 \text{ cm})$ $h < 1.17 \text{ cm}$
The dime does not obscure, so $r > r_{dime} = 0.875 \text{ cm}$.
Therefore, $h > 1.236(0.875 \text{ cm})$
   $h > 1.08 \text{ cm}$.   Hence, we must conclude that $1.08 \text{ cm} < h < 1.17 \text{ cm}$.

24.46 $\sin\theta_c = \dfrac{n_2}{n_1}$

(a) If surrounding medium is air
$$\sin\theta_c = \dfrac{n_2}{n_1} = \dfrac{1.00}{1.53} = 0.6535 \quad \theta_c = 40.8°$$

(b) If surrounding medium is water,
$$\sin\theta_c = \dfrac{n_2}{n_1} = \dfrac{\tfrac{4}{3}}{1.53} = 0.8714 \quad \theta_c = 60.6°$$

24.47 $\sin\theta_c = \dfrac{n_c}{n_p}$, so $n_c = n_p \sin\theta_c = 1.60 \sin 59.5° = 1.379$

24.48 $\sin\theta_c = \dfrac{n_{air}}{n_{pipe}} = \dfrac{1.00}{1.36} = 0.7353 \quad \theta_c = 47.3°$
geometry shows that the angle of refraction at the end is
   $\theta_r = 90° - \theta_c = 90° - 47.3° = 42.7°$
Then, Snell's law at the end,
   $1.00 \sin\theta = 1.36 \sin 42.7°$, gives $\theta = 67.2°$

24.49 The time to go through 1 cm of water $= t_1 = \dfrac{1 \text{ cm}}{v_{water}}$

$= \dfrac{10^{-2} \text{ m}}{v_{water}} = \dfrac{10^{-2} \text{ m}}{c/n_{water}} = \dfrac{n_{water}}{c}(10^{-2} \text{ m})$

and the time to go through 0.5 cm of lucite $= t_2 = \dfrac{0.5 \text{ cm}}{v_{lucite}} = \dfrac{n_{lucite}}{c}(5 \times 10^{-3} \text{ m})$

and the time to go through 1.5 cm of air $= t_3 = \dfrac{1.5 \text{ cm}}{v_{air}} = \dfrac{1}{c}(1.5 \times 10^{-2} \text{ m})$

Thus, $\Delta t = (t_1 + t_2) - t_3 = 2.09 \times 10^{-11} \text{ s}$

24.50 Let us use Snell's law to find the angle of refraction in the ice. We call the air medium 1 and the ice medium 2. We have
   $n_2 \sin\theta_2 = n_1 \sin\theta_1 = (1)\sin 30°$
Thus, $n_2 \sin\theta_2 = \dfrac{1}{2}$
At the second surface, we call medium 3 the water and medium 2 the ice. We have
   $n_3 \sin\theta_3 = n_2 \sin\theta_2$
   $\sin\theta_3 = \dfrac{n_2 \sin\theta_2}{n_3} = \dfrac{\tfrac{1}{2}}{1.333} = 0.375$, and $\theta_3 = 22.0°$

24.51 By use of Snell's law, we find the index of refraction of the material.
   $n_2 \sin\theta_2 = n_1 \sin\theta_1$
   $n_2 \sin 25° = (1)\sin 37°$

CHAPTER TWENTY-FOUR SOLUTIONS

and $n_2 = 1.424$

Thus, $v = \dfrac{c}{n} = \dfrac{3 \times 10^8 \text{ m/s}}{1.424} = 2.11 \times 10^8$ m/s

---

24.52 To have total internal reflection at the diagonal surface, it is necessary that the critical angle be less than the angle of incidence, 45°. From Snell's law, we have
$$n_1 \sin\theta_1 = n_2 \sin\theta_2$$
At the critical angle, $\quad n_1 \sin\theta_c = n_2$
Therefore, if the critical angle is less than or equal to 45°, we have
$$n_1 \sin 45° \geq n_2$$
or, $\quad n_2 \leq 0.707 n_1 = \dfrac{\sqrt{2}}{2} n_1$

where n_1 is the index of refraction of the prism material, and n_2 is the index of refraction of the surrounding fluid.

24.53 The angle of incidence is 80°. We find the angle of refraction from Snell's law.
$$n_2 \sin\theta_2 = n_1 \sin\theta_1$$
$$1.33 \sin\theta_2 = (1)\sin 80°$$
gives, $\quad \theta_2 = 47.6°$
Thus, with 47.6° as the angle between the refracted ray and the normal, we see that the angle between the ray and the horizontal is $90° - 47.6° = 42.4°$.

24.54 We call medium 1 the air and medium 2 the glass. Snell's law is
$$n_2 \sin\theta_2 = n_1 \sin\theta_1$$
$$1.56 \sin\theta_2 = \sin\theta_1$$
But, the conditions of the problem are such that $\theta_1 = 2\theta_2$. Thus, we have
$$1.56 \sin\theta_2 = \sin 2\theta_2$$
We now use the double angle trig identity suggested.
$$1.56 \sin\theta_2 = 2\sin\theta_2 \cos\theta_2$$
or, $\quad \cos\theta_2 = \dfrac{1.56}{2} = 0.78 \quad$ and $\quad \theta_2 = 38.7°$
Thus, $\theta_1 = 2\theta_2 = 77.5°$

24.55 The sketch of the light path is shown. We have
$n_3 \sin\theta_3 = n_1 \sin\theta_4$
$1.5 \sin\theta_3 = (1)\sin 40°$. From which, $\theta_3 = 25.4°$
Then, $n_2 \sin\theta_2 = n_3 \sin\theta_3$
$\quad 1.66 \sin\theta_2 = 1.5 \sin 25.4°$
and $\quad \theta_2 = 22.8°$
Finally, $n_1 \sin\theta_1 = n_2 \sin\theta_2$
$\quad (1)\sin\theta_1 = 1.66 \sin 22.8°$
$\quad \theta_1 = 40°$

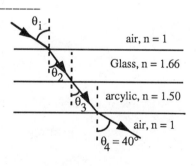

CHAPTER TWENTY-FOUR SOLUTIONS

24.56 We are to show that B = 2A. (See sketch.)
We see that
$$B = A + \beta + \gamma$$
Also, $\alpha + \delta = A$
But, $\beta + \theta_1 = 90°$, and $\alpha + \theta_1 = 90°$.
Therefore, $\beta = \alpha$.
Also, $\gamma + \theta_2 = 90°$, and $\delta + \theta_2 = 90°$
Therefore, $\gamma = \delta$

So,
$$B = A + \alpha + \delta = A + (\alpha + \delta) = A + (A) = 2A$$

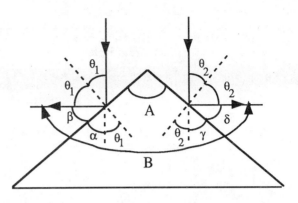

24.57 (See sketch.)

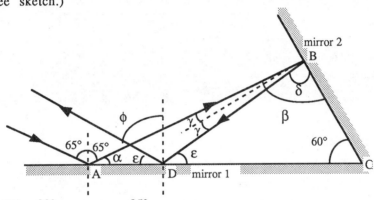

$\alpha + 65° = 90°$, so $\alpha = 25°$
Use triangle ABCA, to find $\alpha + \beta + 60° = 180°$, or $\beta = 120° - \alpha = 95°$
But $\beta = 90° + \gamma$. Therefore, $\gamma = \beta - 90° = 95° - 90° = 5°$
Then, $\delta = 90° - \gamma = 90° - 5° = 85°$

From triangle BCDB, we have $\delta + 60° + \varepsilon = 180°$

Thus, $\varepsilon = 120° - \delta = 120° - 85° = 35°$. Therefore $\phi = 55°$ is the angle the emerging ray makes with the normal to mirror 1. Note that this ray is not parallel to the incident ray.

24.58 (See sketch.)
$$\alpha + (90° - \theta) + 90° = 180°$$
gives, $\alpha - \theta = 0$, or, $\alpha = \theta$
Thus, the angle of incidence at the second mirror is
$$90° - \theta$$
where θ is an arbitrary angle of incidence at the first mirror. Thus, we see that the emerging ray makes an angle θ with the second mirror (a vertical surface which is parallel to the normal line for the first mirror. Therefore, the emerging ray is parallel to the incident ray.

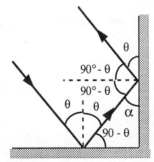

24.59 $\lambda_0 = n_w \lambda_w = n_b \lambda_b$ or $\dfrac{n_w}{n_b} = \dfrac{\lambda_b}{\lambda_w} = \dfrac{390 \text{ nm}}{438 \text{ nm}} = 0.890$

CHAPTER TWENTY-FOUR SOLUTIONS

24.60 (a) For polystyrene surrounded by air internal reflection requires

$$\theta_3 \geq \sin^{-1}\left(\frac{1}{1.49}\right) \geq 42.2°$$

and then from the geometry $\theta_2 = 90° - \theta_3 \leq 47.8°$
From Snell's law, this would require that $\theta_1 \leq 90°$
(b) For polystyrene surrounded by water, we have

$$\theta_3 = \sin^{-1}\left(\frac{1.33}{1.49}\right) = 63.2° \text{ and } \theta_2 = 26.8°,$$

and from Snell's law $\theta_1 = 30.3°$
(c) Total internal reflection is not possible since the beam is initially traveling in a medium of lower index of refraction.

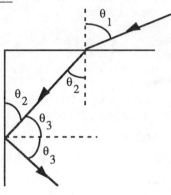

24.61 (See sketch.)
We have, $\delta_{min} = \alpha + \varepsilon$
Now, $\beta + \theta_2 = 90°$ (1)
Use triangle abc.

$$\beta + \frac{A}{2} + 90° = 180°$$

which becomes,

$$\beta + \frac{A}{2} = 90° \quad (2)$$

Comparing (1) and (2), we see that $\theta_2 = \frac{A}{2}$

Using triangle bcd, we have $\frac{A}{2} + 90° + \gamma = 180°$, or $\gamma = 90° - \frac{A}{2} = 90° - \theta_2$
Therefore, $\gamma + \theta_3 = 90°$ or $(90° - \theta_2) + \theta_3 = 90°$
Which gives, $\theta_3 = \theta_2$.
We also have, $(1)\sin\theta_1 = n\sin\theta_2$, and $n\sin\theta_3 = (1)\sin\theta_4$
Since $\theta_3 = \theta_2$ these equations show that $\theta_4 = \theta_1$
Therefore, $\alpha = \theta_1 - \theta_2 = \theta_1 - \frac{A}{2}$, and $\varepsilon = \theta_4 - \theta_3 = \theta_1 - \theta_2 = \theta_1 - \frac{A}{2}$

Then, $\delta_{min} = \alpha + \varepsilon = \theta_1 - \frac{A}{2} + \theta_1 - \frac{A}{2} = 2\theta_1 - A$ or, $2\theta_1 = A + \delta_{min}$

From which, $\theta_1 = \frac{A + \delta_{min}}{2}$. Finally, from (1) $\sin\theta_1 = n\sin\theta_2$, we have

$$n = \frac{\sin\theta_1}{\sin\theta_2} = \frac{\sin\left(\frac{A + \delta_{min}}{2}\right)}{\sin\frac{A}{2}}$$

CHAPTER TWENTY-FOUR SOLUTIONS

24.62 Refer to the sketch.
We see that $\theta_2 + \alpha = 90°$
and $\theta_3 + \beta = 90°$,
so $\theta_2 + \theta_3 + \alpha + \beta = 180°$.
Also, from the figure we see
$\alpha + \beta + \phi = 180°$
Therefore $\phi = \theta_2 + \theta_3$.
By applying Snell's law at the first and second surfaces, we find

$$\theta_2 = \sin^{-1}\left(\frac{\sin\theta_1}{n}\right)$$

and $\theta_3 = \sin^{-1}\left(\frac{\sin\theta_4}{n}\right)$

Substituting these values into the expression for ϕ yields

$$\phi = \sin^{-1}\left(\frac{\sin\theta_1}{n}\right) + \sin^{-1}\left(\frac{\sin\theta_4}{n}\right)$$

The limiting condition for internal reflection at the second surface is θ_4 approaches $90°$. Under these conditions we have $\phi = \sin^{-1}\left(\frac{\sin\theta_1}{n}\right) + \sin^{-1}\left(\frac{1}{n}\right)$,

or $\sin\theta_1 = n\sin\left(\phi - \sin^{-1}\frac{1}{n}\right)$. Using the trigonometry identity for the sine of the difference of two angles, this may be written as

$$\sin\theta_1 = n\left\{\sin\phi \cos[\sin^{-1}\left(\frac{1}{n}\right)] - \cos\phi \sin[\sin^{-1}\left(\frac{1}{n}\right)]\right\}.$$

But $\sin[\sin^{-1}\left(\frac{1}{n}\right)] = \frac{1}{n}$ and $\cos[\sin^{-1}\left(\frac{1}{n}\right)] = \frac{(n^2-1)^{1/2}}{n}$.

Therefore,
$\sin\theta_1 = (n^2-1)^{1/2}\sin\phi - \cos\phi$ or $\theta_1 = \sin^{-1}\left[(n^2-1)^{1/2}\sin\phi - \cos\phi\right]$.

CHAPTER TWENTY-FIVE SOLUTIONS

25.1 (See sketch.)
(1) angle PRQ = angle P'RQ
($90° - \theta = 90° - \theta$)
(2) angle PQR = angle P'QR
(both 90°)
(3) side QR = side QR
Therefore, triangle PQR is congruent to triangle P'QR (angle-side-angle)
Hence, the distance PQ = QP'
or, the image is as far behind the mirror as the object is in front of the mirror.

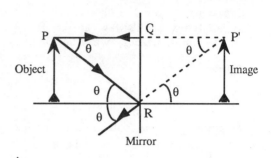

25.2 (a) and (b) From angle-side-angle, we see that triangle FOG is congruent to triangle GOE. Therefore, $h_1 = h_2$
(corresponding sides of congruent triangles)
Also, by angle-side-angle, we have triangle EQD is congruent to triangle DQT. Thus,
$d_1 = d_2$
(corresponding sides of congruent triangles)
Therefore, the height of the man is
$H = h_1 + h_2 + d_1 + d_2 = 2h_2 + 2d_2$
$= 2(h_2 + d_2)$
But $h_2 + d_2$ is the length of the mirror needed, L.
Thus, L = H/2, or the length of the mirror needed is one-half the person's height. The result is independent of distance from the mirror.

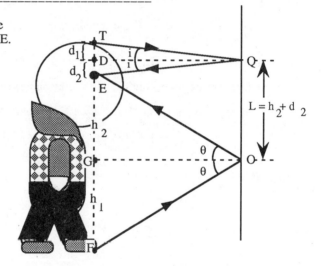

25.3 (1) The first image in the left mirror is 5ft behind the mirror, or 10 ft from the position of the person.
(2) The first image in the right mirror is located 10 ft behind the right mirror, but this location is 25 ft from the left mirror. Thus, the second image in the left mirror is 25 ft behind the mirror, or 30 ft from the person.
(3) The first image in the left mirror forms an image in the right mirror. This first image is 20 ft from the right mirror, and thus, an image 20 ft behind the right mirror is formed. This image in the right mirror also forms an image in the left mirror. The distance from this image in the right mirror to the left mirror is 35 ft. The third image in the left mirror is, thus, 35 ft behind the mirror, or 40 ft from the person.

CHAPTER TWENTY-FIVE SOLUTIONS

25.4 (See sketch.) Note from the sketch that both rays A and B make a 53° angle with the vertical. Thus, rays A and B are parallel to each other.

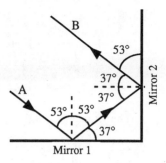

25.5 For a concave mirror both R and f are positive.

(a) $\frac{1}{s} + \frac{1}{s'} = \frac{1}{f}$, gives

$$\frac{1}{s'} = \frac{1}{f} - \frac{1}{s} = \frac{1}{20\text{cm}} - \frac{1}{40\text{ cm}} = \frac{1}{40\text{ cm}}, \text{ and } s' = 40 \text{ cm}$$

$$M = -\frac{s'}{s} = -\frac{40 \text{ cm}}{40 \text{ cm}} = -1$$

Therefore, the image is 40 cm in front of the mirror, real, and is inverted.

(b) $\frac{1}{s'} = \frac{1}{20 \text{ cm}} - \frac{1}{20 \text{ cm}} = 0$, or s' = infinity. No image is formed. The rays are reflected from the mirror parallel to one another.

(c) $\frac{1}{s'} = \frac{1}{f} - \frac{1}{s} = \frac{1}{20\text{cm}} - \frac{1}{10\text{ cm}} = -\frac{1}{20\text{ cm}}$, and s' = -20 cm

$$M = -\frac{s'}{s} = -\frac{-20 \text{ cm}}{10 \text{ cm}} = +2$$

The image is 20 cm behind the mirror, virtual and erect.

25.6 For a concave mirror, both R and f are positive. We also know that $f = \frac{R}{2} = 10$ cm

(a) $\frac{1}{s'} = \frac{1}{f} - \frac{1}{s} = \frac{1}{10\text{cm}} - \frac{1}{40\text{ cm}} = \frac{3}{40\text{ cm}}$, and s' = 13.3 cm

$$M = -\frac{s'}{s} = -\frac{13.3 \text{ cm}}{40 \text{ cm}} = -0.333$$

The image is 13.3 cm in front of the mirror, is real, and inverted.

(b) $\frac{1}{s'} = \frac{1}{f} - \frac{1}{s} = \frac{1}{10\text{cm}} - \frac{1}{20\text{ cm}} = \frac{1}{20\text{ cm}}$, and s' = 20 cm

$$M = -\frac{s'}{s} = -\frac{20 \text{ cm}}{20 \text{ cm}} = -1$$

The image is 20 cm in front of the mirror, is real, and inverted.

(c) $\frac{1}{s'} = \frac{1}{f} - \frac{1}{s} = \frac{1}{10\text{cm}} - \frac{1}{10\text{ cm}} = 0$. Thus, s' = infinity. No image is formed. The rays are reflected parallel to each other.

25.7 For a convex mirror, R and f are negative.

(a) $\frac{1}{s'} = \frac{1}{f} - \frac{1}{s} = \frac{1}{-20\text{cm}} - \frac{1}{40\text{ cm}} = -\frac{3}{40\text{ cm}}$, and s' = -13.3 cm

$$M = -\frac{s'}{s} = -\frac{-13.3 \text{ cm}}{40 \text{ cm}} = \frac{1}{3}$$

The image is 13.3 cm behind mirror, virtual, and erect.

(b) $\frac{1}{s'} = \frac{1}{f} - \frac{1}{s} = \frac{1}{-20\text{cm}} - \frac{1}{20\text{ cm}} = -\frac{1}{10\text{ cm}}$, and s' = -10 cm

CHAPTER TWENTY-FIVE SOLUTIONS

$$M = -\frac{s'}{s} = -\frac{-10 \text{ cm}}{20 \text{ cm}} = \frac{1}{2}$$

The image is 10 cm behind the mirror, virtual, and erect.

(c) $\frac{1}{s'} = \frac{1}{f} - \frac{1}{s} = \frac{1}{-20 \text{cm}} - \frac{1}{10 \text{ cm}} = -\frac{3}{20 \text{ cm}}$, and s' = -6.67 cm

$$M = -\frac{s'}{s} = -\frac{-6.67 \text{ cm}}{10 \text{ cm}} = \frac{2}{3}$$

The image is 6.67 cm behind the mirror, virtual, and erect.

25.8 For a convex mirror, R and f are negative.

(a) $\frac{1}{s'} = \frac{1}{f} - \frac{1}{s} = \frac{1}{-10 \text{cm}} - \frac{1}{40 \text{ cm}} = -\frac{5}{40 \text{ cm}}$, and s' = -8 cm

$$M = -\frac{s'}{s} = -\frac{-8 \text{ cm}}{40 \text{ cm}} = \frac{1}{5}$$

The image is 8 cm behind the mirror, virtual, and erect.

(b) $\frac{1}{s'} = \frac{1}{f} - \frac{1}{s} = \frac{1}{-10 \text{cm}} - \frac{1}{20 \text{ cm}} = -\frac{3}{20 \text{ cm}}$, and s' = -6.67cm

$$M = -\frac{s'}{s} = -\frac{-6.67 \text{ cm}}{20 \text{ cm}} = \frac{1}{3}$$

The image is 6.67 cm behind the mirror, virtual, and erect.

(c) $\frac{1}{s'} = \frac{1}{f} - \frac{1}{s} = \frac{1}{-10 \text{cm}} - \frac{1}{10 \text{ cm}} = -\frac{1}{5 \text{ cm}}$, and s' = -5 cm

$$M = -\frac{s'}{s} = -\frac{-5 \text{ cm}}{10 \text{ cm}} = \frac{1}{2}$$

The image is 5 cm behind the mirror, virtual, and erect.

25.9 (a) $\frac{1}{s} + \frac{1}{s'} = \frac{1}{f}$

$$\frac{1}{f} = \frac{1}{1 \text{ cm}} - \frac{1}{10 \text{ cm}}$$

gives f = 1.11 cm   and   R = 2f = 2.22 cm

(b) $M = -\frac{s'}{s} = \frac{-10}{1} = 10$

25.10 $\frac{1}{s'} = \frac{1}{f} - \frac{1}{s} = -\frac{1}{0.275 \text{ m}} - \frac{1}{10 \text{ m}}$

gives s' = -0.267 m

Thus, the image is virtual.

$$M = -\frac{s'}{s} = -\frac{-0.267}{10 \text{ m}} = 0.027$$

Thus, the image is erect (+ M)

25.11 We are given that s' = -20 cm, and for a concave mirror R and f are positive.

Thus, R = 40 cm, and $f = \frac{R}{2} = 20$ cm

$\frac{1}{s} = \frac{1}{f} - \frac{1}{s'} = \frac{1}{20 \text{ cm}} - \frac{1}{-20 \text{ cm}} = \frac{1}{10 \text{ cm}}$, and   s = 10 cm

The object should be 10 cm in front of the mirror.

25.12 For a convex mirror f is negative.

CHAPTER TWENTY-FIVE SOLUTIONS

We also know that s' = -10 cm because a convex mirror can only form virtual images of real objects. Thus,
$$\frac{1}{s} = \frac{1}{f} - \frac{1}{s'} = \frac{1}{-15 \text{ cm}} - \frac{1}{-10 \text{ cm}} = \frac{1}{30 \text{ cm}}, \text{ and } s = 30 \text{ cm}$$
Thus, the object should be placed 30 cm in front of the mirror.

25.13 From the sign conventions for convex mirrors, we see that R = -3 cm and f = -1.5 cm. Also, s = 10 cm. Thus,
$$\frac{1}{s'} = \frac{1}{f} - \frac{1}{s} = \frac{1}{-1.5 \text{cm}} - \frac{1}{10 \text{ cm}} = -\frac{23}{30 \text{ cm}}, \text{ and } s' = -1.3 \text{ cm}$$
$$M = -\frac{s'}{s} = -\frac{-1.3 \text{ cm}}{10 \text{ cm}} = 0.130$$

25.14 We know that for a convex mirror R and f are negative, and we also are given that $M = \frac{1}{2}$. M is positive because convex mirrors only form erect virtual images of real objects.

Therefore, $M = -\frac{s'}{s} = \frac{1}{2}$, from which, $s' = -\frac{s}{2}$

Thus, $\frac{1}{s} = \frac{1}{f} - \frac{1}{s'} = \frac{1}{-20 \text{ cm}} - \frac{1}{-\frac{s}{2}} = \frac{1}{-20} + \frac{2}{s}$ and s = 20 cm

The object should be 20 cm in front of the mirror.

25.15 We know that R and f are positive for a concave mirror, and we are given that s = 30 cm, and M = -4. The magnification is negative because the only magnified real images formed by a concave mirror of real objects are inverted images. Thus,
$$M = -\frac{s'}{s} = -4, \text{ and } s' = 4s \text{ Since } s = 30 \text{ cm}, \quad s' = 120 \text{ cm}. \text{ Thus,}$$
$$\frac{1}{f} = \frac{1}{s} + \frac{1}{s'} = \frac{1}{30 \text{ cm}} + \frac{1}{120 \text{ cm}} = \frac{5}{120 \text{ cm}} \text{ and } f = 24 \text{ cm}.$$
Therefore, R = 2f = 48 cm

25.16 We know that R and f are positive for a concave mirror. Also, since a concave mirror only forms magnified, erect images when the image is virtual, we know that s' is negative.
Thus, M = 2
and $M = -\frac{s'}{s} = 2$ or, s' = -2(25 cm) = -50 cm Thus,
$$\frac{1}{f} = \frac{1}{s} + \frac{1}{s'} = \frac{1}{25 \text{ cm}} + \frac{1}{-50 \text{ cm}} = \frac{1}{50 \text{ cm}} \text{ and } f = 50 \text{ cm}.$$
But, R = 2f = 100 cm = 1 m

25.17 Since we have a real object and want a magnified image, the mirror must be concave. (A convex mirror forms diminished images of all real objects.) Since image is erect, it must be virtual. (All real images formed by concave mirrors of real objects are inverted.)
Thus, s' < 0, f > 0, s = 10 cm
$$M = \frac{h'}{h} = \frac{4 \text{ cm}}{2 \text{ cm}} = +2 = -\frac{s'}{s}$$
so, s' = -2s = -20 cm

CHAPTER TWENTY-FIVE SOLUTIONS

Thus, $\dfrac{1}{f} = \dfrac{1}{s} + \dfrac{1}{s'} = \dfrac{1}{10 \text{ cm}} + \dfrac{1}{-20 \text{ cm}}$

or, $f = 20$ cm, and $R = 2f = 40$ cm

25.18 $\dfrac{1}{f} = \dfrac{1}{s} + \dfrac{1}{s'} = \dfrac{1}{1.52 \text{ m}} + \dfrac{1}{0.18 \text{ m}}$

or, $f = 0.161$ m $= 16.1$ cm

Now, to get an erect image which is twice the size of the object, we need

$M = -\dfrac{s'}{s} = 2$, so $s' = -2s$

Thus, $\dfrac{1}{s} + \dfrac{1}{s'} = \dfrac{1}{f}$ becomes $\dfrac{1}{s} - \dfrac{1}{2s} = \dfrac{1}{f}$, or $s = \dfrac{f}{2}$

Thus, $s = \dfrac{16.1}{2} = 8.05$ cm is the required object distance.

25.19 A convex mirror has R and f negative. Thus, $f = \dfrac{R}{2} = -5$ cm.

For a virtual image, s' is negative. Therefore,

$M = -\dfrac{s'}{s} = \dfrac{1}{3}$ and $s' = -\dfrac{s}{3}$

Therefore, $\dfrac{1}{f} = \dfrac{1}{s} + \dfrac{1}{s'}$ becomes $\dfrac{1}{-5 \text{ cm}} = \dfrac{1}{s} - \dfrac{3}{s}$ or $s = 10$ cm

25.20 For a convex mirror, R and f are negative. Since we have a real object, the image will be virtual and erect. Thus, s' is negative and the magnification is positive.

$M = -\dfrac{s'}{s} = \dfrac{1}{2}$ and $s' = -\dfrac{s}{2} = -\dfrac{10 \text{ cm}}{2} = -5$ cm

Thus, $\dfrac{1}{f} = \dfrac{1}{s} + \dfrac{1}{s'} = \dfrac{1}{10 \text{ cm}} + \dfrac{1}{-5 \text{ cm}} = -\dfrac{1}{10 \text{ cm}}$ and $f = -10$ cm.

But $R = 2f = -20$ cm

25.21 For a concave mirror, R and f are positive. Also, for an erect image M is positive. Therefore, $M = \dfrac{s'}{s} = 4$ and $s' = -4s$

$\dfrac{1}{f} = \dfrac{1}{s} + \dfrac{1}{s'}$ becomes $\dfrac{1}{40 \text{ cm}} = \dfrac{1}{s} - \dfrac{1}{4s} = \dfrac{3}{4s}$

From which, $s = 30$ cm

25.22 For a concave mirror, R and f are positive. Also,

$M = \dfrac{y'}{y} = -\dfrac{s'}{s}$ We are given that $y' = 5$ cm (erect...thus y' is positive), $y = 2$ cm, and $s = 3$ cm.

Therefore, $s' = -s\dfrac{y'}{y} = -(3 \text{ cm})\dfrac{5 \text{ cm}}{2 \text{ cm}} = -7.5$ cm

And, $\dfrac{1}{f} = \dfrac{1}{s} + \dfrac{1}{s'} = \dfrac{1}{3 \text{ cm}} - \dfrac{1}{7.5 \text{ cm}} = \dfrac{3}{15 \text{ cm}}$ and $f = 5$ cm.

25.23 (a) Since the image is projected on a screen, it is a real image. Therefore, s' is positive. We have $\dfrac{1}{f} = \dfrac{1}{s} + \dfrac{1}{s'} = \dfrac{1}{10 \text{ cm}} + \dfrac{1}{200 \text{ cm}} = \dfrac{21}{200 \text{ cm}}$

and $f = 9.52$ cm.

CHAPTER TWENTY-FIVE SOLUTIONS

Thus, because f is positive, we need a concave mirror.

(b) $\quad M = -\dfrac{s'}{s} = -\dfrac{200 \text{ cm}}{10 \text{ cm}} = -20$

25.24 (a) For a plane mirror, R is infinite.

(b) $\dfrac{1}{s} + \dfrac{1}{s'} = \dfrac{2}{R} = 0$, or $\quad s' = -s$

This says the image is virtual and as far behind the mirror as the object is in front of the mirror.

(c) The results of (b) are in agreement with the results for a plane mirror. Additionally,

$\quad M = -\dfrac{s'}{s} = 1 \quad$ the image is unmagnified and erect.

25.25 (a) For a plane refracting surface, we have

$s' = -\dfrac{n_2}{n_1} s = -\dfrac{1}{\frac{4}{3}} s = -\dfrac{3}{4} s$

Thus, for $s = 2$ m, $\quad s' = -1.5$ m $\quad$ The pool appears to be 1.5 m deep.

(b) If the pool is half filled, $s = 1$ m, and

$s' = -\dfrac{3}{4}(1 \text{ m}) = -0.75$ m.

The bottom surface appears to be 0.75 m below the water surface, or 1.75 m below the top of the pool.

25.26 $s' = -\dfrac{n_2}{n_1} s = -\dfrac{1}{1.309}(50 \text{ cm}) = -38.2$ cm

Thus, the virtual image of the dust speck is 38.2 cm below the top surface of the ice.

25.27 We are given that $s' = -35$ cm. The negative sign arises because we have a virtual image on the same side of the boundary as is the object.

$s = -\dfrac{n_1}{n_2} s' = -\dfrac{1.5}{1.0}(-35 \text{ cm}) = 52.5$ cm

The benzene is 52.5 cm deep in the tank.

25.28 $\dfrac{n_1}{s} + \dfrac{n_2}{s'} = \dfrac{n_2 - n_1}{R}$

But the object is at infinity, and $s' = 2R$,

so $\quad \dfrac{n_2}{2R} = \dfrac{n_2 - 1.0}{R}$

which gives $n_2 = 2$

25.29 $\dfrac{n_1}{s} + \dfrac{n_2}{s'} = \dfrac{n_2 - n_1}{R}$

becomes $\dfrac{1.33}{10 \text{ cm}} + \dfrac{1.0}{s'} = \dfrac{1.0 - 1.33}{-15 \text{ cm}}$

from which $s' = -9.00$ cm

or the fish appears to be 9 cm inside the wall of bowl.

25.30 We have $R = -4$ cm, and $s = 4$ cm.

CHAPTER TWENTY-FIVE SOLUTIONS

$$\frac{n_1}{s} + \frac{n_2}{s'} = \frac{n_2 - n_1}{R}$$

$$\frac{1.5}{4 \text{ cm}} + \frac{1}{s'} = \frac{-0.5}{-4 \text{ cm}}$$

Thus, s' = -4 cm. Note that the location of the virtual image is at the same position as the object.

$$M = \frac{h'}{h} = -\frac{n_1 s'}{n_2 s}$$

$$\frac{h'}{2.5 \text{ mm}} = -\frac{(1.5)(-4 \text{ cm})}{(1.0)(4 \text{ cm})} = 1.5$$

Thus, h' = 3.75 mm

25.31 We use $\quad \dfrac{n_1}{s} + \dfrac{n_2}{s'} = \dfrac{n_2 - n_1}{R}$

(a) $\dfrac{1}{20 \text{ cm}} + \dfrac{1.5}{s'} = \dfrac{0.5}{8 \text{ cm}} \qquad$ gives s' = 120 cm

(b) $\dfrac{1}{8 \text{ cm}} + \dfrac{1.5}{s'} = \dfrac{0.5}{8 \text{ cm}} \qquad$ gives s' = -24 cm

(c) $\dfrac{1}{4 \text{ cm}} + \dfrac{1.5}{s'} = \dfrac{0.5}{8 \text{ cm}} \qquad$ gives s' = -8 cm

(d) $\dfrac{1}{2 \text{ cm}} + \dfrac{1.5}{s'} = \dfrac{0.5}{8 \text{ cm}} \qquad$ gives s' = -3.43 cm

---

25.32 In this case, $\quad \dfrac{n_1}{s} + \dfrac{n_2}{s'} = \dfrac{n_2 - n_1}{R} \quad$ becomes $\quad \dfrac{4}{3s} + \dfrac{1.5}{s'} = \dfrac{1.5 - 1.33}{8 \text{ cm}}$

(a) $\dfrac{1.5}{s'} = \dfrac{0.167}{8 \text{ cm}} - \dfrac{4}{60 \text{ cm}} \qquad$ s' = -32.7 cm

(b) $\dfrac{1.5}{s'} = \dfrac{0.167}{8 \text{ cm}} - \dfrac{1}{6 \text{ cm}} \qquad$ s' = -10.3 cm

(c) $\dfrac{1.5}{s'} = \dfrac{0.167}{8 \text{ cm}} - \dfrac{1}{3 \text{ cm}} \qquad$ s' = -4.8 cm

(d) $\dfrac{1.5}{s'} = \dfrac{0.167}{8 \text{ cm}} - \dfrac{2}{3 \text{ cm}} \qquad$ s' = -2.32 cm

---

25.33 We have R = -8 cm.

(a) $\dfrac{1.5}{s'} = -\dfrac{0.5}{8 \text{ cm}} - \dfrac{1}{20 \text{ cm}} \qquad$ s' = -13.3 cm

(b) $\dfrac{1.5}{s'} = -\dfrac{0.5}{8 \text{ cm}} - \dfrac{1}{8 \text{ cm}} \qquad$ s' = -8 cm

(c) $\dfrac{1.5}{s'} = -\dfrac{0.5}{8 \text{ cm}} - \dfrac{1}{4 \text{ cm}} \qquad$ s' = -4.8 cm

(d) $\dfrac{1.5}{s'} = -\dfrac{0.5}{8 \text{ cm}} - \dfrac{1}{2 \text{ cm}} \qquad$ s' = -2.67 cm

---

25.34 (a) $\dfrac{1.5}{s'} = -\dfrac{0.167}{8 \text{ cm}} - \dfrac{1.33}{20 \text{ cm}} \qquad$ s' = -17.1 cm

(b) $\dfrac{1.5}{s'} = -\dfrac{0.167}{8 \text{ cm}} - \dfrac{1.33}{8 \text{ cm}} \qquad$ s' = -8.00 cm

(c) $\dfrac{1.5}{s'} = -\dfrac{0.167}{8 \text{ cm}} - \dfrac{1.33}{4 \text{ cm}} \qquad$ s' = -4.24 cm

CHAPTER TWENTY-FIVE SOLUTIONS

(d) $\dfrac{1.5}{s'} = -\dfrac{0.167}{8 \text{ cm}} - \dfrac{1.33}{2 \text{ cm}}$ $s' = -2.18$ cm

25.35 $\dfrac{1}{f} = (n-1)\left(\dfrac{1}{R_1} - \dfrac{1}{R_2}\right)$ becomes

$\dfrac{1}{25 \text{ cm}} = 0.58\left(\dfrac{1}{R_1} - \dfrac{1}{1.8 \text{ cm}}\right)$

which gives $R_1 = 1.60$ cm = 16 mm

25.36 $\dfrac{1}{f} = (n-1)\left(\dfrac{1}{R_1} - \dfrac{1}{R_2}\right)$ becomes

$\dfrac{1}{60 \text{ cm}} = (n-1)\left(\dfrac{1}{52.5 \text{ cm}} - \dfrac{1}{-61.9 \text{ cm}}\right)$

which gives $n = 1.473$

25.37 For a converging lens, f is positive. We use $\dfrac{1}{s} + \dfrac{1}{s'} = \dfrac{1}{f}$.

(a) $\dfrac{1}{s'} = \dfrac{1}{f} - \dfrac{1}{s} = \dfrac{1}{20 \text{ cm}} - \dfrac{1}{40 \text{ cm}} = \dfrac{1}{40 \text{ cm}}$ $s' = 40$ cm

$M = -\dfrac{s'}{s} = -\dfrac{40}{40} = -1$

The image is real, inverted, and located 40 cm past the lens.

(b) $\dfrac{1}{s'} = \dfrac{1}{f} - \dfrac{1}{s} = \dfrac{1}{20 \text{ cm}} - \dfrac{1}{20 \text{ cm}} = 0$ $s' = $ infinity

No image is formed. The rays emerging from the lens are parallel to each other.

(c) $\dfrac{1}{s'} = \dfrac{1}{f} - \dfrac{1}{s} = \dfrac{1}{20 \text{ cm}} - \dfrac{1}{10 \text{ cm}} = -\dfrac{1}{20 \text{ cm}}$ $s' = -20$ cm

$M = -\dfrac{s'}{s} = -\dfrac{-20}{10} = 2$

The image is erect, virtual, and 20 cm in front of the lens.

25.38 (a) We find the focal length from

$\dfrac{1}{f} = (n-1)\left(\dfrac{1}{R_1} - \dfrac{1}{R_2}\right) = 0.5\left(\dfrac{1}{15 \text{ cm}} - \dfrac{1}{-10 \text{ cm}}\right)$

Which yields $f = 12$ cm.

(b) When the object is at infinity, the thin lens equation reduces to

$\dfrac{1}{s'} = \dfrac{1}{f}$ Thus, $s' = f = 12$ cm.

(c) $\dfrac{1}{s'} = \dfrac{1}{f} - \dfrac{1}{s} = \dfrac{1}{12 \text{ cm}} - \dfrac{1}{36 \text{ cm}} = \dfrac{2}{36 \text{ cm}}$ $s' = 18$ cm

(d) $\dfrac{1}{s'} = \dfrac{1}{f} - \dfrac{1}{s} = \dfrac{1}{12 \text{ cm}} - \dfrac{1}{12 \text{ cm}} = 0$ $s' = $ infinity (no image formed)

(e) $\dfrac{1}{s'} = \dfrac{1}{f} - \dfrac{1}{s} = \dfrac{1}{12 \text{ cm}} - \dfrac{1}{6 \text{ cm}} = -\dfrac{1}{12 \text{ cm}}$ $s' = -12$ cm

25.39 (a) $\dfrac{1}{s'} = \dfrac{1}{f} - \dfrac{1}{s} = \dfrac{1}{25 \text{ cm}} - \dfrac{1}{26 \text{ cm}}$ $s' = 650$ cm

The image is real, inverted, and enlarged.

CHAPTER TWENTY-FIVE SOLUTIONS

(b) $\dfrac{1}{s'} = \dfrac{1}{f} - \dfrac{1}{s} = \dfrac{1}{25\text{ cm}} - \dfrac{1}{24\text{ cm}}$ $s' = -600$ cm

The image is virtual, erect, and enlarged.

25.40 (a) $\dfrac{1}{s'} = \dfrac{1}{f} - \dfrac{1}{s} = \dfrac{1}{-25\text{ cm}} - \dfrac{1}{26\text{ cm}}$ $s' = -12.75$ cm

The image is erect, virtual, and diminished in size

(b) $\dfrac{1}{s'} = \dfrac{1}{f} - \dfrac{1}{s} = \dfrac{1}{-25\text{ cm}} - \dfrac{1}{24\text{ cm}}$ $s' = -12.24$ cm

The image is erect, virtual, and diminished in size.

25.41 For a diverging lens, f is negative.

(a) $\dfrac{1}{s'} = \dfrac{1}{f} - \dfrac{1}{s} = -\dfrac{1}{20\text{ cm}} - \dfrac{1}{40\text{ cm}} = -\dfrac{3}{40\text{ cm}}$ $s' = -13.3$ cm

$M = -\dfrac{s'}{s} = -\dfrac{-13.3\text{ cm}}{40\text{ cm}} = \dfrac{1}{3}$

The image is virtual, erect, and 13.3 cm in front of the lens.

(b) $\dfrac{1}{s'} = \dfrac{1}{f} - \dfrac{1}{s} = -\dfrac{1}{20\text{ cm}} - \dfrac{1}{20\text{ cm}} = -\dfrac{1}{10\text{ cm}}$ $s' = -10$ cm

$M = -\dfrac{s'}{s} = -\dfrac{-10\text{ cm}}{20\text{ cm}} = \dfrac{1}{2}$

The image is virtual, erect, and 10 cm in front of the lens.

(c) $\dfrac{1}{s'} = \dfrac{1}{f} - \dfrac{1}{s} = -\dfrac{1}{20\text{ cm}} - \dfrac{1}{10\text{ cm}} = -\dfrac{3}{20\text{ cm}}$ $s' = -6.67$ cm

$M = -\dfrac{s'}{s} = -\dfrac{-6.67\text{ cm}}{10\text{ cm}} = 0.667$

The image is virtual, erect, and 6.67 cm in front of the lens.

25.42 (a) $\dfrac{1}{f} = (n - 1)(\dfrac{1}{R_1} - \dfrac{1}{R_2}) = 0.5\,(-\dfrac{1}{15\text{ cm}} - \dfrac{1}{10\text{ cm}})$

Which yields f = -12 cm.

(b) When s = infinity, s' = f = -12 cm

(c) $\dfrac{1}{s'} = \dfrac{1}{f} - \dfrac{1}{s} = -\dfrac{1}{12\text{ cm}} - \dfrac{1}{36\text{ cm}} = -\dfrac{1}{9\text{ cm}}$ $s' = -9$ cm

(d) $\dfrac{1}{s'} = \dfrac{1}{f} - \dfrac{1}{s} = -\dfrac{1}{12\text{ cm}} - \dfrac{1}{12\text{ cm}} = -\dfrac{1}{6\text{ cm}}$ $s' = -6$ cm

(e) $\dfrac{1}{s'} = \dfrac{1}{f} - \dfrac{1}{s} = -\dfrac{1}{12\text{ cm}} - \dfrac{1}{6\text{ cm}} = -\dfrac{1}{4\text{ cm}}$ $s' = -4$ cm

25.43 (a) $\dfrac{1}{f} = (n - 1)(\dfrac{1}{R_1} - \dfrac{1}{R_2}) = 0.5\,(-\dfrac{1}{10\text{ cm}} - \dfrac{1}{15\text{ cm}})$

which yields f = -12 cm

Since f is the same and all the object distances are the same, the image positions for this problem are identical to those for the previous problem (number 42).

CHAPTER TWENTY-FIVE SOLUTIONS

25.44 (a) Note that $s' = 12.9$ cm $- s$
so $\dfrac{1}{s} + \dfrac{1}{12.9 - s} = \dfrac{1}{2.44}$
which yields a quadratic in s as
$-s^2 + 12.9s = 31.48$
which has solutions
$s = 9.63$ cm or $s = 3.27$ cm
both solutions are valid

(b) For a virtual image
$-s' = s + 12.9$ cm
$\dfrac{1}{s} - \dfrac{1}{12.9 + s} = \dfrac{1}{2.44}$ or
$s^2 + 12.9s = 31.8$
from which $s = 2.1$ cm or $s = -15$ cm
We must have a real object, so the -15 cm solution must be rejected.

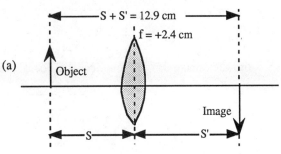

(a)

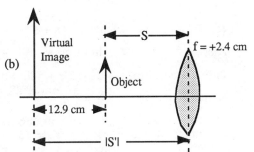

(b)

25.45 We must first realize that we are looking at an erect, magnified, virtual image. Thus, we have a real object located inside the focal point of a converging lens.
Thus, $s > 0$, $s' < 0$, and $f < 0$
$M = 2 = -\dfrac{s'}{s}$, or $s' = -2s = -2(2.84 \text{ cm}) = -5.68$ cm
Thus, $\dfrac{1}{s} + \dfrac{1}{s'} = \dfrac{1}{f}$
$\dfrac{1}{f} = \dfrac{1}{2.84 \text{ cm}} - \dfrac{1}{5.68 \text{ cm}}$ $f = 5.68$ cm

---

25.46 We are given that $s = 30$ cm and $s' = -15$ cm. Thus, from $\dfrac{1}{s} + \dfrac{1}{s'} = \dfrac{1}{f}$.
$\dfrac{1}{f} = \dfrac{1}{s} - \dfrac{1}{s'} = \dfrac{1}{30 \text{ cm}} - \dfrac{1}{15 \text{ cm}} = -\dfrac{1}{30 \text{ cm}}$ and $f = -30$ cm
The lens is diverging.

25.47 We are given that $s = 80$ cm and $s' = -40$ cm. Thus,
$\dfrac{1}{f} = \dfrac{1}{s} + \dfrac{1}{s'} = \dfrac{1}{80 \text{ cm}} - \dfrac{1}{40 \text{ cm}} = -\dfrac{1}{80 \text{ cm}}$ and $f = -80$ cm

---

25.48 Given $f = -40$ cm, and the magnitude of $s' = 30$ cm. But for a real object, a diverging lens can only form virtual images...$s' < 0$
Therefore, $s' = -30$ cm
$\dfrac{1}{s} = \dfrac{1}{f} - \dfrac{1}{s'} = \dfrac{1}{-40 \text{ cm}} - \dfrac{1}{-30 \text{ cm}}$ and $s = 120$ cm
The object should be 120 cm in front of lens.
$M = -\dfrac{s'}{s} = -\dfrac{-30 \text{ cm}}{120 \text{ cm}} = 0.25$

CHAPTER TWENTY-FIVE SOLUTIONS

25.49 Given $f > 0$ and $s = 10f$

Thus, $\dfrac{1}{s'} = \dfrac{1}{f} - \dfrac{1}{s} = \dfrac{1}{f} - \dfrac{1}{10f}$ and $s' = \dfrac{10}{9}f$

or, the image is $\dfrac{1}{9}f$ outside the focal point.

25.50 $\dfrac{1}{s} = \dfrac{1}{f} - \dfrac{1}{s'} = \dfrac{1}{12.5 \text{ cm}} - \dfrac{1}{-30 \text{ cm}}$ and $s = 8.82$ cm

$M = -\dfrac{s'}{s} = -\dfrac{-30 \text{ cm}}{8.82 \text{ cm}} = 3.4$ image is erect

---

25.51 We know that $s + s' = 50$ cm, so $s' = 50$ cm $- s$

Also, $\dfrac{1}{s} + \dfrac{1}{s'} = \dfrac{1}{f}$ becomes $\dfrac{1}{s} + \dfrac{1}{(50 \text{ cm} - s)} = \dfrac{1}{10 \text{ cm}}$

This reduces to $s^2 - 50s + 500 = 0$

Which yields $s = \dfrac{50 \text{ cm} \pm 22.4 \text{ cm}}{2}$

Let us choose the plus sign in the above. We find, $s = 36.2$ cm, which yields an image distance of $s' = 13.8$. The magnification is $M = -0.382$.
If we use the negative sign $s = 13.8$ cm, $s' = 36.2$ cm, and $M = -2.62$

25.52 For a virtual image, the image distance is negative and the magnification is positive.

Thus, $M = -\dfrac{s'}{s} = \dfrac{1}{3}$ and $s' = -\dfrac{s}{3}$

So, $\dfrac{1}{s} + \dfrac{1}{s'} = \dfrac{1}{f}$ becomes $\dfrac{1}{s} - \dfrac{3}{s} = \dfrac{1}{f}$ or, $s = -2f$

25.53 Let us consider the first lens. We find the image position and magnification as

$\dfrac{1}{s_1'} = \dfrac{1}{f_1} - \dfrac{1}{s_1} = \dfrac{1}{15 \text{ cm}} - \dfrac{1}{30 \text{ cm}} = \dfrac{1}{30 \text{ cm}}$ $\quad s_1' = 30$ cm

$M_1 = -\dfrac{s_1'}{s_1} = -\dfrac{30 \text{ cm}}{30 \text{ cm}} = -1$

Now consider the second lens. The image produced by the first lens becomes the object for this second lens. Thus, $s_2 = 40$ cm $- s_1' = 40$ cm $- 30$ cm $= 10$ cm

$\dfrac{1}{s_2'} = \dfrac{1}{f_2} - \dfrac{1}{s_2} = \dfrac{1}{15 \text{ cm}} - \dfrac{1}{10 \text{ cm}} = -\dfrac{1}{30 \text{ cm}}$ $\quad s_2' = -30$ cm

$M_2 = -\dfrac{s_2'}{s_2} = -\dfrac{-30 \text{ cm}}{10 \text{ cm}} = 3$

The overall magnification M is
$M = (M_1)(M_2) (-1)(3) = -3$

25.54 (a) We first find the image position and magnification for the first lens.

$\dfrac{1}{s_1'} = \dfrac{1}{f_1} - \dfrac{1}{s_1} = \dfrac{1}{15 \text{ cm}} - \dfrac{1}{30 \text{ cm}} = \dfrac{1}{30 \text{ cm}}$ $\quad s_1' = 30$ cm

$M_1 = -\dfrac{s_1'}{s_1} = -\dfrac{30 \text{ cm}}{30 \text{ cm}} = -1$

The object position for the second lens is
$s_2 = 20$ cm $- s_1' = 20$ cm $- 30$ cm $= -10$ cm (a virtual object). Thus,

CHAPTER TWENTY-FIVE SOLUTIONS

$$\frac{1}{s_2'} = \frac{1}{f_2} - \frac{1}{s_2} = \frac{1}{15 \text{ cm}} - \frac{1}{-10 \text{ cm}} = \frac{1}{6 \text{ cm}} \qquad s_2' = 6 \text{ cm}$$

$$M_2 = -\frac{s_2'}{s_2} = -\frac{6 \text{ cm}}{-10 \text{ cm}} = 0.6$$

the overall magnification M is
$$M = (M_1)(M_2) = (-1)(0.6) = -0.6$$

(b) For the first lens, $s_1' = 30$ cm and $M_1 = -1$ as before,
The object distance for the second lens is $s_2 = 30$ cm $- s_1' = 30$ cm $- 30$ cm $= 0$

From $\frac{1}{s} + \frac{1}{s'} = \frac{1}{f}$. we see that as s_2 approaches 0 from the positive side, s_2' also approaches 0 but from the negative side so that $\frac{1}{s} + \frac{1}{s'}$ is a constant, namely $\frac{1}{f}$.

We also see that the magnification must be $M_2 = 1$.

Thus, the final image is located at the position of the second lens and the overall magnification is $M = (M_1)(M_2) = (-1)(1.0) = -1$

25.55 Consider the first lens.

$$\frac{1}{s_1'} = \frac{1}{f_1} - \frac{1}{s_1} = \frac{1}{25 \text{ cm}} - \frac{1}{20 \text{ cm}} = -\frac{1}{100 \text{ cm}} \qquad s_1' = -100 \text{ cm}$$

$$M_1 = -\frac{s_1'}{s_1} = -\frac{-100 \text{ cm}}{20 \text{ cm}} = 5$$

The object position for the second lens is $s_2 = 125$ cm. Thus,

$$\frac{1}{s_2'} = \frac{1}{f_2} - \frac{1}{s_2} = -\frac{1}{10 \text{ cm}} - \frac{1}{125 \text{ cm}} \qquad \text{and} \qquad s_2' = -9.26 \text{ cm}$$

$$M_2 = -\frac{s_2'}{s_2} = -\frac{-9.26 \text{ cm}}{125 \text{ cm}} = 0.074$$

the overall magnification M is
$$M = (M_1)(M_2) = (5)(0.074) = 0.370$$

25.56 For the first lens, we have

$$\frac{1}{s_1'} = \frac{1}{f_1} - \frac{1}{s_1} = \frac{1}{-10 \text{ cm}} - \frac{1}{20 \text{ cm}} = -\frac{3}{20 \text{ cm}} \qquad s_1' = -6.67 \text{ cm}$$

$$M_1 = -\frac{s_1'}{s_1} = -\frac{-6.67 \text{ cm}}{20 \text{ cm}} = \frac{1}{3}$$

For the second lens, the object distance is 31.67 cm, and we have

$$\frac{1}{s_2'} = \frac{1}{f_2} - \frac{1}{s_2} = \frac{1}{25 \text{ cm}} - \frac{1}{31.67 \text{ cm}} \qquad \text{and} \qquad s_2' = 119 \text{ cm}$$

$$M_2 = -\frac{s_2'}{s_2} = -\frac{119 \text{ cm}}{31.67 \text{ cm}} = -3.75$$

the overall magnification M is
$$M = (M_1)(M_2) = (\frac{1}{3})(-3.75) = -1.25$$

25.57 The final image will be on the film, so let us start with the second lens and work backward.
$$s_2' = 5 \text{ cm}$$
and $$\frac{1}{s_2} = \frac{1}{f_2} - \frac{1}{s'_2} = \frac{1}{13.3 \text{ cm}} - \frac{1}{5 \text{ cm}} \qquad \text{and} \qquad s_2 = -8.01 \text{ cm}$$

Thus, $s_1' = 7$ cm $+ 8.01$ cm $= 15.01$ cm

CHAPTER TWENTY-FIVE SOLUTIONS

Then, $\dfrac{1}{s_1} = \dfrac{1}{f_1} - \dfrac{1}{s'_1} = \dfrac{1}{15.0 \text{ cm}} - \dfrac{1}{15.01 \text{ cm}}$ and $s_1 = 187$ m (max dist)

Similarly, if $d = 10$ cm, $s_2' = 10$ cm

$\dfrac{1}{s_2} = \dfrac{1}{f_2} - \dfrac{1}{s'_2} = \dfrac{1}{13.3 \text{ cm}} - \dfrac{1}{10 \text{ cm}}$ and $s_2 = -40.3$ cm

Thus, $s_1' = 12$ cm $- 10$ cm $+ 40.3$ cm $= 42.3$ cm

$\dfrac{1}{s_1} = \dfrac{1}{f_1} - \dfrac{1}{s'_1} = \dfrac{1}{15.0 \text{ cm}} - \dfrac{1}{42.3 \text{ cm}}$ and $s_1 = 23.2$ cm (min dist)

25.58 Treating the first lens

$\dfrac{1}{s_1'} = \dfrac{1}{f_1} - \dfrac{1}{s_1} = \dfrac{1}{8 \text{ cm}} - \dfrac{1}{4 \text{ cm}}$ $s_1' = -8.0$ cm

Thus, $s_2 = 14$ cm

$\dfrac{1}{s_2'} = \dfrac{1}{f_2} - \dfrac{1}{s_2} = \dfrac{1}{-16 \text{ cm}} - \dfrac{1}{14 \text{ cm}}$ $s_2' = -7.47$ cm

$M_1 = -\dfrac{s_1'}{s_1} = 2$ and $M_2 = -\dfrac{s_2'}{s_2} = 0.533$

$M = M_1 M_2 = 1.07$

Thus, if the object is 1 cm in height, the image is 1.07 cm tall.

25.59 Start with the second lens and work backward.

$f_2 = 20$ cm, $s_2' = -(50 \text{ cm} - 31 \text{ cm}) = -19$ cm

$\dfrac{1}{s_2} + \dfrac{1}{s'_2} = \dfrac{1}{f}$ becomes, $\dfrac{1}{s_2} - \dfrac{1}{19} = \dfrac{1}{20}$, from which $s_2 = 9.74$ cm

$s_1' = (50 \text{ cm} - 9.74 \text{ cm}) = 40.26$ cm, and $f_1 = 10$ cm

Now, treat the first lens

$\dfrac{1}{s_1} = \dfrac{1}{f_1} - \dfrac{1}{s_1'} = \dfrac{1}{10.0 \text{ cm}} - \dfrac{1}{40.26 \text{ cm}}$ and $s_1 = 13.3$ cm

(b) $M_1 = -\dfrac{s_1'}{s_1} = -\dfrac{40.26 \text{ cm}}{13.3 \text{ cm}} = -3.03$

$M_2 = -\dfrac{s_2'}{s_2} = -\dfrac{-19 \text{ cm}}{9.74 \text{ cm}} = 1.95$

and $M = M_1 M_2 = -5.90$

(c) The overall magnification is less than 0, so the final image is inverted.

25.60 Start with the first pass through the lens

$\dfrac{1}{s_1'} = \dfrac{1}{f_1} - \dfrac{1}{s_1} = \dfrac{1}{80 \text{ cm}} - \dfrac{1}{100 \text{ cm}}$ $s_1' = 400$ cm to right of lens

For the mirror, $s_2 = -300$ cm

so $\dfrac{1}{s_2'} = \dfrac{1}{f_2} - \dfrac{1}{s_2} = \dfrac{1}{-50 \text{ cm}} - \dfrac{1}{-300 \text{ cm}}$ $s_2' = -60$ cm

For the second pass through the lens $s_2 = 160$ cm

$\dfrac{1}{s_3'} = \dfrac{1}{f_1} - \dfrac{1}{s_3} = \dfrac{1}{80 \text{ cm}} - \dfrac{1}{160 \text{ cm}}$ $s_3' = 160$ cm to left of lens

$M_1 = -\dfrac{s_1'}{s_1} = -\dfrac{400 \text{ cm}}{100 \text{ cm}} = -4$

$M_2 = -\dfrac{s_2'}{s_2} = -\dfrac{-60 \text{ cm}}{-300 \text{ cm}} = -\dfrac{1}{5}$

$M_3 = -\dfrac{s_3'}{s_3} = -\dfrac{160 \text{ cm}}{160 \text{ cm}} = -1$

CHAPTER TWENTY-FIVE SOLUTIONS

and $M = M_1 M_2 M_3 = -0.8$
since $M < 0$, the final image is inverted relative to the original object.

25.61 $\quad \dfrac{1}{s_1'} = \dfrac{1}{f_1} - \dfrac{1}{s_1} = \dfrac{1}{10 \text{ cm}} - \dfrac{1}{12.5 \text{ cm}}$

$s_1' = 50$ cm =(to left of mirror)
This serves as an object for the lens (a virtual object), so
$\dfrac{1}{s_2'} = \dfrac{1}{f_2} - \dfrac{1}{s_2} = \dfrac{1}{-16.7 \text{ cm}} - \dfrac{1}{-25 \text{ cm}} \quad s_2' = -50.3$ cm (to right of lens)
Thus, the final image is located 25.3 cm to right of mirror.

$M_1 = -\dfrac{s_1'}{s_1} = -\dfrac{50 \text{ cm}}{12.5 \text{ cm}} = -4$

$M_2 = -\dfrac{s_2'}{s_2} = -\dfrac{-50.3 \text{ cm}}{-25 \text{ cm}} = -2.01$

$M = M_1 M_2 = 8.05$

Thus, the final image is virtual, erect, and 8.05 times the size of object, and 25.3 cm to right of the mirror.

25.62 (a) $R = 2f = 60$ cm

(b) $\dfrac{1}{s'} = \dfrac{1}{f} - \dfrac{1}{s} = \dfrac{1}{30 \text{ cm}} - \dfrac{1}{100 \text{ cm}} = \dfrac{7}{300 \text{ cm}} \quad s' = 42.9$ cm

$M = -\dfrac{s'}{s} = -\dfrac{42.9}{100} = -0.429$

Thus, the image is real, inverted and located 42.9 cm in front of the mirror.

(c) $\dfrac{1}{s'} = \dfrac{1}{f} - \dfrac{1}{s} = \dfrac{1}{30 \text{ cm}} - \dfrac{1}{10 \text{ cm}} = -\dfrac{2}{30 \text{ cm}} \quad s' = -15$ cm

$M = -\dfrac{s'}{s} = -\dfrac{-15}{10} = 1.5$

The image is erect, virtual, and located 15 cm behind the mirror.

25.63 We use $\quad \dfrac{1}{f} = (n - 1)(\dfrac{1}{R_1} - \dfrac{1}{R_2})$

which becomes $\dfrac{1}{f} = (1.5 - 1)(\dfrac{1}{R_1} - 0) = \dfrac{1}{20 \text{ cm}}$ from which, we find $R_1 = 10$ cm

25.64 $\dfrac{1}{f} = (n - 1)(\dfrac{1}{R_1} - \dfrac{1}{R_2}) = (1.5 - 1)(0 - \dfrac{1}{R_2}) = -\dfrac{1}{20 \text{ cm}}$

from which, we find $R_2 = 10$ cm

25.65 For a converging lens, the focal length is positive.

(a) $\dfrac{1}{s'} = \dfrac{1}{f} - \dfrac{1}{s} = \dfrac{1}{20 \text{ cm}} - \dfrac{1}{50 \text{ cm}} = \dfrac{3}{100 \text{ cm}} \quad s' = 33.3$ cm

$M = -\dfrac{s'}{s} = -\dfrac{33.3 \text{ cm}}{50 \text{ cm}} = -0.67$

(b) $\dfrac{1}{s'} = \dfrac{1}{f} - \dfrac{1}{s} = \dfrac{1}{20 \text{ cm}} - \dfrac{1}{30 \text{ cm}} = \dfrac{1}{60 \text{ cm}} \quad s' = 60$ cm

$M = -\dfrac{s'}{s} = -\dfrac{60 \text{ cm}}{30 \text{ cm}} = -2$

(c) $\dfrac{1}{s'} = \dfrac{1}{f} - \dfrac{1}{s} = \dfrac{1}{20 \text{ cm}} - \dfrac{1}{10 \text{ cm}} = -\dfrac{1}{20 \text{ cm}} \quad s' = -20$ cm

CHAPTER TWENTY-FIVE SOLUTIONS

$$M = -\frac{s'}{s} = -\frac{-20 \text{ cm}}{10 \text{ cm}} = 2$$

(d) The magnification has been determined for each case in the above.
For case (a) the image is real, and inverted.
For case (b) the image is real and inverted.
For case (c) the image is erect and virtual.

25.66 We first find the focal length of the mirror.
$$\frac{1}{f} = \frac{1}{s} + \frac{1}{s'} = \frac{1}{10 \text{ cm}} + \frac{1}{8 \text{ cm}} = \frac{9}{40 \text{ cm}} \quad \text{and} \quad f = 4.44 \text{ cm}$$
Hence, if $s = 20$ cm,
$$\frac{1}{s'} = \frac{1}{f} - \frac{1}{s} = \frac{1}{4.44 \text{ cm}} - \frac{1}{20 \text{ cm}} = \frac{15.56}{88.8 \text{ cm}} \quad s' = 5.71 \text{ cm}$$

25.67 (a) We find the image position produced by the first lens.
$$\frac{1}{s_1'} = \frac{1}{f_1} - \frac{1}{s_1} = \frac{1}{20 \text{ cm}} - \frac{1}{40 \text{ cm}} = \frac{1}{40 \text{ cm}} \quad s_1' = 40 \text{ cm}$$
$$M_1 = -\frac{s_1'}{s_1} = -\frac{40 \text{ cm}}{40 \text{ cm}} = -1$$
For the second lens, $s_2 = 10$ cm. Thus,
$$\frac{1}{s_2'} = \frac{1}{f_2} - \frac{1}{s_2} = \frac{1}{5 \text{ cm}} - \frac{1}{10 \text{ cm}} \quad \text{and} \quad s_2' = 10 \text{ cm to right of 2nd lens}$$
$$M_2 = -\frac{s_2'}{s_2} = -\frac{10 \text{ cm}}{10 \text{ cm}} = -1$$
(b) The overall magnification M is
$$M = (M_1)(M_2) = (-1)(-1) = 1$$
Thus, the final image is real, erect, and the same size as the object.
(c) $$\frac{1}{f} = \frac{1}{f_1} + \frac{1}{f_2} = \frac{1}{20 \text{ cm}} + \frac{1}{5 \text{ cm}} \quad \text{and} \quad f = 4 \text{ cm}$$
$$\frac{1}{s'} = \frac{1}{f} - \frac{1}{s} = \frac{1}{4 \text{ cm}} - \frac{1}{5 \text{ cm}} = \frac{1}{20 \text{ cm}} \quad s' = 20 \text{ cm}$$

25.68 The real image formed by the concave mirror serves as a real object for the convex mirror with $s = 50$ cm and $s' = -10$ cm. Therefore,
$$\frac{1}{f} = \frac{1}{s} + \frac{1}{s'}$$
$$\frac{1}{f} = \frac{1}{50 \text{ cm}} + \frac{1}{-10 \text{ cm}} \quad \text{gives } f = -12.5 \text{ cm} \quad \text{and} \quad R = 2f = -25 \text{ cm}.$$

25.69 We use $\quad \frac{1}{f} = (n-1)(\frac{1}{R_1} - \frac{1}{R_2}) = \frac{1}{s} + \frac{1}{s'}$
Substitute the given values and find $R_2 = -30$ cm

25.70 For the first lens, $\frac{1}{f} = \frac{1}{s} + \frac{1}{s'}$ becomes $\frac{1}{-6 \text{ cm}} = \frac{1}{12 \text{ cm}} + \frac{1}{s_1'}$
giving $s_1' = -4$ cm.
When we require that s_2' approach infinity, we see that s_2 must be the focal length of the second lens, and in this case $s_2 = d - (-4 \text{ cm})$.
Therefore $d + 4 \text{ cm} = f_2 = 12$ cm and $d = 8$ cm.

CHAPTER TWENTY-FIVE SOLUTIONS

25.71 We use $\dfrac{n_1}{s} + \dfrac{n_2}{s'} = \dfrac{n_2 - n_1}{R}$ and when R = infinity, we find

$s = -s'\dfrac{n_1}{n_2}$

Thus, $s = -(-35\text{ cm})\dfrac{1.5}{1} = 52.5\text{ cm}$ and $M = -\dfrac{n_1 s'}{n_2 s} = \dfrac{h'}{h}$ or

$h' = -h\dfrac{n_1 s'}{n_2 s} = -(1.5\text{ cm})\dfrac{(1.5)(-35\text{ cm})}{(1)(52.5\text{ cm})} = 1.5\text{ cm}$

25.72 $\dfrac{1}{s_1'} = \dfrac{1}{f_1} - \dfrac{1}{s_1} = \dfrac{1}{10\text{ cm}} - \dfrac{1}{15\text{ cm}}$ $s_1' = 30\text{ cm}$

so $s_2 = -20\text{ cm}$ and $s_2' = -25\text{ cm}$

Thus, $\dfrac{1}{f_2} = \dfrac{1}{-20\text{ cm}} + \dfrac{1}{-25\text{ cm}}$ and $f_2 = -11.1\text{ cm}$

(b) $M_1 = -\dfrac{s_1'}{s_1} = -\dfrac{30\text{ cm}}{15\text{ cm}} = -2$

$M_2 = -\dfrac{s_1'}{s_1} = -\dfrac{-25\text{ cm}}{-20\text{ cm}} = -1.25$

and $M = (M_1)(M_2) = (-2)(-1.25) = 2.5$

(c) The final image is virtual, erect, and enlarged.

25.73 We use $\dfrac{n_1}{s} + \dfrac{n_2}{s'} = \dfrac{n_2 - n_1}{R}$ applied to each surface in turn.

(a) At the first surface, we have

$\dfrac{1}{1\text{ cm}} + \dfrac{1.5}{s_1'} = \dfrac{1.5 - 1}{2\text{ cm}}$ from which, $s_1' = -2\text{ cm}$

Thus, the image is formed 2 cm in front of the surface, a distance of 16 cm from the second surface. This position becomes the object distance for the second surface.

We again use, $\dfrac{n_1}{s} + \dfrac{n_2}{s'} = \dfrac{n_2 - n_1}{R}$

$\dfrac{1.5}{16\text{ cm}} + \dfrac{1}{s_2'} = \dfrac{1.0 - 1.5}{-4\text{ cm}}$ and $s_2' = 32\text{ cm}$

The final image is a real image 32 cm past the second surface.

25.74 At the first surface, we have

$\dfrac{1}{28\text{ cm}} + \dfrac{1.5}{s_1'} = \dfrac{1.5 - 1}{2\text{ cm}}$ from which, $s_1' = 7\text{ cm}$

This image position is 3 cm from the second surface. Thus, the object distance for the second surface is 3 cm.

$\dfrac{1.5}{3\text{ cm}} + \dfrac{1}{s_2'} = \dfrac{1.0 - 1.5}{-4\text{ cm}}$ and $s_2' = -2.67\text{ cm}$

The final image is a virtual image located 2.67 cm to the left of the second surface.

25.75 We solve $\dfrac{1}{f} = \dfrac{1}{s} + \dfrac{1}{s'}$ for s to find $s = \dfrac{fs'}{(s' - f)}$ and we use $M = -\dfrac{s'}{s}$

(a) For $s' = 4f$ $s = \dfrac{f(4f)}{(4f - f)} = \dfrac{4f}{3}$

CHAPTER TWENTY-FIVE SOLUTIONS

(b) For $s' = -3f$, $\quad s = \dfrac{f(-3f)}{(-3f - f)} = \dfrac{3f}{4}$

(c) For case (a), $\quad M = -\dfrac{4f}{\frac{4f}{3}} = -3\;$ and for case b, $\quad M = -\dfrac{-3f}{\frac{3f}{4}} = 4$

CHAPTER TWENTY-SIX SOLUTIONS

26.1 The position of the mth bright fringe is given by $y_m = \frac{m\lambda L}{d}$.

Thus, $\Delta y = y_{m+1} - y_m = \frac{\lambda L}{d}$. From this, we find the wavelength as

$$\lambda = \frac{(\Delta y)d}{L} = \frac{(5 \times 10^{-4} \text{ m})(1.3 \times 10^{-3} \text{ m})}{1.5 \text{ m}} = 433 \text{ nm}$$

26.2 We have $\Delta y_{bright} = \frac{\lambda L}{d}$. (See the solution to problem 26.1)

$$\Delta y = \frac{(6 \times 10^{-7} \text{ m})(2.5 \text{ m})}{5 \times 10^{-5} \text{ m}} = 3 \times 10^{-2} \text{ m} = 3 \text{ cm}$$

26.3 (a) The position of the first bright fringe, m = 1, is given by

$$y_{bright} = \frac{\lambda L}{d} = \frac{(546 \times 10^{-9} \text{ m})(1.2 \text{ m})}{0.25 \times 10^{-3} \text{ m}} = 2.62 \times 10^{-3} \text{ m} = 2.62 \text{ mm}$$

(b) $y_{dark} = (m + \frac{1}{2})\frac{\lambda L}{d}$ from which $\Delta y_{dark} = \frac{\lambda L}{d} = \Delta y_{bright} = 2.62 \text{ mm}$

26.4 (a) $\Delta y = [(y_{max})_{m=1} - (y_{max})_{m=0}] = \frac{\lambda L}{d}$

$$= \frac{(5.7 \times 10^{-7} \text{ m})(2.5 \text{ m})}{8.3 \times 10^{-5} \text{ m}} = 1.72 \text{ cm}$$

(b) $\Delta y = [(y_{max})_{m=4} - (y_{max})_{m=2}] = 4\frac{\lambda L}{d} - 2\frac{\lambda L}{d} = 2\frac{(4.1 \times 10^{-7} \text{ m})(2.5 \text{ m})}{8.3 \times 10^{-5} \text{ m}} =$ 2.47 cm

(c) $\Delta y = [(y_{dark})_{m=1} - (y_{dark})_{m=0}] = \frac{\lambda L}{d}(1 + \frac{1}{2}) - \frac{\lambda L}{d}(0 + \frac{1}{2}) = 1.72$ cm
(calculation same as in part (a)

$\Delta y = [(y_{dark})_{m=4} - (y_{dark})_{m=2}] = \frac{\lambda L}{d}(4 + \frac{1}{2}) - \frac{\lambda L}{d}(2 + \frac{1}{2}) = 2.47$ cm (calculation same as in part (b)

26.5 Location of A = central maximum, and location of B = first minimum. So,
$\Delta y = [(y_{min}) - (y_{max})] = \frac{\lambda L}{d}(0 + \frac{1}{2}) - 0 = \frac{1}{2}\frac{\lambda L}{d} = 20 \text{ m} = \frac{(3 \text{ m})(150 \text{ m})}{d}$
From which, d = 11.3 m

26.6 For bright fringes, the condition is $\delta = d\sin\theta = m\lambda$

Thus, $d = \frac{m\lambda}{\sin\theta} = \frac{(1)(575 \times 10^{-9} \text{ m})}{\sin 16.5°} = 2.02 \text{ μm}$

26.7 From $\Delta y = \frac{\lambda L}{d}$, we have

CHAPTER TWENTY-SIX SOLUTIONS

$$L = \frac{(\Delta y)d}{\lambda} = \frac{(4 \times 10^{-3} \text{ m})(3 \times 10^{-4} \text{ m})}{460 \times 10^{-9} \text{ m}} = 2.61 \text{ m}$$

26.8 For the first order (m = 1) bright fringe, $y = \frac{\lambda L}{d}$.

Thus, for violet light $y_v = \frac{\lambda L}{d} = \frac{(4 \times 10^{-7} \text{ m})(1.5 \text{ m})}{3 \times 10^{-4} \text{ m}} = 2. \times 10^{-3} \text{ m} = 2.0 \text{ mm}$

and for red light $y_r = \frac{\lambda L}{d} = \frac{(7 \times 10^{-7} \text{ m})(1.5 \text{ m})}{3 \times 10^{-4} \text{ m}} = 3.5. \times 10^{-3} \text{ m} = 3.5 \text{ mm}$

Therefore, $\Delta y = y_r - y_v = 3.5 \text{ mm} - 2 \text{ mm} = 1.5 \text{ mm}$

26.9 We have $y_m = \frac{m\lambda L}{d}$, and we are given that $y_2 = 10d = 10(0.2 \text{ mm}) = 2 \text{ mm}$.

Therefore, $2 \times 10^{-3} \text{ m} = \frac{2(587.5 \times 10^{-9} \text{ m})(L)}{2 \times 10^{-4} \text{ m}}$ yields $L = 0.340 \text{ m} = 34 \text{ cm}$

26.10 $\lambda = \frac{(y_m)d}{mL} = \frac{(3.4 \times 10^{-3} \text{ m})(5 \times 10^{-4} \text{ m})}{(1)(3.3 \text{ m})} = 515 \text{ nm}$

26.11 Note, with the conditions given, the small angle approximation does not work well. The approach to be used is outlined below.

(a) at m = 2 maximum, $\tan\theta = \frac{400 \text{ m}}{1000 \text{ m}} = 0.4$ and $\theta = 21.8°$

So, $\lambda = \frac{d\sin\theta}{m} = \frac{(300 \text{ m})\sin 21.8°}{2} = 55.7 \text{ m}$

(b) The next minimum encountered is the m = 2 min, and at that point
$d\sin\theta = (m + \frac{1}{2})\lambda$, which becomes $d \sin\theta = \frac{5}{2}\lambda$

or $\sin\theta = \frac{m\lambda}{2d} = \frac{5(55.7 \text{ m})}{2(300 \text{ m})} = 0.464$, and $\theta = 27.7°$

so, $y = (1000 \text{ m})\tan 27.7° = 524 \text{ m}$

So, the car must travel an additional 124 m

26.12 Note that we are neglecting any phase changes which may occur on reflection from the mountain. If d is the distance to the mountain, we have a path difference of $\delta = 2d$.

For destructive interference, $\delta = (m + \frac{1}{2})\lambda$

Therefore, $2d = (m + \frac{1}{2})\lambda$

For the minimum value of d, we use the smallest value of m, namely m = 0. Thus,

$2d = \frac{\lambda}{2}$ or $d = \frac{\lambda}{4} = \frac{300 \text{ m}}{4} = 75 \text{ m}$

CHAPTER TWENTY-SIX SOLUTIONS

26.13 The wavelength of the wave is
$$\lambda = \frac{c}{f} = \frac{3 \times 10^8 \text{ m/s}}{1.5 \times 10^6 \text{ Hz}} = 200 \text{ m}$$

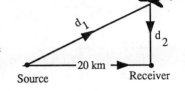

(See sketch.) For destructive interference, we note that
$$\delta = (d_1 + d_2) - 2 \times 10^4 \text{ m} = \frac{\lambda}{2} = 100 \text{ m}$$

Thus, $d_1 = 2.01 \times 10^4 \text{ m} - d_2$ (1)
Also, using the right triangle, we have $d_1^2 = d_2^2 + (2 \times 10^4 \text{ m})^2$ (2)
We solve (1) and (2) simultaneously to find $d_2 = 99.8 \text{ m}$

26.14 (See sketch.) For the minimum height h and hence the minimum path difference to give destructive interference, we have

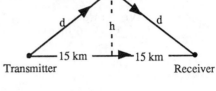

$$\delta = 2d - 3 \times 10^4 \text{ m} = \frac{\lambda}{2} = 175 \text{ m}$$

Therefore, $d = 1.509 \times 10^4 \text{ m}$
Then from the Pythagorean theorem,
$$d^2 = h^2 + (1.5 \times 10^4 \text{ m})^2 = (1.509 \times 10^4 \text{ m})^2$$
and $h = 1.62 \text{ km}$

26.15 (See sketch.) We have
$25° + 90° + \alpha + 25° = 180°$

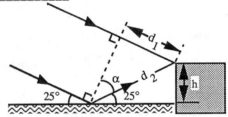

Thus, $\alpha = 40°$
$\delta = d_2 - d_1 = d_2 - d_2 \sin\alpha = d_2(1 - \sin 40°) = (0.357)d_2$
For first order destructive interference,
$$\delta = \frac{\lambda}{2} = 125 \text{ m}$$
Therefore, $(0.357)d_2 = 125 \text{ m}$ and $d_2 = 350 \text{ m}$.
Then, $h = d_2 \sin 25° = (350 \text{ m})\sin 25° = 148 \text{ m}$

26.16 The position of the mth order bright fringe on the screen is given by
$$y_{bright} = \frac{\lambda L}{d} m$$
Thus, if $y = 12 \text{ mm}$, we have
$$m = \frac{yd}{\lambda L} = \frac{(12 \times 10^{-3} \text{ m})(2.41 \times 10^{-4} \text{ m})}{(4.86 \times 10^{-7} \text{ m})(1.19 \text{ m})} = 5$$
Therefore, there must be four other bright fringes between this one and the central maximum.

26.17 The phase difference because of the difference in path lengths is $2n_f t$, where n_f is the index of refraction of the film.
The phase difference due to reflection at the upper surface is $\frac{\lambda}{2}$.
For constructive interference, $\delta = m\lambda$.
Thus, $2n_f t + \frac{\lambda}{2} = m\lambda$.
For the minimum thickness, $m = 1$, and we have

365

CHAPTER TWENTY-SIX SOLUTIONS

$$t = \frac{\lambda}{4n_f} = \frac{500 \text{ nm}}{4(1.36)} = 91.9 \text{ nm}$$

26.18 Both waves undergo a phase shift of half wavelength on reflection. Thus, the difference in phase is determined solely by the path difference. We have $\delta = 2n_f t = m\lambda$
For the smallest thickness, m = 1, and we have
$$t = \frac{\lambda}{2n_f} = \frac{500 \text{ nm}}{2(1.36)} = 184 \text{ nm}$$

26.19 There is a phase shift at the upper surface and none at the lower. We have

$$n_{film}(2t) + \frac{\lambda}{2} = \lambda \text{ as the condition for the thinnest film to give}$$
constructive interference.
Thus, $t = \frac{\lambda}{4n_{film}} = \frac{600 \text{ nm}}{4(1.756)} = 85.4 \text{ nm}$
Other possible thicknesses are odd integral multiples of the above, 256 nm, 427 nm, etc.

26.20 There is a phase shift at the upper surface and none at the lower. We have

$$n_{film}(2t) + \frac{\lambda}{2} = m\lambda \text{ (for constructive interference)}$$
or $(m - \frac{1}{2})\lambda = n_{film}(2t) = 2(1.52)(4.2 \times 10^{-7} \text{ m}) = 1277 \text{ nm}$
or $\lambda = \frac{1277 \text{ nm}}{m - \frac{1}{2}}$ where m = 1,2,3,...
giving λ = 2554 nm, 851 nm, 511 nm, 365 nm (The only visible wavelength is 511 nm.)

26.21 There is a phase difference because of reflection and because of the path difference. We have,
$$2n_f t + \frac{\lambda}{2} = (m + \frac{1}{2})\lambda.$$
or, $t = m\frac{\lambda}{2n_f} = m\frac{580 \text{ nm}}{2(1)}$
which reduces to $t = m(290 \text{ nm})$
When m = 1, t = 290 nm
when m = 2 t = 580 nm
when m = 3 t = 870 nm

26.22 For destructive interference, we have
$$\delta = 2n_f t + \frac{\lambda}{2} = (m + \frac{1}{2})\lambda.$$
or, $2t = m\lambda$, and for the minimum distance, m = 1.
Thus, $t = \frac{\lambda}{2} = 150 \text{ m}$

CHAPTER TWENTY-SIX SOLUTIONS

26.23 Both reflections are of the same type. Thus, there is no phase difference because of the reflection. Thus, for constructive interference, we have
$$\delta = 2n_f d = m\lambda.$$
For the minimum value of d, m = 1
Thus, $d = \frac{\lambda}{2} = \frac{580 \text{ nm}}{2} = 290 \text{ nm}$

26.24 We have
$$\delta = 2n_f t + \frac{\lambda}{2} = (m + \frac{1}{2})\lambda.$$
or, $2n_f t = m\lambda.$
Thus, $t = m\frac{500 \text{ nm}}{2(1)} = m(250 \text{ nm})$
For the 19th order destructive interference,
$t = 19(250 \text{ nm}) = 4750 \text{ nm}$

26.25

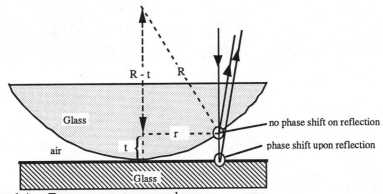

(See sketch.) From geometry, we have
$$R^2 = (R - t)^2 + r^2$$
or $t = R - \sqrt{R^2 - r^2} = 3.0 \text{ m} - \sqrt{(3.0)^2 - (9.8 \times 10^{-3})^2} = 1.6 \times 10^{-5} \text{ m}$
at a bright fringe
$$\text{path difference} = 2n_f t + \frac{\lambda}{2} = m\lambda.$$
or for $n_f = 1$ $(m - \frac{1}{2})\lambda = 2t$, or
$$\lambda = \frac{2t}{m - \frac{1}{2}} = \frac{2(1.6 \times 10^{-5} \text{ m})}{49.5} = 647 \text{ nm (since this is the 50 th (m = 50)}$$
bright fringe.)

26.26 There is a phase shift at both the upper and lower film surfaces. For a bright fringe,
$$\text{path difference} = 2n_f t + 0 = m\lambda$$
$$t = \frac{m\lambda}{2n_f} = \frac{(1)(600 \text{ nm})}{2(1.29)} = 233 \text{ nm}$$

26.27 For constructive interference, we have

CHAPTER TWENTY-SIX SOLUTIONS

$$\delta = 2n_f t + \frac{\lambda}{2} = m\lambda.$$

and $\quad \lambda = \dfrac{2n_f t}{(m - \frac{1}{2})} = \dfrac{2(1.46)(500 \text{ nm})}{(m - \frac{1}{2})}$

For m = 1, 2, 3, 4, etc. we have
λ = 2930 nm, 973 nm, 584 nm, 417 nm, 324 nm, 265 nm etc.
But, between 300 nm and 700 nm, we find
324 nm, 417 nm, and 584 nm

26.28 For destructive interference, we have
$$\delta = 2n_f t + \frac{\lambda}{2} = (m + \frac{1}{2})\lambda.$$
or, $\quad 2n_f t = m\lambda.$

and $\quad t = m\dfrac{500 \text{ nm}}{2(1.50)} = m(193.3 \text{ nm})$

For the smallest thickness other than zero, m = 1, and
t = 193 nm

26.29 There is a phase shift at both the upper and lower surfaces of the film. For a bright fringe
path difference = $2n_f t + 0 = m\lambda.$

or, $\quad \lambda = \dfrac{2n_f t}{m} = \dfrac{2(1.38)(100 \text{ nm})}{m} = \dfrac{276 \text{ nm}}{m}$

which gives λ = 276 nm, 138 nm, 92 nm
Thus, no visible wavelengths produce constructive interference

26.30 There is a phase shift at the upper surface, but none at the lower. For destructive interference,
$$2n_f t + \frac{\lambda}{2} = (2m + 1)\frac{\lambda}{2}$$

or $\quad \lambda = \dfrac{2n_f t}{m} = \dfrac{2(1.55)(177.4 \text{ nm})}{m} = 550$ nm (when m = 1), At the central part of the visible spectrum.

26.31 There is a phase shift at the upper surface, but none at the lower. For destructive interference,
$$2n_f t + \frac{\lambda}{2} = m\lambda.$$

or $\quad t = \dfrac{(m - \frac{1}{2})\lambda}{2n_f} \quad$ and m = 1 for minimum thickness

(a) $\quad t = \dfrac{\frac{1}{2}(656.3 \text{ nm})}{2(1.33)} = 123$ nm

(b) $\quad t = \dfrac{\frac{1}{2}(434.0 \text{ nm})}{2(1.33)} = 81.6$ nm

26.32 There is a phase shift at the first surface and none at the second.
$$2n_f t + \frac{\lambda}{2} = m\lambda. \quad (1)$$

CHAPTER TWENTY-SIX SOLUTIONS

For the same thickness to strongly reflect red ($\lambda = 700$ nm) and green ($\lambda = 500$ nm), we must have
$$(m_R - \tfrac{1}{2})(700 \text{ nm}) = (m_g - \tfrac{1}{2})(500 \text{ nm})$$
which reduces to $\quad 1.4 m_R - m_g = 0.2$
The smallest integer values which satisfy this are $m_R = 3$ and $m_g = 4$. Thus, (1) gives
$$t = \frac{(3 - \tfrac{1}{2})(700 \text{ nm})}{2(1.4)} = 625 \text{ nm}$$

26.33 For destructive interference,
$$\delta = 2n_f t + \frac{\lambda}{2} = (m + \tfrac{1}{2})\lambda.$$
or $\quad 2n_f t = m\lambda.$
Thus, $m = \dfrac{2(1.0)(4 \times 10^{-5} \text{ m})}{546.1 \times 10^{-9} \text{ m}} = 146.5$
Thus, the last dark fringe seen is the 146th order,
and the total number of dark fringes seen (counting the zeroth order) is $146 + 1 = 147$.

26.34 The first dark band occurs for zero separation because of the half-wavelength phase shift upon reflection from the lower plate. Likewise, the second dark band occurs when the separation is $\lambda/2$. Continuing this process, the seventh dark band occurs when the separation is 3λ.
Thus, the thickness of the hair is $3(546.1 \text{ nm}) = 1640$ nm.

26.35 For constructive interference,
$$\delta = 2n_f t + \frac{\lambda}{2} = m\lambda.$$
or $\quad t = \dfrac{(m - \tfrac{1}{2})\lambda}{2 n_f}$

At the 20th bright fringe $t = \dfrac{19.5(434 \text{ nm})}{2(1)} = 4.23 \text{ μm}$
At the 21st bright fringe $t = 4.45$ μm
Thus, to see 20 but not 21 bright fringes, the thickness must lie between 4.23 μm and 4.45 μm.

26.36 For constructive interference
$$\delta = 2n_f t + \frac{\lambda}{2} = m\lambda.$$
$$\delta = 2(1)t + \frac{\lambda}{2} = m\lambda.$$
Thus, the thickness for the mth order bright fringe is
$$t_m = (m - \tfrac{1}{2})\frac{\lambda}{2} = m\frac{\lambda}{2} - \frac{\lambda}{4}$$
and the thickness for the $m - 1$ bright fringe is
$$t_{m-1} = (m - 1)\frac{\lambda}{2} - \frac{\lambda}{4}$$
Thus, $\Delta t = t_m - t_{m-1} = \dfrac{\lambda}{2}$

CHAPTER TWENTY-SIX SOLUTIONS

This is the change in thickness for one fringe shift (bright to dark and back to bright).

Thus, 200 fringe shifts correspond to $200 \frac{\lambda}{2} = 100$ wavelengths in the thickness of the air film.

Thus, the increase in the length of the rod is
$$\Delta L = 100\lambda = (100)(5 \times 10^{-7} \text{ m}) = 5 \times 10^{-5} \text{ m}$$
and from $\Delta L = L_0 \alpha \Delta T$, we have
$$\alpha = \frac{\Delta L}{L_0 \Delta T} = \frac{5 \times 10^{-5} \text{ m}}{(1 \times 10^{-1} \text{ m})(25 \text{ °C})} = 20 \times 10^{-6} \text{ °C}^{-1}$$

26.37 For destructive interference,
$$2n_f t + \frac{\lambda}{2} = (m + \frac{1}{2})\lambda.$$
or, $t = \frac{m\lambda}{2}$

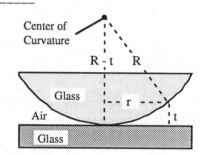

Also, see sketch, we find
$$R^2 = r^2 + (R - t)^2$$
Solving for r^2, we find
$r^2 = 2Rt - t^2$ which is approximately equal to
$r^2 = 2Rt$ if $Rt \gg t^2$
or, $r = \sqrt{2Rt} = \sqrt{m\lambda R}$

26.38 For destructive interference,
$$\sin\theta = m\frac{\lambda}{a} = \frac{\lambda}{a} = \frac{5 \text{ cm}}{36 \text{ cm}} = 0.13889$$
and $\theta = 7.98°$

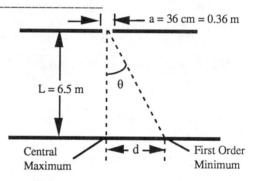

and $\frac{d}{L} = \tan\theta$ gives
$d = L\tan\theta = (6.5 \text{ m})\tan 7.98° = 0.912 \text{ m} = 91.2 \text{ cm}$

26.39 The angles are small enough so that the approximation $\sin\theta = \tan\theta$ holds.

(a) Thus, $\sin\theta = \frac{m\lambda}{a}$, or $a = \frac{\lambda}{\sin\theta} = \frac{12.2 \times 10^{-6} \text{ m}}{2.38 \times 10^{-4}} = 5.12 \times 10^{-2} \text{ m} = 5.12 \text{ cm}$

(b) if $a = \frac{1}{2}$ (old a) = 2.56 cm

then $\sin\theta = \frac{12.2 \times 10^{-6} \text{ m}}{2.56 \times 10^{-2}} = 4.76 \times 10^{-4}$,

and $\tan\theta = \frac{d}{L}$, so $d = (840 \text{ mm})(4.76 \times 10^{-4}) = 0.4 \text{ mm}$

so the spacing between the first order minima = $2d = 0.8$ mm

---

26.40 (a) The dark bands occur where $\sin\theta = m\frac{\lambda}{a}$

CHAPTER TWENTY-SIX SOLUTIONS

For the first dark band $\sin\theta = \frac{\lambda}{a}$

But $\sin\theta \cong \frac{y}{1.5 \text{ m}}$. Thus, $\frac{y}{1.5 \text{ m}} = \frac{\lambda}{a}$

and $y = \frac{(1.5 \text{ m})(600 \times 10^{-9} \text{ m})}{4 \times 10^{-4} \text{ m}} = 2.25$ mm

(b) The width of the central maximum is the distance between the dark lines on each side of the central bright line. This is $2(2.25 \text{ mm}) = 4.50$ mm.

26.41 We have
$$y = mL\frac{\lambda}{a}$$

Thus, for the first dark band, $y_1 = \frac{\lambda}{a}(0.5 \text{ m})$

and the third dark band is at $y_3 = 3\frac{\lambda}{a}(0.5 \text{ m})$

Thus, $\Delta y = y_3 - y_1 = 2\frac{\lambda}{a}(0.5 \text{ m}) = \frac{\lambda}{a}(1 \text{ m})$

and $a = \frac{\lambda}{\Delta y} = \frac{(1 \text{ m})(680 \times 10^{-9} \text{ m})}{3 \times 10^{-3} \text{ m}} = 0.227$ mm

---

26.42 At the first dark band, we have
$$\sin\theta = \frac{\lambda}{a} = \frac{5.5 \times 10^{-7} \text{ m}}{2 \times 10^{-4} \text{ m}} = 2.75 \times 10^{-3} \text{ m}$$

which yields $\theta = 0.158°$

This is the angle between the central maximum and the first dark band. Thus, the angle between the dark bands on each side of the central max is just twice this.
$\alpha = 2\theta = 0.315°$

26.43 (a) At the first dark band, $\sin\theta = \frac{\lambda}{a} = \frac{5.875 \times 10^{-7} \text{ m}}{7.5 \times 10^{-4} \text{ m}} = 7.83 \times 10^{-4}$

But also, $\sin\theta = \frac{y}{L}$, so $L = \frac{y}{\sin\theta}$ thus, $L = \frac{8.5 \times 10^{-4} \text{ m}}{7.83 \times 10^{-4}} = 1.09$ m

(b) The width of the central maximum = $2y = 2(0.85 \text{ mm}) = 1.70$ mm.

26.44 The position of a dark band is given by $y = mL\frac{\lambda}{a}$

Which becomes, $y = m\frac{(1.2 \text{ m})(5 \times 10^{-7} \text{ m})}{5 \times 10^{-4} \text{ m}} = m(1.2 \times 10^{-3} \text{ m}) = m(1.2 \text{ mm})$

Thus, $y_1 = 1.2$ mm, $y_2 = 2.4$ mm, and $y_3 = 3.6$ mm
Thus, the width of the first maximum is $(2.4 - 1.2)$ mm = 1.2 mm
and the width of the second maximum is $(3.6 - 2.4)$ mm = 1.2 mm

---

26.45 We have $\tan\theta_p = \frac{n_2}{n_1} = \frac{1.501}{1} = 1.501$, and $\theta_p = 56.3°$

26.46 $n = \tan\theta_p = \tan(65.6°) = 2.2$

26.47 $n_g = \tan\theta_p = 1.65$

CHAPTER TWENTY-SIX SOLUTIONS

so $\theta_p = 58.8°$
and from Snell's law $\quad n_g \sin\theta_r = n_{air} \sin\theta_p$
which gives $\sin\theta_r = \dfrac{(1)(\sin 58.8°)}{1.65} = 0.5183$, and $\quad \theta_r = 31.2°$

26.48 Starting with Snell's law, we have
$\quad n_1 \sin\theta_1 = n_2 \sin\theta_2$
At Brewster's angle $\quad \theta_1 + \theta_2 + 90° = 180°$, or $\quad \theta_1 + \theta_2 = 90°$
If the material is incident from medium 1, θ_1 is the polarizing angle, θ_p. So,
$\quad \theta_2 = 90° - \theta_p$
Thus, $\sin\theta_2 = \sin(90° - \theta_p) = \cos\theta_p$
and Snell's law becomes, $n_1 \sin\theta_p = n_2 \cos\theta_p$
or, $\quad \dfrac{\sin\theta_p}{\cos\theta_p} = \dfrac{n_2}{n_1}$
and $\quad \tan\theta_p = \dfrac{n_2}{n_1}$

26.49 $\tan\theta_p = \dfrac{n_2}{n_1} = \dfrac{1.50}{1.33} = 1.13$, and $\quad \theta_p = 48.4°$

26.50 The polarizing angle is $\quad \tan\theta_p = \dfrac{n_2}{n_1} = \dfrac{1.33}{1} = 1.33$, and $\quad \theta_p = 53.1°$
The polarizing angle is the angle between the ray of light and the normal to the surface, while the altitude is the angle between the ray of light and the horizontal. This angle is
$\quad \alpha = 90° - \theta_p = 36.9°$

26.51 (a) $\tan\theta_p = \dfrac{n_2}{n_1} = n_2$ because $n_1 = 1$. Thus, $\quad n_2 = \tan 48° = 1.11$
(b) We know that the refracted angle is related to the polarizing angle as
$\quad \theta_2 = 90° - \theta_p = 42°$

26.52 From Snell's law, we find the refracted angle for the ordinary ray.
$\quad \sin\theta_0 = \dfrac{n_1 \sin\theta_1}{n_2} = \dfrac{(1)\sin 20°}{1.66} = 11.9°$ (for ordinary ray)
For the extraordinary ray, the angle of refraction is
$\quad \sin\theta_e = \dfrac{n_1 \sin\theta_1}{n_2} = \dfrac{(1)\sin 20°}{1.49} = 13.3°$ (for extraordinary ray)
Thus, the angle between the two rays is $\quad 13.3° - 11.9° = 1.4°$

26.53 The critical angle for total internal reflection is given by $\sin\theta_c = \dfrac{n_2}{n_1}$,
where n_1 is the index of refraction of the material in which the light travels, and n_2 is the index of the surrounding material. Thus,
$\quad n_1 = \dfrac{n_2}{\sin\theta_c} = \dfrac{1}{\sin 34.4°} = 1.77$ this is the index of refraction of sapphire.
The polarizing angle is given by $\tan\theta_p = \dfrac{n_2}{n_1} = \dfrac{1.77}{1} = 1.77$, and $\theta_p = 60.5°$

CHAPTER TWENTY-SIX SOLUTIONS

26.54 The critical angle is given by $\sin\theta_c = \dfrac{n_{air}}{n_{medium}} = \dfrac{1}{n}$
and the polarizing angle is given by $\tan\theta_p = n$
Thus, $\tan\theta_p = \dfrac{1}{\sin\theta_c}$
or, $\sin\theta_c = \dfrac{1}{\tan\theta_p} = \cot\theta_p$

26.55 We first find the index of refraction of the glass by use of Snell's law.
$n_1\sin\theta_1 = n_g\sin\theta_2$
$(1)\sin 37° = n_g\sin 22°$
From which, $n_g = 1.606$
Then, $\tan\theta_p = \dfrac{n_2}{n_1} = \dfrac{1.606}{1} = 1.606$, and $\theta_p = 58.1°$

26.56 There is a phase shift at both surfaces.
For constructive interference,
$2n_f t + 0 = m\lambda$.
or $t = m\dfrac{\lambda}{2n_f} = m\dfrac{525 \text{ nm}}{2(1.25)} = m(210 \text{ nm})$
Thus, the possible thicknesses are any integral multiples of 210 nm.

26.57 $\sin\theta = \dfrac{\lambda}{a} = \dfrac{632.8 \times 10^{-9} \text{ m}}{3 \times 10^{-4} \text{ m}} = 2.109 \times 10^{-3}$
and $\tan\theta = \dfrac{d}{L}$ gives $d = L\tan\theta = L\sin\theta = (1 \text{ m})(2.109 \times 10^{-3}) = 2.11 \text{ mm}$
Thus, the width of the central maximum = (distance between the first order minimum) = $2d = 4.22$ mm

---

26.58 The wave must strike the water at the polarizing angle.
Thus,
$\tan\theta_p = \dfrac{n_2}{n_1} = \dfrac{1.33}{1} = 1.33$,
But from triangle RST, see sketch, we have
$\tan\theta_p = \dfrac{x}{90 \text{ m}}$, or
$x = (90 \text{ m})\tan\theta_p = (90 \text{ m})(1.33)$
= 120 m
Also, using triangle ABT, we have
$\tan\theta_p = \dfrac{y}{5 \text{ m}}$
and $y = (5 \text{ m})\tan\theta_p = (5 \text{ m})(1.33) = 6.67$ m
The total distance from shore = $x + y = 126.7$ m

26.59 The position of the mth dark fringe is given by $\sin\theta = m\dfrac{\lambda}{a}$
So, $\lambda = \dfrac{a\sin\theta}{m}$. Thus, for the second order and at an angle of 1.56°, we have

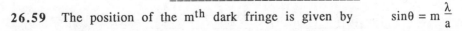

CHAPTER TWENTY-SIX SOLUTIONS

$$\lambda = \frac{(5 \times 10^{-5} \text{ m})\sin 1.56°}{2} = 681 \text{ nm}$$

26.60 For destructive interference, we have
$$(m + \frac{1}{2})\lambda = d\sin\theta \quad m = 0, 1, 2, 3.....$$
The angle θ is 18 min of arc, or 0.3°
At the second minimum, m = 1, and
$$d = \frac{\frac{3}{2}\lambda}{\sin\theta} = \frac{\frac{3}{2}(546 \times 10^{-9} \text{ m})}{\sin 0.3°} = 1.56 \times 10^{-4} \text{ m} = 0.156 \text{ mm}$$

26.61 Bright lines occur when $m\lambda = d\sin\theta$. For the λ_2 light we have,
$$5\lambda_2 = d\sin\theta. \quad (1)$$
For the fourth order bright line of λ_1, we have,
$$4\lambda_1 = d\sin\theta. \quad (2)$$
From (1) and (2), we have $\quad 5\lambda_2 = 4\lambda_1$
Thus, $\lambda_2 = \frac{4}{5}\lambda_1 = 0.8(540 \text{ nm}) = 432 \text{ nm}$.

26.62 (a) We use $I = I_i \cos^2\theta$
If $\theta = 45°$ then $\cos\theta = \frac{1}{\sqrt{2}}$, and $\cos^2\theta = \frac{1}{2}$
Thus, $\frac{I}{I_i} = \frac{1}{2}$
(b) If $\frac{I}{I_i} = \frac{1}{3}$, then $\cos^2\theta = \frac{1}{3}$, and $\cos\theta = \frac{1}{\sqrt{3}} = 0.577$
Which yields, $\quad \theta = 54.7°$

26.63 For $\theta_1 = 20°$, we have $\quad I_2 = I_i\cos^2\theta = 10\cos^2 20° = 8.83$ units
The light emerging from this plate is polarized at 20° with respect to the vertical. At the next plate, we have
$$I_3 = I_2\cos^2(\theta_2 - \theta_1) = 8.83\cos^2 20° = 7.80 \text{ units}$$
The light emerging from the second polarizer has its plane of polarization at 40° from the vertical. At the final plate,
$$I_f = I_3\cos^2(\theta_3 - \theta_2) = 7.80\cos^2 20° = 6.89 \text{ units}$$

26.64 The wavelength of the wave is $\quad \lambda = \frac{c}{f} = \frac{3 \times 10^8 \text{ m/s}}{7.5 \times 10^9 \text{ Hz}} = 4 \times 10^{-2} \text{ m} = 4 \text{ cm}$.
The mth destructive interference occurs at $\quad \sin\theta = m\frac{\lambda}{a}$
or, $\sin\theta = (1)\frac{4 \text{ cm}}{6 \text{ cm}} = 0.667 \quad \theta = 41.8°$

26.65 We are assuming here (not really true) that the first order maximum occurs halfway between the first and second order minima. Thus,
$$[y_{max}]_1 = \frac{[y_{dark}]_1 + [y_{dark}]_2}{2} = \frac{(1)\frac{\lambda}{a}L + (2)\frac{\lambda}{a}L}{2} = \frac{3}{2}\frac{\lambda}{a}L$$

374

CHAPTER TWENTY-SIX SOLUTIONS

Thus, $a = \dfrac{3\lambda L}{2[y_{max}]_1} = \dfrac{3(5 \times 10^{-7}\text{ m})(1.4\text{ m})}{2(3 \times 10^{-3}\text{ m})} = 3.5 \times 10^{-4}\text{ m} = 0.35\text{ mm}$

26.66 (a) The path difference $\delta = d\sin\theta$ and when L is much much greater than y, we have

$$\delta \approx \dfrac{yd}{L} = \dfrac{(1.8 \times 10^{-2}\text{ m})(1.50 \times 10^{-4}\text{ m})}{(1.40\text{ m})} = 1.93 \times 10^{-6}\text{ m}$$

(b) $\dfrac{\delta}{\lambda} = \dfrac{1.93 \times 10^{-6}\text{ m}}{6.43 \times 10^{-7}\text{ m}} = 3$ or $\delta = 3\lambda$

(c) The interference will be a maximum since the path difference is an integer multiple of the wavelength.

26.67 For destructive interference, $2t = \dfrac{\lambda_d}{n}(m + \dfrac{1}{2})$,

and for constructive interference $2t = \dfrac{\lambda_b}{n}(m)$ where n is the index of refraction of the oil. Eliminating m and solving for t, we find

$$t = \dfrac{\lambda_d}{4n(1 - \dfrac{\lambda_d}{\lambda_b})} = \dfrac{500\text{ nm}}{4(1.2)(1 - \dfrac{500}{750})} = 313\text{ nm}$$

26.68 Bright fringes occur when $2t = \dfrac{\lambda}{n}(m + \dfrac{1}{2})$, and dark fringes occur when $2t = \dfrac{\lambda}{n}(m)$ The thickness of the film at x is $t = \dfrac{h}{l}x$.

Therefore, $x_{bright} = \dfrac{\lambda l(m + \dfrac{1}{2})}{2hn}$ and $x_{dark} = \dfrac{\lambda lm}{2hn}$

26.69 In this case constructive interference requires $2x = \dfrac{\lambda}{n}(m + \dfrac{1}{2})$ where 2x is the distance traveled in the film. From the sketch, we have $\cos\theta_2 = \dfrac{t}{x}$.

Therefore, $\dfrac{2nt}{\cos\theta_2} = \lambda(m + \dfrac{1}{2})$

$t = \cos\theta_2 \dfrac{\lambda}{2n}(m + \dfrac{1}{2})$ For minimum thickness, choose m = 0. Also from Snell's law

$\cos\theta_2 = \sqrt{1 - \dfrac{\sin^2\theta_1}{n}}$

Therefore,

$t = \dfrac{\lambda}{4n}\sqrt{1 - \dfrac{\sin^2\theta_1}{n}} = \dfrac{(5.9 \times 10^{-7}\text{ m})}{4(1.38)}\sqrt{1 - \dfrac{\sin^2 30°}{1.38}} = 9.96 \times 10^{-8}\text{ m}$

CHAPTER TWENTY-SIX SOLUTIONS

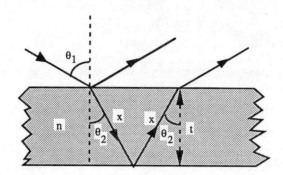

26.70 Let the path difference of two waves be denoted by δ. Then we have
$\sqrt{L^2 + (y-d)^2} - \sqrt{d^2 + (L-x)^2} - \sqrt{y^2 + x^2} = \delta$ where $\dfrac{d}{(L-x)} = \dfrac{y}{x}$, therefore

$\delta = \sqrt{L^2 + (y+d)^2} - \sqrt{L^2 + (y-d)^2}$ $\delta \propto \lambda \ll \sqrt{L^2 + (y-d)^2}$, therefore

$2\delta\sqrt{L^2 + (y-d)^2} = 4yd$ also $(y-d) \ll L$ therefore $y = \dfrac{\delta L}{2d}$ or

$y = \dfrac{m\lambda L}{2d} = \dfrac{(1)(620 \times 10^{-9})(1.2)}{2(2.5 \times 10^{-3})} = 1.49 \times 10^{-4}$ m

CHAPTER TWENTY-SEVEN SOLUTIONS

27.1 From $h' = \dfrac{hs'}{s}$, we have

$$s = \dfrac{h}{h'}f + \dfrac{2.00 \text{ m}}{3.5 \times 10^{-2} \text{ m}}(0.1 \text{ m}) = 5.71 \text{ m}$$

27.2 From $\dfrac{1}{s} + \dfrac{1}{s'} = \dfrac{1}{f}$, we have

$\dfrac{1}{s'} = \dfrac{1}{f} - \dfrac{1}{s} = \dfrac{1}{0.25 \text{ m}} - \dfrac{1}{1.5 \text{ m}}$, and from this $s' = 0.3 \text{ m} = 30 \text{ cm}$

Thus the image is formed 30 cm beyond the lens.

$$M = -\dfrac{s'}{s} = -\dfrac{30 \text{ cm}}{150 \text{ cm}} = -\dfrac{1}{5}$$

27.3 $\theta = \dfrac{h}{100 \text{ m}} = \dfrac{h'}{f}$

so $h = \dfrac{100 \text{ m}(0.092 \text{ m})}{5.2 \times 10^{-2} \text{ m}} = 177 \text{ m}$

27.4 We have $D = \dfrac{f}{f} = \dfrac{12 \text{ cm}}{f}$

f-number = 4 yields $D = \dfrac{12 \text{ cm}}{4} = 3 \text{ cm}$

f-number = 8 yields $D = \dfrac{12 \text{ cm}}{8} = 1.5 \text{ cm}$

f-number = 16 yields $D = \dfrac{12 \text{ cm}}{16} = 0.75 \text{ cm}$

27.5 The focal length is found from the thin lens equation as

$\dfrac{1}{f} = \dfrac{1}{s} + \dfrac{1}{s'} = \dfrac{1}{20 \text{ cm}} + \dfrac{1}{15 \text{ cm}}$, and from this $f = 8.57 \text{ cm}$

$$f\text{-number} = \dfrac{f}{D} = \dfrac{8.57 \text{ cm}}{2 \text{ cm}} = 4.29$$

27.6 Energy = Power(time) = IAt

Thus, $E_1 = IA_1 t$ and $E_2 = IA_2 t$
where I and t are the same for both cases.
Thus, $E_2 = 2E_1$ (film 2 needs twice as much energy)
Then $A_2 = 2A_1$

so $\dfrac{\pi d_2^2}{4} = 2\dfrac{\pi d_1^2}{4}$

or $d_2 = \sqrt{2}\, d_1$

and f-number $= \dfrac{f}{D}$ and f is a constant

so $d_1 f\text{-number})_1 = d_2 f\text{-number})_2$

CHAPTER TWENTY-SEVEN SOLUTIONS

which leads to $\quad f\text{-number})_2 = \dfrac{(f\text{-number})_1}{\sqrt{2}} = \dfrac{11}{\sqrt{2}} = 7.78$

There is no 7.78 setting on a camera, so one should use f/8 setting.

27.7 The f/8 setting uses an aperture opening of twice the diameter of an f/16 setting. We have

$$D_8 = \dfrac{f}{8} \quad \text{and} \quad D_{16} = \dfrac{f}{16}$$

Thus, $\dfrac{D_8}{D_{16}} = \dfrac{16}{8} = 2$

Therefore, since the amount of light entering the lens is proportional to the area of the opening, four times as much light will enter the camera. Thus, you should need $\dfrac{1}{4}$ as much exposure time, or $t = \dfrac{1}{4}(\dfrac{1}{32}\,s) = \dfrac{1}{128}\,s$.

27.8 The exposure time is being reduced by a factor of

$$\dfrac{\frac{1}{256}\,s}{\frac{1}{32}\,s} = \dfrac{1}{8}$$

Thus, you need to get 8 times more light on the film. To do this the aperture opening will have to be 8 times the original area, or $\sqrt{8}$ (original diameter)

Therefore, $D_2 = \sqrt{8}\, D_1$

But, $f\text{-number} = \dfrac{f}{D}$, so $f_2 = \dfrac{f}{D_2} = \dfrac{f}{\sqrt{8}D_1} = \dfrac{1}{\sqrt{8}}\, f_1$

Thus $f_2 = \dfrac{4}{\sqrt{8}} = 1.41$. We must use an f/1.4 setting.

27.9 (a) When an object is at 25 cm in front of the lens (s = 25 cm), the image must be virtual and 100 cm in front of the lens so that the eye can focus on it. (s' = -100 cm)

Thus, $\dfrac{1}{s} + \dfrac{1}{s'} = \dfrac{1}{f}$ becomes $\dfrac{1}{25\text{ cm}} + \dfrac{1}{-100\text{ cm}} = \dfrac{1}{f}$

from which, $f = 33.3$ cm.

(b) $P = \dfrac{1}{f} = \dfrac{1}{0.333\text{ m}} = 3.0$ diopters

---

27.10 $f = \dfrac{1}{P} = \dfrac{1}{2.9} = 0.345$ m $= 34.5$ cm

When the object is at s = 25 cm in front of the lens, the lens should form an erect, virtual image at the near point of the eye. From the thin lens equation,

$\dfrac{1}{s'} = \dfrac{1}{f} - \dfrac{1}{s} = \dfrac{1}{0.345\text{ m}} - \dfrac{1}{0.25\text{ m}}$ which yields s' = -0.909 m = -90.9 cm

Thus, the near point of the eye is at 90.9 cm from the eye.

27.11 The lens should form an erect, virtual image of the most distant objects (s = infinity) at the far point (s' = -100 cm). Thus,

$$\dfrac{1}{s} + \dfrac{1}{s'} = \dfrac{1}{f}$$

CHAPTER TWENTY-SEVEN SOLUTIONS

$$0 + \frac{1}{-100 \text{ cm}} = \frac{1}{f} \quad \text{and} \quad f = -100 \text{ cm} = -1 \text{ m}$$

and the power is $P = \frac{1}{f} = \frac{1}{-1} = -1$ diopter

---

27.12 For the right eye, a virtual image of the most distant object should be 8.44 cm in front of the lens.

$$\frac{1}{s} + \frac{1}{s'} = \frac{1}{f}$$

$$0 + \frac{1}{-8.44 \text{ cm}} = \frac{1}{f}$$

and $f = -8.44$ cm $= -0.0844$ m

so $P = \frac{1}{-0.0844} = -11.8$ diopters

For the left eye, as in the above, the needed lens has a focal length of $f = -$(far point), so

$f = -0.122$ m

and $P = \frac{1}{-0.122} = -8.2$ diopters

27.13 Considering the image formed by the eye as a virtual object for the implanted lens, we have

$s = -(2.53 \text{ cm} + 2.8 \text{ cm}) = -5.33$ cm

and $s' = 2.8$ cm

Thus, $\frac{1}{s} + \frac{1}{s'} = \frac{1}{f}$

$$\frac{1}{-5.33 \text{ cm}} + \frac{1}{2.8 \text{ cm}} = \frac{1}{f}$$

gives $f = 5.90$ cm

and $P = \frac{1}{0.059 \text{ m}} = 17.0$ diopters

27.14 (a) If the far point is at 20 cm, we need an image distance of $s' = -20$ cm when $s =$ infinity. Thus, $\frac{1}{s} + \frac{1}{s'} = \frac{1}{f}$ becomes

$$0 + \frac{1}{-20 \text{ cm}} = \frac{1}{f} \quad \text{and} \quad f = -20 \text{ cm} = -0.2 \text{ m}$$

Thus, the power is $P = \frac{1}{f} = \frac{1}{-0.2} = -5$ diopters

(b) To be seen by the eye, the virtual image cannot be any closer than 13 cm to the lens. Thus, let us find the smallest value the object distance can have.

$$\frac{1}{s} + \frac{1}{s'} = \frac{1}{f}$$

$$\frac{1}{s} + \frac{1}{-0.13 \text{ m}} = \frac{1}{-0.2 \text{ m}} \quad \text{which gives } s = 0.371 \text{ m} = 37.1 \text{ cm}$$

Thus, the near point when glasses are worn is 37.1 cm.

27.15 $f = \frac{1}{P} = \frac{1 \text{ m}}{-3} = -0.333$ m $= -33.3$ cm

The location of the image formed of a very distant object is

$$\frac{1}{s} + \frac{1}{s'} = \frac{1}{f}$$

CHAPTER TWENTY-SEVEN SOLUTIONS

$$0 + \frac{1}{s'} = \frac{1}{-33.3 \text{ cm}} \quad \text{and} \quad s' = -33.3 \text{ cm}$$

The far point of the eye is at 33.3 cm

27.16 (a) To correct nearsightedness, the image of distant objects ($s =$ infinity) should be virtual and located at the far point ($s' = -1.5$ m).

$$\frac{1}{s} + \frac{1}{s'} = \frac{1}{f}$$

$$0 + \frac{1}{-1.5 \text{ m}} = \frac{1}{f} \qquad f = -1.5 \text{ m}$$

Thus, $P = \frac{1}{f} = \frac{1}{-1.5} = -0.67$ diopters

(b) To correct farsightedness, objects at $s = 25$ cm should form a virtual image at the near point ($s' = -0.3$ m)

$$\frac{1}{s} + \frac{1}{s'} = \frac{1}{f}$$

$$\frac{1}{0.25 \text{ m}} + \frac{1}{-0.3 \text{ m}} = \frac{1}{f} \qquad \text{from which} \quad f = 1.5 \text{ m}$$

and $P = \frac{1}{f} = \frac{1}{1.5} = 0.67$ diopters

27.17 The upper portion of the lens is to allow the wearer to see very distant objects. Thus, it must form virtual images of distant objects at the far point of the eye, at $s' = -1.0$ m

$$\frac{1}{s} + \frac{1}{s'} = \frac{1}{f}$$

$$0 + \frac{1}{-1.0 \text{ m}} = \frac{1}{f} \qquad f = -1.0 \text{ m}$$

Thus, $P = \frac{1}{f} = \frac{1}{-1.0} = -1.0$ diopters

The lower portion of the lens is to allow the wearer to view objects at the normal near point (25 cm) clearly. Thus, it must form a virtual image of 25 cm distant objects at the eyes near point.

$$\frac{1}{s} + \frac{1}{s'} = \frac{1}{f}$$

$$\frac{1}{0.25 \text{ m}} + \frac{1}{-0.667 \text{ m}} = \frac{1}{f} \qquad f = 0.40 \text{ m}$$

Thus, $P = \frac{1}{f} = \frac{1}{0.40 \text{ m}} = 2.5$ diopters

---

27.18 The lens is forming a virtual image at the unaided near point of the eye ($s' = -84$ cm) when an object is located at the corrected near point ($s = 50.5$ cm). Thus,

$$\frac{1}{s} + \frac{1}{s'} = \frac{1}{f}$$

$$\frac{1}{50.5 \text{ cm}} + \frac{1}{-84 \text{ cm}} = \frac{1}{f} \qquad f = 127 \text{ cm} = 1.27 \text{ m}$$

27.19 (a) First, locate the image being formed by the lens and the magnification produced by the lens.

$$\frac{1}{s} + \frac{1}{s'} = \frac{1}{f}$$

CHAPTER TWENTY-SEVEN SOLUTIONS

$$\frac{1}{71 \text{ cm}} + \frac{1}{s'} = \frac{1}{39 \text{ cm}} \qquad s' = 86.5 \text{ cm}$$

and $M = -\frac{s'}{s} = -\frac{86.5}{71} = -1.22$, so $h' = -1.22\,h$

(b) We have $\frac{\theta'}{\theta} = \frac{h'/d'}{h/d} = \frac{h'}{h}\frac{d}{d'} = (1.22)\frac{197 \text{ cm}}{39.5 \text{ cm}} = 6.08$

27.20 The maximum magnification is obtained when the virtual image is at the near point of the eye (25 cm in front of the eye).

$$\frac{1}{s} + \frac{1}{-25} = \frac{1}{5 \text{ cm}}$$

$s = 4.17$ cm

and $M = 1 + \frac{25 \text{ cm}}{f} = 1 + \frac{25 \text{ cm}}{5 \text{ cm}} = 6$

27.21 We are given that $s = 3.5$ cm and $s' = -25$ cm.

Thus, $\frac{1}{s} + \frac{1}{s'} = \frac{1}{f}$ becomes $\frac{1}{3.5 \text{ cm}} + \frac{1}{-25 \text{ cm}} = \frac{1}{f}$

which yields $f = 4.07$ cm

27.22 (a) $\frac{1}{s} + \frac{1}{s'} = \frac{1}{f}$ becomes

$$\frac{1}{s} + \frac{1}{-25 \text{ cm}} = \frac{1}{7.5 \text{ cm}} \qquad \text{From which,} \qquad s = 5.77 \text{ cm}$$

(b) $M = -\frac{s'}{s} = -\frac{-25 \text{ cm}}{5.77 \text{ cm}} = 4.33$

27.23 (a) $M = 1 + \frac{25 \text{ cm}}{f} = 1 + \frac{25 \text{ cm}}{25 \text{ cm}} = 2$

(b) When the eye is relaxed $M = \frac{25 \text{ cm}}{f} = \frac{25 \text{ cm}}{25 \text{ cm}} = 1$

27.24 The image of a very distant object is formed at $s' = 5$ cm. Thus, the focal point is

$$\frac{1}{s} + \frac{1}{s'} = \frac{1}{f}$$

$$0 + \frac{1}{5 \text{ cm}} = \frac{1}{f} \quad \text{and} \quad f = 5 \text{ cm}$$

(a) The maximum magnification occurs when the image is at the near point of the eye. Thus, we see that $s' = -15$ cm. We find,

$$\frac{1}{s} + \frac{1}{s'} = \frac{1}{f}$$

$$\frac{1}{s} + \frac{1}{-15 \text{ cm}} = \frac{1}{5 \text{ cm}} \quad \text{yields} \quad s = \frac{15}{4} \text{ cm}$$

The magnification is $M = -\frac{s'}{s} = -\frac{-15 \text{ cm}}{\frac{15}{4} \text{ cm}} = 4$

(b) We have $\theta_0 = \frac{h}{15 \text{ cm}}$ and $\theta = \frac{h}{f}$

CHAPTER TWENTY-SEVEN SOLUTIONS

Therefore, $M = \dfrac{\theta}{\theta_0} = \dfrac{15 \text{ cm}}{f} = \dfrac{15 \text{ cm}}{5 \text{ cm}} = 3$

27.25 The overall magnification is $M = M_1 m_e = -140$ (The negative sign arises because the image is inverted.) Also, we are given that the magnification of the objective is

$M_1 = -12$ (The image is inverted, hence the negative sign.)

Thus, $m_e = \dfrac{-140}{-12} = 11.7$

We now find the object distance for the eyepiece.

$$m_e = -\dfrac{s_e'}{s_e} = -\dfrac{-25 \text{ cm}}{s_e} = 11.7$$

Therefore, $s_e = 2.14$ cm
and the focal length of the eyepiece is

$$\dfrac{1}{s_e} + \dfrac{1}{s_e'} = \dfrac{1}{f_e}$$

$$\dfrac{1}{2.14 \text{ cm}} + \dfrac{1}{-25 \text{ cm}} = \dfrac{1}{f_e} \quad \text{which yields} \quad f_e = 2.34 \text{ cm}$$

(Note: the equation $m_e = \dfrac{25 \text{ cm}}{f_e}$ assumes the final image is at infinity.)

27.26 If the eye is relaxed, the final image is at infinity. Thus the object position for the eyepiece is $s_e = f_e$ and the magnification of the eyepiece is

$$m_e = \dfrac{25 \text{ cm}}{f_e}$$

The image position for the image formed by the objective lens is, with L = the length of the microscope,

$s_0' = L - f_e = 15 \text{ cm} - 2.5 \text{ cm} = 12.5 \text{ cm}$

(Note that the approximation that the the image distance for the objective is nearly equal to the length of the microscope is a poor one in this case. Hence, we should not use $M_1 = \dfrac{L}{f_0}$)

We now find the object position for the objective.

$$\dfrac{1}{s_0} + \dfrac{1}{s_0'} = \dfrac{1}{f_0}$$

$$\dfrac{1}{s_0} + \dfrac{1}{12.5 \text{ cm}} = \dfrac{1}{1 \text{ cm}} \quad \text{which yields} \quad s_0 = 1.09 \text{ cm}$$

Thus, $M_1 = -\dfrac{s_0'}{s_0} = -\dfrac{12.5 \text{ cm}}{1.09 \text{ cm}} = -11.5$

and the magnification of the eyepiece is

$$m_e = \dfrac{25 \text{ cm}}{f_e} = \dfrac{25 \text{ cm}}{2.5 \text{ cm}} = 10$$

Thus, $M = M_1 m_e = (-11.5)(10) = -115$

27.27 $M = -\dfrac{L}{f_0}\left(\dfrac{25 \text{ cm}}{f_e}\right) = -\dfrac{15 \text{ cm}}{0.5 \text{ cm}}\left(\dfrac{25 \text{ cm}}{f_e}\right) = -800$

and $f_e = 0.938$ cm

27.28 (a) $M_1 = \dfrac{L}{f_0} = 50$ Thus, $f_0 = \dfrac{L}{50} = \dfrac{20 \text{ cm}}{50} = 0.4$ cm

CHAPTER TWENTY-SEVEN SOLUTIONS

(b) $m_e = \dfrac{25 \text{ cm}}{f_e} = 20$ and, $f_e = \dfrac{25 \text{ cm}}{20} = 1.25$ cm

(c) $M = M_1 m_e = -(50)(20) = -1000$

27.29 First, find the size of the final image

angular size $= \dfrac{h_e}{|s'_e|} = \dfrac{h_e}{29 \text{ cm}}$

$= 1.43 \times 10^{-3}$ rad

so $h_e = 4.147 \times 10^{-2}$ cm.

Now apply thin lens equation to each lens to find magnification produced by each lens and hence the overall magnification.

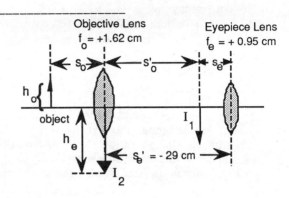

$\dfrac{1}{s_e} + \dfrac{1}{s'_e} = \dfrac{1}{f_e}$

$\dfrac{1}{s_e} + \dfrac{1}{-29 \text{ cm}} = \dfrac{1}{0.95 \text{ cm}}$

which yields $s_e = 0.920$ cm.

Thus, $M_e = -\dfrac{s'_e}{s_e} = -\dfrac{29 \text{ cm}}{0.92 \text{ cm}} = 31.5$

For the objective lens, we have $\dfrac{1}{s_0} + \dfrac{1}{s'_0} = \dfrac{1}{f_0}$, or

$\dfrac{1}{s_0} + \dfrac{1}{-28.08 \text{ cm}} = \dfrac{1}{1.622 \text{ cm}}$ which yields $s_0 = 1.721$ cm.

Thus, the magnitude of the magnification is $M_0 = \dfrac{s'_0}{s_0} = \dfrac{28.08 \text{ cm}}{1.721 \text{ cm}} = 16.3$

$M = M_e M_0 = (31.5)(16.3) = 514$

so $h_0 = \dfrac{h_e}{M} = \dfrac{4.147 \times 10^{-2} \text{ cm}}{514} = 8.07 \times 10^{-2}$ cm $= 0.81$ μm

---

27.30 (a) The diameter of the lens is 5 in = 127 mm

f-number $= \dfrac{f_0}{D} = \dfrac{1250 \text{ mm}}{127 \text{ mm}} = 9.84$

(b) $M = \dfrac{f_0}{f_e} = \dfrac{1250 \text{ mm}}{25 \text{ mm}} = 50$

27.31 $f = \dfrac{1}{P}$, thus $f_0 = \dfrac{1}{2} = 0.50$ m $= 50$ cm

and $f_e = \dfrac{1}{30} = 3.33 \times 10^{-2}$ m $= 3.33$ cm

Therefore, $M = \dfrac{f_0}{f_e} = \dfrac{50 \text{ cm}}{3.33 \text{ cm}} = 15$

27.32 $M = \dfrac{f_0}{f_e} = \dfrac{75 \text{ cm}}{4 \text{ cm}} = 18.8$

27.33 (a) We first find the position of the image formed by the objective lens.

$\dfrac{1}{s_0} + \dfrac{1}{s'_0} = \dfrac{1}{f_0}$

CHAPTER TWENTY-SEVEN SOLUTIONS

$$\frac{1}{150 \text{ cm}} + \frac{1}{s_0'} = \frac{1}{100 \text{ cm}} \quad \text{from which} \quad s_0' = 300 \text{ cm}$$

Then, the object position for the eyepiece is the length of the telescope, 110 cm, minus s_0'.

$$s_e = 110 \text{ cm} - 300 \text{ cm} = -190 \text{ cm} \quad \text{(a virtual object)}$$

The image position for the eyepiece is

$$\frac{1}{s_e} + \frac{1}{s_e'} = \frac{1}{f_e}$$

$$\frac{1}{-190 \text{ cm}} + \frac{1}{s_e'} = \frac{1}{10 \text{ cm}} \quad \text{and} \quad s_e' = 9.5 \text{ cm}$$

(b) An observer would probably not notice the bee because when looking through a telescope, the eye is relaxed (focused at infinity) and would not focus on the image of the bee. Also, one can calculate the magnification of the telescope for this case and see that it is 0.1. Thus the final image of the bee is greatly reduced in size from the actual size of the bee.

27.34 $M = -\frac{f_0}{f_e} = -\frac{175 \text{ cm}}{-9.73 \text{ cm}} = 18$

The negative sign arises here since the Galilean telescope (one with a divergent lens for the eyepiece) forms an erect image rather than an inverted image. Still the basic relationship for the magnification applies.
Thus, the tree appear 18 times larger when the telescope is used.

27.35 (See sketch.) The length, x, of an object on the moon, is found as

$$\frac{x}{3.8 \times 10^8 \text{ m}} = \frac{10^{-2} \text{ m}}{15 \text{ m}}$$

and $x = 2.53 \times 10^5 \text{ m} = 157$ miles

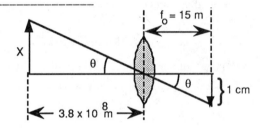

27.36 Consider first the lens used for the left eye. The near point is at 50 cm ($s' = -50$ cm when $s = 25$ cm). Thus,

$$\frac{1}{s} + \frac{1}{s'} = \frac{1}{f} \quad \text{becomes} \quad \frac{1}{25 \text{ cm}} + \frac{1}{-50 \text{ cm}} = \frac{1}{f} \quad \text{and} \quad f = 50 \text{ cm}$$

For the right eye lens, we have a near point of 100 cm ($s' = -100$ cm when $s = 25$ cm).

$$\frac{1}{s} + \frac{1}{s'} = \frac{1}{f} \quad \text{becomes} \quad \frac{1}{25 \text{ cm}} + \frac{1}{-100 \text{ cm}} = \frac{1}{f} \quad \text{and} \quad f = 33.3 \text{ cm}$$

(a) $M = \frac{f_0}{f_e} = \frac{50 \text{ cm}}{33.3 \text{ cm}} = 1.5$

(b) We shall use the 50 cm lens as the objective, and we shall require that a virtual final image be formed at $s_e' = -25$ cm for maximum magnification. The object distance for the eyepiece is

$$\frac{1}{s_e} + \frac{1}{s_e'} = \frac{1}{f_e} \quad \text{or,} \quad \frac{1}{s_e} + \frac{1}{-25 \text{ cm}} = \frac{1}{33.3 \text{ cm}} \quad \text{yielding} \quad s_e = 14.3 \text{ cm}$$

and $m_e = -\frac{s_e'}{s_e} = -\frac{-25 \text{ cm}}{14.3 \text{ cm}} = 1.75$

The image position for the objective is $s_0' = L - s_e = 10 \text{ cm} - 14.3 \text{ cm} = -4.3 \text{ cm}$ and the object position is found as

CHAPTER TWENTY-SEVEN SOLUTIONS

$$\frac{1}{s_0} + \frac{1}{s_0'} = \frac{1}{f_0} \quad \text{or,} \quad \frac{1}{s_0} + \frac{1}{-4.3 \text{ cm}} = \frac{1}{50 \text{ cm}}$$

yielding $s_0 = 3.95$ cm

and $m_0 = -\frac{s_0'}{s_0} = -\frac{-4.3 \text{ cm}}{3.95 \text{ cm}} = 1.09$

The overall magnification is $M = m_0 m_e = 1.90$

These steps can be repeated with the 33.3 cm lens as the objective and the 50 cm as the eyepiece. For this case, the overall magnification will be 1.80. Thus, the arrangement used above gives the greater magnification.

27.37 The angular resolution needed is

$$\theta_m = \frac{s}{r} = \frac{300 \text{ m}}{4.02 \times 10^8 \text{ m}} = 7.46 \times 10^{-7} \text{ radians}$$

For a circular aperture $\theta_m = 1.22 \frac{\lambda}{D}$

or $D = 1.22 \frac{\lambda}{\theta_m} = 1.22 \frac{5 \times 10^{-7} \text{ m}}{7.46 \times 10^{-7} \text{ rad}} = 0.818 \text{ m} = 32.2 \text{ in}$

27.38 (a) For telescope A

$$\theta_m = 1.22 \frac{\lambda}{D} = 1.22 \frac{6 \times 10^{-5} \text{ cm}}{12.7 \text{ cm}} = 5.76 \times 10^{-6} \text{ rad}$$

For B $\theta_m = 1.22 \frac{\lambda}{D} = 1.22 \frac{6 \times 10^{-5} \text{ cm}}{7.62 \text{ cm}} = 9.61 \times 10^{-6} \text{ rad}$

(b) For A $M = \frac{f_0}{f_e} = \frac{120 \text{ cm}}{1.8 \text{ cm}} = 66.7$

For B $M = \frac{f_0}{f_e} = \frac{100 \text{ cm}}{2.5 \text{ cm}} = 40$

27.39 (a) $\theta_m = 1.22 \frac{\lambda}{D} = 1.22 \frac{5.5 \times 10^{-5} \text{ cm}}{6.5 \text{ cm}} = 1.03 \times 10^{-5} \text{ rad}$

Thus, $s = r\theta = (10^3 \text{ m})(1.03 \times 10^{-5} \text{ rad}) = 1.03 \times 10^{-2} \text{ m} = 1.03 \text{ cm}$

(b) $\theta_m = 1.22 \frac{\lambda}{D} = 1.22 \frac{5.5 \times 10^{-5} \text{ cm}}{0.25 \text{ cm}} = 2.68 \times 10^{-4} \text{ rad}$

Thus, $s = r\theta = (10^3 \text{ m})(2.68 \times 10^{-4} \text{ rad}) = 2.68 \times 10^{-1} \text{ m} = 26.8 \text{ cm}$

27.40 If just resolved, $\theta = \theta_m = 1.22 \frac{\lambda}{D} = 1.22 \frac{500 \times 10^{-9} \text{ m}}{0.30 \text{ m}} = 2.03 \times 10^{-6} \text{ rad}$

Thus, $h = \frac{d}{2.03 \times 10^{-6} \text{ rad}} = 4.92 \times 10^5 \text{ m} = 492 \text{ km}$

27.41 If just resolved $\theta = \theta_m = 1.22 \frac{\lambda}{D} = 1.22 \frac{5500 \times 10^{-10} \text{ m}}{0.35 \text{ m}} = 1.92 \times 10^{-6} \text{ rad}$

and $d = \theta h = (1.92 \times 10^{-6} \text{ rad})(2 \times 10^5 \text{ m}) = 0.383 \text{ m} = 38.3 \text{ cm}$

27.42 $\theta_m = 1.22 \frac{\lambda}{D} = 1.22 \frac{5 \times 10^{-7} \text{ m}}{0.508 \text{ m}} = 1.2 \times 10^{-6} \text{ rad}$

Thus, $s = r\theta = (4.70 \times 10^{13} \text{ miles})(1.2 \times 10^{-6} \text{ rad}) = 5.65 \times 10^7 \text{ miles} = 9.09 \times 10^7 \text{ km}$

CHAPTER TWENTY-SEVEN SOLUTIONS

27.43 $\lambda_{medium} = \dfrac{\lambda_{vac}}{n_{medium}} = \dfrac{580 \text{ nm}}{1.33} = 435 \text{ nm}$ (This is the wavelength of the light in water.)

and $\theta_m = 1.22 \dfrac{\lambda}{D} = 1.22 \dfrac{4.35 \times 10^{-7} \text{ m}}{5 \times 10^{-2} \text{ m}} = 1.06 \times 10^{-5} \text{ rad}$

27.44 $r = \dfrac{s}{\theta_m} = \dfrac{s}{1.22 \dfrac{\lambda}{D}} = \dfrac{(4.02 \times 10^8 \text{ m})(1.61 \times 10^6 \text{ m})}{1.22(1 \text{ m})}$

$= 5.31 \times 10^{14} \text{ m} = 5.31 \times 10^{11} \text{ km}$

27.45 (a) $\lambda_{medium} = \dfrac{\lambda_{vac}}{n_{medium}} = \dfrac{500 \text{ nm}}{1.33} = 375 \text{ nm}$

$\theta_m = 1.22 \dfrac{\lambda}{D} = 1.22 \dfrac{3.75 \times 10^{-7} \text{ m}}{2 \times 10^{-3} \text{ m}} = 2.29 \times 10^{-4} \text{ rad}$

(b) $r = \dfrac{s}{\theta_m} = \dfrac{10^{-2} \text{ m}}{2.29 \times 10^{-4} \text{ rad}} = 43.7 \text{ m}$

27.46 $\theta = \dfrac{s}{r} = \dfrac{2 \text{ m}}{10^4 \text{ m}} = 2 \times 10^{-4} \text{ rad}$

But $\theta_m = 1.22 \dfrac{\lambda}{D}$ so $D = \dfrac{1.22 \lambda}{\theta_m} = \dfrac{1.22(8.85 \times 10^{-7} \text{ m})}{2 \times 10^{-4} \text{ rad}} = 5.40 \times 10^{-3} \text{ m} = 5.40 \text{ mm}$

27.47 (a) pixel size $= \dfrac{\text{width of screen}}{\text{number of columns}} = \dfrac{\text{ht of screen}}{\text{number of rows}}$

$= \dfrac{4 \text{ cm}}{516} = 7.75 \times 10^{-3} \text{ cm}$

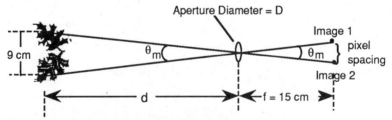

(b) To have 2 images just resolved, the space between the centers of the images must be the same as the space between pixel centers (i.e. 1 pixel size). Therefore, the limiting angle of resolution is

$\theta_{min} = \dfrac{7.75 \times 10^{-3} \text{ cm}}{15 \text{ cm}} = 5.17 \times 10^{-4} \text{ rad}$

But for circular aperture

$\theta_{min} = 1.22 \dfrac{\lambda}{D}$ so $D = \dfrac{1.22 \lambda}{\theta_{min}} = \dfrac{1.22(500 \times 10^{-9} \text{ m})}{5.17 \times 10^{-4} \text{ rad}} = 1.18 \times 10^{-3} \text{ m} = 1.18 \text{ mm}$

(c) $\theta_{min} = \dfrac{9 \text{ cm}}{d}$, so $d = \dfrac{9 \text{ cm}}{5.17 \times 10^{-4} \text{ rad}} = 1.74 \times 10^4 \text{ cm} = 174 \text{ m}$

27.48 $\theta_1 = \theta_m = 1.22 \dfrac{\lambda}{D} = 1.22 \dfrac{500 \times 10^{-9} \text{ m}}{0.10 \text{ m}} = 6.10 \times 10^{-6} \text{ rad}$

$d = \theta_1 f$

CHAPTER TWENTY-SEVEN SOLUTIONS

Also, $\theta_2 = \dfrac{d}{30 \text{ cm}}$, so $\theta_2 = \dfrac{\theta_1 f}{30 \text{ cm}} = (6.10 \times 10^{-6} \text{ rad})\dfrac{140}{30} = 2.85 \times 10^{-5}$ rad = 28.5 μrad

---

27.49 A fringe passes each time the mirror is moved a distance of half a wavelength. Thus, when 481 fringes pass, the mirror is moved a distance of

$$d = 481 \dfrac{\lambda}{2} = 1.2 \times 10^{-4} \text{ m} \quad \text{yielding} \quad \lambda = 499 \text{ nm}$$

27.50 $d = $ (number of fringes) $\dfrac{\lambda}{2} = 155 \dfrac{6.5 \times 10^{-7} \text{ m}}{2} = 50.4$ μm

---

27.51 $d = $ (number of fringes) $\dfrac{\lambda}{2}$

Thus, number of fringes $= \dfrac{d}{\dfrac{\lambda}{2}} = \dfrac{2d}{\lambda} = \dfrac{2(1.8 \times 10^{-4} \text{ m})}{5.5 \times 10^{-7} \text{ m}} = 654$

27.52 The path length of the light through the tube when it is evacuated is $L_1 = L = 5$ cm. When the tube is filled with gas, the effective path length through the tube is changed to
$$L_1' = n_g L = n_g (5 \text{ cm})$$
The increase in the effective path length as the gas is emitted is
$$\delta = L_1' - L_1 = n_g(5 \text{ cm}) - 5 \text{ cm} = (n_g - 1)(5 \text{ cm})$$
If 60 fringe shifts occur, the total increase in the effective path length is $60 \dfrac{\lambda}{2} = 30\lambda$.

Therefore, $(n_g - 1)(5 \text{ cm}) = 30\lambda$

or $n_g = 1 + \dfrac{30\lambda}{5 \text{ cm}}$

27.53 The change in the effective path length is
$$L' - L = nt - t = (n - 1)t$$
Since a fringe shift occurs every time the path length increases by $\dfrac{\lambda}{2}$, the total number of fringes which pass is

$$N = \dfrac{L' - L}{\dfrac{\lambda}{2}} = \dfrac{2(n - 1)t}{\lambda} = \dfrac{2(0.40)(15 \times 10^{-6} \text{ m})}{6 \times 10^{-7} \text{ m}} = 20$$

27.54 $d = \dfrac{1 \text{ cm}}{6000 \text{ lines}} = 1.667 \times 10^{-4}$ cm

and $n\lambda = d \sin\theta$ yields $\lambda = \dfrac{d \sin\theta}{n} = \dfrac{(1.667 \times 10^{-4} \text{ cm}) \sin 29°}{2} = 404$ nm

---

27.55 $d = \dfrac{n\lambda}{\sin\theta} = \dfrac{(1)(546.1 \text{ nm})}{\sin 21°} = 1.524 \times 10^3$ nm

and the number of lines per centimeter is
$\dfrac{1}{d} = \dfrac{1}{1.524 \times 10^{-4} \text{ cm}} = 6562$ lines/cm

CHAPTER TWENTY-SEVEN SOLUTIONS

27.56 (a) $d = \dfrac{1}{1500 \text{ lines/cm}} = 6.667 \times 10^{-4}$ cm $= 6.667 \times 10^{-6}$ m

and $\quad n = \dfrac{d\sin\theta}{\lambda} = \dfrac{(6.667 \times 10^{-6} \text{ m})\sin 90°}{5 \times 10^{-7} \text{ m}} = 13.3$

Thus, 13 complete orders will be observed.

(b) For 15,000 lines/cm $\quad d = 6.667 \times 10^{-7}$ m

and $\quad n = \dfrac{d\sin\theta}{\lambda} = \dfrac{(6.667 \times 10^{-7} \text{ m})\sin 90°}{5 \times 10^{-7} \text{ m}} = 1.33$

Only one order can be seen.

27.57 (a) $d = \dfrac{1}{3660 \text{ lines/cm}} = 2.732 \times 10^{-4}$ cm $= 2.732 \times 10^{-6}$ m $= 2732$ nm

$\lambda = \dfrac{d \sin\theta}{n}$ at $\theta = 10.09°$, $\lambda = 478.7$ nm

at $\theta = 13.71°$, $\lambda = 647.6$ nm, at $\theta = 14.77°$, $\lambda = 696.6$ nm

(b) $d = \dfrac{\lambda}{\sin\theta_1}$

and $\quad (2)\lambda = d\sin\theta_2$

so $\quad \sin\theta_2 = \dfrac{2\lambda}{d} = \dfrac{2\lambda}{\left(\dfrac{\lambda}{\sin\theta_1}\right)} = 2\sin\theta_1$

Therefore, if $\theta_1 = 10.09°$ then $\sin\theta_2 = 2\sin(10.09°)$ gives $\theta_2 = 20.51°$

Similarly, for $\theta_1 = 13.71°$, $\theta_2 = 28.30°$

and for $\theta_1 = 14.77°$, $\theta_2 = 30.66°$

27.58 $n\lambda = d\sin\theta$ becomes

$4\lambda_1 = d\sin\theta_4$

and $\quad 5\lambda_2 = d\sin\theta_5$

but $\quad \theta_5 = \theta_4$ (the fifth order line of the 500 nm wavelength coincides with the fourth order line of λ_1.

Therefore, $4\lambda_1 = 5\lambda_2$ and $\quad \lambda_1 = \dfrac{5}{4}\lambda_2 = \dfrac{5}{4}\,500$ nm $= 625$ nm

---

27.59 (a) $d = \dfrac{1}{410 \text{ lines/mm}} = 2.44 \times 10^{-6}$ m

For the first order ($n = 1$) we have $\sin\theta = \dfrac{\lambda}{d}$.

Which, for λ_1 becomes $\sin\theta_1 = \dfrac{4.10 \times 10^{-7} \text{ m}}{2.44 \times 10^{-6} \text{ m}} = 0.168$ and $\theta_1 = 9.67°$

For λ_4 we have $\quad \theta_4 = 15.6°$

and $\Delta\theta = \theta_4 - \theta_1 = 5.93°$

(b) For the third order for λ_1 we have $\sin\theta_1 = 3\dfrac{\lambda}{d} = \dfrac{3\,(4.10 \times 10^{-7} \text{ m})}{2.44 \times 10^{-6} \text{ m}}$

and $\quad \theta_1 = 30.3°$

Likewise, for λ_3 we find $\quad \theta_3 = 36.7°$

$\Delta\theta = \theta_3 - \theta_1 = 6.42°$

27.60 To obtain an upper limit on the grating spacing, assume that the second order maxima of the shortest wavelength comes exactly at 90°. We have

CHAPTER TWENTY-SEVEN SOLUTIONS

$$d = \frac{n\lambda}{\sin\theta} = \frac{(2)(4 \times 10^{-7} \text{ m})}{\sin 90°} = 800 \text{ nm}$$

Thus, d must be less than 800 nm.
Also, since all of the first order is seen, the lower limit on d is

$$d = \frac{n\lambda}{\sin\theta} = \frac{(1)(7 \times 10^{-7} \text{ m})}{\sin 90°} = 700 \text{ nm}$$

27.61 First, find the longer wavelength,

$$y = \frac{n\lambda}{d} L \text{ and } y_2 - y_1 = \frac{\lambda}{d} L$$

or $\quad 0.844 \text{ cm} = \lambda \dfrac{15 \text{ cm}}{8.333 \times 10^{-4} \text{ cm}}$

where $d = \dfrac{1}{1200 \text{ slits/cm}} = 8.333 \times 10^{-4}$ cm = 8333 nm

from which $\lambda = 4.689 \times 10^{-5}$ cm = 468.9 nm
Now find the other wavelength
The third order of λ_{short} (at θ_3) coincides with the first order of λ_{long}
Thus, $3\lambda_{short} = d\sin\theta_3 = (1) \lambda_{long}$

$$\lambda_{short} = \frac{1}{3} \lambda_{long} = \frac{469 \text{ nm}}{3} = 156 \text{ nm}$$

27.62 $\quad d = \dfrac{1}{5000 \text{ slits/cm}} = 2 \times 10^{-4}$ cm = 2000 nm

In the second order, $\quad 2\lambda = d\sin\theta$

so $\quad \sin\theta_1 = \dfrac{2\lambda_1}{d} = \dfrac{2(480 \text{ nm})}{2000 \text{ nm}} = 0.48 \quad \theta_1 = 28.67°$

and $\quad \sin\theta_2 = \dfrac{2(610 \text{ nm})}{2000 \text{ nm}} = 0.61 \quad \theta_1 = 37.59°$

$y_1 = L\tan\theta_1 = (2 \text{ m}) \tan 28.67° = 1.094$ m
$y_2 = L\tan\theta_2 = (2 \text{ m}) \tan 37.59° = 1.540$ m

and $\quad \Delta s = y_2 - y_1 = 0.445$ m = 44.5 cm

27.63 $\quad N = \dfrac{\text{width}}{d} = \dfrac{3 \times 10^{-2} \text{ m}}{775 \times 10^{-9} \text{ m}} = 3.87 \times 10^4$ slits

Therefore, $\quad R = \dfrac{\lambda}{\Delta\lambda} = mN \quad$ or $\quad \Delta\lambda = \dfrac{\lambda}{mN} = \dfrac{600 \text{ nm}}{(1)(3.87 \times 10^4)} = 0.02$ nm

---

27.64 (a) The resolving power is given by $\quad R = \dfrac{\lambda}{\Delta\lambda} = mN$

Thus, $N = \dfrac{\lambda}{m\Delta\lambda}$

For the first order $N = \dfrac{656.2 \text{ nm}}{(1)(0.18 \text{ nm})} = 3646$ slits

(b) For the second order, $N = \dfrac{656.2 \text{ nm}}{(2)(0.18 \text{ nm})} = 1823$ slits

---

27.65 (a) $L = f_0 + f_e = 101.5$ cm

(b) $M = \dfrac{f_0}{f_e} = \dfrac{100 \text{ cm}}{1.5 \text{ cm}} = 66.7$

CHAPTER TWENTY-SEVEN SOLUTIONS

27.66 (a) Corrective lenses must form a virtual image at s' = -75 cm for objects which are 25 cm in front of the lens (s = 25 cm).

$$\frac{1}{s} + \frac{1}{s'} = \frac{1}{f}$$

$$\frac{1}{25 \text{ cm}} + \frac{1}{-75 \text{ cm}} = \frac{1}{f} \quad \text{and} \quad f = 37.5 \text{ cm} = 0.375 \text{ m}$$

Thus, $P = \frac{1}{f} = \frac{1}{0.375} = 2.67$ diopters

(b) If s' = -75 cm when s = 26 cm rather than 25 cm as assumed in part(a), we have

$$\frac{1}{s} + \frac{1}{s'} = \frac{1}{f}$$

$$\frac{1}{26 \text{ cm}} + \frac{1}{-75 \text{ cm}} = \frac{1}{f'} \quad \text{and} \quad f' = 39.8 \text{ cm} = 0.398 \text{ m}$$

and $P = \frac{1}{f'} = \frac{1}{0.398} = 2.51$ diopters

The miss is 0.16 diopters

27.67 $d = \frac{n\lambda}{\sin\theta} = \frac{(3)(546.1 \text{ nm})}{\sin 81°} = 1.66 \times 10^{-3}$ mm

Thus, lines/mm $= \frac{1}{d} = \frac{1}{1.66 \times 10^{-3} \text{ mm}} = 603$

27.68 The corrective lens must form a virtual image of very distant objects at the far point of the eye (at s' = -75 cm).

Thus, $\frac{1}{s} + \frac{1}{s'} = \frac{1}{f}$

$$0 + \frac{1}{-75 \text{ cm}} = \frac{1}{f} \quad \text{or} \quad f = -75 \text{ cm} = -0.75 \text{ m}$$

and $P = \frac{1}{f} = \frac{1}{-0.75 \text{ m}} = -1.33$ diopters

---

27.69 We first find the focal length of the lens as

$$\frac{1}{s} + \frac{1}{s'} = \frac{1}{f}$$

$$\frac{1}{40 \text{ cm}} + \frac{1}{20 \text{ cm}} = \frac{1}{f} \quad \text{hence,} \quad f = 13.3 \text{ cm}$$

and $M = 1 + \frac{25 \text{ cm}}{f} = 1 + \frac{25 \text{ cm}}{13.3 \text{ cm}} = 2.88$

27.70 (a) The resolving power is given by $R = \frac{\lambda}{\Delta\lambda} = mN$

Thus, $N = \frac{\lambda}{m\Delta\lambda} = \frac{589 \text{ nm}}{(1)(0.6 \text{ nm})} = 982$

(b) $N = \frac{\lambda}{m\Delta\lambda} = \frac{589 \text{ nm}}{(3)(0.6 \text{ nm})} = 327$

27.71 We have,

$n_1\lambda_1 = d\sin\theta = n_2\lambda_2$ because $d\sin\theta$ is the same for both components.

Thus, $\lambda_2 = \frac{n_1}{n_2}\lambda_1 = \frac{3}{2}(440 \text{ nm}) = 660$ nm

CHAPTER TWENTY-SEVEN SOLUTIONS

27.72 $d = \dfrac{n\lambda}{\sin\theta} = \dfrac{(2)(5.77 \times 10^{-5}\text{ cm})}{\sin 41.25°} = 1.75 \times 10^{-4}\text{ cm}$

Thus, the number of lines in a 4 cm width is

$N = \dfrac{4\text{ cm}}{d} = \dfrac{4\text{ cm}}{1.75 \times 10^{-4}\text{ cm}} = 22{,}850$ slits

---

27.73 (a) $\dfrac{1}{s} + \dfrac{1}{s'} = \dfrac{1}{f}$ gives $\dfrac{1}{100\text{ cm}} + \dfrac{1}{2\text{ cm}} = \dfrac{1}{f}$ and $f = 1.96$ cm

(b) and (c) $f_{max} = \dfrac{f}{D_{min}} = \dfrac{1.96\text{ cm}}{0.2\text{ cm}} = 9.8$

and $f_{min} = \dfrac{f}{D_{max}} = \dfrac{1.96\text{ cm}}{0.6\text{ cm}} = 3.27$

27.74 We first find the grating spacing as

$d = \dfrac{n\lambda}{\sin\theta} = \dfrac{(2)(501.5\text{ nm})}{\sin 30°} = 2006$ nm

Then for the 667.8 nm line, we have

$\sin\theta = \dfrac{n\lambda}{d} = \dfrac{(1)(667.8\text{ nm})}{2006\text{ nm}} = 0.333$ From which $\theta = 19.4°$

27.75 The increase in the effective path length is $nt - n = 6\dfrac{\lambda}{2}$

Thus, $(n-1)t = 3\lambda$

and $n = 1 + \dfrac{3\lambda}{t} = 1 + \dfrac{3(5.8 \times 10^{-7}\text{ m})}{2.5 \times 10^{-6}\text{ m}} = 1.696$

27.76 The focal length of the 2 diopter lens is $f = \dfrac{1}{P} = \dfrac{1}{2} = 0.5$ m $= 50$ cm

Since he holds the object at $s = 35$ cm, we have

$\dfrac{1}{s} + \dfrac{1}{s'} = \dfrac{1}{f}$

$\dfrac{1}{35\text{ cm}} + \dfrac{1}{s'} = \dfrac{1}{50\text{ cm}}$ From which, $s' = 117$ cm

Thus, the near point of the reader must now be at 117 cm.
The new corrective lens should produce a virtual image at $s' = -117$ cm when $s = 25$ cm. Thus,

$\dfrac{1}{s} + \dfrac{1}{s'} = \dfrac{1}{f}$

becomes $\dfrac{1}{25\text{ cm}} + \dfrac{1}{-117\text{ cm}} = \dfrac{1}{f}$ yielding $f = 31.8$ cm

and $P = \dfrac{1}{f} = \dfrac{1}{0.318\text{ m}} = 3.14$ diopters

---

27.77 We use $\dfrac{n_1}{s} + \dfrac{n_2}{s'} = \dfrac{n_2 - n_1}{R}$ with $s = $ infinity.

Thus, $R = s'\dfrac{n_2 - n_1}{n_2} = 2\text{ cm}\dfrac{1.34 - 1}{1.34} = 0.507$ cm $= 5.07$ mm

CHAPTER TWENTY-EIGHT SOLUTIONS

28.1 (a) Along path I, the velocity along the ground is
$v_{ground} = 120$ m/.s
Thus, $t_I = \dfrac{200 \times 10^3 \text{ m}}{120 \text{ m/s}} = 1667$ s
Along path II, see sketch, we have
$v_{air}^2 = v_{ground}^2 + v_{wind}^2$
$(100 \text{ m/s})^2 = v_{ground}^2 + (20 \text{ m/s})^2$
gives $v_{ground} = 98$ m/.s

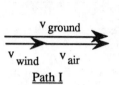

Therefore, $t_{II} = \dfrac{200 \times 10^3 \text{ m}}{98 \text{ m/s}} = 2041$ s

(b) Along path I, $v_{ground} = -100$ m/s $+ 20$ m/s $= -80$ m/s
and $t_I' = \dfrac{-200 \times 10^3 \text{ m}}{-80 \text{ m/s}} = 2500$ s
Along path II, $v_{air}^2 = v_{ground}^2 + v_{wind}^2 = 98$ m/s as before.
Thus, $t_{II}' = \dfrac{200 \times 10^3 \text{ m}}{98 \text{ m/s}} = 2041$ s

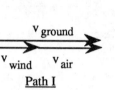

(c) $t_I)_{total} = 1667$ s $+ 2500$ s $= 4167$ s
$t_{II})_{total} = 2041$ s $+ 2041$ s $= 4082$ s
Thus, $\Delta t = 4167$ s $- 4082$ s $= 85$ s

28.2 (a) $\Delta t_{net} = \dfrac{2Lv^2}{c^3} = \dfrac{2(28 \text{ m})(3 \times 10^4 \text{ m/s})^2}{(3 \times 10^8 \text{ m/s})^3} = 1.87 \times 10^{-15}$ s

(b) The path difference is
$\Delta d_{net} = c(\Delta t)_{net} = (3 \times 10^8 \text{ m/s})(1.87 \times 10^{-15} \text{ s}) = 5.60 \times 10^{-7}$ m
This corresponds to a shift of
$N = \dfrac{\Delta d_{net}}{\lambda} = \dfrac{5.60 \times 10^{-7} \text{ m}}{550 \times 10^{-9} \text{ m}} = 1.02$ fringes

28.3 (a) The time for 70 beats $= T = 1$ min. This is the time measured by the astronaut and any observer at rest with respect to the astronaut.
(b) This observer is moving at speed $v = 0.9c$ relative to the astronaut. We have
$\gamma = \dfrac{1}{\sqrt{1 - \dfrac{v^2}{c^2}}} = \dfrac{1}{\sqrt{1 - (0.9)^2}} = \dfrac{1}{\sqrt{0.19}} = \dfrac{1}{0.436} = 2.29$

Thus, $\Delta t = \gamma \Delta t' = 2.29$ min (This is the time for 70 beats as measured by an observer on earth.) Therefore, the rate is
$\dfrac{70 \text{ beats}}{2.29 \text{ min}} = 30.5$ beats/min.

CHAPTER TWENTY-EIGHT SOLUTIONS

28.4 $T' = 2$ s, $\quad T = 7$ s. Therefore, $T = \gamma T'$ becomes $\gamma = \dfrac{T}{T'} = \dfrac{7 \text{ s}}{2 \text{ s}} = 3.5$

But, $\gamma = \dfrac{1}{\sqrt{1 - \dfrac{v^2}{c^2}}}$ becomes $1 - \dfrac{v^2}{c^2} = \dfrac{1}{\gamma^2}$, or $\dfrac{v^2}{c^2} = 1 - \dfrac{1}{\gamma^2}$

Thus, $v = c\sqrt{1 - \dfrac{1}{\gamma^2}}$ Hence, $v = c\sqrt{1 - \dfrac{1}{(3.5)^2}} = c\sqrt{1 - 0.816} = 0.958$ c

---

28.5 If the frequency of a clock is one half its rest frequency, then the period = two times the rest period. Thus, $T = 2T'$, and from $T = \gamma T'$, we see that $\gamma = 2$.

Thus, $v = c\sqrt{1 - \dfrac{1}{\gamma^2}} = c\sqrt{1 - \dfrac{1}{4}} = 0.866$ c

---

28.6 $\gamma = \dfrac{1}{\sqrt{1 - \dfrac{v^2}{c^2}}} = \dfrac{1}{\sqrt{1 - (0.8)^2}} = \dfrac{1}{0.6} = 1.667$

So, $\Delta T = \gamma \Delta T' = 1.667(3.0 \text{ s}) = 5.0$ s

---

28.7 The time required in the Earth's frame of reference = $T_1 = \dfrac{4.2 \text{ light-yr}}{0.95} = 4.42$ yrs

Thus, the time measured in the ships reference frame is $T_2 = \dfrac{T_1}{\gamma} = \dfrac{4.42 \text{ yr}}{3.20} = 1.38$ yrs

(where $\gamma = \dfrac{1}{\sqrt{1 - (0.95)^2}} = 3.20$).

The length between Earth and the star as measured by astronauts is
$L = \dfrac{L'}{\gamma} = \dfrac{4.2 \text{ ly}}{3.20} = 1.313$ ly

---

28.8 (a) $T' = 2.6 \times 10^{-8}$ s, $v = 0.98$ c, and $\gamma = \dfrac{1}{\sqrt{1 - \dfrac{v^2}{c^2}}} = \dfrac{1}{\sqrt{1 - (0.98)^2}} = 5.03$

Thus, $T = \gamma T' = 5.03(2.6 \times 10^{-8} \text{ s}) = 1.31 \times 10^{-7}$ s
(b) $d = vt = (0.98 \text{ c})(1.31 \times 10^{-7} \text{ s}) = 38.5$ m
(c) $d = vt = (0.98 \text{ c})(2.6 \times 10^{-8} \text{ s}) = 7.64$ m

---

28.9 $L = L'\sqrt{1 - \dfrac{v^2}{c^2}}$

We are given that $L = 0.5$ m $= \dfrac{1}{2} L'$. Thus

$\sqrt{1 - \dfrac{v^2}{c^2}} = \dfrac{1}{2}$ From which, $v = 0.866$ c

CHAPTER TWENTY-EIGHT SOLUTIONS

28.10 Length contraction occurs only in the dimension parallel to the motion. Thus, sides b and c are not affected. (See sketch.) However, side a will exhibit length contraction. We have

$$L_a = L_a' \sqrt{1 - \frac{v^2}{c^2}}$$

$$= 2 \text{ m} \sqrt{1 - (0.833)^2} = 1.11 \text{ m}$$

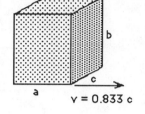

The box looks like a rectangular box with the sides perpendicular to the motion being 2 m long while the sides parallel to the motion appear 1.11 m long.

28.11 $\gamma = \dfrac{1}{\sqrt{1 - \dfrac{v^2}{c^2}}} = \dfrac{1}{\sqrt{1 - (0.9)^2}} = 2.294$

The length measured by an observer in motion relative to the ship = 100 m = L
The length measured by pilot (at rest relative to ship) = L'
$\quad L' = \gamma L = 2.294(100 \text{ m}) = 229.4 \text{ m}$

28.12 The contracted length of the sled (measured by observer at rest on ice) must be less than the proper width of the hole, or $\frac{2}{3}$ L', where L' = proper length of sled.

$\quad L \leq \dfrac{L'}{\gamma} \quad$ becomes $\quad \dfrac{2}{3} L' \leq \dfrac{L'}{\gamma}$

so, $\quad \gamma \leq \dfrac{3}{2}$

Thus, $\dfrac{1}{\sqrt{1 - \dfrac{v^2}{c^2}}} \leq 1.5$, or $\sqrt{1 - \dfrac{v^2}{c^2}} \leq \dfrac{1}{(1.5)^2} = 0.4444$

and $\dfrac{v^2}{c^2} \leq 1 - 0.444 = 0.555$

from which $v \leq 0.745c$ (This is the largest safe speed of the sled.)

28.13 We are given L' = 4 ly. Therefore, if L = 1.5 ly, then

$\quad L = L' \sqrt{1 - \dfrac{v^2}{c^2}} \quad$ becomes $\quad \dfrac{L}{L'} = \dfrac{1.5}{4} = 0.375 = \sqrt{1 - \dfrac{v^2}{c^2}}$

From which, $\dfrac{v}{c} = 0.927$, or $v = 0.927 \, c$

28.14 The lifetime as measured by an observer in motion relative to the particle is

$\quad T = \dfrac{d}{v} = \dfrac{0.15 \text{ m}}{2.2 \times 10^8 \text{ m/s}} = 6.82 \times 10^{-10} \text{ s}$

We also have

$\quad \gamma = \dfrac{1}{\sqrt{1 - \dfrac{v^2}{c^2}}} = \dfrac{1}{\sqrt{1 - \dfrac{(2.2 \times 10^8 \text{ m/s})^2}{(3 \times 10^8 \text{ m/s})^2}}} = \dfrac{1}{0.680} = 1.47$

CHAPTER TWENTY-EIGHT SOLUTIONS

Therefore, $T = \gamma T'$ becomes $T' = \dfrac{T}{\gamma} = \dfrac{6.82 \times 10^{-10} \text{ s}}{1.47} = 4.64 \times 10^{-10}$ s

28.15 (a) Since your ship is identical to his, and you are at rest with respect to your own ship, its length is 20 m
(b) His ship is in motion relative to you, so you see its length contracted to 19 m.

(c) We have, $L = L'\sqrt{1 - \dfrac{v^2}{c^2}}$ from which $\dfrac{L}{L'} = \dfrac{19 \text{ m}}{20 \text{ m}} = 0.95 = \sqrt{1 - \dfrac{v^2}{c^2}}$
and $v = 0.312 c$

28.16 $\gamma = 10.0$.
Given: $L_1 = 2.0$ m, and $\theta_1 = 30°$ (both measured in a reference frame moving relative to the rod). Thus, $L_{1x} = L_1 \cos\theta_1 = (2 \text{ m})(0.867) = 1.734$ m,
and $L_{1y} = L_1 \sin\theta_1 = (2 \text{ m})(0.5) = 1.0$ m.
L_{2x} = a "proper length" is related to L_{1x} by
$L_{1x} = \dfrac{L_{2x}}{\gamma}$.
Therefore, $L_{2x} = 10 L_{1x} = 17.34$ m.
$L_{2y} = L_{1y} = 1.00$ m [lengths perpendicular to the motion are unchanged].
$L_2 = \sqrt{(L_{2x})^2 + (L_{2y})^2}$ gives $L_2 = 17.4$ m
and $\theta_2 = \tan^{-1}\dfrac{L_{2y}}{L_{2x}}$ gives $\theta_2 = 3.3°$.

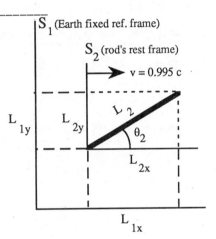

28.17 $\gamma = 3.3715$
Let L_0 = the common length the observer measures for the two rods.
Then since rod A is at rest relative to this observer, the proper length of rod A, $L'_A = L_0$
The proper length of B is given by $L'_B = \gamma L_0 = 3.3715 L_0$
The length the observer in the rest frame of B will measure for rod A is

$$L_{AB} = \dfrac{L'_A}{\gamma} = \dfrac{L_0}{3.3715} = 0.2966 L_0$$

The ratio of lengths measured by observer in B $= \dfrac{\text{measured length of A}}{\text{measured length of B}}$

$= \dfrac{0.2966 L_0}{3.3715 L_0} = 0.088$

28.18 (a) The distance to the star as measured by an observer at rest on earth is
$4 \text{ ly} = 3.78 \times 10^{16}$ m
The time to travel this distance according to an earthbound observer is
$T = \dfrac{d}{v} = \dfrac{3.78 \times 10^{16} \text{ m}}{2.5 \times 10^8 \text{ m/s}} = 1.51 \times 10^8$ s = 4.8 years
(b) This time, measured by the observer in the ship is given by
$T = \gamma T'$ where T' is the time measured by the ship clock, and T is the time measured by an observer on earth (in motion relative to the ship clock). Thus,

CHAPTER TWENTY-EIGHT SOLUTIONS

$$T = \frac{T'}{\gamma} = T'\sqrt{1 - \frac{v^2}{c^2}} = (4.8 \text{ y})\sqrt{1 - \frac{(2.5 \times 10^8 \text{ m/s})^2}{(3 \times 10^8 \text{ m/s})^2}} = 2.65 \text{ years}$$

28.19 The time as measured by an observer in the plane at rest with respect to the clock is
$$T' = 3600 \text{ s}$$

For $v \ll c$, $\gamma = 1 + \frac{1}{2}\frac{v^2}{c^2}$ Therefore,

$$T = (1 + \frac{1}{2}\frac{v^2}{c^2})T' = T' + \frac{1}{2}\frac{v^2}{c^2}T' = 3600 \text{ s} + \frac{1}{2}\frac{(400)^2}{(3 \times 10^8)^2}(3600 \text{ s}) = 3600 \text{ s} + 3.2 \times 10^{-9} \text{ s}$$

28.20 $v = 65$ mi/h $= 29.1$ m/s. We are given $L' = 2$ m.

Thus, $L = L'\sqrt{1 - \frac{v^2}{c^2}} = \frac{L'}{\gamma}$, or $L' = \gamma L = \gamma(L' - \Delta L) = \gamma L' - \gamma \Delta L$

Therefore, $\Delta L = \frac{(\gamma - 1)}{\gamma}L'$

For $v \ll c$, $\gamma = 1 + \frac{1}{2}\frac{v^2}{c^2}$

So, $\Delta L = \dfrac{1 + \frac{1}{2}\frac{v^2}{c^2} - 1}{1 + \frac{1}{2}\frac{v^2}{c^2}} L' = \dfrac{\frac{1}{2}\frac{v^2}{c^2}}{1 + \frac{1}{2}\frac{v^2}{c^2}} L' = \dfrac{\frac{1}{2}\frac{(29.1)^2}{(3 \times 10^8)^2}}{1 + \frac{1}{2}\frac{(29.1)^2}{(3 \times 10^8)^2}} (2 \text{ m})$

$= \dfrac{9.38 \times 10^{-15} \text{ m}}{1 + 4.7 \times 10^{-15}} = 9.38 \times 10^{-15}$ m

28.21 $p = \gamma m_0 v$

(a) $v = 0.01$ c, $\gamma = \dfrac{1}{\sqrt{1 - \frac{v^2}{c^2}}} = \dfrac{1}{\sqrt{1 - (0.01)^2}} = 1.00005 = 1$

Thus, $p = (1)(9.11 \times 10^{-31} \text{ kg})(0.01)(3 \times 10^8 \text{ m/s}) = 2.73 \times 10^{-24}$ kg m/s
(b) Following the same steps as used in part (a), we find
$\gamma = 1.16$ and $p = 1.58 \times 10^{-22}$ kg m/s
(c) $\gamma = 2.29$ and $p = 5.64 \times 10^{-22}$ kg m/s

28.22 The steps are the same here as for problem 28.21 with the exception that the rest mass of the proton is substituted for the mass of the electron. proton rest mass $= 1.67 \times 10^{-27}$ kg.
(a) $\gamma = 1.00$ and $p = 5.0 \times 10^{-21}$ kg m/s
(b) $\gamma = 1.16$ and $p = 2.90 \times 10^{-19}$ kg m/s
(c) $\gamma = 2.29$ and $p = 1.03 \times 10^{-18}$ kg m/s

28.23 (a) We have $p_{rel} = \gamma m_0 v$, and $p_{classical} = m_0 v$. We are given that $p_{rel} = 1.9 p_{classical}$
Thus, $\gamma m_0 v = 1.9 m_0 v$

CHAPTER TWENTY-EIGHT SOLUTIONS

From this, we see $\gamma = 1.9 = \dfrac{1}{\sqrt{1 - \dfrac{v^2}{c^2}}}$ Which yields $v = 0.85\, c$

(b) The result does not change because the mass of the particle cancels from the calculations.

28.24   The momentum of the electron is

$$p_e = \gamma m_{0e} v = \dfrac{9.11 \times 10^{-31}\text{ kg}}{\sqrt{1 - (0.9)^2}} (0.9\, c) = (1.88 \times 10^{-30}\text{ kg})\, c$$

For a proton to have the same momentum as the electron, we have
$p_p = \gamma m_{0p} v = p_e$

or, $\dfrac{1.67 \times 10^{-27}\text{ kg}}{\sqrt{1 - \dfrac{v^2}{c^2}}} (v) = (1.88 \times 10^{-30}\text{ kg})\, c$

This yields, $v = (1.13 \times 10^{-3}\, c) = 3.38 \times 10^5$ m/s

---

28.25   Relativistic momentum must be conserved:
For total momentum to be zero after as it was before, we must have, with subscript 2 referring to the heavier fragment, and subscript 1 to the lighter,

$$p_2 = p_1 \text{ or } \gamma m_{02} v_2 = \gamma m_{01} v_1 = \dfrac{2.5 \times 10^{-28}\text{ kg}}{\sqrt{1 - (0.893)^2}} (0.893\, c)$$

or, $\dfrac{(1.67 \times 10^{-27}\text{ kg}) v_2}{\sqrt{1 - \dfrac{v_2^2}{c^2}}} = (4.960 \times 10^{-28}\text{ kg})\, c$

and $v_2 = 0.285 c$

28.26 $m = \gamma m_0$
Thus, if $m = 10\, m_0$

$\gamma = 10 = \dfrac{1}{\sqrt{1 - \dfrac{v_2^2}{c^2}}}$

gives $v = 0.995\, c$

28.27 $m = \gamma m_0$

Thus, $\dfrac{1}{\gamma} = \dfrac{m_0}{m} = \sqrt{1 - \dfrac{v^2}{c^2}} = \dfrac{0.15\text{ kg}}{0.4\text{ kg}} = 0.375$

From which, $v = 0.927\, c$

28.28   (a) $m = \gamma m_0$

Thus, if $m = 2 m_0$, $\gamma = 2 = \dfrac{1}{\sqrt{1 - \dfrac{v^2}{c^2}}}$

From which, $v = 0.866\, c$
The result is independent of the mass of the particle.

CHAPTER TWENTY-EIGHT SOLUTIONS

28.29 The radius at high speeds is r_2, given by $\quad r_2 = \dfrac{m_2 v_2}{qB}$

At low speeds the radius is r_1, given by $\quad r_1 = \dfrac{m_1 v_1}{qB}$

The ratio of r_2 to r_1 is

$$\frac{r_2}{r_1} = \frac{\gamma_2 m_0 v_2}{\gamma_1 m_0 v_1} =$$

But γ_1 is approximately equal to unity at low speeds. Thus,

$$r_2 = r_1 \frac{\gamma_2 v_2}{v_1} = (0.1 \text{ m}) \frac{\dfrac{1}{\sqrt{1 - (0.96)^2}}(0.96 \times 3 \times 10^8 \text{ m/s})}{(1 \times 10^5 \text{ m/s})} = 1.03 \times 10^3 \text{ m}$$

28.30 $m = \gamma m_0 = \dfrac{1}{\sqrt{1 - (0.66)^2}}(1.67 \times 10^{-27} \text{ kg}) = 2.22 \times 10^{-27} \text{ kg}$

28.31 (See sketch.) The length L_1', and L_2' are not length contracted.
Thus, $L_1 = L_2 = 0.5$ m. However, L_3' is contracted, and its contracted length is

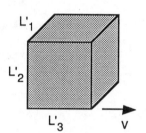

$$L_3 = L_3'\sqrt{1 - \frac{v^2}{c^2}}, \text{ so } \quad \frac{L_3'}{\gamma} = L_3 = \frac{0.5 \text{ m}}{\gamma}$$

Thus, the volume at high speeds is
$$V = L_1 L_2 L_3 = \frac{L_1' L_2' L_3'}{\gamma} = \frac{V_0}{\gamma}$$

Also, the relativistic mass is $\quad m = \gamma m_0$
Thus, the density at high speeds is given by

$$\rho = \frac{m}{V} = \frac{\gamma m_0}{\dfrac{V_0}{\gamma}} = \gamma^2 \frac{m_0}{V_0} \text{ or, } \quad \rho = \gamma^2 \rho_0$$

(a) At low speeds, $\rho_0 = \dfrac{8 \text{ kg}}{0.5 \text{ m}^3} = 64 \text{ kg/m}^3$.

Thus, if $v = 0.9$ c, we find
$\gamma = 2.29$, and $\rho = \gamma^2 \rho_0 = (2.29)^2 (64 \text{ kg/m}^3) = 337 \text{ kg/m}^3$.
(b) If the observer moves with the cube, then he measures the rest density as 64 kg/m^3.

28.32 (a) From $m = \gamma m_0$, we find $\quad \gamma = \dfrac{m}{m_0} = 10^4 = \dfrac{1}{\sqrt{1 - \dfrac{v^2}{c^2}}}$ and from this,

we find $\quad v = 0.999999995$ c
(b) We have $L = \dfrac{L'}{\gamma} = \dfrac{3500 \text{ m}}{10^4} = 0.35 \text{ m} = 35 \text{ cm}$.

28.33 We have $\quad u = \dfrac{u' + v}{1 + \dfrac{u'v}{c^2}} = \dfrac{0.9c + 0.75c}{1 + \dfrac{(0.9c)(0.75c)}{c^2}} = 0.985c$

CHAPTER TWENTY-EIGHT SOLUTIONS

28.34 $u = \dfrac{u' + v}{1 + \dfrac{u'v}{c^2}} = \dfrac{-c + 0.8c}{1 + \dfrac{(-c)(0.8c)}{c^2}} = -c$

28.35 $u = \dfrac{u' + v}{1 + \dfrac{u'v}{c^2}} = 0.8c = \dfrac{-0.95c + v}{1 + \dfrac{(-0.95c)(v)}{c^2}}$

gives $v = 0.994c$

28.36 $u = \dfrac{u' + v}{1 + \dfrac{u'v}{c^2}} = \dfrac{0.95c - 0.92c}{1 + \dfrac{(0.95c)(-0.92c)}{c^2}}$

gives $u = 0.238c$

28.37 $u = \dfrac{u' + v}{1 + \dfrac{u'v}{c^2}} = \dfrac{0.7c + 0.9c}{1 + \dfrac{(0.7c)(0.9c)}{c^2}} = 0.982c$

28.38 We choose a reference frame S to be at rest on ship R, and a frame S' to be at rest on the earth. Then, v (the velocity of the S' frame relative to S) is -0.7c.
Thus,

$u = \dfrac{u' + v}{1 + \dfrac{u'v}{c^2}} = \dfrac{-0.7c - 0.7c}{1 + \dfrac{(-0.7c)(-0.7c)}{c^2}} = -0.94c$

28.39 First, find the velocity of the rocket relative to the pulsar

$u = \dfrac{u' + v}{1 + \dfrac{u'v}{c^2}} = 0.995c \quad \dfrac{-0.95c + v}{1 + \dfrac{(-0.95c)(v)}{c^2}}$

gives $v = 0.99987c$

Proper time = period of pulsar in its own rest frame = $\dfrac{1}{10 \text{ Hz}} = 0.1$ s

so $\Delta t_{rocket} = \dfrac{\Delta t_{proper\ time}}{\sqrt{1 - \dfrac{v^2}{c^2}}} = \dfrac{0.1 \text{ s}}{\sqrt{1 - (0.99987)^2}} = 6.24$ s (period of pulsar as measured in rocket)

or frequency as seen from rocket = $\dfrac{1}{6.24 \text{ s}} = 0.160$ Hz

28.40 $u = \dfrac{u' + v}{1 + \dfrac{u'v}{c^2}} = 0.92c = \dfrac{u' + (-0.95c)}{1 + \dfrac{(-0.95c)(u')}{c^2}}$

gives $u' = 0.998c$

28.41 Let u_A = velocity of rocket relative to A = -0.91c
and u_B = velocity of rocket relative to B

CHAPTER TWENTY-EIGHT SOLUTIONS

Also, $m_A = \dfrac{m_0}{\sqrt{1 - \dfrac{u_A^2}{c^2}}} = 1.2 \times 10^4$ kg

which yields $m_0 = 4975$ kg   (rest mass of rocket)

Then, the mass as detected by B is

$$m_B = \dfrac{m_0}{\sqrt{1 - \dfrac{u_B^2}{c^2}}}$$

becomes $\quad 27{,}000 \text{ kg} = \dfrac{4975 \text{ kg}}{\sqrt{1 - \dfrac{u_B^2}{c^2}}}$

which gives $u_B = \pm\, 0.983c$ (speed of rocket relative to B)

(b) There are two possibilities to explore.

If $u_B = +\, 0.983c$ then

$$u = \dfrac{u' + v}{1 + \dfrac{u'v}{c^2}} = -0.91c = \dfrac{u' + (0.983c)}{1 + \dfrac{(0.983c)(u')}{c^2}}$$

gives $u' = -0.9992c$

if $u_B = -\, 0.983c$

$$u = \dfrac{u' + v}{1 + \dfrac{u'v}{c^2}} = -0.91c = \dfrac{u' - (0.983c)}{1 + \dfrac{(-0.983c)(u')}{c^2}}$$

$u' = 0.695c$

Thus, observer B could be moving at 0.9992c to the left or at 0.695c to the right and obtain the stated mass of the rocket.

28.42 $E_K = E_0$
but $E = E_0 + E_K$
and $E = 2E_0$
becomes $mc^2 = 2m_0c^2 \qquad m = 2m_0$
or $\gamma m_0 = 2m_0$

$$\gamma = 2 = \dfrac{1}{\sqrt{1 - \dfrac{v^2}{c^2}}}$$

from which $v = 0.866c$

28.43 At $v = 0.95c$, $\gamma = 3.20$.
(a) $E_0 = m_0c^2 = (1.67 \times 10^{-27}$ kg$)(3 \times 10^8$ m/s$)^2 = 1.5 \times 10^{-10}$ J = 939 MeV.
(b) $E = mc^2 = \gamma m_0 c^2 = \gamma E_0 = (3.2)(939$ MeV$) = 3006$ MeV.
(c) KE $= E - E_0 = 3006$ MeV $- 939$ MeV $= 2067$ MeV

28.44 $E_0 = m_0c^2 = (1.67 \times 10^{-27}$ kg$)(3 \times 10^8$ m/s$)^2 = 1.5 \times 10^{-10}$ J $= 9.39 \times 10^8$ eV

28.45 We are given $\quad E - E_0 = 0.5 E_0$
Thus, $E = 1.5 E_0$
Also, $E = mc^2 = \gamma m_0 c^2 = \gamma E_0$

CHAPTER TWENTY-EIGHT SOLUTIONS

So, $\gamma = 1.5 = \dfrac{1}{\sqrt{1 - \dfrac{v^2}{c^2}}}$ and from this, $v = 0.745c$

28.46 (a) $E = mc^2 = (0.5 \text{ kg})(3 \times 10^8 \text{ m/s})^2 = 4.5 \times 10^{16}$ J

(b) Power $= \dfrac{E}{t}$

Thus, $t = \dfrac{E}{\text{Power}} = \dfrac{4.5 \times 10^{16} \text{ J}}{100 \text{ J/s}} = 4.5 \times 10^{14}$ s $= 1.43 \times 10^7$ years

---

28.47 (a) From $E = mc^2$, we have $m = \dfrac{E}{c^2} = \dfrac{4 \times 10^{26} \text{ J}}{(3 \times 10^8 \text{ m/s})^2} = 4.44 \times 10^9$ kg

(b) $t = \dfrac{\text{total mass}}{\text{rate of use}} = \dfrac{2 \times 10^{30} \text{ kg}}{4.44 \times 10^9 \text{ kg/s}} = 4.5 \times 10^{20}$ s $= 1.43 \times 10^{13}$ years

---

28.48 The energy equivalent of one electron is
$E_0 = m_0 c^2 = (9.11 \times 10^{-31} \text{ kg})(3 \times 10^8 \text{ m/s})^2 = 8.2 \times 10^{-14}$ J

Thus $E = (\text{power})t = (100 \text{ J/s})(3600 \text{ s}) = 3.6 \times 10^5$ J is the energy required to run the bulb for one hour.

number electrons needed $= \dfrac{\text{total energy needed}}{\text{energy per electron}} = \dfrac{3.6 \times 10^5 \text{ J}}{8.2 \times 10^{-14} \text{ J}} = 4.39 \times 10^{18}$

28.49 $E = 2.86 \times 10^5$ J

Also, the mass-energy relation says that $E = mc^2$

Therefore, $m = \dfrac{E}{c^2} = \dfrac{2.86 \times 10^5 \text{ J}}{(3 \times 10^8 \text{ m/s})^2} = 3.18 \times 10^{-12}$ kg

No, a mass loss of this magnitude (out of a total of 9 g) could not be detected.

28.50 At $0.95c$, $\gamma = 3.20$

$p = \gamma m_0 v = 3.2(1.67 \times 10^{-27} \text{ kg})(0.95 \times 3 \times 10^8 \text{ m/s}) = 1.52 \times 10^{-18}$ kg m/s

$KE = E - E_0 = \gamma E_0 - E_0 = (\gamma - 1)E_0 = (\gamma - 1)m_0 c^2$
$= (3.2 - 1)(1.67 \times 10^{-27} \text{ kg})(3 \times 10^8 \text{ m/s})^2$
$= 3.31 \times 10^{-10}$ J $= 2067$ MeV.

28.51 $KE = E - E_0 = 50$ GeV $= 5.0 \times 10^4$ MeV

Therefore, $E = E_0 + KE = 939$ MeV $+ 5.0 \times 10^4$ MeV $= 50{,}939$ MeV

(a) $E^2 = (pc)^2 + E_0^2$

or, $(pc)^2 = E^2 - E_0^2 = (50{,}939 \text{ MeV})^2 - (939 \text{ MeV})^2$

or, $pc = 5.09 \times 10^4$ MeV, and $p = 5.09 \times 10^4$ MeV/c

(b) From $E = \gamma E_0$ we have $\gamma = \dfrac{E}{E_0} = \dfrac{50939 \text{ MeV}}{939 \text{ MeV}} = 54.2 = \dfrac{1}{\sqrt{1 - \dfrac{v^2}{c^2}}}$

from which, $v = 0.9998c$

28.52 We must conserve both (1) mass-energy and (2) relativistic momentum

CHAPTER TWENTY-EIGHT SOLUTIONS

conservation of mass energy gives, with subscript 1 referring to 0.868c particle and subscript 2 to 0.987c particle,

$E_1 + E_2 = E_{total}$

which is   (1) $\gamma_1 m_{01} c^2 + \gamma_2 m_{02} c^2 = m_{0total} c^2$
and $\gamma_1 = 2.014$ $\gamma_2 = 6.22$

(1) reduces to $m_{01} + 3.09 m_{02} = 1.66 \times 10^{-27}$ kg (A)

Since the momentum after must equal zero, we have

$p_1 = p_2$

gives $\gamma_1 m_{01} v_1 = \gamma_2 m_{02} v_2$
or $(2.014)(m_{01})(.868c) = (6.22)(m_{02})(.987c)$
or $m_{01} = 3.51 m_{02}$ (B)

Solving (A) and (B) simultaneously, we find

$m_{01} = 8.83 \times 10^{-28}$ kg and $m_{02} = 2.52 \times 10^{-28}$ kg

---

28.53 (a) $KE = E - E_0 = 5E_0$

Thus, $E = 6E_0 = 6(9.11 \times 10^{-31} \text{ kg})(3 \times 10^8 \text{ m/s})^2 = 4.92 \times 10^{-13}$ J $= 3.07$ MeV

(b) $E = mc^2 = \gamma m_0 c^2 = \gamma E_0$

Thus, $\gamma = \dfrac{E}{E_0} = 6 = \dfrac{1}{\sqrt{1 - \dfrac{v^2}{c^2}}}$

Which yields, $v = 0.986c$

28.54 $E = E_0 + KE = E_0 + qV = 939$ MeV $+ 500$ eV

Thus, the total energy is still approximately equal to 939 MeV. As a result, $E = E_0$ so $\gamma = 1$ and $m = m_0 = 1.67 \times 10^{-27}$ kg

Classically, $KE = \dfrac{1}{2} m_0 v^2 = qV$, from which $v = 3.1 \times 10^5$ m/s

(b) $E = E_0 + KE = E_0 + qV = 939$ MeV $+ 500$ MeV $= 1439$ MeV, and

$\dfrac{E}{E_0} = \gamma = \dfrac{1439}{939} = 1.53 = \dfrac{1}{\sqrt{1 - \dfrac{v^2}{c^2}}}$ (1)

Therefore, $m = \gamma m_0 = 1.53(1.67 \times 10^{-27} \text{ kg}) = 2.56 \times 10^{-27}$ kg
and, from (1), $v = 0.758c$

28.55 $u = \dfrac{u' + v}{1 + \dfrac{u'v}{c^2}}$ becomes

$0.85c = \dfrac{0.7c + v}{1 + \dfrac{(0.7c)v}{c^2}}$

from which, $v = 0.37c$

28.56 The energy which arrives in one year is

$E = (\text{power})t = (1.79 \times 10^{17} \text{ J/s})(3.156 \times 10^7 \text{ s}) = 5.65 \times 10^{24}$ J

Thus, $m = \dfrac{E}{c^2} = \dfrac{5.65 \times 10^{24} \text{ J}}{(3 \times 10^8 \text{ m/s})^2} = 6.28 \times 10^7$ kg

CHAPTER TWENTY-EIGHT SOLUTIONS

28.57 $E = \gamma E_0 = 5E_0$ Thus, $\gamma = 5$
so, $E^2 = (pc)^2 + E_0^2$ becomes $25E_0^2 = (pc)^2 + E_0^2$
Therefore, $pc = 4.9E_0$
or, $p = \dfrac{4.9E_0}{c}$

(a) For an electron $E_0 = 0.511$ MeV and $p = \dfrac{4.9(0.511 \text{ MeV})}{c} = 2.50$ MeV/c

(b) For a proton $E_0 = 939$ MeV and $p = \dfrac{4.9(939 \text{ MeV})}{c} = 4600$ MeV/c

---

28.58 $E = \gamma E_0 = \gamma(939 \text{ MeV})$
(a) If $v = 0.5c$ $\gamma = 1.16$ and $E = (1.16)(939 \text{ MeV}) = 1089$ MeV
(b) If $v = 0.95c$ $\gamma = 3.20$ and $E = (3.20)(939 \text{ MeV}) = 3005$ MeV

28.59 $\gamma = \dfrac{1}{\sqrt{1 - \dfrac{v^2}{c^2}}} = \dfrac{1}{\sqrt{1 - (0.75)^2}} = 1.51$

Thus, $T = \gamma T' = 1.51(10 \text{ h}) = 15.1$ h
Therefore, $d = vT = (0.75)(3 \times 10^8 \text{ m/s})(15.1 \text{ h})(3600 \text{ s/h}) = 1.22 \times 10^{13}$ m

---

28.60 $\Delta E = (\gamma_1 - \gamma_2)mc^2$ and for an electron $mc^2 = 0.51$ MeV

(a) $\Delta E = \left(\dfrac{1}{\sqrt{1 - (0.75)^2}} - \dfrac{1}{\sqrt{1 - (0.50)^2}} \right)(0.51 \text{ MeV}) = 0.183$ MeV

(b) $\Delta E = \left(\dfrac{1}{\sqrt{1 - (0.99)^2}} - \dfrac{1}{\sqrt{1 - (0.90)^2}} \right)(0.51 \text{ MeV}) = 2.45$ MeV

28.61 (a) $\tau = \gamma \tau' = \dfrac{1}{\sqrt{1 - (0.95)^2}}(2.2 \text{ μs}) = 7.05$ μs

(b) $\Delta t' = \dfrac{d'}{0.95c} = \dfrac{d}{\gamma(0.95c)} = \dfrac{3 \times 10^3 \text{ m}}{3.2(0.95c)} = 3.3$ μs

Therefore, $N = (5 \times 10^4 \text{ muons}) e^{(-3.3 \text{ μs})/(2.2 \text{ μs})} = 11{,}150$ muons

---

28.62 (a) $L_0 = \sqrt{(L_x')^2 + (L_y')^2} = \sqrt{(L_0^2)\cos^2\theta + (L_0^2)\sin^2\theta}$ so that

$L = \sqrt{\dfrac{(L_0^2)\cos^2\theta}{\gamma^2} + (L_0^2)\sin^2\theta} = L_0 \sqrt{1 - \dfrac{v^2}{c^2}\cos^2\theta}$

(b) $\tan\theta_0 = \dfrac{L_y'}{L_x'} = \dfrac{L_y}{\gamma L_x} = \dfrac{1}{\gamma} \tan\theta$

28.63 $f_0 = \dfrac{1}{T_0} = \dfrac{(c - v)}{\lambda_0} = \dfrac{(c - v)}{\dfrac{\lambda}{\gamma}} = \dfrac{(c - v)\gamma}{\dfrac{c}{f}} = \dfrac{(c - v)}{c} \dfrac{1}{\sqrt{1 - \dfrac{v^2}{c^2}}} f$

Therefore $f_0 = \sqrt{\dfrac{c - v}{c + v}} f$ (If source changes direction v becomes -v and the signs are reversed in numerator and denominator.)

CHAPTER TWENTY-NINE SOLUTIONS

29.1 (a) $E = hf = (6.63 \times 10^{-34} \text{ J s})(90 \times 10^6 \text{ Hz}) = 5.97 \times 10^{-26} \text{ J} = 3.73 \times 10^{-7} \text{ eV}$
(b) $E = hf = (6.63 \times 10^{-34} \text{ J s})(10^{13} \text{ Hz}) = 6.63 \times 10^{-21} \text{ J} = 4.14 \times 10^{-2} \text{ eV}$
(c) $E = hf = (6.63 \times 10^{-34} \text{ J s})(10^{16} \text{ Hz}) = 6.63 \times 10^{-18} \text{ J} = 41.4 \text{ eV}$

29.2 $E = hf = h\dfrac{c}{\lambda} = \dfrac{(6.63 \times 10^{-34} \text{ J s})(3 \times 10^8 \text{ m/s})}{\lambda}$
$= \dfrac{1.99 \times 10^{-25} \text{ J m}}{\lambda} = \dfrac{1.243 \times 10^{-6} \text{ eV m}}{\lambda}$
(a) For $\lambda = 5$ cm $= 5 \times 10^{-2}$ m $E = 2.49 \times 10^{-5}$ eV
(b) For $\lambda = 500$ nm $= 5 \times 10^{-7}$ m $E = 2.49$ eV
(c) For $\lambda = 10$ nm $= 10 \times 10^{-9}$ m $E = 124$ eV

29.3 Energy of a single 500 nm photon:
$E_\gamma = hf = \dfrac{hc}{\lambda} = \dfrac{(6.63 \times 10^{-34} \text{ J s})(3 \times 10^8 \text{ m/s})}{500 \times 10^{-9} \text{ m}} = 3.98 \times 10^{-19} \text{ J}$
The energy entering the eye each second
$E = Pt = (IA)t = (4 \times 10^{-11} \text{ W/m}^2)\dfrac{\pi}{4}(8.5 \times 10^{-3} \text{ m})^2 (1 \text{ s}) = 2.27 \times 10^{-15} \text{ J}$
The number of photons required to yield this energy
$n = \dfrac{E}{E_\gamma} = \dfrac{2.27 \times 10^{-15} \text{ J}}{3.98 \times 10^{-19} \text{ J/photon}} = 5.71 \times 10^3$ photons

29.4 From Wien's displacement law
$\lambda_{max} T = 0.2898 \times 10^{-2}$ mK
$T = \dfrac{0.2898 \times 10^{-2} \text{ mK}}{560 \times 10^{-9} \text{ m}} = 5180$ K
Clearly, a firefly is not near this temperature and hence this is not blackbody radiation.

29.5 $\lambda_{max} T = 0.2898 \times 10^{-2}$ m K (Wien's Displacement Law)
Thus, $\lambda_{max} = \dfrac{0.2898 \times 10^{-2} \text{ m K}}{5800 \text{ K}} = 5.00 \times 10^{-7}$ m $= 500$ nm

29.6 $\lambda = \dfrac{c}{f} = \dfrac{3 \times 10^8 \text{ m/s}}{10^{15} \text{ Hz}} = 3 \times 10^{-7}$ m
Thus, $T = \dfrac{0.2898 \times 10^{-2} \text{ m K}}{\lambda_{max}} = \dfrac{0.2898 \times 10^{-2} \text{ m K}}{3 \times 10^{-7} \text{ m}} = 9660$ K

29.7 $E = hf = h\dfrac{c}{\lambda}$

CHAPTER TWENTY-NINE SOLUTIONS

Thus, $\lambda = \dfrac{hc}{E} = \dfrac{(6.63 \times 10^{-34} \text{ J s})(3 \times 10^8 \text{ m/s})}{(2000 \text{ eV})(1.6 \times 10^{-19} \text{ J/eV})} = 6.22 \times 10^{-10}$ m = 0.622 nm

---

29.8 The energy which would be released = mgh = (0.5 kg)(9.8 m/s²)(3 m) = 14.7 J
The energy of a 500 nm photon is

$$E = h\dfrac{c}{\lambda} = \dfrac{(6.63 \times 10^{-34} \text{ J s})(3 \times 10^8 \text{ m/s})}{500 \times 10^{-9} \text{ m}} = 3.98 \times 10^{-19} \text{ J}$$

and number photons released = $\dfrac{14.7 \text{ J}}{3.98 \times 10^{-19} \text{ J/photon}} = 3.7 \times 10^{19}$ photons

---

29.9 The energy equivalent of one photon of 500 nm light is 3.98×10^{-19} J. (See solution to problem 8.)

Thus, mgh = E and $h = \dfrac{E}{mg} = \dfrac{3.98 \times 10^{-19} \text{ J}}{(0.5 \text{ kg})(9.8 \text{ m/s}^2)} = 8.19 \times 10^{-20}$ m

---

29.10 $E = \tfrac{1}{2} kA^2 = \tfrac{1}{2}(20 \text{ N/m})(0.03 \text{ m})^2 = 9 \times 10^{-3}$ J

and $f = \dfrac{1}{2\pi}\sqrt{\dfrac{k}{m}} = \dfrac{1}{2\pi}\sqrt{\dfrac{20 \text{ N/m}}{1.5 \text{ kg}}} = 0.581$ Hz

Therefore, if E = nhf, we have

$$n = \dfrac{E}{hf} = \dfrac{9 \times 10^{-3} \text{ J}}{(6.63 \times 10^{-34} \text{ J s})(0.581 \text{ Hz})} = 2.34 \times 10^{31}$$

(b) If $\Delta n = 1$, then $\Delta E = hf = 3.85 \times 10^{-34}$ J

Thus, $\dfrac{\Delta E}{E} = \dfrac{3.85 \times 10^{-34} \text{ J}}{9 \times 10^{-3} \text{ J}} = 4.28 \times 10^{-32}$

29.11 Energy needed = 1 eV = 1.6×10^{-19} J
The energy absorbed in time t is E = Pt = (IA)t

so, $t = \dfrac{E}{IA} = \dfrac{1.6 \times 10^{-19} \text{ J}}{(500 \text{ J/sm}^2)(\pi(2.82 \times 10^{-15} \text{ m})^2)} = 1.28 \times 10^7$ s = 148 days

---

29.12 $\phi = 3.44$ eV $= 5.51 \times 10^{-19}$ J

$KE = \tfrac{1}{2} mv^2 = \tfrac{1}{2}(9.11 \times 10^{-31} \text{ kg})(4.2 \times 10^5 \text{ m/s})^2 = 8.04 \times 10^{-20}$ J

Energy absorbed = E_γ= KE + ϕ = 6.31×10^{-19} J   (per electron emitted)
Energy incident on one square centimeter of surface each second is
    E = Pt = (IA)t = $(5.5 \times 10^{-2} \text{ J/sm}^2)((10^{-2} \text{ m})^2)(1 \text{ s}) = 5.5 \times 10^{-6}$ J
Therefore, the number of electrons liberated each second, n is

$n = \dfrac{5.5 \times 10^{-6} \text{ J}}{6.31 \times 10^{-19} \text{ J/electron}} = 8.71 \times 10^{12}$ electrons/s (from each square centimeter)

29.13 (a) $KE_{max} = hf - \phi$

$= \dfrac{(6.63 \times 10^{-34} \text{ J s})(3 \times 10^8 \text{ m/s})}{6 \times 10^{-7} \text{ m}} - 2 \times 10^{-19}$ J

$= 1.32 \times 10^{-19}$ J = 0.822 eV

405

CHAPTER TWENTY-NINE SOLUTIONS

(b) $\lambda_c = \dfrac{hc}{\phi} = \dfrac{(6.63 \times 10^{-34} \text{ J s})(3 \times 10^8 \text{ m/s})}{2 \times 10^{-19} \text{ J}} = 9.95 \times 10^{-7}$ m = 995 nm

---

29.14 (a) The photon energy is

$$E = hf = h\dfrac{c}{\lambda} = \dfrac{(6.63 \times 10^{-34} \text{ J s})(3 \times 10^8 \text{ m/s})}{350 \times 10^{-9} \text{ m}} = 5.68 \times 10^{-19} \text{ J} = 3.55 \text{ eV}$$

and $\phi = E - KE_{max} = 3.55$ eV $- 1.31$ eV $= 2.24$ eV

(b) $\lambda_c = \dfrac{hc}{\phi} = \dfrac{(6.63 \times 10^{-34} \text{ J s})(3 \times 10^8 \text{ m/s})}{(2.24 \text{ eV})(1.6 \times 10^{-19} \text{ J/eV})} = 5.55 \times 10^{-7}$ m = 555 nm

(c) $f_c = \dfrac{c}{\lambda_c} = \dfrac{3 \times 10^8 \text{ m/s}}{5.55 \times 10^{-7} \text{ m}} = 5.41 \times 10^{14}$ Hz

---

29.15 (a) The photon energy is

$$E = hf = h\dfrac{c}{\lambda} = \dfrac{(6.63 \times 10^{-34} \text{ J s})(3 \times 10^8 \text{ m/s})}{300 \times 10^{-9} \text{ m}} = 6.63 \times 10^{-19} \text{ J} = 4.14 \text{ eV}$$

and $\phi = E - KE_{max} = 4.14$ eV $- 2.23$ eV $= 1.91$ eV

(b) If $\lambda = 400$ nm, then the photon energy is $E = h\dfrac{c}{\lambda} = 3.11$ eV.

Therefore, $KE_{max} = E - \phi = V_s e$ or $V_s e = 3.11$ eV $- 1.91$ eV $= 1.20$ eV

From which, $V_s = \dfrac{1.20 \text{ eV}}{e} = 1.20$ V.

29.16 $\lambda = \dfrac{c}{f} = \dfrac{3 \times 10^8 \text{ m/s}}{3 \times 10^{15} \text{ Hz}} = 10^{-7}$ m = 100 nm

Thus, the photon energy is $E = h\dfrac{c}{\lambda} = 12.43$ eV

Also, $KE_{max} = V_s e = (7 \text{ V})e = 7$ eV

Therefore, $\phi = E - KE_{max} = 12.43$ eV $- 7$ eV $= 5.43$ eV

29.17 The energy of a 300 nm wavelength photon is $E = h\dfrac{c}{\lambda} = 4.14$ eV

(a) In order for the photoelectric effect to occur, the energy of the photon must be greater than the work function. Thus, the effect will occur in lithium and iron.

(b) For lithium: $KE_{max} = hf - \phi = 4.14$ eV $- 2.3$ eV $= 1.84$ eV

For iron; $KE_{max} = hf - \phi = 4.14$ eV $- 3.9$ eV $= 0.244$ eV

29.18 The frequency of the 253.7 nm wavelength light is

$$f = \dfrac{c}{\lambda} = \dfrac{3 \times 10^8 \text{ m/s}}{253.7 \times 10^{-9} \text{ m}} = 1.18 \times 10^{15} \text{ Hz.}$$

Similarly, the frequency of the 435.8 nm light is 6.88×10^{14} Hz.
The graph you draw should look somewhat like the sketch below. The desired quantities are read from the graph as indicated. You should find that

$f_c = 4.77 \times 10^{14}$ Hz and $\phi = 2.03$ eV.

CHAPTER TWENTY-NINE SOLUTIONS

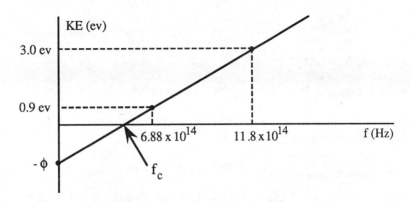

29.19 We have $KE_{max} = \frac{1}{2} m(v_{max})^2 = \frac{1}{2} (9.11 \times 10^{-31} \text{ kg})(10^6 \text{ m/s})^2$
$= 4.56 \times 10^{-19} \text{ J} = 2.85 \text{ eV}$

Then, $E = KE_{max} + \phi = 2.85 \text{ eV} + 2.3 \text{ eV} = 5.15 \text{ eV} = \frac{hc}{\lambda}$
From which, $\lambda = 242$ nm.

29.20 $KE_{max} = hf - \phi = \frac{hc}{\lambda} - \phi$

For wavelength λ_1, $KE_1 = \frac{hc}{\lambda_1} - \phi$

and for wavelength λ_2, $KE_2 = \frac{hc}{\lambda_2} - \phi$

Thus, if $KE_2 = 1.5 KE_1$, we have $\frac{hc}{\lambda_2} - \phi = (1.5)\left(\frac{hc}{\lambda_1} - \phi\right)$

or $\phi = 2hc\left(\frac{1.5}{\lambda_1} - \frac{1}{\lambda_2}\right)$

$= 2(6.63 \times 10^{-34} \text{ J s})(3 \times 10^8 \text{ m/s})\left(\frac{1.5}{670 \times 10^{-9} \text{ m}} - \frac{1}{520 \times 10^{-9} \text{ m}}\right)$

$= 1.26 \times 10^{-19} \text{ J} = 0.784 \text{ eV}$

29.21 $E_{max} = \frac{hc}{\lambda_{min}} = KE = qV$

so $\lambda_{min} = \frac{hc}{qV} = \frac{(6.63 \times 10^{-34} \text{ J s})(3 \times 10^8 \text{ m/s})}{(1.6 \times 10^{-19} \text{ C})(10^4 \text{ J/C})} = 1.24 \times 10^{-10}$ m = 0.124 nm

29.22 $eV = \frac{hc}{\lambda_{min}} = \frac{1243 \text{ nm eV}}{0.03 \text{ nm}} = 4.14 \times 10^4$ eV
Hence, $V = 4.14 \times 10^4$ V

29.23 We have, $\lambda_{min} = \frac{hc}{eV} = \frac{1243 \text{ nm eV}}{eV} = \frac{1243 \text{ nm V}}{V}$

(a) $\lambda_{min} = \frac{1243 \text{ nm V}}{15000 \text{ V}} = 0.083$ nm

(b) $\lambda_{min} = \frac{1243 \text{ nm V}}{1 \times 10^5 \text{ V}} = 0.012$ nm

CHAPTER TWENTY-NINE SOLUTIONS

29.24 From $eV = \dfrac{hc}{\lambda_{min}}$

If $\lambda = 10$ nm, $V = \dfrac{1243 \text{ nm V}}{10 \text{ nm}} = 124$ V

If $\lambda = 10^{-4}$ nm, $V = \dfrac{1243 \text{ nm V}}{10^{-4} \text{ nm}} = 1.24 \times 10^7$ V

29.25 From the grating equation $\sin\theta = \dfrac{n\lambda}{d}$

and $d = \dfrac{1}{25000 \text{ lines/cm}} = 4 \times 10^{-5}$ cm $= 400$ nm

If $n = 1$, $\sin\theta = \dfrac{(1)(0.05 \text{ nm})}{400 \text{ nm}} = 1.25 \times 10^{-4}$ and $\theta = 7.16 \times 10^{-3}$ degrees = 0.430 min of arc = 25.8 sec of arc

---

29.26 $E = \dfrac{hc}{\lambda}$

so $\lambda = \dfrac{(6.63 \times 10^{-34} \text{ J s})(3 \times 10^8 \text{ m/s})}{(11.3 \times 10^3 \text{ eV})(1.60 \times 10^{-19} \text{ J/eV})} = 1.1 \times 10^{-10}$ m $= 0.11$ nm

From Bragg's law

$\sin\theta = \dfrac{m\lambda}{2d}$, so if $m = 2$ and $d = 0.352$ nm

$\sin\theta = \dfrac{2(0.11 \text{ nm})}{2(0.352 \text{ nm})} = 0.312$ and $\theta = 18.2°$

29.27 From $n\lambda = 2d\sin\theta$, we have

$\sin\theta = n\dfrac{\lambda}{2d}$ for constructive interference.

For the minimum angle, we let $n = 1$, the first order. Thus,

$\sin\theta = \dfrac{\lambda}{2d} = \dfrac{0.07 \text{ nm}}{2(0.3 \text{ nm})} = 0.117$ and $\theta = 6.7°$

29.28 From $n\lambda = 2d\sin\theta$, we have

$\lambda = \dfrac{2d\sin\theta}{n} = \dfrac{2(0.296 \text{ nm}) \sin 7.6°}{(1)} = 0.078$ nm

---

29.29 We have $\lambda = \dfrac{2d\sin\theta}{n} = \dfrac{2(0.353 \text{ nm}) \sin 20.5°}{(2)} = 0.124$ nm

---

29.30 $d = \dfrac{n\lambda}{2\sin\theta} = \dfrac{(1)(0.14 \text{ nm})}{2\sin 14.4°} = 0.281$ nm

---

29.31 $\Delta\lambda = \dfrac{h}{m_0 c}(1 - \cos\theta) = \dfrac{hc}{m_0 c^2}(1 - \cos\theta) = \dfrac{hc}{E_0}(1 - \cos\theta)$

At $\theta = 90°$, $\Delta\lambda = \dfrac{1243 \text{ nm eV}}{0.511 \times 10^6 \text{ eV}}(1 - 0) = 2.43 \times 10^{-3}$ nm

---

29.32 $\lambda_c = \dfrac{h}{m_0 c} = \dfrac{6.63 \times 10^{-34} \text{ J s}}{(9.11 \times 10^{-31} \text{ kg})(3 \times 10^8 \text{ m/s})}$

$= 2.43 \times 10^{-12}$ m $= 2.43 \times 10^{-3}$ nm

CHAPTER TWENTY-NINE SOLUTIONS

29.33 $\Delta\lambda = \dfrac{h}{m_0 c}(1 - \cos\theta) = \lambda_c(1 - \cos\theta)$

When $\theta = 90°$, $\Delta\lambda = \lambda_c = 2.43 \times 10^{-3}$ nm (see problem 32)

Thus, $\lambda = \lambda_0 + \Delta\lambda = 0.071$ nm $+ 0.00243$ nm $= 7.34 \times 10^{-2}$ nm

---

29.34 We have $\Delta\lambda = \lambda_c(1 - \cos\theta)$

or, $\cos\theta = 1 - \dfrac{\Delta\lambda}{\lambda_c} = 1 - \dfrac{1.2 \times 10^{-3} \text{ nm}}{2.43 \times 10^{-3} \text{ nm}} = 1 - 0.494 = 0.506$ and $\theta = 59.6°$

29.35 $\Delta\lambda = \lambda_c(1 - \cos\theta)$

becomes $\lambda = \lambda_0 + \lambda_c(1 - \cos\theta) = \lambda_0 + 0.00243(1 - \cos\theta)$

If $\lambda_0 = 0.68$ nm and $\theta = 45°$, the wavelength of the scattered photons is

$\lambda = 0.68$ nm $+ (2.43 \times 10^{-3}$ nm$)(1 - \cos 45°) = 0.6807$ nm

The energy of these photons is

$E = \dfrac{hc}{\lambda} = \dfrac{(6.63 \times 10^{-34} \text{ J s})(3 \times 10^8 \text{ m/s})}{(0.6807 \times 10^{-9} \text{ m})} = 2.92 \times 10^{-16}$ J $= 1.82$ keV

and the momentum is

$p = \dfrac{E}{c} = 9.74 \times 10^{-25}$ kg m/s $= 1.82$ keV/c

---

29.36 $E_{\gamma 0} = \dfrac{hc}{\lambda_0} = \dfrac{(6.63 \times 10^{-34} \text{ J s})(3 \times 10^8 \text{ m/s})}{(0.45 \times 10^{-9} \text{ m})} = 4.42 \times 10^{-16}$ J

$\Delta\lambda = \lambda_c(1 - \cos\theta)$

becomes, for $\theta = 23°$

$\lambda' = \lambda_0 + \lambda_c(1 - \cos\theta) = 0.45$ nm $+ (2.43 \times 10^{-3}$ nm$)(1 - \cos 23°) = 0.45019$ nm

$E_\gamma = \dfrac{hc}{\lambda'} = \dfrac{(6.63 \times 10^{-34} \text{ J s})(3 \times 10^8 \text{ m/s})}{(0.45019 \times 10^{-9} \text{ m})} = 4.418 \times 10^{-16}$ J

(a) Energy given the electron $= E_{\gamma 0} - E_\gamma = 1.90 \times 10^{-19}$ J $= 1.18$ eV

(b) from $KE = \dfrac{1}{2} mv^2$

$v = \sqrt{\dfrac{2(KE)}{m}} = \sqrt{\dfrac{2(1.90 \times 10^{-19} \text{ J})}{(9.11 \times 10^{-31} \text{ kg})}} = 6.45 \times 10^5$ m/s

---

29.37 $KE = \dfrac{1}{2} mv^2 = \dfrac{1}{2}(9.11 \times 10^{-31}$ kg$)(1.4 \times 10^6$ m/s$)^2 = 8.93 \times 10^{-19}$ J

but $KE = E_\gamma - E'_\gamma = \dfrac{hc}{\lambda} - \dfrac{hc}{\lambda'} = hc\left(\dfrac{\lambda' - \lambda}{\lambda'\lambda}\right) = hc\dfrac{\Delta\lambda}{\lambda^2}$

(a) so $\Delta\lambda = \dfrac{\lambda^2(KE)}{hc} = \dfrac{(8. \times 10^{-10} \text{ m})^2(8.93 \times 10^{-19} \text{ J})}{(6.63 \times 10^{-34} \text{ J s})(3 \times 10^8 \text{ m/s})}$

$= 2.87 \times 10^{-12}$ m $= 2.87 \times 10^{-3}$ nm

(b) From the Compton equation

$1 - \cos\theta = \dfrac{\Delta\lambda}{\lambda_c} = \dfrac{2.87 \times 10^{-3} \text{ nm}}{2.43 \times 10^{-3} \text{ nm}} = 1.182$

from which $\theta = 100.5°$

CHAPTER TWENTY-NINE SOLUTIONS

29.38 For the recoiling electron to have a kinetic energy, equal to the energy of the scattered photon, the energy of the incident photon must be split equally between the electron and the scattered photon.
Therefore, $E'_\gamma = \frac{1}{2} E_\gamma$
or $\frac{hc}{\lambda'} = \frac{1}{2} \frac{hc}{\lambda}$ so $\lambda' = 2\lambda$
and from the Compton equation
$$\cos\theta = 1 - \frac{\lambda}{\lambda_c} = 1 - \frac{0.0016 \text{ nm}}{0.00243 \text{ nm}} = 0.3416 \quad \text{and } \theta = 70.0°$$

29.39 (a) The minimum energy, E, of the photon required is twice the rest energy of a proton. Thus, $E = 2E_0 = 2m_0 c^2$
Therefore, $E = 2(1.67 \times 10^{-27} \text{ kg})(3 \times 10^8 \text{ m/s})^2 = 3.006 \times 10^{-10} \text{ J} = 1878 \text{ MeV}$
(b) $E = hf = \frac{hc}{\lambda}$
or, $\lambda = \frac{hc}{E} = \frac{(6.63 \times 10^{-34} \text{ J s})(3 \times 10^8 \text{ m/s})}{3.006 \times 10^{-10} \text{ J}} = 6.62 \times 10^{-16} \text{ m}$

29.40 (a) $E = 2(1.67495 \times 10^{-27} \text{ kg})(3 \times 10^8 \text{ m/s})^2 = 3.0149 \times 10^{-10} \text{ J} = 1880 \text{ MeV}$
(b) $\lambda = \frac{hc}{E} = \frac{(6.63 \times 10^{-34} \text{ J s})(3 \times 10^8 \text{ m/s})}{3.0149 \times 10^{-10} \text{ J}} = 6.60 \times 10^{-16} \text{ m}$

29.41 The kinetic energy of the pair will be the energy of the photon used to produce them less the combined rest energy of the two. Thus,
$KE = E - 2E_0 = 3 \text{ MeV} - 2(0.511 \text{ MeV}) = 1.98 \text{ MeV}$

29.42 (a) We have $E = KE + 2E_0 = 2.5 \text{ MeV} + 2(0.511 \text{ MeV}) = 3.52 \text{ MeV}$.
(b) $f = \frac{E}{h} = \frac{3.52 \text{ MeV}(1.6 \times 10^{-13} \text{ J/MeV})}{6.63 \times 10^{-34} \text{ J s}} = 8.50 \times 10^{20} \text{ Hz}$

29.43 Assuming we are in the rest frame of the particles before annihilation, we have $p_{before} = 0$, and as a result, $p_{after} = 0$.
$\frac{E_\gamma}{c} = \frac{E'_\gamma}{c}$, or $E_\gamma = E'_\gamma$
The two photons have equal energy
From the energy-mass relationship,
$2E_0 = 2E_\gamma$ or $E_\gamma = E_0$
$hf = m_p c^2$
gives $f = \frac{m_p c^2}{h} = \frac{(1.673 \times 10^{-27} \text{ kg})(3 \times 10^8 \text{ m/s})^2}{6.63 \times 10^{-34} \text{ J s}} = 2.27 \times 10^{23} \text{ Hz}$
$\lambda = \frac{c}{f} = \frac{3 \times 10^8 \text{ m/s}}{2.27 \times 10^{23} \text{ Hz}} = 1.32 \times 10^{-15} \text{ m}$

29.44 (a) $\lambda = \frac{h}{p} = \frac{h}{mv}$, so $v = \frac{h}{m\lambda} = \frac{6.63 \times 10^{-34} \text{ J s}}{(9.11 \times 10^{-31} \text{ kg})(500 \times 10^{-9} \text{ m})} = 1456 \text{ m/s}$
(b) $p = mv = (9.11 \times 10^{-31} \text{ kg})(10^7 \text{ m/s}) = 9.11 \times 10^{-24} \text{ kg m/s}$
Thus, $\lambda = \frac{h}{p} = \frac{6.63 \times 10^{-34} \text{ J s}}{9.11 \times 10^{-24} \text{ kg m/s}} = 7.28 \times 10^{-11} \text{ m}$

CHAPTER TWENTY-NINE SOLUTIONS

29.45 $\lambda = \dfrac{h}{p} = \dfrac{6.63 \times 10^{-34} \text{ J s}}{(2 \times 10^3 \text{ kg})(65 \text{ mi/h})(0.447 \text{ m/s/mi h})} = 1.14 \times 10^{-38}$ m

29.46 (a) At 10^4 m/s, the proton is non-relativistic. Thus,
$p = m_0 v = (1.67 \times 10^{-27} \text{ kg})(10^4 \text{ m/s}) = 1.67 \times 10^{-23}$ kg m/s

So, $\lambda = \dfrac{h}{p} = \dfrac{6.63 \times 10^{-34} \text{ J s}}{1.67 \times 10^{-23} \text{ kg m/s}} = 3.97 \times 10^{-11}$ m

(b) At 10^7 m/s $\gamma = 1.0006$

Thus, $p = \gamma m_0 v = (1.0006)(1.67 \times 10^{-27} \text{ kg})(10^7 \text{ m/s}) = 1.67 \times 10^{-20}$ kg m/s

and $\lambda = \dfrac{h}{p} = \dfrac{6.63 \times 10^{-34} \text{ J s}}{1.67 \times 10^{-20} \text{ kg m/s}} = 3.97 \times 10^{-14}$ m

29.47 (a) $KE = \dfrac{1}{2} mv^2 = \dfrac{m^2 v^2}{2m} = \dfrac{p^2}{2m}$ or, $p^2 = 2m(KE)$

Thus, $p = \sqrt{2(9.11 \times 10^{-31} \text{ kg})(50 \times 1.6 \times 10^{-19} \text{ J})} = 3.82 \times 10^{-24}$ kg m/s

and $\lambda = \dfrac{h}{p} = \dfrac{6.63 \times 10^{-34} \text{ J s}}{3.82 \times 10^{-24} \text{ kg m/s}} = 1.74 \times 10^{-10}$ m $= 0.174$ nm

(b) Consider relativistic effects.

$KE = 50$ keV $= E - E_0$

or, $E = E_0 + 50$ keV $= 511$ keV $+ 50$ keV $= 561$ keV

and $E^2 = p^2 c^2 + E_0^2$

or, $p = \dfrac{1}{c}\sqrt{(561 \text{ keV})^2 - (511 \text{ keV})^2} = 232$ keV/c

$= 232 \text{ keV/c} \left(\dfrac{c}{3 \times 10^8 \text{ m/s}}\right)\left(\dfrac{1.6 \times 10^{-16} \text{ J}}{1 \text{ keV}}\right) = 1.23 \times 10^{-22}$ kg m/s

Thus, $\lambda = \dfrac{h}{p} = \dfrac{6.63 \times 10^{-34} \text{ J s}}{1.23 \times 10^{-22} \text{ kg m/s}} = 5.37 \times 10^{-12}$ m

29.48 $p = \dfrac{h}{\lambda} = \dfrac{6.63 \times 10^{-34} \text{ J s}}{10^{-11} \text{ m}} = 6.63 \times 10^{-23}$ kg m/s

Relativistically, $E^2 = (pc)^2 + E_0^2$
$= [(6.63 \times 10^{-23} \text{ kg m/s})(3 \times 10^8 \text{ m/s})]^2 + [(9.11 \times 10^{-31} \text{ kg})(3 \times 10^8 \text{ m/s})^2]^2$

gives $E = 8.44 \times 10^{-14}$ J

and $E_K = E - E_0$
$= 8.44 \times 10^{-14}$ J $- (9.11 \times 10^{-31} \text{ J})(3 \times 10^8 \text{ m/s})^2 = 2.38 \times 10^{-15}$ J $= 14.8$ keV

If done classically,

$E_K = \dfrac{p^2}{2m_0} = 2.41 \times 10^{-15}$ J $= 15.1$ keV

(b) $E_\gamma = \dfrac{hc}{\lambda} = \dfrac{(6.63 \times 10^{-34} \text{ J s})(3 \times 10^8 \text{ m/s})}{10^{-11} \text{ m}} = 1.99 \times 10^{-14}$ J $= 124$ keV

29.49 From conservation of energy, $\dfrac{1}{2} mv^2 = mgh$,

or $v = \sqrt{2gh} = \sqrt{2(9.8 \text{ m/s}^2)(50 \text{ m})} = 31.3$ m/s

Thus, $p = mv = (0.2 \text{ kg})(31.3 \text{ m/s}) = 6.26$ kg m/s

CHAPTER TWENTY-NINE SOLUTIONS

and $\lambda = \dfrac{h}{p} = \dfrac{6.63 \times 10^{-34} \text{ J s}}{6.26 \text{ kg m/s}} = 1.06 \times 10^{-34} \text{ m}$

29.50 $KE = eV = 10^7 \text{ eV} = 10 \text{ MeV}$
Thus, $E = E_0 + KE = 939 \text{ MeV} + 10 \text{ MeV} = 949 \text{ MeV}$
and $\gamma = \dfrac{E}{E_0} = \dfrac{949}{939} = 1.01 = \dfrac{1}{\sqrt{1 - \dfrac{v^2}{c^2}}}$ yields $\dfrac{v}{c} = 0.145$

Therefore,
$p = \gamma m_0 v = (1.01)(1.67 \times 10^{-27} \text{ kg})(0.145)(3 \times 10^8 \text{ m/s}) = 7.33 \times 10^{-20} \text{ kg m/s}$
Finally, $\lambda = \dfrac{h}{p} = \dfrac{6.63 \times 10^{-34} \text{ J s}}{7.33 \times 10^{-20} \text{ kg m/s}} = 9.05 \times 10^{-15} \text{ m}$

29.51 $p = \dfrac{h}{\lambda} = \dfrac{6.63 \times 10^{-34} \text{ J s}}{10^{-10} \text{ m}} = 6.63 \times 10^{-24} \text{ kg m/s}$

and $pc = (6.63 \times 10^{-24} \text{ kg m/s})(3 \times 10^8 \text{ m/s}) \dfrac{1 \text{ MeV}}{1.6 \times 10^{-13} \text{ J}} = 1.24 \times 10^{-2} \text{ MeV}$

Thus, $E = \sqrt{p^2 c^2 + E_0^2} = \sqrt{(1.24 \times 10^{-2} \text{ MeV})^2 + (0.511 \text{ MeV})^2} = 0.5115 \text{ MeV}$,
and $\gamma = \dfrac{E}{E_0} = 1.0003$ (Thus, classical calculations suffice.)

We have, $v = \dfrac{p}{m_0} = \dfrac{6.63 \times 10^{-24} \text{ J s}}{9.11 \times 10^{-31} \text{ kg}} = 7.28 \times 10^6 \text{ m/s}$

and $V = \dfrac{KE}{e} = \dfrac{(9.11 \times 10^{-31} \text{ kg})(7.28 \times 10^6 \text{ m/s})^2}{2(1.6 \times 10^{-19} \text{ C})} = 151 \text{ V}$

29.52 For a non-relativistic electron, conservation of energy
$\dfrac{1}{2} m_0 v^2 = eV$ yields $v = \sqrt{\dfrac{2eV}{m_0}}$

Thus, $p = m_0 v = \sqrt{2 m_0 eV}$ and $\lambda = \dfrac{h}{p} = \dfrac{h}{\sqrt{2 m_0 e} \sqrt{V}}$

$= \dfrac{6.63 \times 10^{-34} \text{ J s}}{\sqrt{2(9.11 \times 10^{-31} \text{ kg})(1.6 \times 10^{-19} \text{ C})}\sqrt{V}} = \dfrac{1.228 \text{ nm}}{\sqrt{V}}$

29.53 $\dfrac{y}{L} = m \dfrac{\lambda}{a}$ gives $y_2 - y_1 = 2\dfrac{L\lambda}{a} - \dfrac{L\lambda}{a} = \dfrac{L\lambda}{a}$

or $\lambda = \Delta y \dfrac{a}{L} = \dfrac{(2.1 \text{ cm})(0.5 \text{ nm})}{20 \text{ cm}} = 0.0525 \text{ nm}$

Thus, $p = \dfrac{h}{\lambda} = \dfrac{6.63 \times 10^{-34} \text{ J s}}{(5.25 \times 10^{-11} \text{ kg})} = 1.26 \times 10^{-23} \text{ kg m/s}$

Classically, $KE = \dfrac{p^2}{2 m_0} = \dfrac{(1.26 \times 10^{-23} \text{ kg m/s})^2}{2(9.11 \times 10^{-31} \text{ kg})} = 8.75 \times 10^{-17} \text{ J} = 546 \text{ eV}$

Relativistically, $KE = \sqrt{p^2 c^2 + E_0^2} - E_0 = 8.748 \times 10^{-17} \text{ J} = 546 \text{ eV}$

29.54 A relativistic electron, so $E = E_0 + KE$
and $E_0 = m_0 c^2 = (9.11 \times 10^{-31} \text{ kg})(3 \times 10^8 \text{ m/s})^2 = 8.2 \times 10^{-14} \text{ J}$

CHAPTER TWENTY-NINE SOLUTIONS

and $KE = 3$ MeV $= 4.806 \times 10^{-13}$ J
so $E = E_0 + KE = 5.626 \times 10^{-13}$ J

Also, $p = \frac{1}{c}\sqrt{E^2 - E_0^2}$

$= \frac{1}{3 \times 10^8 \text{ m/s}} \sqrt{(5.626 \times 10^{-13} \text{ J})^2 - (8.2 \times 10^{-14} \text{ J})^2} = 1.855 \times 10^{-21}$ kg m/s

Then, $\lambda = \frac{h}{p} = \frac{6.63 \times 10^{-34} \text{ J s}}{1.855 \times 10^{-21} \text{ kg m/s}} = 3.57 \times 10^{-13}$ m

---

29.55 $(\Delta x)(\Delta p) \geq \frac{h}{4\pi}$ We have $v = 30$ m/s, and $\Delta v = 0.1\%(v) = 3 \times 10^{-2}$ m/s

Thus, $\Delta p = (50 \times 10^{-3} \text{ kg})(3 \times 10^{-2} \text{ m/s}) = 1.5 \times 10^{-3}$ kg m/s

and $\Delta x \geq \frac{h}{4\pi(\Delta p)} = \frac{6.63 \times 10^{-34} \text{ J s}}{4\pi(1.5 \times 10^{-3} \text{ kg m/s})} = 3.50 \times 10^{-32}$ m

---

29.56 We have $\Delta x = 0.5$ cm $= 5 \times 10^{-3}$ m

Thus, $\Delta p \geq \frac{h}{4\pi(\Delta x)} = \frac{6.63 \times 10^{-34} \text{ J s}}{4\pi(5 \times 10^{-3} \text{ m})} = 1.05 \times 10^{-32}$ kg m/s

and $\Delta p = \Delta(mv) = m(\Delta v)$

so $\Delta v = \frac{\Delta p}{m} \geq \frac{1.05 \times 10^{-32} \text{ kg m/s}}{0.5 \text{ kg}} = 2.10 \times 10^{-32}$ m/s

Thus v is of the order of the uncertainty, and equal to $v = 2.10 \times 10^{-32}$ m/s

---

29.57 $\Delta p \geq \frac{h}{4\pi(\Delta x)} = \frac{6.63 \times 10^{-34} \text{ J s}}{4\pi(10^{-10} \text{ m})} = 5.28 \times 10^{-25}$ kg m/s

so $\Delta v = \frac{\Delta p}{m} \geq \frac{5.28 \times 10^{-25} \text{ kg m/s}}{9.11 \times 10^{-31} \text{ kg}} = 5.79 \times 10^5$ m/s

---

29.58 $\Delta p = \frac{h}{4\pi(\Delta x)} = \frac{6.63 \times 10^{-34} \text{ J s}}{4\pi(10^{-10} \text{ m})} = 5.28 \times 10^{-25}$ kg m/s

$\Delta v = \frac{\Delta p}{m} = \frac{5.28 \times 10^{-25} \text{ kg m/s}}{9.11 \times 10^{-31} \text{ kg}} = 5.8 \times 10^5$ m/s

Thus, v is of the order of 10^5 m/s

29.59 $\frac{\Delta p}{p} = 10^{-3}$ (i.e. one part in 1000)

Thus, if $v = 6.4 \times 10^6$ m/s, then $p = mv = (9.11 \times 10^{-31} \text{ kg})(6.4 \times 10^6 \text{ m/s})$
$= 5.83 \times 10^{-24}$ kg m/s,

and $\Delta p = (10^{-3})p = 5.83 \times 10^{-27}$ kg m/s

Then, $\Delta x = \frac{h}{4\pi \Delta p} \geq \frac{6.63 \times 10^{-34} \text{ J s}}{4\pi(5.83 \times 10^{-27} \text{ kg m/s})} = 9.05 \times 10^{-9}$ m

or $\frac{\Delta x}{L} = \frac{9.05 \times 10^{-9} \text{ m}}{1 \text{ m}} = 9.05 \times 10^{-9}$ m approximately 10^{-8} m, or 1 part in 10^8

29.60 $E = 2$ eV $= 3.2 \times 10^{-19}$ J   Thus, one percent of $E = \frac{E}{100} = 3.2 \times 10^{-21}$ J

CHAPTER TWENTY-NINE SOLUTIONS

and from $(\Delta E)(\Delta t) \geq \dfrac{h}{4\pi}$

we have $\Delta t \geq \dfrac{h}{4\pi \Delta E} = \dfrac{6.63 \times 10^{-34} \text{ J s}}{4\pi(3.2 \times 10^{-21} \text{ J})} = 1.64 \times 10^{-14}$ s

---

29.61 (a) $KE = \dfrac{1}{2}mv^2 = \dfrac{(mv)^2}{2m} = \dfrac{p^2}{2m}$

(b) $\Delta p \geq \dfrac{h}{4\pi(\Delta x)} = \dfrac{6.63 \times 10^{-34} \text{ J s}}{4\pi(10^{-15} \text{ m})} = 5.25 \times 10^{-20}$ kg m/s

The smallest momentum is also of the order of Δp. Thus, $p = 5.28 \times 10^{-20}$ kg m/s.

Thus, $KE_{min} = \dfrac{p_{min}^2}{2m} = \dfrac{(5.28 \times 10^{-20} \text{ kg m/s})^2}{2(1.67 \times 10^{-27} \text{ kg})} = 8.34 \times 10^{-13}$ J = 5.21 MeV

29.62 $KE_{max} = eV = 0.92$ eV, and the photon energy is $E = hf = \dfrac{hc}{\lambda} = 4.97$ eV

Thus from, $KE_{max} = E - \phi$, we find

$\phi = E - KE_{max} = 4.97$ eV $- 0.92$ eV $= 4.05$ eV.

29.63 The retarding potential is just sufficient to stop the most energetic electrons liberated by the 350 nm light. Thus, $eV = KE_{max} = E - \phi$

So, we use $V = \dfrac{E - \phi}{e}$ with $E = hf = \dfrac{hc}{\lambda} = 3.55$ eV, and $\phi = 2$ eV

$V = \dfrac{3.55 \text{ eV} - 2 \text{ eV}}{e} = 3.55$ V $- 2$ V $= 1.55$ V

29.64 $E = KE)_{max} = \dfrac{1}{2}mv^2 = \dfrac{1}{2}(70 \text{ kg})(2 \text{ m/s})^2 = 140$ J

If $E = nhf$, then $n = \dfrac{E}{hf} = \dfrac{140 \text{ J}}{(6.63 \times 10^{-34} \text{ J s})(0.5 \text{ Hz})} = 4.22 \times 10^{35}$

(b) If $\Delta n = 1$, then $\Delta E = hf = (6.63 \times 10^{-34}$ J s$)(0.5$ Hz$) = 3.32 \times 10^{-34}$ J

29.65 The energy of one photon is $E = hf = \dfrac{hc}{\lambda} = 3.37 \times 10^{-19}$ J

The energy emitted per second by the lamp = (power)t = (100 J/s)(1 s) = 100 J

Thus, the number of photons emitted = $\dfrac{\text{energy from lamp}}{\text{energy per photon}}$

$= \dfrac{100 \text{ J}}{3.37 \times 10^{-19} \text{ J}} = 2.96 \times 10^{20}$

29.66 (a) $\lambda = \dfrac{hc}{eV} = \dfrac{1243 \text{ nm eV}}{e(5 \times 10^4 \text{ V})} = 2.49 \times 10^{-2}$ nm

(b) From $n\lambda = 2d \sin\theta$, we have $d = \dfrac{n\lambda}{2\sin\theta} = \dfrac{(1)2.49 \times 10^{-2} \text{ nm}}{2\sin 2.5°} = 0.285$ nm

---

29.67 We use $KE_{max} = E - \phi$ (1)

The energy per photon of the first source is $E = \dfrac{hc}{\lambda}$

And the energy per photon for the second is $E' = \dfrac{hc}{\lambda'} = \dfrac{hc}{\frac{\lambda}{2}} = 2\dfrac{hc}{\lambda} = 2E$

414

CHAPTER TWENTY-NINE SOLUTIONS

Thus, for case 1, (1) becomes $\quad 1\text{ eV} = E - \phi \quad$ (2)
For case 2, (1) becomes $\quad 4\text{ eV} = 2E - \phi \quad$ (3)
Solving (2) and (3) simultaneously, we find $\quad \phi = 2\text{ eV}$

29.68 $KE = \frac{3}{2}kT = \frac{3}{2}(1.38 \times 10^{-23}\text{ J/K})(2000\text{ K}) = 4.14 \times 10^{-20}\text{ J}$. For energies this low, we do not have to consider relativistic effects. Thus, for the electron we have

$$p = \sqrt{2m(KE)} = \sqrt{2(9.11 \times 10^{-31}\text{ kg})(4.14 \times 10^{-20}\text{ J})} = 2.75 \times 10^{-25}\text{ kg m/s}$$

and $\lambda = \dfrac{h}{p} = \dfrac{6.63 \times 10^{-34}\text{ J s}}{2.75 \times 10^{-25}\text{ kg m/s}} = 2.41\text{ nm}$

and for the proton

$$p = \sqrt{2m(KE)} = \sqrt{2(1.67 \times 10^{-27}\text{ kg})(4.14 \times 10^{-20}\text{ J})} = 1.18 \times 10^{-23}\text{ kg m/s}$$

and $\lambda = \dfrac{h}{p} = \dfrac{6.63 \times 10^{-34}\text{ J s}}{1.18 \times 10^{-23}\text{ kg m/s}} = 0.056\text{ nm}$

29.69 We recall that the orbit of a charged particle in a magnetic field is given by

$$r = \frac{mv}{qB} \quad \text{or,} \quad mv = qrB$$

Now, $p_{max} = mv_{max} = (1.6 \times 10^{-19}\text{ C})(0.2\text{ m})(2 \times 10^{-5}\text{ T}) = 6.4 \times 10^{-25}\text{ kg m/s}$

and $KE_{max} = \dfrac{p_{max}^2}{2m} = \dfrac{(6.4 \times 10^{-25}\text{ kg m/s})^2}{2(9.11 \times 10^{-31}\text{ kg})} = 2.25 \times 10^{-19}\text{ J} = 1.41\text{ eV}$

The photon energy is $\quad E = hf = \dfrac{hc}{\lambda} = 2.76\text{ eV}$

So, $\quad \phi = E - KE_{max} = 2.76\text{ eV} - 1.41\text{ eV} = 1.35\text{ eV}$

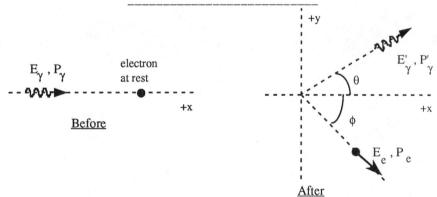

29.70 (a) Conserving mass-energy,

$E_\gamma + E_0 = E'_\gamma + (E_0 + KE)$ where E'_γ = the energy of the scattered photon, E_γ = is the energy of the incident photon, and $(E_0 + KE)$ is the energy of the scattered electron.

Thus, $E_\gamma = 120\text{ keV} + 40\text{ keV} = 160\text{ keV}$, and

$$\lambda = \frac{hc}{E_\gamma} \quad \text{becomes} \quad \lambda = \frac{(6.63 \times 10^{-34}\text{ J s})(3 \times 10^8\text{ m/s})}{(160\text{ keV})1.6 \times 10^{-16}\text{ J/keV}} = 7.77 \times 10^{-12}\text{ m}$$

(b) $\lambda' = \dfrac{hc}{E'_\gamma} = \dfrac{(6.63 \times 10^{-34}\text{ J s})(3 \times 10^8\text{ m/s})}{(120\text{ keV})1.6 \times 10^{-16}\text{ J/keV}} = 1.04 \times 10^{-11}\text{ m}$

CHAPTER TWENTY-NINE SOLUTIONS

Therefore, $\Delta\lambda = \lambda' - \lambda = \lambda_c(1 - \cos\theta)$
$= 1.04 \times 10^{-11}$ m $- 7.77 \times 10^{-12}$ m $= (2.43 \times 10^{-12}$ m$)(1 - \cos\theta)$
From which $\theta = 93.8°$

(c) $p_\gamma = \dfrac{E_\gamma}{c} = \dfrac{h}{\lambda} = \dfrac{(6.63 \times 10^{-34} \text{ J s})}{7.77 \times 10^{-12} \text{ m}} = 8.53 \times 10^{-23}$ kg m/s

and $p'_\gamma = \dfrac{E'_\gamma}{c} = \dfrac{h}{\lambda'} = \dfrac{(6.63 \times 10^{-34} \text{ J s})}{1.04 \times 10^{-11} \text{ m}} = 6.38 \times 10^{-23}$ kg m/s

so $p_x)_{\text{before}} = p_x)_{\text{after}}$ (where θ = scattering angle of photon, and ϕ = scattering angle of electron)
$p_\gamma = p_e \cos\phi + p'_\gamma \cos\theta$ gives

$p_e \cos\phi = p_\gamma - p'_\gamma \cos\theta = 8.53 \times 10^{-23}$ kg m/s $- (6.38 \times 10^{-23}$ kg m/s$)\cos(93.8°)$, or
$p_e \cos\phi = 8.95 \times 10^{-23}$ kg m/s (1)

Also, $p_y)_{\text{before}} = p_y)_{\text{after}}$
$0 = p'_\gamma \sin\theta - p_e \sin\phi$

gives $p_e \sin\phi = (6.38 \times 10^{-23}$ kg m/s$)\sin 93.8° = 6.37 \times 10^{-23}$ kg m/s (2)
Solving (1) and (2) simultaneously gives $\phi = 35.4°$

CHAPTER THIRTY SOLUTIONS

30.1 We use $\quad \frac{1}{\lambda} = R\left(1 - \frac{1}{n^2}\right)$

When $n = 2$, $\frac{1}{\lambda} = (1.0974 \times 10^7 \text{ m}^{-1})\left(1 - \frac{1}{4}\right) = 8.23 \times 10^6 \text{ m}^{-1}$, and from this,

$\lambda = 1.215 \times 10^{-7} \text{ m} = 121.5 \text{ nm}$

When $n = 3$, $\frac{1}{\lambda} = (1.0974 \times 10^7 \text{ m}^{-1})\left(1 - \frac{1}{9}\right) = 9.75 \times 10^6 \text{ m}^{-1}$, and from this,

$\lambda = 1.025 \times 10^{-7} \text{ m} = 102.5 \text{ nm}$

When $n = 4$, $\frac{1}{\lambda} = (1.0974 \times 10^7 \text{ m}^{-1})\left(1 - \frac{1}{16}\right) = 1.029 \times 10^7 \text{ m}^{-1}$, and from this,

$\lambda = 9.72 \times 10^{-8} \text{ m} = 97.2 \text{ nm}$

These lines are in the far ultraviolet

30.2 $\frac{1}{\lambda} = R\left(\frac{1}{2^2} - \frac{1}{n^2}\right)$

When $n = 3$, $\frac{1}{\lambda} = (1.0974 \times 10^7 \text{ m}^{-1})\left(\frac{1}{4} - \frac{1}{9}\right) = 1.524 \times 10^6 \text{ m}^{-1}$, and from this,

$\lambda = 6.56 \times 10^{-7} \text{ m} = 656 \text{ nm}$

When $n = 4$, $\frac{1}{\lambda} = (1.0974 \times 10^7 \text{ m}^{-1})\left(\frac{1}{4} - \frac{1}{16}\right) = 2.058 \times 10^6 \text{ m}^{-1}$, and from this,

$\lambda = 4.86 \times 10^{-7} \text{ m} = 486 \text{ nm}$

When $n = 5$, $\frac{1}{\lambda} = (1.0974 \times 10^7 \text{ m}^{-1})\left(\frac{1}{4} - \frac{1}{25}\right) = 2.304 \times 10^6 \text{ m}^{-1}$, and from this,

$\lambda = 4.34 \times 10^{-7} \text{ m} = 434 \text{ nm}$

These lines are in the visible part of the spectrum

30.3 $\frac{1}{\lambda} = R\left(\frac{1}{3^2} - \frac{1}{n^2}\right)$

When $n = 4$ $\frac{1}{\lambda} = (1.0974 \times 10^7 \text{ m}^{-1})\left(\frac{1}{9} - \frac{1}{16}\right) = 5.33 \times 10^5 \text{ m}^{-1}$, and from this,

$\lambda = 1.875 \times 10^{-6} \text{ m} = 1875 \text{ nm}$ and $f = \frac{c}{\lambda} = 1.60 \times 10^{14} \text{ Hz}$

When $n = 5$, $\frac{1}{\lambda} = (1.0974 \times 10^7 \text{ m}^{-1})\left(\frac{1}{9} - \frac{1}{25}\right) = 7.804 \times 10^5 \text{ m}^{-1}$, and from this,

$\lambda = 1.281 \times 10^{-6} \text{ m} = 1281 \text{ nm}$ and $f = \frac{c}{\lambda} = 2.46 \times 10^{14} \text{ Hz}$

These lines are in the infrared.

CHAPTER THIRTY SOLUTIONS

30.4 The wavelengths, and hence frequencies, which can be absorbed are the same wavelengths which the gas is capable of emitting. Assuming the gas is heated sufficiently to have the n = 2 level populated, we have (See problem 30.2 for wavelengths.)

for $\lambda_1 = 656$ nm, $\quad f_1 = \dfrac{c}{\lambda_1} = \dfrac{3 \times 10^8 \text{ m/s}}{6.56 \times 10^{-7} \text{ m}} = 4.57 \times 10^{14}$ Hz

for $\lambda_2 = 486$ nm, $\quad f_2 = \dfrac{c}{\lambda_2} = \dfrac{3 \times 10^8 \text{ m/s}}{4.86 \times 10^{-7} \text{ m}} = 6.17 \times 10^{14}$ Hz

for $\lambda_3 = 434$ nm, $\quad f_3 = \dfrac{c}{\lambda_3} = \dfrac{3 \times 10^8 \text{ m/s}}{4.34 \times 10^{-7} \text{ m}} = 6.91 \times 10^{14}$ Hz

---

30.5 $\quad \dfrac{1}{\lambda} = R\left(\dfrac{1}{m^2} - \dfrac{1}{n^2}\right)$

m = 1 = Lyman series
m = 2 = Balmer series
m = 3 = Paschen series

(a) When n = 6 $\quad \dfrac{1}{\lambda} = (1.0974 \times 10^7 \text{ m}^{-1})\left(\dfrac{1}{9} - \dfrac{1}{36}\right)$, and from this,

$\lambda = 1.094 \times 10^{-6}$ m = 1094 nm

(b) When n = 3 $\quad \dfrac{1}{\lambda} = (1.0974 \times 10^7 \text{ m}^{-1})\left(\dfrac{1}{4} - \dfrac{1}{9}\right)$, and from this,

$\lambda = 656.1$ nm

(c) When n = ∞ $\quad \dfrac{1}{\lambda} = (1.0974 \times 10^7 \text{ m}^{-1})\left(\dfrac{1}{9} - \dfrac{1}{\infty}\right)$, and from this,

$\lambda = 820.1$ nm

(d) When n = 3 $\quad \dfrac{1}{\lambda} = (1.0974 \times 10^7 \text{ m}^{-1})\left(\dfrac{1}{1} - \dfrac{1}{9}\right)$, and from this,

$\lambda = 102.5$ nm

30.6 $\quad \dfrac{1}{\lambda} = R\left(\dfrac{1}{m^2} - \dfrac{1}{n^2}\right)$

m = 1 = Lyman series
m = 2 = Balmer series
m = 3 = Paschen series

(a) For series limit (shortest wavelength), n = ∞

Lyman series $\quad \dfrac{1}{\lambda} = (1.0974 \times 10^7 \text{ m}^{-1})\left(\dfrac{1}{1} - 0\right)$, and from this,

$\lambda = \dfrac{1}{R} = 91.13$ nm

Balmer series $\quad \dfrac{1}{\lambda} = (1.0974 \times 10^7 \text{ m}^{-1})\left(\dfrac{1}{4} - 0\right)$, and from this,

$\lambda = \dfrac{4}{R} = 364.5$ nm

Paschen series $\quad \dfrac{1}{\lambda} = (1.0974 \times 10^7 \text{ m}^{-1})\left(\dfrac{1}{9} - 0\right)$, and from this,

$\lambda = \dfrac{9}{R} = 820.1$ nm

CHAPTER THIRTY SOLUTIONS

Brackett series $\frac{1}{\lambda} = (1.0974 \times 10^7 \text{ m}^{-1})\left(\frac{1}{16} - 0\right)$, and from this,

$\lambda = \frac{16}{R} = 1458$ nm

(b) $E_g = \frac{hc}{\lambda} = \frac{1241.6 \text{ nm eV}}{\lambda}$ yields

$E_g = 13.6$ eV Lyman
$E_g = 3.41$ eV Balmer
$E_g = 1.51$ eV Paschen
$E_g = 0.85$ eV Brackett

30.7 $\frac{1}{\lambda} = R\left(1 - \frac{1}{n^2}\right)$ for Lyman series

Thus, if $\lambda = 94.96$ nm, we have

$\frac{1}{94.96 \times 10^{-9} \text{ m}} = (1.0974 \times 10^7 \text{ m}^{-1})\left(1 - \frac{1}{n^2}\right)$

and $n = 5$

(b) This spectral line could not be associated with the Paschen or Brackett series because it is shorter than the series limit of either of these series. (See problem 30.6.)

30.8 $\frac{mk^2e^4}{2\hbar^2} = \frac{(9.11 \times 10^{-31} \text{ kg})(9 \times 10^9 \text{ N m}^2/\text{C}^2)^2(1.6 \times 10^{-19} \text{ C})^4}{2(1.055 \times 10^{-34} \text{ J s})^2}$

$= 2.173 \times 10^{-18}$ J $= 13.6$ eV

30.9 Starting with $\frac{1}{2}mv^2 = \frac{ke^2}{2r}$

we have $v^2 = \frac{ke^2}{mr}$ and using $r_n = \frac{n^2\hbar^2}{mke^2}$

gives $v_n^2 = \frac{ke^2}{m\frac{n^2\hbar^2}{mke^2}}$ or $v_n = \frac{ke^2}{n\hbar}$, which for $n = 1$ becomes

$\frac{v}{c} = \frac{ke^2}{\hbar c} = \frac{(9 \times 10^9 \text{ N m}^2/\text{C}^2)(1.6 \times 10^{-19} \text{ C})^2}{(1.055 \times 10^{-34} \text{ J s})(3 \times 10^8 \text{ m/s})} = 7.2796 \times 10^{-3} = \frac{1}{137}$

30.10 Using the energy level diagram of Fig. 30.6, we have
(a) six different transitions, and hence wavelengths, possible. The transitions are
$n = 4$ to $n = 1$, $n = 4$ to $n = 2$, $n = 4$ to $n = 3$, $n = 3$ to $n = 1$, $n = 3$ to $n = 2$, $n = 2$ to $n = 1$.
(b) Transition $n = 4$ to $n = 3$ yields the smallest energy photon and hence the longest wavelength:

$E = \frac{hc}{\lambda}$ gives $\lambda = \frac{hc}{E} = \frac{1241.6 \text{ nm eV}}{(-0.85 \text{ eV} - (-1.51 \text{ eV}))} = 1880$ nm (a member of the Paschen family)

30.11 (a) The energy of the photon is found as

$E = E_i - E_f = \frac{-13.6 \text{ eV}}{n_i^2} - \frac{(-13.6 \text{ eV})}{n_f^2} = 13.6 \text{ eV}\left(\frac{1}{n_f^2} - \frac{1}{n_i^2}\right)$

CHAPTER THIRTY SOLUTIONS

Thus, for n = 3 to n = 2 transition, $\quad E = 13.6 \text{ eV}\left(\frac{1}{4} - \frac{1}{9}\right) = 1.89 \text{ eV}$

(b) $E = \frac{hc}{\lambda}$ and $\quad \lambda = \frac{1243 \text{ nm eV}}{1.89 \text{ eV}} = 658 \text{ nm}$

(c) $f = \frac{c}{\lambda} = \frac{3 \times 10^8 \text{ m/s}}{6.58 \times 10^{-7} \text{ m}} = 4.56 \times 10^{14} \text{ Hz}$

30.12 We use $E_n = \frac{-13.6 \text{ eV}}{n^2}$

To ionize the atom when the electron is in the nth level, it is necessary to add an amount of energy given by $\quad E = -E_n = \frac{13.6 \text{ eV}}{n^2}$

(a) Thus, in the ground state where n = 1, we have $\quad E = 13.6 \text{ eV}$

(b) In the n = 3 level, $\quad E = \frac{13.6 \text{ eV}}{9} = 1.51 \text{ eV}$

30.13 From the energy level sketch shown, we find the energy of the absorbed photon, E, is

$E = E_3 - E_1 = -1.51 \text{ eV} - (-13.6 \text{ eV}) = 12.09 \text{ eV}$.

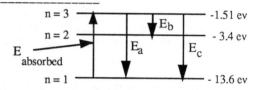

(b) The three possible transitions are shown in the energy level diagram. The energies are

$E_a = E_i - E_f = -1.51 \text{ eV} - (-13.6 \text{ eV}) = 12.09 \text{ eV}$
$E_b = E_i - E_f = -1.51 \text{ eV} - (-3.4 \text{ eV}) = 1.89 \text{ eV}$
$E_c = E_i - E_f = -3.4 \text{ eV} - (-13.6 \text{ eV}) = 10.2 \text{ eV}$

30.14 $r_0 = \frac{\hbar^2}{mke^2} = \frac{(1.055 \times 10^{-34} \text{ J s})^2}{(9.11 \times 10^{-31} \text{ kg})(9 \times 10^9 \text{ N m}^2/\text{C}^2)(1.6 \times 10^{-19} \text{ C})^2}$
$= 0.529 \times 10^{-10} \text{ m}$

30.15 $R = \frac{mk^2e^4}{4\pi c\hbar^3} = \frac{(9.11 \times 10^{-31} \text{ kg})(9 \times 10^9 \text{ N m}^2/\text{C}^2)^2(1.6 \times 10^{-19} \text{ C})^4}{4\pi(3 \times 10^8 \text{ m/s})(1.055 \times 10^{-34} \text{ J s})^3}$
$= 1.09243 \times 10^7 \text{ m}^{-1}$

30.16 $r_n = n^2 r_0 = n^2(0.0529 \text{ nm})$
For n = 1, $\quad r_1 = r_0 = 0.0529 \text{ nm}$
For n = 2, $\quad r_2 = 4r_0 = 4(0.0529 \text{ nm}) = 0.212 \text{ nm}$
For n = 3, $\quad r_3 = 9r_0 = 9(0.0529 \text{ nm}) = 0.476 \text{ nm}$

30.17 Starting with $\frac{1}{2}mv^2 = \frac{ke^2}{2r}$

we have $\quad v^2 = \frac{ke^2}{mr}$ and using $r_n = \frac{n^2\hbar^2}{mke^2}$

gives $v_n^2 = \frac{ke^2}{m\frac{n^2\hbar^2}{mke^2}}$ or $\quad v_n = \frac{ke^2}{n\hbar}$

420

CHAPTER THIRTY SOLUTIONS

30.18 (a) $r_n = n^2 r_0 = (2)^2 (0.0529 \text{ nm}) = 0.212 \text{ nm}$

(b) $p = mv = m\dfrac{ke^2}{n\hbar}$ (see problem 30.17) gives

$$p = \dfrac{(9.11 \times 10^{-31} \text{ kg})(9 \times 10^9 \text{ N m}^2/\text{C}^2)(1.6 \times 10^{-19} \text{ C})^2}{2(1.05 \times 10^{-34} \text{ J s})} = 9.995 \times 10^{-25} \text{ kg m/s}$$

(c) $L = n\hbar = 2(1.05 \times 10^{-34} \text{ J s}) = 2.10 \times 10^{-34} \text{ J s}$

(d) $KE = \dfrac{ke^2}{2r} = \dfrac{(9 \times 10^9 \text{ N m}^2/\text{C}^2)(1.6 \times 10^{-19} \text{ C})^2}{2(0.212 \times 10^{-9} \text{ m})} = 5.434 \times 10^{-19} \text{ J} = 3.39 \text{ eV}$

(e) $PE = -\dfrac{ke^2}{r} = -\dfrac{(9 \times 10^9 \text{ N m}^2/\text{C}^2)(1.6 \times 10^{-19} \text{ C})^2}{(0.212 \times 10^{-9} \text{ m})}$
$= -1.0868 \times 10^{-18} \text{ J} = -6.78 \text{ eV}$

(f) $E = KE + PE = 3.39 \text{ eV} - 6.78 \text{ eV} = -3.39 \text{ eV}$

30.19 $E = \dfrac{13.6}{n^2}$ becomes for (n = 1) 13.6 eV,

and $\lambda = \dfrac{hc}{E} = \dfrac{1243 \text{ nm eV}}{13.6 \text{ eV}} = 91.4 \text{ nm}$

30.20 $r = r_0 = 0.529 \times 10^{-10} \text{ m}$

Thus, $F = \dfrac{kq_1 q_2}{r^2} = \dfrac{(9 \times 10^9 \text{ N m}^2/\text{C}^2)(1.6 \times 10^{-19} \text{ C})^2}{(0.529 \times 10^{-10} \text{ m})^2} = 8.23 \times 10^{-8} \text{ N}$

30.21 $L = n\hbar = 3(1.05 \times 10^{-34} \text{ J s}) = 3.15 \times 10^{-34} \text{ J s}$

30.22 For absorption, $E = E_f - E_i = -\dfrac{13.6 \text{ eV}}{n^2_f} - \left(-\dfrac{13.6 \text{ eV}}{n^2_i}\right) = 13.6\left(\dfrac{1}{n^2_i} - \dfrac{1}{n^2_f}\right)$

(a) $E = 13.6\left(\dfrac{1}{9} - \dfrac{1}{25}\right) = 0.97 \text{ eV}$

(b) $E = 13.6\left(\dfrac{1}{25} - \dfrac{1}{49}\right) = 0.27 \text{ eV}$

30.23 $\Delta E = 13.6\left(\dfrac{1}{n^2_i} - \dfrac{1}{n^2_f}\right)$ (see problem 30.22) Where for $\Delta E > 0$ we have absorption and for $\Delta E < 0$ we have emission.
(A) for $n_i = 2$ and $n_f = 5$ $\Delta E = 2.86 \text{ eV}$ (absorption)
(B) for $n_i = 5$ and $n_f = 3$ $\Delta E = -0.97 \text{ eV}$ (emission)
(C) for $n_i = 7$ and $n_f = 4$ $\Delta E = -0.57 \text{ eV}$ (emission)
(D) for $n_i = 4$ and $n_f = 7$ $\Delta E = 0.57 \text{ eV}$ (absorption)

(a) $E = \dfrac{hc}{\lambda}$ so the shortest wavelength is emitted in transition B

(b) atom gains most energy in transition A

(c) atom loses energy in transitions B and C

30.24 The shortest wavelength corresponds to the highest energy emitted. For the Lyman series, the energy of the photon is given by

CHAPTER THIRTY SOLUTIONS

$E = 13.6 \text{ eV}\left(1 - \dfrac{1}{n^2}\right)$ and E is a max when n = infinity. Thus, $E_{max} = 13.6$ eV

and $\lambda_{min} = \dfrac{hc}{E_{max}} = \dfrac{1243 \text{ nm eV}}{13.6 \text{ eV}} = 91.4$ nm

---

30.25 (a) In the first Bohr orbit, we have n = 1 and r = 0.529 X 10^{-10} m. From the quantization of angular momentum, we have

$v_n = \dfrac{n\hbar}{mr_n}$, or $v_1 = \dfrac{n\hbar}{mr_0} = \dfrac{1.055 \times 10^{-34} \text{ J s}}{(9.11 \times 10^{-31} \text{ kg})(0.529 \times 10^{-10})} = 2.19 \times 10^6$ m/s

(b) $t = \dfrac{2\pi r}{v} = \dfrac{2\pi(0.529 \times 10^{-10} \text{ m})}{2.19 \times 10^6 \text{ m/s}} = 1.52 \times 10^{-16}$ s

(c) $I = \dfrac{\Delta Q}{\Delta t} = \dfrac{e}{t} = \dfrac{1.6 \times 10^{-19} \text{ C}}{1.52 \times 10^{-16} \text{ s}} = 1.05 \times 10^{-3}$ A

30.26 (a) The velocity of the moon in its orbit is

$v = \dfrac{2\pi r}{T} = \dfrac{2\pi(3.84 \times 10^8 \text{ m})}{2.36 \times 10^6 \text{ s}} = 1.02 \times 10^3$ m/s

So,

$L = mvr = (7.36 \times 10^{22} \text{ kg})(1.02 \times 10^3 \text{ m/s})(3.84 \times 10^8 \text{ m}) = 2.89 \times 10^{34}$ kg m²/s

(b) We have $L = n\hbar$, or $n = \dfrac{L}{\hbar} = \dfrac{2.89 \times 10^{34} \text{ kg m}^2/\text{s}}{1.055 \times 10^{-34} \text{ J s}} = 2.74 \times 10^{68}$

(c) We have $\dfrac{\Delta r}{r} = \dfrac{(n+1)^2 R - n^2 R}{n^2 R} = \dfrac{2n+1}{n^2}$

which is approximately equal to $\dfrac{2}{n} = 7.30 \times 10^{-69}$

30.27 $E = 13.6\left(\dfrac{1}{n^2_i} - \dfrac{1}{n^2_f}\right)$

(a) $E = 13.6\left(\dfrac{1}{1} - \dfrac{1}{4}\right) = 10.2$ eV

and $f = \dfrac{E}{h} = \dfrac{10.2 \text{ eV}(1.602 \times 10^{-19} \text{ J/eV})}{(6.63 \times 10^{-34} \text{ J s})} = 2.46 \times 10^{15}$ Hz

Period $= \dfrac{1}{f_{orb}} = \dfrac{2\pi r_n}{v_n}$, or $f_{orb} = \dfrac{v_n}{2\pi r_n}$ Now substitute for v_n using the expression in problem 30.17 and for r_n from the text, to find

$f = \dfrac{mk^2e^4}{2\pi n^3 \hbar^3} = \dfrac{(9.11 \times 10^{-31} \text{ kg})(9 \times 10^9 \text{ Nm}^2/\text{C}^2)^2(1.6 \times 10^{-19} \text{ C})^4}{2\pi(2)^3(1.05 \times 10^{-34} \text{ J s})^3}$

$= 8.31 \times 10^{14}$ Hz

(b) $E = 13.6\left(\dfrac{1}{(9999)^2} - \dfrac{1}{(10000)^2}\right) = 2.72 \times 10^{-11}$ eV

and $f = \dfrac{E}{h}$ gives $f = 6565$ Hz

and $f_{orb} = \dfrac{mk^2e^4}{2\pi n^3 \hbar^3}$ yields 6649 Hz

For small n, significant differences between classical and quantum results appear. However, as n becomes large, classical theory and quantum theory approach one another in their results. (correspondence principle)

CHAPTER THIRTY SOLUTIONS

30.28 The plot originates from the equation $E_n = -\dfrac{Z^2(13.6\text{ eV})}{n^2}$

with $Z = 2$, $E_n = -\dfrac{4(13.6\text{ eV})}{n^2} = -\dfrac{54.4\text{ eV}}{n^2}$

and the diagram should look somewhat like that sketched below.

```
n = ∞ ─────────────────────  E = 0
n = 5 ─────────────────────  -2.18 ev
n = 4 ─────────────────────  -3.40 ev

n = 3 ─────────────────────  -6.04 ev


n = 2 ─────────────────────  -13.6 ev




n = 1 ─────────────────────  -54.4 ev
```

30.29 (a) From $E_n = -\dfrac{Z^2(13.6\text{ eV})}{n^2}$

we have, with $Z = 3$ and $n = 1$, $\quad E_1 = -9(13.6\text{ eV}) = -122.4\text{ eV}$

(b) From $r_n = \dfrac{n^2 r_0}{Z}$ we have $\quad r_1 = \dfrac{0.529 \times 10^{-10}\text{ m}}{3} = 1.76 \times 10^{-11}\text{ m}$

30.30 We use $E = -\dfrac{Z^2(13.6\text{ eV})}{n^2}$ with $Z = 3$ to give $E = -\dfrac{122.4\text{ eV}}{n^2}$ to give

```
n = ∞ ─────────────────────  E = 0
n = 5 ─────────────────────  -4.90 ev
n = 4 ─────────────────────  -7.65 ev

n = 3 ─────────────────────  -13.6 ev


n = 2 ─────────────────────  -30.6 ev




n = 1 ─────────────────────  -122.4 ev
```

30.31 $r_n = \dfrac{n^2 r_0}{Z}$ we have $\quad r_n = \dfrac{0.529 \times 10^{-10}\text{ m}}{Z}$

(a) for He^+ $r_n = \dfrac{0.529 \times 10^{-10}\text{ m}}{2} = 0.265 \times 10^{-10}\text{ m} = 0.265\text{ Å}$

(b) for Be^{3+} $r_n = \dfrac{0.529 \times 10^{-10}\text{ m}}{4} = 0.132 \times 10^{-10}\text{ m} = 0.132\text{ Å}$

CHAPTER THIRTY SOLUTIONS

30.32 $E = -\dfrac{Z^2(13.6 \text{ eV})}{n^2}$ with $Z = 2$ for helium to give $E_n = -\dfrac{54.4 \text{ eV}}{n^2}$

For the Lyman series $n_f = 1$ and $E = -54.4$ eV

The longest wavelength occurs when $n_i = 2$, so $E_i = -\dfrac{54.4 \text{ eV}}{4} = -13.6$ eV

Thus, $E = E_i - E_f = 40.8 \text{ eV} = \dfrac{hc}{\lambda}$

so $\lambda_{long} = \dfrac{(6.63 \times 10^{-34} \text{ J s})(3 \times 10^8 \text{ m/s})}{(40.8 \text{ eV})(1.6 \times 10^{-19} \text{ J/eV})} = 30.4$ nm

The shortest wavelength occurs when $n_i = \infty$
so $E = 54.4$ eV

and $\lambda_{short} = \dfrac{(6.63 \times 10^{-34} \text{ J s})(3 \times 10^8 \text{ m/s})}{(54.4 \text{ eV})(1.6 \times 10^{-19} \text{ J/eV})} = 22.8$ nm

---

30.33 We have $R = \dfrac{Z^2 m k^2 e^4}{r \pi c \hbar^2} = Z^2 R_{hydrogen}$

For singly ionized helium $Z = 2$, and we have
$R = 4 R_{hydrogen} = 4(1.0974 \times 10^7 \text{ m}^{-1}) = 4.39 \times 10^7 \text{ m}^{-1}$

(b) $\dfrac{1}{\lambda} = R\left(\dfrac{1}{n_f^2} - \dfrac{1}{n_i^2}\right) = 4.39 \times 10^7 \text{ m}^{-1}\left(\dfrac{1}{1} - \dfrac{1}{4}\right) = 3.29 \times 10^7 \text{ m}^{-1}$

or, $\lambda = 3.03 \times 10^{-8}$ m $= 30.3$ nm

(c) This wavelength is in the deep ultraviolet region.

30.34 In the P shell, $n = 6$ and l goes 0, 1, 2, 3, 4, 5

subshell	notation	number of electrons
$n = 6, l = 0$	6s	$2(2l + 1) = 2(1) = 2$
$n = 6, l = 1$	6p	$2(2l + 1) = 2(3) = 6$
$n = 6, l = 2$	6d	$2(2l + 1) = 2(5) = 10$
$n = 6, l = 3$	6f	$2(2l + 1) = 2(7) = 14$
$n = 6, l = 4$	6g	$2(2l + 1) = 2(9) = 18$
$n = 6, l = 5$	6h	$2(2l + 1) = 2(11) = 22$
(a)		(b)

(c) The total number of electrons that can exist in the P shell
$= 2 + 6 + 10 + 14 + 18 + 22 = 72$

CHAPTER THIRTY SOLUTIONS

30.35 In the 3d subshell, n = 3 and $l = 2$, We have

n	l	m_l	m_s
3	2	+2	+1/2
3	2	+2	−1/2
3	2	+1	+1/2
3	2	+1	−1/2
3	2	0	+1/2
3	2	0	−1/2
3	2	−1	+1/2
3	2	−1	−1/2
3	2	−2	+1/2
3	2	−2	−1/2

(A total of 10 states.)

30.36 In the 30 subshell, n = 3 and $l = 1$. We have

n	l	m_l	m_s
3	1	+1	+1/2
3	1	+1	−1/2
3	1	0	+1/2
3	1	0	−1/2
3	1	−1	+1/2
3	1	−1	−1/2

(A total of 6 states.)

30.37 We list the quantum numbers as
n = 4 $l = 3$ $m_l = \pm 3, \pm 2, \pm 1, 0$
          $l = 2$ $m_l = \pm 2, \pm 1, 0$
          $l = 1$ $m_l = \pm 1, 0$
          $l = 0$ $m_l = 0$

There are (a) 4 different values for l, and (b) there are 16 different combinations, but only 7 different (or distinct) values for m_l.

30.38 (a) For helium, Z = 2, so there are 2 electrons. Their quantum numbers are
n = 1, $l = 0$, $m_l = 0$, and $m_s = \pm 1/2$ Thus, the electronic configuration is $1s^2$
(b) For neon, Z = 10. Two of these electrons are in the 1s level, two are in the 2s level, and the remaining 6 are in the 2p level. The configuration is
$1s^2 2s^2 2p^6$
(c) For argon, Z = 18. Ten of the electrons complete the neon configuration above. Two of the remaining eight are in the 3s level, and the last six are in the 3p level. The complete configuration is written as $1s^2 2s^2 2p^6 3s^2 3p^6$

30.39 (a) The electronic configuration for oxygen (Z = 8) is $1s^2 2s^2 2p^4$
(b) The quantum numbers are

$1s^2$ state   n = 1 $l = 0$ $m_l = 0$ $m_s = \pm 1/2$
$2s^2$ state   n = 2 $l = 0$ $m_l = 0$ $m_s = \pm 1/2$
$2p^4$ state   n = 2 $l = 1$ $m_l = -1$ $m_s = \pm 1/2$
                              $m_l = 0$ $m_s = \pm 1/2$

CHAPTER THIRTY SOLUTIONS

30.40 (a) For n = 1, $l = 0$, and there are $2(2l + 1)$ states = $2(1) = 2$ sets of quantum numbers
(b) For n = 2, $l = 0$ for $2(2l + 1)$ states = $2(1) = 2$ sets
     and $l = 1$ for $2(2l + 1)$ states = $2(3) = 6$ sets
     total number of sets = 8
(c) For n = 3 $l = 0$ for $2(2l + 1)$ states = $2(1) = 2$ sets
     and $l = 1$ for $2(2l + 1)$ states = $2(3) = 6$ sets
     and $l = 2$ for $2(2l + 1)$ states = $2(5) = 10$ sets
     total number of sets = 18
(d) For n = 4 $l = 0$ for $2(2l + 1)$ states = $2(1) = 2$ sets
     and $l = 1$ for $2(2l + 1)$ states = $2(3) = 6$ sets
     and $l = 2$ for $2(2l + 1)$ states = $2(5) = 10$ sets
     and $l = 3$ for $2(2l + 1)$ states = $2(7) = 14$ sets
     total number of sets = 32
(e) For n = 5 $l = 0$ for $2(2l + 1)$ states = $2(1) = 2$ sets
     and $l = 1$ for $2(2l + 1)$ states = $2(3) = 6$ sets
     and $l = 2$ for $2(2l + 1)$ states = $2(5) = 10$ sets
     and $l = 3$ for $2(2l + 1)$ states = $2(7) = 14$ sets
     and $l = 4$ for $2(2l + 1)$ states = $2(9) = 18$ sets
     total number of sets = 50

For n = 1, $2n^2 = 2$
For n = 2 $2n^2 = 8$
For n = 3 $2n^2 = 18$
For n = 4 $2n^2 = 32$
For n = 5 $2n^2 = 50$

Thus, the number of sets of quantum states agrees with the $2n^2$ rule.

30.41 $n = 3, l = 0, m_l = 0$, and $m_s = \pm 1/2$
(a) If the exclusion principle were inoperative, so that electrons could be in identical states, there would be 4 possible states for the system as shown:

state	electron 1 (n, l, m_l, m_s)	electron 2 (n, l, m_l, m_s)
1	$3, 0, 0, +\frac{1}{2}$	$3, 0, 0, +\frac{1}{2}$
2	$3, 0, 0, +\frac{1}{2}$	$3, 0, 0, -\frac{1}{2}$
3	$3, 0, 0, -\frac{1}{2}$	$3, 0, 0, +\frac{1}{2}$
4	$3, 0, 0, -\frac{1}{2}$	$3, 0, 0, -\frac{1}{2}$

(b) Taking the exclusion principle into account, it is seen that states 1 and 4 above are not allowed, leaving only the remaining two states.

30.42 $E = E_L - E_K = -951 \text{ eV} - (-8979 \text{ eV}) = 8028 \text{ eV}$

and $\lambda = \dfrac{(6.63 \times 10^{-34} \text{ J s})(3 \times 10^8 \text{ m/s})}{(8028 \text{ eV})(1.6 \times 10^{-19} \text{ J/eV})} = 0.155 \text{ nm}$

To produce a K_α line, an electron must be excited from the K shell to the L shell and must give 8028 eV to the atom. Thus, the target must be bombarded with 8028 eV electrons, so an accelerating voltage of 8.03 kV is needed.

CHAPTER THIRTY SOLUTIONS

30.43 $E_K = -(Z-1)^2(13.6 \text{ eV}) = -(28-1)^2(13.6 \text{ eV}) = -(27)^2(13.6 \text{ eV}) = -9914 \text{ eV}$

and $E_L = -(Z-3)^2 \dfrac{(13.6 \text{ eV})}{2^2} = -(28-3)^2 \dfrac{(13.6 \text{ eV})}{4} = -(25)^2 \dfrac{(13.6 \text{ eV})}{4} = -2125 \text{ eV}$

Thus, $\Delta E = E_L - E_K = -2125 \text{ eV} - (-9914 \text{ eV}) = 7789 \text{ eV}$ = the energy, E, of the x-ray emitted.

and $f = \dfrac{E}{h} = \dfrac{(7789 \text{ eV})(1.6 \times 10^{-19} \text{ J/eV})}{6.63 \times 10^{-34} \text{ J s}} = 1.88 \times 10^{18}$ Hz

---

30.44 $E_K = -(Z-1)^2(13.6 \text{ eV}) = -(42-1)^2(13.6 \text{ eV}) = -(41)^2(13.6 \text{ eV}) = -22{,}862 \text{ eV}$

and $E_L = -(Z-3)^2 \dfrac{(13.6 \text{ eV})}{2^2} = -(42-3)^2 \dfrac{(13.6 \text{ eV})}{4} = -(39)^2 \dfrac{(13.6 \text{ eV})}{4} = -5171 \text{ eV}$

Thus, $\Delta E = E_L - E_K = -5171 \text{ eV} - (-22{,}862 \text{ eV}) = 17{,}691 \text{ eV}$ = the energy, E, of the x-ray emitted.

so, $\lambda = \dfrac{hc}{E} = \dfrac{1243 \text{ nm eV}}{17691 \text{ eV}} = 7.03 \times 10^{-2}$ nm

---

30.45 $E = \dfrac{hc}{\lambda}$

$= \dfrac{1241.6 \text{ eV nm}}{\lambda}$

$= \dfrac{1.2416 \text{ keV nm}}{\lambda}$

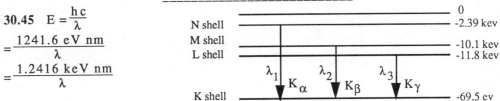

for $\lambda_1 = 0.0185$ nm, E = 67.11 keV

$\lambda_2 = 0.0209$ nm, E = 59.4 keV

$\lambda_3 = 0.0215$ nm, E = 57.7 keV

The ionization energy for K shell = 69.5 keV, so, the ionization energies for the other shells are L shell = 11.8 keV : M shell = 10.1 keV : N shell = 2.39 keV

30.46 $E = \dfrac{hc}{\lambda} = \dfrac{1241.6 \text{ eV nm}}{\lambda} = \dfrac{1241.6 \text{ eV nm}}{97.3 \text{ nm}} = 12.76$ eV

$E = E_f - E_i$ gives $E_f = E + E_i = 12.76 - 13.6 \text{ eV} = -0.84$ eV

and $E_f = -\dfrac{13.6 \text{ eV}}{n_f^2}$ -0.84 eV, gives $n_f = 4$

The sketch and calculation of wavelengths are given below:

(1) $\lambda = \dfrac{1241.6 \text{ nm eV}}{E_i - E_f} = \dfrac{1241.6 \text{ nmeV}}{-0.84 \text{ eV} - (-13.6 \text{ eV})}$

$= 97.3$ nm

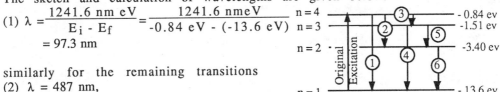

similarly for the remaining transitions
(2) $\lambda = 487$ nm,
(3) $\lambda = 1881$ nm,
(4) $\lambda = 103$ nm,
(5) $\lambda = 657$ nm, (6) $\lambda = 1217$ nm

30.47 The energy emitted in 1 s is $E = (\text{power})t = (0.5 \text{ J/s})(1 \text{ s}) = 0.5$ J

and the energy per photon is $E' = \dfrac{hc}{\lambda} = \dfrac{1243 \text{ nm eV}}{640 \text{ nm}} = 1.94 \text{ eV} = 3.11 \times 10^{-19}$ J

CHAPTER THIRTY SOLUTIONS

Thus, the number of photons per second $= \dfrac{E}{E'}$

$= \dfrac{5 \times 10^{-1} \text{ J}}{3.11 \times 10^{-19} \text{ J/photon}} = 1.61 \times 10^{18}$ photons

30.48 (a) The energy difference between these two states is equal to the energy that is absorbed.

Thus, $E = E_2 - E_1 = \dfrac{-13.6}{4} - \dfrac{(-13.6 \text{ eV})}{1} = 10.2 \text{ eV} = 1.63 \times 10^{-18}$ J

(b) We have $E = \dfrac{3}{2} kT$, or $T = \dfrac{2}{3k} E = \dfrac{2(1.63 \times 10^{-18} \text{ J})}{3(1.38 \times 10^{-23} \text{ J/K})} = 7.88 \times 10^4$ K

30.49 (a) $r_n = n^2 r_0 = (3)^2 (0.0529 \text{ nm}) = 0.476$ nm

(b) From the quantization of angular momentum, we have

$L_n = mv_n r_n = n\hbar$, or $v_n = \dfrac{n\hbar}{mr_n}$

Thus, $v_3 = \dfrac{3(1.055 \times 10^{-34} \text{ J s})}{(9.11 \times 10^{-31} \text{ kg})(4.76 \times 10^{-10} \text{ m})} = 7.30 \times 10^5$ m/s

(c) $\lambda = \dfrac{h}{p} = \dfrac{h}{mv} = \dfrac{6.63 \times 10^{-34} \text{ J s}}{(9.11 \times 10^{-31} \text{ kg})(7.30 \times 10^5 \text{ m/s})} = 0.997$ nm

(d) The ratio of the circumference divided by the wavelength is

$\dfrac{\text{circumference}}{\lambda} = \dfrac{2\pi r}{\lambda} = \dfrac{2\pi (4.76 \times 10^{-10} \text{ m})}{9.97 \times 10^{-10} \text{ m}} = 3$

30.50 (a) The energy is found as,

$E = (\text{power})t = (5 \times 10^{-3} \text{ J/s})(25 \times 10^{-3} \text{ s}) = 1.25 \times 10^{-4}$ J

(b) The energy of a single photon is

$E' = \dfrac{hc}{\lambda} = \dfrac{(6.63 \times 10^{-34} \text{ J s})(3 \times 10^8 \text{ m/s})}{632.8 \times 10^{-9} \text{ m}} = 3.14 \times 10^{-19}$ J/photon

Thus, the number of photons per pulse is

$\dfrac{E}{E'} = \dfrac{1.25 \times 10^{-4} \text{ J}}{3.14 \times 10^{-19} \text{ J/photon}} = 3.98 \times 10^{14}$ photons

---

30.51 $E = KE + PE = \dfrac{1}{2} mv^2 - \dfrac{ke^2}{r}$ but $\dfrac{1}{2} mv^2 = \dfrac{1}{2} \dfrac{ke^2}{r}$

Thus, $E = -\dfrac{1}{2} \dfrac{ke^2}{r} = \dfrac{PE}{2}$, so $PE = 2E = 2(-13.6 \text{ eV}) = -27.2$ eV

and $KE = E - PE = -13.6 \text{ eV} - (-27.2 \text{ eV}) = 13.6$ eV

30.52 (a) $r_1 = (0.0529 \text{ nm})n^2 = 0.0529$ nm (when $n = 1$)

(b) $mv = m\sqrt{\dfrac{ke^2}{mr}} = \sqrt{\dfrac{(9.1 \times 10^{-31} \text{ kg})(9 \times 10^9 \text{ N m}^2/\text{C}^2)}{5.29 \times 10^{-11} \text{ m}}} (1.6 \times 10^{-19} \text{ C})$

$= 1.99 \times 10^{-24}$ kg m/s

(c) $L = mvr = (1.99 \times 10^{-24} \text{ kg m/s})(5.29 \times 10^{-11} \text{ m}) = 1.05 \times 10^{-34}$ kg m²/s

(d) $KE = 13.6$ eV (see preceding problem)

(e) $PE = -27.2$ eV (see preceding problem)

(f) $E = K + PE = -13.6$ eV

CHAPTER THIRTY SOLUTIONS

30.53 (a & b) $E = \dfrac{hc}{\lambda}$

$= \dfrac{1241.6 \text{ eV nm}}{\lambda}$

we have
$\lambda = 253.7$ nm, $E = 4.894$ eV
$\lambda = 185.0$ nm, $E = 6.711$ eV
$\lambda = 158.5$ nm, $E = 7.833$ eV

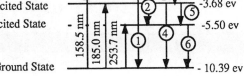

The wavelengths of the emission lines shown are:
$\lambda_1 = 158.5$ nm, $\lambda_2 = 422$ nm, $\lambda_3 = 1109$ nm, $\lambda_4 = 185$ nm, $\lambda_5 = 682$ nm, $\lambda_6 = 253.7$ nm

(c) To have an inelastic collision, we must excite the atom from ground state to the first excited state, and electron must have a minimum kinetic energy of 10.39 eV $- 5.50$ eV $= 4.89$ eV

so $v = \sqrt{\dfrac{2(4.89 \text{ eV})(1.6 \times 10^{-19} \text{ J/eV})}{9.11 \times 10^{-31} \text{ kg}}} = 1.31 \times 10^6$ m/s

30.54 $E = \dfrac{hc}{\lambda} = \dfrac{1241.6 \text{ eV nm}}{\lambda}$

$= \Delta E$
$\lambda_1 = 310$ nm, so $\Delta E_1 = 4.005$ eV
$\lambda_2 = 400$ nm, $\Delta E_2 = 3.104$ eV
$\lambda_3 = 1378$ nm, $\Delta E_3 = 0.901$ eV
and the ionization energy = 4.10 eV

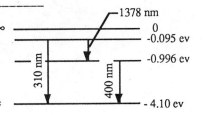

The energy level diagram having the fewest number of levels and consistent with these energies is shown above.

30.55 Since $\lambda = d = 2r_0 = 2(0.0529 \text{ nm}) = 0.1058$ nm, the energy of the electron is nonrelativistic, so we can use $p = \dfrac{h}{\lambda}$ and $KE = \dfrac{p^2}{2m}$

$KE = \dfrac{h^2}{2m\lambda^2} = \dfrac{(6.626 \times 10^{-34} \text{ J s})^2}{2(9.1 \times 10^{-31} \text{ kg})(0.1058 \times 10^{-9})^2} = 2.155 \times 10^{-17}$ J $= 135$ eV This is about 10 times as large as the ground-state energy of hydrogen, which is 13.6 eV.

30.56 (a) In this problem, the electron must be treated relativistically. The momentum of the electron is
$p = \dfrac{h}{\lambda} = \dfrac{6.626 \times 10^{-34} \text{ J s}}{10^{-14} \text{ m}} = 6.626 \times 10^{-20}$ kg m/s
The energy of the electron is
$E = \sqrt{p^2 c^2 + m^2 c^4}$
$= \sqrt{(6.626 \times 10^{-20})^2 (3 \times 10^8)^2 + (0.511 \times 10^6)^2 (1.6 \times 10^{-19})^2}$
$E = 1.99 \times 10^{-11}$ J $= 1.24 \times 10^8$ eV   so that $K = E - mc^2 \approx 124$ MeV
(b) The kinetic energy is too large to expect that the electron could be confined to a region the size of the nucleus.

CHAPTER THIRTY SOLUTIONS

30.57 $hf = \Delta E = \dfrac{4\pi^2 mk^2 e^4}{2h^2}\left(\dfrac{1}{(n-1)^2} - \dfrac{1}{n^2}\right)$

$f = \dfrac{2\pi^2 mk^2 e^4}{h^3}\left(\dfrac{2n-1}{(n-1)^2 n^2}\right)$

30.58 As n approaches infinity, we have f approaching $\dfrac{2\pi^2 mk^2 e^4}{h^3}\dfrac{2}{n^3}$

The classical frequency is $f = \dfrac{v}{2\pi r} = \dfrac{1}{2\pi}\sqrt{\dfrac{ke^2}{m}\dfrac{1}{r^{3/2}}}$ where $r = \dfrac{n^2 h^2}{4\pi mke^2}$

Using this equation to eliminate r from the expression for f, we find

$f = \dfrac{2\pi^2 mk^2 e^4}{h^3}\dfrac{2}{n^3}$

30.59 (a) $\Delta x \Delta p = \hbar$ so if $\Delta x = r$, $\Delta p \approx \dfrac{\hbar}{r}$

(b) $KE = \dfrac{p^2}{2m} \approx \dfrac{(\Delta p)^2}{2m} = \dfrac{\hbar^2}{2mr^2}$, $\qquad PE = -\dfrac{ke^2}{r} \qquad E = \dfrac{\hbar^2}{2mr^2} - \dfrac{ke^2}{r}$

30.60 (a) $\dfrac{1}{\alpha} = \dfrac{\hbar c}{ke^2} = \dfrac{(6.63 \times 10^{-34})(3 \times 10^8)}{2\pi(9 \times 10^9)(1.6 \times 10^{-19})^2} = 137.036$

(b) $\dfrac{r_0}{\lambda} = \dfrac{\hbar^2}{mke^2}\dfrac{\hbar}{mc} = \dfrac{1}{2\pi}\dfrac{\hbar c}{ke^2} = \dfrac{137}{2\pi} = \dfrac{1}{2\pi\alpha}$

(c) $\dfrac{1}{Rr_0} = \dfrac{mke^2}{\hbar^2}\dfrac{4\pi c \hbar^3}{mk^2 e^4} = \dfrac{4\pi}{\alpha}$

CHAPTER THIRTY-ONE SOLUTIONS

31.1 The expression for the potential energy is $PE = \frac{1}{2}kx^2$, or specifically, one should plot the graph of $PE = (75 \text{ N/m})x^2$. The result should look like the sketch below.

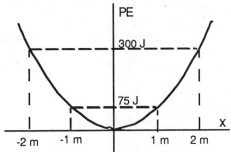

31.2 The potential energy function that you should plot is of the form, $PE = -\frac{\text{const}}{r}$, and should look like the sketch below.

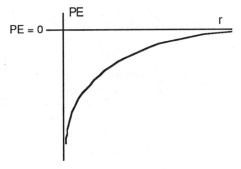

31.3 (a) Assuming the ions act as point charges, the force is the Coulomb force between two point charges:
$$F = \frac{kq_1q_2}{r^2} = \frac{(9 \times 10^9 \text{ Nm}^2/\text{C}^2)(1.6 \times 10^{-19} \text{ C})^2}{(2.8 \times 10^{-10} \text{ m})^2} = 2.94 \times 10^{-9} \text{ N}$$

(b) The potential energy between the charges is
$$PE = \frac{kq_1q_2}{r} = -\frac{(9 \times 10^9 \text{ Nm}^2/\text{C}^2)(1.6 \times 10^{-19} \text{ C})^2}{(2.8 \times 10^{-10} \text{ m})} = -8.23 \times 10^{-19} \text{ J} = -5.14 \text{ eV}$$

31.4 At point A, the slope is zero, and the force is zero.
At point B, the slope is less than zero, so the force is in the positive x direction.
At point C, the slope is zero, so the force is also zero.
At point D, the slope is positive, so the force is in the negative x direction.
At point E, the slope and the force are again zero.

CHAPTER THIRTY-ONE SOLUTIONS

31.5 $PE = 0.15$ eV $= 2.4 \times 10^{-20}$ J and $x = 0.01$ nm $= 10^{-11}$ m

(a) $PE = \frac{1}{2} kx^2$, so $k = \frac{2(PE)}{x^2} = \frac{2(2.4 \times 10^{-20} \text{ J})}{(10^{-11} \text{ m})^2} = 480$ N/m

(b) $\mu = \frac{35}{36}\mu = 1.614 \times 10^{-27}$ kg

$$f = \frac{1}{2\pi}\sqrt{\frac{k}{\mu}} = \frac{1}{2\pi}\sqrt{\frac{480 \text{ N/m}}{1.614 \times 10^{-27} \text{ kg}}} = 8.68 \times 10^{13} \text{ Hz}$$

31.6 The energy required to form the ionic bond in KCl equals the energy required to form K^- (ionize a neutral K atom) less the energy obtained when a Cl^- ion forms from a free electron and a neutral Cl atom.
Thus, Energy required = ionization energy of potassium - electron affinity of Cl

$$= 4.34 \text{ eV} - 3.7 \text{ eV} = 0.64 \text{ eV}$$

31.7 Energy required to form KF = ionization energy of neutral K atom - electron affinity of F.

$$= 4.34 \text{ eV} - 4.1 \text{ eV} = 0.24 \text{ eV}$$

31.8 The heat of vaporization of water $= 2.26 \times 10^6$ J/kg, and the molecular weight of water $= 18$. Thus, 18 kg of water has 6.02×10^{26} molecules. Therefore, in 1 kg of water there are

$$\frac{6.02 \times 10^{26}}{18} \text{ molecules} = 3.34 \times 10^{25} \text{ molecules}$$

The number of bonds $= \left(\frac{3.34 \times 10^{25} \text{ molecules/kg}}{2}\right)\left(\frac{6 \text{ bonds}}{\text{molecule}}\right)$

$$= 1.00 \times 10^{26} \text{ bonds/kg}$$

(The number of molecules is divided by two in the above because it takes two molecules to form each bond.)
Thus, to evaporate 1 kg of water, 2.26×10^6 J are required to break all of the 10^{26} bonds involved. Therefore, the energy required to break each bond is

$$\text{energy per bond in water} = \frac{2.26 \times 10^6 \text{ J}}{1.00 \times 10^{26} \text{ bonds}}$$

$$= 2.26 \times 10^{-20} \text{ J/bond}\left(\frac{1 \text{ eV}}{1.6 \times 10^{-19} \text{ J}}\right) = 0.141 \text{ eV}$$

31.9 The energy required to break all bonds in methane is 1.03×10^8 J/kg. (See example 31.3) Therefore, the amount of methane which could be broken apart by an input of 3.63×10^8 J of energy is

$$\text{Methane broken} = \frac{\text{energy input}}{\text{energy input per kg}} = \frac{3.63 \times 10^8 \text{ J}}{1.03 \times 10^8 \text{ J/kg}} = 3.52 \text{ kg}$$

31.10 The molecular weight of ethane (C_2H_6) = 30. Thus, the number of molecules in one kg of ethane equals

$$\text{number of molecules} = \frac{1}{30}(6.02 \times 10^{26} \text{ molecules}) = 2.01 \times 10^{25} \text{ molecules}$$

In each molecule there are 1 C-C bond and 6 C-H bonds. Thus, in 1 kg of ethane, there are

CHAPTER THIRTY-ONE SOLUTIONS

2.01×10^{25} C-C bonds, and

$6 \times (2.01 \times 10^{25}) = 1.20 \times 10^{26}$ C-H bonds

The energy required to break the C-H bonds is

1.20×10^{26} bonds$(4.28$ eV/bond$) = 5.15 \times 10^{26}$ eV

and the energy required to break the C-C bonds is

2.01×10^{25} bonds$(3.60$ eV/bond$) = 7.22 \times 10^{25}$ eV

Thus, the total energy required to break all the bonds in 1 kg of ethane is

Total energy $= 5.15 \times 10^{26}$ eV $+ 7.22 \times 10^{25}$ eV $= 5.88 \times 10^{26}$ eV $= 9.40 \times 10^{7}$ J

31.11 We place the fluorine atom at the origin of the coordinate system and the hydrogen atom at 0.11 nm on the x axis. The x coordinate of the center of mass is

$$x_{cm} = \frac{\sum m_i x_i}{\sum m_i} = \frac{(3.15 \times 10^{-26} \text{ kg})(0) + (1.67 \times 10^{-27} \text{ kg})(0.11 \text{ nm})}{3.15 \times 10^{-26} \text{ kg} + 1.67 \times 10^{-27} \text{ kg}}$$

$= 5.54 \times 10^{-3}$ nm

31.12 We place the oxygen atom at the origin of the coordinate system and the nitrogen atom at x = 0.114 nm.

$$x_{cm} = \frac{\sum m_i x_i}{\sum m_i} = \frac{(2.66 \times 10^{-26} \text{ kg})(0) + (2.33 \times 10^{-26} \text{ kg})(0.114 \times 10^{-9} \text{ m})}{2.33 \times 10^{-26} \text{ kg} + 2.66 \times 10^{-26} \text{ kg}}$$

$= 5.32 \times 10^{-11}$ m

31.13 The Fluorine atom is located at a distance of 5.54×10^{-12} m from the center of mass, and the hydrogen atom is at 1.04×10^{-10} m from the center of mass. Thus, the moment of inertia about the center of mass is

$I = \Sigma mr^2 = (3.15 \times 10^{-26}$ kg$)(5.54 \times 10^{-12}$ m$)^2$

$\quad + (1.67 \times 10^{-27}$ kg$)(1.045 \times 10^{-10}$ m$)^2$

$= 1.92 \times 10^{-47}$ kg m^2

31.14 The oxygen atom is located at a distance of 5.32×10^{-11} m from the center of mass and the nitrogen atom is at 6.08×10^{-11} m from the center of mass. The moment of inertia about the center of mass is

$I = \Sigma mr^2 = (2.66 \times 10^{-26}$ kg$)(5.32 \times 10^{-11}$ m$)^2 + (2.33 \times 10^{-26}$ kg$)(6.08 \times 10^{-11}$ m$)^2$

$= 1.61 \times 10^{-46}$ kg m^2

31.15 $E_{rot} = \dfrac{\hbar}{2I} J'(J' + 1)$

$\Delta E = \dfrac{\hbar^2}{2I}(3)(3+1) - \dfrac{\hbar^2}{2I}(2)(2+1) = 6\dfrac{\hbar^2}{2I}$

For HF molecules, $I = 1.92 \times 10^{-47}$ kg m^2 (see problem 13)

Thus, $E = 6\dfrac{(1.05 \times 10^{-34} \text{ J s})^2}{2(1.92 \times 10^{-47} \text{ kg m}^2)} = 1.72 \times 10^{-21}$ J

and $\lambda = \dfrac{hc}{E} = \dfrac{(6.63 \times 10^{-34} \text{ J s})(3 \times 10^{-8} \text{ m/s})}{1.72 \times 10^{-21} \text{ J}} = 1.15 \times 10^{-4}$ m $= 0.115$ mm

CHAPTER THIRTY-ONE SOLUTIONS

31.16 $E_{rot} = \dfrac{\hbar^2}{2I} J'(J' + 1)$

For the NO molecule $I = 1.61 \times 10^{-46}$ kg m^2 (see problem 14)
Thus,

$$E = \dfrac{\hbar^2}{2I}(4)(4+1) - \dfrac{\hbar^2}{2I}(3)(3+1) = 8\dfrac{\hbar^2}{2I} = 4\dfrac{\hbar^2}{I} = \dfrac{4(1.05 \times 10^{-34} \text{ J s})^2}{(1.61 \times 10^{-46} \text{ kg m}^2)}$$

$= 2.74 \times 10^{-22}$ J

and $f = \dfrac{E}{h} = \dfrac{2.74 \times 10^{-22} \text{ J}}{6.63 \times 10^{-34} \text{ J s}} = 4.13 \times 10^{11}$ Hz

and $\lambda = \dfrac{c}{f} = 7.26 \times 10^{-4}$ m $= 0.726$ mm

---

31.17 Since both constituents of the molecule have the same mass, the center of mass is located halfway between them. Thus, the moment of inertia about the center of mass is

$I = \Sigma mr^2 = 2(2.33 \times 10^{-26}$ kg$)(5.5 \times 10^{-11}$ m$)^2 = 1.41 \times 10^{-46}$ kg m^2

Thus, the energy in a state is given by

$$E_{J'} = \dfrac{\hbar^2}{2I} J'(J' + 1) = \dfrac{(1.05 \times 10^{-34} \text{ J s})^2}{2(1.41 \times 10^{-46} \text{ kg m}^2)} J'(J' + 1) = (3.91 \times 10^{-23} \text{ J})J'(J' + 1)$$

Therefore, $E_0 = 0$, and $E_1 = (3.91 \times 10^{-23}$ J$)(2) = 7.82 \times 10^{-23}$ J

So, the frequency of the photon that would produce the transition is

$f = \dfrac{\Delta E}{h} = \dfrac{E_1 - E_0}{h} = \dfrac{7.82 \times 10^{-23} \text{ J}}{6.63 \times 10^{-34} \text{ J s}} = 1.18 \times 10^{11}$ Hz

---

31.18 We locate the Br atom at the origin of the coordinate system and the hydrogen atom at the position 1.1×10^{-10} m along the x axis. The x coordinate of the center of mass is

$$x_{cm} = \dfrac{\Sigma m_i x_i}{\Sigma m_i} = \dfrac{(1.33 \times 10^{-25} \text{ kg})(0) + (1.67 \times 10^{-27} \text{ kg})(1.1 \times 10^{-10} \text{ m})}{1.33 \times 10^{-25} \text{ kg} + 1.67 \times 10^{-27} \text{ kg}}$$

$= 1.37 \times 10^{-12}$ m

The distance from the Br atom to the system's center of mass is 1.37×10^{-12} m. The distance from the hydrogen atom to the center of mass is 1.087×10^{-10} m. The moment of inertia about the center of mass is

$I = \Sigma mr^2 = (1.33 \times 10^{-25}$ kg$)(1.37 \times 10^{-12}$ m$)^2$
$\qquad\qquad + (1.67 \times 10^{-27}$ kg$)(1.087 \times 10^{-10}$ m$)^2$
$= 2.00 \times 10^{-47}$ kg m^2

The energy in a state J' is given by

$$E_{J'} = \dfrac{\hbar^2}{2I} J'(J' + 1) = \dfrac{(1.05 \times 10^{-34} \text{ J s})^2}{2(2.00 \times 10^{-47} \text{ kg m}^2)} J'(J' + 1) = (2.76 \times 10^{-22} \text{ J})J'(J' + 1) \text{ J}$$

$= 1.73 \times 10^{-3}$ eV$)J'(J' + 1)$

Thus, $E_0 = 0$, and $E_1 = 3.45 \times 10^{-3}$ eV

The energy carried away by the photon in a $J' = 1$ to a $J' = 0$ transition is

$E = E_1 - E_0 = 3.45 \times 10^{-3}$ eV

31.19 We locate the Cl atom at the origin of the coordinate system. Then,

434

CHAPTER THIRTY-ONE SOLUTIONS

$$x_{cm} = \frac{\sum m_i x_i}{\sum m_i} = \frac{(35\,\mu)(0) + (1\,\mu)(0.128\,nm)}{36\,\mu}$$

$$= \frac{1}{36}(0.128\,nm) = 3.56 \times 10^{-3}\,nm$$

(a) and the distance of the H atom from the center of mass = 0.1244 nm.

(b) $I = \Sigma mr^2 = (35\,\mu)(3.56 \times 10^{-3}\,nm)^2 + (1\,\mu)(0.1244\,nm)^2$

gives $I = 2.64 \times 10^{-47}\,kg\,m^2$

(c) For $J' = 0$, $E_{rot} = 0$, and for $J' = 1$,

$$E_{rot} = \frac{\hbar^2}{2I}(2) = \frac{\hbar^2}{I} = \frac{(1.05 \times 10^{-34}\,J\,s)^2}{2.64 \times 10^{-47}\,kg\,m^2} = 4.17 \times 10^{-22}\,J$$

(d) $\lambda = \frac{hc}{\Delta E} = \frac{(6.63 \times 10^{-34}\,J\,s)(3 \times 10^8\,Jm/s)}{4.17 \times 10^{-22}\,J} = 4.77 \times 10^{-4}\,m$

31.20 $\Delta E = \frac{\hbar^2}{2I}(6)(6+1) - \frac{\hbar^2}{2I}(5)(5+1) = 6\frac{\hbar^2}{I}$ (for $J' = 5$ to $J' = 6$)

$\Delta E = \frac{\hbar^2}{2I}(1)(1+1) - 0 = \frac{\hbar^2}{I}$ (for $J' = 0$ to $J' = 1$)

Thus, $\frac{hc}{\lambda_1} = 6\frac{hc}{\lambda_2}$ gives $\lambda_2 = 6\lambda_1 = 6(1.35 \times 10^{-2}\,m) = 8.1 \times 10^{-2}\,m = 8.1\,cm$

and $f = \frac{c}{\lambda}$ gives $f = 3.70 \times 10^9\,Hz$

(b) $\frac{hc}{\lambda_1} = 6\frac{\hbar^2}{I}$ gives $I = \frac{3h\lambda_1}{2\pi^2 c}$ gives $I = 4.53 \times 10^{-45}\,kg\,m^2$

31.21 The energy level diagram should look like the one sketched below. The energies calculated from the given information are shown.

⋮ ⋮ ⋮

$v = 5$ ——— $\frac{9}{2}hf = 1.62\,ev$

$v = 4$ ——— $\frac{7}{2}hf = 1.26\,ev$

$v = 3$ ——— $\frac{5}{2}hf = 0.903\,ev$

$v = 2$ ——— $\frac{3}{2}hf = 0.542\,ev$

$v = 1$ ——— $\frac{1}{2}hf = 0.18\,ev$

31.22 We have $E_0 = \frac{1}{2}hf$, where $E = hf$ is the energy of the photon which must be absorbed to move the molecule from one vibrational state to the next higher one. We have

$E = hf = (6.63 \times 10^{-34}\,J\,s)(8.66 \times 10^{13}\,Hz) = 5.74 \times 10^{-20}\,J = 0.359\,eV$

Thus, $E_0 = \frac{1}{2}hf = \frac{0.359\,eV}{2} = 0.179\,eV$

CHAPTER THIRTY-ONE SOLUTIONS

31.23 $\mu = \frac{35}{36}\mu = 1.614 \times 10^{-27}$ kg

$$f = \frac{1}{2\pi}\sqrt{\frac{k}{\mu}} = \frac{1}{2\pi}\sqrt{\frac{480 \text{ N/m}}{1.614 \times 10^{-27} \text{ kg}}} = 8.68 \times 10^{13} \text{ Hz}$$

$E_{vib} = (v + \frac{1}{2})hf$, so

$\Delta E_{vib} = hf = (6.63 \times 10^{-34} \text{ J s})(8.68 \times 10^{13} \text{ Hz}) = 5.75 \times 10^{-20} \text{ J} = 0.359$ eV

$$\lambda = \frac{hc}{\Delta E_{vib}} = \frac{(6.63 \times 10^{-34} \text{ J s})(3 \times 10^8 \text{ m/s})}{5.75 \times 10^{-20} \text{ J}} = 3.46 \times 10^{-6} \text{ m} = 3.46 \text{ }\mu\text{m}$$

31.24 $\mu = \frac{126.9}{127.9}\mu = 1.647 \times 10^{-27}$ kg

$k = 4\pi^2 f^2 \mu = 4\pi^2 (6.69 \times 10^{13} \text{ Hz})^2 (1.647 \times 10^{-27} \text{ kg}) = 291$ N/m

Similarly for the NO molecule

$\mu = 1.239 \times 10^{-26}$ kg

and $k = 4\pi^2 f^2 \mu = 4\pi^2 (5.63 \times 10^{13} \text{ Hz})^2 (1.239 \times 10^{-26} \text{ kg}) = 1550$ N/m

(b) k is much larger for the NO molecule because a larger number of electrons participate in the molecular bond.

31.25 The photon must give the electron at least 1.14 eV in order to allow the electron to cross the gap to the conduction band. Since the energy of a photon is given by $E = hf$, we find the minimum frequency from

$$f = \frac{E}{h} = \frac{1.14 \text{ eV}(1.6 \times 10^{-19} \text{ J/eV})}{6.63 \times 10^{-34} \text{ J s}} = 2.75 \times 10^{14} \text{ Hz}$$

This corresponds to a wavelength of

$$\lambda = \frac{c}{f} = \frac{3 \times 10^8 \text{ m/s}}{2.75 \times 10^{14} \text{ Hz}} = 1.09 \times 10^{-6} \text{ m} = 1.09 \text{ }\mu\text{m}$$

31.26 The width of the gap is equal to the minimum energy of a photon that will promote an electron across the gap. Thus, the energy of the gap is

$$E_{gap} = \frac{hc}{\lambda_{max}} = \frac{(6.63 \times 10^{-34} \text{ J s})(3 \times 10^8 \text{ m/s})}{1.85 \times 10^{-6} \text{ m}} = 1.075 \times 10^{-19} \text{ J} = 0.672 \text{ eV}$$

31.27 Aluminum has 3 valence electrons which can form covalent bonds with 3 of the 4 nearest neighbor silicon atoms, leaving an electron deficiency or "hole" in the bond with the fourth neighbor silicon atom. This hole can migrate through the lattice structure and act in the same way as would a positive charge carrier.
The resulting semiconductor is p-type.

31.28 Arsenic has 5 valence electrons. Four of these participate in covalent bonds with the four nearest neighbor silicon atoms and one electron is left over. This loosely bound valence electron is easily removed from the atom, allowing it to migrate through the lattice structure as a negative charge carrier.
This is an n-type semiconductor.

31.29 (a) $kT = 0.01$ eV $= 1.6 \times 10^{-21}$ J.

CHAPTER THIRTY-ONE SOLUTIONS

Thus, $T = \dfrac{1.6 \times 10^{-21} \text{ J}}{1.38 \times 10^{-23} \text{ J/K}} = 116 \text{ K} = -157°C$

(b) $kT = 7 \text{ eV} = 1.12 \times 10^{-18}$ J

and $T = \dfrac{1.12 \times 10^{-18} \text{ J}}{1.38 \times 10^{-23} \text{ J/K}} = 8.12 \times 10^4 \text{ K} = 8.09 \times 10^4$ °C

31.30 The band energy gap = energy of photon emitted when an electron crosses the gap

$E_{gap} = \dfrac{hc}{\lambda} = \dfrac{(6.63 \times 10^{-34} \text{ J s})(3 \times 10^8 \text{ m/s})}{650 \times 10^{-9} \text{ m}} = 3.06 \times 10^{-19} \text{ J} = 1.91 \text{ eV}$

31.31 The gap should be narrow enough so that the longest wavelength photon present could excite an electron across the gap. Thus,

$E_{gap} < E_{min} = \dfrac{hc}{\lambda} = \dfrac{(6.63 \times 10^{-34} \text{ J s})(3 \times 10^8 \text{ m/s})}{10^{-6} \text{ m}} = 1.99 \times 10^{-19} \text{ J} = 1.24 \text{ eV}$

31.32 The graph should look like the one sketched below.

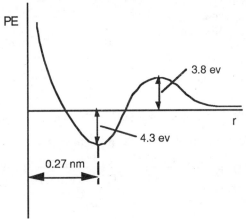

31.33 The molecular weight of propane = $3(12) + 8(1) = 44$

Thus, 1 kg mol = 44 kg or, 1 kg = $\dfrac{1}{44}$ kg mol

In 1 kg of propane, there are $\dfrac{1}{44}$ (6.02 × 10^{26} molecules) = 1.37×10^{25} molecules.

Thus, there are $2(1.37 \times 10^{25}) = 2.74 \times 10^{25}$ C-C bonds
and $8(1.37 \times 10^{25}) = 1.09 \times 10^{26}$ C-H bonds

The total energy to break the C-C bonds = $(2.74 \times 10^{25})(3.6 \text{ eV/bond})$
$= 9.85 \times 10^{25}$ eV $= 1.58 \times 10^7$ J

The total energy to break the C-H bonds = $(1.09 \times 10^{26})(4.28 \text{ eV/bond})$
$= 4.68 \times 10^{26}$ eV $= 7.5 \times 10^7$ J

The total energy to break all the bonds = 9.07×10^7 J

31.34 (a) We place the Cl ion at the origin of our coordinate system and the hydrogen ion at the x axis at the 0.12 nm position. The x location of the center of mass of the system is

CHAPTER THIRTY-ONE SOLUTIONS

$$x_{cm} = \frac{\sum m_i x_i}{\sum m_i} = \frac{(5.88 \times 10^{-26} \text{ kg})(0) + (1.67 \times 10^{-27} \text{ kg})(1.2 \times 10^{-10} \text{ m})}{5.88 \times 10^{-26} \text{ kg} + 1.67 \times 10^{-27} \text{ kg}}$$

$$= 3.32 \times 10^{-12} \text{ nm}$$

Thus, the distance of the Cl ion from the center of mass is 3.32×10^{-12} m, and the distance of the hydrogen ion from the center of mass is 1.17×10^{-10} m. The moment of inertia about the center of mass is

$$I = \Sigma mr^2 = (5.88 \times 10^{-26} \text{ kg})(3.32 \times 10^{-12} \text{ m})^2 + (1.67 \times 10^{-27} \text{ kg})(1.17 \times 10^{-10} \text{ m})^2$$

$$= 2.34 \times 10^{-47} \text{ kg m}^2$$

Thus, the energy in the J' state is given by

$$E_{J'} = \frac{\hbar^2}{2I} J'(J'+1) = \frac{(1.05 \times 10^{-34} \text{ J s})^2}{2(2.34 \times 10^{-47} \text{ kg m}^2)} J'(J'+1) = (2.36 \times 10^{-22} \text{ J})J'(J'+1)$$

Thus, $E_0 = 0$, and $E_1 = 4.71 \times 10^{-23}$ J

The energy of the photon that would cause a transition from J' = 0 to J' = 1 is

$$E = hf = E_1 - E_0 = 4.71 \times 10^{-23} \text{ J}$$

and $f = \frac{E}{h} = \frac{4.71 \times 10^{-23} \text{ J}}{6.63 \times 10^{-34} \text{ J s}} = 7.11 \times 10^{11}$ Hz

This photon has a wavelength of

$$\lambda = \frac{c}{f} = \frac{3 \times 10^8 \text{ m/s}}{7.11 \times 10^{11} \text{ Hz}} = 4.22 \times 10^{-4} \text{ m} = 0.422 \text{ mm}$$

31.35 For an oxygen molecule, the center of mass is halfway between the two atoms

$r_1 = r_2 = \frac{d}{2} = 6 \times 10^{-11}$ m and $m_o = 2.66 \times 10^{-26}$ kg

and $I = \Sigma mr^2 = 2(2.66 \times 10^{-26} \text{ kg})(6 \times 10^{-11} \text{ m})^2 = 1.91 \times 10^{-46}$ kg m^2

and $E_{rot} = \frac{\hbar^2}{2I} J'(J'+1) = \frac{(1.05 \times 10^{-34} \text{ J s})^2}{2(1.91 \times 10^{-46} \text{ kg m}^2)} (1)(2) = 5.77 \times 10^{-23}$ J (for J' = 1)

and for J' = 2

$E_{rot} = \frac{\hbar^2}{2I} (2)(3) = (3) \frac{(1.05 \times 10^{-34} \text{ J s})^2}{(1.91 \times 10^{-46} \text{ kg m}^2)} = 1.73 \times 10^{-22}$ J

31.36 $\lambda = 4.67$ μm, so $f = 6.42 \times 10^{13}$ Hz

for CO, $\mu = 1.138 \times 10^{-26}$ kg, so

(a) $k = 4\pi^2 f^2 \mu = 4\pi^2 (6.42 \times 10^{13} \text{ Hz})^2 (1.138 \times 10^{-26} \text{ kg}) = 1850$ N/m

(b) and from $E = \frac{1}{2} kA^2$

$$A = \sqrt{\frac{2hf}{k}} = \sqrt{\frac{2(6.63 \times 10^{-34} \text{ Js})(6.42 \times 10^{13} \text{ Hz})}{1850 \text{ N/m}}} = 6.78 \times 10^{-12} \text{ m}$$

(c) $\frac{A}{d} 100\% = \frac{6.78 \times 10^{-12} \text{ m}}{0.113 \times 10^{-9} \text{ m}} 100 = 6\%$

31.37 For the hydrogen molecule, the center of mass is midway between the two protons, or 5.3×10^{-11} m from each atom.

$I = \Sigma mr^2 = 2(1.67 \times 10^{-27} \text{ kg})(5.3 \times 10^{-11} \text{ m})^2 = 9.38 \times 10^{-48}$ kg m^2

CHAPTER THIRTY-ONE SOLUTIONS

(a) $E_{rot} = \dfrac{\hbar^2}{2I} J'(J' + 1) = \dfrac{(1.05 \times 10^{-34} \text{ J s})^2}{2(9.38 \times 10^{-48} \text{ kg m}^2)} J'(J' + 1)$

$= (5.88 \times 10^{-22} \text{ J})(J')(J' + 1) = (3.67 \times 10^{-3} \text{ eV})(J')(J' + 1)$

(b) $\mu = 8.35 \times 10^{-28}$ kg, so

$$f = \dfrac{1}{2\pi}\sqrt{\dfrac{k}{\mu}} = \dfrac{1}{2\pi}\sqrt{\dfrac{40 \text{ N/m}}{8.35 \times 10^{-28} \text{ kg}}} = 3.48 \times 10^{13} \text{ Hz}$$

and

$$E_{vib} = (v + \tfrac{1}{2})hf = (v + \tfrac{1}{2})(6.63 \times 10^{-34} \text{ J s})(3.48 \times 10^{13} \text{ Hz})$$

$$= (v + \tfrac{1}{2})2.31 \times 10^{-20} \text{ J} = (v + \tfrac{1}{2})0.144 \text{ eV}$$

31.38 For the chlorine molecule, the center of mass is midway between the two atoms, or 1.25×10^{-10} m from each atom.

$I = \Sigma mr^2 = 2(5.81 \times 10^{-26} \text{ kg})(1.25 \times 10^{-10} \text{ m})^2 = 1.82 \times 10^{-45} \text{ kg m}^2$

$\mu = 2.905 \times 10^{-26}$ kg, so

$$f = \dfrac{1}{2\pi}\sqrt{\dfrac{k}{\mu}} = \dfrac{1}{2\pi}\sqrt{\dfrac{50 \text{ N/m}}{2.905 \times 10^{-26} \text{ kg}}} = 6.60 \times 10^{12} \text{ Hz}$$

For T = 4.5 K, $kT = (1.38 \times 10^{-23}$ J/mol K$)(4.5$ K$) = 6.21 \times 10^{-23}$ J

If $E_{rot} = \dfrac{\hbar^2}{2I} J'(J' + 1) = kT$

then $J'(J' + 1) = \dfrac{2I}{\hbar^2}(kT) = \dfrac{2(1.82 \times 10^{-45} \text{ kg m}^2)(6.21 \times 10^{-23} \text{ J})}{(1.05 \times 10^{-34} \text{ J s})^2}$

gives a quadratic in J' which can be solved to show J' = 4

Similarly, $E_{vib} = (v + \tfrac{1}{2})hf = kT$

gives $v = -0.485$, but v must be greater than zero, so at T = 4.5 K, only the zero point energy level (v = 0) would be excited in vibrational states.
At T = 450 K, the corresponding calculations are $kT = 6.21 \times 10^{-21}$ J and J' = 44 and $v = 0.92$. Thus, at 450 K, still only the v = 0 vibrational state is populated.
At T = 4500 K, $kT = 6.21 \times 10^{-20}$ J, J' = 142, and v = 13.

31.39 (a) The lowest energy photon must have an energy of 0.7 eV = 1.12×10^{-19} J

Thus, since E = hf, we have $f = \dfrac{E}{h} = \dfrac{1.12 \times 10^{-19} \text{ J}}{6.63 \times 10^{-34} \text{ J s}} = 1.69 \times 10^{14}$ Hz

(b) $\lambda = \dfrac{c}{f} = \dfrac{3 \times 10^8 \text{ m/s}}{1.69 \times 10^{14} \text{ Hz}} = 1.78 \times 10^{-6}$ m = 1.78 µm

---

31.40 (a) Phosphorus has 5 valence electrons; thus a n-type material is produced.
(b) Gallium has 3 valence electrons; thus the material produced is p-type.
(c) Indium has 3 valence electrons, so the material is p-type.

CHAPTER THIRTY-TWO SOLUTIONS

32.1 (a) $r = r_0 A^{1/3} = (1.2 \times 10^{-15} \text{ m})(4)^{1/3} = 1.90 \times 10^{-15}$ m
(b) $r = r_0 A^{1/3} = (1.2 \times 10^{-15} \text{ m})(238)^{1/3} = 7.4 \times 10^{-15}$ m

---

32.2 From $r = r_0 A^{1/3}$, we have $A = \dfrac{r^3}{r_0^3} = \dfrac{(4.36 \times 10^{-15} \text{ m})^3}{(1.2 \times 10^{-15} \text{ m})^3} = 48$

32.3 From $r = r_0 A^{1/3}$, the radius of uranium is $r_u = r_0(238)^{1/3}$
Thus, if $r = \dfrac{1}{2} r_u$ then $r_0 A^{1/3} = \dfrac{1}{2} r_0 (238)^{1/3}$
From which, $A = 30$

32.4 $r = r_0 A^{1/3}$ for $A = 2$ gives $r = (1.2 \times 10^{-15} \text{ m})(2)^{1/3} = 1.51 \times 10^{-15}$ m
for $A = 60$, $r = (1.2 \times 10^{-15} \text{ m})(60)^{1/3} = 4.70 \times 10^{-15}$ m
for $A = 197$, $r = (1.2 \times 10^{-15} \text{ m})(197)^{1/3} = 6.98 \times 10^{-15}$ m
for $A = 239$, $r = (1.2 \times 10^{-15} \text{ m})(239)^{1/3} = 7.45 \times 10^{-15}$ m

---

32.5 (a) The mass density in the specified system of units is
$$\rho_n = 2.3 \times 10^{17} \text{ kg/m}^3 \left(\dfrac{1 \text{ slug}}{14.6 \text{ kg}}\right)\left(\dfrac{2.83 \times 10^{-2} \text{ m}^3}{1 \text{ ft}^3}\right) = 4.5 \times 10^{14} \text{ slugs/ft}^3.$$
Therefore, the weight density is
$$\rho_n g = (4.5 \times 10^{14} \text{ slugs/ft}^3)(32 \text{ ft/s}^2) = 1.4 \times 10^{16} \text{ lbs/ft}^3$$
(b) The volume of the sphere would be
$$V = \dfrac{4}{3}\pi r^3 = \dfrac{4}{3}\pi (5 \times 10^{-2} \text{ m})^3 \left(\dfrac{1 \text{ ft}^3}{2.83 \times 10^{-2} \text{ m}^3}\right) = 1.85 \times 10^{-2} \text{ ft}^3$$
and the weight is
$$w = (\text{weight per unit volume})(V) = (1.4 \times 10^{16} \text{ lbs/ft}^3)(1.85 \times 10^{-2} \text{ ft}^3)$$
$$= 2.6 \times 10^{14} \text{ lbs}$$

32.6 We have
mass of sphere = mass of earth
$(\text{Volume})_{\text{sphere}}(\text{density})_{\text{sphere}} = (\text{Volume})_{\text{earth}}(\text{density})_{\text{earth}}$
Thus, $(\dfrac{4}{3}\pi r_s^3)(\rho_n) = (\dfrac{4}{3}\pi R_E^3)(\rho_E)$
or, $r_s = \dfrac{(\rho_E)^{1/3}}{(\rho_n)^{1/3}} R_E = \dfrac{(5.52 \times 10^3 \text{ kg/m}^3)^{1/3}}{(2.3 \times 10^{17} \text{ kg/m}^3)^{1/3}} (6.37 \times 10^6 \text{ m}) = 1.84 \times 10^2$ m
Thus, the diameter = $2r = 368$ m

32.7 The mass of the hydrogen atom is approximately equal to the mass of the proton which is 1.67×10^{-27} kg. If the radius of the atom is 0.53×10^{-10} m, the volume is $V = \dfrac{4}{3}\pi r^3 = 6.24 \times 10^{-31}$ m^3.
Therefore, the density of the atom is

440

CHAPTER THIRTY-TWO SOLUTIONS

$$\rho_a = \frac{m}{V} = \frac{1.67 \times 10^{-27} \text{ kg}}{6.24 \times 10^{-31} \text{ kg/m}^3} = 2.68 \times 10^3 \text{ kg/m}^3$$

and the ratio of the nuclear density to the atomic density is

$$\frac{\rho_n}{\rho_a} = \frac{2.3 \times 10^{17} \text{ kg/m}^3}{2.68 \times 10^3 \text{ kg/m}^3} = 8.6 \times 10^{13}$$

32.8 $V = \frac{4}{3}\pi r^3 = 4.163 \times 10^{-5} \text{ m}^3$.

$m = \rho V = (2.3 \times 10^{17} \text{ kg/m}^3)(4.163 \times 10^{-5} \text{ m}^3) = 9.575 \times 10^{12} \text{ kg}$

and $F = G\frac{m_1 m_2}{r^2} = (6.67 \times 10^{-11} \text{ N m}^2/\text{kg}^2)\frac{(9.575 \times 10^{12} \text{ kg})^2}{(1 \text{ m})^2} = 6.11 \times 10^{15} \text{ N}$

32.9 We have $PE)_f = KE)_i$

or, $\frac{kQ_1 Q_2}{r_{min}} = 0.5 \text{ MeV}$

For an alpha particle, $Q = 2e$, and for gold $Q = 79e$

Thus, $\frac{k(2e)(79e)}{r_{min}} = (0.5)(1.6 \times 10^{-13} \text{ J})$

or, $r_{min} = \frac{(9 \times 10^9 \text{ N m}^2/\text{C}^2)(158)(1.6 \times 10^{-19} \text{ C})^2}{(0.5)(1.6 \times 10^{-13} \text{ J})} = 4.55 \times 10^{-13} \text{ m}$

32.10 The potential energy of an alpha particle at a distance of 3.2×10^{-14} m from a gold nucleus is

$PE = \frac{kQ_1 Q_2}{r_{min}} = \frac{(9 \times 10^9 \text{ N m}^2/\text{C}^2)(2e)(79e)}{3.2 \times 10^{-14} \text{ m}} = 1.14 \times 10^{-12} \text{ J}$

Thus, $\frac{1}{2}mv^2 = 1.14 \times 10^{-12} \text{ J} = \frac{1}{2}(6.64 \times 10^{-27} \text{ kg})v^2$

From which, $v = 1.85 \times 10^7$ m/s

(b) $KE = 1.14 \times 10^{-12} \text{ J}(\frac{1 \text{ MeV}}{1.6 \times 10^{-13} \text{ J}}) = 7.11 \text{ MeV}$

32.11 The kinetic energy of the proton is $KE = 0.5 \text{ MeV} = 8 \times 10^{-14} \text{ J}$

Thus, $\frac{kQ_1 Q_2}{r_{min}} = 8 \times 10^{-14} \text{ J}$

Also, $Q_1 = e \quad Q_2 = 79e$

Thus, $r_{min} = \frac{(9 \times 10^9 \text{ N m}^2/\text{C}^2)(79)(1.6 \times 10^{-19} \text{ C})^2}{(8 \times 10^{-14} \text{ J})} = 2.28 \times 10^{-13} \text{ m}$

32.12 Assume a head-on collision for the minimum speed situation.

$\frac{1}{2}mv^2 = \frac{kQ_1 Q_2}{r_{min}}$, thus

$v = \sqrt{\frac{2(9 \times 10^9 \text{ N m}^2/\text{C}^2)(26)(2)(1.6 \times 10^{-19} \text{ C})^2}{4(1.66 \times 10^{-27} \text{ kg})(4.6 \times 10^{-15} \text{ m})}} = 2.80 \times 10^7 \text{ m/s}$

32.13 $F = k\frac{Q_1 Q_2}{r^2} = (9 \times 10^9 \text{ N m}^2/\text{C}^2)\frac{(2)(6)(1.6 \times 10^{-19} \text{ C})^2}{(1 \times 10^{-14} \text{ m})^2} = 27.6 \text{ N}$

441

CHAPTER THIRTY-TWO SOLUTIONS

(b) $a = \dfrac{F}{m} = \dfrac{27.6 \text{ N}}{6.64 \times 10^{-27} \text{ kg}} = 4.16 \times 10^{27} \text{ m/s}^2$

(c) $PE = k \dfrac{Q_1 Q_2}{r} = (9 \times 10^9 \text{ N m}^2/\text{C}^2) \dfrac{(2)(6)(1.6 \times 10^{-19} \text{ C})^2}{(1 \times 10^{-14} \text{ m})}$

$= 2.765 \times 10^{-13} \text{ J} = 1.73 \text{ MeV}$

32.14 mass difference = (total mass of component parts) - (mass of nucleus)

$\Delta m = (Z m_H + (A - Z) m_n) - m_N$

and the binding energy equals $E_b = (\Delta m) c^2$

Therefore, for $^{20}_{10}\text{Ne}$

$\Delta m = 10(1.007825 \text{ } \mu) + 10(1.008665 \text{ } \mu) - 19.992439 \text{ } \mu = 0.172461 \text{ } \mu$.

Therefore, $E_b = (\Delta m)c^2 = (0.172461 \text{ } \mu)(931.5 \text{ MeV}/\mu) = 161 \text{ MeV}$

32.15 $\Delta m = 20 m_H + 20 m_n - m_{Ca} = 20(1.007825 \text{ } \mu) + 20(1.008665 \text{ } \mu) - 39.96259 \text{ } \mu$
$= 0.367209 \text{ } \mu$.

Therefore, $E_b = (\Delta m)c^2 = (0.367209 \text{ } \mu)(931.5 \text{ MeV}/\mu) = 342 \text{ MeV}$

32.16 $\Delta m = 41 m_H + 52 m_n - m_{Nb} = 41(1.007825 \text{ } \mu) + 52(1.008665 \text{ } \mu) - 92.906378 \text{ } \mu$
$= 0.865027 \text{ } \mu$.

Therefore, $E_b = (\Delta m)c^2 = (0.865027 \text{ } \mu)(931.5 \text{ MeV}/\mu) = 805.8 \text{ MeV}$

and $\dfrac{E_b}{A} = \dfrac{805.8 \text{ MeV}}{93} = 8.66 \text{ MeV/nucleon}$

32.17 $\Delta m = 79 m_H + 118 m_n - m_{Au} = 79(1.007825 \text{ } \mu) + 118(1.008665 \text{ } \mu) - 196.96656 \text{ } \mu$
$= 1.674085 \text{ } \mu$.

Therefore, $E_b = (\Delta m)c^2 = (1.674085 \text{ } \mu)(931.5 \text{ MeV}/\mu) = 1559.4 \text{ MeV}$

and $\dfrac{E_b}{A} = \dfrac{1559.4 \text{ MeV}}{197} = 7.92 \text{ MeV/nucleon}$

32.18 $\Delta m = 12 m_H + 12 m_n - m_{mg} = 12(1.007825 \text{ } \mu) + 12(1.008665 \text{ } \mu) - 23.985045 \text{ } \mu$
$= 0.212835 \text{ } \mu$

and $E_b = (\Delta m)c^2 = 198.25 \text{ MeV}$

$\dfrac{E_b}{A} = \dfrac{198.25 \text{ MeV}}{24} = 8.26 \text{ MeV/nucleon (for } ^{24}_{12}\text{Mg)}$

For $^{85}_{37}\text{Rb}$, $\Delta m = 37 m_H + 48 m_n - m_{Rb}$

$= 37(1.007825 \text{ } \mu) + 48(1.008665 \text{ } \mu) - 84.911800 \text{ } \mu$
$= 0.793645 \text{ } \mu$

and $E_b = (\Delta m)c^2 = 739.3 \text{ MeV}$

$\dfrac{E_b}{A} = \dfrac{739.3 \text{ MeV}}{85} = 8.70 \text{ MeV/nucleon}$

32.19 For $^{15}_{8}\text{O}$ $Z = 8$, and $A - Z = 7$

Thus, $E_b = (8 m_H + 7 m_n - 15.003065 \text{ } \mu)(931.5 \text{ MeV}/\mu) = (0.12019 \text{ } \mu)(931.5 \text{ MeV}/\mu)$
$= 111.96 \text{ MeV}$

CHAPTER THIRTY-TWO SOLUTIONS

For For $^{15}_{7}N$ $Z = 7$, and $A - Z = 8$

Thus, $E_b = (7m_H + 8m_n - 15.000109\ \mu)(931.5\ \text{MeV}/\mu) = (0.123986\ \mu)(931.5\ \text{MeV}/\mu)$
$= 115.49$ MeV
So, $\Delta E_b = 3.54$ MeV

32.20   For $^{3}_{1}H$ $\Delta m = 2m_n + m_H - m_{tritium}$

$= (1.007825\ \mu) + 2(1.008665\ \mu) - 3.016049\ \mu$
$= 0.009106\ \mu$

and $E_b = (\Delta m)c^2 = 8.482$ MeV

For $^{3}_{2}He$, $\Delta m = m_n + 2m_H - m_{helium}$ $= 2(1.007825\ \mu) + (1.008665\ \mu) - 3.016029\ \mu$

$= 0.008286\ \mu$

and $E_b = (\Delta m)c^2 = 7.718$ MeV

The difference in binding energy = 0.764 MeV, with the decreased binding energy in helium being mainly due to the increased Coulomb repulsion present.

32.21   For $^{23}_{11}Na$ $Z = 11$, and $A - Z = 12$

Thus, $E_b = (11m_H + 12m_n - 22.989770\ \mu)(931.5\ \text{MeV}/\mu)$
$= (0.200285\ \mu)(931.5\ \text{MeV}/\mu)$
$= 186.6$ MeV

and $\dfrac{E_b}{A} = 8.112$ MeV/nucleon

For $^{23}_{12}Mg$ $Z = 12$, and $A - Z = 11$

Thus, $E_b = (12m_H + 11m_n - 22.994127\ \mu)(931.5\ \text{MeV}/\mu)$
$= (0.195088\ \mu)(931.5\ \text{MeV}/\mu)$
$= 181.7$ MeV

and $\dfrac{E_b}{A} = 7.901$ MeV/nucleon

Thus, $\Delta\left(\dfrac{E_b}{A}\right) = 0.211$ MeV/nucleon

The difference is mostly due to increased Coulomb repulsion due to the extra proton in $^{23}_{12}Mg$.

32.22   The energy to remove the last neutron in $^{43}_{20}Ca$ is

$E = [(M_{42Ca} + m_n) - M_{43Ca}]\ (931.5\ \text{MeV}/\mu)$
$= [41.958622\ \mu + 1.008665\ \mu - 42.958770\ \mu]\ (931.5\ \text{MeV/nucleon})$
$= (0.008517\ \mu)(931.5\ \text{MeV}/\mu) = 7.93$ MeV

32.23   (a) The surface-effect term must have a negative sign because the volume term over-approximates the binding energy by treating all nucleons as if they were completely surrounded by other nucleons. The nucleons on the surface of the "drop" are not totally surrounded and the surface term

CHAPTER THIRTY-TWO SOLUTIONS

attempts to correct for this effect by subtracting off the excess which was included in the volume term.

(b) First note that for a sphere, $V = \frac{4}{3}\pi r^3$ and from this, $r = \frac{(3V)^{1/3}}{(4\pi)^{1/3}}$

We have $\frac{V}{A} = \frac{\frac{4}{3}\pi r^3}{4\pi r^2} = \frac{r}{3} = \frac{(3V)^{1/3}}{3(4\pi)^{1/3}} = 0.207 V^{1/3}$.

For a cube $V = L^3$ and $L = V^{1/3}$.

Thus, $\frac{V}{A} = \frac{L^3}{6L^2} = \frac{L}{6} = \frac{V^{1/3}}{6} = 0.167 V^{1/3}$

Hence, for a given enclosed volume, a sphere has a larger volume to surface ratio than a cube. Thus, a spherical nucleus is more tightly bound and stable than a cubical one.

32.24 $E_b = C_1 A - C_2 A^{2/3} - C_3 \frac{(Z)(Z-1)}{A^{1/3}}$

$C_1 = 15.7$ MeV, $C_2 = 17.8$ MeV, and $C_3 = 0.71$ MeV

For $^{64}_{29}Cu$, $A = 64$, and $Z = 29$

Thus, $E_b = (15.7 \text{ MeV})(64) - (17.8 \text{ MeV})(64)^{2/3} - (0.71 \text{ MeV})\frac{(29)(28)}{(64)^{1/3}} = 576$ MeV

32.25 The decay constant is $\lambda = \frac{0.693}{T_{1/2}} = \frac{0.693}{30 \text{ min}} = 2.31 \times 10^{-2}$ min^{-1}.

$N = N_0 e^{-\lambda t} = (3 \times 10^{16}) e^{-(2.31 \times 10^{-2} \text{ min}^{-1})(10 \text{ min})} = 3 \times 10^{16} e^{-0.231}$
$= 3 \times 10^{16}(0.794) = 2.38 \times 10^{16}$ nuclei

32.26 We use $N = N_0 e^{-\lambda t}$ with $N = \frac{1}{2} N_0$, or $\frac{N}{N_0} = \frac{1}{10}$

Thus, $\frac{1}{10} = e^{-\lambda t}$ which becomes $t = \frac{\ln(10)}{\lambda}$

But, $\lambda = \frac{0.693}{T_{1/2}}$. So, $t = T_{1/2}\frac{\ln(10)}{0.693} = (140 \text{ days})\frac{2.30}{0.693} = 465$ days

---

32.27 We start with $N = N_0 e^{-\lambda t}$ and multiply both sides by m the mass of a nucleus of the substance. Thus, $mN = mN_0 e^{-\lambda t}$
But $mN = M =$ the current mass of the sample, and $mN_0 = M_0$ the original mass of the sample.

So, $M = M_0 e^{-\lambda t}$ or $\frac{M}{M_0} = \frac{0.25 \times 10^{-3} \text{ g}}{10^{-3} \text{ g}} = 0.25 = e^{-\lambda(2 \text{ h})}$

From which, $\lambda = 0.693$ h^{-1} and $T_{1/2} = \frac{0.693}{\lambda} = \frac{0.693}{0.693 \text{ h}^{-1}} = 1$ h

---

32.28 $T_{1/2} = 14$ days $= 1.21 \times 10^6$ s.

and $\lambda = \frac{0.693}{T_{1/2}} = \frac{0.693}{1.21 \times 10^6 \text{ s}} = 5.73 \times 10^{-7}$ s^{-1}.

The activity is $R = \lambda N = (5.73 \times 10^{-7} \text{ s}^{-1})(3 \times 10^{16}) = 1.72 \times 10^{10}$ decays/s
But using the conversion 1 curie $= 3.7 \times 10^{10}$ decays/s, we find

CHAPTER THIRTY-TWO SOLUTIONS

R = 0.465 curies

32.29 $R = \lambda N = \lambda(N_0 e^{-\lambda t}) = R_0 e^{-\lambda t}$

$\lambda = \dfrac{0.693}{T_{1/2}} = \dfrac{0.693}{6.05 \text{ h}} = 0.11455 \text{ /h}$

Thus, if $R_0 = 1.1 \times 10^4$ Bq and $t = 2$ h

$R = (1.1 \times 10^4 \text{ Bq}) e^{-(0.11455/h)(2.0 \text{ h})} = 8.75 \times 10^3 \text{ Bq} = 8.75 \times 10^3$ decays/s

---

32.30 (a) $\lambda = \dfrac{0.693}{T_{1/2}} = \dfrac{0.693}{8.04 \text{ days}} = 8.619 \times 10^{-2}$ days^{-1} = 9.98×10^{-7} s^{-1}

(b) $N = \dfrac{R}{\lambda}$ Thus, if $R = 0.5 \, \mu\text{Ci} = (0.5 \times 10^{-6})(3.7 \times 10^{10} \text{ 1/s}) = 1.85 \times 10^4$ decays/s

then $N = \dfrac{1.85 \times 10^4 \text{ decays/s}}{9.98 \times 10^{-7} \text{ s}^{-1}} = 1.85 \times 10^{10}$ nuclei

---

32.31 $T_{1/2} = 8.1$ days $= 7 \times 10^5$ s. and $\lambda = \dfrac{0.693}{T_{1/2}} = \dfrac{0.693}{7 \times 10^5 \text{ s}} = 9.90 \times 10^{-7}$ s^{-1}.

From $R = \lambda N$, $N = \dfrac{R}{\lambda}$

If $R = 0.2 \, \mu\text{Ci} = 7.40 \times 10^3$ decays/s then

$N = \dfrac{7.40 \times 10^3 \text{ decays/s}}{9.90 \times 10^{-7} \text{ s}^{-1}} = 7.47 \times 10^9$ nuclei

---

32.32 (a) $R_0 = 10$ mCi $= 3.7 \times 10^8$ decays/s.

We have $R = \lambda N$ and $R_0 = \lambda N_0$

Thus, $\dfrac{R}{R_0} = \dfrac{\lambda N}{\lambda N_0} = \dfrac{N}{N_0} = e^{-\lambda t}$

If $R_0 = 10$ mCi initially, and $R = 8$ mCi after 4 hours, we have

$\dfrac{8}{10} = e^{-\lambda(4 \text{ h})}$ and from this $\lambda = 5.58 \times 10^{-2}$ h^{-1}

Now, $T_{1/2} = \dfrac{0.693}{\lambda} = \dfrac{0.693}{5.58 \times 10^{-2} \text{ h}^{-1}} = 12.4$ h

(b) $N_0 = \dfrac{R_0}{\lambda} = \dfrac{3.7 \times 10^8 \text{ decays/s}}{(5.58 \times 10^{-2} \text{ h}^{-1})(1 \text{ h}/3600 \text{ s})} = 2.39 \times 10^{13}$ nuclei

(c) $R = R_0 e^{-\lambda t} = (10 \text{ mCi}) e^{-(5.58 \times 10^{-2} \text{ h}^{-1})(30 \text{ h})} = 1.88$ mCi

---

32.33 $\lambda = \dfrac{0.693}{T_{1/2}} = \dfrac{0.693}{12.33 \text{ y}} = 5.62 \times 10^{-2}$ y^{-1}.

At $t = 5$ y; $\dfrac{N}{N_0} = e^{-\lambda t} = e^{-(5.62 \times 10^{-2} \text{ y}^{-1})(5 \text{ y})} = 0.755$ \qquad (1)

(1) is the fraction of the original which will remain after 5 y. Thus, the fraction which has decayed = 1 - 0.755 = 0.245.
Thus, the % decayed = 24.5 %

32.34 $R = \lambda N = \lambda(N_0 e^{-\lambda t}) = R_0 e^{-\lambda t}$
with $R_0 = \lambda N_0$

$N_0 = \dfrac{30 \times 10^{-15} \text{ kg.}}{108(1.66 \times 10^{-27} \text{ kg})} = 1.6734 \times 10^{11}$ nuclei

CHAPTER THIRTY-TWO SOLUTIONS

$$\lambda = \frac{0.693}{T_{1/2}} = \frac{0.693}{(2.42 \text{ min})(60 \text{ s/min})} = 4.773 \times 10^{-3} \text{ s}^{-1}.$$

Thus, $R_0 = \lambda N_0 = (4.773 \times 10^{-3} \text{ s}^{-1}.)(1.6734 \times 10^{11}) = 7.986 \times 10^8 \text{ s}^{-1} = 21.6 \text{ mCi}$

32.35 $N_0 = \dfrac{\text{mass present}}{\text{mass of nucleus}} = \dfrac{5.0 \text{ kg}}{89.9077 \, \mu (1.66 \times 10^{-27} \text{ kg}/\mu)}$

$= 3.35 \times 10^{25}$ nuclei

$\lambda = \dfrac{0.693}{T_{1/2}} = \dfrac{0.693}{(28.8 \text{ y})} = 2.4063 \times 10^{-2} \text{ y}^{-1} .= 4.575 \times 10^{-8} \text{ min}^{-1}$.(half-life is taken from appendix B)

$R_0 = \lambda N_0 = (4.575 \times 10^{-8} \text{ min}^{-1}.)(3.35 \times 10^{25}) = 1.533 \times 10^{18}$ counts/min

$\dfrac{R}{R_0} = e^{-\lambda t} = \dfrac{10}{1.533 \times 10^{18} \text{ counts/min}} = 6.525 \times 10^{-18}$

and, $\lambda t = -\ln(6.525 \times 10^{-18}) = 39.57$
giving $t = 1645$ y

32.36 $^{212}_{83}\text{Bi} \rightarrow \,^{208}_{81}\text{Tl} + \,^{4}_{2}\text{He}$

$^{95}_{36}\text{Kr} \rightarrow \,^{95}_{37}\text{Rb} + \,^{0}_{-1}e$

$^{144}_{60}\text{Nd} \rightarrow \,^{4}_{2}\text{He} + \,^{140}_{58}\text{Ce}$

32.37 $^{12}_{5}\text{B} \rightarrow \,^{12}_{6}\text{C} + \,^{0}_{-1}e$

$^{234}_{90}\text{Th} \rightarrow \,^{230}_{88}\text{Ra} + \,^{4}_{2}\text{He}$

$^{14}_{6}\text{C} \rightarrow \,^{14}_{7}\text{N} + \,^{0}_{-1}e$

32.38 $^{232}_{92}\text{U} \rightarrow \,^{4}_{2}\text{He} + \,^{228}_{90}\text{Th}$

$E = (\Delta m)c^2 = [(M_{232U}) - (M_{4He} + M_{228Th})] \, (931.5 \text{ Mev}/\mu)$
$= [232.03714 \, \mu - (4.002602 \, \mu + 228.02873 \, \mu)] \, (931.5 \text{ Mev}/\mu)$
$E = (0.005807 \, \mu)(931.5 \text{ Mev}/\mu) = 5.41 \text{ MeV}$

32.39 $^{238}_{92}\text{U} \rightarrow \,^{4}_{2}\text{He} + \,^{234}_{90}\text{Th}$

$E = (\Delta m)c^2 = [(M_{238U}) - (M_{4He} + M_{234Th})] \, (931.5 \text{ Mev}/\mu)$
$= [238.050786 \, \mu - (4.002602 \, \mu + 234.043583 \, \mu)] \, (931.5 \text{ Mev}/\mu)$
$E = (0.0046 \, \mu)(931.5 \text{ Mev}/\mu) = 4.28 \text{ MeV}$

32.40 $^{8}_{4}\text{Be} \rightarrow \,^{4}_{2}\text{He} + \,^{4}_{2}\text{He}$

CHAPTER THIRTY-TWO SOLUTIONS

First, assume the decay will occur and compute the energy balance.
$E = (\Delta m)c^2 = [(M_{8Be}) - 2(M_{4He})]$ (931.5 Mev/μ)
$= [8.005305 \text{ μ} - 2(4.002603) \text{ μ}]$ (931.5 Mev/μ)
$E = 92.2$ keV
Since $E > 0$ this means that the decay can occur spontaneously with an energy release.

32.41 (a) $^{40}_{20}Ca \rightarrow ^{0}_{+1}e + ^{40}_{19}K$

For positron decay,
$\Delta m = M_p - M_d - 2m_e = 39.962591 \text{ μ} - 39.96400 \text{ μ} - 2(0.000549 \text{ μ})$
$\Delta m < 0$ Hence, the decay cannot occur spontaneously.

(b) $^{144}_{60}Nd \rightarrow ^{4}_{2}He + ^{140}_{58}Ce$

For alpha decay,
$\Delta m = M_p - (M_d + m_\alpha) = 143.910096 \text{ μ} - (139.90544 \text{ μ} + 4.002603 \text{ μ})$
$= 2.053 \times 10^{-3} \text{ μ}$
The mass difference is greater than zero, so the decay can occur spontaneously.

32.42

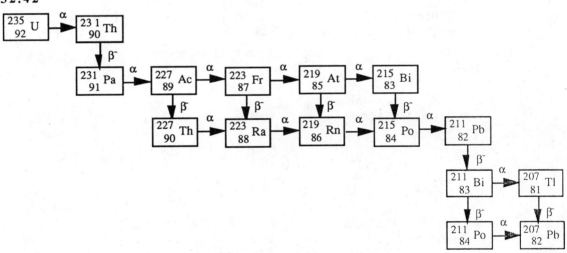

32.43 (a) $^{66}_{28}Ni \rightarrow ^{66}_{29}Cu + ^{0}_{-1}e + \bar{\nu}$

(b) $KE_{max} = (M_{ni} - M_{cu})c^2$
$= (65.929 \text{ μ} - 65.9289 \text{ μ})(1.66 \times 10^{-27} \text{ kg/μ})(3 \times 10^8 \text{ m/s})^2$
$= 2.988 \times 10^{-14}$ J $= 0.187$ MeV $= 187$ keV

32.44 $R = 0.6R_0$
and $\dfrac{R}{R_0} = e^{-\lambda t} = 0.6$
Thus, $\lambda t = -\ln(0.6) = 0.5108$

CHAPTER THIRTY-TWO SOLUTIONS

and for carbon-14, $\lambda = \dfrac{0.693}{T_{1/2}} = \dfrac{0.693}{5730 \text{ y}} = 1.2094 \times 10^{-4} \text{ y}^{-1}$.

Thus, $t = \dfrac{0.5108}{\lambda} = \dfrac{0.5108}{1.2094 \times 10^{-4} \text{ y}^{-1}} = 4220 \text{ y}$

32.45 The original activity of this 1000 g sample of carbon:
R_0 = m(15 decays/min g) = 1000 g(15 decays/min g) = 15000 decays/min
The current activity = 2000 decays/min.
The halflife of the sample is 3.01×10^9 min,
and the decay constant = 2.30×10^{-10} min^{-1}
We have $R = R_0 e^{-\lambda t}$
Thus, 2000 decays/min = (15000 decays/min)$e^{-(2.3 \times 10^{-10})t}$
From which, $t = 8.75 \times 10^9$ min = 1.66×10^4 y = 16,600 y

---

32.46 $^4_2\text{He} + ^9_4\text{Be} \rightarrow ^{12}_6\text{C} + ^1_0\text{n}$

$Q = [(M_{4He} + M_{9Be}) - (M_{12C} + M_n)]$ (931.5 Mev/μ)
 = [(4.002603 μ + 9.012183 μ) - (12.00000 μ + 1.008665 μ)] (931.5 MeV/μ)
 = (6.121 × 10^{-3} μ)(931.5 MeV/μ) = 5.70 MeV

32.47 $Q = [(M_{7Li} + M_H) - (M_{He} + M_{He})]$ (931.5 Mev/μ)
 = [(7.016005 μ + 1.007825 μ) - (4.002603 μ + 4.002603 μ)] (931.5 MeV/μ)
 = (0.018624 μ)(931.5 MeV/μ) = 17.35 MeV

32.48 $^{27}_{13}\text{Al} + ^1_0\text{n} \rightarrow ^{24}_{11}\text{Na} + ^4_2\text{He}$

$\Delta m = (M_{Al} + m_n) - (M_{Na} + m_\alpha)$
= 26.981541 μ + 1.008665 μ) - (23.9909646μ + 4.002603 μ)
     = -3.361 × 10^{-3} μ
$Q = \Delta mc^2 = (-3.361 \times 10^{-3}$ μ)(1.66 × 10^{-27}/μ)(3 × 10^8 m/s)^2
     = -5.02 × 10^{-13} J = -3.14 MeV

(b) $^{11}_5\text{Be} + ^1_1\text{H} \rightarrow ^{11}_6\text{C} + ^1_0\text{n}$

Δm = (11.009305 μ + 1.007825 μ) - (11.011433 μ + 1.008665 μ)
 = -2.968 × 10^{-3} μ
Thus, $Q = \Delta mc^2$ = -4.434 × 10^{-13} J = -2.77 MeV

32.49 $^7_3\text{Li} + ^4_2\text{He} \rightarrow ^{10}_5\text{B} + ^1_0\text{n}$

(b) $\Delta m = (M_{Li} + m_\alpha) - (M_B + m_n)$
= (7.016005 μ + 4.002603 μ) - (10.012938 μ + 1.008665 μ)
     = - 0.002995 μ = - 4.97 × 10^{-30} kg
and $Q = \Delta mc^2$ = -4.475 × 10^{-13} J = -2.80 MeV

32.50 $^4_2\text{He} + ^{14}_7\text{N} \rightarrow ^1_1\text{H} + ^{17}_8\text{O}$

CHAPTER THIRTY-TWO SOLUTIONS

$${}_{3}^{7}Li + {}_{1}^{1}H \rightarrow {}_{2}^{4}He + {}_{2}^{4}He$$

32.51 $\quad {}_{13}^{27}Al + {}_{2}^{4}He \rightarrow {}_{0}^{1}n + {}_{15}^{30}P$

$${}_{0}^{1}n + {}_{5}^{10}B \rightarrow {}_{2}^{4}He + {}_{3}^{7}Li$$

32.52 (a) $\quad {}_{13}^{27}Al + {}_{2}^{4}He \rightarrow {}_{15}^{30}P + {}_{0}^{1}n$

The product nucleus is ${}_{15}^{30}P$

(b) $Q = [(M_{Al} + M_{He}) - (M_P + M_n)]$ (931.5 Mev/μ)
$= [(26.981541\ \mu + 4.002603\ \mu) - (29.978310\ \mu + 1.008665\ \mu)]$ (931.5 MeV/μ)
$= (-2.831 \times 10^{-3}\ \mu)(931.5\ \text{MeV}/\mu) = -2.64\ \text{MeV}$

32.53 (a) $\quad {}_{3}^{6}Li + {}_{1}^{1}H \rightarrow {}_{2}^{3}He + {}_{2}^{4}He$

The product nucleus is ${}_{2}^{3}He$

(b) $Q = [(M_{Li} + M_H) - (M_{3He} + M_{4He})]$ (931.5 Mev/μ)
$= [(6.015123\ \mu + 1.007825\ \mu) - (3.016029\ \mu + 4.002603\ \mu)]$ (931.5 MeV/μ)
$= (4.316 \times 10^{-3}\ \mu)(931.5\ \text{MeV}/\mu) = 4.02\ \text{MeV}$

32.54 $\Delta m = (m_n + M_U) - (M_{Zr} + M_{Te} + 3m_n)$
$= (1.008665\ \mu + 235.043925\ \mu) - (97.9120\ \mu + 134.9087\ \mu + 3(1.008665\ \mu)$
$= 0.20589\ \mu = 3.418 \times 10^{-28}\ \text{kg}$
and $Q = \Delta mc^2 = 3.076 \times 10^{-11}\ \text{J} = 192\ \text{MeV}$

32.55 Neglect recoil of product nucleus. (i.e. do not require momentum conservation) Then energy balance gives
$KE_{emerging} = KE_{incident} + Q$
To find Q,
$\Delta m = (m_p + M_{Al}) - (M_{Si} + m_n)$
$= (1.007825\ \mu + 26.981541\ \mu) - (26.986721\ \mu + (1.008665\ \mu)$
$= -6.02 \times 10^{-3}\ \mu = -9.993 \times 10^{-30}\ \text{kg}$
and $Q = \Delta mc^2 = -8.994 \times 10^{-13}\ \text{J} = -5.621\ \text{MeV}$
Thus, $KE_{emerging} = 6.61\ \text{MeV} - 5.621\ \text{MeV} = 0.989\ \text{MeV}$

32.56 (a) $\quad {}_{5}^{10}B + {}_{2}^{4}He \rightarrow {}_{6}^{13}C + {}_{1}^{1}H$

The product nucleus is ${}_{6}^{13}C$

CHAPTER THIRTY-TWO SOLUTIONS

(b) $_{6}^{13}C + _{1}^{1}H \rightarrow _{5}^{10}B + _{2}^{4}He$

The product nucleus is $_{5}^{10}B$

32.57 (a) The second reaction has as its target and incident projectile the products formed in the first reaction.

The Q value for the first reaction, $_{5}^{10}B + _{2}^{4}He \rightarrow _{6}^{13}C + _{1}^{1}H$ is

$Q_1 = [(M_B + M_{He}) - (M_C + M_H)]$ (931.5 Mev/μ)

and the Q value for the second reaction, $_{6}^{13}C + _{1}^{1}H \rightarrow _{5}^{10}B + _{2}^{4}He$ is

$Q_2 = [(M_C + M_H) - (M_B + M_{He})]$ (931.5 Mev/μ)

Note that $Q_2 = -Q_1$ Thus, the absolute values of Q for the two reactions are equal.

32.58 (a) $_{8}^{18}O + _{1}^{1}H \rightarrow _{9}^{18}F + _{0}^{1}n$

The other particle is a neutron.
(b) $Q = [(M_O + M_H) - (M_F + M_n)]$ (931.5 Mev/μ)
$= [(17.999160 \mu + 1.007825 \mu) - (M_F + 1.008665 \mu)]$ (931.5 MeV/μ) = -2.453 MeV
From which, we find
$M_F = 18.000953 \mu$

32.59 $_{0}^{1}n + _{2}^{4}He \rightarrow _{1}^{2}H + _{1}^{3}H$

The Q value is
$Q = [(M_n + M_{He}) - (M_{2H} + M_{3H})]$ (931.5 Mev/μ)
$= [(1.008665 \mu + 4.002603 \mu) - (2.014102 \mu + 3.016049 \mu)]$ (931.5 MeV/μ)
$= (-0.018883 \mu)(931.5$ MeV/$\mu) = -17.6$ MeV
Thus,
$$KE_{min} = (1 + \frac{m_{incident\ projectile}}{m_{target\ nucleus}})|Q| = (1 + \frac{1.008665}{4.002603})(17.6\ MeV) = 22.0\ MeV$$

32.60 The original activity of the sample is
$R_0 = 18$ g(15 decays/min g) = 270 decays/min
The decay constant is $\lambda = 1.209 \times 10^{-4}$ y^{-1}.
and $R = R_0 e^{-\lambda t} = (270$ decays/min$)e^{-(1.209 \times 10^{-4}\ y^{-1})(2 \times 10^4\ y)}$
From which, $R = 24.0$ decays/min

32.61 If the decay $_{6}^{12}C \rightarrow _{2}^{4}He + _{2}^{4}He + _{2}^{4}He$ occurs, the Q value is

$Q = [M_C - 3(M_{He})]$ (931.5 Mev/μ)
$= [12.00000 \mu - 3(4.002603 \mu)]$ (931.5 MeV/μ)
$= (-7.81 \times 10^{-3} \mu)(931.5$ MeV/$\mu) = -7.27$ MeV
Since the Q value is less than zero, the decay cannot occur spontaneously.

CHAPTER THIRTY-TWO SOLUTIONS

32.62 The decay constant is $\lambda = \dfrac{0.693}{T_{1/2}} = \dfrac{0.693}{14 \text{ days}} = 4.95 \times 10^{-2} \text{ day}^{-1}$

Therefore, from $R = R_0 e^{-\lambda t}$ we have $\dfrac{R}{R_0} = e^{-\lambda t}$

Thus, $\dfrac{20 \text{ mCi}}{200 \text{ mCi}} = e^{-(4.95 \times 10^{-2} \text{ day}^{-1})t}$

From which, $t = 46.5$ days

32.63 $^{1}_{1}H + ^{7}_{3}Li \rightarrow ^{7}_{4}Be + ^{1}_{0}n$

(b) $Q = [(M_H + M_{Li}) - (M_{Be} + M_n)](931.5 \text{ MeV}/\mu)$
$= [(1.007825\ \mu + 7.016005\ \mu) - (7.016930\ \mu + 1.008665\ \mu)](931.5 \text{ MeV}/\mu)$
$= (-1.765 \times 10^{-3}\ \mu)(931.5 \text{ MeV}/\mu) = -1.644 \text{ MeV}$

Thus,
$KE_{min} = \left(1 + \dfrac{m_{\text{incident projectile}}}{m_{\text{target nucleus}}}\right)|Q| = \left(1 + \dfrac{1.007825}{7.016005}\right)(1.644 \text{ MeV}) = 1.88 \text{ MeV}$

32.64 The Q value of the reaction is
$Q = [(2M_{2H}) - (M_{3He} + M_n)](931.5 \text{ MeV}/\mu)$
$= [2(2.014102\ \mu) - (3.016029\ \mu + 1.008665\ \mu)](931.5 \text{ MeV}/\mu)$
$= (3.51 \times 10^{-3}\ \mu)(931.5 \text{ MeV}/\mu) = 3.27 \text{ MeV}$

Because the Q value is greater than zero, no threshold energy is required.

32.65 Let R_0 equal the total activity withdrawn from the stock solution.
$R_0 = (25 \text{ mCi/ml})(10 \text{ ml}) = 25 \text{ mCi}$.
Let R_0' equal the initial activity of the working solution.
$R_0' = \dfrac{25 \text{ mCi}}{250 \text{ ml}} = 0.1 \text{ mCi/ml}$

After 48 hours the activity of the working solution will be
$R' = R_0' e^{-\lambda t} = (0.1 \text{ mCi/ml}) e^{-(0.693/15 \text{ h})(48 \text{ h})} = 0.011 \text{ mCi/ml}$

and the activity in the sample will be
$R = (0.011 \text{ mCi/ml})(5 \text{ ml}) = 0.055 \text{ mCi}$

32.66 (a) # nuclei $= \dfrac{\text{mass parent}}{\text{mass of 1 nucleus}} = \dfrac{1 \text{ kg}}{239\ \mu(1.66 \times 10^{-27} \text{ kg}/\mu)}$
$= 2.52 \times 10^{24}$

(b) $\lambda = \dfrac{0.693}{T_{1/2}} = \dfrac{0.693}{(2.4 \times 10^4 \text{ y})(3.156 \times 10^7 \text{ s/yr})} = 9.149 \times 10^{-13} \text{ s}^{-1}$.

$R_0 = \lambda N_0 = (9.149 \times 10^{-13} \text{ s}^{-1})(2.52 \times 10^{24}) = 2.306 \times 10^{12}$ decays/s
$= 2.306 \times 10^{12}$ Bq

(c) $R = R_0 e^{-\lambda t}$ or $e^{-\lambda t} = \dfrac{R}{R_0} = \dfrac{0.1 \text{ Bq}}{2.306 \times 10^{12} \text{ Bq}} = 4.336 \times 10^{-14}$ Bq

$\lambda t = -\ln(4.336 \times 10^{-14}) = 30.77$

and $t = \dfrac{30.77}{\lambda} = \dfrac{30.77}{9.149 \times 10^{-13} \text{ s}^{-1}} = 3.363 \times 10^{13}$ s $= 1.07 \times 10^6$ y $= 1.07$ million years.

CHAPTER THIRTY-TWO SOLUTIONS

32.67 The original activity per unit area is

$$R_0 = \frac{5 \times 10^6 \text{ Ci}}{10^{10} \text{ m}^2} = 5 \times 10^{-4} \text{ Ci/m}^2.$$

and the final desired activity is $R = 2 \times 10^{-6}$ Ci/m^2.

The decay constant is $\lambda = \frac{0.693}{T_{1/2}} = \frac{0.693}{28.7 \text{ y}} = 2.414 \times 10^{-2} \text{ y}^{-1}$.

From $R = R_0 e^{-\lambda t}$

we have $\quad \dfrac{R}{R_0} = \dfrac{2 \times 10^{-6} \text{ Ci/m}^2}{5 \times 10^{-4} \text{ Ci/m}^2} = .004 = e^{-(2.414 \times 10^{-2} \text{ y}^{-1})t}$

From which, $\quad t = 228$ y

32.68 The kinetic energy of the neutrons is

$$\tfrac{1}{2} mv^2 = 0.04 \text{ eV} = 6.4 \times 10^{-21} \text{ J}$$

and from this, the velocity of the neutrons (mass = 1.675×10^{-27} kg) is

$$v = 2.76 \times 10^3 \text{ m/s}$$

The time for these neutrons to move 10 km is

$$t = \frac{L}{v} = \frac{10 \times 10^3 \text{ m}}{2.76 \times 10^3 \text{ m/s}} = 3.62 \text{ s}$$

The decay constant for the neutron is

$$\lambda = \frac{0.693}{T_{1/2}} = \frac{0.693}{12 \text{ min}(60 \text{ s/min})} = 9.63 \times 10^{-4} \text{ s}^{-1}.$$

and from $N = N_0 e^{-\lambda t}$ we have $\quad \dfrac{N}{N_0} = e^{-\lambda t} = e^{-(9.63 \times 10^{-4} \text{ s}^{-1})(3.62 \text{ s})} = 0.9965$

So, $\dfrac{N}{N_0} = 0.9965$ is the fraction remaining, and the fraction having decayed is

$$1 - \frac{N}{N_0} = 0.0035, \text{ or } 0.35 \%$$

32.69 The decay constant of the isotope is

$$\lambda = \frac{0.693}{T_{1/2}} = \frac{0.693}{64.8 \text{ h}} = 1.069 \times 10^{-2} \text{ h}^{-1} = 1.78 \times 10^{-4} \text{ s}^{-1}$$

The initial number of nuclei is

$$N_0 = \frac{R_0}{\lambda} = \frac{(40 \times 10^{-6})(3.7 \times 10^{10} \text{ decays/s})}{1.78 \times 10^{-4} \text{ s}^{-1}} = 8.30 \times 10^9 \text{ nuclei}$$

Now use $N = N_0 e^{-\lambda t}$ to find the number present at $t = 10$ h.

$$N_{10} = (8.30 \times 10^9 \text{ nuclei}) e^{-(1.069 \times 10^{-2} \text{ h}^{-1})(10 \text{ h})} = 7.465 \times 10^9 \text{ nuclei}$$

Similarly, the number present at 12 h can be found to be

$$N_{12} = 7.306 \times 10^9 \text{ nuclei}$$

Thus, the number which have decayed between $t = 10$ h and $t = 12$ h is

$$\Delta N = N_{10} - N_{12} = 1.60 \times 10^8 \text{ nuclei}$$

32.70 (a) $R = R_0 e^{-\lambda t}$,

$$R_0 = N_0 \lambda = 1.3 \times 10^{-12} N_0(^{12}C)\lambda = (1.3 \times 10^{-12} \times 25 \times \frac{6.02 \times 10^{23}}{12}) \lambda$$

where $\lambda = \dfrac{0.693}{5730 \times 3.15 \times 10^7} = 3.84 \times 10^{-12}$ decays/s

So $R_0 = 376$ decays/min

CHAPTER THIRTY-TWO SOLUTIONS

$R = (3.76 \times 10^2)e^{(-3.84 \times 10^{-12} \text{ s}^{-1})(2.5 \times 10^4 \text{ y})(3.15 \times 10^7 \text{ s/y})} = 18.3$ counts/min
(b) The observed count rate is slightly less than the average background and would be difficult to measure accurately within reasonable counting times.

32.71 $R = 10$ Ci $= 10(3.7 \times 10^{10}$ decays/s) $= 3.7 \times 10^{11}$ decays/s

$$\lambda = \frac{0.693}{T_{1/2}} = \frac{0.693}{5.2 \text{ y}} = 0.1332 \text{ y}^{-1} = 4.2227 \times 10^{-9} \text{ s}^{-1}$$

Also, $R = R_0 e^{-\lambda t}$, so $R_0 = Re^{\lambda t} = (3.7 \times 10^{11}$ decays/s$)e^{(0.13327 \text{ yr}^{-1})(2.5 \text{ y})}$
$= 5.163 \times 10^{11}$ decays/s

so $N_0 = \frac{R_0}{\lambda} = \frac{5.163 \times 10^{11} \text{ decays/s}}{4.2227 \times 10^{-9} \text{ s}^{-1}} = 1.223 \times 10^{20}$ (number of original nuclei needed)

Therefore, the initial mass of ^{60}Co must be
$mN_0 = (60 \text{ }\mu)(1.66 \times 10^{-27} \text{ kg/}\mu)(1.223 \times 10^{20}) = 1.217 \times 10^{-5}$ kg $= 12.2$ mg

32.72 The intial specific activity of ^{59}Fe in the steel,

$(R/m)_0 = \frac{20 \text{ }\mu\text{Ci}}{0.2 \text{ kg}} = \frac{100 \text{ }\mu\text{Ci}}{\text{kg}} \frac{3.7 \times 10^4 \text{ Bq}}{\mu\text{Ci}} = 3.7 \times 10^6$ Bq/kg

After 1000 h,

$\frac{R}{m} = (R/m)_0 \, e^{-\lambda t} = (3.7 \times 10^6 \text{ Bq/kg}) \, e^{-(6.4 \times 10^{-4} \text{ h}^{-1})(1000 \text{ h})} = 1.95 \times 10^6$ Bq/kg

The activity of the oil, $R_{oil} = (\frac{800}{60}$ Bq/liter$)(6.5$ liters$) = 86.7$ Bq. Therefore,

$m_{\text{in oil}} = \frac{R_{oil}}{(R/m)} = \frac{86.7 \text{ Bq}}{1.95 \times 10^6 \text{ Bq/kg}} = 4.45 \times 10^{-5}$ kg

So that   Wear Rate $= \frac{4.45 \times 10^{-5} \text{ kg}}{1000 \text{ h}} = 4.45 \times 10^{-8}$ kg/h

CHAPTER THIRTY-THREE SOLUTIONS

33.1 $\; {}^{1}_{0}n + {}^{235}_{92}U \rightarrow {}^{144}_{56}Ba + {}^{89}_{36}Kr + 3{}^{1}_{0}n$

$Q = (\Delta m)c^2 = [(M_n + M_U) - (M_{Ba} + M_{Kr} + 3m_n)]\,(931.5\text{ MeV}/\mu)$
$= [235.043925\,\mu - 143.922673\,\mu - 88.917563\,\mu - 2(1.008665\,\mu)](931.5\text{ MeV}/c)$
$= (0.186359\,\mu)(931.5\text{ MeV}/c) = 174\text{ MeV}$

33.2 $\; {}^{1}_{0}n + {}^{235}_{92}U \rightarrow {}^{141}_{56}Ba + {}^{92}_{36}Kr + 3{}^{1}_{0}n$

3 neutrons are produced

33.3 $\; {}^{1}_{0}n + {}^{235}_{92}U \rightarrow {}^{90}_{38}Sr + {}^{144}_{54}Xe + 2{}^{1}_{0}n$

or, $\quad {}^{1}_{0}n + {}^{235}_{92}U \rightarrow {}^{90}_{38}Sr + {}^{143}_{54}Xe + 3{}^{1}_{0}n$

or, $\quad {}^{1}_{0}n + {}^{235}_{92}U \rightarrow {}^{90}_{38}Sr + {}^{142}_{54}Xe + 4{}^{1}_{0}n$

33.4 $\; {}^{1}_{0}n + {}^{235}_{92}U \rightarrow {}^{88}_{38}Sr + {}^{136}_{54}Xe + 12{}^{1}_{0}n$

$Q = (\Delta m)c^2 = [(M_n + M_U) - (M_{Sr} + M_{Xe} + 12m_n)]\,(931.5\text{ MeV}/\mu)$
$= [235.043925\,\mu - 87.905625\,\mu - 135.90722\,\mu - 11(1.008665\,\mu)](931.5\text{ MeV}/\mu)$
$= (0.135765\,\mu)(931.5\text{ MeV}/\mu) = 126\text{ MeV}$

33.5 The energy needed by the bulb in one hour is
$\quad E = (\text{power})t = (100\text{ J/s})(3600\text{ s}) = 3.60 \times 10^5\text{ J} = 2.25 \times 10^{18}\text{ MeV}$
Thus, the number of fission events required is

$$N = \frac{\text{total energy needed}}{\text{energy per event}} = \frac{2.25 \times 10^{18}\text{ MeV}}{208\text{ MeV/fission}} = 1.08 \times 10^{16}\text{ fissions}$$

33.6 We are given $V_{\text{sphere}} = V_{\text{cube}}$
If the sphere has a radius a and the cube has a length L on each side, we have
$\quad \frac{4}{3}\pi a^3 = L^3, \quad \text{or,} \quad L = (4\pi/3)^{1/3}(a)$

(a) Sphere: $\dfrac{A}{V} = \dfrac{4\pi a^2}{\frac{4}{3}\pi a^3} = \dfrac{3}{a}$

(b) Cube: $\dfrac{A}{V} = \dfrac{6L^2}{L^3} = \dfrac{6}{L} = \dfrac{6}{(4\pi/3)^{1/3}(a)} = \dfrac{3.72}{a}$

CHAPTER THIRTY-THREE SOLUTIONS

(c) $\frac{A}{V}$ is lowest for the sphere. Thus, the sphere has a better shape to minimize leakage.

33.7 The total energy output of the plant is
$$E = (\text{power})t = (1000 \times 10^6 \text{ J/s})(86,400 \text{ s}) = 8.64 \times 10^{13} \text{ J}$$
And the total energy which must be released is
$$E_r = \frac{E}{0.3} = \frac{8.64 \times 10^{13} \text{ J}}{0.3} = 2.88 \times 10^{14} \text{ J}$$
The number of events required is
$$N = \frac{E_r}{Q} = \frac{2.88 \times 10^{14} \text{ J}}{(208 \text{ MeV/fission})(1.6 \times 10^{-13} \text{ J/MeV})} = 8.65 \times 10^{24} \text{ events}$$
The number of moles of uranium required is
$$8.65 \times 10^{24} \text{ nuclei} \frac{1 \text{ mol}}{6.02 \times 10^{23} \text{ atoms}} = 14.4 \text{ mol}$$
and the mass required is
$$m = (14.4 \text{ mol})\frac{235 \text{ g}}{\text{mol}} = 3.38 \times 10^3 \text{ g} = 3.38 \text{ kg}$$

33.8 The energy used in one year = $(2000 \text{ kWh/mo})(12 \text{ mo/y})$
$$= 2.4 \times 10^4 \text{ kWh/y} = 8.64 \times 10^{10} \text{ J/y}$$
Assuming 208 MeV of energy released per fission event, the number of nuclei which must fission to yield this energy is
$$N = \frac{8.64 \times 10^{10} \text{ J}}{(208 \text{ MeV/fission})(1.6 \times 10^{-13} \text{ J/MeV})} = 2.60 \times 10^{21} \text{ nuclei}$$
This represents a mass of
$$M = Nm_{nucleus} = (2.60 \times 10^{21} \text{ nuclei})(235 \text{ }\mu/\text{nucleus})(1.66 \times 10^{-27} \text{ kg/}\mu)$$
$$= 1.01 \times 10^{-3} \text{ kg} = 1.01 \text{ g}$$

33.9 The weight of $^{235}_{92}\text{U}$ present is

$w = 10^9 \text{ tons}(7 \times 10^{-3}) = 7 \times 10^6 \text{ tons}$, equivalent to a mass of 6.36×10^9 kg.
The number of moles of U-235 is
$$m = (6.36 \times 10^{12} \text{ g})\frac{1 \text{ mol}}{235 \text{ g}} = 2.71 \times 10^{10} \text{ mol}$$
And the total number of U-235 nuclei is
$$N = (2.71 \times 10^{10} \text{ mol})(6.02 \times 10^{23} \text{ nuclei/mol}) = 1.63 \times 10^{34} \text{ nuclei}$$
Therefore, assuming 208 MeV released per fission, the energy available is
$$E = (208 \text{ MeV/fission})(1.63 \times 10^{34} \text{ nuclei}) = 3.39 \times 10^{36} \text{ MeV} = 5.43 \times 10^{23} \text{ J}$$
(The calculation of E assumes that all the nuclei fission and there is 100% conversion efficiency in the production of power.)
The length of time is
$$t = \frac{E}{\text{rate}} = \frac{5.43 \times 10^{23} \text{ J}}{7 \times 10^{12} \text{ J/s}} = 7.75 \times 10^{10} \text{ s} = 2.46 \times 10^3 \text{ y}$$

33.10 The energy released was $E = (20 \times 10^3 \text{ ton})(4 \times 10^9 \text{ J/ton}) = 8 \times 10^{13} \text{ J}$
The number of nuclei required at 208 MeV per fission is
$$N = \frac{8 \times 10^{13} \text{ J}}{(208 \text{ MeV})(1.6 \times 10^{-13} \text{ J/MeV})} = 2.40 \times 10^{24} \text{ nuclei}$$

CHAPTER THIRTY-THREE SOLUTIONS

The number of moles used is
$$n_{moles} = \frac{2.40 \times 10^{24} \text{ nuclei}}{6.02 \times 10^{23} \text{ nuclei/mol}} = 3.99 \text{ mol}$$
and the mass of U-235 used is
$$M = (3.99 \text{ mol})(235 \text{ g/mol}) = 938 \text{ g}$$

33.11 The mass of U-235 present in 1 kg of fuel = $0.017(1 \text{ kg}) = 1.7 \times 10^{-2}$ kg = 17.0 g.
The number of U-235 atoms (and hence the number of nuclei) present
$$= 17 \text{ g}\left(\frac{6.02 \times 10^{23} \text{ atoms/mol}}{235 \text{ g/mol}}\right) = 4.355 \times 10^{22} \text{ nuclei}$$
Thus, in using 1 kg of fuel, 4.355×10^{22} fissions occur with the release of 208 MeV per fission.
Energy released = 4.355×10^{22} (208 MeV)(1.6×10^{-13} J/MeV) = 1.45×10^{12} J
Energy used effectively = (energy released)(efficiency)
$$= (1.45 \times 10^{12} \text{ J})(0.2) = 2.90 \times 10^{11} \text{ J}$$
Thus, the work done = $(10^5 \text{ N})s = 2.9 \times 10^{11}$ J
and $s = 2.9 \times 10^6$ m = 2900 km = 1800 miles

33.12 $\ _1^2H + \ _1^2H \rightarrow \ _1^3H + \ _1^1H$

$Q = (\Delta m)c^2 = [2M_{2H} - M_{3H} - M_H]$ (931.5 MeV/µ)
$\quad = [2(2.014102 \text{ µ}) - 3.016049 \text{ µ} - 1.007825 \text{ µ}](931.5 \text{ MeV/µ})$
$\quad = (4.33 \times 10^{-3} \text{ µ})(931.5 \text{ MeV/µ}) = 4.03$ MeV

33.13 $\ _1^2H + \ _1^3H \rightarrow \ _2^4He + \ _0^1n$

$Q = (\Delta m)c^2 = [M_{2H} + M_{3H} - M_{He} - M_n]$ (931.5 MeV/µ)
$\quad = [2.014102 \text{ µ}) + 3.016049 \text{ µ} - 4.002603 \text{ µ} - 1.008665 \text{ µ}]$ (931.5 MeV/µ)
$\quad = (1.8883 \times 10^{-2} \text{ µ})(931.5 \text{ MeV/µ}) = 17.6$ MeV

33.14 The total energy required is
$\quad E = (2000 \text{ kWh/mo})(12 \text{ mo/y})(3.6 \times 10^6 \text{ J/kWh}) = 8.64 \times 10^{10}$ J/y
The number of fusion events required at 17.6 MeV released per event is
$$N = \frac{8.64 \times 10^{10} \text{ J/y}}{(17.6 \text{ MeV/event})(1.6 \times 10^{-13} \text{ J/MeV})} = 3.07 \times 10^{22} \text{ events/y}$$

33.15 (a) $\ _1^1H + \ _6^{12}C \rightarrow \ _7^{13}N + \gamma \quad (A = \ _7^{13}N)$

(b) $\ _7^{13}N \rightarrow \ _{+1}^0e + \ _6^{13}C \quad (B = \ _6^{13}C)$

(c) $\ _1^1H + \ _6^{13}C \rightarrow \ _7^{14}N + \gamma \quad (C = \ _7^{14}N)$

(d) $\ _1^1H + \ _7^{14}N \rightarrow \ _8^{15}O + \gamma \quad (D = \ _8^{15}O)$

CHAPTER THIRTY-THREE SOLUTIONS

(e) $^{15}_{8}O \rightarrow\ ^{0}_{+1}e + ^{15}_{7}N$ $(E = ^{15}_{7}N)$

(f) $^{15}_{7}N + ^{1}_{1}H \rightarrow\ ^{4}_{2}He + ^{12}_{6}C$ $(F = ^{12}_{6}C)$

33.16 (a) $^{4}_{2}He + ^{4}_{2}He \rightarrow\ ^{8}_{4}Be + \gamma$ $(A = ^{8}_{4}Be)$

(b) $^{8}_{4}Be + ^{4}_{2}He \rightarrow\ ^{12}_{6}C + \gamma$ $(B = ^{12}_{6}C)$

(c) Consider the first reaction,
$Q_1 = (\Delta m)c^2 = [2(4.002603\ \mu) - 8.00305\ \mu](931.5\ MeV/\mu) = 2.01\ MeV$
and the second reaction yields
$Q_2 = (\Delta m)c^2 = [8.00305\ \mu + 4.002603\ \mu - 12.00000\ \mu](931.5\ MeV/\mu) = 5.27\ MeV$
The total energy released = 7.27 MeV

33.17 (a) 1 gallon = 3786 cm^3 = 3786 g of water
The number of moles of water present is

$$n = \frac{3786\ g}{18\ g/mol} = 210.3\ mol$$

The number of water molecules present = 210.3 mol(6.02 X 10^{23} molecules/mol)
= 1.27 X 10^{26} molecules
The number of hydrogen atoms = 2(1.27 X 10^{26}) = 2.53 X 10^{26} hydrogen atoms
The number of deuterium nuclei = 1.56 X 10^{-4}(2.53 X 10^{26} hydrogen atoms)
= 3.95 X 10^{22} deuterium nuclei

(b) The reaction $^{2}_{1}H + ^{2}_{1}H \rightarrow\ ^{3}_{2}He + ^{1}_{0}n$ has a Q value of 3.27 MeV

The number of reactions which can occur = $\frac{1}{2}$ (number of deuterium nuclei)
= 1.98 X 10^{22} reactions

Thus, the total energy released is
E = (3.27 MeV/reaction)(1.98 X 10^{22} reactions)(1.6 X 10^{-13} J/MeV) = 1.03 X 10^{10} J

(c) Number of gallons of gasoline required = $\frac{1.03\ X\ 10^{10}\ J}{2\ X\ 10^8\ J/gal}$ = 51.7 gallons

---

33.18 (a) $\langle KE \rangle = \frac{1}{2} m v_{av}^2 = \frac{3}{2} kT$,

or, $\bar{v} = \sqrt{\frac{3kT}{m}} = \sqrt{\frac{3(1.38\ X\ 10^{-23}\ J/K)(10^8\ K)}{(2.014\ \mu)(1.66\ X\ 10^{-27}\ kg/\mu)}} = 1.11\ X\ 10^6\ m/s$

(b) The average time for one of these deuterons to cross a 10 cm cube is
$t = \frac{d}{v} = \frac{10^{-1}\ m}{1.11\ X\ 10^6\ m/s} = 9\ X\ 10^{-8}\ s$ (about 10^{-7} s)

33.19 The # of water molecules present

= 1.32 X 10^{21} kg $\left(\frac{6.02\ X\ 10^{26}\ molecules/kg\ mol}{18\ kg/kg\ mol} \right)$ = 4.415 X 10^{46} molecules.

CHAPTER THIRTY-THREE SOLUTIONS

The number of hydrogen nuclei = 2 per molecule = $2(4.415 \times 10^{46})$
= 8.83×10^{46} nucleons

The number of deuterons = 1.56×10^{-4}(number of hydrogen nuclei)
= 1.377×10^{43} deuterons

The number of fusions possible = $\dfrac{\text{number of deuterons}}{2 \text{ deuterons/fusion}}$ = 6.89×10^{42} fusions

(a) Energy released = $(6.89 \times 10^{42} \text{ events})(3.27 \text{ MeV/event})(1.6 \times 10^{-13} \text{ J/MeV})$
= 3.60×10^{30} J

(b) If rate of consumption = 100(present rate) = 7×10^{14} J/s, the time this will last is

$$t = \dfrac{\text{total energy available}}{\text{rate of use}} = \dfrac{3.60 \times 10^{30} \text{ J}}{7 \times 10^{14} \text{ J/s}} = 5.15 \times 10^{15} \text{ s} = 1.63 \times 10^{8} \text{ yr} =$$

163 million years.

33.20 The rad is a measure of the total energy absorbed. The second worker is absorbing energy at twice the rate of the first worker and continues the exposure for the same amount of time as the first worker. Thus, the second worker receives twice the dose received by the first worker.

33.21 dosage rate = 100 mrad = 0.1 rad
dosage rate in rem = (dosage rate in rad)RBE
and the RBE factor for gamma rays is 1.
Thus, dosage rate in rem = 0.1 rem/h

(a) time to receive total dose of 1 rem = $\dfrac{1 \text{ rem}}{0.1 \text{ rem/h}} = 10$ h

(b) Since isotropic $I = \dfrac{I_0}{r^2}$ or $I r^2 = I_0$

$I_2 r_2^2 = I_1 r_1^2$ or $r_2^2 = \dfrac{I_1 r_1^2}{I_2} = \dfrac{(0.1 \text{ rad/h})(1 \text{ m})^2}{0.01 \text{ rad/h}}$

gives $r_2 = 3.16$ m

33.22 (a) The total number of x-rays taken per year
= (8 per day)(5 days/week)(52 weeks/y) = 2080 x-rays/y.

Thus, the average dose per x-ray = $\dfrac{5 \text{ rem/y}}{2080 \text{ x-rays/y}} = 2.40 \times 10^{-3}$ rem/x-ray

(b) The average dose from low-level background is 0.13 rem/y

Therefore, $\dfrac{5 \text{ rem/y}}{0.13 \text{ rem/y}} = 38$ times the background level.

33.23 Energy absorbed = (dose in rad)(energy absorbed per rad per unit of mass)(mass)

$$= (25 \text{ rad})(10^{-2} \dfrac{\text{J/rad}}{\text{kg}})(75 \text{ kg}) = 18.8 \text{ J}$$

33.24 For a 100 rad dose of x-rays,
dose in rem = (100 rad)(1) = 100 rem   (The RBE factor for x-rays = 1)
For heavy ions: dose in rem = (dose in rad)(20), where the RBE factor for heavy ions is 20.
Thus, for the 2 doses in rem to be equivalent, we must have
(heavy ion dose in rad)(20) = 100 rem = (100 rad)(1)

CHAPTER THIRTY-THREE SOLUTIONS

or, Heavy ion dose in rad = $\frac{100 \text{ rad}}{20}$ = 5 rad

---

33.25 (a) Energy delivered per unit of mass = (200 rad)(10^{-2} J/kg rad) = 2 J/kg
(b) $mc\Delta T$ = total energy absorbed = m(2 J/kg)
or, $c\Delta T$ = 2 J/kg
Thus, $\Delta T = \frac{2 \text{ J/kg}}{4184 \text{ J/kg °C}}$ = 4.78 X 10^{-4} °C

33.26 The total energy absorbed = (10^3 rad)(10^{-2} J/rad kg)M = (10 J/kg)M
But $Mc\Delta T$ = total energy absorbed = M(10 J/kg)
and $\Delta T = \frac{10 \text{ J/kg}}{4184 \text{ J/kg °C}}$ = 2.39 X 10^{-3} °C

33.27 rate of depositing energy = 10 rad/s(10^{-2} J/ rad kg) = 0.1 J/kg s
energy needed per kg of water heated = $mc\Delta T$ = (1 kg)(4186 J/kg °C)(50 °C) = 2.09 X 10^5 J/kg
time required = $\frac{\text{energy needed}}{\text{rate of energy input}} = \frac{2.09 \times 10^5 \text{ J/kg}}{0.1 \text{ J/kg s}}$ = 2.09 X 10^6 s = 24.2 days

---

33.28 (a) absorbed energy per kg = $\frac{\text{rest energy}}{\text{mass}} = \frac{m_0 c^2}{m_0} = c^2$ = 9 X 10^{16} J/kg

absorbed dose = $\frac{\text{absorbed energy/kg}}{10^{-2} \text{ J/ rad kg}}$ = 9 X 10^{18} rad

(b) absorbed energy per kg = $\frac{\text{heat required}}{m} = \frac{mL_v}{m} = L_v$ = 2.26 X 10^6 J/kg

absorbed dose = $\frac{\text{absorbed energy/kg}}{10^{-2} \text{ J/ rad kg}} = \frac{2.26 \times 10^6 \text{ J/kg}}{10^{-2} \text{ J/ rad kg}}$ = 2.26 X 10^8 rad

---

33.29 decay constant = $\lambda = \frac{0.693}{14.3 \text{ days}}$ = 4.846 X 10^{-2} day^{-1} = 5.61 X 10^{-7} s^{-1}

initial number of radioactive nuclei present = $N_0 = \frac{R_0}{\lambda}$

= $\frac{1.31 \times 10^6 \text{ s}^{-1}}{5.61 \times 10^{-7} \text{ s}^{-1}}$ = 2.34 X 10^{12}

The number remaining after 10 days = $N = N_0 e^{-\lambda t}$
= (2.34 X 10^{12})$e^{-(4.846 \times 10^{-2} \text{ day}^{-1})(10 \text{ days})}$ = 1.44 X 10^{12} (after 10 days)
(a) number of electrons emitted = number of decays that have occurred
= $N_0 - N$ = 9.00 X 10^{11}
(b) energy deposited = (700 keV/decay)(9 X 10^{11} decays)(1.6 X 10^{-16} J/keV) = 1.01 X 10^{-1} J
(c) aborbed dose = $\frac{\text{energy deposited}}{\text{mass}} = \frac{0.101 \text{ J}}{0.1 \text{ kg}} \frac{1 \text{ rad}}{10^{-2} \text{ J/kg}}$ = 101 rad

---

33.30 total energy released = energy stored in capacitor = $\frac{1}{2}CV^2$

= $\frac{1}{2}$ (5 X 10^{-12} F)(1000 V)^2 J = 2.5 X 10^{-6} J

CHAPTER THIRTY-THREE SOLUTIONS

energy input = 0.5 MeV(1.6 X 10^{-13} J/MeV) = 8 X 10^{-14} J

(a) energy amplification = $\dfrac{2.5 \times 10^{-6} \text{ J}}{8 \times 10^{-14} \text{ J}}$ = 3.13 X 10^7

(b) number of electrons avalanched = number of electrons required to neutralize the positive plate of the capacitor = $\dfrac{Q}{e}$

$\quad$ Q = CV = (5 X 10^{-12} F)(1000 V) = 5 X 10^{-9} C

number of electrons = $\dfrac{5 \times 10^{-9} \text{ C}}{1.6 \times 10^{-19} \text{ C/electron}}$ = 3.13 X 10^{10}

33.31 (a) Assume the incident electron started from rest at zero potential. When it arrives at the first dynode it has 100 eV of energy. Therefore, it can free 10 electrons.

(b) Each of these freed electrons gains 100 eV of energy as they accelerate to the second dynode, where each will free 10 electrons, and so forth. Thus, the multiplication factor is 10 at each dynode. As a result, the total multiplication factor is

$\quad$ (10)(10)(10)(10)(10)(10)(10) = 10^7

33.32 The energy gained by an electron as it travels s cm in the field is given by

$\quad$ W = Fs = qEs = e(1000 V/cm)s = (10^3 eV/cm)s

Thus, if W = 40 eV $\quad$ 40 eV = (10^3 eV/cm)s

and $\quad$ s = 4 X 10^{-2} cm

33.33 The number of electrons in the avalanche doubles in a characteristic distance of

$\quad$ 4 X 10^{-2} cm. (See previous problem.)

Thus, if the total distance is 0.5 cm, the number will double

$\quad \dfrac{0.5 \text{ cm}}{4 \times 10^{-2} \text{ cm}}$ = 12.5 (or 12 times)

Thus, the final number of electrons in the avalanche = 2^{12} = 4096

33.34 $\langle KE \rangle = \dfrac{1}{2} m\bar{v}^2 = \dfrac{3}{2} kT$,

or, $\quad \bar{v} = \sqrt{\dfrac{3kT}{m}} = \sqrt{\dfrac{3(1.38 \times 10^{-23} \text{ J/K})(3 \times 10^7 \text{ K})}{(1.67 \times 10^{-27} \text{ kg})}}$ = 8.62 X 10^5 m/s

---

33.35 (a) $\dfrac{1}{2} mv^2 = kT$,

or $\quad v = \sqrt{\dfrac{2kT}{m}} = \sqrt{\dfrac{2(1.38 \times 10^{-23} \text{ J/K})(300 \text{ K})}{(1.675 \times 10^{-27} \text{ kg})}}$ = 2.22 X 10^3 m/s

Thus, p = mv = (1.675 X 10^{-27} kg)(2.22 X 10^3 m/s) = 3.72 X 10^{-24} kg m/s

(b) $\lambda = \dfrac{h}{p} = \dfrac{6.63 \times 10^{-34} \text{ J s}}{3.72 \times 10^{-24} \text{ kg m/s}}$ = 1.78 X 10^{-10} m

The de Broglie wavelength is about the size of an atom (10^{-10} m), and about 10^5 times the size of a nucleus (10^{-15} m).

CHAPTER THIRTY-THREE SOLUTIONS

33.36 The energy output required = (power)t
= $(10^8 \text{ J/s})(100 \text{ days})(86,400 \text{ s/day})$
= 8.64×10^{14} J

The total energy input required is $\dfrac{8.64 \times 10^{14} \text{ J}}{0.3} = 2.88 \times 10^{15}$ J

Assuming an average of 208 MeV per fission of U-235, the number of fission events, or nuclei required, is

$$N = \frac{2.88 \times 10^{15} \text{ J}}{(208 \text{ MeV/fission})(1.6 \times 10^{-13} \text{ J/MeV})} = 8.65 \times 10^{25} \text{ nuclei}$$

The number of moles used = $\dfrac{8.65 \times 10^{25} \text{ atoms}}{6.02 \times 10^{23} \text{ atoms/mol}} = 143.8$ mol

and the mass of this much uranium is $M = (143.8 \text{ mol})(0.235 \text{ kg/mol}) = 33.8$ kg

---

33.37 KE after a collision = $\dfrac{1}{2}$ (KE before collision)

Thus, after N collisions

$KE = \dfrac{1}{2^N}$ (original KE)

or $2^N = \dfrac{KE_{original}}{KE_{final}} = \dfrac{2 \times 10^6 \text{ eV}}{0.039 \text{ eV}} = 5.128 \times 10^7$

Thus, $N \ln 2 = \ln(5.128 \times 10^7) = 17.75$
from which $N = 25.6$
Thus, it requires 26 collisions

33.38 (Volume of air present) = $\dfrac{\text{mass}}{\text{density}} = \dfrac{2.20 \times 10^{-3} \text{ kg}}{1.29 \text{ kg/m}^3} = 1.705 \times 10^{-3}$ m^3

or volume = 1705 cm^3

exposure in Roentgens = $\dfrac{\dfrac{1.65 \times 10^9 \text{ ion pairs}}{1705 \text{ cm}^3}}{2.08 \times 10^9 \text{ ion pairs/cm}^3/\text{R}} = 4.65 \times 10^{-4}$ R

exposure rate = $\dfrac{4.65 \times 10^{-4} \text{ R}}{2.4 \text{ s}} = 1.939 \times 10^{-4}$ R/s = 0.698 Roentgens/h

---

33.39 The number of moles in one kg of water = $\dfrac{1000 \text{ g}}{18 \text{ g/mol}} = 55.6$ mol

and the number of molecules in one kg = $(55.6 \text{ mol})(6.02 \times 10^{23} \text{ molecules/mol})$
= 3.34×10^{25} molecules

The thermal energy = (number of molecules)(kT)
= $(3.34 \times 10^{25} \text{ molecules})(1.38 \times 10^{-23} \text{ J/K})(300 \text{ K}) = 1.38 \times 10^5$ J

The energy delivered by the radiation = (dose in rad)(10^{-2} J/kg rad)(1 kg)
In order for this to equal the thermal energy,
(dose in rad)(10^{-2} J/rad) = 1.38×10^5 J

and dose in rad = $\dfrac{1.38 \times 10^5 \text{ J}}{10^{-2} \text{ J/rad}} = 1.38 \times 10^7$ rad

---

33.40 (a) 1 gallon = 3786 cm^3 which is equivalent to a mass of 3786 g.

The number of moles present = $\dfrac{3786 \text{ g}}{18 \text{ g/mol}} = 210.3$ mol

CHAPTER THIRTY-THREE SOLUTIONS

The number of water molecules present = 210.3 mol(6.02 X 10^{23} molecules/mol)
$$= 1.27 \times 10^{26} \text{ molecules}$$
The number of hydrogen atoms present = 2(1.27 X 10^{26}) = 2.53 X 10^{26} atoms

and the number of deuterons present = $\dfrac{2.53 \times 10^{26} \text{ atoms}}{6500 \text{ atoms/deuteron}}$

= 3.90 X 10^{22} deuterons

Thus, the energy available, assuming Q = 3.27 MeV per fusion event (which consumes 2 deuterons) is

E = (3.90 X 10^{22} deuterons)(1.64 MeV/deuteron) = 6.39 X 10^{22} MeV
= 1.02 X 10^{10} J

(b) t = $\dfrac{\text{energy available}}{\text{rate of consumption}} = \dfrac{1.02 \times 10^{10} \text{ J}}{10^4 \text{ J/s}}$ = 1.02 X 10^6 s = 11.8 days

CHAPTER THIRTY-FOUR SOLUTIONS

34.1 The minimum energy is released, and hence the minimum frequency photons are produced, when the proton and antiproton are at rest when they annihilate. That is, $E = E_0$, and $E_K = 0$. To conserve momentum, each photon must carry away one-half the energy. Thus,

$$E_{min} = hf_{min} = \frac{1}{2}(2E_0) = E_0 = 938.3 \text{ MeV}.$$

Thus, $f_{min} = \dfrac{(938.3 \text{ MeV})(1.6 \times 10^{-13} \text{ J/MeV})}{6.63 \times 10^{-34} \text{ J s}} = 2.26 \times 10^{23}$ Hz

and $\lambda = \dfrac{c}{f_{min}} = \dfrac{3 \times 10^8 \text{ m/s}}{2.26 \times 10^{23} \text{ Hz}} = 1.32 \times 10^{-15}$ m

---

34.2 Assuming that the proton and antiproton are left at rest after they are produced, the energy of the photon, E, must be

$E = 2E_0 = 2(938.3 \text{ MeV}) = 1876.6 \text{ MeV} = 3.00 \times 10^{-10}$ J

Thus, $E = hf = 3.00 \times 10^{-10}$ J and $f = \dfrac{3.00 \times 10^{-10} \text{ J}}{6.63 \times 10^{-34} \text{ J s}} = 4.53 \times 10^{23}$ Hz

and $\lambda = \dfrac{c}{f} = \dfrac{3 \times 10^8 \text{ m/s}}{4.53 \times 10^{23} \text{ Hz}} = 6.62 \times 10^{-16}$ m

---

34.3 The rest energy of the Z^0 boson is $E_0 = 96$ GeV.

The maximum time a virtual Z^0 boson can exist is found from $\Delta E \Delta t = \hbar$,

or $\Delta t = \dfrac{\hbar}{\Delta E} = \dfrac{1.055 \times 10^{-34} \text{ J s}}{(96 \text{ GeV})(1.6 \times 10^{-10} \text{ J/GeV})} = 6.87 \times 10^{-27}$ s

The maximum distance it can travel in this time is

$d = c(\Delta t) = (3 \times 10^8 \text{ m/s})(6.87 \times 10^{-27} \text{ s}) = 2.06 \times 10^{-18}$ m

The distance d is an approximate value for the range of the weak interaction.

34.4 $\mu^+ + e \rightarrow \nu + \nu$

muon-lepton number before = $(-1) + (0)$
electron-lepton number before = $(0) + (1) = 1$
Therefore, after the reaction, the muon-lepton number must be -1. Thus, one of the neutrinos must be the anti-neutrino associated with muons or $\overline{\nu}_\mu$. Also, after the reaction, the electron-lepton number must be 1. Thus, one of the neutrinos must be the neutrino associated with electrons, or ν_e.

Thus, $\mu^+ + e \rightarrow \overline{\nu}_\mu + \nu_e$

34.5 The time for a particle traveling with the speed of light to travel a distance of 3×10^{-15} m is $\Delta t = \dfrac{d}{v} = \dfrac{3 \times 10^{-15} \text{ m}}{3 \times 10^8 \text{ m/s}} = 10^{-23}$ s

CHAPTER THIRTY-FOUR SOLUTIONS

34.6 Total momentum before = 0 = total momentum after. Thus, the photons must go in opposite directions and have equal momenta, and hence lower energies since $E_g = p_g c$.

Thus, from mass energy conservation,

$E_0 = 2E_g$ (where E_0 is the energy of the π^0)

or $\quad E_g = \frac{1}{2}E_0 = \frac{1}{2}$ (135 MeV) = 67.5 MeV for each gamma ray

---

34.7 $? + p \rightarrow n + \mu^+$

Conservation of charge yields $Q + e = 0 + e$, so $Q = 0$
Conservation of Baryon number yields, $B + 1 = 1 + 0$, so $B = 0$
Conservation of Lepton number yields, $L_e + 0 = 0 + 0$, so $L_e = 0$
and $\quad L_\mu + 0 = 0 - 1$, gives $L_\mu = -1$
$\quad L_\tau + 0 = 0 + 0$, so $L_\tau = 0$

Thus, the particle must be an antilepton. Since the μ^+ is on the right, the particle on the left must be the $\overline{\nu}_\mu$

34.8 The $\rho^0 \rightarrow \pi^+ + \pi^-$ decay must occur via the strong interaction.

The $K^0 \rightarrow \pi^+ + \pi^-$ decay must occur via the weak interaction.

34.9 (a) $\pi^- + p \rightarrow 2\eta^0$

violates conservation of Baryon number as $0 + 1 \rightarrow 0$

(b) $K^- + n \rightarrow \Lambda^0 + \pi^-$

Baryon number = $0 + 1 \rightarrow 1 + 0$
charge = $-1 + 0 \rightarrow 0 - 1$
Strangeness, $-1 + 0 \rightarrow -1 + 0$
Lepton number, $0 \rightarrow 0$

All are conserved and the interaction may occur via the strong interaction

(c) $K^- \rightarrow \pi^- + \pi^0$

Strangeness, $-1 \rightarrow 0 + 0$
Baryon number, $0 \rightarrow 0$
Lepton number, $0 \rightarrow 0$
charge, $-1 \rightarrow -1 + 0$

Thus, strangeness is not conserved, but everything else is. Thus, the reaction can occur via the weak interaction, but not the strong or electromagnetic.

(d) $\Omega^- \rightarrow \Xi^- + \pi^0$

Baryon number, $1 \rightarrow 1 + 0$
Lepton number, $0 \rightarrow 0$
Charge, $-1 \rightarrow -1 + 0$
Strangeness, $-3 \rightarrow -2 + 0$

May occur by weak interaction, but not by strong or electromagnetic

(e) $\eta^0 \rightarrow 2\gamma$

CHAPTER THIRTY-FOUR SOLUTIONS

Baryon number, $0 \to 0$

Lepton number, $0 \to 0$

Charge, $0 \to 0$

Strangeness, $0 \to 0$

No conservation laws are violated, but photons are the mediators of the electromagnetic interaction. Also, the lifetime of the η^0 is consistent with the electromagnetic interaction.

34.10 $K^0 \to \pi^+ + \pi^-$

Strangeness, $+1 \to 0 + 0$

Baryon number, $0 \to 0 + 0$

Lepton number, $0 \to 0$

charge, $0 \to +1 - 1$

Does not violate any absolute conservation law, and violates strangeness by only 1 unit. Thus, it can occur via the weak interaction.

(b) $\Lambda^0 \to \pi^+ + \pi^-$

Baryon number, $-1 \to 0 + 0$

Since all interactions conserve Baryon number, this process cannot occur.

34.11 (a) $\Lambda^0 \to p + \pi^-$

Strangeness: $-1 \to 0 + 0$ (strangeness is not conserved)

(b) $\pi^- + p \to \Lambda^0 + K^0$

Strangeness: $0 + 0 \to -1 + 1$ (0 = 0 and strangeness is conserved)

(c) $\bar{p} + p \to \bar{\Lambda}^0 + \Lambda^0$

Strangeness: $0 + 0 \to +1 - 1$ (0 = 0 and strangeness is conserved)

(d) $\pi^- + p \to \pi^- + \Sigma^+$

Strangeness: $0 + 0 \to 0 - 1$ (0 does not equal -1 so strangeness is not conserved)

(e) $\Xi^- \to \Lambda^0 + \pi^-$

Strangeness: $-2 \to -1 + 0$ (-2 does not equal -1 so strangeness is not conserved)

(f) $\Xi^0 \to p + \pi^-$

Strangeness: $-2 \to 0 + 0$ (-2 does not equal 0 so strangeness is not conserved)

34.12 (a) $\mu^- \to e + \gamma$
L_e goes $0 \to 1 + 0$ and L_μ goes $1 \to 0 + 0$
(b) $n \to p + e + \nu_e$
L_e goes $0 \to 0 + 1 + 1$
(c) $\Lambda^0 \to p + \pi^0$
Strangeness: $-1 \to 0 + 0$, and charge $0 \to +1 + 0$
(d) $p \to e^+ + \pi^0$

CHAPTER THIRTY-FOUR SOLUTIONS

Baryon number; $+1 \to 0 + 0$
(e) $\Xi^0 \to n + \pi^0$
Strangeness: $-2 \to 0 + 0$

34.13 (a) $\Xi^- \to \Lambda^0 + \mu^- + \nu_\mu$
Baryon number, $+1 \to +1 + 0 + 0$
Lepton number, $0 \to 0 + 0 + 0$
Charge, $-1 \to 0 - 1 + 0$
Strangeness, $-2 \to -1 + 0 + 0$
$L_e, 0 \to 0 + 0 + 0$
$L_\mu, 0 \to 0 + 1 + 1$
$L_\tau, 0 \to 0 + 0 + 0$
Conserved quantities are B, charge, L_e, L_τ,

(b) $K^0 \to 2\pi^0$
Baryon number, $0 \to 0$
charge, $0 \to 0$
$L_e, 0 \to 0$
$L_\mu, 0 \to 0$
$L_\tau, 0 \to 0$
Strangeness, $+1 \to 0$
Conserved quantities are B, charge, L_e, L_μ, L_τ,

(c) $K^- + p \to \Sigma^0 + n$
Baryon number, $0 + 1 \to 1 + 1$
charge, $-1 + 1 \to 0 + 0$
$L_e, 0 + 0 \to 0 + 0$
$L_\mu, 0 + 0 \to 0 + 0$
$L_\tau, 0 + 0 \to 0 + 0$
Strangeness, $-1 + 0 \to -1 + 0$
Conserved quantities are S, charge, L_e, L_μ, L_τ,

(d) $\Sigma^0 \to \Lambda^0 + \gamma$
Baryon number, $+1 \to 1 + 0$
charge, $0 \to 0$
$L_e, 0 \to 0 + 0$
$L_\mu, 0 \to 0 + 0$
$L_\tau, 0 \to 0 + 0$
Strangeness, $-1 \to -1 + 0$
Conserved quantities are B, S, charge, L_e, L_μ, L_τ,

(e) $e^+ + e^- \to \mu^+ + \mu^-$

CHAPTER THIRTY-FOUR SOLUTIONS

Baryon number, $0 + 0 \to 0 + 0$

charge, $+1 -1 \to +1 - 1$

L_e, $-1 + 1 \to 0 + 0$

L_μ, $0 + 0 \to +1 - 1$

L_τ, $0 + 0 \to 0 + 0$

Strangeness, $0 + 0 \to 0 + 0$

Conserved quantities are B, S, charge, L_e, L_μ, L_τ,

(f) $\bar{p} + n \to \Lambda^+ + \Sigma^-$

Baryon number, $-1 + 1 \to -1 + 1$

charge, $-1 + 0 \to 0 - 1$

L_e, $0 + 0 \to 0 + 0$

L_μ, $0 + 0 \to 0 + 0$

L_τ, $0 + 0 \to 0 + 0$

Strangeness, $0 + 0 \to +1 - 1$

Conserved quantities are B, S, charge, L_e, L_μ, L_τ,

34.14 (a) $K^+ + p \to \underline{} + p$

The strong interaction conserves everything.

Baryon number, $0 + 1 \to B + 1$, so $B = 0$

charge, $+1 + 1 \to Q + 1$, so $Q = +1$

Lepton numbers, $0 + 0 \to L + 0$, so $L_e = L_\mu = L_\tau = 0$

Strangeness, $+1 + 0 \to S + 0$, so $S = 1$

The conclusion is that the particle must be positively charged, a non-Baryon, with strangeness of +1. Of particles in Table 34.2, it can only be the K^+. Thus, this is an elastic scattering process.

(b) The weak interaction conserves everything but strangeness, and there $\Delta S = \pm 1$.

$\Omega^- \to \underline{} + \pi^-$

Baryon number, $+ 1 \to B + 0$, so $B = 1$

charge, $- 1 \to Q - 1$, so $Q = 0$

Lepton numbers, $0 \to L + 0$, so $L_e = L_\mu = L_\tau = 0$

Strangeness, $-3 \to S + 0$, so $\Delta S = 1$, and $S = -2$

The particle must be a neutral baryon with strangeness of -2. Thus, it is the Ξ^0

(c) $K^+ \to \underline{} + \mu^+ + \nu_\mu$

Baryon number, $0 \to B + 0 + 0$, so $B = 0$

charge, $+ 1 \to Q + 1 + 0$, so $Q = 0$

L_e, $0 \to L_e + 0 + 0$, so $L_e = 0$

L_μ, $0 \to L_\mu - 1 + 1$, so $L_\mu = 0$

L_τ, $0 \to L_\tau + 0 + 0$, so $L_\tau = 0$

Strangeness, $1 \to S + 0 + 0$, so $\Delta S = \pm 1$ (for weak interaction), and $S = 0$

CHAPTER THIRTY-FOUR SOLUTIONS

Thus, the particle must be a neutral meson with strangeness = 0. Thus, it is the π^0

34.15 (a) $p + \bar{p} \rightarrow \mu^+ + e$
L_e goes $0 + 0 \rightarrow 0 + 1$ and L_μ goes $0 + 0 \rightarrow -1 + 0$
(b) $\pi^- + p \rightarrow p + \pi^+$
charge goes $-1 + 1 \rightarrow +1 + 1$
(c) $p + p \rightarrow p + \pi^+$
baryon number: $1 + 1 \rightarrow 1 + 0$
(d) $p + p \rightarrow p + p + n$
baryon number: $1 + 1 \rightarrow 1 + 1 + 1$
(e) $\gamma + p \rightarrow n + \pi^0$
charge: $0 + 1 \rightarrow 0 + 0$

34.16 $\pi^+ + p \rightarrow K^+ + \Sigma^+$
A strong interaction. Charge, baryon number, and strangeness are conserved.
$\pi^+ + p \rightarrow \pi^+ + \Sigma^-$
A strong interaction. Charge and baryon number are conserved, but strangeness is not conserved.

34.17 (a) $\pi^- \rightarrow \mu^- + \bar{\nu}_\mu$
L_μ: $0 \rightarrow 1 - 1$
(b) $K^+ \rightarrow \mu^+ + \nu_\mu$
L_μ: $0 \rightarrow -1 + 1$
(c) $\bar{\nu}_e + p \rightarrow n + e^+$
L_e: $-1 + 0 \rightarrow 0 - 1$
(d) $\nu_e + p \rightarrow n + e$
L_e: $1 + 0 \rightarrow 0 + 1$
(e) $\nu_\mu + n \rightarrow p + \mu^-$
L_μ: $1 + 0 \rightarrow 0 + 1$
(f) $\mu^- \rightarrow e + \bar{\nu}_e + \nu_\mu$
L_μ: $1 \rightarrow 0 + 0 + 1$, and L_e: $0 \rightarrow 1 - 1 + 0$

34.18 The relevant conservation laws are those involving L_e, L_μ, and L_τ
(a) $\pi^+ \rightarrow \pi^0 + e^+ + ?$
L_e: $0 \rightarrow 0 - 1 + L_e$ so $L_e = 1$, so we have a ν_e
(b) $? + p \rightarrow \mu^- + p + \pi^+$
L_μ: $L_\mu + 0 \rightarrow +1 + 0 + 0$, so $L_\mu = 1$, so we have a ν_μ
(c) $\Lambda^0 \rightarrow p + \mu^- + ?$
L_μ: $0 \rightarrow 0 + 1 + L_\mu$, so $L_\mu = -1$, so we have a $\bar{\nu}_\mu$
(d) $\tau^+ \rightarrow \mu^+ + ? + ?$
L_μ: $0 \rightarrow -1 + L_\mu$, so $L_\mu = 1$, so we have a ν_μ
L_τ: $+1 \rightarrow 0 + L_\tau$, so $L_\tau = 1$, so we have a $\bar{\nu}_\tau$

CHAPTER THIRTY-FOUR SOLUTIONS

So, $L_\mu = 1$ for one particle, and $L_\tau = 1$ for the other particle. We have ν_μ and $\bar{\nu}_\tau$

34.19 (a) $p \to \pi^+ + \pi^0$
baryon number is violated; $1 \to 0 + 0$
(b) $p + p \to p + p + \pi^0$
This reaction can occur.
(c) $p + p \to p + \pi^+$
baryon number is violated: $1 + 1 \to 1 + 0$
(d) $\pi^+ \to \mu^+ + \nu_\mu$
This reaction can occur.
(e) $n \to p + e + \bar{\nu}_e$
This reaction can occur
(f) $\pi^+ \to \mu^+ + n$
violates baryon number: $0 \to 0 + 1$,
and violates muon-lepton number: $0 \to -1 + 0$

34.20 Total momentum before = 0 and they must have zero momentum after. Also, they have equal original energies. Since the π^+ is at rest, the p and n must go in opposite directions with equal magnitude momenta. For minimum energy of incident protons, the p and n will be at rest afterward. Mass-energy conservation gives

$2E_p = E_{0p} + E_{0n} + E_{0\pi} = 938.3 \text{ MeV} + 939.6 \text{ MeV} + 139.6 \text{ MeV}$
or $E_p = 1008.75$ MeV
But, $E_p = KE + E_0$
and $KE = E_p - E_0 = 1008.75 - 938.3 \text{ MeV} = 70.5 \text{ MeV}$

34.21 We must conserve both mass-energy and momentum.
$p_{before} = 0$, so $p_\Lambda = p_\gamma$
or $E^2_\Lambda - E^2_{0\Lambda} = E^2_\gamma$ or $E^2_\Lambda - (1115.6 \text{ MeV})^2 = E^2_\gamma$ (1)
From mass-energy conservation, $E_{0\Sigma} = E_\Lambda + E_\gamma$
which gives $E_\gamma = (1192.5 \text{ MeV}) - E_\Lambda$ (2)
Solving (1) and (2) simultaneously gives
$E_\Lambda = 1118.08$ MeV, and $E_\gamma = 74.4$ MeV

34.22 (a)

	K^0	d	$\bar{s}$	total
strangeness	1	0	1	1
baryon number	0	$\frac{1}{3}$	$-\frac{1}{3}$	0
charge	0	$-\frac{1}{3}e$	$\frac{1}{3}e$	0

CHAPTER THIRTY-FOUR SOLUTIONS

(b)

	Λ^0	u	d	s	total
strangeness	-1	0	0	-1	-1
baryon number	1	$\frac{1}{3}$	$\frac{1}{3}$	$\frac{1}{3}$	1
charge	0	$\frac{2}{3}e$	$-\frac{1}{3}e$	$-\frac{1}{3}e$	0

34.23 (a)

	proton	u	u	d	total
strangeness	0	0	0	0	0
baryon number	1	$\frac{1}{3}$	$\frac{1}{3}$	$\frac{1}{3}$	1
charge	e	$\frac{2}{3}e$	$\frac{2}{3}e$	$-\frac{1}{3}e$	e

(b)

	neutron	u	d	d	total
strangeness	0	0	0	0	0
baryon number	1	$\frac{1}{3}$	$\frac{1}{3}$	$\frac{1}{3}$	1
charge	0	$\frac{2}{3}e$	$-\frac{1}{3}e$	$-\frac{1}{3}e$	0

34.24 (a) $\bar{u}\,\bar{u}\,\bar{d}$

charge = $(-\frac{2}{3}e) + (-\frac{2}{3}e) + (\frac{1}{3}e) = -e$ This is the antiproton.

(b) $\bar{u}\,\bar{d}\,\bar{d}$

charge = $(-\frac{2}{3}e) + (\frac{1}{3}e) + (\frac{1}{3}e) = 0$ This is the antineutron.

34.25 (a) $\pi^- + p \rightarrow K^+ + \Lambda^0$

In terms of constituent quarks: $\bar{u}d + uud \rightarrow d\bar{s} + uds$
up quarks: $-1 + 2 \rightarrow 0 + 1$
down quarks: $1 + 1 \rightarrow 1 + 1$
strange quarks: $0 + 0 \rightarrow -1 + 1$

(b) $\pi^+ + p \rightarrow K^+ + \Sigma^+$

$u\bar{d} + uud \rightarrow u\bar{s} + uus$
up quarks: $1 + 2 \rightarrow 1 + 2$
down quarks: $-1 + 1 \rightarrow 0 + 0$
strange quarks: $0 + 0 \rightarrow -1 + 1$

(c) $K^- + p \rightarrow K^+ + K^0 + \Omega^-$

$\bar{u}s + uud \rightarrow u\bar{s} + d\bar{s} + sss$
up quarks: $-1 + 2 \rightarrow 1 + 0 + 0$
down quarks: $0 + 1 \rightarrow 0 + 1 + 1$
strange quarks: $1 + 0 \rightarrow -1 - 1 + 3$

(d) $p + p \rightarrow K^0 + p + \pi^+ + ?$

$uud + uud \rightarrow d\bar{s} + uud + u\bar{d} + ?$

The quark combination of ? must be such as to balance the last equation for u, d, and s quarks.
u: $2 + 2 = 0 + 2 + 1 + ?$ (has 1 u quark)
d: $1 + 1 = 1 + 1 - 1 + ?$ (has 1 d quark)

470

CHAPTER THIRTY-FOUR SOLUTIONS

s: 0 + 0 = -1 + 0 + 0 + ? (has 1 s quark)
quark composite = uds = Λ^0

34.26 Compare the given quark states to the entries in Table 34.4

(a) uds = Λ^0 or Σ^0

(b) $\bar{u}\,d = \pi^-$

(c) $\bar{s}\,d = K^0$

(d) ssd = Ξ^-

34.27 Quark composition of proton = uud
and of neutron = udd
Thus, if we neglect binding energies, we may write
$m_p = 2m_u + m_d$ (1)
and $m_n = m_u + 2m_d$ (2)
Solving simultaneously, we find

$$m_u = \frac{1}{3}(2m_p - m_n) = \frac{1}{3}(2(938.3 \text{ MeV}/c^2) - 939.6 \text{ MeV}/c^2) = 312.3 \text{ MeV}/c^2$$

and from either (1) or (2), $m_d = 313.6$ MeV/c^2

CHAPTER THIRTY-FOUR SOLUTIONS

EVEN-NUMBERED ANSWERS

ANSWERS TO EVEN-NUMBERED PROBLEMS

CHAPTER ONE

2. (a) $\frac{L}{T^2}$, (b) L

4. (a) $\frac{L}{T}$ on the left side does not equal $\frac{L}{T} + \frac{L^2}{T^2}$ on the right side. Thus, the equation is not dimensionally correct. (b) Both sides have units of L, so the equation is dimensionally correct.

8. (a) 3 significant figures, (b) 4 significant figures, (c) 3 significant figures, (d) 2 significant figures.

10. (a) 719, (b) 2.2

12. 115.9 m

14. 9.82 cm

16. 10^{17} ft

18. (a) 3.16×10^7 s, (b) 6.05×10^{10} yr

20. 10^6 m^2 = 0.39 sq. miles

22. (a) 1000 kg, (b) $m_{cell} = 5.2 \times 10^{-16}$ kg, $m_{kidney} = 0.27$ kg, (c) $m_{fly} = 1.3 \times 10^{-5}$ kg

24. 6.71×10^8 mi/h

26. 2.57×10^6 m^3

28. 7300 balls, (assumes 81 games per season, 9 innings per game, and an average of 10 hitters per inning)

30. 750 tuners, (assumes 1 tuner per 10,000 residents and a population of 7.5 million)

32. (2.05, 1.43

34. r = 2.24 m, θ = 26.6°

36. (a) 3, (b) 3, (c) 4/5, (d) 4/5, (e) 4/3

38. 35.5°

40. (a) 6.71 m, (b) 0.894, (c) 0.746

42. (a) 0.677 g/cm^3, (b) 4.30×10^{16} m^2

44. 2×10^5 tons/yr, (assumes an average of 0.5 oz of aluminum per can)

EVEN-NUMBERED ANSWERS

CHAPTER TWO

2. 1.26 h (75.5 min)
4. (a) 52.9 km/h, (b) 90 km
6. 218 km
8. 0.182 mi west of the flagpole
10. (a) 1.17 m/s, (b) 1.40 m/s
12. (a) 6.00 m/s, (b) 8.00 m/s
14. (a) 4.00 m/s, (b) -4.00 m/s, (c) 0, (d) 2.00 m/s
16. 0.75 m/s^2
18. -1500 m/s^2
20. (a) 0, 1.60 m/s^2, 0.80 m/s^2, (b) 0, 1.60 m/s^2, 0
22. 3.73 s
24. (a) 1.25 m/s^2, (b) 8.00 s
26. (a) 107 m, (b) 1.49 m/s^2
28. (a) -8.00 m/s^2, (b) 100 m
30. (a) 3 X 10^{-10} s, (b) 1.26 X 10^{-4} m
32. 29.1 s
34. Yes, the necessary acceleration = 0.032 m/s^2
36. (a) 31.9 m, (b) 2.55 s, (c) 2.55 s, (d) -25 m/s
38. 0.82 m/s
40. (a) 6.26 m/s, (b) 1.28 s
42. (a) 2.33 s, (b) -32.9 m/s
44. (a) 113.5 s, (b) -423 m/s
46. (a) -3.5 X 10^5 m/s^2, (b) 2.86 X 10^{-4} s
48. (a) -3 m/s^2, (b) 3 m/s^2, (c) 0, (d) -3 m/s^2, 3 m/s^2, 0
50. (a) -29.4 m/s, (b) 939 m/s^2, (c) 3.13 X 10^{-2} s
52. 1.05 s
54. (a) 5500 ft, (b) 367 ft/s
56. 96 m

EVEN-NUMBERED ANSWERS

CHAPTER THREE

2. (a) 205 m eastward, (b) 45 m westward
4. (a) 5.2 m at 60° above x axis, (b) 3.0 m at 30° below x axis, (c) 3.0 m at 150° with the x axis, (d) 5.2 m at 60° below x axis.
6. 7.92 m at 4.34° north of west
8. 15.3 m at 58° south of east
10. 25 m
12. 240 m at 237°
14. 1320 mi at 17.0° north of east
16. 196 cm at 14.7° below x axis
18. 2.77 m from base of table, $v_x = 5$ m/s, $v_y = -5.42$ m/s
20. 25.1 m
22. 5.05 m
24. (a) 24.5 m horizontally from base of cliff, (b) 1.53 s
26. (a) -15 km/h, (b) 15 km/h
28. 61.4 s
30. (a) 0.85 m/s, (b) $v_{1w} = 2.05$ m/s, $v_{2w} = -2.05$ m/s
32. 249 ft upstream
34. (a) $v_{wc} = 57.7$ km/h at 60° west of vertical, (b) $v_{we} = 28.9$ km/h downward
36. (a) 16.4 m/s, (b) 11.5 m/s
38. 12 s
40. approximately 2.3 m/s horizontal velocity
42. 37.2 m/s
44. 10.8 m
46. (a) 1522 m, (b) 36.1 s, (c) 4045 m
48. (a) 22.9 m/s, (b) 360 m horizontally from base of cliff
52. (a) 50, 86.6, (b) -64.3, 76.6, (c) -94.9, -34.2, (d) 34.2, -94.0
54. (a) 131.8 cm at 69.6° (male), 110.9 cm at 70° (female), (b) 146.4 cm at 69.6° (male), 132.0 cm at 70° (female), $\Delta d = 2.24$ cm at 65.9°
56. 1.97 m, 1.98 m, 83.6°

EVEN-NUMBERED ANSWERS

CHAPTER FOUR

2. (a) 0, (b) 0
4. 3.71 N, 58.7 N, 2.27 kg
6. See solution to this problem in this manual.
8. 2.45 m/s^2
10. 9.6 N
12. 1.59 m/s^2 at 65.2° north of east
14. (a) 0.20 m/s^2, (b) 10.0 m, (c) 2.00 m/s
16. 1080 N at 204°
18. 75 N, 130 N
20. 171.4 N at 60.55°
22. 1.36 m
24. (a) 14.3 m/s, (b) 589 m
26. 12.8 N down the incline
28. 154 N
30. 64 N
32. (a) 7250 N, (b) 4.57 m/s^2
34. (a) 2150 N, (b) 645 N, (c) 645 N to rear, (d) 10190 N at 15.9° left of vertical
36. 32.7 N, 6.53 m/s^2
38. (a) 39.2 N, (b) 0.78 m/s^2
40. $\mu_s = 0.383$, $\mu_k = 0.306$
42. $\mu_k = 0.229$
44. 894 N
46. (a) $\mu_s = 0.404$, (b) 45.8 lb
48. $\mu_k = 0.436$
50. $\mu = 0.456$
52. 4130 N
54. $\mu_s = 0.727$, $\mu_k = 0.577$
56. 3.92 m/s^2
58. 700 N, 700 N
60. 60.6 N, 35.0 N
62. (a) 84.9 N vertically, (b) 84.9 N
64. 99.9 N
66. (a) 3.35 m/s^2, (b) 2.44 s

EVEN-NUMBERED ANSWERS

68. 50 m
70. 10.9 N, (b) 2.73 m/s^2
72. 515 N
74. 1.155Mg, 0.5775Mg
76. (a) 50 N, (b) μ_s = 0.5, (c) 25 N
78. 1.77 m/s^2
80. (a) 1.78 m/s^2, (b) μ_k = 0.368, (c) 2.67 m/s
82. 0.685 m/s^2
84. (a) 1.67 m/s^2, 16.7 N, (b) 0.687 m/s^2, 16.7 N
86. (a) 0.682 m, (b) 3.2 m/s^2, (c) 2.09 m/s
88. $a_1 = \dfrac{2F}{(4m_1 + m_2)}$, $a_2 = \dfrac{F}{(4m_1 + m_2)}$, $T = \dfrac{2m_1 F}{(4m_1 + m_2)}$
90. (a) 1.02 m/s^2, (b) 2.04 N, 3.06 N, 4.08 N, (c) F_{12} = 14 N, F_{23} = 8 N
92. (a) 78 N, 35.9 N, (b) μ = 0.655

CHAPTER FIVE

2. 700 J
4. 0.675 J
6. 8.75 M
8. (a) -560 J, (b) 1.17 m
10. (a) 5.14 X 10^3 J, - 5.14 X 10^3 J, μ_k = 0.653
12. 150 J (fast), 96.0 J (slower)
14. 0.265 m/s
16. (a) 90.0 J, (b) 180 N
18. 2.00 m/s, (b) 200 N
20. 1.00 m/s
22. 147 J
24. (a) 2.59 X 10^5 J, (b) 2.59 X 10^5 J
26. (a) 80.0 J, (b) 10.7 J, (c) 0
28. (a) -30.0 J, (b) -51.2 J, (c) -42.4 J, (d) The results show that friction forces are non-conservative.
30. 0.459 m
32. (a) 20.4 m, (b) 14.1 m/s
34. (a) KE$_{javelin}$ = 349 J, KE$_{discus}$ = 676 J, KE$_{shot}$ = 741 J, (b) F$_{javelin}$ = 175 N, F$_{discus}$ = 338 N, F$_{shot}$ = 371 N, (c) Yes, if 371 N can be exerted on the shot, a

EVEN-NUMBERED ANSWERS

similar force should be exerted on the others. This would lead to a larger range than given above.

36. 3.68 m/s
38. 2060 N
40. 104 m/s
42. (a) 0.415 m, (b) -4.94 J, (c) 2.94 J, -6.51 J
44. 289 m
46. (a) 2.39×10^4 W, (b) 4.77×10^4 W
48. 0.315 hp
50. (a) 2.06×10^4 J, (b) 0.919 hp
52. 8.73 hp
54. 6.47×10^3 N
56. (a) 7.50 J, (b) 15.0 J, (c) 7.50 J, (d) 30.0 J
58. 5.10 m
60. 2.59×10^6 J
62. 5.33×10^3 hp
64. 563 lb
66. (a) 2.34×10^3 N, (b) 469 N
68. $\mu_k = 0.306$
70. 2.50 m
72. 0.70 m/s
74. (a) 51.0 J, (b) 69.0 J
76. 3914 J
78. (a) 63.9 J, (b) -35.4 J, (c) -9.51 J, (d) 19.0 J
80. (a) 0.588 J, (b) 0.588 J, (c) 2.42 m/s, (d) 0.196 J, 0.392 J

CHAPTER SIX

2. (a) momentum is doubled, (b) kinetic energy is quadrupled
4. (a) 0, (b) 1.06 kg m/s
6. 7.50×10^4 N
8. (a) -7.50 kg m/s, (b) 375 N
10. (a) 12.0 N s, (b) 6.00 m/s, (c) 4.00 m/s
12. (a) -6.30 kg m/s, (b) -3.15×10^3 N
14. 6530 N downward
16. (a) 0.49 m/s, (b) 2.01×10^{-2} m/s

EVEN-NUMBERED ANSWERS

18. 2.36 cm
20. 62.1 s
22. $v_{thrower}$ = 2.48 m/s, $v_{catcher}$ = 2.25 cm/s
24. 0.30 m/s
26. (a) 1.80 m/s, (b) 2.16 X 10^4 J
28. (a) -6.70 cm/s, 13.3 cm/s, (b) 0.889
30. 17.1 cm/s, 22.1 cm/s
32. 3.40 X 10^2 m/s
34. (a) 2.88 m/s at 32.3° with respect to initial direction of travel of fullback, (b) 783 J
36. v_{white} = 7.07 m/s, v_{black} = 5.89 m/s
38. (a) $\frac{KE_{carbon}}{KE_{neutron}}$ = 0.284, (b) $KE_{neutron}$ = 1.15 X 10^{-13} J, KE_{carbon} = 4.54 X 10^{-14} J
40. (a) 60.0°, (b) 3.46 m/s, 2.00 m/s
42. -2.3 m
44. (0,-12.0 in)
46. 4.67 X 10^6 m
48. (0.333 m, 1.67 m)
50. 14.8 kg m/s opposite to initial velocity
52. 0.267 m/s eastward
54. (a) 537 kg m/s, (b) 380 kg m/s
56. (a) $-\frac{1}{5}$ m/s, (b) $-\frac{8}{3}$ m/s
58. 56.7 m
60. 91.2 m/s
62. 0.96 m
64. 2.78 X 10^3 N
66. (a) 0.556 m/s, (b) 11.1 J
68. (a) 12.0 N s, (b) 8.00 N s, (c) 8.00 m/s, 5.33 m/s
70. 152 m/s
72. 1.25 X 10^7 m/s at 41.8°
74. 0.40 N
76. $D = \frac{2v_0^2}{9\mu g} - \frac{4d}{9}$

EVEN-NUMBERED ANSWERS

CHAPTER SEVEN

2. 60°, 216°, 540°
4. 7.27×10^{-5} rad/s
6. (a) 3.46 rad/s, (b) 5.19 rad
8. 1.67 rad, 95.7°
10. (a) 5.24 s, (b) 27.4 rad
12. 41.0 rad/s^2
14. -12.7 rad/s^2, 3.14 s
16. 3.20 rad
18. 1.02 m
20. (a) 7.5×10^{-3} rad/s^2, (b) 4.38 rad
22. 147 rev
24. 10.5 m/s, 218 m/s^2
26. 4.94×10^{-2} rad/s
28. (a) 8.0 rad/s, (b) 2.4 m/s, 1.2 m/s^2, (c) 516°, or 156° counterclockwise from a horizontal reference line
30. The required tension in the vine is 1377 N. He does not make it.
32. 2.69×10^3 N at 56.9° from the vertical
34. 6.56×10^{15} rev/s
36. (a) 18.0 m/s^2, (b) 900 N, (c) 1.84, a coefficient of friction greater than 1 is unreasonable. She will not stay on the carousel.
38. (b) 20.1°
40. (a) 0, (b) 1287 N, (c) 2.06×10^3 N
42. 321 N toward Earth
44. 1.05×10^{-10} N at 71.5° with respect to x axis
46. 6.01×10^{24} kg. The estimate is high because the Moon actually orbits about the center of mass of the Earth-Moon system, not about the center of the Earth.
48. 0.184 m/s^2
50. (a) 5.58×10^3 m/s, (b) 240 min, (c) 1.46×10^3 N
52. 1.90×10^{27} kg
54. See solution to this problem in this manual.
56. (a) 2.38×10^3 m/s (5300 mph), (b) 4.18×10^3 m/s (9350 mph),

EVEN-NUMBERED ANSWERS

(c) 6.02×10^4 m/s (135,000 mph)

58. (a) 2.51 m/s, (b) 7.90 m/s^2, (c) 4.00 m/s
60. (a) 7.76×10^3 m/s, (b) 5.36×10^3 s (89.3 min)
62. (a) 31.4 rad/s, (b) 2.09 m/s
64. (a) 2.34×10^{-10} N (in -x direction), (b) 1.00×10^{-10} N (in + x direction)
66. $\mu = 0.218$
68. 108.8 N (upper), 56.5 N (lower)
70. (a) $N = mg - \dfrac{mv^2}{r}$, (b) 17.1 m/s
74. (a) $v_{min} = \left(\sqrt{Rg\dfrac{(\tan\theta - \mu)}{(1 + \mu\tan\theta)}}\right)$ $v_{max} = \left(\sqrt{Rg\dfrac{(\tan\theta + \mu)}{(1 - \mu\tan\theta)}}\right)$,

   8.57 m/s to 16.6 m/s

CHAPTER EIGHT

2. 68.4 N m clockwise
4. 705 N m clockwise
6. $\tau_A = 207$ N m (clockwise), $\tau_B = 145$ N m (clockwise), $\tau_C = 95.7$ N m (CW)
8. 1200 N
10. (a) 400 N, (b) H = 346 N (to right), V = 0
12. $T_1 = 501$ N, $T_2 = 672$ N, $T_3 = 384$ N
14. (a) $\Sigma\tau_R = (m_1 g)(x_1) + (m_2 g)(x_2) + \ldots$, (b) $Wx_{cg} = (Mg)(x_{cg})$
16. 209 N
18. T(left string) = $\dfrac{w}{3}$, T(right string) = $\dfrac{2}{3}$ w
20. 6.15 m
22. $\dfrac{w}{3}$ on each front tire, $\dfrac{w}{6}$ on each rear tire
24. 28.1 kg m^2
26. (a) 99.0 kg m^2, (b) 44.0 kg m^2, (c) 143 kg m^2
28. 1.36 rad/s
30. -5.65×10^{-2} N m
32. $\mu_k = 0.312$
34. (a) 5.35 m/s^2 downward, (b) 42.8 m, (c) 8.91 rad/s^2
36. 1.41 m/s^2
38. (a) 1.27 N, (b) 3.18 N
40. 29.0 J

EVEN-NUMBERED ANSWERS

42. (a) $I = 92.0$ kg m^2, KE = 184 J, (b) $v_4 = 6.0$ m/s, $v_2 = 4.0$ m/s, $v_3 = 8.0$ m/s, $KE_t = 184$ J
44. 276 J
46. (a) 500 J, (b) 250 J, (c) 750 J
48. $a_{sphere} > a_{cylinder} > a_{ring}$ Thus, sphere wins and ring comes in last.
50. 149 rad/s
52. 24.2 m
54. (a) 7.08×10^{33} J s, (b) 2.66×10^{40} J s
56. Days would be four times longer.
58. 11.1 %, Kinetic energy has increased because she must do work to pull her arms inward.
60. (a) 1.91 rad/s, (b) $KE_i = 2.53$ J, $KE_f = 6.44$ J
62. (a) 3.58 rad/s, (b) 539 J (Difference results from work done by man as he walks inward.)
64. 5.99×10^{-2} J
66. $\Delta t_{sphere} = 1.44 \times 10^{-3}$ s, $\Delta t_{disk} = 1.80 \times 10^{-3}$ s, $\Delta t_{shell} = 3.60 \times 10^{-3}$ s
68. (a) -3.32 rad/s^2, (b) 29.2 rev, (c) -1.31×10^{-3} N m
70. 0.167 rev/s
72. 35.6 rad/s
74. $T_1 = 11.2$ N, $T_2 = 1.39$ N, $F = 7.23$ N
76. 2000 N
78. $F_B = 6.47 \times 10^5$ N, horizontal toward right, $F_A = 6.59 \times 10^5$ N, at 78.9° to left of vertical
80. 8 rev/s
84. (a) 3.24 J, (b) 1.44 s, (c) Yes, 2.59 m is required.
86. (a) 4.5 m/s, (b) 10.1 N
88. 3.22×10^3 W (4.32 hp)
90. (a) 3.12 m/s^2, (b) $T_1 = 26.7$ N, $T_2 = 9.37$ N
92. $h = 2.7(R - r)$
94. (a) 1.09 m/s^2, $T_1 = 21.8$ N, $T_2 = 43.6$ N

CHAPTER NINE

2. 6.93 μm
4. 1.18×10^3 N

EVEN-NUMBERED ANSWERS

6. 1.32 mm
8. 80 people
10. 2.1×10^7 Pa
12. 6.28×10^4 N (14,100 lb)
14. 1.95 cm
16. $\rho = 2.70 \times 10^3$ kg/m^3, (crown is aluminum)
18. 6.28 N
20. 3.58×10^6 Pa
22. 1.13 atm
24. 9.86×10^4 Pa
26. 10.2 m
28. 2.04 N m
30. (a) 1.28×10^5 Pa, (b) 2.67×10^4 Pa
32. (a) 1.96×10^3 N, (b) remains the same
34. 5.60 N
36. 5.10 cm
38. 9.41×10^3 N
40. 77.7 kg
42. 1.25 cm
44. 7.32×10^{-2} N/m
46. 2.27 cm
48. 5.56×10^{-2} N/m
50. 29.8 cm
52. 140°
54. 2.04×10^6 Pa
56. 1.03×10^4 Pa
58. 150 cm^2
60. (a) 1.46×10^{-2} m^3, (b) 2.10×10^3 kg/m^3
62. (a) 3.57×10^3 kg/m^3, (b) 643 kg/m^3
64. 4.14×10^3 m^3
66. (a) 1.160×10^5 Pa, (b) 52.0 Pa
68. 833 kg/m^3
70. 15 m
72. 1.71 cm
74. 1.07×10^{-2} N

EVEN-NUMBERED ANSWERS

76. (a) $\dfrac{\rho_1 h_1 + \rho_2 h_2}{h_1 + h_2}$, (b) $d = \dfrac{\rho_1 h_1 + \rho_2 h_2}{\rho_w}$ (c) Same as above, $d' = d$

CHAPTER TEN

2. 12.7 min
4. (a) 11.1 m/s, (b) 50.9 s
6. 8.85 m/s
8. 4.35×10^{-2} Pa
10. (a) 585 Pa, (b) 1.02×10^5 N upward
12. (a) 7.86×10^3 Pa, (b) 2.64×10^4 Pa
14. (a) 80 g/s, (b) 2.67×10^{-2} cm/s
16. 9.0 cm
18. (a) 2.65 m/s, (b) 2.31×10^4 Pa
20. (a) 17.7 m/s, (b) 1.73 mm
22. (a) the larger pipe. See solution in this manual. (b) 296 cm^3/s
24. 7.92×10^{-2} Ns/m^2
26. 20.7 atm
28. (a) 2.73×10^{-1} cm^3/s, (b) 30.6 min
30. 0.412 mm
32. 8 cm/s
34. 0.6 m/s
36. RN = 2890, The flow is unstable, but not necessarily turbulent.
38. 0.10 kg/m^4
40. 4.00×10^{-13} kg
42. 1.43×10^4 s (3.99 h)
44. 1.41×10^{-5} N s/m^2
46. 2.82 microns
48. (a) 16.0 m/s, (b) 1.73×10^5 Pa
50. 1.08×10^4 W (14.4 hp)
52. 3 cm/s
54. (a) 0.318 m, (b) 1.34×10^5 Pa

EVEN-NUMBERED ANSWERS

CHAPTER ELEVEN

2. (a) -273.5°C, (b) $P_f = 1.272$ atm, $P_b = 1.737$ atm
4. (a) -423°F, 20 K, (b) 68°F, 293 K
6. (a) 30°C, (b) 30 K
8. (a) 119°C, (b) 246.2°F, 832.3°F
12. 2.0034 m
14. (a) 1.2995 m, (b) fast
16. 55°C
18. (a) 437°C, (b) 2099°C, aluminum melts at 660°C
20. 641 N
22. 18.702 m
24. 1020 gallons
26. (a) 1.0001 cm^3, (b) 1.0000 cm^3
28. (a) 4.16 X 10^{-5} mol, (b) 4.11 X 10^{-21} mol
30. 5.87 atm
32. (a) 666 g, (b) 83.2 g
34. (a) 627°C, (b) 927°C
36. 0.417 liters
38. 109°C
40. (a) 468 kg (assumes a 1500 ft^2 house with 8 ft ceilings), (b) 84 kg
42. $\rho = \dfrac{PM}{RT}$
44. (a) 2.45 X 10^{25} molecules, (b) 2.42 X 10^9 molecules
46. 6.64 X 10^{-24} g/molecule
48. 16.0 N, 3.20 Pa
50. 8.0 N, 1.6 Pa
52. 6.21 X 10^{-21} J
54. (a) 8.76 X 10^{-21} J/molecule, (b) v_{helium} = 1620 m/s, v_{argon} = 514 m/s
56. 3.34 X 10^5 Pa
58. 86.2 kg
62. P_N = 0.533 atm, P_O = 0.467 atm
64. -48°C
66. 1.63 X 10^{-1} kg/m^3
68. 4.04 X 10^{25} molecules/m^3
70. 3.28 cm

EVEN-NUMBERED ANSWERS

72. 28.4 m
74. 1.15 atm = 1.17 X 10^5 Pa
76. See solution in this manual. (b) 3.66 X 10^{-3} C^{-1}
78. See solution in this manual.

CHAPTER TWELVE

2. 1030 J
4. 176 °C
6. (2.34 X 10^{-2} Cal/kg)m
8. 14.2 m
10. 1056 J
12. 88.2 W
14. 1215 J/kg °C
16. 29.5 °C
18. 59.4°C
20. 34.7 °C
22. (a) 84.44 °C, (b) Q_{Hg} = 2330 J, Q_{alc} = 4184 J, Q_w = 6514 J
24. 32.4 g
26. 20.7 g
28. 6.66 g of steam is condensed, and we are left with a mixture of 3.34 g of steam and 22.6 g of liquid water.
30. 27.8 g
32. 15.5 °C
34. 2.27 X 10^3 m
36. (a) 474 J/s, (b) 1.66 X 10^3 J/s
38. 3.6 X 10^7 J (8.6 X 10^6 cal)
40. (a) 100°C/m, (b) 1.59 J/s, (c) 60°C
42. Answers will vary depending on details of construction.
44. T_{inside} = 9.02 °C, $T_{outside}$ = 8.98 °C
46. 109 W
48. 2.63 X 10^3 °C
50. 89.8 °C (silver), 89.7°C (copper), copper wins
52. (a) 9.2 X 10^6 J/°C, (b) 5.18 X 10^7 J/day, (c) 20.6°C
54. 1.25 X 10^4 J/s

EVEN-NUMBERED ANSWERS

56. 2.35 kg
58. 28.7°C
60. 404 cm^3/h
62. (a) 74 stops, (b) Assumes no heat loss to surroundings and that all heat stays with brakes until next application of the brakes.
64. 12.2 h
66. 28.1°C
68. (a) 75.0°C, (b) 3.58 X 10^4 J

CHAPTER THIRTEEN

2. 1.13 X 10^4 J
4. (a) 810 J, (b) 507 J, (c) 203 J
6. (a) 6.08 X 10^5 J, (b) -4.56 X 10^5 J
8. 800 cm^3
10. 3.04 X 10^{-2} J
12. (a) -89 J, (b) 721 J
14. (a) 8.24 J, (b) 28.8 J, (c) 20.5 J
16. (a) 338 J, (b) 4520 J, (c) 4182 J
18. $W_{BC} = 0$, $W_{CA} < 0$, $W_{AB} > 0$, $Q_{BC} < 0$, $Q_{CA} < 0$, $Q_{AB} > 0$, $\Delta U_{AB} > 0$, $\Delta U_{BC} < 0$, $\Delta U_{CA} < 0$
20. (a) -6000 J, (b) 0, (c) -6000 J
22. (a) 4.86 X 10^{-2} J, (b) 1.62 X 10^4 J, (c) 1.62 X 10^4 J
24. 2.76 liters
26. (a) 7.65 liters, (b) 32°C
28. (a) $W_{IAF} = 76$ J, $W_{IBF} = 101$ J, $W_{IF} = 88.7$ J, (b) $Q_{IAF} = 167$ J, $Q_{IBF} = 192$ J, $Q_{IF} = 180$ J
30. (a) Eff = 0.294, (b) 500 J, (c) 1667 W
32. (a) Eff = 0.333, (b) $\frac{Q_c}{Q_h} = \frac{2}{3}$
34. (a) 560 J, (b) 350 K
36. Eff = 19.7%
38. $\frac{T_c}{T_h} = \frac{1}{3}$
40. 5.12 m^3/h
42. $\frac{Q_h}{W} = 1.17$
44. 6.06 X 10^3 J/K
46. 2.70 X 10^3 J/K

EVEN-NUMBERED ANSWERS

48. (a) Refer to solution section for table...Result = 2H and 2T. (b) all H or all T, (c) 2H and 2T.

50. (a) one way, (b) six ways

52. 1730 J

54. (a) 251 J, (b) 314 J, (c) 104 J, (d) -104 J, (e) 0 in both cases

56. (a) 0.945 J, (b) 3.15×10^5 J, (c) 3.15×10^5 J

58. (a) 5.23×10^3 J, (b) 3.17×10^3 J

60. 78.8 W (0.106 hp)

62. (a) 1.76×10^3 W, (b) COP = 3.57

64. 4.11×10^3 J

66. (a) $\frac{21}{2} RT_0$, (b) $\frac{17}{2} RT_0$, (c) Eff = $\frac{4}{21}$ (about 19%), (d) Eff$_{Carnot}$ = $\frac{5}{6}$ (about 83.3%)

68. (b) 864 J

CHAPTER FOURTEEN

2. (a) 24.0 N, 60.0 m/s^2

4. 58.8 N/m

6. (a) 575 N/m, (b) 46.0 J

8. (a) 2.12 m, (b) 1.90 m/s

10. (a) 0.938 cm, (b) 1.25 J

12. (a) 11.0 cm/s, (b) 6.32 cm/s, (c) 3.0 N

14. 2.61 m/s

16. (a) 28.0 cm/s, (b) 26.0 cm/s, (c) 26.0 cm/s, (d) 3.46 cm

18. 39.2 N

20. (a) 126 N/m, (b) 17.8 cm

22. (a) 1.99 Hz, (b) 0.503 s

24. (a) 0.628 m/s, (b) 0.50 Hz, (c) 3.14 rad/s

26. 0.627 s

28. 2.23 Hz

30. (a) at t = 0, x = 0.300 m, at t = 0.2 s, x = 0.293 m, (b) 0.3 m, (c) $\frac{1}{6}$ Hz, (d) 6.00 s

32. (a) 0.50 s, (b) 1.0 s, (c) 0.75 s

34. 105 oscillations

36. 0.248 m

EVEN-NUMBERED ANSWERS

38. $\frac{g_c}{g_t} = 1.0015$

40. 58.8 s

42. 2.40 m/s

44. 31.9 cm

46. 0.80 m/s

48. 9.47 X 10^{15} m

50. 219 N

52. 2.61 X 10^{-1} kg

54. 586 m/s

56. 40.0 m/s

58. (a) constructive interference yields A = 0.5 m, (b) destructive interference yields A = 0.1 m

60. (a) 0.25 m, (b) 0.474 N/m, (c) 0.232 m, (d) -0.116 m/s

62. 7.07 m/s

64. 0.75 J

66. 12.2 cm/s

70. 1.25 cm/s

72. See the solution section of this manual.

74. See the solution section of this manual.

76. (a) 6.93 m/s, (b) 1.14 m

78. 1.07 m/s

CHAPTER FIFTEEN

2. 1.73 X 10^{-2} m to 17.3 m

4. 0.196 s

6. 5.0 X 10^{-7} m

8. 1.04 X 10^9 Pa

10. (a) 10^{-8} W/m^2, 10^{-2} W/m^2

12. 10 machines

14. 64 dB

16. $\frac{I_A}{I_B} = 2, \frac{I_A}{I_C} = 5$

18. 3.14 X 10^{-3} W

20. (a) 94 dB, (b) 90.5 dB, (c) 88 dB

22. (a) 1095 Hz, (b) 920 Hz

EVEN-NUMBERED ANSWERS

24. 480 Hz
26. 0.391 m/s
28. The velocity of the medium has no effect.
30. 41.8°
32. 690 Hz
34. 5.17 m
36. 800 m
38. 824 N
40. 845 Hz, 1690 Hz, 2535 Hz
42. Nodes at 0, 2.67 m, 5.33 m, and 8.00 m. Antinodes at 1.33 m, 4.00 m, and 6.67 m, (b) 18.6 Hz
44. 378 Hz
46. 3450 Hz
48. 349 m/s
50. Resonance occurs at frequencies given by $f_n = n(28.8 \text{ Hz})$, where n is an odd integer between 1 and 694.
52. (a) 0.40 m, (b) 1080 Hz
54. (a) 5.72 mm to 21.4 m, (b) 8.58 mm to 32.1 m
56. (a) 0.655 m, (b) 11.7°C
58. 2.94 cm
60. 21.5 Hz
62. 3.79 m/s toward station, 3.88 m/s away from station
64. (a) 120 cm, (b) 30.0 Hz
66. 3.01 dB
68. 10 mosquitoes
70. 100 dB
72. 4 Hz
74. 1.34×10^4 N
76. 1.93 m/s
78. $\dfrac{L_1}{L_2} = 6.5$

CHAPTER SIXTEEN

2. 1.80×10^{-6} N
4. (a) 2.16×10^{-5} N, (b) 9.00×10^{-7} N

EVEN-NUMBERED ANSWERS

6. 3.72×10^{-9} kg
8. 5.08 m
10. $F_6 = 46.8$ N (left), $F_{1.5} = 157.5$ N (right), $F_{-2} = 110.7$ N (left)
12. 1.38×10^{-5} N at 77.5° below -x axis
14. (a) 8.86×10^{-8} N, (b) 2.23×10^6 m/s
16. (a) 2.75×10^{23} electrons, (b) -4.40×10^4 C
18. 7.22×10^{-9} C
20. 4.53×10^{-6} C and 5.95×10^{-4} C
22. (a) 8×10^{-17} N (westward), (b) 8×10^{-17} N (east)
24. 1.42×10^{-8} C
26. (a) 1.20×10^4 N/C toward 30×10^{-9} C charge, (b) 3.60×10^4 N/C toward the -60×10^{-9} C charge.
28. (a) 5.58×10^{-11} N/C (downward), (b) 1.02×10^{-7} N/C (upward)
30. 1.49 g
32. (a) 3.20×10^{-16} N (in + x direction), (b) 1.91×10^{11} m/s^2, (c) 5.23×10^{-6} s
34. (a) 1.28×10^4 N/C, (b) 4.24×10^6 m/s
36. 2.76×10^3 N/C at 77.5° below - x axis
38. 641 N/C at 86.1° above - x axis
40. (a) at center of triangle, (b) 3.44×10^5 N/C (upward)
42. 1.02×10^{-7} N/C
44. 0.392 m
46. See solution section of this manual.
48. See solution section of this manual.
50. See solution section of this manual.
52. (a) 0, (b) +5 µC, -5 µC, (c) 0, - 5 µC, (d) 0, 5 µC
54. 1.33×10^{-3} C
56. (a) 4.8×10^{-15} N, (b) 2.87×10^{12} m/s^2
58. (a) 3.83×10^{10} m/s^2, (b) 383 m/s, (c) 1.23×10^{-22} J
60. at y = 0.853 m
62. (a) 1.20×10^{-14} N m (CCW), (b) 1.04×10^{-14} Nm (CCW)
64. 41.8 cm
66. $1.91 \frac{kq^2}{a^2}$ at 45° below x axis
68. See solution section of this manual.
70. See solution section of this manual.

EVEN-NUMBERED ANSWERS

CHAPTER SEVENTEEN

2. See solution section of this manual.
4. 1.67×10^6 N/C
6. (a) 1.13×10^5 V/m, (b) 1.80×10^{-14} N, (c) 4.38×10^{-17} J
8. 1.44×10^{-20} J
10. (a) 4.0×10^{-14} N, (b) 2.40×10^{13} m/s^2
12. -0.502 V
14. (a) 2.65×10^7 m/s, (b) 6.19×10^5 m/s
16. (a) 1.44×10^{-7} V, (b) -7.2×10^{-8} V, (c) -1.44×10^{-7} V, 7.2×10^{-8} V
18. (a) 225 V, (b) 105 V, (c) -135 V, -15 V
20. 0.546 m, -1.20 m
22. (a) 103 V, (b) -3.86×10^{-7} J, (c) Positive work must be done to separate the charges.
24. 2.30×10^{-28} J
26. (a) 2.5×10^4 eV, (b) 2.19×10^6 m/s
28. 1.13×10^8 m^2 (43.6 sq miles)
30. 49 V
32. (a) 800 V, (b) $\dfrac{Q_2}{Q_1} = \dfrac{1}{2}$
34. C is quadrupled
36. (a) 111 μF, (b) 17.8 C
38. (a) 9 V, (b) $Q_5 = 45$ μC, $Q_{12} = 108$ μC,
40. (a) 13.3 μC on each, (b) 20 μC, 40 μC
42. (a) 12 V across each, $Q_5 = 60$ μC, $Q_4 = 48$ μC, $Q_9 = 108$ μC, (b) 21.4 μC on each, $V_5 = 4.28$ V, $V_4 = 5.35$ V, $V_9 = 2.38$ V
44. (a) 2 μF, (b) $Q_3 = 24$ μC, $Q_4 = 16$ μC, $Q_2 = 8$ μC, $V_2 = V_4 = 4$ V, $V_3 = 8$ V
46. 1.83 C
48. (a) All four should be connected in parallel. (b) Two in parallel followed by another group of two in parallel, or two in series which are in parallel with another group of two in series. (c) One in series with a group of three in parallel. (d) All four in series.
50. 12 μF, (b) $Q_4 = 144$ μC, $Q_2 = 72$ μC, $Q_{24} = Q_8 = 216$ μC,
52. 2.16×10^{-4} J, (b) 5.40×10^{-5} J
54. doubles
56. $W_4 = 3.2 \times 10^{-5}$ J, $W_2 = 1.6 \times 10^{-5}$ J, $W_3 = 9.6 \times 10^{-5}$ J

EVEN-NUMBERED ANSWERS

58. (a) 4.00 μF, (b) 48 μC, 5.71 V, (c) 8.40 μF
60. κ = 4.00
62. (a) 1.33×10^{-8} C, (b) 2.72×10^{-7} C
64. 3 m, 2×10^{-7} C
66. -7.84×10^3 V
68. 3.11 μF
70. 2.40×10^{-5} J
72. 0.301 m
74. (a) 1.73 m^2, (b) 3.60 J
76. (a) 800 μC, (b) 200 V
78. $Q_4 = 1280$ μC, $Q_6 = 1920$ μC
80. See solution section of this manual.

CHAPTER EIGHTEEN

2. 2.81×10^{20} electrons
4. 9.89 mA
6. 48 C
8. 3.64 h
10. 1.3×10^{-4} m/s
12. (a) 4.5×10^{18} electrons/s, (b) 0.72 A
14. 24 Ω
16. 0.310 Ω
18. 0.50 A
20. (a) 0.375 A, (b) 0.542 A
22. (a) 1.50 Ω, (b) 2.0 A
24. 4.45 m
26. 8.20 m
28. 1.57×10^{-3} Ωm
30. (a) 5.89×10^{-2} Ω, (b) 5.45×10^{-2} Ω
32. 25.8 mA
34. -7%
36. 1435°C
38. 67.6°C
40. (a) 3.24×10^5 J, (b) 1080 s (18 min)
42. 2.70 W

EVEN-NUMBERED ANSWERS

44. 8.67 W
46. 18 bulbs
48. 34.4 Ω
50. $\dfrac{P_A}{P_B} = 2$
52. 768 kg
54. (a) 5.04 cents, (b) 71%
56. 48.2 Ω
58. 558 W
60. 8.64×10^5 J
62. 13.5 h
64. 2.24×10^{-5} V
66. See solution section of this manual.
68. 2 Ω
70. 37.4 MΩ
72. 256 Ω
74. (a) 2.55×10^{-5} Ω, (b) 76.5 kg

CHAPTER NINETEEN

2. 5 V
4. Connect the three in parallel.
6. 1.20 V
8. See solution section of this manual.
10. (a) 0.75 A, $V_{18} = 13.5$ V, $V_6 = 4.5$ V, (b) $V_{18} = V_6 = 18$ V, $I_{18} = 1$ A, $I_6 = 3$ A
12. (a) 30 V, (b) 2.25 V
14. 4.53 V
16. $R_A = 6$ Ω, $R_B = 3$ Ω,
18. (a) $V_A = 0$, $V_B = 10$ V, $V_C = 5$ V, (b) $V_A = -10$ V, $V_B = 0$, $V_C = -5$ V, (c) $V_A = -5$ V, $V_B = 5$ V, $V_C = 0$,
20. (a) 4.59 Ω, (b) fraction dissipated = 0.082
22. 30 mA, 12 V
24. 0.353 A, 0.118 A, 0.471 A
26. 5.4 V with point a at higher potential than b
28. $I_1 = 3.5$ A, $I_2 = 2.5$ A, $I_3 = 1$ A,

EVEN-NUMBERED ANSWERS

30. $V_2 = 3.05$ V, $V_3 = 4.57$ V, $V_4 = 7.38$ V, $V_5 = 1.62$ V
32. (a) 2×10^{-3} s, (b) 180 μC, (c) 0.09 A, (d) 113 μC, (e) 3.33×10^{-2} A
34. 4.1 μA
36. (a) 14.6 μC, (b) 0.567 μA
38. (a) 91.7 V, (b) 27.6 kΩ, (c) 75.4 V
40. (a) 2 A, (b) 0.113 A, (c) 1.97 A
42. $R_x = R_3 \dfrac{L_1}{L_2}$
44. (a) Toaster = 8.33 A, Heater = 10.8 A, Grill = 12.5 A, (b) I_{total} = 31.6 A, 30 A circuit is not sufficient.
46. (a) 6.25 A, (b) 750 W
48. 15.5 A
50. total distinct values = 13. See solution section of this manual.
52. (a) 0.0999 Ω, (b) current in R_1 = 50 A, current in 100 Ω, R_2 and R_3 = 0.045 A, (b) R = 1.09 Ω, current in R_1 and R_2 = 4.55 A, current in 100 Ω and R_3 = 0.045 A, (c) R = 9.991 Ω, current in R_1, R_2, and R_3 = 0.45 A; current in 100 Ω = 0.05 A
54. I_1 = 1 A, I_2 = I_3 = 0.5 A, I_4 = 0.3 A, I_5 = 0.2 A, (b) V_{ac} = 6 V, V_{ce} = 1.2 V, V_{ed} = 1.8 V, V_{fd} = 1.8 V, V_{cd} = 3 V, (c) P_{ac} = 6 W, P_{ce} = 0.6 W, P_{ed} = 0.54 W, P_{fd} = 0.36 W, P_{cd} = 1.5 W, P_{db} = 6 W
56. (a) 10^{-2} F, (b) 0.414 C
58. I_1 = 0, I_2 = I_3 = 0.5 A
60. (a) 7 Ω, (b) $I_8 = \dfrac{3}{8}$ A, $I_6 = \dfrac{4}{3}$ A
62. (a) 14 Ω, (b) 56 W, (c) 2 A
64. (a) 0.5 A, (b) 3 W, (c) 1.06 W
66. 12 Ω
68. 0.39 A, 1.5 V

CHAPTER TWENTY

2. 1.92×10^{14} m/s^2
4. (a) 4.09×10^{-14} T, (b) horizontal
6. (a) westward, (b) no deflection, (c) upward, (d) downward
8. 806 N
10. (a) 7.90×10^{-12} N, (b) 0
12. See solution section of this manual.
14. 7.50 N

EVEN-NUMBERED ANSWERS

16. 8×10^{-3} T in +z direction
18. (a) into page, (b) toward right, (c) toward bottom of page
20. 0.245 T eastward
22. 1.96 T out of page
24. 9.98 N m clockwise (as viewed from above the loop)
26. (a) 0.628 N m, (b) 30°
28. 5.00×10^{-3} Ω
30. 74, 960 Ω
32. $R_1 = 2.99 \times 10^4$ Ω, $R_2 = 2.70 \times 10^5$ Ω, $R_3 = 2.70 \times 10^6$ Ω,
34. 1.04 cm
36. 0.885 m
38. See solution section of this manual.
40. 1.77 cm
42. (a) 2.08×10^{-1} m (singly charged), 1.04×10^{-1} m (doubly charged), (b) 20.8 cm
44. 4.97×10^{-27} kg
46. 2 cm
48. (a) right to left, (b) out of page, (c) lower left to upper right
50. (a) 4×10^{-6} T (downward), (b) 6.67×10^{-6} T at 77° to the left of vertical
52. 5.33×10^{-5} N directed toward other wire
54. 4.5 mm
56. 300 turns
58. See solution section of this manual.
60. 1.73×10^{-11} N + z direction
62. 1.74 N m
64. 0.109 A (toward the right)
66. (a) 2.22×10^{-4} A, (b) required current would be smaller by a factor of 150.
68. 1.10×10^{-2} T
70. 1.6×10^{-5} T (toward top of page)
72. See solution section of this manual.
74. 0.588 T

CHAPTER TWENTY-ONE

2. (a) 10^{-7} T m^2, (b) 8.66×10^{-8} T m^2, (c) 0
4. (a) 0, (b) 0

EVEN-NUMBERED ANSWERS

6. 2.96×10^{-5} T m^2
8. (a) $\phi = B(\pi r^2)\cos\omega t$, (b) 0
10. 83.8 mV
12. 160 A
14. 94.2 mV
16. 0.18 T
18. 2.67 T/s
20. (a) 4.0×10^{-6} T m^2, (b) 1.78×10^{-5} V
22. 272 m
24. 625 m/s (1400 mi/h), This exceeds the normal speed of a plane, so the induced voltage will not reach 1.5 V.
26. 1.00 m/s
28. (a) 2 nW, (b) 2 nW, (c) 4×10^{-10} N, (d) 2 nW
30. (a) clockwise as viewed from above, (b) counterclockwise as viewed from above
32. (a) left to right, (b) no current is present, (c) right to left
34. left to right
36. from top toward bottom
38. See solution section of this manual.
40. 1.52 T
42. 3.18×10^{-4} rad/s
44. (a) 8 A, (b) 3.17 A, (c) 60 V
46. 0.64 N m
48. 19.5 mV
50. (a) 1.97 mH, (b) 38.1 A/s
52. 1.92×10^{-5} T m^2
54. See solution section of this manual.
56. (a) 0, (b) 2.22 V, (c) 6 V, (d) 3.78 V
58. 3.01 A
60. (a) 18 J, (b) 7.14 J
62. 6.48×10^{-2} J
64. 20 mH
66. 2.21×10^{-3} C
68. 5.55×10^{-4} J
70. (a) 1.27×10^{-7} J, (b) 2.38×10^{-7} J

EVEN-NUMBERED ANSWERS

72. $\dfrac{L_A}{L_B} = \dfrac{1}{2}$

74. 3.45 μs

76. $v_t = \dfrac{mgR}{L^2 B^2}$

CHAPTER TWENTY-TWO

2. 8.49 A, 339 V
4. (a) 192 Ω, (b) 144 Ω
6. 6.76 W
8. 20.1 Hz
10. 0.36 Ω
12. 398 Hz
14. 16.6 μF
16. See solution section of this manual.
18. L > 7.03 H
20. (a) 12.5 A, (b) 0.125 A
22. 3.14 A
24. (a) 0.361 A, (b) 18.1 V, (c) 24.0 V, (d) -53°
26. 2.79 kHz
28. (a) 4.68 A, (b) 35.3 V, (c) 93.6 V, (d) 20.7°
30. (a) 138 V, (b) 104 V, (c) 729 V, (d) 640 V
32. 1.88 V
34. (a) 653 mA, (b) 1.44 H
36. (a) 1.94 A, (b) 149 V
38. (a) power factor = 0.627, (b) 0.69 A, (c) 4.22 W
40. (a) power factor = 0.492, 48.3 W, (b) power factor = 0.404, 32.6 W
42. (a) 24.1 Ω, (b) 31.9 Ω, (c) 867 W
44. (a) 179 Ω, (b) 0.709 H
46. I_{10} = 0.5 mA, I_{100} = 6.32 mA, I_{1000} = 2.35 mA, $I_{10,000}$ = 0.225 mA, (b) ϕ_{10} = -89.2°, ϕ_{100} = -79.7°, ϕ_{1000} = 86.2°, $\phi_{10,000}$ = 89.6°, (c) 212 Hz
48. 2.05 MHz, 146 m
50. 18.9 pF to 930 pF
52. (a) 480 W, (b) 0.192 W, (c) 0.03 W, (d) 0.192 W, (e) 0.03 W, The power delivered to the circuit is maximum when the frequency of the source equals the resonant frequency of the circuit.

EVEN-NUMBERED ANSWERS

54. (a) 1.59 Hz, (b) 15.9 Hz
56. (a) 100 turns, (b) 6 A
58. (a) 1110 kW, (b) 309 A, (c) 8.33 × 10^3 A
60. (a) 0.155 A, (b) 57 V, (c) 7.22 W
62. (a) 28.1 µF, (c) 0
64. (a) 0.116 A, (b) 1.38 A, (c) for (a) P = 1.08 W, for (b) 151 W
66. (a) 15 turns, (b) 9 V, all 15 turns; 6 V, 10 turns; 3 V, 5 turns
68. 0.429 A
70. 1.72 cents
72. 41.1 Hz
74. (a) 2.65 cm, 26.5 turns, (b) 9.41 × 10^{-5} H, (c) 0.108 Ω, (d) 176 A
76. 2.9 × 10^4 W, (b) $\frac{P_{loss}}{P}$ = 0.0058, (c) See solution section of this manual.
78. See solution section of this manual.

CHAPTER TWENTY-THREE

2. 75 MHz
4. See solution section of this manual.
6. 29.8 W/m^2
8. 3.74 × 10^{26} W
10. (a) 0.949 N/C, (b) 0.0949 N/C, (c) 9.49 × 10^{-3} N/C
12. (a) 10^{14} Hz, (b) north and south in a horizontal plane
14. 2.32 × 10^{10} J
16. 720 J
18. (a) 300 m, (AM radio), (b) 3 m, (TV, FM radio), (c) 3 cm, (microwaves), (d) 30 microns, (infrared), (e) 300 nm, (ultraviolet), (f) 3 nm, (x-rays), (g) 0.003 Å (γ rays)
20. 5.45 × 10^{14} Hz
22. (a) 6.25 × 10^{14} Hz, (b) 480 nm
24. (a) 3.33 × 10^{-4} s, (b) 11.5 cm
26. 7.71 m
28. (a) 4.0 × 10^{-5} s, (b) 2 mm
30. (a) 3 × 10^{14} Hz, (b) 3 × 10^{17} Hz, (c) 3 × 10^{18} Hz
32. See solution section of this manual.
34. 3.79 × 10^{16} m
36. 5.16 m

EVEN-NUMBERED ANSWERS

38. (a) 2×10^{-8} T, (b) 600 kW
40. 4.47×10^{-9} J

CHAPTER TWENTY-FOUR

2. 1.98×10^{11} m
4. (a) 536 rev/s, (b) 1070 rev/s
6. 75°
8. 5 reflections
10. (a) 1.81×10^8 m/s, (b) 2.25×10^8 m/s, (c) 1.56×10^8 m/s
12. (a) 327 nm, (b) 287 nm
14. 16.5°
16. 43.4°
18. 110.6°
20. 0.387 cm
22. 1.06×10^{-10} s
24. $\theta = 30.4°$, $\theta' = 23.3°$
26. 106 m
28. 6.39 ns
30. 2.37 cm
32. 0.39°
34. (a) 41.6°, (b) 18.4°, (c) 27.3°, (d) 42.3°
36. (a) 38.5°, (b) n = 1.44
38. (a) 43.9°, (b) 68.4°, (c) No total internal reflection can occur.
40. 4.54 m
42. 27.5°
44. $n \geq 1.89$
46. (a) 40.8°, (b) 60.6°
48. 67.2°
50. 22.0°
52. $0.707\ n_{prism}$
54. 77.5°
56. See solution section of this manual.
58. See solution section of this manual.
60. (a) 90°, (b) 30.3°, (c) Not possible since the beam is initially traveling in a medium of lower index of refraction.

EVEN-NUMBERED ANSWERS

62. $\theta_1 = \sin^{-1}[(n^2-1)^{1/2}\sin\phi - \cos\phi]$

CHAPTER TWENTY-FIVE

2. (a) 3 ft, (b) The result is independent of the distance from the mirror.

4. They are parallel to each other.

6. (a) 13.3 cm in front of mirror, real, inverted, M = -0.333, (b) 20 cm in front of mirror, real, inverted, M = -1, (c) No image is formed.

8. (a) 8 cm behind mirror, virtual, erect, $M = \frac{1}{5}$, (b) 6.67 cm behind mirror, virtual, erect, $M = \frac{1}{3}$, (c) 5 cm behind mirror, virtual, erect, $M = \frac{1}{2}$

10. 26.7 cm behind mirror, erect, M = 0.027

12. 30 cm in front of mirror

14. 20 cm in front of mirror

16. 1.0 m

18. 8.05 cm

20. -20 cm

22. 5 cm

24. (a) R is infinite, (b) s' = -s, (c) See solution section of this manual.

26. 38.2 cm below top surface of ice.

28. n = 2.00

30. 3.75 mm

32. (a) -32.7 cm, (b) -10.3 cm, (c) -4.8 cm, (d) -2.32 cm

34. (a) -17.1 cm, (b) -8.00 cm, (c) -4.24 cm, (d) -2.18 cm

36. n = 1.47

38. (a) 12 cm, (b) 12 cm, (c) 18 cm, (d) no image is formed, (e) -12 cm

40. (a) 12.8 cm behind lens, erect, virtual, M = 0.490, (b) 12.2 cm behind lens, erect, virtual, M = 0.510

42. (a) -12 cm, (b) -12 cm, (c) -9 cm, (d) -6 cm, (e) -4 cm

44. (a) 9.63 cm or 3.27 cm, (b) 2.1 cm

46. (a) -30 cm, (b) diverging

48. 120 cm in front of lens, M = 0.25

50. M = 3.4, erect

52. at s = -2f

54. (a) 6 cm in front of second lens, M = -0.6, (b) at the position of the second lens, M = -1

EVEN-NUMBERED ANSWERS

56. 119 cm beyond second lens, M = -1.25
58. 7.74 cm to left of second lens, 1.07 cm tall, erect, virtual
60. 160 cm to left of lens, inverted, M = -0.8
62. (a) 60 cm, (b) 42.9 cm in front of mirror, real, inverted, 0.429 times the size of the object, (c) 15 cm behind mirror, erect, virtual, 1.5 times the size of the object
64. 10 cm
66. real, inverted image, 5.71 cm in front of mirror
68. -25 cm
70. 8 cm
72. (a) -11.1 cm, (b) M = 2.5, (c) virtual, erect
74. 2.67 cm to left of second surface, (virtual image)

CHAPTER TWENTY-SIX

2. 3.0 cm
4. (a) 1.72 cm, (b) 2.47 cm, (c) 1.72 cm, (d) 2.47 cm
6. 2.02 μm
8. 1.5 mm
10. 515 nm
12. 75 m
14. 1.62 km
16. 4 lines
18. 184 nm
20. 511 nm
22. 150 m
24. 4750 nm
26. 233 nm
28. 193 nm
30. 550 nm
32. 625 nm
34. 1640 nm
36. 20×10^{-6} °C^{-1}
38. 91.2 cm

EVEN-NUMBERED ANSWERS

40. (a) 2.25 mm, (b) 4.50 mm

42. 0.32°

44. 1.2 mm, 1.2 mm

46. n = 2.2

48. See solution section of this manual.

50. 36.9°

52. 1.4°

54. See solution section of this manual.

56. 210 nm or any integral multiple of 210 nm

58. 127 m

60. 0.156 mm

62. $\frac{I}{I_i} = \frac{1}{2}$, (b) 54.7°

64. 41.8°

66. (a) 1.93 μm, (b) δ = 3λ, (c) maximum

68. See solution section of this manual.

70. 1.49 × 10⁻⁴ m

CHAPTER TWENTY-SEVEN

2. 30 cm beyond the lens, M = $-\frac{1}{5}$

4. 3 cm, 1.5 cm, 0.75 cm, respectively

6. f/8

8. f/1.4

10. 90.9 cm

12. For right eye, P = - 11.8 diopters, for left eye, P = - 8.2 diopters

14. (a) -5.0 diopters, (b) 37.1 cm

16. (a) -0.67 diopters, (b) 0.67 diopters

18. 1.27 m

20. (a) 4.17 cm in front of eye, (b) M = 6

22. (a) 5.77 cm, (b) M = 4.33

24. (a) M = 4, (b) M = 3

26. M = - 115

28. (a) 0.4 cm, (b) 1.25 cm, (c) M = - 1000

30. (a) f/9.84, (b) M = 50

32. M = 18.8

EVEN-NUMBERED ANSWERS

34. 18 times larger
36. (a) M = 1.5, (b) M = 1.90
38. (a) For A 5.76×10^{-6} rad, For B 9.61×10^{-6} rad, (b) For A, M = 66.7, for B, M =
40. 492 km
42. 5.65×10^7 miles (9.09×10^7 km)
44. 5.31×10^{11} km (3.30×10^{11} miles)
46. 5.40 mm
48. 28.5 μradians
50. 50.4 μm
52. $n_g = 1 + \dfrac{30\lambda}{5\ \text{cm}}$
54. 404 nm
56. (a) 13 orders, (b) 1 order
58. 625 nm
60. between 700 nm and 800 nm
62. 44.5 cm
64. (a) 3646 slits, (b) 1823 slits
66. (a) 2.67 diopters, (b) 0.16 diopters too low
68. -1.33 diopters
70. (a) 982 lines, (b) 327 lines
72. 22,800 slits
74. 19.4°
76. 3.14 diopters

CHAPTER TWENTY-EIGHT

2. (a) 1.87×10^{-15} s, (b) 1.02 fringes
4. 0.958c
6. 5.0 s
8. (a) 1.31×10^{-7} s, (b) 38.5 m, (c) 7.64 m
10. (a) a rectangular box, (b) sides perpendicular to velocity are 2 m long, sides parallel to velocity are 1.11 m long.
12. $v \leq 0.745c$
14. 4.64×10^{-10} s
16. (a) 17.4 m, (b) 3.3°

EVEN-NUMBERED ANSWERS

18. 4.8 y, (b) 2.65 y

20. 9.38×10^{-15} m

22. (a) 5.00×10^{-21} kg m/s, (b) 2.90×10^{-19} kg m/s, (c) 1.03×10^{-18} kg m/s

24. 3.38×10^5 m/s

26. 0.995c

28. 0.866c. The result is independent of the mass of the particle.

30. 2.22×10^{-27} kg

32. (a) 0.999999995c, (b) 35 cm

34. - c

36. 0.238c to the right

38. -0.94c

40. 0.998c

42. 0.866c

44. 9.39×10^8 eV

46. (a) 4.5×10^{16} J, (b) 1.43×10^7 y

48. 4.39×10^{18} electrons

50. 1.52×10^{-18} kg m/s (2850 MeV/c), 2070 MeV

52. m(faster) = 2.52×10^{-28} kg, m(slower) = 8.83×10^{-28} kg

54. (a) 1.67×10^{-27} kg, 3.1×10^5 m/s, (b) 2.56×10^{-27} kg, 0.758c

56. 6.28×10^7 kg

58. (a) 1090 MeV, (b) 3000 MeV

60. (a) 0.183 MeV, (b) 2.45 MeV

62. See solution section of this manual.

CHAPTER TWENTY-NINE

2. (a) 2.49×10^{-5} eV, (b) 2.49 eV, (c) 124 eV

4. 5180 K, Clearly, a firefly is not at this temperature, so this is not blackbody radiation.

6. 9660 K

8. 3.7×10^{19} photons

10. (a) n = 2.34×10^{31}, (b) $\frac{\Delta E}{E} = 4.28 \times 10^{-32}$

12. 8.71×10^{12} electrons

14. (a) 2.24 eV, (b) 555 nm, (c) 5.41×10^{14} Hz

16. 5.43 eV

18. 4.77×10^{14} Hz, 2.03 eV

EVEN-NUMBERED ANSWERS

20. 0.784 eV
22. 4.14×10^4 V
24. 124 V to 1.24×10^7 V
26. 18.2°
28. 0.078 nm
30. 0.281 nm
32. See solution section of this manual.
34. 59.6°
36. (a) 1.18 eV, (b) 6.45×10^5 m/s
38. 70.0°
40. (a) 1880 MeV, (b) 6.60×10^{-16} m
42. (a) 3.52 MeV, (b) 8.50×10^{20} Hz
44. (a) 1460 m/s, (b) 7.28×10^{-11} m
46. (a) 3.97×10^{-11} m, (b) 3.97×10^{-14} m
48. (a) 14.8 keV (if done relativistically), 15.1 keV (if done classically), (b) 124 keV
50. 9.05×10^{-15} m
52. See solution section of this manual.
54. 3.57×10^{-13} m
56. 2.10×10^{-32} m/s
58. order of magnitude = 10^5 m/s
60. 1.64×10^{-14} s
62. 4.05 eV
64. (a) n = 4.22×10^{35}, (b) 3.32×10^{-34} J
66. (a) 2.49×10^{-2} nm, (b) 0.285 nm
68. 2.41 nm (electron), 0.056 nm (proton)
70. (a) 7.77×10^{-12} m, (b) 93.8°, (c) 35.4°

CHAPTER THIRTY

2. 656 nm, 486 nm, 434 nm, visible region
4. 4.57×10^{14} Hz, 6.17×10^{14} Hz, 6.91×10^{14} Hz
6. (a) 91.13 nm (Lyman), 364.5 nm (Balmer), 820.1 nm (Paschen), 1458 nm (Brackett), (b) 13.6 eV (Lyman), 3.41 eV (Balmer), 1.51 eV (Paschen), 0.85 eV (Brackett)
8. See solution section of this manual.

EVEN-NUMBERED ANSWERS

10. (a) six transitions, n = 4 to 1, 4 to 2, 4 to 3, 3 to 1, 3 to 2, and 2 to 1. (b) 1880 nm (Paschen series)
12. (a) 13.6 eV, (b) 1.51 eV
14. See solution section of this manual.
16. 0.053 nm, 0.212 nm, 0.476 nm
18. (a) 0.212 nm, (b) 10^{-24} kg m/s, (c) 2.10×10^{-34} J s, (d) 3.39 eV, (e) -6.78 eV, (f) -3.39 eV
20. 8.23×10^{-8} N
22. (a) 0.97 eV, (b) 0.27 eV
24. 91.4 nm
26. (a) 2.89×10^{34} kg m²/s, (b) $n = 2.74 \times 10^{68}$, $\frac{\Delta r}{r} = 7.30 \times 10^{-69}$
28. See solution section of this manual.
30. See solution section of this manual.
32. 30.4 nm, 22.8 nm
34. (a) and (b) See solution section of this manual. (c) 72 electrons
36. See solution section of this manual.
38. See solution section of this manual.
40. (a) 2, (b) 8, (c) 18, (d) 32, (e) 50
42. 0.155 nm, 8.03 kV
44. 7.03×10^{-2} nm
46. 97.3 nm, 103 nm, 487 nm, 657 nm, 1217 nm, 1881 nm
48. (a) 10.2 eV, (b) 7.88×10^4 K
50. 1.25×10^{-4} J, (b) 3.98×10^{14} photons
52. (a) 0.0529 nm, (b) 1.99×10^{-24} kgm/s, (c) 1.05×10^{-34} kg m²/s, (d) 13.6 eV, (e) -27.2 eV, (f) -13.6 eV
54. See solution section of this manual.
56. (a) 124 MeV, (b) No, its energy is too large to allow it to be confined.
58. See solution section section of this manual.
60. (a) $\frac{1}{\alpha} = 137.036$, (b) $\frac{r_0}{\lambda} = \frac{1}{2\pi\alpha}$, (c) $\frac{1}{Rr_0} = \frac{4\pi}{\alpha}$

CHAPTER THIRTY-ONE

2. See solution section of this manual.
4. point A, F = 0; point B, F is in positive x direction; point C, F = 0; point D, F is in negative x direction; point E, F = 0

EVEN-NUMBERED ANSWERS

6. 0.64 eV
8. 0.141 eV/bond
10. 9.40×10^7 J
12. 5.32×10^{-11} m from the oxygen atom
14. 1.61×10^{-46} kg m^2
16. 0.726 mm, 4.13×10^{11} Hz
18. 3.45×10^{-3} eV
20. (a) 8.1 cm, 3.70×10^9 Hz, (b) 4.53×10^{-45} kg m^2
22. 0.179 eV
24. (a) 291 N/m, (HI molecule), 1550 N/m (NO molecule), (b) A larger number of electrons participate in the molecular bond.
26. 0.672 eV
28. n-type, See solution section of this manual for explanation.
30. 1.91 eV
32. See solution section of this manual.
34. 7.11×10^{11} Hz
36. (a) 1850 N/m, (b) 6.78×10^{-12} m, (c) The vibration amplitude is about 6% of the separation distance.
38. J' = 4 and ν = 0 for 4.5 K, J' = 44 and ν = 0 for 450 K, J' = 142 and ν = 13 for 4500 K
40. (a) n-type, (b) p-type, (c) p-type

CHAPTER THIRTY-TWO

2. A = 48
4. A = 2, r = 1.51×10^{-15} m, A = 60, r = 4.70×10^{-15} m, A = 197, r = 6.98×10^{-15} m, A = 239, r = 7.45×10^{-15} m,
6. 368 m
8. 6.11×10^{15} N
10. (a) 1.85×10^7 m/s, (b) 7.11 MeV
12. 2.80×10^7 m/s
14. 161 MeV
16. 8.66 MeV/nucleon
18. 8.26 MeV/nucleon for $^{24}_{12}$M, 8.70 MeV/nucleon for $^{85}_{37}$Rb
20. 0.764 MeV

EVEN-NUMBERED ANSWERS

22. 7.93 MeV
24. 576 MeV
26. 465 days
28. 0.465 curies
30. (a) 9.98×10^{-7} s^{-1}, (b) 1.85×10^{10} nuclei
32. 5.58×10^{-2} h^{-1}, 12.4 h, (b) 2.39×10^{13} nuclei, (c) 1.88 mCi
34. 21.6 mCi
36. $^{208}_{81}$Tl, $^{95}_{37}$Rb, $^{144}_{60}$Nd
38. 5.41 MeV
40. Yes, the mass of Be is greater than the mass of two alpha particles, so the decay can occur spontaneously with the release of 92.2 keV of energy.
42. See solution section of this manual.
44. 4220 y
46. 5.70 MeV
48. (a) -3.14 MeV, (b) -2.77 MeV
50. $^{4}_{2}$He, $^{4}_{2}$He
52. (a) $^{30}_{15}$Po, (b) -2.64 MeV
54. 192 MeV
56. (a) $^{13}_{6}$C, (b) $^{10}_{5}$B
58. (a) $^{1}_{0}$n, (b) Fluorine mass = 18.000953 μ
60. 24.0 decays/min
62. 46.5 days
64. Q > 0, no threshold energy is required
66. (a) 2.52×10^{24} nuclei, (b) 2.306×10^{12} Bq, (c) 1.07×10^{6} y
68. fraction = 0.35%
70. 18.3 counts/min, (b) The observed count rate is slightly less than the average background and would be difficult to measure accurately using reasonable counting times.
72. 4.45×10^{-8} kg/h

EVEN-NUMBERED ANSWERS

CHAPTER THIRTY-THREE

2. $^{1}_{0}n + ^{235}_{92}U \rightarrow ^{141}_{56}Ba + ^{92}_{36}Kr + 3\,^{1}_{0}n$

4. 126 MeV

6. (a) For sphere, $\frac{A}{V} = \frac{3}{a}$, (b) For cube, $\frac{A}{V} = \frac{3.72}{a}$, (c) $\frac{A}{V}$ is lowest for sphere

8. 1.01 g

10. 938 g

12. 4.03 MeV

14. 3.07×10^{22} events/y

16. (a) $^{8}_{4}Be$, (b) $^{12}_{6}C$, (c) 7.27 MeV

18. (a) 1.11×10^{6} m/s, (b) about 10^{-7} s

20. The second worker receives twice the dose.

22. (a) 2.40×10^{-3} rem/x-ray, (b) This is about 38 times background.

24. 5 rad

26. 2.39×10^{-3} °C

28. (a) 9×10^{18} rad, (b) 2.26×10^{8} rad

30. (a) Energy amplification = 3.13×10^{7}, (b) 3.13×10^{10} electrons

32. 4×10^{-2} cm

34. 8.62×10^{5} m/s

36. 33.8 kg

38. 4.65×10^{-4} R, 0.698 Roentgens/h

40. (a) See solution section in this manual. (b) 11.8 days

CHAPTER THIRTY-FOUR

2. 4.53×10^{23} Hz, 6.62×10^{-16} m

4. $\bar{\nu}_\mu$, and ν_e.

6. 67.5 MeV

8. The first decay is via the strong interaction; the second via the weak interaction.

10. First can occur via the weak interaction; the second violates conservation of baryon number.

12. See solution section of this manual.

14. K^+, Ξ^0, π^0,

EVEN-NUMBERED ANSWERS

16. The second reaction does not conserve strangeness

18. (a) ν_e, (b) ν_μ, (c) $\bar{\nu}_\mu$, (d) ν_μ and $\bar{\nu}_\tau$,

20. 70.5 MeV

22. (a)

	K^0	d	$\bar{s}$	total
strangeness	1	0	1	1
baryon number	0	$\frac{1}{3}$	$-\frac{1}{3}$	0
charge	0	$-\frac{1}{3}e$	$\frac{1}{3}e$	0

(b)

	Λ^0	u	d	s	total
strangeness	-1	0	0	-1	-1
baryon number	1	$\frac{1}{3}$	$\frac{1}{3}$	$\frac{1}{3}$	1
charge	0	$\frac{2}{3}e$	$-\frac{1}{3}e$	$-\frac{1}{3}e$	0

24. (a) $-e$ (antiproton), (b) 0, antineutron

26. (a) Λ^0 or Σ^0, (b) π^-, (c) K^0, (d) Ξ^-